Models in Geography

THE SECOND MADINGLEY LECTURES

Models in Geography

Edited by

RICHARD J. CHORLEY
PETER HAGGETT

METHUEN & CO LTD

First published in 1967
by Methuen & Co, Ltd
11 *New Fetter Lane, London EC*4
© 1967 *Methuen & Co Ltd*
Reprinted 1968
Printed in Great Britain
by Ebenezer Baylis & Son, Ltd
The Trinity Press, Worcester, and London
SBN 416 29020 5
1.2

Distributed in the USA by Barnes & Noble Inc

Contents

ACKNOWLEDGMENTS   13

I THE ROLE OF MODELS

1   MODELS, PARADIGMS AND THE NEW GEOGRAPHY   19
Peter Haggett and Richard J. Chorley
Departments of Geography, Bristol and Cambridge Universities
Facts, models and paradigms   19
Classificatory paradigms in geography   28
Towards a model-based paradigm of geography   33
Epilogue   38
References   39

2   THE USE OF MODELS IN SCIENCE   43
F. H. George
Educational and Scientific Developments Ltd, and Bristol University
Introduction   43
Scientific method   45
Static models   48
Dynamic models   49
Methods used in preparing dynamic models as programmes   52
References   56

II MODELS OF PHYSICAL SYSTEMS

3   MODELS IN GEOMORPHOLOGY   59
Richard J. Chorley
Department of Geography, Cambridge University

Introduction	59
Natural analogue systems	60
Physical systems	63
General systems	76
Testing the model	89
References	90

4 MODELS IN METEOROLOGY AND CLIMATOLOGY — 97
R. G. Barry
Department of Geography, Southampton University

Introduction	97
The general circulation	99
The Westerlies	107
Synoptic-scale models	114
Tropical models	118
Meso-scale models	127
Climatological implications	129
Prediction models	133
References	135

5 HYDROLOGICAL MODELS AND GEOGRAPHY — 145
Rosemary J. More
Department of Civil Engineering, Imperial College London

Introduction	145
Function of models in hydrology	148
Physical and systems hydrology	153
Models in physical hydrology	154
Overall catchment models	165
Stochastic models	172
Hydro-economic models	174
Hydrological models and geography	179
References	179

III MODELS OF SOCIO-ECONOMIC SYSTEMS

6 DEMOGRAPHIC MODELS AND GEOGRAPHY — 189
E. A. Wrigley
Department of Geography, Cambridge University

Introduction	189
Animal population behaviour	191
A simple demographic model	193
Demographic characteristics and geographical conditions	197
Fertility levels in pre-industrial societies	198
A modified demographic model	200
Demographic models and post-industrial societies	207
The developing countries	209
Conclusion	212
References	213

7 SOCIOLOGICAL MODELS IN GEOGRAPHY 217
R. E. Pahl
  Faculty of Social Sciences, University of Kent at Canterbury

Internalized models	218
The myth of a value-free geography	219
Sociological models in geography	220
The normative orientation of action systems	223
Weber's 'ideal type' models	225
Functionalism as a model	227
Functionalism and the problem of change	230
Sociological models and urban geography	237
References	240

8 MODELS OF ECONOMIC DEVELOPMENT 243
D. E. Keeble
  Department of Geography, Cambridge University

Geography, models and economic development	243
Non-spatial models of economic development	248
Models of spatial distribution of economic development	257
Conclusion	287
References	287

9 MODELS OF URBAN GEOGRAPHY AND
  SETTLEMENT LOCATION 303
B. Garner
  Department of Geography, Bristol University

Introduction	303
Settlement patterns	306
The internal structure of cities	335

	Conclusion	354
	References	355

10   MODELS OF INDUSTRIAL LOCATION   361
F. E. Ian Hamilton
Department of Geography, London School of Economics and Political Science, and School of Slavonic and East European Studies, London University

The changing character of model techniques	362
In search of reality	364
Early models of industrial location	369
The Weber model: merits and demerits	370
Weber's model reformed: an approach to reality	374
Industrial location under capitalism	377
Location policy in the socialist world	381
Allocation-location models	386
Industrial location and settlement hierarchies	389
Models of an ideal spatial dispersion of industry	393
Industry through history	395
Some observations on Monte-Carlo and Markov chain models	401
Structure, process and stage	402
The international distribution of industry	403
Industry in the national setting	405
Regional industrial *'climaxes'*	406
Industrial inertia and migration	410
Intra-regional change and the multiplier model	413
An industrial example: iron and steel	415
Conclusion	416
References	417

11   MODELS OF AGRICULTURAL ACTIVITY   425
Janet D. Henshall
Department of Geography, King's College, London University

A general model	425
Experimental models	426
Conceptual models	429
Taxonomic models	437
Models of the location of agricultural activity	443
Land potential models	449
Future trends	452
References	452

IV MODELS OF INTEGRATED SYSTEMS

12 REGIONS, MODELS AND CLASSES 461
 David Grigg
  Department of Geography, Sheffield University
 Introduction 461
 The development of the regional concept in geography 464
 Is the region a concrete object? 472
 Ecological regions 476
 Cores and boundaries 478
 Regionalization and classification 479
 Some principles of regionalization 485
 Statistics and regional delimitation 489
 Regions as models 494
 References 501

13 ORGANISM AND ECOSYSTEM AS GEOGRAPHICAL MODELS 511
 D. R. Stoddart
  Department of Geography, Cambridge University
 Introduction 511
 The organic analogy 512
 The organic analogy in geography 514
 The organic analogy: components and criticism 518
 Human ecology and the urban sociologists 521
 The ecosystem as a model of reality 522
 Geography, the ecosystem and general systems 537
 References 538

14 MODELS OF THE EVOLUTION OF SPATIAL PATTERNS IN HUMAN GEOGRAPHY 549
 D. Harvey
  Department of Geography, Bristol University
 Theories and metaphors 549
 Models 552
 Models of the evolution of spatial patterns – operational considerations 561
 The application of models of spatial evolution in geographic research 588
 References 597

15 NETWORK MODELS IN GEOGRAPHY — 609
Peter Haggett
Department of Geography, Bristol University

Path geometry — 610
Tree geometry — 624
Circuit geometry — 632
Cell geometry — 646
Network transformations — 654
Conclusion — 664
References — 664

V INFORMATION MODELS

16 MAPS AS MODELS — 671
C. Board
Department of Geography, London School of Economics and Political Science

The map-model cycle: the argument — 672
Making the model — 675
Testing the model — 713
Conclusion — 719
References — 719

17 HARDWARE MODELS IN GEOGRAPHY — 727
M. A. Morgan
Department of Geography, Bristol University

Introduction — 727
Static models — 728
Dynamic models in physical geography — 732
Conducting sheet analogues — 746
Magnetic and electro-magnetic analogues — 756
Blotting paper analogues — 759
Location models — 762
Networks — 768
References — 771

18 MODELS OF GEOGRAPHICAL TEACHING — 775
S. G. Harries
Department of Education, Cambridge University

Introduction	775
Models in geography teaching	776
Science and teaching models	779
Learning models	785
Conclusion	790
References	792

INDEX 793
   D. R. Stoddart

Acknowledgments

The editors and contributors would like to thank the following learned societies, editors, publishers, universities, organizations and individuals for permission to reproduce figures and tables:

Learned Societies
American Geophysical Union for fig. 3.12 from the *Transactions*, fig. 5.14 from *Water Resources Research* (and M. B. Fiering) and fig. 15.15 from the *Journal of Geophysical Research*: American Meteorological Society for fig. 4.8 (and D. Fultz) from the *Journal of Meteorology*: American Society of Civil Engineers for fig. 5.12A from the *Journal of the Hydraulics Division*: American Sociological Association for fig. 7.1 from the *American Sociological Review*: Association of American Geographers for figs. 3.2A, 3.2B, 12.1 and 16.14 from the *Annals*; and fig. 15.17 from the *Professional Geographer*: Geological Society of America for figs. 3.4A, 3.4B, 3.4D, 15.11, 17.2 and 17.5 from the *Bulletin*: Geologists' Association of London for fig. 17.6 (and M. J. Kenn) from the *Proceedings*: Glaciological Society for fig. 3.3C from the *Journal of Glaciology*: Institution of Civil Engineers for figs. 5.3A, 5.3B, 5.3C and 5.3D from the *Proceedings*: Institution of Water Engineers for fig. 5.2 from the *Journal*: Operations Research Society of America for figs. 15.2 (and W. Miehle) and 15.4 (and S. B. Akers, Jr.) from *Operations Research*: Pacific Science Association for fig. 13.1 from the *Proceedings*: Regional Science Association for fig. 9.27 from the *Papers and Proceedings*: Royal Geographical Society for fig. 16.3 from the *Geographical Journal*: Society of Economic Paleontologists and Mineralogists for figs. 3.5D, 3.5E and 17.3 from the *Journal of Sedimentary Petrology*: Society of Sigma XI for figs. 13.6 and 13.7 from the *American Scientist*.

Editors
American Journal of Psychology for fig. 1.1: *American Journal of Science* for figs. 3.5A, 3.5B and 15.14: *Australian Geographical Studies* for fig. 5.3E: *Cartographic Journal* for fig. 16.8: *Econometrica* for fig. 17.25 (and S. Enke): *Economic Geography* for figs. 9.19 and 13.3: *The Economist* for fig. 8.3:

[14] ACKNOWLEDGMENTS

Geografiska Annaler for fig. 14.4 (and E. Bylund): *Geography* for figs. 13.2, 13.4 and 13.5: *Gerlands Beitrage zur geophysik* for figs. 17.10 and 17.11: *Publications in Geography, State University of Iowa* for fig. 16.21: *Journal of Asian Studies* for fig. 9.12: *Journal of Hydrology* for fig. 15.8: *The Times* for fig. 16.19B: *Tijdschrift voor Economisch en Sociale Geografie* for Fig.16. 13: *Weather* for fig. 4.16 (and F. H. Ludlam).

Publishers
John Bartholomew, London for figs. 16.10 and 16.18: B.B.C. Publications for fig. 17.9: Cambridge University Press for figs. 15.23 and 15.24: Cassell Ltd., London for fig. 16.18A: Clarendon Press, Oxford for fig. 16.18C: The Controller, Her Majesty's Stationery Office (Crown Copyright Reserved) for figs. 3.3B and 4.11: Edward Arnold Ltd., London for figs. 9.3, 9.5, 9.8, 9.9, 9.14, 9.15, 9.16, 9.17, 9.22 and 9.24; also for table 9.1: Elsevier, Amsterdam for fig. 15.6: Faber and Faber Ltd., London for fig. 16.18E: C. W. K. Gleerup Publishers, Sweden for figs. 9.20, 14.8 (and R. L. Morrill), 14.9 (and T. Hägerstrand), 14.10 (and T. Hägerstrand), 15.20 (and R. L. Morrill) and 15.21 (and R. L. Morrill) from *Lund Studies in Geography*: Harvard University Press for fig. 17.1 (and L. W. Pollack): Longmans, Green and Co. Ltd., London for figs. 17.7 and 17.8 (and J. Allen): The Macmillan Co., New York for fig. 5.4B: McGraw-Hill Book Co., New York for fig. 4.7 (and J. Namias); for figs. 5.4A, 5.4D and 5.5A from *Applied Hydrology* by R. K. Linsley, M. A. Kohler and L. H. Paulhus; for fig. 5.5B from *Hydrology* by O. E. Meinzer (Ed.); and for figs. 5.10A and 5.10B from *Handbook of Applied Hydrology* by V. T. Chow (Ed.): Pergamon Press Ltd., London for fig. 4.6: George Philip and Son Ltd., London for figs. 10.13 (and J. E. Martin) and 16.18B: Prentice-Hall Inc., New Jersey for figs. 5.10E and 5.10F: Princeton University Press for fig. 15.5: Random House Inc., New York for fig. 15.3: Scientific American Inc., New York for fig. 5.13 by R. Revelle (All rights reserved): University of Chicago Press for figs. 3.4C (by W. E. H. Culling), 3.8 (by M. A. Melton), 3.10 (by M. A. Melton) and 3.11 (by R. L. Miller and J. M. Zeigler) reprinted from the *Journal of Geology*; and for figs. 9.21, 9.25 and 9.28 reprinted from *Commercial Structure and Commercial Blight* by B. J. L. Berry *et al*: University of Washington Press for fig. 9.29 reprinted from *Studies of the Central Business District and Urban Freeway Development* by E. M. Horwood and R. Boyce: John Wiley and Sons Inc., New York for figs. 5.4C, 9.7 and 9.10: Yale University Press for figs. 9.6 and 15.22.

Universities
Massachusetts Institute of Technology, Department of Civil Engineering, Soil Mechanics Division, for figs. 3.3D and 3.3E: Northwestern University, Department of Geography, for figs. 9.26 and 9.30; and the Transportation Center for fig. 15.18: Stanford University, Department of Civil Engineering,

for fig. 5.11A: University of Michigan for fig. 15.7 from *Michigan Inter-University Community of Mathematical Geographers, Discussion Papers*: University of Oregon, School of Business Administration, for fig. 9.11 from *Urban Systems and Economic Development*.

Organizations
The Director General, Ordnance Survey for figs. 16.9 and 16.11: The Director, Trigsurvey, Pretoria, for fig. 16.5: The Director, United States Geological Survey for figs. 3.4E, 3.4F, 3.4G, 3.5C, 5.10C and 5.10D: London Transport for fig. 16.20: Ministry of Housing and Local Government (Crown Copyright Reserved) for fig. 16.19E: Provinces of Alberta and British Columbia Highways Departments for fig. 16.7: Société Nationale des Chemins de fer français for fig. 16.4: Standard Oil Co. and State of Illinois Highways Department for fig. 16.6: Thames Conservancy for fig. 5.9A: United Nations Economic Commission for Europe for fig. 8.6: United States Department of the Army, Waterways Experiment Station, Corps of Engineers, Vicksburg, Mississippi, for figs. 3.2C, 3.2D and 3.3A: Water Research Association, Medmenham, for figs. 5.7 and 5.8.

Individuals
H. Beguin for fig. 11.1: C. Board of the London School of Economics for fig. 11.3: H. Flohn of the University of Bonn for fig. 4.14: B. A. Kennedy of the Department of Geography, University of British Columbia for fig. 3.7: T. Mrzygłód for fig. 10.5: D. R. Stoddart of the Department of Geography, Cambridge University for fig. 1.2.

Finally, the following thanks are also due:
Chapter 3. The author acknowledges his debt to his association with L. B. Leopold, M. A. Melton, S. A. Schumm, A. N. Strahler and, particularly, W. C. Krumbein: Chapter 5. The author would like to thank T. O'Donnell and F. V. Appleby for reading the manuscript and making some suggestions: Chapter 7. The author is grateful to Geoffrey Hawthorn of the Department of Sociology at the University of Essex and to his fellow sociologist Rex Taylor of the University of Kent at Canterbury for their helpful comments: Chapter 9. The author is grateful to A. E. Frey and W. Campbell of the Department of Geography, Bristol University, for their help in clarifying the presentation: Chapter 10. The author thanks R. C. Estall for kindly reading and commenting on the chapter: Chapter 11. The author and editors are grateful to D. W. Harvey of the Department of Geography, Bristol University, for making available an unpublished review (subsequently Harvey, 1966) of theoretical models of agricultural land-use patterns: Chapter 13. The author would like to thank P. Haggett and R. J. Chorley for discussion of several points. Parts of this chapter have appeared in different form in

[16] ACKNOWLEDGMENTS

articles in *Geography* and the *Annals of the Association of American Geographers* (Stoddart, 1965 and 1966). The author also thanks Yu. G. Saushkin of the University of Moscow for his comments on the paper by Polonskiy. Dr. S. A. Wainwright of Duke University and Dr. P. E. Gibbs of Queen Mary College read the manuscript: Chapter 16. The author acknowledges the assistance of the Staff of the Drawing Office and Dark Room of the London School of Economics and, particularly, the opportunity of discussing some of the matters covered in the chapter with W. C. Krumbein and with his colleagues at the Department of Geography, London School of Economics, and at the Ohio State University: Index. The editors would like to thank D. R. Stoddart of the Department of Geography, Cambridge University, for his painstaking preparation of the index, and also for employing his rare bibliographical talents in assisting with checking the proofs.

The idea for the jacket design was by M. A. Morgan.

PART I

The Role of Models

CHAPTER ONE

Models, Paradigms and the New Geography

P. HAGGETT and R. J. CHORLEY

Models are undeniably beautiful, and a man may justly be proud to be seen in their company. But they may have their hidden vices. The question is, after all, not only whether they are good to look at, but whether we can live happily with them. KAPLAN, 1964, p. 288.

In concluding the previous volume in this series, we attempted to review the paths taken by various workers in moving towards what they saw to be the 'frontier' in geographical research. We argued there that the quest for a model or models was a recurrent theme in their search. This volume is a direct outcome of that conclusion in that specific workers were asked to discuss the role of model-building within their own special fields of geographical research. While we would not wish to pre-judge their findings, it will be evident from the format and arrangement of the chapters that there is: (1) some measurable contrast between their approaches to geography, various as they are, and those that characterize the great part of established geographical patterns of thinking, as evidenced by existing textbooks and syllabuses; and (2) a community of common ideas that link all contributors into what Price (1963, p. 62) would characterize as an 'invisible college' of geographical practitioners. Whether this communality is sufficient to form the basis of what Manley (1966) has termed a 'New Geography' is not for us to judge. However, it is perhaps significant that the greater part of the volume is based on work produced since 1945, and much of it since 1960. In this opening chapter we discuss what we believe to be the significance of this new search for a model-based geography.

FACTS, MODELS AND PARADIGMS
The nature of facts
Information in geography is capable of treatment in terms of general information theory. In this context factual information only has relevance within

[20] MODELS IN GEOGRAPHY

some more general frame of reference, and such a basic operation as the definition of a relevant fact can only be made on the basis of some theoretical framework. There are also different levels of organization of relevant information. Some information can be relevantly organized only at a small scale, whereas the orderly large-scale patterns of other information are blurred or swamped altogether on the local scale. One can therefore view geographical information registration and analysis, from one point of view at any rate, as a problem in the separation of regional and local information patterns from the more randomly-organized information which, as 'noise', obscures them (Chorley and Haggett, 1965). Of course, one may choose to regard the noise

1.1 A photograph of melting snow taken on impulse by a photographer in China just before the last war. The pattern makes no sense until it is organized as a full-face and shoulders, similar in style to a late-medieval representation of Christ; the upper margin cutting the brow and illuminated from the right (*Source: Partly from Porter, 1954*).

as the more significant element and to ask whether it is useful to try to recognize any order in reality. This results in the stress being placed on the variety of geographical information available and in attempts to subdivide information. However, it is becoming increasingly popular to ask what kinds of order are exhibited by geographical information and on what scales of space and time each operates. In short, the 'simple' registration of facts is being recognized not only as unsatisfactory but as an impossibility. Hanson (1958, pp. 8–19) has pointed out that what is observed depends not only on the context in which a particular phenomenon is set, but in the manner in which one is prepared to view it. In the words of Sigwart: 'That there is more order in the world than appears at first sight is not discovered *till the order is looked for*' (Quoted by Hanson, 1958, p. 204). Figure 1.1 gives a striking illustration both of the close apparent relationship between order and disorder, and of the subjective approach necessary to identify what it believed to be orderly.

The distinction between the idiographic and nomothetic approaches to the real world was recognized by Aristotle, although not in the terms which we currently employ, when he pointed out that poetry is more philosophical and of graver import than history because it is concerned with what is pervasive and universal, whereas history is addressed to what is special and singular (Nagel, 1961, p. 547). Today the distinction is made commonly between the 'humanities' which are primarily concerned with the unique and non-recurrent, and the 'sciences' which seek to establish general statements for repeatable events and process. Contemporary geography obviously lies athwart this apparent gulf, which must either be bridged or must lead to the dismemberment of the existing discipline. The dichotomy between the general and the particular was clearly stated by Francis Bacon in his *Maxims of the Law*; 'For there be two contrary faults and extremities in the debating and sifting of the law, which may be noted in two several manner of arguments: some argue upon general grounds, and come not near the point in question; others, without laying any foundation of a ground or difference of reason, do loosely put cases, which, though they go near the point, yet being put so scattered, prove not, but rather serve to make the law appear more doubtful than to make it more plain'. Indeed, the distinction between the idiographic and nomothetic views of geography, so strongly put by Bunge (1962), may be useful in highlighting many of the current shortcomings in the subject, but is less valuable from the more purely philosophical standpoint. Bambrough (1964, p. 100), for example, points out that all reasoning is ultimately concerned with particular cases, and that laws, rules and principles are merely devices for bringing particular cases to bear on other particular cases. 'The ideal limiting case of representation is reduplication, and a duplicate is too true to be useful. Anything that falls short of the ideal limit of reduplication is too useful to be altogether true' (Bambrough, 1964, p. 98). In short, every individual is, by definition, different, but the most significant statement which can be made about modern scholarship in general is that it has been found to be intellectually more profitable, satisfying and productive to view the phenomena of the real world in terms of their 'set characteristics', rather than to concentrate upon their individual deviations from one another.

The nature of models

The catholic view of models taken in this volume derives largely from Skilling (1964). He argued that a model can be a theory or a law or an hypothesis or a structured idea. It can be a role, a relation or an equation. It can be a synthesis of data. Most important from the geographical viewpoint, it can also include reasoning about the real world by means of translations in space (to give spatial models) or in time (to give historical models).

The need for idealization. The traditional reaction of man to the apparent complexity of the world around him has been to make for himself a simplified and intelligible picture of the world. 'He then tries to substitute this cosmos of his own for the world of experience, and thus to overcome it' (Chorafas, 1965, p. 1). The mind decomposes the real world into a series of simplified systems and thus achieves in one act 'an overview of the essential characteristics of a domain' (Apostel, 1961, p. 15). This simplification requires both sensual and intellectual creativity (Keipers, 1961, p. 132). 'The mind needs to see the system in opposition and distinction to all others; therefore the separation of the system from others is made more complete than it is in reality. The system is viewed from a certain scale; details that are too microscopical or too global are of no interest to us. Therefore they are left out. The system is known or controlled within certain limits of approximation. Therefore effects that do not reach this level of approximation are neglected. The system is studied with a certain purpose in mind; everything that does not affect this purpose is eliminated. The various features of the system need to be known as aspects of one identical whole; therefore their unity is exaggerated' (Apostel, 1961, pp. 15–16). According to this view, reality exists as a patterned and bounded connexity which has been explored by the use of simplified patterns of symbols, rules and processes (Meadows, 1957, pp. 3-4). The simplified statements of this structural interdependence have been termed 'models'. A model is thus a simplified structuring of reality which presents supposedly significant features or relationships in a generalized form. Models are highly subjective approximations in that they do not include all associated observations or measurements, but as such they are valuable in obscuring incidental detail and in allowing fundamental aspects of reality to appear. This selectivity means that models have varying degrees of probability and a limited range of conditions over which they apply. The most successful models possess a high probability of application and a wide range of conditions in which they seem appropriate. Indeed, the value of a model is often directly related to its level of abstraction. However, all models are constantly in need of improvement as new information or vistas of reality appear, and the more successfully the model was originally structured the more likely it seems that such improvement must involve the construction of a different model.

Characteristics of models. The term 'model' is conventionally employed in a number of different ways. It is used as a noun implying a representation, as an adjective implying a degree of perfection, or as a verb implying to demonstrate or to show what something is like (Ackoff, Gupta and Minas, 1962, p. 108). In fact models possess all these properties.

The most fundamental feature of models is that their construction has involved a highly *selective* attitude to information, wherein not only noise but less important signals have been eliminated to enable one to see something of the heart of things. Models can be viewed as selective approximations which, by the elimination of incidental detail, allow some fundamental, relevant or interesting aspects of the real world to appear in some generalized form. Thus models can be thought of as selective pictures and 'a direct description of the logical characteristics of our knowledge of the external world shows that each of these pictures gives undue prominance to some features of our knowledge and obscures and distorts the other features that rival pictures emphasize. Each of them directs such a bright light on one part of the scene that it obscures other parts in a dark shadow' (Bambrough, 1964, p. 102). As Black (1962, p. 220) wrote of scale models, '... only by being unfaithful in *some* respect can a model represent its original'.

Another important model characteristic is that models are *structured*, in the sense that the selected significant aspects of the 'web of reality' are exploited in terms of their connections. It is interesting that what is often termed a model by logicians is called by econometricians a 'structure' (Suppes, 1961, p. 165; Kaplan, 1964, p. 267). Science has profited greatly from this *pattern seeking*, in which phenomena are viewed in terms of a kind of organic relationship.

This model feature leads immediately to the *suggestive* nature of models, in that a successful model contains suggestions for its own extension and generalization (Hesse, 1953-54, pp. 213-214). This implies, firstly, that the whole model structure has greater implications than a study of its individual parts might lead one to suppose (Deutsch, 1948-49), and, secondly, that predictions can be made about the real world from the model. Models have thus been termed 'speculative instruments', and Black (1962, pp. 232-233) has described a promising model as 'one with implications rich enough to suggest novel hypotheses and speculations in the primary field of investigation'. Similarly, Toulmin (1953, pp. 38-39) regards a good model as experimentally fertile, suggesting further questions, taking us beyond the phenomena from which we began, and tempting us to formulate hypotheses. The 'intuitive grasp' (*Gestalt knowledge*) of the capacities and implications of a model is thus the key to the exploitation of its suggestive character.

Selectivity implies that models are different from reality in that they are *approximations* of it. A model must be simple enough for manipulation and understanding by its users, representative enough in the total range of the implications it may have, yet complex enough to represent accurately the system under study (Chorafas, 1965, p. 31). In another sense, too, models represent compromises in that each has a circumscribed range of conditions within which it has relevance (Skilling, 1964, p. 389A).

Because models are different from the real world they are *analogies*. The use of hardware models is an obvious example of the general aim of the model builder to reformulate some features of the real world into a more familiar, simplified, accessible, observable, easily-formulated or controllable form, from which conclusions can be deduced, which, in turn, can be reapplied to the real world (Chorley, 1964, pp. 127–128).

Reapplication is a prerequisite for models in the empirical sciences. Although some mathematical model builders disclaim responsibility for the degree to which their idealizations may represent the real world, claiming that their responsibility is discharged completely and with honour if they avoid internal error (Camp, 1961, p. 22); most geographical model builders would judge the value of a model almost entirely in terms of its reapplicability to the real world.

The functions of models. Models are necessary, therefore, to constitute a bridge between the observational and theoretical levels; and are concerned with simplification, reduction, concretization, experimentation, action, extension, globalization, theory formation and explanation (Apostel, 1961, p. 3). One of their main functions is *psychological* in enabling some group of phenomena to be visualized and comprehended which could otherwise not be because of its magnitude or complexity. Another is *acquisitive*, in that the model provides a framework wherein information may be defined, collected and ordered. Models have not only an *organizational* function with respect to data, but also a *fertility* in allowing the maximum amount of information to be squeezed out of the data (see the 'statistical models' of Krumbein and Graybill, 1965). Models also perform a *logical* function by helping to explain how a particular phenomenon comes about. The question as to what constitutes a satisfactory explanation is a complex one, but Bridgman (1936, p. 63) put it in model terms when he wrote; 'Explanation consists of analysing our complicated systems into simpler systems in such a way that we recognize in the complicated systems the interplay of elements already so familiar to us that we accept them as not needing explanation'. Models also perform a *normative* function by comparing some phenomenon with a more familiar one (Hutton, 1953–54, pp. 285–286). The *systematic* function of model building has already been stressed in which reality is viewed in terms of interlocking systems, such that one view of the history of science is that it represents the construction of a succession of models by which systems have been explored and tested (Meadows, 1957, p. 3). This leads to the *constructional* function of models in that they form stepping stones to the building of theories and laws. Models and theories are very closely linked (Theobald, 1964, p. 260), perhaps differing only in the degree of probability with which they can predict reality. The terms 'true' or 'false' cannot usefully be applied in the evaluation of models, however, and must be replaced by ones like 'appropriate', 'stimulating' or 'significant'. Laws are statements of very high probability and, as such, all laws are models, but not all models are laws. Finally there is the *cognative* function of models, promoting the communication of scientific ideas. This communication 'is not a matter merely of the sociology of science, but is intrinsic to its logic; as in art, the idea is nothing till it has found expression' (Kaplan, 1964, p. 269).

Types of models. Chorley (1964) provided an initial structure for the classification of models currently used in geography and this 'model of models' has been expanded and revised with special reference to geomorphology in Chapter 3 (Fig. 3.1).

The term 'model' has been used, however, in such a wide variety of contexts that it is difficult to define even the broad types of usage without ambiguity. One division is between the *descriptive* and the *normative*; the former concerned with some stylistic description of reality and the latter with what might be expected to occur under certain stated conditions. Descriptive models can be dominantly *static*, concentrating on equilibrium structural features, or *dynamic*, concentrating on processes and functions through time. Where the time element is particularly stressed *historical* models result. Descriptive models may be concerned with the organization of empirical information, and be termed *data, classificatory (taxonomic),* or *experimental design* models (Suppes, 1962). Normative models often involve the use of a more familiar situation as a model for a less familiar one, either in a time (*historical*) or a *spatial* sense, and have a strongly *predictive* connotation.

Models can also be classed according to the stuff from which they are made, into, firstly, *hardware, physical* or *experimental* constructions, and, secondly, into *theoretical, symbolic, conceptual* or *mental* models. The former can either be *iconic* (Ackoff, Gupta and Minas, 1962), wherein the relevant properties of the real world are represented by the same properties with only a change in *scale,* or *analogue (simulation)* models, having real-world properties represented by different properties. The latter are concerned with symbolic or *formal* assertions of a *verbal* or *mathematical* kind in *logical* terms (Rosenblueth and Wiener, 1944–45, p. 317; Beament, 1960). Mathematical models can be further classed according to the degree of probability associated with their prediction into *deterministic* and *stochastic.*

Another view of models concentrates upon them as *systems* which can be defined on the basis of the relative interest of the model builder in the input/output variables, as distinct from the internal status variables. In order of decreasing interest in the status variables, many models can be viewed as *synthetic systems, partial systems* and *black boxes.*

The scale on which models are valuable and the standpoint from which they are constructed allow further distinctions, notably into *internalized* models which give a very parochial view of reality, and *paradigms* which are broadly significant models of value to a wide community of scholars.

Pitfalls in model building. The characteristics of models imply the existence of many dangers to which the model builder may fall prey. Simplification might lead to 'throwing the baby out with the bath water'; structuring to spurious correlation; suggestiveness to improper prediction; approximation to unreality; and analogy to unjustifiable leaps into different domains. Kaplan (1964, pp. 275–288) has summed up many of the dangers as problems of *overemphasis* on symbols, form, simplification, rigor and prediction. According to this view, a bad model would be heavily symbolic, present an overly-formalized view of reality, be much over-simplified, represent an attempt to erect a more exact structure than the data allows, and be used for inappropriate prediction.

Many philosophers have pointed to the dangers of craving for generality and of adopting a contemptuous attitude towards the particular case. They have often considered reality to be of too complex and multivariate a character to be susceptible to reasoning by analogy, and have asked whether the use of models introduces too great a detour into the reasoning process. In short, some hold that we should take heed of the Second Commandment: 'Thou shalt not make unto thee any graven image, or any likeness of anything that is in the heaven above, or that is in the earth beneath, or that is in the water under the earth'. In reply to this view of model building Ubbink (1961, p. 178) asks 'if this should be the case: in what sense can knowledge be true?'. Model building and reasoning are indissoluble, but 'the price of the employment of models is eternal vigilance' (Braithwaite, 1953, p. 93). Kaplan (1964, p. 276) believes that such vigilance is all the more necessary when model building is currently fashionable: 'The danger is all the greater with respect to model building because so much else in our culture conspires to make of it the glass of fashion and the mould of form. Models seem peculiarly appropriate to a brave new world of computers, automation and space technology, and to the astonishing status suddenly accorded to the scientist in government, industry and the military. It is easy to feel drawn to the wave of the future, and such tides are flowing strong today.'

The nature of paradigms

Paradigms may be regarded as stable patterns of scientific activity. They are in a sense large-scale models, but differ from models in the sense used above in that: (1) they are rarely so specifically formulated; and (2) they refer to patterns of searching the real world rather than to the real world itself. Scientists whose research is based on shared paradigms are committed to the same problems, rules and standards, i.e. they form a continuing community devoted to a particular research tradition. In a sense then, paradigms may be regarded here as 'super models' within which the smaller-scale models are set. As such, Thomas Kuhn in his *Structure of Scientific Revolutions* (1962), has assigned to the origin, continuance, and obsolescence of paradigms a prior place in the history of the evolution of science. Progress in research requires the continual discarding of outdated models, and subsequent remodelling. The more internally consistent the original paradigm or model, the more difficult it may be to remodel an existing structure in step with changing notions and increasing data. It is usual, therefore, for the most significant intellectual steps to be marked by the emergence of completely new models (Skilling, 1964, p. 389A). As Kuhn (1962, p. 17) has argued '... no natural history can be interpreted in the absence of at least some implicit body of intertwined theoretical and methodological belief that permits selection, evaluation, and criticism'. Without such paradigms all the available facts may seem equally-likely candidates for inclusion. As a direct consequence there is: (1) no case for the highly-defined, fact-gathering so typical of the exact sciences; and (2) a tendency to restrict fact-gathering to the wealth of available data which comes easily to hand. The fact that much of this data is a secondary by-product of administrative systems adds further to the massive data-handling problem. Certainly most geographical accounts are strongly 'circumstantial', juxtaposing facts of theoretical interest with others so unrelated or so complex as to be outside the bounds of available explanatory models.

The importance of the paradigm lies then, in Kuhn's terms, in providing rules that: (1) tell us what both the world – and our science – are like; and (2) allow us to concentrate on the esoteric problems that these rules together with existing knowledge define. Paradigms tend to be, by nature, highly restrictive. They focus attention upon a small range of problems, often enough somewhat esoteric problems, to allow the concentration of investigation on some part of the man-environment system in a detail and depth that might otherwise prove unlikely, if not inconceivable. This concentration appears to have been a necessary part of scientific advance, allowing the solution of puzzles outside the limits of pre-paradigm thinking.

In practice such 'rules' are acquired through one's education and subsequent exposure to the literature, rather than being formally taught. Indeed a concern about them only comes to the fore when there is a deep and recurrent insecurity about the nature of the existing paradigm. Methodological debates, concern over 'legitimate' problems or appropriate methods of analysis are symptomatic of the pre-paradigm period in the evolution of a science. Once the paradigm is fully established the debate languishes through lack of interest or lack of need. Thus in contemporary economics, the most successful and sophisticated of social sciences, the early debates over the nature of economics have been replaced by rather stable – but largely invisible – rules as to what problems and methods economic science should cultivate.

CLASSIFICATORY PARADIGMS IN GEOGRAPHY

Whatever the range of debate over the purpose and nature of geography, there is considerable communality of practice in the ways in which geographers have tackled their problems. Berry (1964) has analysed this paradigm of practice in terms of alternative approaches to a 'geographical data-matrix'. Here we look at his findings and attempt to diagnose the widespread unease generated by the continued use of this classificatory approach.

The geographical data matrix

Although regional geography, systematic geography, and historical geography are regarded as being quite distinct types of geographical study, Berry (1964) has deftly illustrated that each may be regarded merely as a different axis of approach to the same basic geographical data-matrix.

If a matrix has only *one* column, it is commonly called a 'column vector' (Krumbein and Graybill, 1965, p. 251), in which may be stored a series of j bits of information:

$$\begin{pmatrix} a_{11} \\ a_{12} \\ a_{13} \\ a_{14} \\ \cdot \\ a_{1j} \end{pmatrix}$$

Similarly we may store information about j elements (i.e. temperatures, elevations, population densities, etc.) in a regional column, to give an inventory of all the available characteristics of a given location.

A matrix with only *one* row is termed a 'row vector'. Here we may store a series of i bits of information:

$$(a_{11} \quad a_{21} \quad a_{31} \quad a_{41} \quad \cdot \quad a_{i1})$$

In this approach we store information about the same element but we vary the location to give the standard pattern of systematic geography, i.e. the mapping of a single feature (e.g. population densities).

By combining both the set of regions ($1 \ldots i$) and the set of elements ($1 \ldots j$) we have a rectangular array of the form:

$$\begin{pmatrix} a_{11} & a_{21} & a_{31} & a_{41} & \cdot & a_{i1} \\ a_{12} & a_{22} & a_{32} & a_{42} & \cdot & a_{i2} \\ a_{13} & a_{23} & a_{33} & a_{43} & \cdot & a_{i3} \\ a_{14} & a_{24} & a_{34} & a_{44} & \cdot & a_{i4} \\ \cdot & \cdot & \cdot & \cdot & & \cdot \\ a_{1j} & a_{2j} & a_{3j} & a_{4j} & \cdot & a_{ij} \end{pmatrix}$$

This matrix or box is termed by Berry the *geographical data-matrix* in that items containing information about the earth's surface may be stored in terms of their *regional* (or locational) characteristics and their *elemental* (or substantive) characteristics. Table 1.1A gives a formal example of this sort of matrix, and Grigg (Chap. 12, below) discusses its logical basis in terms of regional models.

TABLE 1.1

Transformation of Vectors in Geographical Data Matrices*

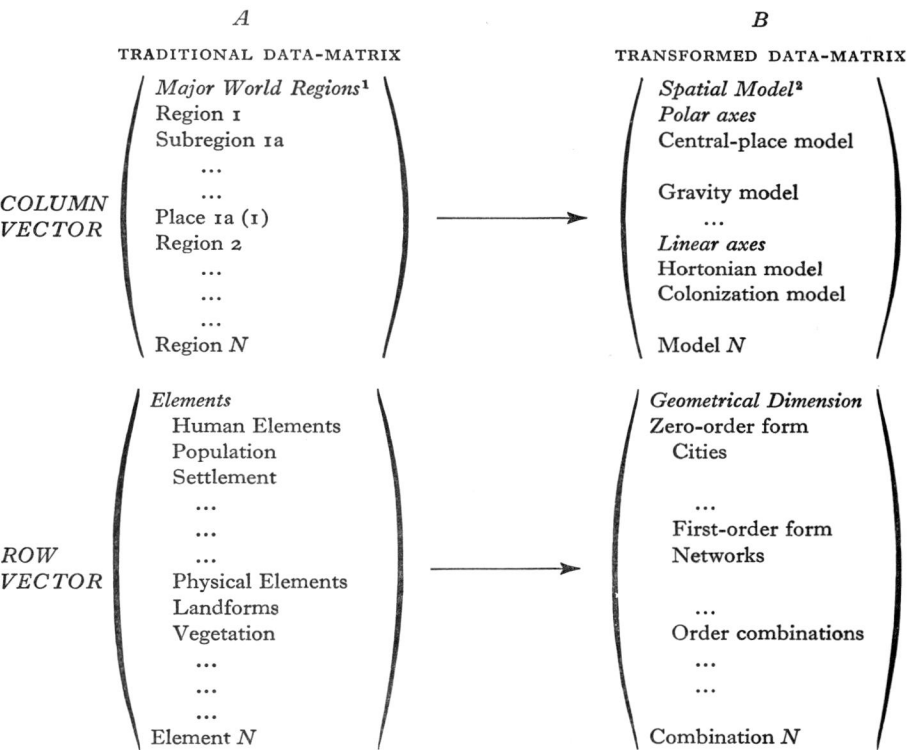

By an ingenious series of row and column comparisons and by the addition of a third (time) dimension, Berry is able to reduce the great part of conventional geographical study to ten basic operations on the matrix. For example, areal differentiation is seen as column-vector comparisons and spatial covariance studies as row-vector comparisons. Comparison of a column over

* Adapted from Berry (1964, pp. 6–8).
[1] Information located with respect to *absolute* location (X, Y co-ordinates measured from a common base, e.g. latitude and longitude).
[2] Information plotted with respect to relative *location* (x, y co-ordinates measured from a variable base, e.g. distance and direction from a diffusion hearth).

time becomes sequent occupance, while other manipulations give the major modes of historical geography distinguished by Darby (1953). Indeed the major – and apparently fundamental – contrasts between regional, systematic and historical geography are seen by Berry (1964, p. 9) largely as a function of the relative length, breadth and depth of the study in terms of the three axes of the matrix.[1]

Difficulties of conventional matrix operations

The need for the information in the data matrix to be structured, given coherence, generalized and made intelligible has long been recognized, and as Wrigley (1965) points out, both Von Humboldt and Ritter scorned the attitude of their predecessors which had reduced geographical studies in the eighteenth century to mere 'pigeon-holing'. There is growing evidence however that neither of the two major solutions – the study of column vectors and of row vectors – effectively meet present-day demands.

(a) *The explosion of the data matrix.* The exact rate of growth of 'information' that could conceivably be stored in a geographical data-matrix is difficult to assess exactly. We have useful figures on the speed of mapping for selected areas (Langbein and Hoyt, 1959), and are familiar with the rapid accelerations caused by the development of air-photography after World War I and of satellite scanning and remote sensing since World War II. If we add to this the growing volume of statistical material collected by international agencies, national governments, state and local administrators, then the size of the potential world 'data bank' becomes staggering. If we restrict our data bank to one fact about each square-mile unit of the earth's surface then we have a basic store of 10^9 bits of information. If we relax this assumption to include the range of possible parameters for each unit (ranging, say, from a minimum of 1 to a maximum of 10^{10}) and a more finely-divided world grid (say 10^{11} units), then we may well have a storage problem of the order of 10^{11} to 10^{21} bits on our hands.

This is of course a speculative calculation, but what gives any such figure added point is the general rate of growth of information. Price (1963, pp. 4–13), after a wide-ranging survey of the growth of scientific information, found evidence of an exponential growth of 'impressive consistency and regularity' – that is, the more information that exists the faster it grows. Depend-

---

[1] In more recent, and as yet unpublished work, Berry has developed a second data-matrix (an interaction matrix) in which pairs of locations (*dyads*) occupy the rows and interactions occupy the columns. By reducing the two matrices through factor analysis, the general relationships between spatial structure (Matrix I) and spatial interaction or behaviour (Matrix II) may be explored. Through this extension Berry has not only clarified the relationships between formal and functional regions, but has also laid the basis for a much more general 'field-theory' of spatial behaviour.

ing upon what we measure, it is possible to estimate that the amount of information tends to double within a period of 10 to 15 years – probably slightly shorter. If one accepts the general form of the curve then the problem facing Humboldt and Ritter was about 1/1,000 times as small as that facing the current generation of geographers. The fact that large areas of the world were still then 'blank on the map' suggests this is in fact a sizeable overestimate; that is, the rate of growth of locationally 'storable' information has been *more* rapid than that of scientific information as a whole as the ecumene itself has expanded. Stoddart (1967) has suggested that, although the intensity of geographical data generation and handling over the past 200 years has been roughly half that of science as a whole (Fig. 1.2), there are indications of a current increase of activity.

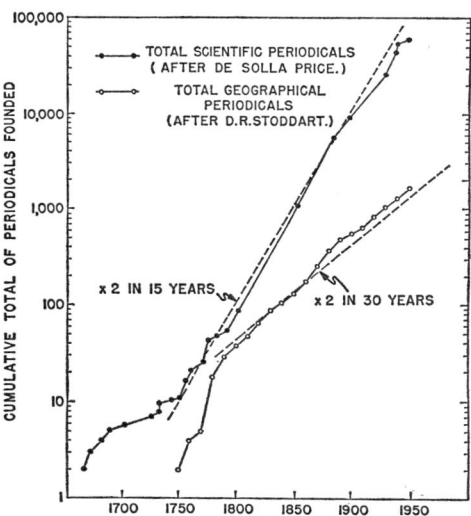

1.2 Cumulative totals of scientific and geographical periodicals founded. The dashed lines represent doubling in 15 and 30 years, respectively. (*Source: D. R. Stoddart. See also Stoddart, 1967*).

(b) *The weakening of conventional vector analysis.* Continuing Berry's (1964) matrix algebra analogy, we may trace the problems in columnar analyses to: (1) the expanding number of locational columns as old areas were ever more finely divided and new areas added; and (2) the proven failure of local 'column' factors to explain the juxtaposition of other variables in the same column. Wrigley (1965) has shown, for example, how the development of general production techniques during the Industrial Revolution tended to destroy unique regional features and to make the geographical characteristics of these regions less explainable in purely local terms. Analysis of row vectors (i.e. systematic distribution studies) has been weakened by the tendency for individual systematic sciences to 'split off' their own rows for separate analyses. It is significant, for example, that systematic sciences are increasingly taking on their own mapping analysis – and in some cases (e.g. Perring and Walters, 1962) are themselves making substantial contributions to cartographic technique.

Computer-orientated analysis of data matrices

One trend which runs counter to the information explosion is the increasing adaptation of digital computers to this problem. The whole field of data storage, data retrieval, data analysis, and data display is expanding rapidly and we may expect that geography, as a special type of *regional data-storage system*, will reap considerable benefits from this field of electronic technology.

Strides at the moment are being made in reducing the size and complexity of the data matrix through general data-evaluation and pattern-search programmes, some of which involve factor-analysis, cluster analysis and related multivariate techniques (Krumbein and Graybill, 1965, Chap. 15). WHIRLPOOL is a typical example of a powerful sequential multiple linear regression programme (Krumbein, Benson and Hempkins, 1964) for 'sorting out' significant factors in the series of elements in the matrix. An alternative approach to the regional columns of the data matrix is being led by Berry (1961; Steiner, 1965) in using a linked series of factor analysis, D^2 analysis, and discriminant analysis to 'optimize' regional divisions; thereby 'collapsing' the number of regional vectors into the smallest or most convenient number consistent with a given level of information (Haggett, 1965, p. 256). Other approaches have concentrated on the collapsing of the detailed isarithmic map into a series of *regional trends*. Chorley and Haggett (1965) have shown how orthogonal polynomials may be used to store vast amounts of regional information on the undulating isarithmic surface in terms of a very few basic equations, while Tobler (1966) has analysed other forms of numerical mapping and rapid map display.

It is clear then that technological improvements, largely in computer engineering and programming logic, are at last beginning to restore to the greatly-strained geographical data-matrix some of the order that its rapid expansion threatened to destroy. While empirical approaches to the stored material are likely to throw up a good deal of unexpected patterning, we should recall that these programmes are '. . . greatly facilitated when preceded by analysis of the problem in terms of some conceptual model – a mental picture – that defines the class of objects or events to be studied, the kinds of measurements to be made, and the properties or attributes of these measurements' (Krumbein and Graybill, 1965, p. viii). It is to these 'mental pictures' that we turn again in our last section.

TOWARDS A MODEL-BASED PARADIGM OF GEOGRAPHY

The continuing problems of the traditional class-based paradigm of geography suggest that experimentation with alternative approaches might prove fruitful. Here we: (1) attempt to isolate some of the components in a model-based paradigm; (2) discuss the impact it might have on research; and (3) inquire whether it meets certain basic requirements that a new 'candidate' should possess.

A model-based paradigm: a proposal

We do not propose to alter the basic Hartshorne (1959, p. 11) definition of geography's prime task, nor challenge the appropriateness of the matrix concept. We suggest, however, that it may be possible to derive a *second* matrix from the first by transforming the two basic vectors in such a way as to throw emphasis away from that of classification towards model-building. Table 1.1B gives an example of such a transformed matrix.

(a) *Vector I: locational relativity.* The search for greater accuracy along the absolute locational vector (e.g. exact latitude and longitude), and the search for significant 'fixes' in this absolute regional space (e.g. the location of the source of the Nile) is now largely centred in specialized government mapping agencies outside the university research world. However, location in an absolute sense on an isotropic surface is clearly only one way of viewing space. Bunge (1962) has urged that quite different and alternative ways are both possible and common, e.g., most ideas on *accessibility* or *isolation* refer to distance measured in a specialized way (usually in terms of *energy* translated into cost terms) and from specialized origins or axes. The selection of these *axes of symmetry* is largely determined by the spatial model being adopted: in a diffusion model the appropriate axis might be the diffusion *hearth*, and the appropriate parameter distance (accessibility) from that axis. Many of the models featured in later chapters use just such relativistic ideas of location with space measured in metric terms (see Henshall's discussion on Von Thünen's model in Chap. 11, below) or topological terms (see Haggett's review of Pitt's accessibility model in Chap. 15, below), with the measurements taken from single points, sets of points, or linear axes (e.g. Horton's model of erosion in terms of distance from the watershed). Attempts to use *relative* location rather than absolute location are a feature of spatial model-building. Although no formal attempts have yet been made to build up a 'general theory of locational relativity', the multivariate analysis of distance (measured in 'real' energy terms rather than 'neutral' *mileage* terms) and their reduction to suitable display maps pose fascinating and complex research problems. A whole new family of atlases might be envisaged showing completely different spatial patterns than the familiar absolute patterns of traditional maps. As Bunge (1964) has pointed out, on a cost-transformed surface pre-railway England reduces to coalfields grouped around an *inland* sea!

(b) *Vector II: topological-geometrical form.* The successive sub-division of the 'elements' vector has seen the constant departure of newly fledged sciences with their own geographical interpretations. Here we seem to have missed a fundamental point that these elements (e.g. vegetation, population) are not *in themselves* the objects of geography. Sten de Geer (1923, p. 2) argued that '... it is certain abstract qualities in the objects which are studied by Geography, and not the objects themselves'. We propose therefore that certain 'abstract properties', – the topological and geometrical form of that object or objects – be substituted for the standard 'element-class' vector. Cole (1966) has extended point-set theory to suggest that using a simple dimensional classification into zero-order (points, cities), first-order (lines, networks), second-order (areas, states), and third-order (surfaces, terrain) provides such an appropriate topological vector. Bunge (1962, p. 197) concludes his *Theoretical Geography* on a similar vein: 'the profession might, as a matter of efficiency, start dividing itself into various theoretical spatial fields, such as point problems, area problems, description of mathematical surfaces, and central place problems, rather than the current arrangement of climatology, population geography, landforms, etc.' Despite the apparent simplicity of this schema it can be rigorously extended (by set combinations) to include complex combinations of the basic topological forms: thus the four basic dimensions give fifteen combinations (Cole, 1966; see his Fig. 3.1). If we divide each of the four into only two time states (*stable* and *unstable*) then the number of possible combinations goes up to over 250. We have argued elsewhere that '.... many of the more successful attempts at geographical models have stemmed from this (geometrical) type of analysis' (Chorley and Haggett, 1965, p. 376) and will not repeat those arguments in detail. One such topological class – the linear network – is currently being investigated (see Haggett, Chap. 15, below; Haggett and Chorley, In preparation). In general we feel that geometrical analysis offers a logical, consistent and geographically more relevant alternative to the 'element-orientated' approach with its inevitable tendency to sub-divide geography and force it outwards towards the relevant external systematic disciplines. It not only offers a chance to weld human and physical geography into a new working partnership, but revives the central role of cartography (see Board, Chap. 16, below) in relation to the two.

Bernal (1965, p. 97) traces the change in biology through phases when it was 'primarily a descriptive science, more like geography, dealing with the structure and working of a number of peculiarly organized entities, at a particular moment of time on a particular planet', through its fission into zoology, botany and their subdivisions, and its present convergence towards common concern with chemistry and physics in molecular biology, cell biology, biophysics, biochemistry, etc. Geography has likewise shown a tendency to replace its first phase of fact gathering with a fission into physical and human geography (and their subdivisions). Whether there are signs within contemporary geography of a third converging stage are matters of debate. Certainly the general concept of systems analysis (Chorley, 1962) has been put forward by Ackerman (1963) as a fundamental integrating concept for geography, and Stoddart (1965; see also Chap. 13, below) has analysed its particular role in relation to ecological systems (ecosystems). Whatever the problems that remain to be solved, particularly problems of system identification and energy monitoring, we may expect that regional systems analysis will emerge as a major theme in geographical work over the next decade. It is not without interest that two of the most tractable systems identified and analysed by geographers – the watershed system (see Chorley, Chap. 3, below: and More, Chap. 4, below) and the nodal 'city region' (Garner, Chap. 9) – are distinctive geometrical forms with particular mathematical properties (e.g. nesting) and are organized with respect to specific axes from which relative distances are measured. That is, we may argue that distinctive *geosystems* (Stoddart, Chap. 13, below) are likely to be found at the intersections of distinctive vectors within our transformed data matrix.

It would be irrelevant here to pursue the applications of our transformation model to detailed cases. Table 1.2 (overleaf) is included to show the organization used by one of the writers in dealing with a typical nodal-system in human geography: it is based largely on topological distinctions (i.e. Vector I organization).

Implications for research

Kuhn (1962) has shown that the period which follows the wholesale adoption of a paradigm is commonly devoted to three main classes of research problem: (1) the determination of significant facts; (2) the matching of facts with theory; and (3) the further articulation of theory.

(1) One of the curious effects of paradigm adoption is to replace a period of data abundance with one of data scarcity. The very definition of the major research problem highlights the paucity of *relevant* data and great attention is paid to the most accurate measurement of constants that would have seemed, in the pre-paradigm period, to be either too recondite or too elusive to measure. Thus a study of urban-centred fields has led to the most careful study of city sizes using stringent operational criteria, and the need for observational networks of a new rigour and accuracy makes increasing demands on research time. With further concentration we might expect that problems

TABLE 1.2

Three-Stage Model for the Analysis of Regional Systems*

	STAGE I *System Identification*	STAGE II A. *Form Differentiation (Static)*			B. *Form Differentiation (Dynamic)*	STAGE III *System Integration*
Dimensional Number	$\{0, 2\}$	0	1	3	4	$\int_4^0$
Geographical Form	City (*polar axis*) City region (*boundaries*)	Cities Settlements Urban hierarchies	Transport networks Communication systems	Urban fields Density gradients Land-use intensity	Innovation waves Frontier movements Sequent occupance Colonization	Regional systems Internal feedbacks Interregional systems External feedbacks
Analytical Techniques	Numerical taxonomy Local residuals Regional analogues	Rank-size analysis Nearest-neighbour analysis Quadrat analysis	Graph-theoretic analysis Connectivity Network geometry	Trend-surface analysis Harmonic analysis Fourier analysis	Physical simulation Monte Carlo models Markov-chain models	Matrix analysis, factor analysis Input-output analysis Interregional linear programming
Spatial Model	Regional hierarchies Formal, functional regions	Central-place theory Gravity models Weberian models Basic non-basic models	Network models Random graph models Geodesic models	Gravity models Absorption models Intervening-opportunity models Von Thünen models Potential models	Diffusion models Migration models Colonization models	Regional climax models Regional multipliers Growth poles
Major Sources for Spatial Models	Decision theory (Psychol.) Taxonomy (Biol.) Discriminant analysis (Stats.)	Point set theory (Math.) Organization models (Manag.) Packing theory (Math.)	Graph theory (Math.) Circuit design (Electr.) Search theory (Math., Psychol., Zool.)	Least-effort models (Sociol.) Minimum-energy models (Phys.) Potential models (Phys.) Game theory (Psychol.)	Epidemic theory (Medic.) Diffusion theory (Fluid dyn.) Rumour theory (Sociol.) Colonization & succession models (Bot.)	General Systems theory Ecosystems (Biol.) Interregional trade theory (Econ.) Multiplier models (Econ.)

* The table is schematic only: it follows the general arrangement of topics in Haggett (1965), Chaps. 2–6 inclusive.

of interregional flow and internal energy flux would put data collection and data standardization as major consumers of research time. In the case of physical sciences such problems are commonly accompanied by the design and construction of measuring apparatus of increasing precision and complexity. The analogy in non-experimental sciences may well be the development of computer programmes for squeezing the greatest amount of information from limited data.

(2) Determination of facts that can be directly related to paradigm models form a second class of activity (Kuhn, 1962, p. 33). Although such facts cannot be said to test the model in any strict sense, they provide a focus for research areas. Thus Gunawardena (1964) directed attention at the central-place structure of southern Ceylon in an attempt to test existing central-place models. Much of the stimulus that a model provides is the drive to test (and possibly overthrow) existing models. The strength of the paradigm is that it allows the progressive evolution of models within its general terms. Kuhn argues that only when no new models (or wholly incompatible new models) can be produced is the paradigm itself likely to shift.

(3) Extension and articulation of theory therefore provides the third class of research. This may take a number of forms but will probably include: (a) the determination of constants in existing predictive equations; (b) the quantification or further mathematization of existing qualitative models; and (c) the speculative and exploratory extension of extant models into areas of unproven and even unlikely application. Olsson's (1965) study of migration rates might fall into the first category, Dacey's (1965) elegant extension of Christaller-Lösch settlement models through point-set algebra into the second, and Gould's (1965) extension of search theory to the extension of transport networks in East Africa into the third.

According to the Kuhn thesis none of these major types of research is designed to produce results entirely outside the paradigm's limits, and the range of acceptable results from the studies is small – certainly small in relation to the results that could be conceived. The internal discipline of the paradigm, its unwritten rules and traditions, guides the pattern of research and ensures by the successive cumulation of small highly-limited advances that the science as a whole will progress.

Needed characteristics of a new paradigm

How should we recognize an efficient paradigm if we saw one? Kuhn (1962, pp. 152–158) has studied the introduction of new paradigms in fields as unlike as chemistry and electricity, and suggests three minimum ingredients for success.

(1) Firstly, the new paradigm must be able to solve at least some of the problems that have brought the old one to crisis point. We have argued above

that the most sorely-troubling feature of present geography is the explosion of the traditional data-matrix, and the forcing of geographers to study both areas and topics less and less relevant to the general shaping of the earth surface they profess to study. A new paradigm in geography must be able to rise above this flood-tide of information and push out confidently and rapidly into new data-territories. It must possess the scientific habit of seeking for relevant pattern and order in information, and the related ability to rapidly discard irrelevant information: '. . . it is the capacity for pattern-seeing and not the actual surveying of the landscape which explains this rapidity' (of scientific development). 'It explains why scientific activity remains a mystery for those devoid of theoretical insight, who see only facts' (Van Duijn, 1961, p. 67).

(2) Secondly, the new paradigm must appeal to the workers' sense of what is elegant, appropriate and simple. This somewhat aesthetic characteristic is difficult to define in specific terms and is most clearly seen in the mathematician's attraction towards elegant rather than inelegant mathematical proofs. At the smaller scale, the demonstration that Horton's law of stream numbers was a simple combinatorial system applicable to a much wider range of phenomena (Haggett: Chap. 15, below) represents an appropriate and economical simplification. Similarly Hägerstrand's (1953) overview of diffusion waves replaced many separate, clumsy, and individually articulated 'frontier' and 'sequent occupance' studies. A new paradigm for geography needs to provide a similar economical and elegant simplification for the *whole* field.

(3) Thirdly, the new paradigm must contain more 'potential for expansion' than the old. This characteristic is believed by Kuhn to be often the decisive one, albeit its adoption is based on faith in the new rather than its proven ability. Geography, coming late to the paradigm race, has the compensating advantage that it can study at leisure the 'take-off' paradigms of other sciences. There is good reason to think that those subjects which have modelled their forms on mathematics and physics – themselves 'leading sectors' in the scientific community (to continue Rostow's language) – have climbed considerably more rapidly than those which have attempted to build internal or idiographic structures. Not a little of this success stems from the great elasticity of mathematical analysis and the hierarchy of ever-more-complex equations that can be derived for observed patterns. As Kaplan (1964, p. 262) remarked: 'The use of mathematics and the construction of logical systems marks a certain coming of age'.

EPILOGUE

In this introductory chapter we have set out to examine the nature of models and their relation to facts on the one hand and to paradigms on the other. We have looked at the traditional paradigmatic model of geography and suggested that it is largely classificatory and that it is under severe stress. We have tentatively suggested an alternative model-based approach.

In judging the success of this approach and of the models discussed in the chapters which follow, we should recall that geography must measure its progress by the number of puzzles it has effectively solved, not by the magnitude of those that remain unsolved. In welcoming Ackerman's (1963, p. 435) reminder that the philosophical goal of geography is '... nothing less than an understanding of the vast, interacting system comprising all humanity and its natural environment on the surface of the earth' we should recall, with Humboldt, that such a goal is utterly unattainable in any complete sense – either now or in the future. Successful application of models in geography ensures no teleological progress towards full understanding, for scientific effort does not reduce the sum total of problems to be solved – it rather increases them.

REFERENCES

ACKERMAN, E. A., [1963], Where is a research frontier?; *Annals of the Association of American Geographers*, 53, 429-440.
ACKOFF, R. L., GUPTA, S. K. and MINAS, J. S., [1962], *Scientific Method: Optimizing Research Decisions*, (New York), 464 pp.
APOSTEL, L., [1961], Towards the formal study of models in the non-formal sciences; In Freudenthal, H., (Ed.), *The Concept and the Role of the Model in Mathematics and Natural and Social Sciences*, (Dordtrecht, Holland), 1-37.
BAMBROUGH, R., [1964], Principia Metaphysica; *Philosophy*, 39, 97-109.
BEAMENT, J. W. L., (Ed.), [1960], *Models and Analogues in Biology*; Symposium No. 14 of the Society for Experimental Biology, (Cambridge), 255 pp.
BERNAL, J. D., [1965], Molecular structure, biochemical function, and evolution; In Waterman, T. H. and Morowitz, H. J., (Eds.), *Theoretical and Mathematical Biology*, (New York), 96-135.
BERRY, B. J. L., [1961], A method for deriving multifactor uniform regions; *Przeglad Geograficzny*, 33, 263-282.
BERRY, B. J. L., [1964], Approaches to regional analysis: a synthesis; *Annals of the Association of American Geography*, 54, 2-11.
BLACK, M., [1962], *Models and Metaphors*, (Ithaca, New York), 267 pp.
BRAITHWAITE, R. B. [1953] *Scientific Explanation*, (Cambridge).
BRAITHWAITE, R. B., [1962], Models in the empirical sciences; In Nagel, E., Suppes, P. and Tarski, A., (Eds.), *Logic, Methodology and Philosophy of Science*, (Stanford), 224-231.
BRIDGMAN, P. W., [1936], *The Nature of Physical Theory*, (Princeton).
BROWN, L., [1965], Models for spatial diffusion research: a review; *Office of Naval Research, Geography Branch, Contract Nonr* 1288 (33), *Technical Report*, 3.

BUNGE, W., [1962], Theoretical geography; *Lund Studies in Geography, Series C, General and Mathematical Geography*, 1.
BUNGE, W., [1964], Geographical dialectics; *Professional Geographer*, 16 (4), 28–29.
CAMP, G. D., [1961], Models as approximations; In Banbury, J. and Maitland, J., (Eds.), *Proceedings of the Second International Conference on Operational Research*, (Aix-en-Provence), 20–25.
CAWS, P., [1965], *The Philosophy of Science*, (Princeton), 354 pp.
CHORAFAS, D. N., [1965], *Systems and Stimulation*, (New York), 503 pp.
CHORLEY, R. J., [1962], Geomorphology and general systems theory; *U.S. Geological Survey, Professional Paper* 500-B, 10 pp.
CHORLEY, R. J., [1964], Geography and analogue theory; *Annals of the Association of American Geographers*, 54, 127–137.
CHORLEY, R. J. and HAGGETT, P., [1965], Trend-surface mapping in geographical research; *Transactions of the Institute of British Geographers*, No. 37, 47–67.
COLE, J. P., [1966], Set theory and geography; *Nottingham University, Department of Geography, Bulletin of Quantitative Data*, 2.
DACEY, M. F., [1965], The geometry of central place theory; *Geografiska Annaler*, 47B, 111–124.
DARBY, H. C., [1953], On the relations of geography and history; *Transactions of the Institute of British Geographers*, No. 19, 1–13.
DEUTSCH, K. W., [1948–49], Some notes on research on the role of models in the natural and social sciences; *Synthèse*, 7, 506–533.
GEER, S. DE, [1923], On the definition, methods and classification of geography; *Geografiska Annaler*, 5, 1–37.
GOULD, P., [1965], *A bibliography of space-searching procedures for geographers*; Pennsylvania State University, Department of Geography (Mimeographed).
GUNAWARDENA, K. A., [1964], Service centres in southern Ceylon; *University of Cambridge, Ph.D. Thesis*.
HÄGERSTRAND, T., [1953], *Innovationsförloppet ur korologisk synpunkt*, (Lund).
HAGGETT, P., [1965], *Locational Analysis in Human Geography*, (London).
HAGGETT, P. and CHORLEY, R. J., [1965], Frontier movements and the geographical tradition; in Chorley, R. J. and Haggett, P., (Eds.), *Frontiers in Geographical Teaching*, (London), 358–378.
HAGGETT, P. and CHORLEY, R. J., (In preparation), *Network models in geography*, (London).
HANSON, N. R., [1958], *Patterns of Discovery*, (Cambridge), 241 pp.
HARTSHORNE, R., [1959], *Perspective on the Nature of Geography*, (London).
HUTTON, E. H., [1953–54], The role of models in physics; *British Journal of the Philosophy of Science*, 4, 284–301.
HESSE, M., [1953–54], Models in physics; *British Journal of the Philosophy of Science*, 4, 198–214.
KAPLAN, A., [1964], *The Conduct of Inquiry*, (San Francisco), 428 pp.
KRUMBEIN, W. C. and GRAYBILL, F. A., [1965], *An Introduction to Statistical Models in Geology*, (New York).
KRUMBEIN, W. C., BENSON, B. and HEMPKINS, W. B., [1964], WHIRLPOOL: a computer programme for 'sorting out' independent variables by sequential

multiple linear regression; *Office of Naval Research, Geography Branch, Technical Report* 14, Task No. 389–135.
KUHN, T. S., [1962], *The Structure of Scientific Revolutions*, (Chicago).
KUIPERS, A., [1961], Model and insight; in Freudenthal, H., (Ed.), *The Concept and the Role of the Model in Mathematics and Natural and Social Sciences*, (Dordrecht, Holland), 125–132.
LANGBEIN, W. B. and HOYT, W. G., [1959], *Water Facts for the Nation's Future*, (New York), 228 pp.
LEWONTIN, R. C., [1963], Models, mathematics and metaphors; *Synthèse*, 15, 222–244.
MANLEY, G., [1966], A new geography; *The Guardian*, March 17th, 1966.
MEADOWS, P., [1957], Models, system and science; *American Sociological Review*, 22, 3–9.
NAGEL, E., [1961], *The Structure of Science*, (London), 618 pp.
OLSSON, G., [1965], Distance and human interaction: a bibliography and review; *Regional Science Research Institute, Bibliography Series*, 2.
PERRING, F. H. and WALTERS, S. M., [1962], *Atlas of the British Flora*, (London).
PORTER, P. B., [1954], Another puzzle-picture; *American Journal of Psychology*, 67, 550–551.
PRICE, D. J. DE SOLLA, [1963], *Little Science, Big Science*, (New York).
ROSENBLUETH, A. and WIENER, N., [1944–45], The role of models in science; *Philosophy of Science*, 11–12, 316–321.
SKILLING, H., [1964], An operational view; *American Scientist*, 52, 388A–396A.
STEINER, D., [1965], A multivariate statistical approach to climatic regionalization and classification; *Tijdschrift van het Koninklijk Nederlandsch Aardrijkskundig Genootschap*, 82, 329–347.
STODDART, D. R., [1965], Geography and the ecological approach: the ecosystem as a geographic principle and method; *Geography*, 50, 242–251.
STODDART, D. R., [1967], Growth and structure of geography; *Transactions of the Institute of British Geographers*, 41.
SUPPES, P., [1961], A comparison of the meaning and uses of models in mathematical and empirical sciences; In Freudenthal, H., (Ed.), *The Concept and the Role of the Model in Mathematics and Natural and Social Sciences*, (Dordrecht, Holland), 163–177.
SUPPES, P., [1962], Models of data; In Nagel, E., Suppes, P. and Tarski, A., (Eds.), *Logic, Methodology and Philosophy of Science*, (Stanford), 252–261.
THEOBALD, D. W., [1964], Models and method; *Philosophy*, 39, 260–267.
TOBLER, W., [1966]. Numerical map generalization; *Michigan Inter-University Community of Mathematical Geographers, Discussion Papers*, 8.
TOULMIN, S., [1953], *The Philosophy of Science*, (London).
UBBINK, J. B., [1961], Model, description and knowledge; In Freudenthal, H., (Ed.), *The Concept and the Role of the Model in Mathematics and Natural and Social Sciences*, (Dordrecht, Holland), 178–194.
VAN DUIJN, P., [1961], A model for theory finding in science; *Synthèse*, 13, 61–67.
WRIGLEY, E. A., [1965], Changes in the philosophy of geography; In Chorley, R. J. and Haggett, P., (Eds.), *Frontiers in Geographical Teaching*, (London), 3–20.

CHAPTER TWO

The Use of Models in Science

F. H. GEORGE

INTRODUCTION

The idea of using models in science is by no means new, and Braithwaite, among many others, has analysed the use of such models and their relation to theories in science.

There is a sense in which almost anything can be used as a model for almost anything else, but as in the use of analogies or metaphors, so models, to have predictive value, must bear some measure of similarity to the structure or process being modelled.

We can classify models in various ways into at least those which model structure, those which model function and those that model both. These are *static* (for structure) and *dynamic* models (for processes or functions). We also have models which are obviously physical (e.g. a wind tunnel model) and those which are obviously symbolic (e.g. mathematical and statistical models for simulation), and so on and so forth.

Following the line of thought of Braithwaite, we might regard scientific theories as capable of being *formalized*. This process of formalization is essentially one of stripping down theories in ordinary language and showing the underlying logic, it is almost indeed a matter of making the original theory more precise, where we take some theory and rewrite in greater detail and in more 'molecular' fashion.

We can also argue that a scientific theory is made up of empirical statements which use terms capable of operational definition, whereas the model uses logical statements, which are not in themselves verifiable.

We can illustrate our general meaning by a simple example. This example shows one sort of formalization:

In Hull's theory of learning, we find a statement to the effect that:

> *the greater the similarity between the conditioned stimulus and the unconditioned stimulus, the greater the absolute value of the increment of tendency to respond to the conditioned stimulus.* (1)

To formalize this needs a lot of detailed definitions, but if s is a stimulus and r is a response, then the strength of association $_sH_r$ can be defined:

$$_sH_r(t_2) = {_sH_r(t_0)} + \Sigma\delta\ _sH_r(t_2, s', r') \qquad (2)$$

where (t_0, t_2) is a time interval, and $\Sigma\delta\ _sH_r$ is the change of strength of association during the interval (t_0, t_2), s' and r' are sets of s's and r's that occur in the interval (t_0, t_2).

We now define a function $F(x, y, z, w)$ such that:

$$F(x, y, z, w) = \delta\ _sH_r(t_2, s', r') \qquad (3)$$

and where $x = \mathcal{J}(t_1, t_2)$, $t_1 = T(r')$, $y = S(s, s')$, $z = S(r, r')$ and $w = T(r') - T(s')$.

Suffice it for our purposes that the variables x, y, z and w can be precisely defined, but roughly speaking x is concerned with drive reduction, and y and z are concerned with stimulus and response similarity, and finally w is a function of time delay between the occurrence of the stimulus and the response.

This leads to a statement (or axiom):

$$\frac{\partial F}{\partial y}(x, y, z, w) > 0 \qquad (4)$$

and this total argument, given here in skeleton form, is a *formalization* of Hull's original statement (1). This is formalized in (2), (3) and (4), although even this is barely adequate as an approximation to the much more precise statement still required.

Another kind of example we might mention is that of the Russell and Whitehead's formalization of number in *Principia Mathematica*.

By 'formalizing' then, we mean 'making more precise', or 'reducing statements' to their underlying logic, and clearly 'formalizing' is itself a vague word, which could well mean 'to expose the model underlying the theory'. Not only is formalizing vague, but it is also a matter of degree.

Similarly we can look at the matter the other way around, and *interpret* models in many different ways to get many different theories. Models and theories are connected by formalizing *and* by interpretation, and theories are linked directly with emperical descriptions. Let us now look at an example of interpretation e.g. the well-formed formula

$$p \supset \cdot p \supset q$$

or

$$p \supset (p \supset q)$$

is from an axiomatic system P and can be interpreted so that p, q are propositions, and $\supset$ is the logical connective called 'material implication'. The dot is a bracket as illustrated by our alternative rendering of the formula. This interpretation is a part of what is called the propositional calculus.

But the identical statement $p \supset .p \supset q$ could be interpreted as a statement in a Boolean Algebra B. When the second interpretation is adopted, a *conventionally* different notation is used. So we have:

$$A \rightarrow (A \rightarrow B)$$

or since

$$A \rightarrow B =\text{df. } A^1 \cup B$$

where $A^1 \equiv \text{df. not} - A$
and $A \cup B = \text{df.}$ the sum of the two classes A and B, then

$$A \rightarrow (A \rightarrow B)$$

becomes

$$A^1 \cup (A^1 \cup B)$$

The fact is that the same model is being given two different interpretations.

Here the process could be described as taking a model or structure, which by itself has no meaning or reference, and supplying the model with the meaning. The word (name) 'Chicago' means to refer to the city Chicago, only by common agreement that 'Chicago' is the name for Chicago. The process of naming is like tying labels on physical objects.

With all this in mind, let us now turn to scientific method.

SCIENTIFIC METHOD

The next thing that must be done is to outline the principles of scientific method. Scientific method is naturally vital to any scientific investigation of the use of models, since, as we have seen, models and scientific theories are so closely connected.

A large number of books have been written on scientific method, logic, and philosophy of science, so there will be no attempt made here to describe the whole of science and the full complications of scientific method. We shall though, try to give an abbreviated account of what scientists do, or what they should ideally do.

Let us suppose a man is walking down the street, and he sees a shop which is open. He notices the shop is selling sweets and chocolates, and from this he might infer that all sweets and chocolates shops are open for the sale of sweets and chocolates at that moment, but will this be a reasonable thing to infer? Obviously not, as he will surely be the first to admit. It could happen that the shop he has seen is open at exceptional hours or on unusual days. On the other hand, if he remembers that he has often bought sweets at that time of day before, and indeed on that particular day – a Wednesday let us say, then he will feel some added strength from this in the belief that all sweets and chocolates shops are in fact open.

In the same way, we note at what shops we can buy what sort of things, so that by and large, it is true that one buys one's vegetables at the greengrocer's, and one's groceries at the grocer's shop, and so on. We may know roughly when they open and when they shut and apply our previous experience to the present. This enables us to make predictive theories.

Science is precisely the process of using one's previous experience as a basis for prediction. The prediction is important, the whole point of having a science is to be able to make (accurate) predictions about the future state of some system, and we can only do that if we have had some previous experience of that system, or some similar systems. Not that prediction is in fact everything, we want also to understand the system, and we want furthermore to be able to control the system, usually our surroundings, to some extent. This ability to control depends at the very least on being able to make a prediction of some sort.

Our own experience sensibly used in our everyday life can be said to be identical with science and scientific method. The only differences we find are in the language used, which, for the scientist becomes more complicated, and there is the means of making the observations, which may be far more detailed than in everyday life.

Language and science then are much alike. We want to build a sort of *map* (see next section) of our surroundings and consult it when we are going somewhere to make sure we take the right route. The map of course is only one of a number of things we use – we also use time-tables, and dictionaries, and whole sets of principles that we have learned for ourselves, and we may have couched in ordinary language. Thus, 'The pub never opens until 12 o'clock on Sunday', 'The only place you can get fresh meat is the butchers'; and 'X's is the best butcher in town'. Of course we can apply deductive logic too. From the last two sentences we have written it is easy to see that we have to go to X's if we are in town and we want some fresh meat, which is also the best.

In fact we shall want to make a distinction between learning, and having learned, the language of science or any other language. Thus, for example, the learning part is to come to realize that in fact X's does have fresh meat, and then perhaps also that Y has fresh meat too, and then finally to realize that a certain type of meat shop has fresh meat and a certain type does not. This learning is what is called 'induction' or the process of making inductions or inductive generalizations.

Inductive generalizations are statements of the kind 'all . . . are –' and examples would be 'All men are featherless bipeds'. 'All birds fly', 'All cats are black' and so on, but of course often we have to be satisfied with some weaker generalization such as 'Almost all cats are black'.

In principle, a science would be ideally made up of a set of inductive generalizations which form the basis of a deductive argument. The laws of nature would be our base and then anything we wanted to know would follow from applying deductive logic to these laws. In terms of our example, it would mean being able to say that 'Charlie is black' because 'Charlie is a cat', and we know that 'All cats are black'.

Science is the business of trying to discover all the necessary inductive generalizations, and the search is likely to go on for ever. One difficulty is knowing whether a generalization is true or not. If you say 'All cats are black', then you are faced with the problem of tracking down every cat, and not only that, logicians are inclined to say 'All cats are black' means now, not only that they are black, but they have always been, and always will be, black. But how can we be sure that cats will not one day become white? One way of answering this would be to say that if they become white you should no longer call them 'cats', but if you did that, then the whole game is farcical because cats are, as logicians say, 'necessarily black', if indeed, because their being cats depends in part on their being black; so there can be no possible reason for looking around the world as they are, to put it another way, 'black by definition'.

Our learning process leads us to generalizations that are probabilistic; they cannot be known to be certain because we cannot fully confirm them. Once set up though, the inductive generalizations are precise enough for making completely certain deductions. But the certainty derived is dependent on clarity over language.

Complete certainty is here meant to mean of course that the certainty depends on our *complete understanding* of the language. Let us illustrate the point by an example: 'If there are three hundred and sixty-seven persons in a room', then deductively it follows that 'at least two persons must have the same birthday'. Of course the extra person is included in case someone mentions Leap Year. But this sort of argument is easy enough to follow, and after a little thought no one would doubt that the force of the argument quite clearly depends on our understanding of what the words involved mean. If the meaning of the word 'birthday' for example is un-clear, then the whole argument will fail to persuade. So also with a vague word, this may be the key to the whole success or failure of deductive argument.

We shall now summarize our argument; we assume that the scientist, and that means every 'rational' person, shows behaviour that is made up of learning and the application of what has been learned. This means making inductive inferences, and then arguing deductively from these inductions. Of course most people learn from other people's as well as from their own direct experience, and we cater for this by saying that some learning is by acquaintance and some is by description.

We can easily distinguish between learning by acquaintance and learning by description, if we remember that we may have learned of the existence of China by description, since we have never been there. But we may have learned of the effect of the sun on the skin by acquaintance – through actually trying to sunbathe.

The method of argument that scientists ideally use is called axiomatic, and involves the restatement of the generalizations as careful statements or axioms ('hypotheses' or postulates, are other words meaning much the same). There is of course a lot more to say on this subject but we have said enough to relate the idea of models to language and to science, and that was the main point of our current discussion of scientific method.

STATIC MODELS

Of the many possible models that can be used to illustrate the static class of models, we shall select that of *maps*.

We can say that maps are like pictures of territory, although they are also different from pictures because they use abstractions in symbolic form of the territory itself.

It is characteristic that maps should be likened to languages and scientific theories. This is itself a reminder that languages and scientific theories are themselves related closely to each other. As a result, in view of what we said about model-theory relationships, we shall sometimes think of maps as models for languages and scientific theories. Let us look carefully at the idea of maps.

A map is any form of contour or line drawing, with or without colours, of any shape or shapes whatever. A map may be on any scale, using any projection, and may represent any detail we choose.

If we made abstract maps this would be part of a sort of universal geometry, although restricted – if we so constrict maps – to two dimensions.

Given such a set (clearly an infinite set) of maps, we could then, if we chose, interpret them as actual territories. The most obvious way of carrying out this interpretation is to choose a map with a *shape* similar to the territory. Because of projecting spherical areas onto a plane, many distortions actually occur, and in any case, we often use highly formalized maps which make no attempt to retain topographical details, only topological ones.

Maps, and we can think of them now as 'real' (i.e. ones that are used for some purpose relative to some actual territory), are to territory like scientific theories or scientific statements are to 'reality'. They either allow you to make *predictions* from territory-to-map and map-to-territory or they do not. If they do allow predictions (assuming no errors in map reading) this is because the structure of the map is in some way similar to the structure of the territory.

The map though is an abstraction – it represents roads, railways, or counties, etc. and it does so symbolically. The symbols of the map refer to abstracted bits of territory like words may refer to parts of reality.

Statements, particularly empirical statements, are highly confirmed (and are therefore probably true) if they allow accurate prediction. So the similarity between language and maps is clear enough. But words are generally non-pictorial and do not have topographical or topological significance in themselves, so in some respects maps are more like pictures. This underlines a point that is continually arising. If we use analogies (models are like analogies) then we must bear in mind *both the similarities and the differences*.

Now language is also sometimes generalized so we can refer to concepts like *reality, democracy*, etc. which cannot be derived from maps alone. A shape is specific or particular, a word may be a universal. So that part of language that is removed from immediate descriptions of the environment ('reality'), or sense-data statements, do not find any equivalent among maps.

A map which contained everything in the reality it mapped, would be indistinguishable from the reality itself. This is why we talk of maps as representing abstractions from the reality. If we now go back to *abstract maps* we can in some measure mimic the equivalent of universals in language.

Strictly speaking, we should not wish to choose maps as models of language because one is symbolic and not pictorial and the other is mainly pictorial with symbols attached. The correct model for language would be logic or some symbolic model which makes no specific appeal to pictorial concepts.

The usefulness of maps as models is of considerable pedagogical value, but it is also of conceptual value, and may even be of literal value. Language and maps come together in some measure in the use of graphs and algebraic geometry generally. Perhaps the most important aspect of the use of maps lies in its more direct appeal to the eye. The use of words is secondary rather than primary, and the two can clearly be used together, and when this is done, we can place the interpretation of 'scientific theory' on the result.

Maps normally do not change much, if at all, because territory does not change much. If maps were changing then they would be dynamic models, but for our dynamic models we shall look at the field of digital computer programmes.

DYNAMIC MODELS

In using computers (or computer programmes) as dynamic models, we are using them to simulate or synthesize a certain type of problem.

The central problem we are concerned with is the programming of a computer to demonstrate various aspects of artificial intelligence. We are primarily concerned with the synthesis of artificial intelligence, but also, inescapably, we are, in doing so, bound up with the problem of human intelligence. This connection is thought to be inescapable, since any form of synthesis must, it seems, at least be suggestive of simulation, however the fact remains that synthesis as such is our primary object. Our problem is to programme a computer so that it can accept and understand questions in a natural language such as English, or possibly in a form of 'empirical' logic. It then searches through its own store to find the answer to these questions and, in general, it will state the answers if they are known to the computer, (Simmonds, 1963, Green *et al.*, 1963). This is not essential to a dynamic model as such, but essential to our use of the computer as a dynamic model of high-grade artificial intelligence.

We want also to programme a computer so that it asks questions of any 'human-like' source of information in its environment. The reason for asking these questions will, in general, result from the need for information to complete some strategy, or act as a basis of information for some decision, whether or not this decision leads to action. It should be made clear that the computer is to be thought of as acting in an environment, which contains at least one human-like source of information and also contains events which occur, some of which may be under the control of the computer, and are also capable of being described by it. The computer's problem is to learn which aspects of the environment it can control, depending on information gained by direct experience of that environment on one hand, or by a question-and-answer interplay, through language, with whatever human-like source is available on the other. It should be said that apart from the ability to collect information and store that information, the computer must have the capacity to draw inferences, of both an inductive and a deductive kind, and take whatever steps are necessary to seek information which may be needed by some other source of information.

The first such environment we are concerned with is that of a well-defined game, such as noughts and crosses (G1), or some other similar such game. The game itself is obviously trivial, and it is only used because it is one which has a decision procedure, and is also one which illustrates the way in which either decision procedure can be discovered or heuristics supplied, without obscuring the process of that derivation because of the complexity of the game.

The position is that the computer both plays the game and makes inferences about it, based on the rules it has been told, at the same time. By the phrase 'at the same time' we mean that the computer must go on-line then off-line, playing, then arguing about or reasoning about or asking questions about, the game, and then be playing the game again, and then making inferences again, and so on. This clearly implies the capacity in the computer to make inductive and deductive inferences, where the deductive process at least is one similar to that described by such writers as Newell, Shaw and Simon (1963), Gelerntner (1963) and others.

The computer must be capable of describing what it is doing and answering questions it is asked with respect to what it is doing. The second stage of this operation is to show that the computer can be given a sufficiently general vocabulary and the formal rules of some further game, G2 (e.g. checkers), so that without any more special programming technique being involved, the reasoning capacity derived from the first game, G1, plus the actual playing of the game in an environment, is sufficient to allow it to improve its performance. Initially we shall expect this to be little more than a definite improvement, without expecting it either to derive an *algorithm*, or without it necessarily being able to play a highly-skilled game.

The same technique can now be generalized and applied to any games whatever, G3, G4, . . . , Gm, whether these games be well defined such as noughts and crosses and checkers, or whether they be ill-defined, requiring optimization techniques, as well as requiring the conditional probabilities needed to play well-defined games. The computer must in the next stage of operations be able to generate new words *and* be able to understand new words which are supplied to it by the other computer or human-like sources of information in its environment. We now would like, of course, also to generalize this and say there may be many human-like sources in its environment, and to this end, we shall have to have a model not only of the environment in which the computer operates but also a model of all the features of that environment which are what we have called 'human-like' sources. This means the computer must keep a model of each human-like source and must assess the likely reliability, in the light of experience, of that model and also the likely motivation that that human-like source may have in making the particular statement it has made (Maron, *et al.*, 1964). The next stage in this programming undertaking (or dynamic modelling), surrounds the need that we do not merely describe the process in general, but supply either compiler codes or machine codes, which show precisely how this whole undertaking can actually be programmed. The next problem is to show how the computer under these circumstances of being exposed to other human-like sources, must guard itself against the possibility of being bluffed or in other ways deceived. This, in common sense terms, is easy to understand. It must depend, of course, on the nature of the ascribed motivation to the utterer of the source of information, and must depend in some respects on the reliability of that source of information, and must be a matter for processing before the computer can decide whether this source of information is a reliable one or not. The difficulty here is to show precisely what instructions you might put into a computer to show how this type of undertaking is carried out.

It is clear from what has been said already that, although the aim is to provide a synthetic system (or model) of artificial intelligence, depending on the ability to learn, to use language, and to develop and learn new languages, both from understanding external linguistic sources and from manufacturing new words internally, the overall picture is one which comes near to being a simulation of the problem solving, decision taking and planning activities of human behaviour. It should be added that a large number of programmes for a variety of different computers, including computer programmes written in ALGOL, have already been written in pursuance of the plan laid out in this section, and we will explain a little bit more of the methods in the next section of this article.

METHODS USED IN PREPARING DYNAMIC MODELS AS PROGRAMMES

In the last section we gave an overall glimpse of the plan on which the whole of our artificial intelligence undertaking is based. We gave a few references to previous work that fitted in to this general picture, and we can remind you again at this point of some of this source work:

1. Logic Theorem Proving and Geometry Theorem Proving by Newell, Shaw and Simon (1963) on one hand and Gelerntner (1963) on the other.
2. Pattern Recognition Problems, not primarily in the field of visual pattern recognition, but as the basis for inductive inference. Here, the work of Minsky and Selfridge springs directly to mind.
3. Problems in Data Retrieval.
4. Concept Formation, which suggests especially the work of Kochen, Bannerji and his associates, among others.
5. The Confirmation of Hypotheses and degrees of factual support. This is a field which has so far impinged on artificial intelligence only very little, and one thinks in this context of the work of Carnap, Reichenbach, Popper, Hempel and Oppenheim, etc.
6. Risk analysis, where the work has been well summarized by Thrall, Coombs and Davis.

The above sample list of activities which are involved in the total process of artificial intelligence are more or less representative of the stages through which the process must go. Figure 2.2 shows this process in a very simple and generalized block diagram:

The first stage in the incoming process from the computer point of view involves the ability to recognize the nature of the input. One decision the computer must make is as to whether the input is linguistic or non-linguistic, i.e. whether it is from a human-like source in language form or whether it is from a non-human like source and represents the occurrence of events in the environment. As far as the computer is concerned, we have agreed that without the need for specific description of pattern recognition systems, a merely conventional distinction needs to be made between the linguistic and the non-linguistic inputs.

The recognition process clearly involves, among other things, comparison

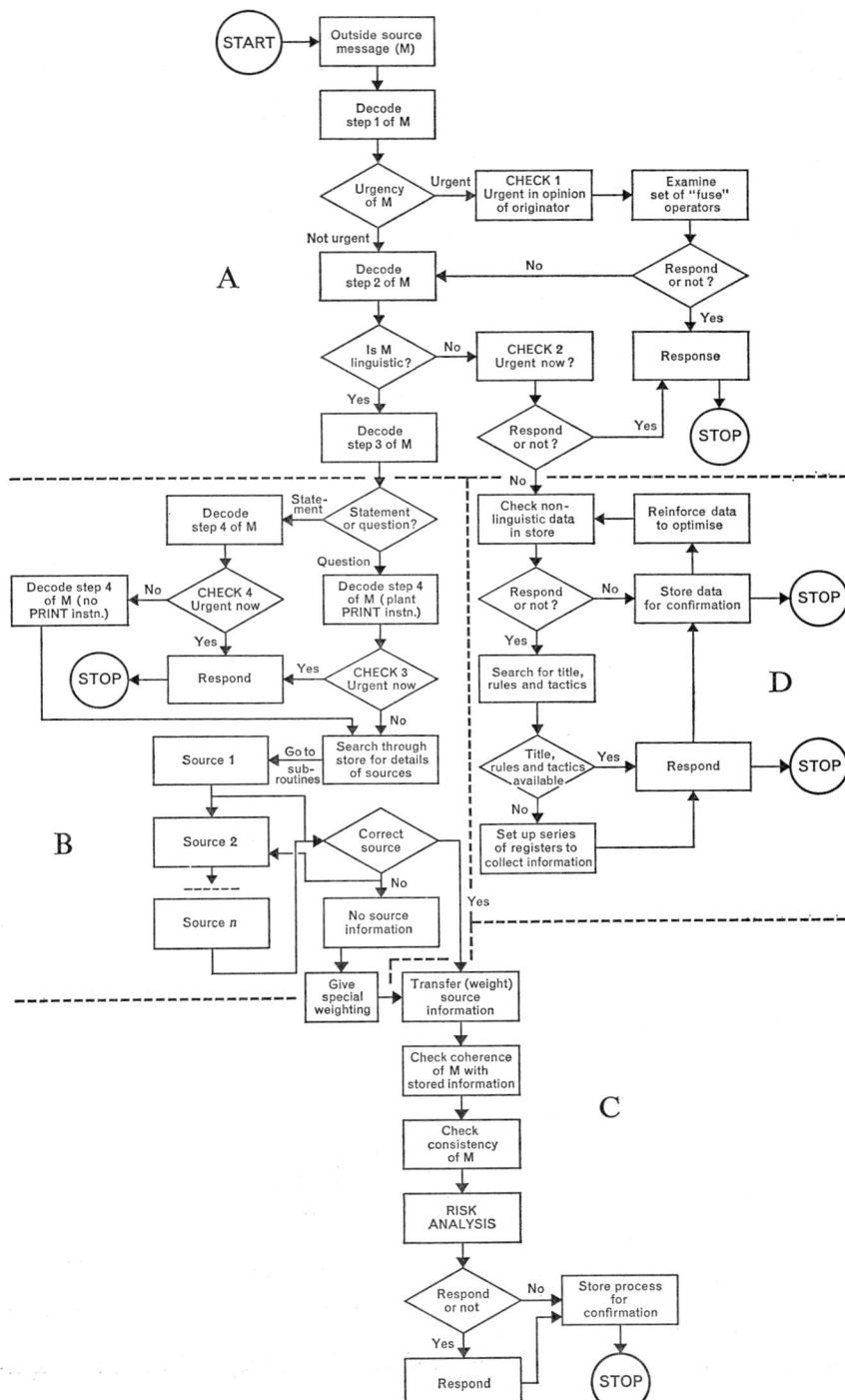

2.1

with existing information in store and, though there may be many stages in recognition, which associate it with recall and other forms of remembering, it also specifically associates it with the problem of data retrieval. It is at this stage of recognition that we have to specify the organization of the storage system, the cross-indexing and cross-referencing, so that an input is appropriately associated with existing information in store. The second stage of the operation is the processing of the information once it has been suitably recognized. Information which is not recognized, of course, may be either discarded or treated by trial-and-error type responses. The internal processing of information at the simplest level is event learning. Thus it is relatively easy to show that the computer can be programmed to collect information, and simulate a conditional probability computer, and learn to optimize output provided always that the game is either well-defined or reasonably well-defined, and there is a degree of periodicity of events which allows some sort of predictability to take place. This simple method, which may be applied to trivial games like noughts and crosses, leads to a decision procedure, without too much difficulty. The need, of course, is to show how to generalize over these simple games so that transfer can be made to more complicated games. The need is also to show that *heuristics* can be developed to deal with games where no decision procedure exists or where the game is far too complex for a decision procedure to be arrived at or even used in economical time. This requires the ability to make inductive inferences which depend in turn on the ability to categorize in a theoretic way. This draws attention in turn to the fact that the ability to make generalizations depends primarily on language, and draws attention further to the fact that the integral organization of the computer must include relationships between the event occurrences which are stored and the verbal occurrences which are stored. The relationship between these two sets of stored occurrences are what we might call *semantic rules*. The semantic rules must themselves be capable of being learnt by the computer in the course of transacting with its environment. We might say at this stage that the computer has organized in hierarchical fashion, sets of hypotheses or beliefs, ranging from more or less trivial ones which require more or less simple response to a stimulus, up to very much more complicated ones which are contingent on the occurrence or non-occurrence of various features in a very complex input. We can liken the process of confirming these hypotheses to those set down in the field of philosophy of science, and say that the next stage of proceedings which will interact between on-line and off-line learning by transaction with the environment, will be the process of confirming hypotheses. This is where the work on degree of factual support, credibility, inverse probability, Bayes Rule, and all the procedures of induction must be used to lend weight and credence, or the opposite, to the hypotheses stored inside the computer.

If we were primarily concerned with simulation here, and even as a synthetic undertaking, we must give some consideration to matters of priority and urgency. Thus, if our artificially intelligent system is to make decisions for us over a domain where urgency may be a prime factor, then some sort of recognition of the urgency, and this will occur right in the first stage of operations, must be brought into the picture. The fact that language is critical to this whole undertaking is a reminder in the present context that natural language will normally be transposed or translated into the logical form (functional calculi will be the general form of language used but with associated probabilities which give it the power of empirical description) and the ability to make deductive inferences on the part of the computer will result from the computer's ability to translate natural language statements not, of course, only into machine code, which is necessary for manufacturing the necessary computer orders, but also into logical language in order to make the

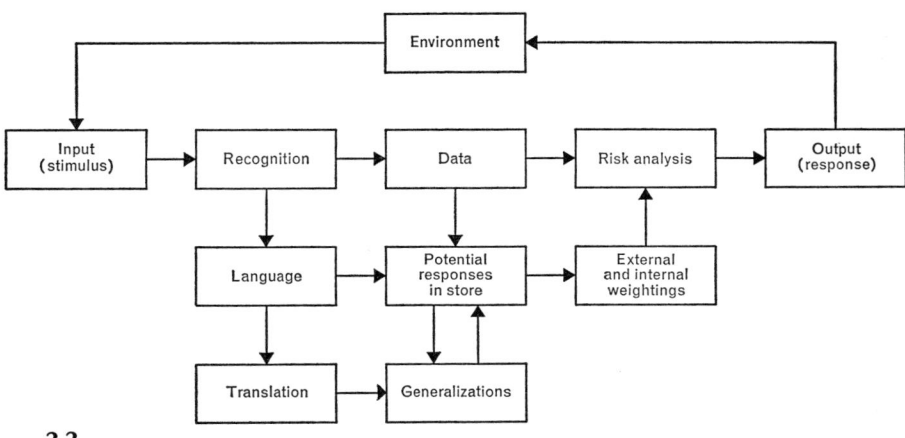

2.2

inferences with maximum convenience. At the output end it must be remembered the computer is capable of performing acts, as well as making statements or asking questions. The difference between performing an act and asking a question or making a statement is critically important to the whole success of the undertaking. Wherever an act or a statement is required as the final output on the behalf of the computer, a prior analysis of risk must be undertaken. The methods cannot be described here, but they are very closely associated with theory of games and so-called 'games against nature', and they involve the manifestly important point of assessing the risk for any particular decision. It may be that the same odds, in probability terms, in one context will be a totally different risk from the same odds in another context. It is obvious that taking a risk over a game of cards in the family context is quite different from taking a risk over the use of a nuclear deterrent, even though the probabilities of success or failure may be the same for each.

The relationship between scientific method and dynamic models should now be fairly clear, since our attempts at computer programming described above could readily be interpreted as an attempt to simulate the scientist. This all means that maps, and static models in general, are limiting special cases of the more general dynamic system.

We have lengthily described our dynamic model on the computer and we have gone well beyond the minimum requirements for a dynamic model.

It should be remembered that many programmes have already been written along the lines suggested (George, 1962, George and Gill, 1964; George, 1965), and finally we will consider a flow chart (Fig. 2.1, p. 53) of the general processes outlined in this last section.

REFERENCES

GELERNTER, H., [1963], Realization of a Geometry-Theorem Proving Machine; In *Computers and Thought*, by Fiegenbaum, A. and Feldman, J. (Eds.), (New York).

GEORGE, F. H., [1962], Simple Adaptive Programs for Computers; Paper read at *Conference on Cybernetics*, University of California, (Los Angeles), *October*.

GEORGE, F. H., [1965], Computer Applications in Decision Taking and Process Control; Paper read at *International Symposium on Long Range Planning for Management*, (Paris), *September*.

GEORGE, F. H. and GILL, R., [1964], A Computer Model for Planning; *Memo ESD*/10.3.64 *Educational & Scientific Developments Limited*.

GREEN, B. F., WOLF, A. K., CHOMSKY, C. and LAUGHERY, K., [1963], Baseball: An Automatic Question Answerer; In *Computers and Thought*, by Fiegenbaum, A. and Feldman, J. (Eds.), (New York).

MARON, M. E., *Computers and Comprehension* Memorandum R.M.-40650-PR.

NEWELL, A., SHAW, J. C. and SIMON, H. A., [1963], The Theorem Proving Machine; In *Computers and Thought*, by Fiegenbaum, A. and Feldman, J. (Eds.), (New York).

SIMMONDS, R., [1962], Synthex; In *Computer Applications in Behavioural Sciences*, by Borko, H. (Ed.), Systems Development Corporation Publication.

PART II

Models of Physical Systems

CHAPTER THREE

Models in Geomorphology

RICHARD J. CHORLEY

INTRODUCTION

The changes which have taken place in geomorphology since the Second World War have been profound. This profundity does not lie so much in the flood of new factual information about the surface of the earth, its processes and its transformations, as in our basic attitudes to the subject as a whole. However, the new mass data-generation methods of remote terrain sensing (e.g. air photography, infra-red photography, airborne radar, as well as other sophisticated military techniques linked with satellites) (McCauley, 1964; Ellermeier and Simonett, 1965; Heacock *et al*, 1966; Latham, 1966; Moore, 1966; Office of Naval Research, 1966; Rouse *et al*, 1966; Rowan and McCauley, 1966; Simpson, 1966), combined with computer methods of data processing and information extraction (Krumbein and Graybill, 1965, Chaps. 13–15), are also beginning to impose the need for radical methodological and conceptual rethinking in geomorphology. Indeed, it may be asked legitimately whether the study of landforms still exists as a discrete scholarly entity, for the most important recent changes in the subject have tended to impress upon scholars the disparate character of modern researches, together with the inability of workers to identify broad common objectives of even the most general character, or even to communicate to one another their mutual objections. Although geomorphology has always been a discipline of fine diversity (Chorley, Dunn and Beckinsale, 1964), many scholars now feel that the decline in the popularity of the Davisian basis for the subject (Chorley, 1965) has produced a conceptual vacuum which has not yet been reoccupied by any comparably broad systematic approach. This deficiency has served to highlight many national preoccupations, some of long standing, with particular geomorphic objectives. Thus the development of the American style of 'dynamic-process' geomorphology, the Franco-German climatic geomorphology, the British denudation chronology/geological approach, the Polish Pleistocene-dominated geomorphology, the Russian applied geomorphology, the Swedish studies of process almost *per se*, the Eastern European morphological mapping, and the Central European tectonic bases have created a

Godot-like atmosphere of articulate introspection. Perhaps this is why many geomorphologists have become increasingly concerned with the basic structure of the subject (e.g. Strahler, 1952; Dylik, 1957; Chorley, 1962 and 1966; Schumm and Lichty, 1965; Howard, 1965), and the relationships of its parts. The present volume, with its theme of model building, seems to provide both an opportunity and a kindly environment for taking such a broad structural view of the subject.

The value of model theory in geographical studies as a whole has been stated briefly and inadequately by the present author (1964) and this paper may serve as an introduction to the more comprehensive model for geomorphological work presented in Figure 3.1. In this the thought processes of abstraction and decision-making lead to the identification of three distinct, if marginally-interlocking, systematic views of geomorphology in which all types of past and present work find conveniently-linked places. Thus the simplified conceptual model system can either be approached by being translated in time and/or space to produce a *natural analogue system*, by being dissected into some supposedly integral parts which are examined in terms of a *physical system*, or by some broad conception the phenomena being structured into a complete *general system* from the outset. Obviously these systematic approaches grade into one another and can (electrically speaking) operate either in 'parallel' or in 'series', but the characteristic members of each seem to be reasonably distinct. Such a view allows apparently very different types of geomorphic work to fall into some more general pattern, and enables us to see more clearly from this *map of geomorphic activity* the areas of previous effort and neglect, so that we may more surely chart our respective future paths through regions the opportunities and challenges of which are unfolding with awe-inspiring speed.

NATURAL ANALOGUE SYSTEMS

One of the most common methods of illuminating a given geomorphological phenomenon is to translate its supposedly-important or characteristic features into some analogous natural system believed to be simpler, better known or in some respect more readily observable than the original. By such reasoning we see how classifications (i.e. like objects grouped together for the purpose of making some general statements about all of them) form an integral part of one kind of model building. Obviously such translations involve large intuitive leaps, for assumptions of natural analogy are often made on the most subjective bases; however, large sections of geomorphic work traditionally lie in this area. One may recognize two classes of natural analogue systems – the historical and the spatial.

Historical analogues group together geomorphic phenomena with regard

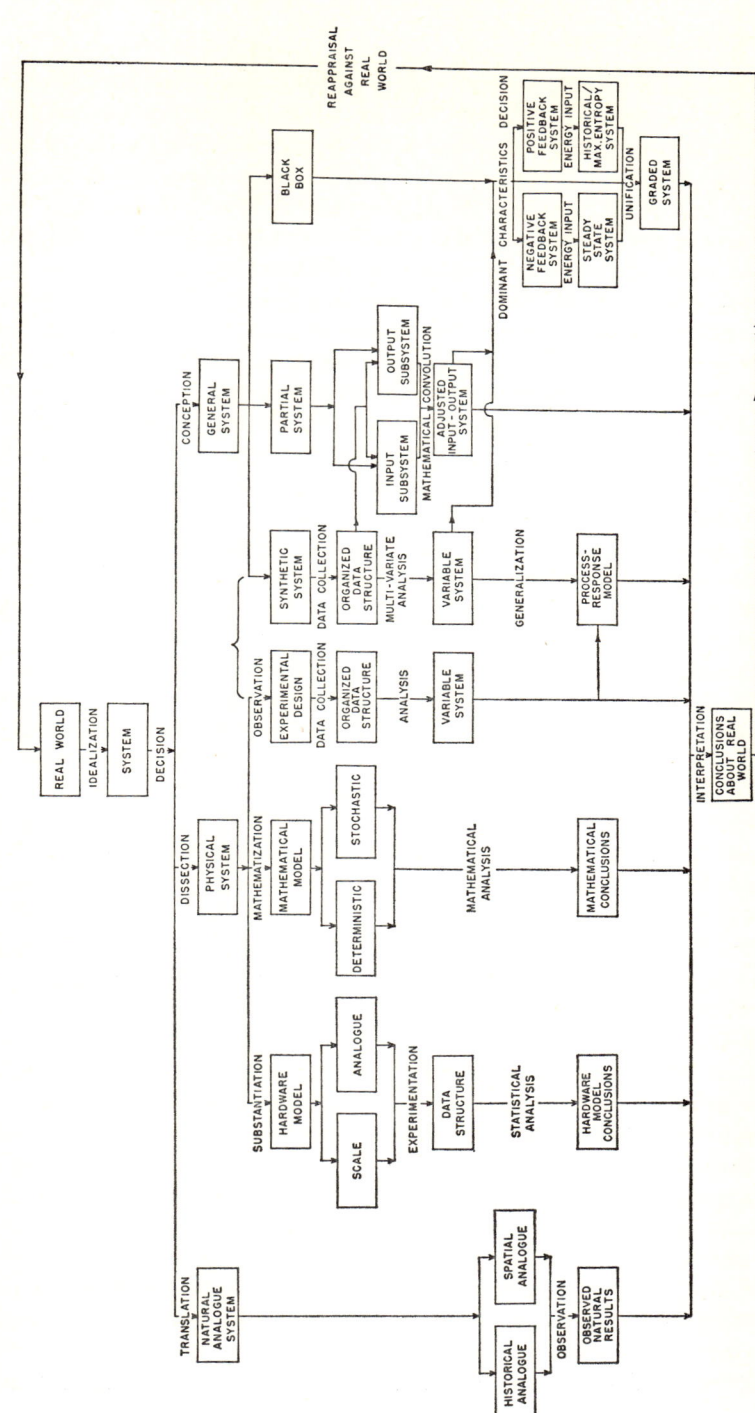

3.1 A map of geomorphic activity.

to their assumed positions in time-controlled sequences, on the assumption that what has happened before will happen again, or that what existed in the past has relevance to what exists now. Thus the phenomenon under consideration is viewed as part of 'a sequence of real, individual but interrelated events' (Simpson, 1963, p. 25; see also Kitts, 1963). It is, after all, the most fundamental canon of geomorphology that past landscapes can be most completely understood with reference to present ones (Hutton, 1795) and, for example, Wills (1929, Chap. 7) has effectively used analogies with present deserts to describe and explain the late-Carboniferous and Permian landscapes of Britain. From another point of view, it may be argued that present landscapes can be better understood with reference to past ones, and recently Ambrose (1964, p. 850) has suggested that the sedimentary stripping of the 'sub-Ordovician paleoplain' of the Canadian shield is reactivating an ancient topography 'virtually intact'. Much of denudation chronology (e.g. Johnson, 1931; Wooldridge and Linton, 1955) depends on this latter reasoning (Chorley, 1965) and it is interesting how this historical preoccupation of the extreme 'westernmost' part of the map of geomorphic activity (Fig. 3.1) reappears in the 'easternmost' (see later Section on The Black Box), suggesting that our map 'closes' to form a globe.

Spatial analogues associate one set of phenomena with others, on the assumption that observations at another place are easier to make or simpler in character than those of the original, or that comparison with other areas believed to be in some way similar will enable one to make more confident and meaningful generalizations about the original area. It is in this last sense that the classificatory character of this phase of model building appears most clearly (Sokal, 1965), although historical (genetic) classifications on the basis of assumed common history have also been abundant in geomorphology (e.g. Johnson's (1919) classification of coasts). The most common form of spatial analogue model is that in which adjacent contiguous areas are grouped together on the assumption that each unit can be better understood in terms of generalizations about some larger region of which it forms a part. Thus on the continental scale of 'morphotectonics' (Hills, 1961 and 1963) and on the regional scale of 'structural geomorphology' (Melton, 1959) individual landforms are grouped into tectonic or structural provinces. The concept of physiographic regions, at least as understood by Fenneman (1914), has a similarly strong structural-geological basis. Under the concept of 'morphological mapping' (Tricart, 1964; Savigear, 1965), individually-mapped landscape 'facets' (e.g. arbitrary land slope angle classes, changes of slope of different character, etc.) are regionally associated (Figs. 3.2A and B). The belief that total landscapes can be adequately or usefully defined in terms of such facets springs very largely from the concepts of denudation chronology (see, for example, Wooldridge and Linton, 1955, p. 56).

Another type of geomorphic spatial analogue is the so-called 'natural model', in which what are believed to be characteristic assemblages of landform units are identified and presented as type assemblages. Beckett and Webster (1962), for example, have proposed such facet assemblages, or 'patterns', for an area around Oxford. In a somewhat similar American approach, Van Lopik and Kolb (1959) have employed the idea that one varied and accessible region (or 'gross landscape') can be divided into 'component landscapes' which are each defined in terms of a characteristic association of four measurable terrain factors (characteristic slope, relief, plan profile and occurrence of steep slopes – Figure 3.2C), to form 'terrain types' with which other regions can be compared and classified. This scheme of desert 'terrain analogues' is based on the landscape around the Yuma Test Station, part of which is shown in Figure 3.2D.

PHYSICAL SYSTEMS

The physical systems approach is the one most obviously associated with conventional 'scientific method' and was the first to be applied to quantitative data in the earth sciences, particularly in the 1930's and 1940's. This approach is based on the view that research can best be pursued by dissecting the geomorphic problem structure into its supposedly components parts, such that the operation of each part and the interactions between the parts can be conveniently examined, leading (it is hoped) to a full synthesis of the components into a working whole. Amorocho and Hart (1964) have pointed to the dichotomy in hydrological research between what are here called the physical systems and general systems approaches, and the pivotal geomorphic work of the hydrologist Robert E. Horton (1945) may be held to belong largely to the former. However, it will be argued here that this distinction does not appear as clearly cut in geomorphology and that there is a considerable identity between experimental designs and synthetic systems, leading to the construction of 'process-response models' (Fig. 3.1). It is convenient to divide physical system investigations into three, often interrelated types: those wherein the important structural elements are substantiated into a hardware model, mathematized into a mathematical model, or subjected to field observation under some convenient experimental design.

Hardware Models

It is a constant source of surprise that hardware models have not hitherto proved to be of greater value in geomorphic research. Indeed, with one or two notable exceptions (e.g. Friedkin, 1945), it is probably true to say that the most valuable hardware models have been those which were basically parts of unscaled reality, closely circumscribed and examined in great detail.

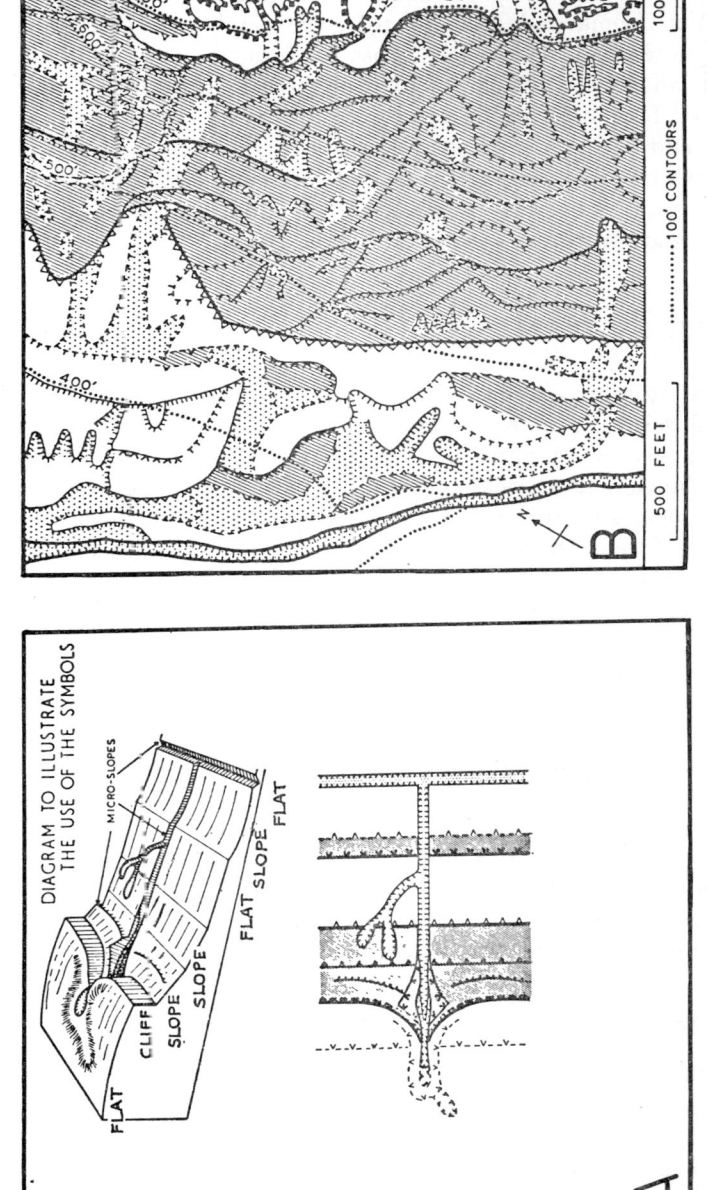

3.2 Spatial analogues in geomorphology.
A Examples of one system of symbols used in morphological mapping (*From Savigear, 1965*).
B An area in West Cornwall, England, mapped by use of the symbols depicted in Figure 3.2A (*From Savigear, 1965*).

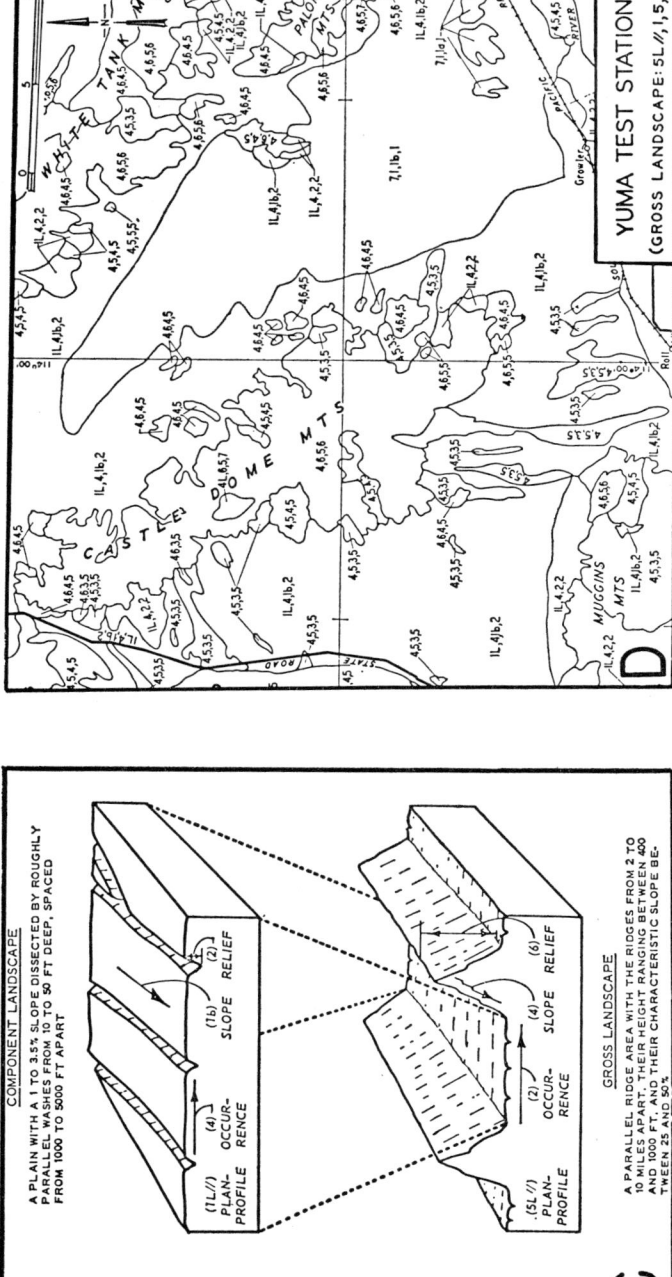

3.2 (*continued*)
c Example of a component landscape defined in terms of four terrain factors, and the relation between a component and a gross landscape (*From Van Lopik and Kolb, 1959*).
d Part of the Yuma Test Station, Arizona, divided into characteristic component landscapes (*From Van Lopik and Kolb, 1959*).

The two most obvious instances of this are Bagnold's (1941) wind tunnel observations on sand movement and Schumm's (1956) investigation of the erosional forms and transformations of the Perth Amboy badlands. This, of course, gives the clue to the basic difficulty of the construction of hardware models in geomorphology; that the complexity of nature imposes scaling or changes of media requirements of a very high order of sophistication. It is consequently easier for an engineer to simulate a life-size man-made structure than for an earth scientist to simulate a natural complex.

Scale models are closely imitative of a segment of the real world, which they resemble in some very obvious respects (i.e. being composed mostly of the same types of materials), and the resemblance may sometimes be so close that the scale model becomes merely a suitably-controlled portion of the real world (e.g. Schumm, 1956). The most obvious geomorphic advantages of the use of scale models are the high degree of control which can be achieved over the simplified experimental conditions and the manner in which time can be compressed. The general problems of scaling and of dimensional similarity have been comprehensively treated by Murphy (1949), Langhaar (1951), Wilson (1952, pp. 317–330) and Duncan (1953), and with particular reference to the earth sciences by Hubbert (1937) and Strahler (1958). Of these problems, the most difficult one is that changes of scale affect the relationships between certain properties of the model and the real world (e.g. scale ratios) in different ways, such that, for example, the kinematic scale ratios (i.e. those involving velocities and accelerations) behave differently from linear scale ratios (i.e. those involving lengths and shapes). Similar difficulties are involved with attempts to produce meaningful dynamic scale ratios (i.e. those involving gravity forces, such as mass and inertia). Such discrepancies can be commonly circumvented in any of three interrelated ways. First, a distortion of one important attribute can (usually by rule of thumb) be reduced or eliminated by the distortion of another attribute – for example, a distortion of the vertical linear scale of river models enables the effects due to turbulence (e.g. a kinematic ratio) to be more or less faithfully reproduced. The second, and most important, way in which analogous model ratios can be produced is by dimensionless combinations of attributes. Thus a combination of density, velocity, depth and viscosity (combined in the Reynolds' Number) enables viscous effects to be accurately reproduced; and a combination of velocity, length and the acceleration of gravity (e.g. the Froude Number) is important where gravity effects need to be accurately scaled in the model. Third, one or more of the media can be changed in the model to assist the truer scaling of other effects; but such considerations naturally lead one into the second type of hardware model – the analogue model. Difficulties of scaling natural geomorphic phenomena explain the failure of attempts to reproduce whole fluvial landform associations and their transformations (e.g. Wurm, 1935),

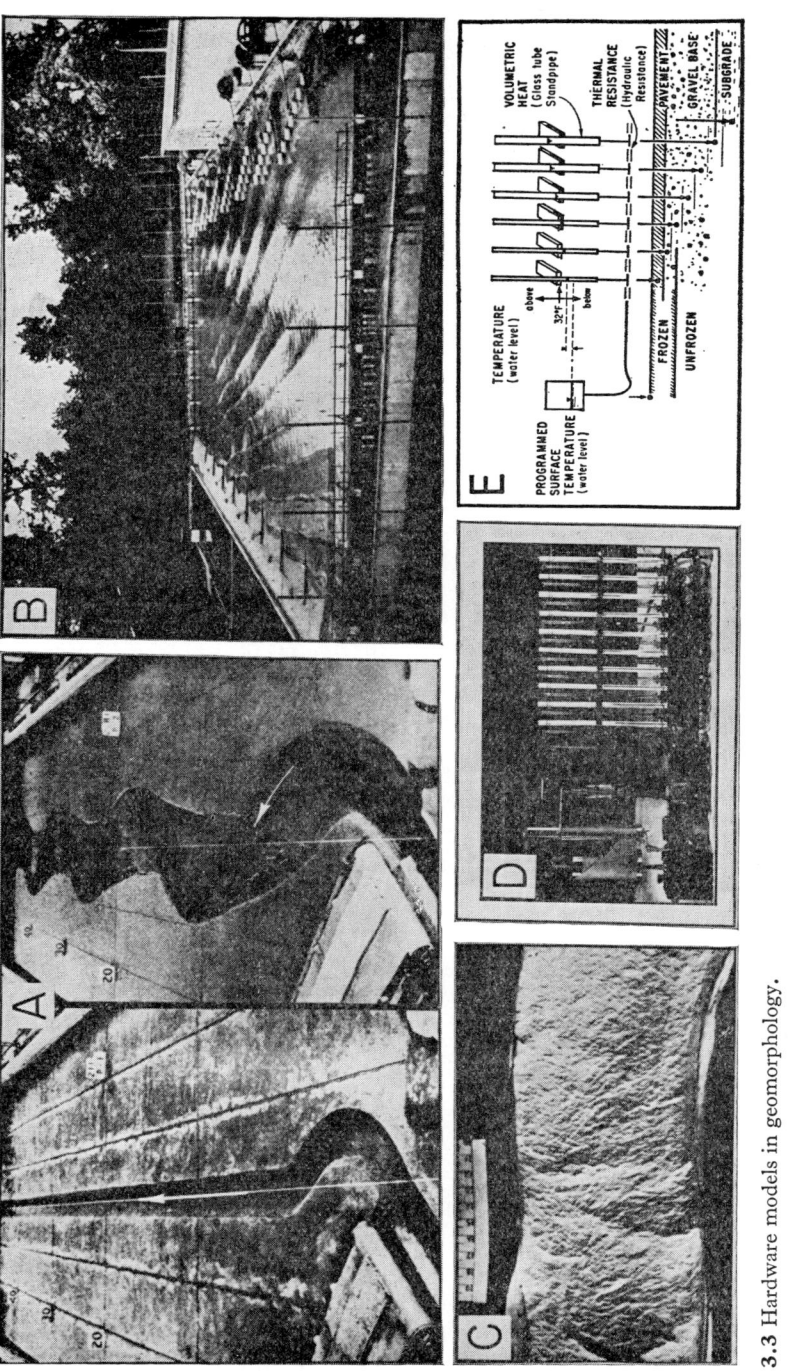

3.3 Hardware models in geomorphology.

A The development of meanders in a laboratory tank: Left, the initial channel and, right, after 3 hours (*From Friedkin, 1945*).
B An open-air wave tank at the Hydraulics Experiment Station, Wallingford, England (*From Hydraulics Research, 1956*) (*Crown Copyright Reserved*).
C Detail of the surface features at a bend in a kaolin glacier (scale in cms.) (*From Lewis and Miller, 1955*).
D Photograph of an hydraulic analogue computer for the study of soil freezing (*From Massachusetts Institute of Technology, 1956*).
E Diagram of the equipment shown in Figure 3.3D (*From Massachusetts Institute of Technology, 1956*).

and why the most successful work has involved attempts to reproduce more limited features like river meander bends (Friedkin, 1945) (Fig. 3.3A), nick points (Brush and Wolman, 1960) and beach segments (Saville, 1950; King, 1966A, Chap. 4) (Fig. 3.3B).

Analogue models involve radical changes in the media of which the model is constructed. They have much more limited aims than scale models in that they are intended to reproduce only some aspects of the structure or web of relationships recognized in the simplified model or idealized system of the real world segment. Such transformations are obviously rather difficult and great potential sources of 'noise' (i.e. extraneous confusion), in that great and often questionable assumptions must be made regarding the appropriateness of the changes of media involved. One example of such an analogue hardware model is Lewis and Miller's (1955) use of a kaolin mixture to simulate some features of the deformation and crevassing of a valley glacier (Fig. 3.3C). A more elaborate geomorphological analogue model is that constructed at the Massachusetts Institute of Technology (1956) to simulate the freezing and thawing of soil layers (Figs. 3.3D and E). The ground surface and successive soil layers are represented by a reservoir and glass tubes, in which the water level is programmed to correspond to the temperatures of the soil layers, the flow of heat in the soil is duplicated by the flow of water, and latent heat by suitably-placed expansion wells. Such analogue models have recently tended to be replaced by electronic computers operating in association with mathematical models or experimental designs.

Often the use of hardware models ends with the presentation of some simple visual impression of the geomorphic phenomenon (e.g. Lewis and Miller, 1955), but where quantitative experimental data results from the operation of the model (e.g. Saville, 1950) it can be built into a data structure and analysed by conventional statistical techniques (see sections on Experimental Design and Synthetic Systems) to produce hardware model conclusions.

Mathematical Models

A mathematical model is an abstraction in that it replaces objects, forces, events, etc., by an expression containing mathematical variables, parameters and constants (Krumbein and Graybill, 1965, p. 15) involving the adoption of 'a number of idealizations of the various phenomena studied and in ascribing to the various entities involved some strictly defined properties' (Neyman and Scott, 1957, p. 109). The essential features of the phenomena are then 'analogous to the relationship between certain abstract symbols, which we can write down. The observed phenomena resemble closely something extremely simple, with very few attributes. The resemblance is so close that the equations are a kind of working model, from which we can predict features of the real thing which we have never observed' (Daniel, 1955, p. 34). The common type of geomorphic mathematical model is concerned with some simplified statement of certain important features of the real world (usually geometrical ones) which can be transformed according to assumptions regarding the basic operation of the system (usually related to changes through time), yielding, by checking the model predictions against the appropriate real world situations, some information about the basic mechanisms involved and the succession of geometric changes to which the earth's surface is subject through time. Thus mathematical analysis follows the symbolic statement of the assumed basic static and dynamic relationships, and the logical mathematical conclusions then are checked against the real world; such that the correspondence or divergence between the real world and the effects predicted by the model indicate the success which we have had in building the original model. 'The differences revealed may then lead to the discovery of further causes, and these observed facts may gradually become understood in greater completeness and detail' (Jeffreys, 1918, p. 179). Mathematical models can be commonly divided into deterministic and stochastic.

Deterministic mathematical models are based on classic mathematical notions of exactly predictable relationships between independent and dependent variables (i.e. between cause and effect), and consist of a set of exactly specified mathematical assertions (derived from experience or intuition) from which unique consequences can be derived by logical mathematical argumentation. Such models are thus intimately concerned with relationships and 'driving forces' between the factors identified in the simplified model. Jeffreys (1918), for example, developed such a model for the denudation of a land surface by runoff, and by this means deduced theoretically the form of the resulting peneplain. Strahler (1952) produced a similar model involving the time relationship governing the elevation of a given point on a graded stream (Fig. 3.4D2), assuming an exponential longitudinal stream profile (Fig. 3.4D1) and a rate of stream-bed lowering always proportional to the gradient at that point. Such assumptions are basic to the construction of mathematical models and it is these which have to be tested according to the accuracy of predictability of the model. The most common type of deterministic geomorphic model involves the transformation of slope profiles under various assumptions (Scheidegger, 1961). Here assumptions are similarly made regarding the original slope geometry and the manner of its transformation. Figure 3.4A shows diagrammatically the geometry of a subtalus rock slope as a steeply-inclined original face retreats by uniform weathering and debris removal such that the debris wedge (of greater volume than the original rock slice) accumulates at a repose angle at the slope base. Obviously a vast number of combinations of assumptions can be made, and Figure 3.4B shows a case when the depth of weathering on the original slope increases with height.

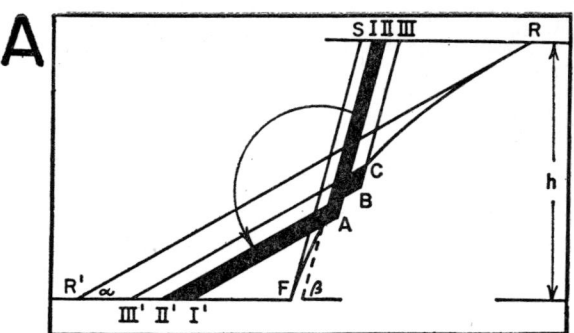
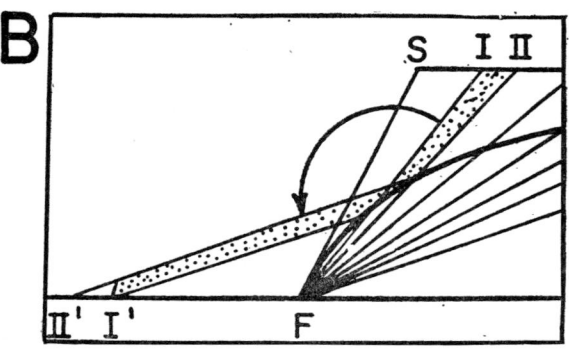
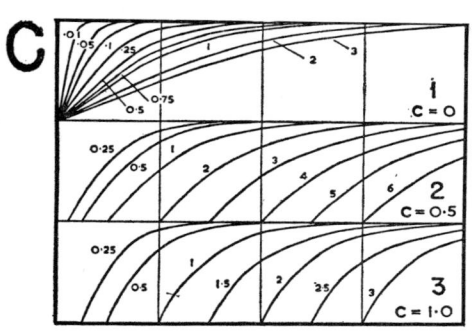
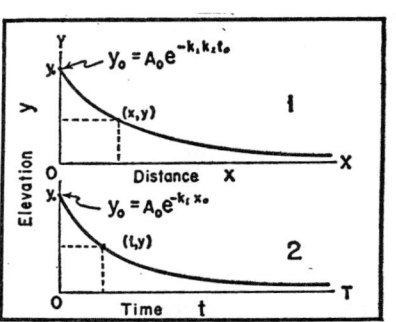
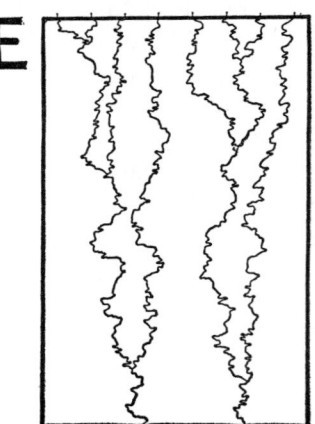

3.4 Mathematical models in geomorphology

A Parallel weathering of a steep slope with screes collecting at the base (*After Lehmann; From Scheidegger, 1961*).

B Central rectilinear slope recession, with weathering rate increasing with elevation (*After Bakker and Le Heux; From Scheidegger, 1961*).

C The progressive degradation of hill slopes by creep under three different rates (1, 2, 3) of under-cutting (*c*) (*From Culling, 1963*).

D A mathematical model of a longitudinal stream profile (1), together with the degradation through time of any given point on it (2) (*From Strahler, 1952*).

E A stream-like network generated from points equally spaced at the origin line by a random-walk procedure wherein uphill (backward) steps were excluded (*From Leopold and Langbein, 1962*).

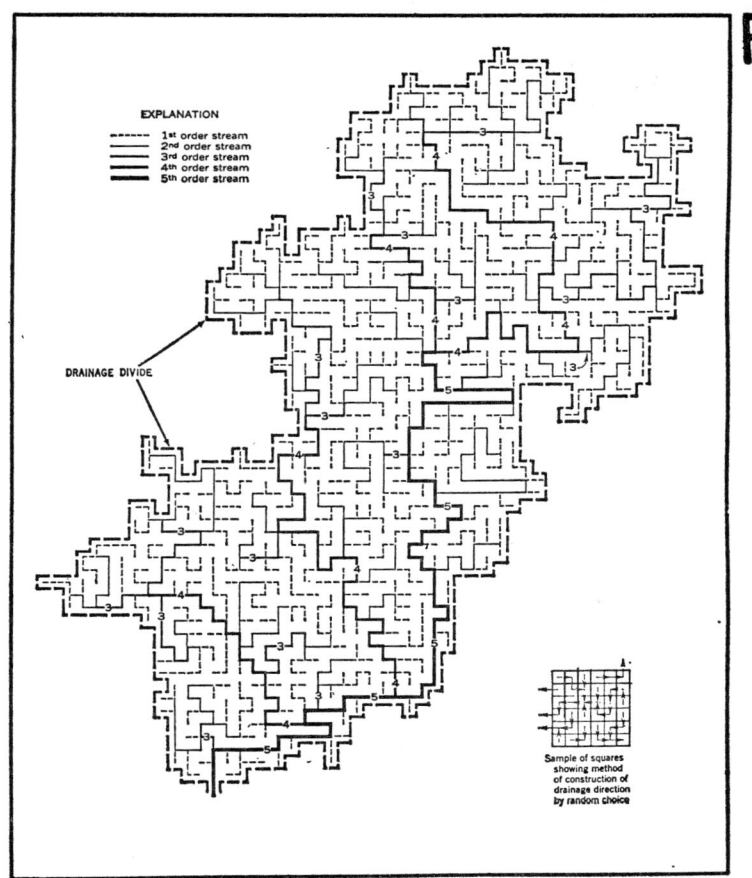

3.4 (*continued*)

F Development of a 5th-order drainage basin network by a random-walk method from a series of squares each of which has an equal chance of draining in any of the 4 cardinal directions (*From Leopold and Langbein, 1962*).

G Plots of stream number and mean length for the network shown in Figure 3.4F, indicating their adherence to the laws of morphometry for a homogeneous region (*From Leopold and Langbein, 1962*).

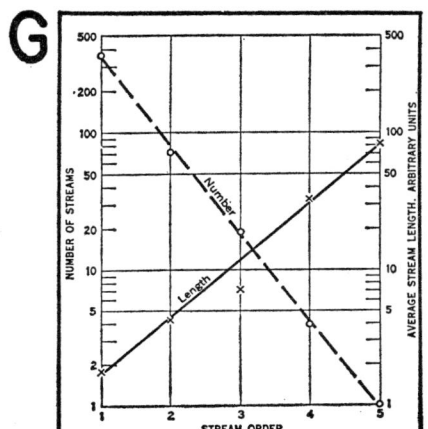

However, few deterministic statements can completely specify all the variables included in a complex natural situation, so that discrepancies occur which, together with the random unpredictable effects inherent in natural processes, combine to produce 'noise' which tends to obscure the simpler deterministic relationships. Often these random effects are so important in determining the results of natural processes that partly or wholly statistical (stochastic) models have to be constructed to take account of them. Culling (1963 and 1965) has proposed a slope model based upon certain assumptions that individual soil particle movements are influenced by pore-space characteristics that are partly random, producing a macroscopic soil creep the rate of which is proportional to the surface gradient. Figure 3.4C gives the two-dimensional development of such a creep model through time with three different rates ($c=0$, $0\cdot5$ and $1\cdot0$) of basal lateral stream undercutting. It is interesting that the model predicts a tendency of slope form towards a stationary state dependent upon the relationship between the rates of undercutting and denudation, and that, once this stationary state has been reached, the slope should retreat parallel.

Stochastic mathematical models are thus expressions involving mathematical variables, parameters, and constants together with one or more random components (Krumbein and Graybill, 1965, p. 17), the latter arising from unpredictable fluctuations in observational or experimental data. A simple example of a stochastic model is a 'Markov process' expressed by the equation (Amorocho and Hart, 1964, p. 318):

$$y_{t+1} = \rho y_t + \eta \bar{E}$$

where: y_t=the situation at time or place t; y_{t+1}=the situation at time or place $t+1$; ρ=a constant; η=a constant dependent upon the distribution of y in time or space; and $\bar{E}$=a random variable. It can be seen that the Markov-type phenomenon (y_{t+1}) is dependent partly upon a previous state of the same phenomenon, whereas in the 'Monte Carlo' model (the other main type of stochastic model) it does not depend on its previous states. The Markov-type model has found more favour with geomorphologists, Curl (1959) applying it to limestone cavern development and Melton (1962) used it to distinguish between regional and local random variations in longitudinal river profiles in Arizona. On the same subject, Leopold and Langbein (1962, pp. 4–8) adapted the statistical thermodynamic model to longitudinal stream profile development on the assumptions that absolute temperature and height above base level can be interchanged, and that there is a continuity of entropy (i.e. rate of increase of entropy in the system+rate of outflow of entropy=rate of internal generation of entropy). The first assumption has been criticized in the grounds, among others, that absolute temperature (measured on a ratio scale) is not analogous to elevation (interval scale), being defended by Scheidegger (1964), and has been subsequently elaborated by the same authors (Langbein and Leopold, 1964) to rationalize the typical longitudinal stream profile in terms of the most probable compromise between the dual tendencies towards a minimum total rate of work in the whole fluvial system and towards a uniform distribution of energy expenditure throughout the system. The same authors (Leopold and Langbein, 1962, pp. 14–17) again presented a more obvious Markov-type random-walk model to demonstrate how a combination of deterministic and stochastic terms can produce a bifurcating drainage-like network (Fig. 3.4E) which, when elaborated into a basin system (Fig. 3.4F), produces relationships akin to those found in natural stream systems (Fig. 3.4G). The possibilities of this technique have been investigated by Schenck (1963) with the aid of a computer. One of the few examples of the geomorphic application of the type of Monte Carlo model so extensively used in diffusion models of human geography has been by McConnell (1965) who examined the spacing and arrangement of topographic elements (e.g. terrain summits) in terms of Poisson and binomial distributions.

Experimental design models

In terms of the contributions made to geomorphology, the most important aspect of the physical systems approach involves the recognition that within a given range of observational data exist certain meaningful component parts which can be identified by employing a suitable experimental design (Fig. 3.1) (Krumbein and Miller, 1953; Krumbein, 1955; Melton, 1960; Krumbein and Graybill, 1965, Chap. 9). This design, which is derived from past observation, logical deduction, intuition, or a combination of these, provides a structure within which other data are collected and then analysed by conventional statistical means (Strahler, 1954; Miller and Kahn, 1962; Krumbein and Graybill, 1965; Chorley, 1966) to produce some generalization which overcomes the inherent variance of data in the earth sciences and gives some overall statement regarding the general data properties. Such statistical generalization commonly involves the fitting to the data of simple, multiple or 3-dimensional regressions (Krumbein and Graybill, 1965, Chaps. 10, 12 and 13, respectively). In a consideration of this approach to geomorphic problems it must be recognized that we are dealing with a very border-line type of model building which, in a rather more complex and sophisticated form, also appears under the category of synthetic systems (Fig. 3.1). The latter represent more ambitious experimental designs which lead to the construction of general process-response models.

On the basis of the experimental design, constructed with reference to some conceptual model of the nature of the problem and suitable operational definitions of its component parts, the numerical data are collected, checked as regards their scalar and number-system characteristics (Krumbein and

Graybill, 1965, Chap. 3), and a 'data matrix' (*organized data structure*, or 'data model') produced. This data structure is then commonly analysed by regression-type techniques to produce a simple variable system, in which workable correlations are identified involving the direction and intensity of assumed causation. With simple experimental designs, however, the design may be so restricted, or the sources of data so constrained, that it is not possible to extend the general conclusions far beyond some comments on the original data, although indications may be given which may lead ultimately to the more general statements characteristic of synthetic-system reasoning. The following examples will suffice to show both the features of experimental design models and this gradational character with synthetic systems.

Strahler (1950) set up a simple experimental design to test the effect of basal stream corrasion on the angles of the associated valley-side slopes in the Verdugo Hills, California, and established a significant control (Fig. 3.5A). Leopold and Maddock (1953) sampled the width, depth and mean velocity of the Powder River, at Locate, Montana, and produced logarithmic relationships between discharge and these three dependent variables (Fig. 3.5C). Geomorphic periodicities have been examined for topographic cross-sections by Piexoto *et al* (1964) by means of harmonic analysis, and for meander patterns by Speight (1964) using spectral analysis. A simple 2-kilometre, rectangular sampling design was employed by Chorley, Stoddart, Haggett and Slaymaker (1966) to examine the variation of surface sand-size facies in the Breckland, Eastern England, and the third-order polynomial surface fitted to the data (Fig. 3.5D) was subsequently broadly substantiated by independent sampling. Such 'trend-surface' treatment of areal data (Krumbein and Graybill, 1965, Chap. 13; Krumbein, 1966) illustrates well the relationships between regression-type analogues and information theory, involving as it does the separation of regional and local 'signals' from 'noise' (Chorley and Haggett, 1965). A more ambitious experimental design model was adopted by Krumbein (1959) in investigating the multiple controls exercised by geometric mean diameter, phi standard deviation, moisture content and porosity over the firmness of beach sand (expressed by 'ball penetration' – Fig. 3.5E). The use of the multiple regression technique enabled reductions in the sums of squares effected by all combinations of the four independent variables (i.e. some of which are given as percentages in Figure 3.5E) to be obtained, indicating the dominance of moisture, with diameter second in individual importance. All four independent variables seemed to explain just over three-quarters of the variation in beach firmness.

With the last example it can be seen readily that we are passing into the extended range of generalization characteristic of synthetic systems, and Melton's (1957) ambitious multiple-correlation investigation of the factors controlling slope steepness and drainage density even more obviously extends

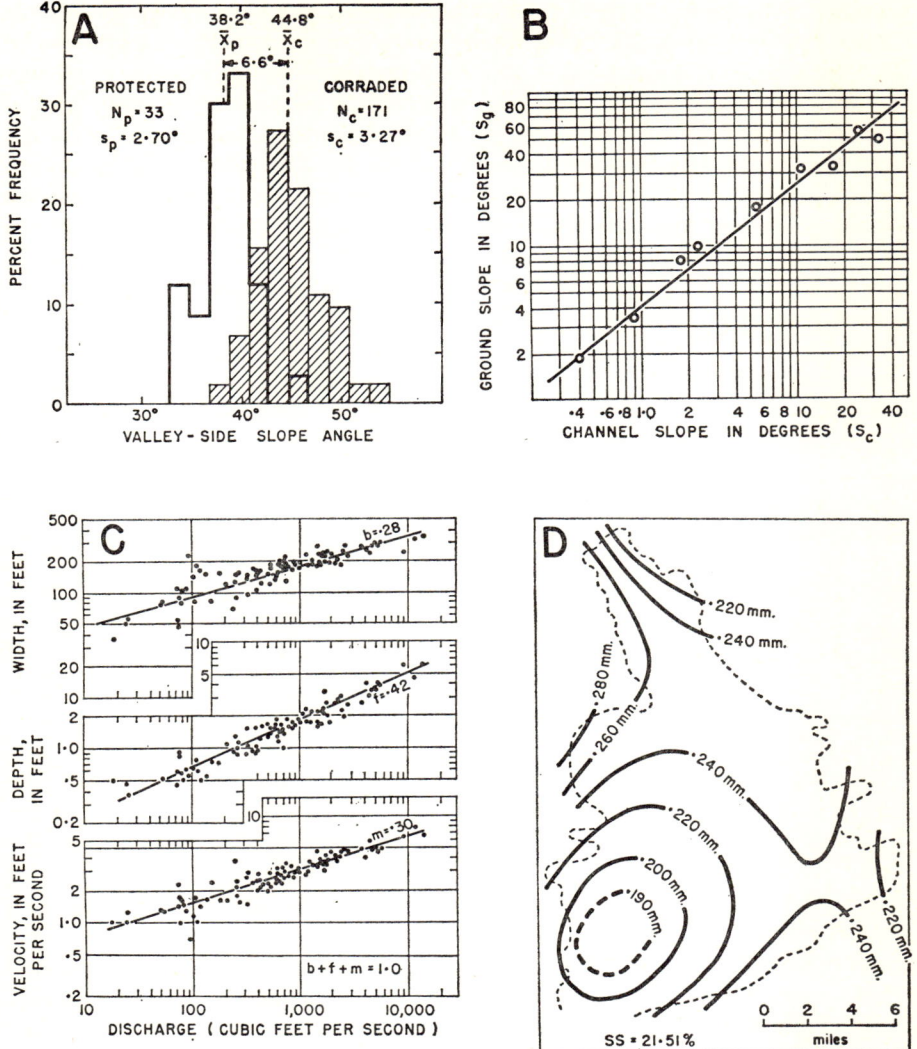

3.5 Experimental design models in geomorphology.
A Histograms of two samples of maximum valley-side slope angles in the Verdugo Hills, California for protected and basally-corraded slopes (*From Strahler, 1950*).
B General relationship between basal channel slope and valley-side (ground) slope for nine maturely-dissected regions (*From Strahler, 1950*).
C Relation of stream width, depth and velocity to discharge for the Powder River at Locate, Montana (*From Leopold and Maddock, 1953*).
D Best-fit cubic trend surface fitted to surface sand facies in the Breckland, Eastern England (*From Chorley* et al., 1966).

beyond the simple experimental design model into the synthetic system. Indeed, all simple experimental designs are capable, through an extension of the sampling, of expansion into synthetic systems to produce process-response models (Whitten, 1964) of quite general application. Strahler (1950) extended his valley-side slope analysis to suggest broad regional relationships between these slopes and the gradients of the associated basal stream channels (Figure 3.5B); Leopold and Maddock's (1953) observations have been widely confirmed and now form one of the bases for the science of 'hydraulic

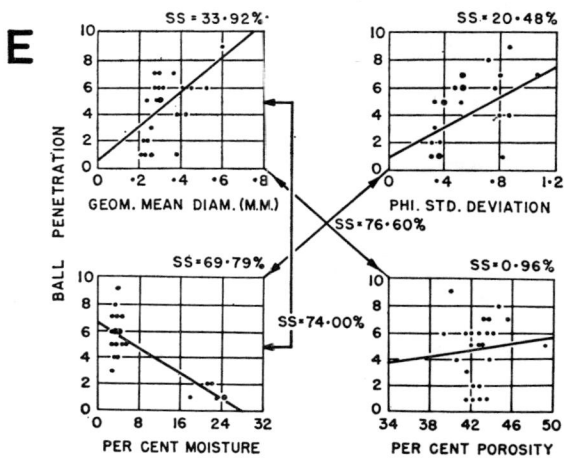

3.5 (continued)

E Relations between geometric mean diameter, phi standard deviation, moisture and porosity, and beach firmness for Wilmette beach, Illinois. Per cent 'explanations' are given for the four individual independent variables, for the most important pair, and for the four together (*From Krumbein*, 1959).

geometry' (Leopold, Wolman and Miller, 1964); and Chorley and Haggett (1965) have drawn attention to the extension of trend-surface analysis to produce generalized process-response models.

GENERAL SYSTEMS

The general systematic approach to the study of landforms is based upon some broad attitude to groups of geomorphic phenomena which is obtained as the result of experience (perhaps from some other type of analysis – e.g. experimental design models) or intuition. Here the emphasis lies in the organization and operation of the system *as a whole or as linked components*, rather than in detailed study of individual system elements (Von Bertalanffy, 1962). However, detailed knowledge of the internal operations of parts of the system (perhaps gained from experimental design models) is useful in the selection of an appropriate general system model.

A geomorphic 'system' is an integrated complex of landforms which operate together according to some discernible pattern (e.g. a drainage basin); energy and matter input into the system giving rise to a predictable system response in terms of internal organization and the resulting energy

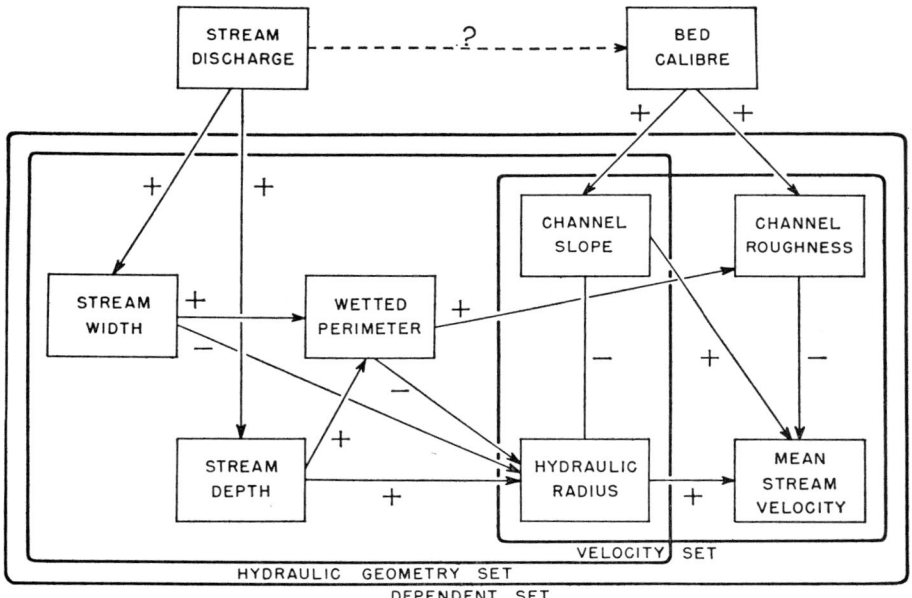

3.6 General relationships between the major factors involved in hydraulic geometry. Arrows suggest independence and dependence; the link without an arrow yoked interaction of two variables; and the mathematical signs whether variations are direct (+) or inverse (−).

and matter output. The instantaneous condition of the system is termed its 'state', which is characterized by its composition, organization, and flow of energy and mass (Howard, 1965), and defined by system parameters. State may be steady or variable through time, and the complexity of the system is expressed by the number of dimensions in the system 'phase space' (Melton, 1958A). The systems approach emphasizes an overall operational process, the operations performed by a system being dependent on the system state, which may be externally or historically determined. A description of such a system thus involves the specification of: (1) the nature of inputs; (2) the nature of outputs; (3) the system phase space; and (4) the model relating inputs, outputs and system states in time.

[78] MODELS IN GEOGRAPHY

Geomorphic systems can all be considered part of 'supersystems' (e.g. whole landform assemblages) and as being composed of 'subsystems' (e.g. slope or channel segments). Subsystems are thus the basic components of a system, and can be identified as distinct input-output linkages (Amorocho, 1965). In geomorphology subsystems are commonly combined by 'cascading' the output of one subsystem into another to form its input (i.e. the output of slope segments forms part of the input of the basal stream segments). When interest in the subsystem operations is very detailed, the systems approach is replaced (at least in the initial stages of the investigation) by the experimental design model. Two systems are said to have 'identity' when there is exact equality of all components, and to have 'equivalence' when they

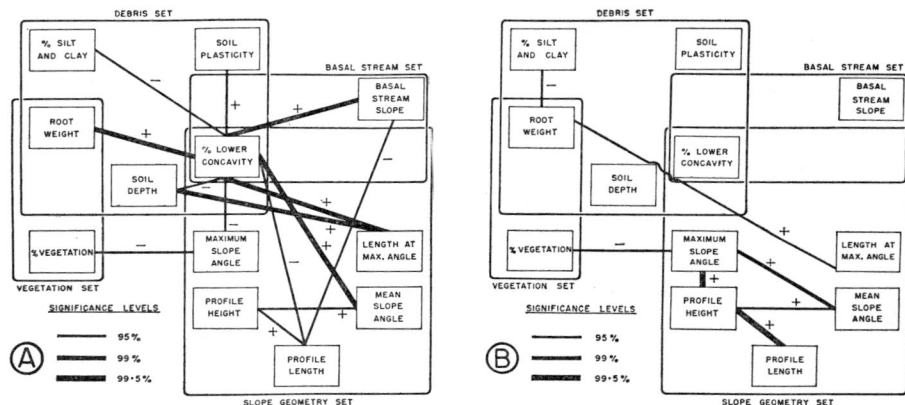

3.7 The interlocking of parameters relating to slope geometry, debris, vegetation and basal stream activity for slopes on the Charmouthien Limestone, Plateau de Bassigny, Northern France.
A Where the basal stream is moving towards the slope base.
B Where the stream is moving away from it. (*After Kennedy*, 1965).

transform the same inputs into identical outputs. The latter is much more common in geomorphology (i.e. two basins being of similar order in a homogeneous region). The internal organization of systems can be profitably viewed in terms of 'feedback' with reference to changes in the controlling external variables (i.e. Leopold and Maddock (1953) consider discharge and sediment characteristics to be the most important external variables in stream system operations) (Fig. 3.6). 'Positive feedback' occurs whenever externally-induced changes of input produce changes in the same direction as the input change (i.e. lead to a progressively-changing, 'timebound' state). 'Negative feedback' operates when changes in the system input result in changes in other system components which regulate the effect of the changed input such as to bring about a new 'timeless' equilibrium or 'steady state'. A steady state usually manifests itself in terms of high interlocking correlations between observed subsystems (e.g. parameters) (Fig. 3.7). Langbein and Leopold (1964) have defined it in terms of balanced tendencies towards an equal areal distribution of energy expenditure and towards minimum total work. Self-regulation is thus the diagnostic feature of open systems with negative feedback. However, this regulation to changes in external variables is often complicated by: (1) 'secondary responses' which may eventually result from primary changes (e.g. a change in precipitation may change discharge which may immediately alter the hydraulic geometry; subsequently vegetational changes may further alter discharge and channel characteristics); (2) 'thresholds', the passage of the system through which involves drastic

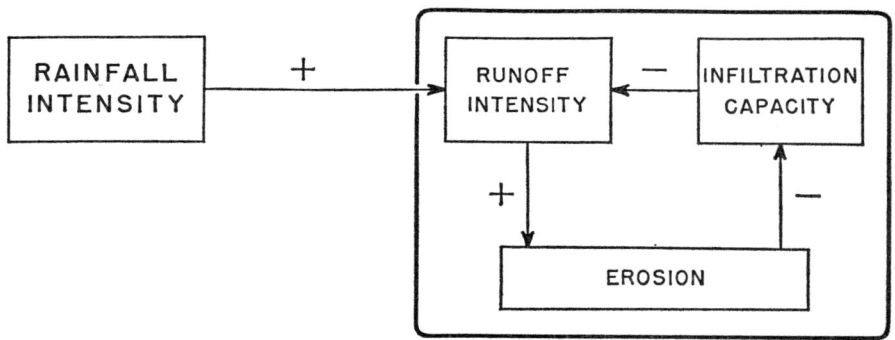

3.8 Correlation structure of a system involving rainfall intensity, runoff intensity, erosion and infiltration capacity exhibiting dominantly positive-feedback characteristics, at least over limited timespans; in this instance as long as erosional decreases of infiltration capacity continue (*From Melton, 1958B*).

changes in the system state (Howard, 1965). It is profitable to view certain geomorphic systems as exhibiting dominantly positive feedback characteristics for limited periods, in particular certain drainage basin subsystems (Melton, 1958B, p. 454) (Fig. 3.8). However, most geomorphic systems operate dominantly as negative-feedback open systems of the self-regulatory type. This self-regulation involves internal reorganization of the system and is accomplished in a time period referred to as 'relaxation time' (Langbein and Leopold, 1964, p. 782; Howard, 1965). Relaxation time depends on: (1) the resistance to input changes of the individual system components; (2) the complexity of the system (i.e. the phase space or number and linkages of variables involved); and (3) the magnitude and direction of the change of input. The relaxation time of some geomorphic systems is short (i.e. the channel adjustments to changing discharge which are the subject of hydraulic geometry), but that of others is long (e.g. changes of erosion induced on granite terrains by climatic change). Relaxation time is therefore some measure of the resistance of the system to external changes. The concept is complicated in that different system parameters have different relaxation times, and that some landform elements carry a longer-term historical record than others. Just because some landscape features (e.g. those related to hydraulic geometry) are in equilibrium, this does not necessarily imply that all other features are as well (Howard, 1965, pp. 309–310). Under such a systems view, however, it is possible to associate apparently very different geomorphic phenomena by examining their tendencies towards the achievement of the steady state. Where the relaxation time is short the timeless character of the system is dominant, where it is long the timebound or historical features preoccupy one (Chorley, 1966). Many of the methodological confusions of Pleistocene and Recent geomorphology are involved with systems where the relaxation time is long or where changes of energy input are so rapid as to produce the coalescing of energy pulses and the blurring together of partial form responses. The common result of such confusion is to produce the general impression that some general changes are taking place through time.

Few geomorphic systems are solely of the negative feedback type when considered through long time spans. If the mere flow of energy through the system induces its progressive internal reorganization (i.e. the removal of mass and the reduction of drainage basin relief (Chorley, 1962, p. 3)) or if the system input suffers a progressive change through time, then some features of the internal form or organization of the system will be susceptible of sequential change through time and an historical parameter introduced (Chorley, 1962; Schumm and Lichty, 1965). Where these changes are especially important they may mask the self-regulatory characteristics of the system and it may be more convenient to view it in some other terms – i.e. as an historically-oriented sequence of operations.

It is convenient to divide general systems into *synthetic systems, partial systems* and '*black boxes*'.

Synthetic systems

As pointed out earlier, synthetic systems in their initial phases bear a striking identity with experimental design models, being concerned with recognizing supposedly crucial features of the structure of geomorphic phenomena, together with their sampling and analysis. The aim of synthetic systems research, however, goes beyond this in that its goals lie in the synthesis of the analysed structure and the extension and generalization of conclusions to the point of producing process-response models.

The procedure for building up synthetic systems begins with the identification of certain key structural elements in a given geomorphic complex, together perhaps with some views as to the possible relationship between these elements (e.g. direction of causation ('impulse'); 'yoked' variables which change together; feedback from effects to assumed causes; the strength (or statistical confidence) of the bonds between elements; etc.) (Fig. 3.9). The succeeding stages are well exemplified by the work of Melton (1957 and 1958B), who used non-parametric methods (Siegel, 1956; for the parametric methods see Krumbein and Graybill, 1965, Chap. 15) to construct a 'variable system' – the mapping of highly-correlated variables into plane sets, in which the directions of causality are stated and with one or more of the

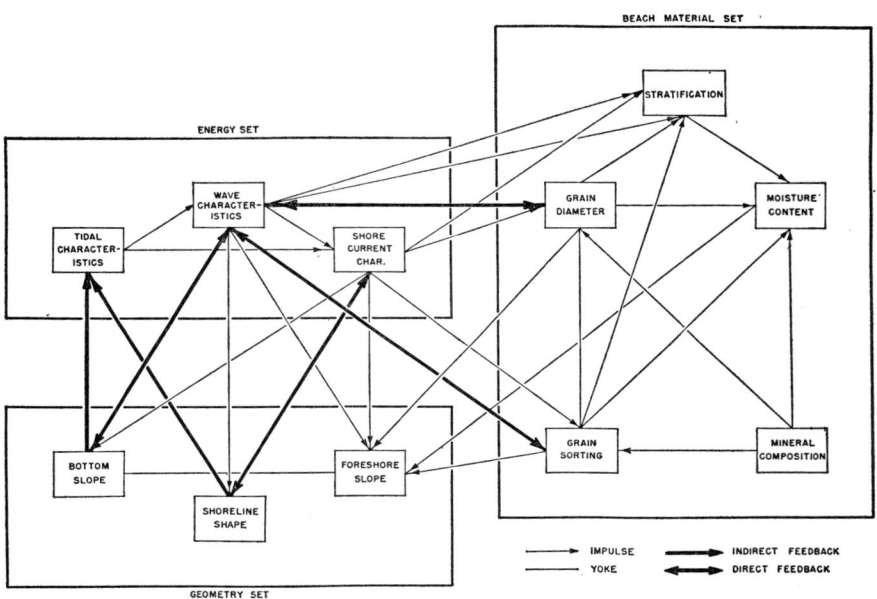

3.9 Tentative relationships between parameters relating to wave, tide, or current characteristics, beach material and shore geometry. Feedbacks are suggested wherever parameters of beach material or shore geometry might influence the wave, tide or current energy. (*Suggested by Krumbein*, Ann. Bull. Beach Erosion Board, *17, 1964*).

variables correlated with variables outside the sets, so that the systems are 'open' in that well-defined relationships exist between their members and the surrounding environments (Melton, 1958B, pp. 443–446) (Fig. 3.10).

Blalock (1964) has outlined certain principles for examining these correlation structures of variable systems, in which regression equations are employed as causal models. As Melton (1958B) has pointed out, the building up of correlation links between variables by techniques such as cluster analysis (Krumbein and Graybill, 1965, pp. 406–408) can give important indications as to the relationships between geomorphic variables without any preconceptions regarding the assignment of cause and effect. A difficulty, however, is that one commonly wishes to evaluate the composite effects of groups of variables, and factor analysis (Imbrie, 1963; Wong, 1963; King, 1966B) provides a means of collapsing the data matrix into a small number of

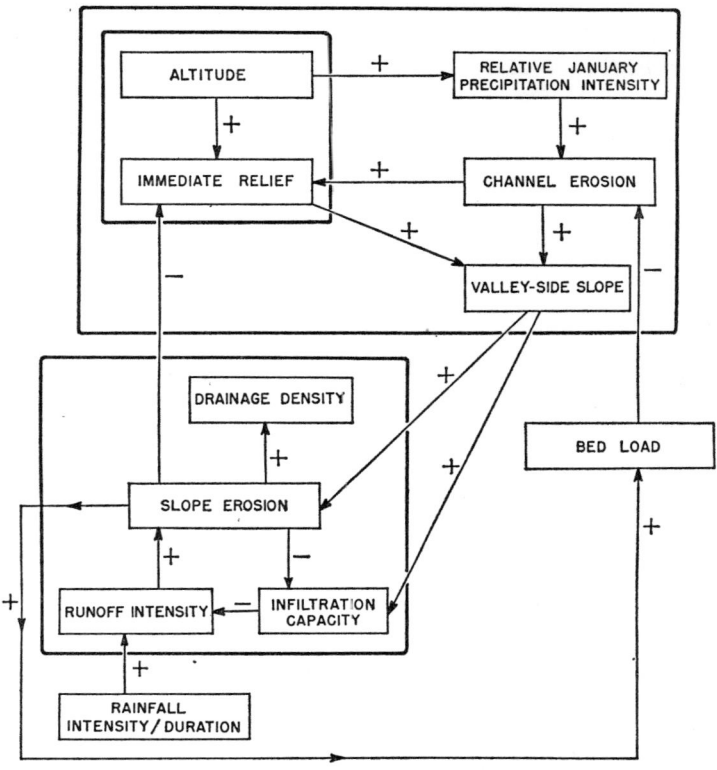

3.10 Correlation structure of a system involving morphometric, climatic and surficial variables, emphasizing the negative-feedback loop: valley-side slope, slope erosion, immediate relief (*From Melton, 1958B*).

idealized variables which account for most of the observed variability of the data. Another means of circumventing the difficulty of handling a large number of combined effects of variables is to maintain their original identity but to eliminate all except the most predictive ones, so that their combined effects can be examined in detail. Sequential multiple linear regression, 'WHIRLPOOL' (Krumbein *et al*, 1964), is such a data-search procedure for sifting out low-correlation elements, retaining a minimum number of variables to achieve some required level of prediction, and to examine completely their interrelationships (Harrison and Krumbein, 1964).

MODELS IN GEOMORPHOLOGY [83]

In their most sophisticated forms such correlation structures provide process-response models (Whitten, 1964; Krumbein and Graybill, 1965, pp. 19–27; Harrison and Krumbein, 1964), which are highly-generalized frameworks of impulses and responses to which further observations can

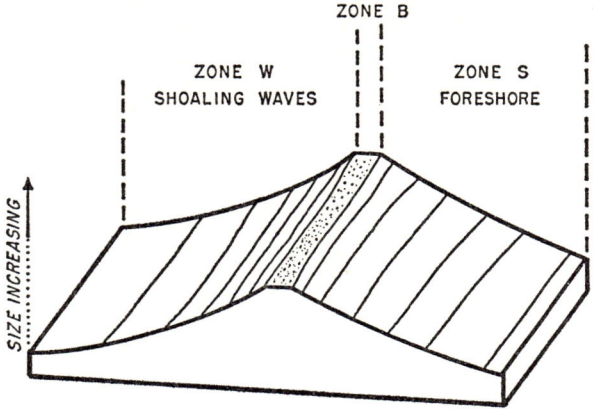

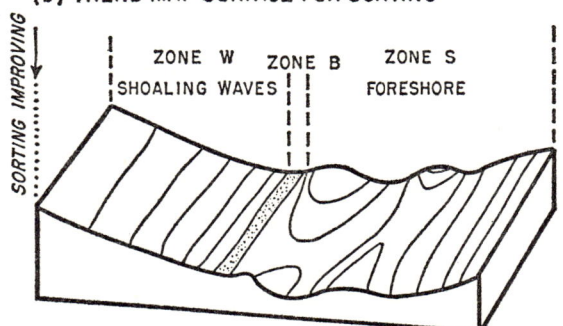

3.11 Trend maps for nearshore breaker zone and foreshore distribution of sediment size (a) and sorting (b) characteristics representing simple process-response models (*From Miller and Zeigler, 1958*).

be referred as an aid in identification of phenomena and as a basis for prediction. Trend surfaces, considered as 'response surfaces', are proving especially valuable in suggesting process-response models (Chorley and Haggett, 1965) and Miller and Zeigler (1958) have suggested a near-shore model for sediment characteristics (Fig. 3.11).

Partial systems

The partial systems approach is one which has been primarily associated with the solving of essentially practical problems in the earth sciences. It has been concerned with the establishment of *workable relationships* between often coarsely-grouped sets of factors or subsystems. Detailed knowledge regarding the internal functioning of these subsystems is not regarded as necessary, but the acquiring of specific information about the interrelationships between these coarse subsystems enables one to identify and predict the behaviour of the whole system under different input conditions. This information may come from a variety of sources (commonly from the Synthetic

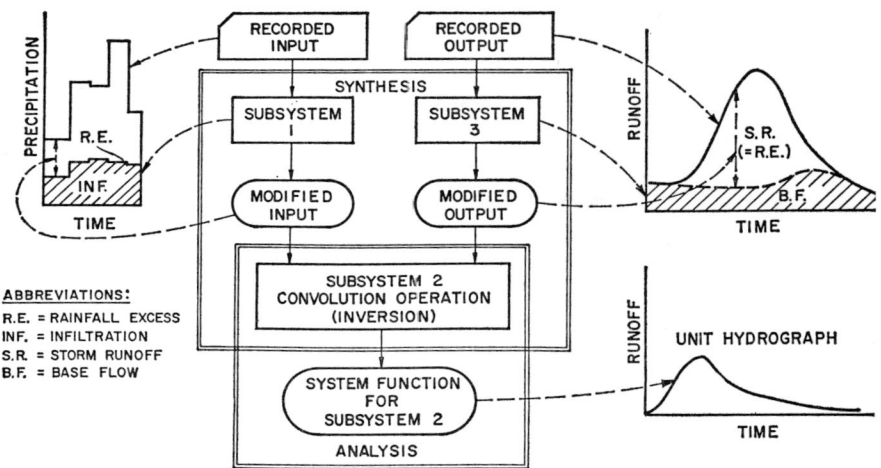

3.12 Flow chart for partial system synthesis involving the prediction of runoff from precipitation characteristics, with a minimum knowledge of the internal operations of the system components (*From Amorocho and Hart, 1964*).

Systems or Experimental Design approaches), but the object is always to establish mathematical relationships between the system inputs and outputs, without any attempt to describe explicitly the internal mechanisms of the system. The resulting behaviour patterns are then used to predict, and to a lesser extent to understand, the behaviour patterns in other similar systems or in the same system at different times (Amorocho and Hart, 1964).

The best example of the partial systems approach is given by Amorocho and Hart (1964) (Fig. 3.12) to predict runoff patterns for different storms over a stream basin. All that is known about the complex operation of the basin system hydrological cycle is; (1) the apportionment of the precipitation between infiltration and rainfall excess, and (2) the apportionment of the resulting hydrograph between base flow and storm runoff. On the basis of past input/output relationships (i.e. storm precipitation/runoff relationships), modified inputs (rainfall excess) are related to modified outputs (storm runoff) by purely mathematical processes to produce an adjusted input-output system which, from a given pattern of storm rainfall, will predict a pattern of storm runoff from which a unit hydrograph can be obtained.

Needless to say, such a fundamentally practical partial systems approach to groups of phenomena does not find many counterparts in geomorphology. A reasonably close equivalent might be provided, however, by attempts to establish workable relationships between precipitation and drainage density based upon certain gross assumptions regarding the factors controlling drainage density and, in particular, regarding the characteristics of surface runoff (Chorley, 1957; Strahler, 1958; Chorley and Morgan, 1962; Carlston, 1963).

The black box

The most extreme application of general systems models involves the concept of the 'black box'. This requires little or no detailed information regarding the system components or subsystems within the black box system, interest being focused upon the nature of the outputs which result from differing inputs. Thus the black box is analogous to the 'grey box' of the partial system and the 'white box' of the synthetic system.

In geomorphology the black box approach has nourished much of the important work of the last one hundred years. Such work has been characterized by broad intuitive leaps wherein decisions are taken (sometimes explicitly, but more usually implicitly) regarding the supposed dominant characteristics of broad landform assemblages. Although there has been much guesswork and speculation regarding the detailed nature of the processes at work, the characteristic patterns of landform transformations through time, and the rates of operation of geomorphic processes, we are only now making a beginning in answering these and other similar questions. Indeed, for most of the general systematic models of geomorphology the lack of such information has not been a disadvantage, its very deficiency allowing the theoretical model builders much greater scope than they would have otherwise have had. So we find that, on closer examination, such apparently disparate approaches to geomorphology as Gilbert's 'dynamic equilibrium' model (Gilbert, 1877; Chorley, 1962), climatic geomorphology, the Davis cycle, and the Penck geomorphic system are all based upon gross intuitive assumptions regarding the significant behavioural patterns of landform assemblages, all falsely give the impression that they are based upon detailed knowledge of geomorphological processes, and are all simply concerned with the landform outputs which are supposed to result from combinations of process and tectonic inputs. Not only does attention to general systems theory enable these similarities and limitations to emerge, but also makes possible the recognition that these mainstream approaches to the study of landforms can be divided into two broad groups – those wherein negative and positive feedback are dominant, respectively (Fig. 3.1).

G. K. Gilbert (1877; Chorley, Dunn and Beckinsale, 1964, pp. 546–572) has provided the most important negative-feedback black-box model for landforms for, despite his later preoccupation with fluvial processes (Gilbert, 1914), he was broadly concerned with the gross adjustments of form to process and of form with form. More modern work (especially Horton, 1945 and Strahler, 1950) has revived interest in this 'dynamic equilibrium' attitude to landforms, with the concept of 'grade' providing some clues as to the feedback mechanism. However, the work of Gilbert is mainly important in that it provided a general attitude for the study of landforms as open systems in a steady state (Chorley, 1962), through his laws of declivities, structure and divides, rather than offering a complete geomorphic framework. By stressing the timeless, equilibrium aspects of landforms Gilbert's ideas provided an important antidote to those of Davis and have also stimulated much of the modern work in dynamic geomorphology. More recently the development of climatic geomorphology is expressive of the same type of attitude, although the historical or even cyclic overtones of much of this work often tend to obscure the central input (process)/output (form) relationships. This attitude, foreshadowed by the recognition of the arid cycle by Davis (1905), has found particular fruition among Continental and associated workers (Birot, 1960; Büdel, 1963; Butzer, 1964; Tricart, 1957). Peltier (1950) and Holzner and Weaver (1965) have given statements of the underlying principle that each climatic regime possesses essentially characteristic geomorphic process assemblages which find unique expression in regional topographic distinctions which transcend structural regions. Thus gross climatic process inputs are viewed as resulting in gross geomorphic form outputs, usually irrespective of any considerations of the detailed investigation of the nature of the processes involved. Of course, considerations of climatic changes have caused many present climatic landform assemblages to be viewed as currently adjusting to recently-changed climatic inputs (Büdel, 1959) in which the assumed long relaxation times mean that 'almost everywhere, in particular in the extratropical zone, fossil landforms occupy a far greater area than the landforms developed by currently active processes' (Büdel, 1963). It is therefore partly through cyclical considerations and partly through those of climatic change that the timebound aspects of climatic geomorphology appear.

The other type of black-box system approach to geomorphic phenomena has proved extremely popular and, indeed, contains most of what has come to be regarded as mainstream geomorphology. This approach, too, is significantly vague as to the precise details of form and process relationships within landform assemblages, and is based upon a general attitude that the important and worthwhile aspects of geomorphology involve considerations of the geometrical transformations of landforms through time (i.e. the output) in response to changes of input (usually baselevel changes). These time-bound black-box models stress progressive changes rather than steady-state adjustments and, for this reason, have been termed positive-feedback systems (Fig. 3.1). With these dominantly historical systems we arrive at the 'east' of our *map of geomorphic activity* and find that it has acquired the features of a *globe*, with the two ends joining to form a zone of common historical endeavour, the segment of the real world forming one pole, the model conclusions about the real world the other pole, and the basic reappraisal of the model results being the central axis of the globe. The features of the Davis geographical cycle (Davis, 1909) conforming to this systems approach have been treated previously in some detail (Chorley, 1962 and 1965), with particular reference to the progressive, sequential and irreversible landform transformations attendant upon the input of one large energy pulse (i.e. uplift). Indeed, the cycle seems to operate in some respects like a closed system tending towards maximum entropy. Studies of denudation chronology (e.g. Baulig, 1928; Johnson, 1931; Wooldridge and Linton, 1955) have been dominantly concerned with interpreting geomorphic outputs (peneplain remnants, breaks of slope, etc.) in terms of a discontinuous sequence of supposed input pulses (changes of baselevel, either of diastrophic or eustatic character). The Walther Penck geomorphic system (Penck, 1924 and 1953), superficially so different from the other two, in reality simply carries this positive-feedback black-box model a step further by trying to explain the landform assemblage outputs in terms of assumed continuous inputs of varying patterns of tectonic baselevel changes. In none of the above geomorphic models was there any real need for detailed internal examination of the systems (e.g. of processes) once the initial decision had been taken regarding the supposed dominant system characteristics.

Naturally the more extreme advocates of both the negative- and positive-feedback black-box approaches have encountered criticism (see, for example Bretz's (1962) comments on Hack (1960), and Chorley's (1962 and 1965) opinions regarding the Davisian cycle), but one of the most heartening features of geomorphic work during the past three years or so has been the attempt on the part of several workers to reconcile the two approaches. Thus, Chorley (1966), Holmes (1964) and Howard (1965), among others, have been concerned with the possibility of the construction of a new model of geomophology combining the important elements of both the timeless and timebound models, such that considerations of steady-state adjustments can find a place within a time-directed framework. This attempt has been termed, perhaps not too appropriately, the 'graded system' in Figure 3.1. It has been left to Schumm and Lichty (1965), however, to make what is perhaps the most significant contribution in this connexion. They believe that distinctions

[88] MODELS IN GEOGRAPHY

between cause and effect in the moulding of landforms depend upon the span of time under consideration. During a long period of time ('cyclic time' – Fig. 3.13A) the most important feature of the landform assemblage (or many individual morphometric features – in this case stream channel gradient) is the continual change of form attendant upon the decrease of potential energy through erosion, such that there may be no time-constant

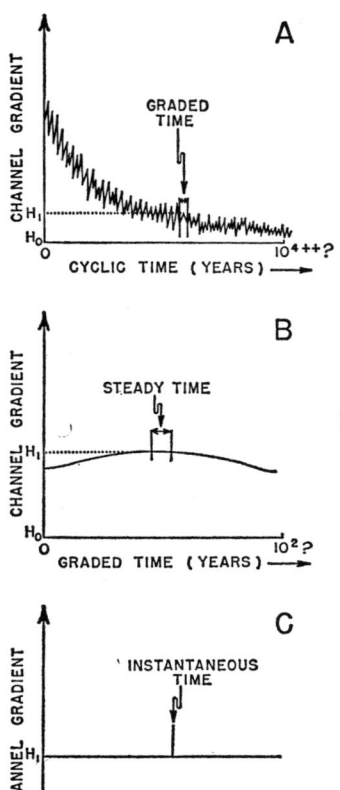

3.13 Diagrams illustrating changes of stream channel gradient at a point through (A) cyclic, (B) graded, and (C) steady time (*Adapted from Schumm and Lichty, 1965*).

relations between many of the dependent and independent variables. Within this progressive change, however, shorter time spans ('graded time' – Fig. 3.13B) can be identified wherein a dynamic equilibrium exists between form and process, largely time-independent and dominated by negative-feedback mechanisms. This graded condition does not apply to all components of a landform system at any one time, but it might be expected that the proportion of graded features would increase with time, and that the periods of temporary graded condition for any one feature would become more frequent with the passage of time. This implies that many correlations between process and response can be fruitfully investigated at this level. During even shorter time spans ('steady time' – Fig. 3.13C) true steady states often exist with very highly-correlated relationships between form and process. Here we are well within the realm of interest of the engineer and the hydraulic geometry relationships of a stream reach having a steady and equal import and export of water and sediment provide an example of this approach. Although this fragmentation of the time-continuum does not provide a completely-integrated model, it does advance us towards a fusion of the timebound and timeless approaches to geomorphology.

TESTING THE MODEL

The testing of the patterns, relationships and process/responses predicted by geomorphic models involves reappraisal against the real world conditions. Although such testing is undoubtedly the most important single step in successful model building, certain types of models have proved more susceptible to reappraisal than others.

The most readily testable models are found among the hardware, partial system and experimental design models. Engineering-type hardware models are constantly checked against reality during construction to allow for adequate representation of present conditions; they are often 'moulded' to known past conditions and tested as to their reproduction of known historical sequences of events; and, finally, predictions from the model usually form the basis for engineering works which themselves form costly tests for the appropriateness of the model (Price and Kendrick, 1963). Similarly, the building of partial system models involves the constant checking of real-world inputs and outputs so that the final model can usually reproduce a limited aspect of reality with reasonable faithfulness. Experimental design models can also be readily tested by collecting new data which are statistically checked against the relationships derived from the original model. The dominant feature of the simple trend-surface model of Breckland sand facies was supported by the results of subsequent sampling design of quite different character (Fig. 3.5D) (Chorley, Stoddart, Haggett and Slaymaker, 1966).

Rather more difficult to test are the process-response models resulting from the synthetic system approach, and the mathematical models. This is because of the higher level of generality presumed by these models and the difficulty of disentangling local complications (i.e. 'noise') from genuine errors in the model. An interesting example of a test of a process-response model has been given by Miller and Zeigler (1964) who tested their small-scale beach sediment model (originally developed for an area of 75 by 26 feet) (Fig. 3.11) over a larger (1,800 by 700 feet) and more complex area. They concluded that these complexities (e.g. the oblique wave attack) merely superimposed other areal sediment features which obscured the simple predictions of the original model. With regard to deterministic mathematical models few concerted attempts have been made to check their implications in the field, but recent measurements of slope transformations in the field (e.g. Schumm, 1964) are providing the kind of information to make this possible. Stochastic models (e.g. Leopold and Langbein, 1962) (Figs. 3.4F and 3.4G), however, seem to have been developed with a view to direct testing.

The most difficult types of model to test are some of the natural analogue system and the black-box general system models because their construction in the past has involved such great leaps into generality based upon decisions regarding the dominant system characteristics the origins of which may be obscure. Thus, for example, certain denudation chronology models and the cycle of erosion involve so many built-in assumptions that any testing to which they have been subjected usually develops into circular reasoning (Chorley, 1965). However, an example of a simpler type of black-box model which received striking field support after a hundred years was Darwin's theory of the development of atolls from fringing and barrier reefs by the slow subsidence of reef foundations. This model was deductively supported by Davis and, after several abortive attempts to drill deep ocean atolls, cores at Eniwetok Atoll in 1951 showed that basalt underlies shallow reef limestones dating from the Eocene at a depth of over 4,000 feet (Stoddart, Personal communication).

The reappraisal of geomorphic models is seldom entirely unequivocal, but testing, discarding and remodelling must become the centre of geomorphic interest if the subject is to develop from a subjective catalogue of phenomena into a coherent and rational discipline.

REFERENCES

AMBROSE, J. W., [1964], Exhumed paleoplains of the pre-Cambrian shield of North America; *American Journal of Science*, 262, 817–857.

AMOROCHO, J., [1965], *Glossary on parametric hydrology* (Tentative); (Mimeo.).

AMOROCHO, J. and HART, W. E., [1964], A critique of current methods in hydrologic systems investigation; *Transactions of the American Geophysical Union*, 45, 307–321.

BAGNOLD, R. A., [1941], *The Physics of Blown Sand and Desert Dunes*, (London), 265 pp.

BAULIG, H., [1928], *Le Plateau Central de la France et sa bordure Mediterranéenne*, (Paris), 590 pp.

BECKETT, P. H. T. and WEBSTER, R., [1962], The storage and collection of information on terrain (An interim report); *Military Engineering Experimental Establishment, Christchurch, Hampshire*, 39 pp., (Mimeo.).

BIROT, P., [1960], Le cycle d'érosion sous les differents climats; *Curso de Altos Estudos Geograficos I, Centro des Pesquisas de Geografico do Brazil*, 137 pp.
BLALOCK, H. M., [1964], *Causal Inferences in Nonexperimental Research*, (Univ. of North Carolina), 200 pp.
BRETZ, J. H., [1962], Dynamic equilibrium and the Ozark land forms; *American Journal of Science*, 260, 427–438.
BRUSH, L. M., and WOLMAN, M. G., [1960], Knickpoint behavior in noncohesive material: A laboratory study; *Bulletin of the Geological Society of America*, 71, 59–74.
BÜDEL, J., [1959], The periglacial-morphologic effects of the Pleistocene climate over the entire world; *International Geology Review*, 1, 1–16.
BÜDEL, J., [1963], Klima-genetische geomorphologie; *Geographische Rundschau*, 15, 269–285.
BUTZER, K. W., [1964], *Environment and Archaeology*, (London), 524 pp.
CARLSTON, C. A., [1963], Drainage density and streamflow; *U.S. Geological Survey, Professional Paper* 422-C, 8 pp.
CHORLEY, R. J., [1957], Climate and morphometry; *Journal of Geology*, 65, 628–638.
CHORLEY, R. J., [1962], Geomorphology and general systems theory; *U.S. Geological Survey, Professional Paper* 500-B, 10 pp.
CHORLEY, R. J., [1964], Geography and analogue theory; *Annals of the Association of American Geographers*, 54, 127–137.
CHORLEY, R. J., [1965], A re-evaluation of the geomorphic system of W. M. Davis; Ch. 2 in Chorley, R. J. and Haggett, P. (Eds.), *Frontiers in Geographical Teaching*, (London).
CHORLEY, R. J., [1966], The application of statistical methods to geomorphology; In Dury, G. H. (Ed.), *Essays in Geomorphology*, (London).
CHORLEY, R. J., DUNN, A. J. and BECKINSALE, R. P., [1964], *The History of the Study of Landforms*, Vol. I, (London), 678 pp.
CHORLEY, R. J. and HAGGETT, P., [1965], Trend-surface mapping in geographical research; *Transactions of the Institute of British Geographers*, 37, 47–67.
CHORLEY, R. J. and MORGAN, M. A., [1962], Comparison of morphometric features, Unaka Mountains, Tennessee and North Carolina, and Dartmoor, England; *Bulletin of the Geological Society of America*, 73, 17–34.
CHORLEY, R. J., STODDART, D. R., HAGGETT, P. and SLAYMAKER, H. O., (1966), Regional and local components in the areal distribution of surface sand facies in, the Breckland, Eastern England; *Journal of Sedimentary Petrology*, 36, 209–220.
CULLING, W. E. H., [1963], Soil creep and the development of hillside slopes; *Journal of Geology*, 71, 127–161.
CULLING, W. E. H., [1965], Theory of erosion on soil-covered slopes; *Journal of Geology*, 73, 230–254.
CURL, R. L., [1959], Stochastic models of cavern development; *Bulletin of the Geological Society of America*, 70, 1802, (Abstract).
DANIEL, V., [1955], The uses and abuses of analogy; *Operations Research Quarterly*, 6, 32–46.
DAVIS, W. M., [1905], The geographical cycle in an arid climate; *Journal of Geology*, 13, 381–407.
DAVIS, W. M., [1909], *Geographical Essays*, (Boston), 777 pp.

DUNCAN, W. J., [1953], *Physical Similarity and Dimensional Analysis*, (London), 156 pp.
DYLIK, J., [1957], Dynamical geomorphology, its nature and methods; *Bulletin de la Société des Sciences et des Lettres de Lódz*, Classe III, Vol. VIII, 12, 1–42.
ELLERMEIER, R. D. and SIMONETT, D. S., [1965], Imaging radars on spacecraft as a tool for studying the earth; *Center for Research Inc., Engineering Science Division, University of Kansas, Lawrence, Kansas, CRES Rept.* No. 61–6, 25 pp.
FENNEMAN, N. M., [1914], Physiographic boundaries within the United States; *Annals of the Association of American Geographers*, 4, 84–134.
FRIEDKIN, J. F., [1945], A laboratory study of the meandering of alluvial rivers; *U.S. Waterways Experiment Station, Vicksburg, Mississippi*, 40 pp.
GILBERT, G. K., [1877], Land sculpture; Ch. 5 in *The Geology of the Henry Mountains*; U.S. Department of the Interior, Washington.
GILBERT, G. K., [1914], The transportation of debris by running water; *U.S. Geological Survey, Professional Paper* 86, 263 pp.
HACK, J. T., [1960], Interpretation of erosional topography in humid temperate regions; *American Journal of Science*, 258A, 80–97.
HARRISON, W. and KRUMBEIN, W. C., [1964], Interactions of the beach-ocean-atmosphere system at Virginia Beach, Va.; *U.S. Army Coastal Engineer Research Center, Technical Memo.* No. 7.
HEACOCK, R. L., KUIPER, G. P., SHOEMAKER, E. M., UREY, H. C. and WHITAKER, E. A. (Eds.), [1966], Ranger VIII and IX: Part II. Experimenters' Analyses and Interpretations; *National Aeronautics and Space Administration, Technical Rept.* No. 32–800, (Jet Propulsion Lab., Pasadena, Calif.), 382 pp.
HILLS, E. S., [1961], Morphotectonics and the geomorphological sciences, with special reference to Australia; *Quarterly Journal of the Geological Society of London*, 117, 77–89.
HILLS, E. S., [1963], Geomorphology and structure morphotectonics; Ch. 14 in *Elements of Structural Geology*, (London), 483 pp.
HOLMES, C. D., [1964], Equilibrium in humid-climate physiographic processes; *American Journal of Science*, 262, 436–445.
HOLZNER, L. and WEAVER, G. D., [1965], Geographic Evaluation of climatic and climato-genetic geomorphology; *Annals of the Association of American Geographers*, 55, 592–602.
HORTON, R. E., [1945], Erosional development of streams and their drainage basins; Hydrophysical approach to quantitative morphology; *Bulletin of the Geological Society of America*, 56, 275–370.
HOWARD, A. D., [1965], Geomorphological systems – equilibrium and dynamics; *American Journal of Science*, 263, 302–312.
HUBBERT, M. K., (1937), Theory of scale models as applied to the study of geological structures; *Bulletin of the Geological Society of America*, 48, 1459–1520.
HUTTON, J., [1795], *Theory of the Earth*, 2 Vols., (Edinburgh).
IMBRIE, J., [1963], Factor and vector analysis programs for analysing geologic data; *Office of Naval Research Project* 389–135, *Technical Report 6, Department of Geology, Columbia University, New York*.
JEFFREYS, H., [1918], Problems of denudation; *Philosophical Magazine*, 6th Series, 36, 179–190.

JOHNSON, D. W., [1919], *Shore Processes and Shoreline Development*, (New York), 584 pp.
JOHNSON, D. W., [1931], *Stream Sculpture on the Atlantic Slope*, (New York), 142 pp.
KENNEDY, B. A., [1965], *An Analysis of the Factors Influencing Slope Development on the Charmouthien Limestone of the Plateau de Bassigny, Haute-Marne, France*, B.A. Thesis, Department of Geography, Cambridge University.
KING, C. A. M., [1966A], *Techniques in Geomorphology*, (London), 342 pp.
KING, C. A. M., [1966B], An introduction to factor analysis with a geomorphological example from Northern England; *Bulletin of Quantitative Data for Geographers No. 6, Department of Geography, Nottingham University*.
KITTS, D. B., [1963], Historical explanation in geology; *Journal of Geology*, 71, 297–313.
KRUMBEIN, W. C., [1955], Experimental design in the earth sciences; *Transactions of the American Geophysical Union*, 36, 1–11.
KRUMBEIN, W. C., [1959], The 'sorting out' of geological variables illustrated by regression analysis of factors controlling beach firmness; *Journal of Sedimentary Petrology*, 29, 575–587.
KRUMBEIN, W. C., [1966], A comparison of polynomial and Fourier models in map analysis; *Office of Naval Research Project 388-078, Technical Report 2, Department of Geology, Northwestern University, Evanston*.
KRUMBEIN, W. C., BENSON, B. T. and HEMPKINS, W. B., [1964], WHIRLPOOL, a computer program for 'sorting out' independent variables by sequential multiple linear regression; *Office of Naval Research Project 398-135, Technical Report 14, Department of Geology, Northwestern University, Evanston*, 49 pp.
KRUMBEIN, W. C. and GRAYBILL, F. A., [1965], *An Introduction to Statistical Models in Geology*, (New York), 475 pp.
KRUMBEIN, W. C. and MILLER, R. L., [1953], Design of experiments for statistical analysis of geological data; *Journal of Geology*, 61, 510–532.
LANGBEIN, W. B. and LEOPOLD, L. B., [1964], Quasi-equilibrium states in channel morphology; *American Journal of Science*, 262, 782–794.
LANGHAAR, H. L., [1951], *Dimensional Analysis and Theory of Models*, (New York), 166 pp.
LATHAM, J. P., [1966], Remote sensing of the environment; *Geographical Review*, 56, 288–291.
LEOPOLD, L. B. and LANGBEIN, W. B., [1962], The concept of entropy in landscape evolution; *U.S. Geological Survey, Professional Paper* 500-A, 20 pp.
LEOPOLD, L. B. and MADDOCK, T., [1953], The hydraulic geometry of stream channels and some physiographic implications; *U.S. Geological Survey, Professional Paper* 252, 57 pp.
LEOPOLD, L. B., WOLMAN, M. G. and MILLER, J. P., [1964], *Fluvial Processes in Geomorphology*, (San Francisco), 522 pp.
LEWIS, W. V. and MILLER, M. M., [1955], Kaolin model glaciers; *Journal of Glaciology*, 2, 535–538.
MASSACHUSETTS INSTITUTE OF TECHNOLOGY, [1956], Design and operation of an hydraulic analogue computer for studies of freezing and thawing of soils; *Soil Engineering Division, Department of Civil Engineering, Technical Report* No. 62, Corps of Engineers, U.S. Army, 38 pp.

MCCAULEY, J. F., [1964], A preliminary report on the terrain analysis of the lunar equatorial belt; *U.S. Geological Survey (Preliminary), NASA Contract*, 44 pp.
MCCONNELL, H., [1965], Randomness in spatial distributions of terrain summits; Department of Geography, Northern Illinois University, DeKalb, Illinois, (Mimeo.).
MELTON, F. A., [1959], Aerial photographs and structural geomorphology; *Journal of Geology*, 67, 355–370.
MELTON, M. A., [1957], An analysis of the relations among elements of climate, surface properties and geomorphology; *Office of Naval Research Project NR 389-042, Technical Report* 11, Department of Geology, Columbia University, New York, 102 pp.
MELTON, M. A., [1958A], Geometric properties of mature drainage systems and their representation in an E_4 phase space; *Journal of Geology*, 66, 35–54.
MELTON, M. A., [1958B], Correlation structure of morphometric properties of drainage systems and their controlling agents; *Journal of Geology*, 66, 442–460.
MELTON, M. A., [1960], Intravalley variation in slope angles related to microclimate and erosional environment; *Bulletin of the Geological Society of America*, 71, 133–144.
MELTON, M. A., [1962], Methods for measuring the effect of environmental factors on channel properties; *Journal of Geophysical Research*, 67, 1485–1490.
MILLER, R. L. and KAHN, S. J., [1962], *Statistical Analysis in the Geological Sciences*, (New York), 357 pp.
MILLER, R. L. and ZEIGLER, J. M., [1958], A model relating dynamics and sediment pattern in equilibrium in the region of shoaling waves, breaker zone and foreshore; *Journal of Geology*, 66, 417–441.
MILLER, R. L. and ZEIGLER, J. M., [1964], A study of sediment distribution in the zone of shoaling waves over complicated bottom topography; Ch. 8 in Miller, R. L. (Ed.), *Papers in Marine Geology: Shepard Commemoration Volume*, (New York).
MOORE, R. K., [1966], Radar as a remote sensor; *Center for Research Inc., Engineering Science Division, University of Kansas, Lawrence, Kansas, CRES Rept. No. 61-67*, 55 pp.
MURPHY, N. F., [1949], Dimensional analysis; *Bulletin of the Virginia Polytechnic Institute, Engineering Experiment Station Series* No. 73, Vol. XLII (6), 41 pp.
NEYMAN, J. and SCOTT, E. L., [1957], On a mathematical theory of populations conceived as a conglomeration of clusters; *Cold Spring Harbor Symposia on Quantitative Biology*, 22, 109–120.
OFFICE OF NAVAL RESEARCH, [1966], Proceedings of the Fourth Symposium on Remote Sensing of Environment (12–14 April, 1966); *Infrared Physics Laboratory, Willow Run Laboratories, Institute of Science and Technology, University of Michigan*, Ann Arbor, Michigan, 871 pp.
PELTIER, L. C., [1950], The geomorphic cycle in periglacial regions as it is related to climatic geomorphology; *Annals of the Association of American Geographers*, 40, 214–236.
PENCK, W., [1924], *Die morphologische Analyse*, (Stuttgart), 283 pp.
PENCK, W., [1953], *Morphological Analysis of Landforms*, Translated by H. Czech and K. C. Boswell, (London), 429 pp.

PIEXOTO, J. P., SALTZMAN, B. and TEWELES, S., [1964], Harmonic analysis of the topography along parallels of the earth; *Journal of Geophysical Research*, 69, 1501–1505.
PRICE, W. A. and KENDRICK, M. P., [1963], Field and model investigation into the reasons for siltation in the Mersey Estuary; *Proceedings of the Institution of Civil Engineers*, 24, 473–517.
ROUSE, J. W., WAITE, W. P. and WALTERS, R. L., [1966], Use of orbital radars for geoscience investigations; *Center for Research Inc., Engineering Science Division, University of Kansas, Lawrence, Kansas, CRES Rept. No. 61-8*, 31 pp.
ROWAN, L. C. and MCCAULEY, J. F., [1966], Lunar terrain analysis; In *Lunar Orbiter-Image Analysis Studies Report, U.S. Geological Survey, NASA Contract W-12*, 123, 89–129.
SAVIGEAR, R. A. G., [1965], A technique of morphological mapping; *Annals of the Association of American Geographers*, 55, 514–538.
SAVILLE, T., [1950], Model study of sand transport along an infinitely long straight beach; *Transactions of the American Geophysical Union*, 31 (4), 555–565.
SCHEIDEGGER, A. E., [1961], Mathematical models of slope development; *Bulletin of the Geological Society of America*, 72, 37–49.
SCHEIDEGGER, A. E., [1964], Some implications of statistical mechanics in geomorphology; *Bulletin of the International Association of Scientific Hydrology*, Year 9, 12–16.
SCHENCK, H., [1963], Simulation of the evolution of drainage-basin networks with a digital computer; *Journal of Geophysical Research*, 68, 5739–5745.
SCHUMM, S. A., [1956], The evolution of drainage systems and slopes in badlands at Perth Amboy, New Jersey; *Bulletin of the Geological Society of America*, 67, 597–646.
SCHUMM, S. A., [1964], Seasonal variations of erosion rates and processes on hillslopes in western Colorado; *Zeitschrift für Geomorphologie*, Supplement 5, 215–238.
SCHUMM, S. A. and LICHTY, R. W., [1965], Time, space, and causality in geomorphology; *American Journal of Science*, 263, 110–119.
SIEGEL, S., [1956], *Nonparametric Statistics for the Behavioral Sciences*, (New York), 312 pp.
SIMPSON, G. G., [1963], Historical science; In Albritton, C. C. (Ed.), *The Fabric of Geology*, (Reading, Mass.), 24–48.
SIMPSON, R. B., [1966], Radar, Geographical tool; *Annals of the Association of American Geographers*, 56, 80–89.
SOKAL, R. R., [1965], Statistical methods in systematics; *Biological Reviews of the Cambridge Philosophical Society*, 40, 337–391.
SPEIGHT, J. G., [1965], Meander spectra of the Angabunga River, Papua; *Journal of Hydrology*, 3, 1–15.
STODDART, D. R., (Personal Communication), *Darwin's coral reef theory and deep drilling through reefs.*
STRAHLER, A. N., [1950], Equilibrium theory of erosional slopes approached by frequency distribution analysis; *American Journal of Science*, 248, 673–696 and 800–814.

STRAHLER, A. N., [1952], Dynamic basis of geomorphology; *Bulletin of the Geological Society of America*, 63, 923–938.
STRAHLER, A. N., [1954], Statistical analysis in geomorphic research; *Journal of Geology*, 62, 1–25.
STRAHLER, A. N., [1958], Dimensional analysis applied to fluvially eroded landforms; *Bulletin of the Geological Society of America*, 69, 279–300.
TRICART, J., [1957], Application du concept de zonalité à la géomorphologie; *Tijdschrift van Het Koninklijk Nederlandsch Aardrijkskundig Genootschap*, 74, 423–430.
TRICART, J., [1964], Geomorphological mapping; *Arid Zone*, No. 24, 12–14.
VAN LOPIK, J. R. and KOLB, C. R., [1959], A technique for preparing desert terrain analogs; *U.S. Army Engineer Waterways Experiment Station, Vicksburg, Mississippi, Technical Report* 3-506, 70 pp.
VON BERTALANFFY, L., [1962], General system theory – A critical review; *General Systems*, 7, 1–20.
WHITTEN, E. H. T., [1964], Process-response models in geology; *Bulletin of the Geological Society of America*, 75, 455–464.
WILLS, L. J., [1929], *The Physiographical Evolution of Britain*, (London), 376 pp.
WILSON, E. B., [1952], *An Introduction to Scientific Research*, (New York).
WONG, S. T., [1963], A multivariate statistical model for predicting mean annual flood in New England; *Annals of the Association of American Geographers*, 53, 298–311.
WOOLDRIDGE, S. W. and LINTON, D. L., [1955], *Structure, Surface and Drainage in South East England*, 2nd Ed., (London), 176 pp.
WURM, A., [1935], Morphologische Analyse und Experiment Hangentwicklung, Einebnung Piedmontreppen; *Zeitschrift für Geomorphologie*, 9, 57–87.

CHAPTER FOUR

Models in Meteorology and Climatology
R. G. BARRY

INTRODUCTION

Meteorological theory rests upon the basic laws of physics and hydrodynamics, and climatology, treating the seasonal characteristics of weather phenomena over specific areas in terms of general relationships between the atmosphere and the earth's surface, is dependent on the fundamentals of meteorology. In consequence climatological models are primarily statistical while those of meteorology are physical-mathematical, although the two approaches are by no means mutually exclusive. Since much research effort during the past twenty years has been devoted to problems of the global air circulation the principal themes of this chapter are 'dynamic climatology' (Bergeron, 1930; Court, 1957; Huschke, 1959) which deals with large-scale atmospheric circulations and synoptic-scale models which are a necessary adjunct for the understanding of global patterns. Models equally have their place, however, in bioclimatology and physical climatology and some important aspects of the latter are discussed in Chapter 5.

It is intended first to summarize some of the important laws and relationships of thermodynamics and hydrodynamics indicating sources which provide a fuller treatment of them. It should be noted that several of these basic principles apply to idealized (model) conditions.

(1) *The first law of thermodynamics.* This refers to the conservation of energy within a thermodynamic system. The law states that the heat supplied to a gas is equal to the increase in its internal energy plus the work carried out in expansion against its surroundings. When for example differential heating sets up a thermally direct circulation, such as a sea-breeze, kinetic energy is produced at the expense of internal and potential energy by the rising of warmer air and sinking of cooler air. (Willett and Sanders, 1959, p. 144; Hare, 1965.)

(2) *The second law of thermodynamics.* The entropy of a closed system either increases or remains constant during any process operating within the system. In meteorology, it is convenient to use potential temperature (the temperature of unsaturated air brought adiabatically to 1,000 mb pressure)

which is a direct function of entropy (Hess, 1959, p. 20; Petterssen, 1956, p. 11). Entropy and potential temperature are constant for dry adiabatic processes.

(3) *The geostrophic wind equation.* This is a model for wind flow based on Newton's laws of motion. It expresses the wind velocity as a balance between the horizontal pressure force and the horizontal Coriolis deflection due to the earth's rotation when there is no friction or curvature of the isobars. The velocity is proportional to the pressure gradient. The Coriolis parameter ($2\Omega \sin \varphi$) is twice the angular velocity of the earth's rotation times the sine of the latitude angle (Hess, 1959, p. 175; Willett and Sanders, 1959, p. 118; Crowe, 1965). The gradient wind equation which includes the centripetal (or cyclostrophic) term for curved isobars is usually an unnecessary refinement. Atmospheric models commonly use the geostrophic approximation which relates the wind and pressure fields and acts as a 'noise filter' removing small-scale ageostrophic components.

(4) *The thermal wind equation.* This relates the geostrophic wind at an upper level to the low-level geostrophic wind and the mean temperature in the intervening tropospheric layer. The thickness of this layer is proportional to its mean temperature (Sutcliffe and Forsdyke, 1951; Hess, 1959, p. 195). In analogy to Buys Ballot's wind law the thermal wind in the northern hemisphere is parallel to the thickness lines with low thickness (cold air) to the left.

(5) *The continuity equation* (the principle of conservation of mass). This states that for an incompressible atmosphere, where density is unchanged by motion, the horizontal divergence of the wind $\left(\dfrac{\partial u}{\partial x}+\dfrac{\partial v}{\partial y}\right)$ is balanced by downward vertical motion $\left(-\dfrac{\partial w}{\partial z}\right)$. u, v and w are the velocity components along the axes x (eastward), y (northward) and z (vertical), respectively. It may be noted that geostrophic winds are non-divergent and therefore cannot be weather-producing. The complete equation of continuity shows that the local rate of change of density is determined by the net rate of mass inflow (Hess, 1959, p. 212; Petterssen, 1956, p. 7; Sutton, 1962, p. 83).

(6) *Pressure tendency.* Combining the continuity equation with the hydrostatic equation $\left(\dfrac{dp}{dz}=-g\rho,\text{ where }\dfrac{dp}{dz}\text{ is the pressure change with height, }g\right.$ gravity, ρ density), the rate of pressure change at sea-level which is termed the pressure tendency is shown to depend on the vertically integrated net mass convergence (Hess, 1959, p. 219).

(7) *The vorticity equation.* Vorticity is a measure of the rotation of an infinitely small fluid element. It may arise through curvature of the streamlines

or wind shear across a current. Absolute vorticity about a vertical axis (Q) is the sum of the vorticity due to the earth's rotation ($f=2\Omega \sin \varphi$) plus the vorticity of motion about a vertical axis relative to the earth's surface

$$\left(\zeta = \frac{\partial v}{\partial x} - \frac{\partial u}{\partial y}, \text{ positive when cyclonic}\right)$$

$$\text{i.e. } Q = f + \zeta$$

The rate of change of absolute vorticity $\left(\frac{dQ}{dt}\right)$ is proportional to the convergence (negative divergence) if the motion is frictionless and there is no lateral gradient of vertical velocity.

$$\frac{dQ}{dt} = -\left(\frac{\partial u}{\partial x} + \frac{\partial v}{\partial y}\right) Q, \text{ or } \frac{dQ}{dt} = -DQ \text{ (The vorticity equation)}$$

For large-scale horizontal motion the vertical component of absolute vorticity is approximately conserved

$$\frac{d(f+\zeta)}{dt} = 0$$

(Hess, 1959, p. 247; Petterssen, 1956, p. 125; Scorer, 1958, p. 49; Sutton, 1962, p. 80; Willett and Sanders, 1959, p. 148).

These fundamental principles provide the basis for a great range of models characterizing various aspects of the complex processes in the atmosphere. The construction of conceptual, mathematical and experimental models is essential throughout the whole field of atmospheric science and the treatment in this chapter is necessarily selective. The approach is to examine first the large-scale models for the general circulation and the westerlies, then synoptic models in mid-latitudes and the quite different ones developed for the tropics, specialized meso-scale models, followed by a section on the application of models in synoptic climatology and palaeoclimatology and finally operational forecasting models.

THE GENERAL CIRCULATION

The aim of dynamic climatology is ultimately to provide a comprehensive explanation of the general circulation; that is, the large-scale motion of the atmosphere in time and space. The behaviour of the atmosphere follows the classical laws of thermodynamics and hydrodynamics yet the complexities introduced by surface effects, transformations of energy and interacting scales of behaviour, together with observational limitations, have so far precluded the development of an adequate theoretical model. The first real progress in this direction has only recently been achieved by Smagorinsky (1963, 1964) and his associates in the United States Weather Bureau (Manabe et al, 1965) with a programme of successively more sophisticated models.

The time-scale appropriate to the problem of the general circulation balance of heat and momentum is often regarded as being of the order of months or years and major wind and pressure systems are treated statistically. Indeed, Mörth (1964) suggests that the mean fields of temperature, pressure and wind along a meridional cross-section can be explained in terms of radiative control by ozone in the lower stratosphere – upper troposphere and water vapour in the middle and lower troposphere. However, many circulation models represent attempts to explain those elements of individual synoptic flow patterns which occur regularly. The principal features of the circulation may be summarized following Fleagle (1957):

(1) large amplitude unstable disturbances in the westerlies in middle and high latitudes. In winter there are about eight to ten at 45° latitude and two to three near the pole.
(2) mainly slow, steady easterly flow in low latitudes, but including intense vortices.
(3) the frequent presence of jet streams in the upper troposphere over middle latitudes.

Jeffreys (1926) recognized that the mid-latitude westerlies would cease due to friction in about ten days if their momentum source were cut off and similarly Sutcliffe (1949) noted that the general dynamic and thermodynamic balance of the atmosphere appears to be maintained over periods of a few days rather than over a month or year. In view of this it seems unlikely that a satisfactory model of the general circulation can be based solely on a simple climatological view of its characteristics. In fact there are no *a priori* reasons for supposing that space – time averages of the circulation are easier to comprehend than synoptic patterns since 'filtering' by taking averages may create its own special problems of interpretation.

Conceptual models of the general circulation

The history of attempts to characterize the essential structure of the general circulation provides illustrations of all the major categories of model building although until recently the majority have been conceptual. As early as 1686 Edmund Halley outlined a thermal circulation model with maximum heating in low latitudes and a thermally direct cell accounting for the equatorward flow of the trade winds. On a stationary earth a direct cell would produce a meridional-plane circulation as shown in Figure 4.1. This scheme was improved by Hadley (1735) who incorporated the effects of the earth's rotation into the model to explain the north-easterly and south-easterly trades (cf. Crowe,

1965). Hadley envisaged a compensatory south-westerly counter-current above the trades (Fig. 4.2) and this meridional-plane toroidal circulation is still referred to as the 'Hadley cell'.

These models overlooked the westerly wind belt and the first reasonably complete picture of the major wind systems was not produced until 1856 by Ferrel. He postulated three meridional cells in a scheme not unlike that of Rossby in 1941 (Fig. 4.3). Ferrel outlined the concept of the conservation of angular momentum and attempted to explain the location of the sub-tropical anticyclones in these terms. The concept implies that zonal rings of

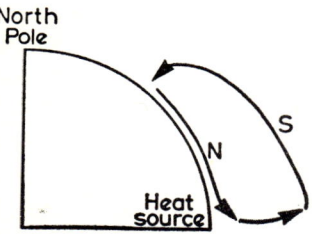

4.1 Thermally-direct cell on a stationary earth (*Halley model, 1686*).

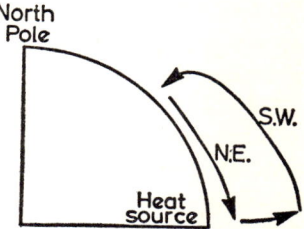

4.2 Thermally-direct cell on a rotating earth (*Hadley model, (1735)*).

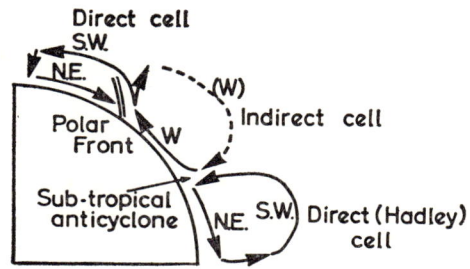

4.3 Three-cell model of the mean circulation (*after Rossby, 1941*).

air (between two latitude circles) develop westerly winds (westerly relative momentum) if displaced polewards and easterly winds if displaced equatorwards. Ferrel's ideas did not affect the mainstream of meteorological study, though in recognition of his work the mid-latitudes westerlies are sometimes referred to as the Ferrel westerlies.

The three cell model with mean meridional-plane circulations was again put forward by Bergeron (1928) and later by Rossby (1941) in modified form. Basically, Rossby assumed that zonal circulation rings conserve their absolute angular momentum. Nevertheless, he recognized the existence of strong upper westerlies in mid-latitudes and suggested that these could be accounted for if westerly momentum from the upper branches of the tropical and polar cells (Fig. 4.3) were spread as a result of lateral mixing (in the horizontal plane) by tropospheric wave patterns. Rossby considered that the frictionally-driven mid-latitude cell was in this way obscured by the upper westerlies.

Rossby's introduction of longitudinal variations into the model had in fact been anticipated by Jeffreys who in 1926 demonstrated that the poleward transport of angular momentum required to maintain the westerlies could be carried out solely by eddy circulations in the horizontal plane. However, his ideas lay dormant until upper data became readily available after World War II. Then Priestley (1949; and 1951) calculated the transport of momentum and heat by mean winds and eddy circulations and a major research programme along the same lines was undertaken at Massachusetts Institute of Technology (Starr, 1948; Starr and White, 1951, 1952 and 1954).

The method of analysing these transports derives from studies of fluid flow in pipes by O. Reynolds in the late nineteenth century. He regarded the instantaneous total transport (or flux) as consisting of a time-mean flow plus a superimposed eddy component. In the same way if we consider the local meridional flow across a latitude circle at a given pressure level then this flow v can be separated into the space-mean flow across the whole latitude circle $[v]$ and the deviation from the mean v'

$$v = [v] + v'$$

The poleward transport of momentum is expressed by uv where u and v are respectively the eastward and northward wind components. Thus, the space-mean poleward momentum flux is

$$[uv] = [u][v] + [u'v']$$

Extending the method to the space-time average, where the bar denotes a time average and [ ]' is the deviation of a space average from a space-time mean, we obtain following Starr and White (1952)

$$[\overline{uv}] = [\overline{u}][\overline{v}] + [\overline{u'v'}]$$
$$\text{and } [\overline{uv}] = [\overline{u}][\overline{v}] + [\overline{[u]'[v]'}] + [\overline{u'v'}]$$

$[\overline{u}][\overline{v}]$ is the transport due to mean meridional circulations
$\overline{[u]'[v]'} = \overline{[u][v]} - [\overline{u}][\overline{v}]$ represents instantaneous net meridional circulations due to time correlations between space means.
$[\overline{u'v'}] = [\overline{uv}] - \overline{[u][v]}$ is the transport by large-scale horizontal eddies arising from spatial correlations between the instantaneous wind components u and v.

Starr and White (1954) used this technique to examine the meridional transports of momentum, heat and water vapour across selected latitude circles in the northern hemisphere during the year 1950. The results of their calculations showed that the horizontal eddy component (surface cyclones and anticyclones and tropospheric troughs and ridges) accounted for a large proportion of the balance requirements of energy and momentum determined independently by other workers. In spite of these results, however, it would be incorrect to formulate a circulation model with only horizontal eddy transports. Palmén's model (1951B) of the mean winter circulation in the northern hemisphere (Fig. 4.4) has been verified by Tucker (1959) who finds evidence of a weak, indirect mid-latitude cell and an important low latitude Hadley cell for momentum transport. The role of the Hadley cell has also been emphasized in terms of the low-level equatorward transport of vapour in the tropics by Palmén and Vuorela (1963).

Finally, some of the complexities of the low latitude circulation problem may be mentioned. The gross over-simplification of the atmosphere's heat sources and sinks has been a common feature of circulation models until quite recently. In particular, the fact that the earth does not possess a simple

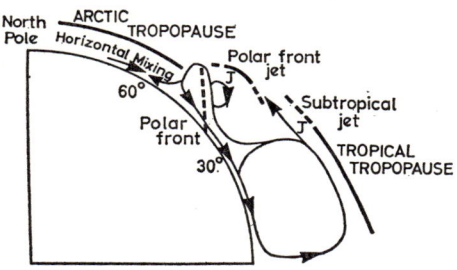

4.4 The mean meridional circulation in winter (*after Palmén, 1951*). J indicates the positions of the jet streams. The latitude of the polar front is very variable.

heat source at the equator is usually insufficiently taken into account. Riehl (1926B), for example, shows that upward heat transfer in the equatorial zone is localized within large cumuli rather than occurring generally. The concept of a general anti-trade is also erroneous as the poleward heat flux is confined to limited longitudinal sectors. In addition to these points there is increasing evidence (Tucker, 1965) that inter-hemispheric exchanges cannot be neglected in hemispheric circulation models.

Sufficient illustration has been given of the complexity of the general circulation to show the difficulties facing those workers who seek to develop adequate models of its behaviour. The need to isolate the really fundamental processes at work in the atmosphere has encouraged the application of mathematical and experimental methods which will now be discussed.

Mathematical models

Much current research into the operation of the general circulation uses mathematical models based on simplified hydrodynamical equations. Certain atmospheric properties are specified, many of them parametrically (by reference to empirical results or semi-intuitive ideas), appropriate boundary conditions are assumed and calculations are made of the resultant circulations. There are two basic atmospheric models. An atmosphere where isobaric surfaces and constant density (isosteric) or constant temperature surfaces are parallel is referred to as *barotropic*, while one in which these surfaces intersect is termed *baroclinic*. A frontal zone with sharp intersection of temperature and pressure surfaces is hyper-baroclinic (Fig. 4.9). If we visualize sets of isobaric-isosteric surfaces intersecting in space they form tightly packed quadrilateral tubes termed solenoids.

Some of the most significant contributions to general circulation theory have come from attempts to explain the development of unstable waves in a baroclinic current (Eliassen, 1956; Fleagle, 1957). The model assumes that potential temperature increases with height (a stable stratification) and that there is a poleward gradient of potential temperature. In a wave disturbance superimposed on a basic westerly current air moving north ahead of the trough is rising while that moving south in the rear is sinking. The north-south slope of the moving air (δ) is determined by

$$\delta = \frac{\mu - \beta f / 2HS\alpha^2}{1 + f^2 / 2H^2 S\alpha^2} \qquad \text{(Fleagle, 1960)}$$

where μ = the slope of the potential isotherms (isentropes)
$\beta = \partial f / \partial y$, the variation of the Coriolis parameter (f) with latitude
H = the depth of atmosphere (10 km.)
S = the static stability
α = the wave number = $2\pi/L$ (L = wavelength).

For short-waves, where α is large, the slope δ is much greater than μ as shown in Figure 4.5A and since the rising air has a lower potential temperature

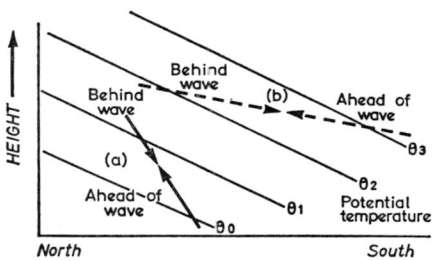

4.5 Baroclinic instability model.
(a) short wavelength case
(b) medium wavelength case
The potential isotherms refer to arbitrary values.

than the subsiding air the potential energy is increased. This means that no energy is available for growth of the disturbance. With medium wavelengths, still with moderate vertical motion, the potential energy is being decreased by the southward movement of sinking potentially cold air and northward movement of rising potentially warm air shown in Figure 4.5B. The maximum conversion of potential energy to kinetic energy occurs for $\delta=\mu/2$ when $L=2\pi H\sqrt{(2S/f)}$. For very long waves the vertical motion is too slight to produce significant energy exchanges.

The maximum slope of potential temperature surfaces is between 30-60° latitude and since β/f increased southwards, causing δ to approach μ, the maximum baroclinic instability is found in higher mid-latitudes. Theoretically, there is a critical latitude at 35° with an essentially stable atmosphere equatorward of this. These deductions, go some way towards accounting for the observed circulation features.

One of the most interesting mathematical experiments is due to Phillips (1956) who used a two-layer baroclinic model of the atmosphere (see p. 135). It was assumed that non-adiabatic heating occurred at low latitudes with a high-latitude sink so that no net heating resulted. Starting from an atmosphere at rest with uniform temperature gradient, zonal flow developed after 130 'days' with a slow meridional cell and equatorward flow at low levels. However, after the introduction of random disturbances and longitudinal variations into the model three meridional cells developed and a marked westerly jet stream appeared. Energy calculations showed that the eddies supplied kinetic energy to the mean zonal flow, a result anticipated by Rossby (1949) and by other theoretical investigations of Kuo and Fjortoft, but contrary to the 'classical' view that the perturbations are driven by the kinetic energy of the mean flow. Smagorinsky (1963) elaborated these techniques using a two-level model with the 'primitive equations of motion' which allow for ageostrophic components and thus provide a proper representation of wind flow in low latitudes. Subsequent work is based on a nine-level model incorporating the effects of surface boundary layer fluxes, computed radiative effects and the moisture cycle (Manabe et al, 1965). The reduced dependence on parametric representation of such factors is a major advance on previous models. The computations provide a realistic simulation of many atmospheric features in at least qualitative terms and the quantitative agreement with a number of observed characteristics is also good. Ultimately it should be possible to describe the entire climatology of atmospheric motions by means of a time-dependent, deterministic model.

Experimental Models

Important contributions to general circulation theory have also been made by laboratory studies of rotating fluids. The crucial problem in developing useful experimental models is a knowledge of the similarity conditions between the model and physical reality. The quantities normally requiring description are lengths, flow characteristics and fluid properties such as density and viscosity. In the case of the atmosphere scale presents a problem since the depth of the model layer cannot be reduced in the same proportion as the horizontal dimension, but such difficulties are overcome if the equations of motion for the atmosphere and the model are made directly comparable through non-dimensional parameters. An important parameter is the Rossby number (Ro), the ratio of the velocity relative to the absolute speed of the point on the equator.

$$Ro = u/Ce$$

When Ro is small (approximately <0.2) the motion is quasi-geostrophic. One intractable problem of working models is that gravity is always a force acting in the same direction.

There have been three major types of experimental model (see Fultz, 1960). They are (1) the Benard cell in which a non-rotating fluid is either heated from below or cooled from above. Cellular convection develops when the temperature gradient exceeds the critical value necessary for instability. A large critical temperature gradient is required if the fluid is rotated; (2) the Hide annulus (Hide, 1953; Riehl and Fultz, 1957 and 1958) in which a tall column of water is rotated between two concentric cylinders. The temperature of the outer wall is gradually raised and that of the inner wall lowered, keeping a constant temperature in the liquid. Beyond a critical temperature gradient the initial slow, symmetrical, easterly flow (Hadley regime) breaks down first into a seven-wave westerly pattern, then with systematically fewer waves down to an eccentric westerly vortex as the temperature gradient is increased (Fig. 4.6); this sequence is referred to as the Rossby regime. With still greater temperature gradients a further Hadley regime appears with a symmetrical westerly current at the upper surface. The transition from the Hadley to the Rossby regimes also depends on the rate of rotation (Fig. 4.6). When the fluid is close to a transition between different wave numbers it is found that the pattern commonly shows a vacillation similar to the index cycle (see p. 111); (3) the 'rotating dishpan' containing two immiscible fluids with mechanically induced motion. The lower layer can be induced to form a 'polar cap' with eastward-moving waves along the fluid interface, simulating the polar front (Fultz, 1960 and 1961).

More specific studies include those relating to the effect of different shaped barriers on fluid flow in a rotating dishpan (Long, 1952 and 1955) or in a rotating hemispheric shell (Frenzen, 1955).

Energy balance calculations for experiments in a steady rotating annulus (Riehl and Fultz, 1958) produced some results which disagreed with the numerical experiment of Phillips. When calculations were made in curvilinear co-ordinates related to the jet stream axis it was found that the kinetic energy production was maintained by a single direct cell and was transferred from the mean circulation to the eddies. This type of approach is valuable in that it suggests ideas which may be tested with observations from the atmosphere. Indeed Krishnamurti (1961) subsequently demonstrated that there is a Hadley circulation around the axis of the subtropical jet stream of the northern hemisphere in winter.

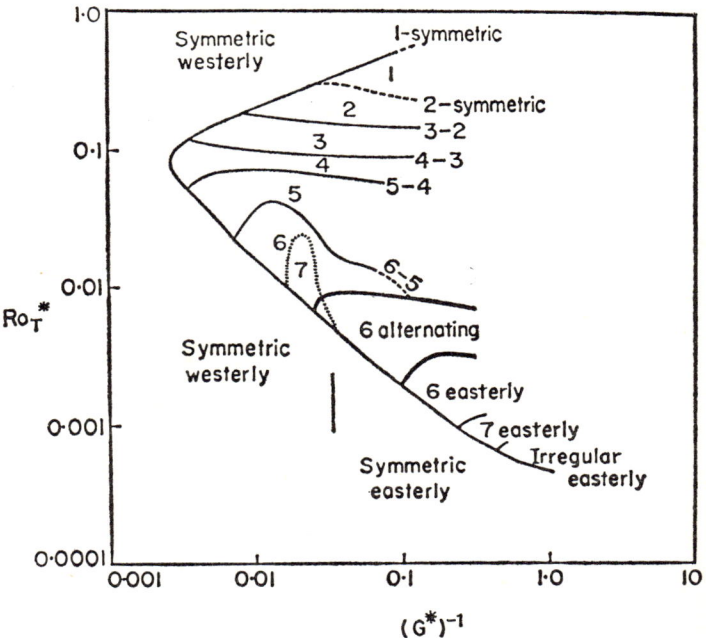

4.6 Circulation regimes in an annulus experiment (*Fultz, 1960*). $Ro_T{}^*$ is proportional to the radial temperature gradient, $(G^*)^{-1}$ is proportional to the square of the rate of rotation of the annulus. Hadley regimes lie to the left of the main curve, Rossby regimes to the right. The numbers indicate the number of waves in the flow pattern.

THE WESTERLIES

Theoretical models of the long waves

Investigation of the middle and upper troposphere has demonstrated the existence of quasi-permanent long waves in the westerlies. Mean contour maps for the 500 mb surface in winter (Sutcliffe, 1951) show deep troughs over eastern North America and eastern Siberia and a minor one at about

60° E., while in summer, although the westerlies are much weaker, the first two troughs remain prominent in almost the same locations and the other is replaced by a weak ridge.

Rossby (1939 and 1940) developed the first theory of wave patterns on the global scale by assuming horizontal motion in a barotropic atmosphere. The crux of his argument was that since the Coriolis parameter increases with latitude then if absolute vorticity is to be conserved a current moving poleward must reduce its relative vorticity, thus becoming more anticyclonic and eventually swinging equatorward. He related the zonal wind velocity (u), the velocity (c) of a simple harmonic wave superimposed on the zonal current, wavelength (L) and a parameter β ($=\partial f/\partial y$), the latitudinal variation of the Coriolis parameter (f), in the following equation

$$u-c=\frac{\beta L^2}{4\pi^2}$$

For stationary waves $L=2\pi\sqrt{\frac{u}{\beta}}$ and approximate values of these wavelengths in km. for different latitudes and zonal velocities are

Zonal Velocity (m/s)	Latitude 30°	45°	60°
4	2820	3120	3710
12	4890	5400	6430

The wave equation applies equally to the 'free', travelling waves and the quasi-permanent ones. Rossby suggested that the latter were anchored by the combined effects of topography and the solenoidal field at coastlines. Mountain barriers modify the depth of the air column and thereby change the vorticity producing a ridge over the mountains and a downstream trough. This effect is easily appreciated if we consider that vertical shrinking of the column over the barrier leads to lateral expansion (divergence) and anticyclonic vorticity (see the vorticity equation, p. 99), with the reverse as the column expands vertically downstream of the barrier. Thermal effects are due to energy sources such as the Gulf Stream and Kuro Shio encouraging low-level convergence (high-level divergence) in winter months while over the continents there is surface divergence (high-level convergence). This pattern results in upper troughs in the westerlies over the eastern continents (Palmen, 1951A).

Despite Rossby's views, two schools of thought developed in the 1950's as to the origin of these 'Rossby waves'. Charney and Eliassen (1949) and Bolin (1951) proposed that mountain barriers were primarily responsible since the troughs appeared to indicate no seasonal variation in position, although it can be shown that different interpretations result in part from different series of average contour maps. The Rockies were visualized as the primary control in the northern hemisphere, giving rise to the eastern North America trough, while Bolin found a similar pattern downstream of the Andes. Theoretical considerations (Queney, 1948) and physical models (Long, 1952 and 1955) also suggested the importance of such barriers but it must be emphasized that the calculations of Queney and Charney and Eliassen were based upon the simplified assumptions of a barotropic-model atmosphere.

Sutcliffe (1951) raised qualitative objections to these ideas and argued that since the atmosphere is baroclinic the role of solenoidal fields cannot be neglected. He presented 1,000–500 mb thickness maps and 500 mb contour charts to demonstrate the close correlation between their patterns in both January and July.

Investigations by Smagorinsky (1953) and Academica Sinica (1957 and 1958) using a variety of baroclinic models suggest that both factors need to be taken into account, although their relative importance has not been fully established. Gilchrist (1954) showed with a baroclinic model that in winter there can be a surface trough over a heat source with the upper trough to the west as a result of high zonal velocities, but in summer with low zonal velocities there is a surface trough over a heat sink with the upper trough to the east. Such changes are observed over Asia and the Pacific and may depend on wavelength adjustments due to different zonal velocities or on a different pattern of heat sources. The distribution of the latter, however, is still the subject of considerable uncertainty. For example, Clapp (1961) calculated the heat sources and sinks for the northern hemisphere in winter by two independent methods and obtained heating patterns which in one case were positively correlated with temperature and in the other negatively. Other studies indicate that the relation is positive, implying that the heat sources do generate the mean motion of the atmosphere (Sheppard, 1962). The effect of heat sources is further complicated by the fact that the heating of the atmosphere is a function of the motion of the air itself and models must make allowance for this non-linearity (Sutcliffe, 1950; Döös, 1962).

Analogue studies of the wave patterns in the two hemispheres may help to resolve some of the present differences of opinion. Lamb (1959) considers that south of 45° S. the dominant control is thermal, but data for the southern westerlies are at present too incomplete for final conclusions to be drawn.

Harmonic analysis – a mathematical model of wave patterns

Recent studies of the wave patterns in the westerlies use the technique of harmonic (Fourier) analysis which allows a complex curve to be resolved into simple component parts comprising only sine and cosine functions of the form $\sin x$, $\sin 2x$ ... $\sin nx$ and $\cos x$, $\cos 2x$ ... $\cos nx$. The wave number (n) as specified by these functions, indicates the number of troughs along a selected latitude circle. Wave number 1 represents an eccentricity of the polar vortex, number 2 refers to two troughs 180 degrees apart and so on (Hare, 1960). The technique also allows the specification of the phase difference (longitudinal distance) between each wave, the amplitude (latitudinal extent) of the wave and the percentage contribution of each wave number to the total variance.[1]

Wave numbers 1, 2 and 3 are dominant at high latitudes (Eliassen, 1958; Saltzman and Peixoto, 1957) and in middle latitudes waves of number 5, 6 and 7 are superimposed upon the longest waves and travel almost continuously eastwards. Wave numbers 11 and 12 are effectively 'noise' as a result of the coarse data network and uncertainties in the original contour analyses. It must be noted that wave number 7 at 30° N. is equivalent to number 3 at 60° N. in terms of actual wavelength. A weakness of the method is that the wave numbers are discrete (Boville and Kwizak, 1958). Thus the dominance of a wave number n, accompanied by high percentages of wave numbers $n-1$ and $n+1$ may indicate that the spacing of the 'n' troughs is not consistently $2\pi/n$.

In addition to analytical studies of the eddy pattern in the zonal circulation the technique can also be used to investigate the partition of energy between the various wave spectra at different latitudes and isobaric surfaces. For example, Wiin Nielsen et al. (1964) show that a large proportion of the total transport of heat and momentum in winter is due to wave numbers 1 to 4 whereas in other months these long waves are less important.

Zonal indices and models of circulation patterns

Several attempts have been made to formulate models of hemispheric circulation patterns by reference to the overall strength and average latitude of mid-latitude westerlies. The basis of these ideas, the zonal wind velocity, is only an analytical tool, but it is necessary first to outline the method before discussing the models which derive from it.

The overall strength of the westerlies can be assessed by measuring the average hemispheric pressure gradient between 35° and 55° N. and converting this into geostrophic west wind. This zonal index may be determined for MSL isobars or upper air contours. The method was first applied by Rossby et al. (1939) in relation to the long-wave equation and subsequently developed in detail by Allen et al. (1940). The MSL index is termed high when the average pressure difference between 35°–55° N. exceeds about eight mb and low when it is less than about three mb. There is a seasonal trend, however, in response to the weaker westerlies in summer.

The belt 35°–55° N. is the mean zone of strongest westerlies though often in winter their circulation pole is displaced from the geographical pole to-

---

[1] Other applications include the resolution of cyclical time series into their principal harmonic components. An example is the analysis of annual precipitation profiles for North America (Lyle and Bryson, 1960; Sabbagh and Bryson, 1962).

wards 170° W. Such displacements are closely related to the strength of the zonal flow (La Seur, 1954) and since the index only reflects activity within the stated latitude belt the precise interpretation of index fluctuations may be difficult. For this reason some studies also give indices for 20°–35° N. and 55°–70° N., but La Seur advocated the use of a 'moving co-ordinate system' related to the circulation pole of the westerlies. An alternative solution is to plot a time profile of the average zonal west wind component for 5° latitude belts. Such zonal profiles reveal northward and southward trends of relative maxima and minima within the mean westerlies which may persist several weeks (Riehl et al., 1952).

Simple models of the synoptic patterns associated with high and low index were indicated by Rossby et al. (1939) and Allen et al. (1940). With high index there is rapid eastward movement of cyclones and little meridional exchange of air masses despite the strong meridional temperature gradient. The subtropical anticyclones and the Aleutian and Icelandic lows are intense features and the latter tend to be eastward of their normal position. With low index the pattern is strongly cellular and longitudinal temperature contrasts are marked. The Aleutian and Icelandic lows tend to be westward of their normal position. Rossby and Willett (1948) and later Namias (1950) elaborated these ideas and showed that over a period of four to six weeks 'index cycles' occur, during which the westerlies increase in strength as the Ferrel vortex expands and the circulation breaks down into cellular patterns (Fig. 4.7).

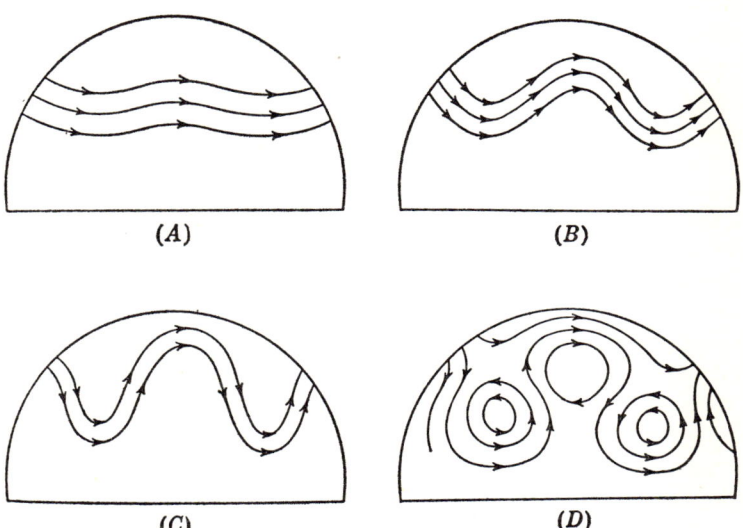

4.7 The index cycle (*Namias, 1950*). Four stages in the change-over from zonal flow (*A*) to a meridional cellular pattern (*D*).

A more detailed model of the index cycle was put forward by Riehl *et al.* (1952) on the basis of the stages in the northward or southward drifts of relative westerly maxima and minima on the zonal profile. This more complex scheme shows that the previous model might prove a dangerous oversimplification. There are for instance four stages which broadly correspond to 'low index' but only one is closely similar to 'high index'. Each stage is characterized by a particular set of synoptic patterns at the surface and in the upper air. In spite of this clarification of the complex nature of index fluctuations and the expression of doubts about its usefulness (Forsdyke 1951), the simplified terminology of high and low index continues to be used. These extreme states are also employed in palaeoclimatology (see p. 131) and in analytical studies of the circulation. For example, Bradbury (1958) examined the behaviour patterns of cyclones and anticyclones for strong and weak circulations.

Jet streams and general circulation models

Narrow bands of very strong winds or jet streams are regularly observed in the upper tropospheric westerlies. In winter there is a subtropical jet stream about 200 mb near the poleward limit of the trade wind cell and there may be one or more jet streams at about 300 mb associated with the major frontal zones in middle and high latitudes (Newton and Persson, 1962; Reiter, 1963; Riehl, 1954 and 1962A). Recent research has also shown the existence of a 'polar night jet' in the winter stratosphere over the arctic and antarctic (Hare, 1962). The tropospheric pattern is more complex in summer with a tropical easterly jet stream across Afro-Asia (see p. 124) and intermittent westerly jets in middle latitudes.

The similarity of jet stream structure to narrow, high velocity ocean currents such as the Gulf Stream was recognized by Rossby and the analogy has been further developed by his associates (Newton, 1959). The fundamental role of jet streams in the circulation has also been emphasized by numerical models (Phillips, 1956) and dishpan experiments (Riehl and Fultz, 1958), as well as by analytical studies (Krishnamurti, 1961), but the occurrence of jet streams still remains to be accounted for by a satisfactory theory. The most comprehensive model so far prepared is due to Rossby (1949) who suggested that tropospheric wave patterns promote lateral mixing through airmass interchange. This equalizes temperatures in higher latitudes and creates a strong mid-latitude temperature gradient. Fleagle (1957 and 1960) has also shown theoretically how this temperature gradient is built-up to the rear of a mid-latitude wave trough. The thermal wind equation (p. 98) implies that strong zonal flow and vertical shear develop with such a meridional temperature gradient, but it must be stressed that lateral mixing is the simultaneous cause of both the wind and temperature fields.

Rossby demonstrated that the mixing leads to constant absolute vorticity in higher mid-latitudes and a latitudinal profile of zonal wind similar to that observed in the northern hemisphere as shown in Fig. 4.8 (Fultz, 1960; Tucker, 1962). At the equatorial limit of this region of constant absolute vorticity (about 35° latitude) there is strong lateral shear such that on the equator-

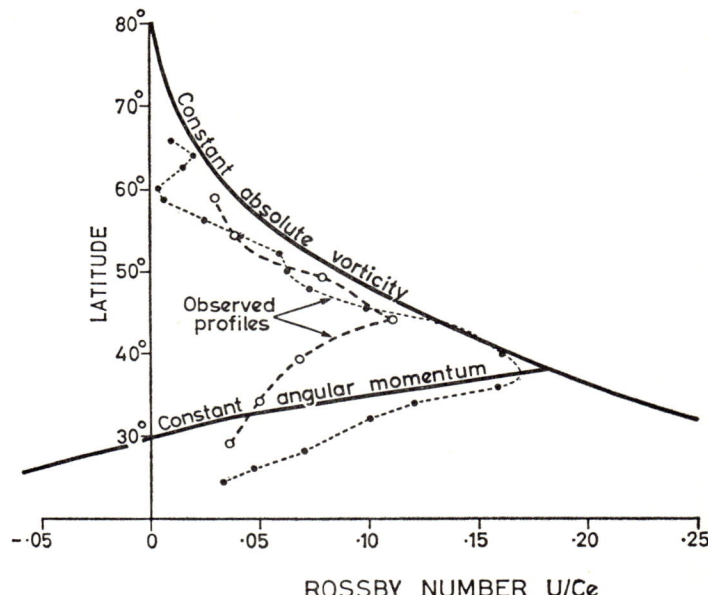

4.8 Theoretical profiles of wind velocity according to the principles of the conservation of absolute vorticity and the conservation of absolute angular momentum compared with observed profiles near the tropopause (*Staff Members, University of Chicago, 1947*). The profile shown with open circles refers to a nine-day average over North America. Wind velocity (u), is expressed as a non-dimensional ratio by dividing by the linear equatorial velocity of the earth Ce.

ward side the profile of wind velocity approximates to that for constant absolute angular momentum (zero absolute vorticity).

This model is still far from accounting for many observed features, particularly the existence of double westerly jet streams in winter. Palmén (1951) regards the subtropical jet as related to the high-level poleward flux of momentum associated with the Hadley cell (Figure 4.4) and it may be that Rossby's scheme is more appropriate for the polar-front jet.

SYNOPTIC-SCALE MODELS

A synoptic model is essentially a standard configuration of the field of pressure or flow with which is associated a characteristic weather pattern.

FORSDYKE, 1960

Extra-tropical depression models

From the earliest synoptic charts Fitzroy (1863) recognized certain essential features of mid-latitude depressions. He pointed out the occurrence of cyclogenesis along the boundary between contrasting air masses, but despite reiteration of similar views by a number of workers (Von Helmholtz, 1888; Shaw and Lempfert, 1906) this elementary model of cyclone development received little attention. A detailed model of the life cycle of a wave disturbance was not put forward until 1922 by Bjerknes and Solberg. Reference need only be made to the full treatment of this model in Petterssen (1956, p. 217) or the simpler account in Pedgley (1961) since its characteristics are widely known.

Refinement of the frontal depression model has followed with more intensive upper air soundings. Godson (1950) for example indicated typical configurations of upper frontal boundaries by means of frontal contour charts and these ideas were further developed by Anderson, Boville and McClellan (1955) and Penner (1955) in the three-front model (Galloway, 1958). The nature of the front, which is not generally a sharp temperature discontinuity, has also been clarified. Figure 4.9 illustrates the structure of a typical frontal zone in winter. The meridional temperature gradient is most intense in the frontal zone but is not confined to it. It may be noted that the presence of sharp frontal zones in the middle and upper troposphere is commonly due to subsidence and not to air mass origin (Sawyer, 1958). Detailed models of frontal structure and air motion relative to the frontal surface are given by Sansom (1951) and Miles (1962).

Important improvements of the model of a developing frontal wave are due to the work of Holmboe and Bjerknes (1944) on the relationships between upper tropospheric waves and surface waves. They demonstrate that the continued fall of pressure at the centre of the depression is made possible, despite active low-level convergence during the deepening stage, by active divergence aloft. Upper troughs have gradient winds of sub-geostrophic value whereas in ridges they are super-geostrophic as a result of the opposite effect of the centripetal term. As a result the transport of air in the contour channels of troughs is less than in ridges causing divergence to be most commonly located ahead of upper troughs in the westerlies. The divergence due to this curvature effect generally exceeds the convergence in poleward flow which results from the increasing Coriolis parameter. The theory of cyclogenesis has been greatly extended by the work of Sutcliffe and others on divergence – vorticity relationships (Sutcliffe, 1947; Sutcliffe and Forsdyke, 1951; Petterssen, 1956, p. 320 *et seq.*). Sutcliffe analysed the vorticity equation (p. 99)

$$\frac{dQ}{dt} = -DQ$$

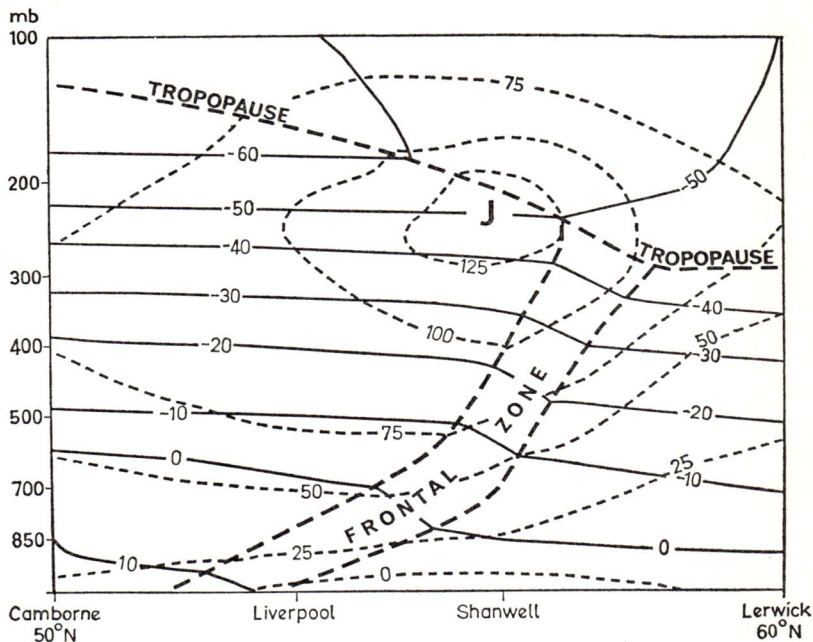

4.9 Model of a frontal zone (based on 26 September 1963)
Full lines – isotherms (°C.).
Dashed lines – isotachs of the westerly wind component (knots)

and showed that cyclonic development $(-D)$ comprises two major components – the advection of vorticity in the middle troposphere and thermal advection. Their role can be illustrated with reference to the model of a developing depression in Figure 4.10. In the initial stages (*a* and *b*) the primary control is considered to be the positive vorticity advection ahead of an advancing upper trough. Subsequent intensification at the centre of the surface wave is mainly dependent on thermal advection (shown by the schematic thickness lines) in the warm sector strengthening the production of vorticity. The original Bjerknes – Solberg model emphasized the upgliding of warm sector air at the warm front but this is now regarded as of secondary importance. Development may also be affected by non-adiabatic heating or

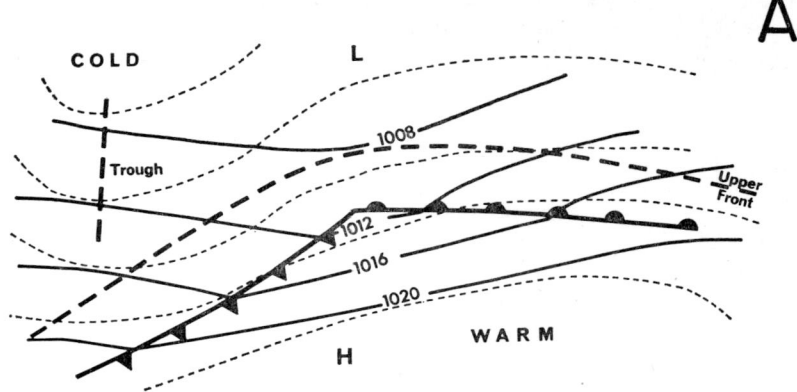

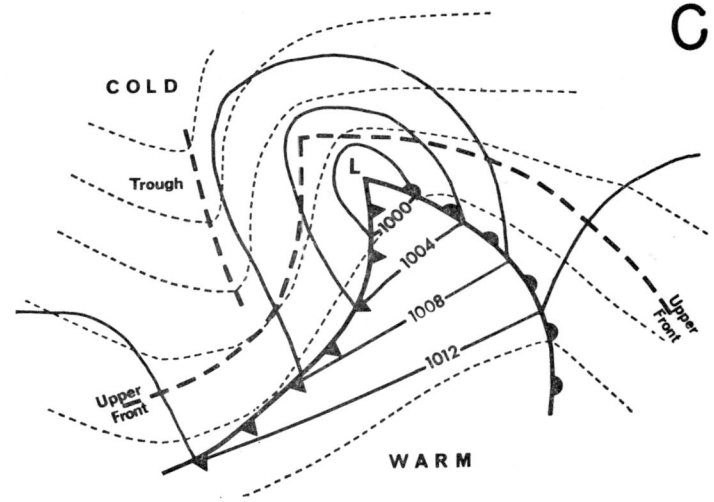

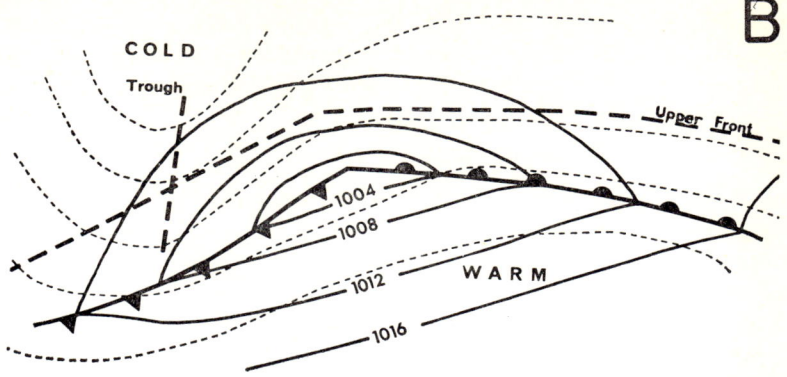

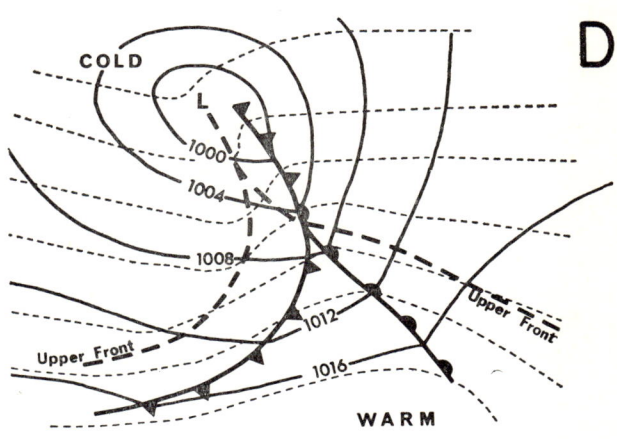

4.10 Idealized frontal wave development. The full lines are surface isobars, the pecked lines relate to schematic 1000–500 mb thickness lines and the dashed lines show the axis of the upper cold trough and the position of the upper front at about 500 mb.

cooling due to air moving over heterogeneous surfaces and by adiabatic influences such as stability and adiabatic cooling with ascending air. The detailed evaluation of these four terms in individual cases is highly complex but this model provides the basis for routine forecasting.

Synoptic models of the jet stream

The most recent addition to the depression model is the linking with it of polar-front jet streams. Figure 4.11 illustrates a typical pattern with the jet axis roughly parallel to the mid-tropospheric position of the frontal zone. The vertical structure is shown in Figure 4.9. The jet core (maximum winds) is generally situated to the rear of the cold front.

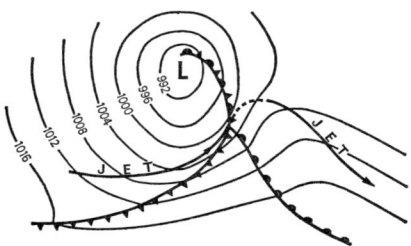

4.11 Jet stream and depression model (*Boyden, 1963*). The arrowed lines show the idealized location of the upper tropospheric polar front jet stream, the broken line indicates a weak or discontinuous jet stream.

As might be expected, active depressions and bad weather in mid-latitudes are frequently associated with an upper jet, but the converse is not necessarily true. The general relationships between jet streams and precipitation patterns have been demonstrated in several models (Riehl *et al.*, 1952; Johnson and Daniels, 1954). These indicate a tendency for ascending air and more precipitation to occur on the right of the jet entrance zone and on the left of the exit looking downstream. The actual distribution of convergence depends on both the contour curvature and the lateral wind shear so that different patterns ensue when the jet maximum is located in a trough or (more commonly) in a ridge.

TROPICAL MODELS

Intensive study of the wind systems and disturbances of the Tropics dates only from about 1940. The earlier literature is for the most part characterized by faulty concepts based on inadequate observations and frequently biased by ideas developed in middle latitudes. Only Riehl's unsurpassed text on tropical meteorology (1954) and Trewartha's recent book (1961) provide any comprehensive coverage of modern views on the tropical atmosphere.

Palmer (1951) refers to the early workers as mainly belonging either to the 'climatological school' or the 'air mass – frontal school'. The former considered the weather in terms of diurnal and seasonal controls plus local effects while the latter transferred the models of Bjerknes and Bergeron to low latitudes, (for example Garbell, 1947). A specific illustration may be cited. In 1936 the trade wind inversion was explained by Ficker in terms of two distinct air masses, a view unfortunately repeated by Pédelaborde (1958). However, Riehl et al. (1951) demonstrated that in the north-east Pacific the inversion is due to subsidence and that particle trajectories cross the inversion boundary. More serious problems appeared when attempts were made to reconcile the Bjerknes model of the frontal depression with observations of tropical disturbances and the polar front with the inter-tropical front. These problems are by no means solved at the present time, but it is possible to indicate the more realistic models which have been proposed during the last twenty-five years.

The inter-tropical discontinuity, also termed the inter-tropical convergence zone (ITCZ) or, to avoid genetic implications, the equatorial low pressure trough, is generally not a frontal discontinuity in the mid-latitude sense since air mass differences in the tropics are frequently insignificant. This is particularly so with regard to density and it is mainly in west Africa and over the Indian sub-continent that the term inter-tropical front (ITF) has some validity. Equally, convergence of the two trade wind systems is far from continuous in space or time and the synoptic occurrence of convergence along the ITCZ is to a significant extent controlled by perturbations in the easterlies (Crowe, 1949 and 1950; Palmer, 1952).

Wave disturbances

Attention was first drawn to the occurrence of wave disturbances in the trades by Dunn (1940) working in the Caribbean. He referred to these features as isallobaric waves, but this term is not now used. Dunn showed that bad weather develops behind the low pressure troughs, the converse of extratropical depressions. He noted that the great majority of the waves were stable and that development appeared to depend on the breaking-down of the trade wind inversion. Riehl (1954) was the first to examine these phenomena in detail by means of streamline analysis. This is necessary in view of the (assumed) non-geostrophic balance, and hence ambiguous relationships between wind and pressure fields, in low latitudes. The classical model of the easterly wave is shown in Figure 4.12 with its associated pattern of convergence and divergence when the easterlies are travelling faster than the wave form, a frequent but not habitual case. The explanation of the configuration rests upon the equation for the conservation of potential vorticity assuming adiabatic motion due to C-G. Rossby (1940):

$$\frac{(\zeta+f)}{\Delta p} = k$$

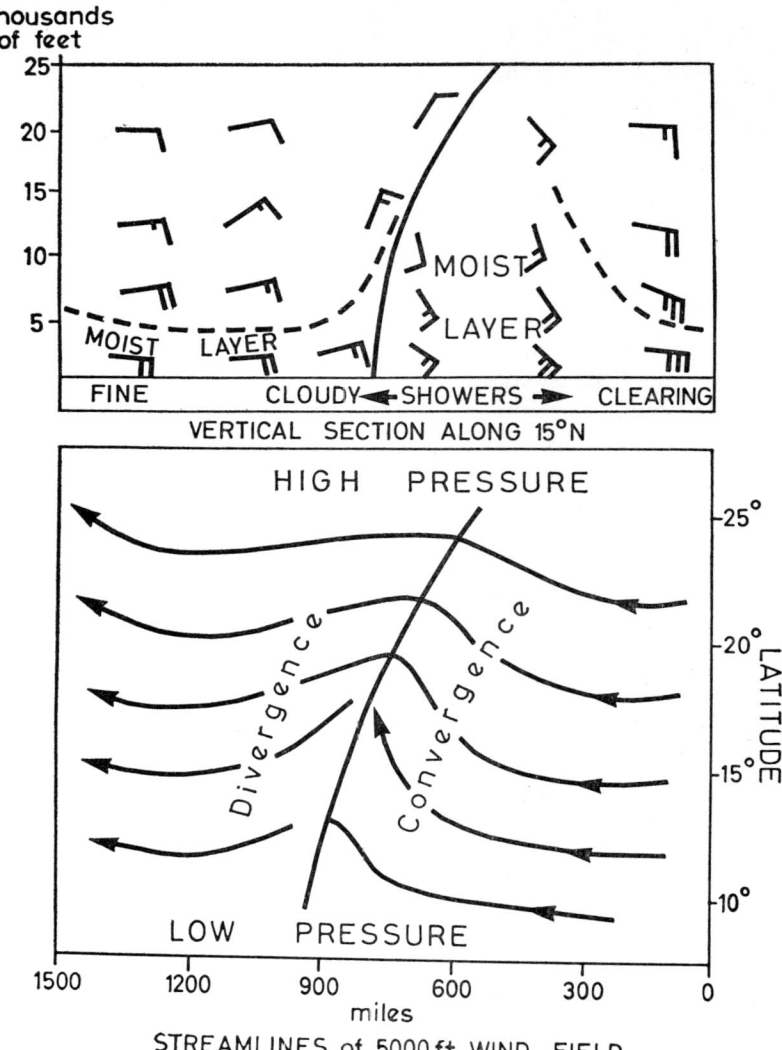

4.12 Model of an easterly wave (*after Riehl and Malkus*). The lower diagram shows the low-level wind flow in relation to the trough line. The vertical section shows the wind direction (where the plotting convention assumes north is at the top of the diagram) and speed, the depth of the moist layer (dashed line) and the general weather sequence.

where k = a constant
  f = the Coriolis parameter
  ζ = relative vorticity (for cyclonic vorticity ζ is positive)
  Δp = the thickness of the layer between two isentropic surfaces.

The expression implies that for air moving northward from the rear of the wave towards the trough and acquiring greater absolute vorticity $(\zeta+f)$ due to increase of f and ζ, there must be concomitant increase in Δp. Such vertical expansion necessitates horizontal contraction i.e. convergence. Conversely there is divergence for air moving southward ahead of the wave and undergoing anticyclonic curvature. It must be noted that this argument assumes that curvature is more important for vorticity than lateral shear.

If the wave is travelling through the wind field it can be shown that the distribution of convergence is reversed. Thus on occasions when the easterlies decrease with height it is possible for divergence to overlie low-level convergence to the rear of the wave trough, creating optimum conditions for the development of bad weather as a result of vertical motion. Recent aerial surveys (Malkus and Riehl, 1964) emphasize the importance of upper tropospheric divergence for cloud build-up, especially when it is accompanied by a low-level wave disturbance.

Similar perturbations are observed in the central and western equatorial Pacific (Palmer, 1952), although there, as noted by La Seur (1960), the equatorial easterlies are convergent in the mean whereas the trades are divergent in the mean. Thus, when waves are embedded in the equatorial easterlies there are only isolated areas of divergence. A further point is that a double system of cyclonic circulation develops when a wave in the equatorial easterlies straddles the equator.

Freeman (1948) pointed out that the easterly wave theory fails at the equator, since there is no horizontal Coriolis component, yet disturbances still occur. He developed a physical-mathematical analogy between incompressible airflow beneath an inversion and the high-speed flow of a compressible fluid. Freeman suggested that jump-type waves were possible in the easterlies and gave an example from the New Guinea area. These ideas have not been further applied in this particular context, mainly perhaps as a result of the difficulty of using them for an operational model in areas of generally sparse data. However, the analogy has been developed with reference to meso-scale squall-lines and planetary blocking waves (Freeman, 1951; Fujita, 1955). The 'disturbance lines' of West Africa (Eldridge, 1957) are probably one such category of squall-line. They travel westwards at between 20–30 kt. (about the speed of the upper easterlies at 300 mb) and may be several hundred miles in length roughly from north to south.

It is probable that a wide range of models will be required for the varied perturbation types in the tropics (Forsdyke, 1960; La Seur, 1960). In this connection the use of satellite cloud pictures is proving a useful analytical tool and should eventually provide information on the more frequent patterns. On this basis for example Merritt (1964) suggests that Riehl's classical model of the easterly wave occurs less frequently than previously supposed and he illustrates a range of linear and vortical perturbations. A frequent disturbance type has a wave at low levels with a closed cyclonic circulation about the 600 mb level.

Vortices

The development of tropical cyclones is a crucial problem for tropical meteorologists, but no satisfactory forecast model has yet been put forward. Early theoretical models based on convection cells or frontal concepts are now regarded as erroneous, and it is generally considered that the storms originate from pre-existing wave disturbances. Indeed in a few cases the development of Atlantic hurricanes has been successfully traced back to west Africa. Erickson (1963) reports that Hurricane Debbie of the 1961 season originated in this area and deepened after moving over the Cape Verde Islands.

The main problem is the mechanism responsible for triggering full hurricane development since many tropical vortices fail to attain such intensity (Malkus, 1958). The essential characteristic of a hurricane model in this respect appears to be the presence of a warm core with upper tropospheric temperatures $5°-8°$ C. warmer in the core than the surroundings. Without this feature the storm cannot develop very low surface pressure and the accompanying high wind speeds nor high level outflow. Two factors are thought to be necessary for the growth of the warm core. One is the release of latent heat by rising air, undiluted by entrainment of drier air of external origin. The 'hot towers' of cumulonimbus clouds may occupy less than one per cent of the disturbance and Scorer (1966) considers that a persistent shower area some 100 to 200 miles across is required to maintain an adequate latent heat supply. The second is that the rising air must have a higher heat content than normal tropical air in order to generate the requisite surface pressure decrease. The additional heat energy (approximately $2 \cdot 5$ cal/g) is acquired in the form of sensible and latent heat as the air spirals inward at high speed. Hence the whole process is dependent upon a feedback mechanism (Malkus, 1962, pp. 231–253; and Yanai, 1964). In the opinion of some authorities, moist adiabatic cooling is by itself insufficient to explain the observed warmth of the core and the effect of descending air in the eye from the ring of cloud forming the 'eye wall' probably accounts for the additional warmth. It may be noted that the warming process appears to operate from the upper troposphere downwards (Estoque, 1964). The inception of the eye is clearly a vital feature of any hurricane model but details of this process are still largely unknown.

The initial development of the cyclone prior to eye formation is dependent on interaction between divergence in the upper troposphere and a pre-existing low-level wave (Riehl, 1951 and 1954) but no complete model of this stage is available. One hypothesis put forward by Sawyer (1947) which linked high level outflow with negative absolute vorticity ('dynamic instability') as a starting mechanism is not now considered to be applicable. Baroclinic instability (see p. 104) is a more probable source of the initial growth of cyclones (Yanai, 1964). General requirements for development include sea-surface temperatures exceeding about 27° C. and a location several degrees from the equator to ensure cyclonic rotation. Once developed the system is self-maintaining through release of latent heat as long as there is a moisture supply and the air diverging aloft descends well away from the centre so as not to destroy the temperature difference between the core and its surroundings. In these respects the mature hurricane is a prime example of an open system maintaining dynamic equilibrium. When the storm moves over land or cool seas it decays rapidly.

Monsoons

The classical model of the monsoon depicts the regime as a large-scale land-sea breeze phenomenon. Text-book statements about the Asian monsoon refer to it typically as follows:

> In January, decidedly high pressure develops over central Siberia, and the doldrums are somewhat south of the equator in the Indian Ocean. Steady north-east winds therefore blow out of Asia in an anticyclonic circulation across the Indian Ocean to the equator (Blair, 1942, p. 65),

although later in the book (p. 327) Blair noted the effect of the mountains separating the cold interior of Asia from the plains of India. However, as recent studies demonstrate, even a modified thermal model of summer heating over northern India leading to monsoonal inflow, and local outflow from the subcontinent in winter, differs considerably from reality.

The construction of three-dimensional models using upper air data is now beginning to clarify the process of seasonal change-over. According to Yin (1949) the winter flow pattern in the middle troposphere characteristically features a westerly jet stream both north and south of the Tibetan Plateau owing to the mechanical effect of the barrier. Downstream over China these currents re-unite. Analogues of this pattern have been obtained with laboratory models by Long (1952 and 1955). Subsidence tends to occur beneath the southern side of the branch of the westerlies over northern India, creating persistent dryness in the low-level north-easterlies. In early summer the southerly branch weakens and disintegrates leaving the main westerlies now over central Asia. This breakdown of the middle and upper tropospheric flow is linked in time with the onset of the south-west monsoon over India and a close causal relationship between these events was at first assumed.

Subsequent investigations cast doubt upon the barrier effect of the Tibetan Plateau. Chang Chia-Ch'eng (1960) shows that in fact there is a low frequency of wind maxima just north of the Tibetan Plateau and that two jet streams are also commonly observed to the west of the Plateau. Nevertheless, the branch over northern India is remarkably stable and Chia-Ch'eng demonstrates the existence of a strong latitudinal gradient of solar radiation between 20°–40° N. over 35°–80° E. from November to April which he holds responsible for the presence of the westerly maximum.

The breakdown in May–June is undoubtedly linked with the seasonal readjustment of the global circulation but Flohn (1960A) postulates a new role for the Tibetan Plateau. Much of the area is close to the 600 mb level with mountain peaks well above this. Absorption of solar radiation by this elevated source raises the air temperature considerably above that of the surrounding free air (Academica Sinica 1957, 1958 and Ramakrishnam et al. 1960). Since a warm air column has a less rapid decrease of pressure with height than a cold one high pressure should form over the area and Flohn (1960A) and Pisharoty and Asnani (1960) amply demonstrate the existence of an anticyclone at the 500 mb level. Paradoxically, the solenoidal field produces at the same time descent away from the plateau and rising air above the Tibetan anticyclone (Lockwood, 1965). The inception of this summer pattern undoubtedly contributes to the disruption of the upper westerly flow over northern India, since the anticyclonic circulation assists the development of an easterly jet, with a core about 150 mb, over southern Asia between 10°–15° N. as shown in Figure 4.13 (Koteswaram, 1958). The high level (Pacific) easterlies (above 8 km.) appear first over Indo-China in May, as a col develops about 95° E. in the upper troposphere between the western Pacific anticylonic cell and that over Africa-western Asia, and they push northwestward towards northern India in June and July (Ramanthan, 1960) aided by the easterly flow round the southern side of the Tibetan high at 500 mb. This model has recently been criticized, however, by Frost and Stephenson (1965) who find no evidence of a trough in the upper westerlies.

Breaks in the summer monsoon may occur even when it is apparently fully established. Ramaswamy (1962) shows that they are associated with the temporary redevelopment of a westerly jet stream over northern India and the removal of the Tibetan high by deep troughs during low index circulation. His results lend considerable weight to the foregoing models stressing the role of the upper tropospheric circulation.

A more general model of the circulation reversal over southern Asia in summer can now be examined. The reversal, which usually takes place in early June, is primarily a manifestation of the seasonal shift of the equatorial low pressure trough. Its movement permits the rapid northward expansion of the equatorial westerlies to produce the south-west monsoon. These westerlies are commonly regarded as south-east trades of the southern hemisphere deflected by the changed sign of the Coriolis parameter at the equator, but as Flohn (1960B) points out Halley in 1686 recognized that in the Indian Ocean the change of direction occurs at 2–3° South as a discontinuity not as a gradual transition. Equally, over Indonesia in the southern hemisphere summer the change is at 2–3° North. In Flohn's view (1960B, C and D) the

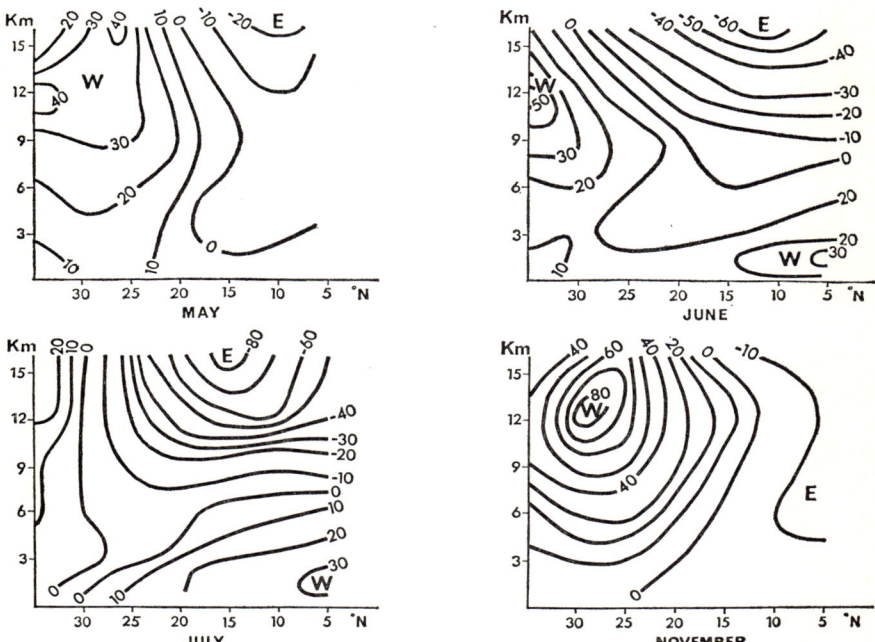

4.13 Average zonal wind components (knots) for longitudes 75°–80° E. (*after Ramakrishnan, Sreenivasaiah and Venkiteshwaran, 1960*). June and July relate to the south-west monsoon season, November represents the north-east monsoon and May the transition to the south-west monsoon.

equatorial westerlies represent a planetary wind belt of the summer hemisphere, which would be continuous on a continental globe. This wind system results from the poleward displacement of the equatorial trough in summer, a movement which is demonstrably large over the land masses of the northern hemisphere. The strong pressure gradient between the equator and the trough induces deflection of the westerlies before they cross the equator due to the inertia of the predominant meridional current near the equator. On an oceanic globe the trough would move only a few degrees of latitude and no westerlies would intervene between the two Trade systems, due to the small

Coriolis component near the Equator. The circulation models for hypothetical extremes are illustrated in Figure 4.14. In a sense therefore, the thermal low still enters into the explanation but the new models differ sharply from the old in that they link the surface and upper tropospheric circulations over southern Asia and take account of global as well as regional factors.

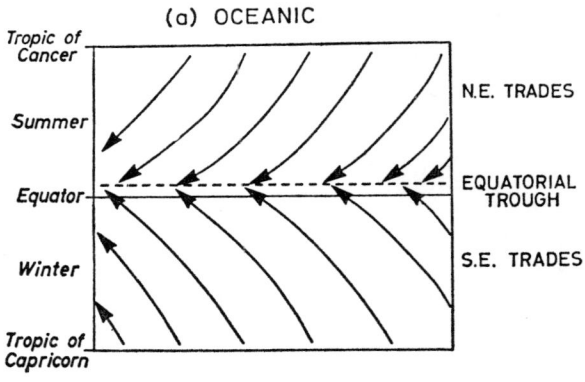

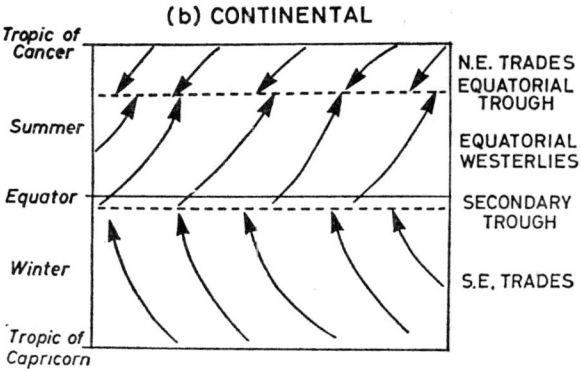

4.14 Schematic circulation models (a) for an oceanic globe, (b) for a continental globe (*Flohn, 1960*).

Equatorial circulation models

Attempts to apply the existing models of wave disturbances in African forecasting have generally failed and this has led to quite different ideas from those developed in other sectors of the tropics. A study of Johnson and Mörth (1960) provides a valuable illustration of model building apart from yielding new insight into equatorial flow patterns. The occurrence of daily rainfall over east Africa was shown not to be closely related to advective changes of moisture or stability conditions, but study of contour charts during periods of steady pressure fields suggested certain relationships between pressure, wind and weather. Hypothetical flow models and deduced patterns of convergence and divergence were found to show reasonable agreement with actual situations and the large-scale distribution of precipitation.

Johnson and Mörth term the two basic types of flow pattern the 'duct', where ridges in each hemisphere lead to easterly flow, and the 'drift' where there is trans-equatorial flow from high to low pressure (Fig. 4.15). The fundamental question is how does the air north of the equator 'know' whether pressure is high or low to the south of the equator, since a particle responds to instantaneous forces and in each case the contours parallel the equator. In view of the negligible curvature term it is concluded that the pressure force must be balanced by the Coriolis deflection even near the equator. Acceleration of air in the confluence zone of the duct entrance leads to convergence near the equator and this is most effective if the anticyclones are intensifying. Ageostrophic components may also encourage convergence through friction or the inability of the Coriolis term to maintain geostrophic balance within about three degrees of the equator (Johnson, 1963). In the drift pattern subgeostrophic velocity causes equatorward airflow and recurvature takes place at several degrees south as geostrophic equilibrium is recovered. This pattern regularly occurs over east Africa between January–March with fine weather over Kenya but convergence and rain in the westerlies over southern Tanzania.

Other patterns discussed by Johnson and Mörth are the 'bridge', with stable quasi-geostrophic westerly flow in both hemispheres and convergence near the equator, 'zonal downgradient flow' with meridional orientation of the pressure cells (Fig. 4.15) and the 'step' with geostrophic westerly flow in one hemisphere and geostrophic easterly flow in the other. The latter is more common in the upper troposphere.

MESO-SCALE MODELS

The development of models for systems with dimensions of the order of 10–100 miles requires the assistance of radar meteorology and a close network of stations. Consequently this field of study dates mainly from post-World War II. The idea that localized surface heating controls all convective storms is still a common misconception although it has long been recognized that many thunderstorms follow clearly defined paths usually in association with cold front zones.

Investigations by Byers and Braham (Byers, 1959, Chap. 19) and especially Fujita (1955) have produced much information about the structure and life

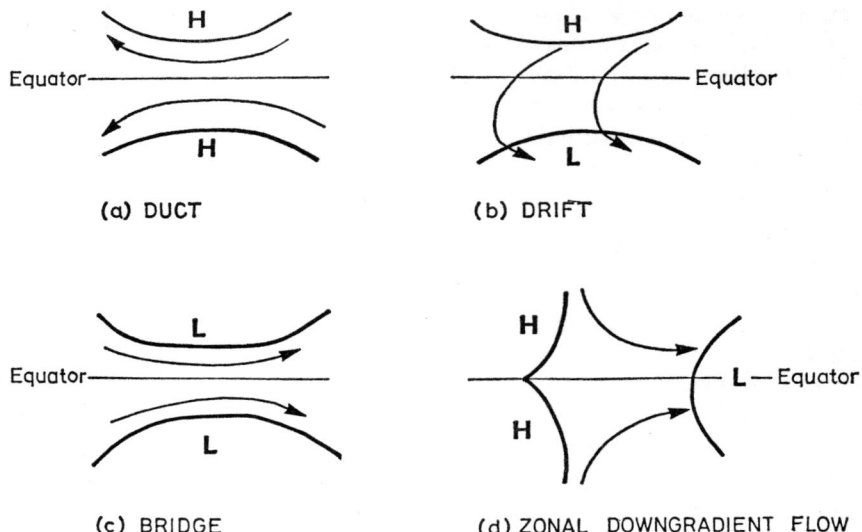

4.15 Equatorial circulation models (*after Johnson and Mörth, 1960*).

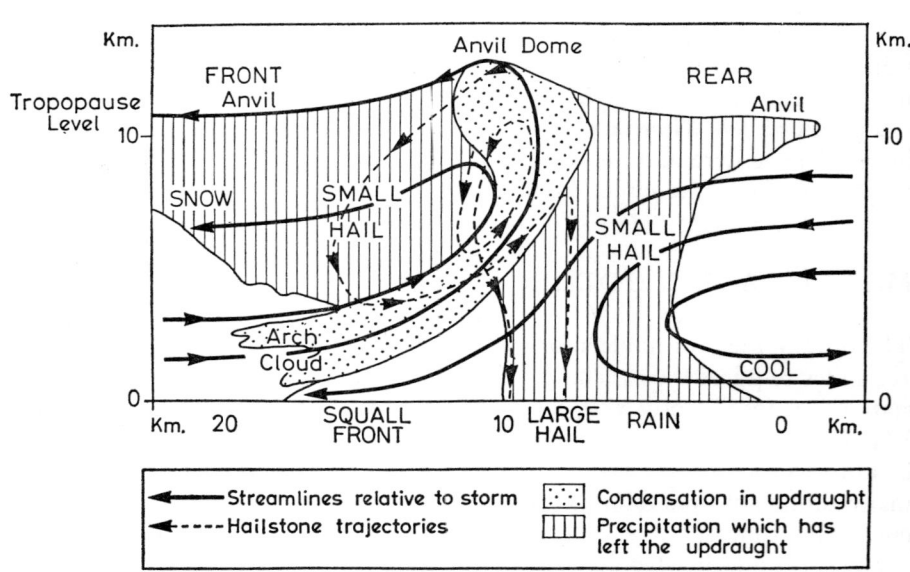

4.16 An idealized hailstorm (*Ludlam, 1961*). The storm is travelling from right to left.

cycle of storms. Models of cold front storms show that in front of the storm there is a line of pressure surge behind which is a pressure excess of about $+3$ mb with a 'wake depression' (a deficit of about -2 mb) in the rear. The thunderstorm high is associated with a cold downdraught initiated by the drag of precipitation drops and the low is induced in the wake of the moving high, unless the general low-level winds are travelling faster than the pressure surge line. More detailed models including a three-dimensional one of airflow within a severe hailstorm have been presented by Ludlam (1961) and Browning and Ludlam (1962) on the basis of radar studies. The essential feature of their model (Fig. 4.16) is an updraught which is tilted backwards with low-level air ahead of the squall-line being shovelled upwards by undercutting (potentially) cold air moving from higher levels at the rear of the storm. At low levels flow relative to the storm is towards the squall-line. Given potentially cold air overlying potentially warm air and strong vertical wind shear the system, once organized, is self-maintaining.

The storm model readily accounts for the growth of large hail (an inch or more in diameter). The concentric structure of clear and opaque layers of ice was previously explained in terms of a fluctuating updraught but the idea of very unsteady currents with velocities of 30-40 m/sec able to carry large hailstones was never satisfactory. The new model, on the other hand, shows that hailstone embryos are blown ahead of the storm by strong upper tropospheric winds after their initial passage through the updraught and again fall back on to the scoop. Some small hailstones may now grow sufficiently quickly for their increasing fall-speed to slow down the rate of lift while small fluctuations in cloud water content and the transverse path of stones through the upcurrent are considered to be the means of growth of clear and opaque ice (Ludlam 1961). Large stones eventually fall from the updraught in a narrow belt to the rear of the storm.

CLIMATOLOGICAL IMPLICATIONS

Air masses and synoptic climatology

The classical studies of Bergeron (1928 and 1930) on air masses and their dynamic relationships with the major frontal systems were a logical development of the synoptic concepts of J. Bjerknes. The ideas provided a new basis for regional climatology removed from the earlier statistical abstractions and consequently air mass study has been applied in many parts of the world (Willett, 1933; Belasco, 1952). Nevertheless, the occurrence of fronts makes a complete climatological analysis using air masses impossible. Frisby and Green (1949) for example omitted consideration of 60 per cent of days with frontal systems in a study of air mass properties over Britain.

The air mass approach overlooks the fact that the air in any vertical column comprises particles which at different levels may have followed quite different trajectories. Furthermore, vertical stability and vapour content are far from being conservative properties. The concept remains of value for descriptive purposes if applied to the lowest 10,000 feet or so of the atmosphere but a more flexible approach is to analyse airflow pattern types. These may refer only to the pressure pattern and general airflow (Abercromby, 1883; Lamb, 1950) or may also take some account of fronts (Barry, 1960). Large-scale tropospheric steering patterns are referred to as *Grosswetter* (Baur, 1951). The complexities of weather-map pressure patterns raise their own classification problems, but synoptic-climatological models allow climatological averages to be calculated on a realistic synoptic basis rather than for arbitrary time periods (Barry, 1963).

The classification of synoptic maps involves the recognition of analogues, but most of the subjectivity involved in this procedure can now be eliminated by one of several mathematical specification techniques. Lund (1963) illustrated the use of correlation coefficients for this purpose. The correlation of pressure values over the north-eastern United States was calculated by computer for 445 days giving a (445×445) matrix. The pattern of the day which had the largest number of maps correlated with it (for a coefficient of at least $+0.7$) was selected as type A. After abstraction of these cases the day with next highest number of related maps was designated type B and so on. Ten such types accounted for 89 per cent of the period. The degree of similarity within each type is specified by the size of the correlation coefficient; a coefficient of 0.7 implies that 49 per cent of the variance of the pattern is common to a pair of maps.

A more elegant method involves the fitting of trend surfaces to the pressure map to obtain the maximum reduction of variance with a limited number of elementary surfaces (see Chap. 3). Examples of this approach are given for polar pressure patterns by Hare (1958).

The selection of synoptic patterns is only a first step towards analysis of the synoptic climatology but a more direct approach has been indicated by Grimmer (1963). By means of factor analysis he abstracted the major components of patterns of mean monthly temperature anomalies over Europe and thereby indicated the influence of continentality and blocking patterns. The technique generates the patterns of each component, thus avoiding the inherent tendency of trend surface analysis to predetermine the answer. At the same time it must be emphasized that the derived 'factors', in this case component patterns, are mathematical abstractions and are not necessarily uniquely related to an individual physical factor. Nevertheless, used with care, this technique holds considerable promise for study of the bases underlying climatic anomaly fields. A recent study by Steiner (1965) illustrates the application of this statistical model to the climatic regionalization of the United States.

Models in palaeoclimatology

The key problem in palaeoclimatology is the relationship between the meridional temperature gradient and the intensity of the atmospheric circulation with its consequent effects upon the evaporation-condensation cycle. There are, for example, various models of the effect of solar radiation changes on the circulation. Simpson (1957) argues for an increase of solar radiation producing a steeper temperature gradient and a stronger circulation. Increased cloud and precipitation imply lower summer temperatures in higher latitudes and this may allow the persistence of snow-cover and eventually ice-caps. However, calculations for a model atmosphere by Sawyer (1963) indicate that increased solar radiation can give greater temperature rise near the poles and weakened extratropical circulation.

At present most theories of climatic change are unverifiable. For this reason, and because these ideas are treated in several standard texts, attention is concentrated on the more immediate question of the range of behaviour of the circulation, since we assume that changes in the mean circulation represent differing frequencies of various circulation types. The present as the key to the past is probably a quite inadequate guide here as we have no knowledge of the pattern of circulation on a globe without polar ice-caps and this situation appears to have been 'normal' throughout most of geological time. Furthermore our period of instrumental records covers only very minor climatic fluctuations during the last hundred years, so that these minor changes assume an importance disproportionate to their magnitude and duration. Recent energy budget calculations for an ice-free Arctic (Donn and Shaw, 1966; Fletcher, 1965) may provide a solution to the first of these problems but the data shortage cannot be fully remedied.

The improved models of the general circulation discussed earlier provide an initial basis for postulating the probable circulation patterns during recent palaeoclimatic epochs. Interest has centred primarily on glacial-interglacial conditions and much use has been made of zonal index concepts, particularly the simple high-low index model. Many workers hold the view that strong westerly flow in higher middle latitudes (high index) is an interglacial and interpluvial pattern (Butzer 1957A) while the circulation responsible for the glacial maxima resembled the low index model with an expanded circumpolar vortex and well-developed meridional components (Leighly, 1949; Willett, 1950). Butzer (1957B) considers that the retreat stages of the glacials were also characterized by high index circulations. More recently Willett and Sanders (1959, pp. 194–5) suggested that the type of circulation most conducive to glaciation is the low-latitude phase of the zonal pattern which in the index cycle (Riehl et al., 1952) commonly precedes the northward shift of the westerlies and the establishment of high index. The extreme cellular type is thought to produce only localized extremes of temperature and precipitation.

Contrary views have been expressed by Flint and Dorsey (1945). They held that high index conditions were responsible for the inception of glaciation in eastern North America and similar suggestions have been advanced by Rex (1950) for Scandinavia. Rex argued that increased storminess associated with strong zonal flow aloft over Europe is required for glacierization and showed that blocking anticyclones over Scandinavia are unfavourable for glaciation. However, the high-latitude blocks described by Rex are associated with intensified zonal flow upstream and the 500 mb zonal index for $35°-55°$ N. remains above normal during the blocking period. As Willett (1950) pointed out this type of Scandinavian blocking needs to be differentiated from that in lower middle latitudes.

Many of the apparently conflicting views could probably be resolved if the circulation models, particularly index phases, employed in palaeoclimatological studies were carefully defined. Models of the circulation changes during the Pleistocene also need to take account of the time-lag between primary circulation changes and the development of a major ice-sheet over a period of approximately 20,000 years and secondary circulation changes consequent upon the presence of an ice-sheet. Again, these stages are not always clearly differentiated. A further consideration is that longitudinal differences of Pleistocene climates must have been considerable (Mather, 1954). For this reason over-simplified hemispheric models may be more misleading than specialized regional ones such as Butzer (1958 and 1961) has tried to develop for the Mediterranean area.

Another type of palaeoclimatic model concerns the relative chronology of the mid-latitude glacials and the lower latitude pluvials. Simpson (1957) and Butzer (1958) envisaged quite different relationships as shown in Table 4.1 and necessarily this affects the related circulation models which they proposed.

TABLE 4. I

Simpson (1957)		Butzer (1958)	
		Late Würm	Post-Pluvial
		Main Würm	Minor Pluvial
Würm	⎫	Early Würm	Major Pluvial
Warm wet Inter-Glacial	⎬ Pluvial	Inter-Glacial	Inter-Pluvial
Riss	⎭	Riss	Pluvial
Cold dry Inter-Glacial	Inter-Pluvial	Inter-Glacial	Inter-Pluvial
Mindel	⎫	Mindel	Pluvial
Warm wet Inter-Glacial	⎬ Pluvial		
Günz	⎭		
Cold dry Inter-Glacial	Inter-Pluvial		
Glacial	⎫		
Warm wet Inter-Glacial	⎬ Pluvial		
Glacial	⎭		

A very different approach to climatic change has been outlined by Curry (1962) who uses stochastic (probability) models based on queue theory to suggest that the atmospheric circulation could shift from its present state of energy balance to a different equilibrium without there necessarily being any significant changes of energy input. The crucial factor is the amount of energy stored at any given time in the oceans and ice-caps. Over the short-term of months or a few years 'feed-back' processes may operate to affect the circulation through such physical factors as snow-cover, sea-surface, temperatures and soil moisture (Namias, 1963). Persistent changes in the zonal or meriodional flow components over longer time scales are likely to be controlled by the degree of anchoring of the long waves by solenoidal fields and orographic barriers. Models of this type are only just beginning to be developed but they appear in the present state of knowledge to provide a more realistic approach to the question of climatic changes than the numerous theories invoking unique causative agents.

PREDICTION MODELS

Operational models for weather prediction (Bergeron, 1959) must be considered in terms of the particular time-scale involved and it is pertinent in this connection to summarize the relative time-orders of the major weather systems. These are:

Mesoscale features	1– 6 hours
Synoptic systems	1– 3 days
Long waves	7–10 days
Index cycle	4– 6 weeks
Monsoon circulation	3– 6 months

The general categories of forecast are short-period ones covering 12–24 hours, extended-range for 5–7 days ahead and long-range outlooks for about a month. Naturally the type of model which can be applied is greatly controlled by the essential predictability of the atmosphere's behaviour over these different periods.

Short-range forecasts were originally based on rule-of-thumb extrapolation estimates of the movement of pressure systems and synoptic models of characteristic weather patterns of the type outlined by Abercromby and Marriott (1883) for the depression. The major developments up to the 1940's were improved synoptic models such as the Bjerknes-Solberg frontal wave. Even when numerical methods are fully developed there will undoubtedly remain a need for specialized models in short-range forecasting particularly in regions of sparse data, and in this respect the models of the easterly wave, the three-front depression and the jet stream are invaluable. Such models also help to maintain continuity of map analysis.

Different approaches are necessary in extended and long-range forecasting. Considerable stimulus was provided after about 1940 by the Rossby long-wave model and the zonal index concept. During the same period also analogue methods of synoptic climatology were being developed in search of operational models (Jacobs, 1946; Baur, 1951). Their application to long-range outlooks has been assisted by the introduction of physical principles including the specification of anomalies of surface temperature and ice distribution (Willett, 1951), but such forecasts remain primarily statistical. Selection of suitable map analogues on which to base forecasts has, however, been greatly facilitated by computer methods. Thirty-day forecasts of temperature and precipitation anomalies in the United States, published since 1947, are derived from calculated changes of the 700 mb contour pattern using kinematic methods and synoptic-climatological models (Namias, 1953; Klein, 1965). Attempts have been made to detect decadal trends on the basis of time-series analysis, including cycles of solar activity (Willet, 1961), but there is at present little prospect of developing a satisfactory method for such very long-range predictions (Craddock, 1964).

The most significant advances in forecasting are due to the development of numerical models based on the fundamental hydrodynamical equations. The methods demand accurate knowledge of the initial state of the atmosphere and equations which adequately predict the atmospheric motion while remaining tractable. In 1922 Richardson tried to use this approach to calculate sea-level pressure tendencies by applying the equations of horizontal motion and the continuity equation at short time intervals over a spatial (finite difference) grid. However, the real changes were swamped by cumulative errors due to 'noise' resulting from the effect of sound and gravity waves on the equations.

Post-World War II developments in this field initially used very simplified structures. One of these was the barotropic 'one-parameter' model (Eliassen, 1956; Haltiner and Martin, 1957; Sutton, 1960). The principle of the conservation of absolute vorticity is applied to the 500 mb surface which is approximately the mean level of non-divergence such that divergence (convergence) below this level is compensated by convergence (divergence) above. It is assumed that the flow is geostrophic and that the motion at 500 mb is independent of that at other levels. Under these assumptions the vorticity equation (p. 99) becomes

$$\frac{\partial Q}{\partial t} + u\frac{\partial Q}{\partial x} + v\frac{\partial Q}{\partial y} = 0$$

where $f + \zeta = Q$ (the absolute vorticity)

$$\zeta = \frac{g}{f}\left(\frac{\partial^2 h}{\partial x^2} + \frac{\partial^2 h}{\partial y^2}\right), u = -\frac{g}{f}\frac{\partial h}{\partial y} \text{ and } v = \frac{g}{f}\frac{\partial h}{\partial x}$$

where h = the height of the 500 mb surface
g = gravity

The forecast equation

$$\left(\frac{\partial^2}{\partial x^2} + \frac{\partial^2}{\partial y^2}\right)\frac{\partial h}{\partial t} = -u\frac{\partial Q}{\partial x} - v\frac{\partial Q}{\partial y}$$

must be solved numerically from grid values of contour heights to obtain the pattern of the height tendency over the grid (the left hand side of the equation). At each grid point finite difference approximations are used to evaluate the terms on the right (Hess, 1959; Knighting, 1958; Sutton, 1960). The height tendency for a given time interval may then be added to the initial contour field and the computations are repeated. This process is carried out until the period of the forecast has been covered.

Bolin (1955) compares barotropic forecasts and actual 500 mb contours for 24, 48 and 72 hour periods. The model is shown to be reasonably satisfactory for the short periods as long as the patterns are displaced with little development because the model specifies that the vorticity is only advected by the geostrophic wind. The results are also affected by the restrictions imposed by the size of the grid area.

Subsequent models include consideration of two or more levels and more realistic atmospheres with additional parameters. For example, height changes at the 750 mb and 250 mb levels allow the calculation of vertical motion at 500 mb and determination of thickness changes in the 750–250 mb layer introduces baroclinicity into the model (Eliassen, 1956; Haltiner and Martin, 1957). The use of three levels provides better estimates of vertical motion and allows the effects of horizontal variations of static stability to be included. The results of such studies indicate that operational numerical forecasting has a bright future and although there is a long way to go before the weather elements themselves can be predicted directly progress is being made along these lines.

REFERENCES

ABERCROMBY, R., [1883], On certain types of British weather; *Quarterly Journal of the Royal Meteorological Society*, 9, 1–23.
ABERCROMBY, R. and MARRIOTT, W., [1883], Popular weather prognostics; *Quarterly Journal of the Royal Meteorological Society*, 9, 27–43.
ACADEMICA SINICA, [1957 and 1958], On the general circulation over eastern Asia; *Tellus*, 9, 432–446; 10, 58–75 and 299–312.

ALLEN, R. A., FLETCHER, J., HOLMBOE, J., NAMIAS, J. and WILLETT, H. C., [1940], Report on an experiment in five-day weather forecasting; *Papers in Physical Oceanography and Meteorology*, 8, No. 3 (Massachusetts Institute of Technology and Woods Hole), 94 pp.

ANDERSON, R., BOVILLE, B. W. and MCCLELLAN, D. E., [1955], An operational frontal contour-analysis model; *Quarterly Journal of the Royal Meteorological Society*, 81, 588–599.

BARRY, R. G., [1960], A note on the synoptic climatology of Labrador-Ungava; *Quarterly Journal of the Royal Meteorological Society*, 86, 557–565.

BARRY, R. G., [1963], Aspects of the synoptic climatology of central southern England; *Meteorological Magazine*, 92, 300–308.

BAUR, F., [1951], Extended-range weather forecasting; in *Compendium of Meteorology*; (Ed., T. F. Malone), American Meteorological Society, (Boston, Mass.), 814–833.

BELASCO, J. E., [1952], Characteristics of air masses over the British Isles; *Geophysical Memoirs, Meteorological Office*, 11 (No. 87), 34 pp.

BERGERON, T., [1928], Uber die dreidimensional verknüpfende Wetteranalyse; *Geofysiske Publikationer*, 5, No. 6 (Oslo), 111 pp.

BERGERON, T., [1930], Richtlinien einer dynamischen Klimatologie; *Meteorologische Zeitschrift*, 47, 246–262.

BERGERON, T., [1954], The problem of tropical hurricanes; *Quarterly Journal of the Royal Meteorological Society*, 80, 131–164.

BERGERON, T., [1959], Methods in scientific weather analysis and forecasting; in *The atmosphere and the sea in motion*; (Ed., B. Bolin), (Rockefeller Institute Press, New York), 440–470.

BJERKNES, J. and SOLBERG, H., [1922], Life cycle of cyclones and the polar front theory of atmospheric circulation; *Geofysiske Publikationer*, 3, No. 1 (Oslo), 18 pp.

BLAIR, T. A., [1942], *Climatology, general and regional*, (New York), 484 pp.

BOLIN, B., [1950], On the influence of the earth's orography on the general characteristics of the westerlies, *Tellus*, 2, 184–195.

BOLIN, B., [1952], Studies of the general circulation of the atmosphere; *Advances in Geophysics*, 1, 87–118.

BOLIN, B., [1955], Numerical forecasting with the barotropic model; *Tellus*, 7, 27–49.

BOVILLE, B. W. and KWIZAK, M., [1959], *Fourier analysis applied to hemispheric waves of the atmosphere*; CIR-3155, TEC-292, Meteorological Branch, Department of Transport, Canada, 21 pp.

BOYDEN, C. J., [1963], Jet streams in relation to fronts and the flow at low levels; *Meteorological Magazine*, 92, 319–328.

BRADBURY, D. L., [1958], On the behaviour of cyclones and anticyclones as related to zonal index; *Bulletin of the American Meteorological Society*, 39, 149–151.

BROWNING, K. A. and LUDLAM, F. H., [1962], Airflow in convective storms; *Quarterly Journal of the Royal Meteorological Society*, 88, 117–135.

BUTZER, K. W., [1957A], Mediterranean pluvials and the general circulation of the Pleistocene; *Geografiska Annaler*; 39, 48–53.

BUTZER, K. W., [1957B], The recent climatic fluctuation in lower latitudes and the general circulation of the Pleistocene; *Geografiska Annaler*; 39, 105–113.

BUTZER, K. W., [1958], Quaternary stratigraphy and climate in the Near East; *Bonner Geographische Abhandlungen*, Nr. 24, 157 pp.
BUTZER, K. W., [1961], Climatic change in arid regions since the Pliocene; in *A history of land use in arid regions*; (Ed., L. D. Stamp), Arid Zone Research, UNESCO, (Paris), 31–56.
BYERS, H. R., [1959], *General Meteorology*, (New York), 3rd Edn., 540 pp.
CHANG CHIA-CH'ENG, [1960], Some views on the nature of the China monsoon; *Trudy Glavnoi Geofizicheskoi Observatorii, Leningrad*, Vyp. 90, 24–42 (translated 1961 by Office of Technical Services, Washington, D.C.).
CHARNEY, J. G. and ELIASSEN, A., [1949], A numerical method for predicting the perturbations of the middle westerlies; *Tellus*, 1 (2), 38–54.
CLAPP, P. F., [1961], Normal heat sources and sinks in the lower troposphere in winter; *Monthly Weather Review*, 89, 147–162.
COURT, A., [1957], Climatology: complex, dynamic and synoptic; *Annals of the Association of American Geographers*, 47, 125–136.
CRADDOCK, J. M., [1964], The analysis of time series for use in forecasting; *The Statistician*, 15, 167–190.
CROWE, P. R., [1949], The trade-wind circulation of the world; *Transactions of the Institute of British Geographers*, No. 15, 37–56.
CROWE, P. R., [1950], The seasonal variation in the strength of the trades; *Transactions of the Institute of British Geographers*, No. 16, 23–47.
CROWE, P. R., [1965], The geographer and the atmosphere; *Transactions of the Institute of British Geographers*, No. 36, 1–19.
CURRY, L., [1962], Climatic change as a random series; *Annals of the Association of American Geographers*, 52, 21–31.
DONN, W. L. and SHAW, D. M., [1966], The heat budgets of an ice-free and an ice-covered Arctic Ocean; *Journal of Geophysical Research*, 71, 1087–93.
DÖÖS, B. R., [1962], The influence of exchange of sensible heat with the earth's surface on the planetary flow; *Tellus*, 14, 133–147.
DUNN, G. E., [1940], Cyclogenesis in the tropical Atlantic; *Bulletin of the American Meteorological Society*, 21, 215–229.
EADY, E. T. and SAWYER, J. S., [1951], Dynamics of flow patterns in extra-tropical regions; *Quarterly Journal of the Royal Meteorological Society*, 77, 531–551.
ELDRIDGE, R. H., [1957], A synoptic study of West African disturbance lines; *Quarterly Journal of the Royal Meteorological Society*, 83, 303–314.
ELIASSEN, A., [1956], Instability theories of cyclone formation; Numerical forecasting; Chapters 15 and 18 in *Weather Analysis and Forecasting* Volume 1, S. Petterssen, (New York), 305–319 and 371–387.
ELIASSEN, A., [1958], A study of the atmospheric waves on the basis of zonal harmonic analysis; *Tellus*, 10, 206–215.
ERICKSON, C. O., [1963], An incipient hurricane near the West African coast; *Monthly Weather Review*, 91, 61–68.
ESTOQUE, M. A., [1964], Formation and structure of tropical hurricanes; in *Proceedings of the Symposium on Tropical Meteorology, Rotorua, New Zealand, 1963*, (Ed., J. W. Hutchings), (Wellington, New Zealand), 595–613.
FERREL, W., [1856], An essay on the winds and currents of the ocean; *Nashville Journal of Medicine and Surgery*, 11(4), 287–301.

FITZROY, R., [1863], *The weather book. A manual of practical meteorology*, (London), 480 pp.
FLEAGLE, R. G., [1957], On the dynamics of the general circulation; *Quarterly Journal of the Royal Meteorological Society*, 83, 1–20.
FLEAGLE, R. G., [1960], The general circulation; *Science Progress*, 48, 72–81.
FLETCHER, J. O., [1965], *Climate and the heat budget of the central Arctic*; The Rand Corporation, (Santa Monica, California), 22 pp.
FLINT, R. F. and DORSEY, JR., H. G., [1945], Iowan and Tazewell drifts and the North American ice sheet; *American Journal of Science*, 243, 615–635.
FLOHN, H., [1960A], Recent investigations on the mechanism of the 'summer monsoon' of southern and eastern Asia; in *Monsoons of the World*, India Meteorological Department, (Delhi), 75–88.
FLOHN, H., [1960B], Monsoon winds and the general circulation; in *Monsoons of the World*, India Meteorological Department, Delhi, 65–74.
FLOHN, H., [1960C], The structure of the inter-tropical convergence zone; in *Tropical Meteorology in Africa*, (Ed., D. J. Bargman), Munitalp Foundation, (Nairobi), 244–251.
FLOHN, H., [1960D], Equatorial westerlies over Africa, their extension and significance; in *Tropical Meteorology in Africa*, (Ed., D. J. Bargman), Munitalp Foundation, (Nairobi), 253–264.
FORSDYKE, A. G., [1951], On zonal and other indices; *Meteorological Magazine*, 80, 151–160.
FORSDYKE, A. G., [1960], Synoptic models of the tropics; in *Tropical Meteorology in Africa*, (Ed., D. J. Bargman), Munitalp Foundation, (Nairobi), 14–19.
FREEMAN JR., J. C., [1948], An analogy between the equatorial easterlies and supersonic gas flow; *Journal of Meteorology*, 5, 138–146.
FREEMAN JR., J. C., [1951], The solution of nonlinear meteorological problems by the method of characteristics; in *Compendium of Meteorology*, (Ed., T. F. Malone), American Meteorological Society, (Boston, Mass.), 421–433.
FRENZEN, P., [1955], Westerly flow past an obstacle in a rotating hemispheric shell; *Bulletin of the American Meteorological Society*, 36, 204–210.
FRISBY, E. M. and GREEN, F. W. H., [1949], Further notes on comparative regional climatology; *Transactions of the Institute of British Geographers*, No. 15, 143–151.
FROST, R. and STEPHENSON, P. H., [1965], Mean streamlines and isotachs at standard pressure levels over the Indian and West Pacific Oceans and adjacent land areas; *Geophysical Memoirs, Meteorological Office*, 14 (No. 109), 24 pp.
FUJITA, T., [1955], Results of detailed synoptic studies of squall-lines; *Tellus*, 7, 405–436.
FULTZ, D., [1951A], Non-dimensional equations and modelling criteria for the atmosphere; *Journal of Meteorology*, 8, 262–267.
FULTZ, D., [1951B], Experimental analogies to atmospheric motions; in *Compendium of Meteorology*, (Ed., T. F. Malone), American Meteorological Society, (Boston, Mass.), 1235–1248.
FULTZ, D., [1960], Experimental models of rotating fluids and possible avenues for future research; in *Dynamics of Climate*, (Ed., R. L. Pfeffer), (Pergamon Press), 71–77.

FULTZ, D., [1961], Developments in controlled experiments on larger scale geophysical problems; *Advances in Geophysics*, 7, 1–103.

GALLOWAY, J. L., [1958A], The three-front model; its philosophy, nature, construction and use; *Weather*, 13, 3–10.

GALLOWAY, J. L., [1958B], The three-front model, the tropopause and the jet stream; *Weather*, 13, 395–403.

GARBELL, H. A., [1947], *Tropical and equatorial meteorology*, (London), 237 pp.

GILCHRIST, B., [1954], The seasonal phase changes of thermally produced perturbations in the westerlies; *Proceedings of the Toronto Meteorological Conference 1953*, 129–131.

GODSON, W. L., [1950], The structure of North American weather systems; *Centenary Proceedings of the Royal Meteorological Society*, 89–106.

GRIMMER, M., [1963], The space-filtering of monthly surface anomaly data in terms of pattern, using empirical orthogonal functions; *Quarterly Journal of the Royal Meteorological Society*, 89, 395–408.

HADLEY, G., [1735], Concerning the cause of the general tradewinds; *Philosophical Transactions, London*, 29, 58–62.

HALTINER, G. J. and MARTIN, F. L., [1957], *Dynamical and physical meteorology*, (New York), 470 pp.

HARE, F. K., [1958], The quantitative representation of the north polar pressure field; in *The Polar Atmosphere Symposium*, Part I, (Ed., R. C. Sutcliffe), (Pergamon Press), 137–150.

HARE, F. K., [1960], The westerlies; *Geographical Review*; 50, 345–367.

HARE, F. K., [1962], The stratosphere; *Geographical Review*; 52, 525–547.

HARE, F. K., [1965], Energy exchanges and the general circulation; *Geography*, 50, 229–241.

HESS, S. L., [1959], *Introduction to theoretical meteorology*, (New York), 362 pp.

HIDE, R., [1953], Some experiments on thermal convection in a rotating liquid; *Quarterly Journal of the Royal Meteorological Society*, 79, 161.

HOLMBOE, J. and BJERKNES, J., [1944], On the theory of cyclones; *Journal of Meteorology*, 1, 1–22.

HUSCHKE, R. E. (Ed.), [1959], *Glossary of meteorology*; American Meteorological Society, (Boston, Mass.), 638 pp.

JACOBS, W. C., [1946], Synoptic climatology; *Bulletin of the American Meteorological Society*, 27, 306–311.

JEFFREYS, H., [1926], On the dynamics of geostrophic winds; *Quarterly Journal of the Royal Meteorological Society*, 52, 85–104.

JOHNSON, D. H., [1962], Tropical meteorology; hurricane and typhoons; *Science Progress*, 50, 403–419.

JOHNSON, D. H., [1963], Tropical meteorology; other weather systems; *Science Progress*, 51, 587–601.

JOHNSON, D. H. and DANIELS, S. M., [1954], Rainfall in relation to the jet stream; *Quarterly Journal of the Royal Meteorological Society*, 80, 212–217.

JOHNSON, D. H. and MÖRTH, H. T., [1960], Forecasting research in East Africa; in *Tropical meteorology in Africa*, (Ed., D. J. Bargman), Munitalp Foundation, (Nairobi), 56–132.

KOTESWARAM, P., [1958], The easterly jet stream in the tropics; *Tellus*, 10, 43–57.

KLEIN, W. H., [1965], Synoptic climatological models for the United States; *Weatherwise*, 18, 252–259.
KNIGHTING, E., [1958], Numerical weather forecasting; *Weather*, 13, 39–50.
KRISHNAMURTI, T. N., [1961], On the role of the subtropical jet stream of winter in the atmospheric general circulation; *Journal of Meteorology*, 18, 657–670.
LAMB, H. H., [1950], Types and spells of weather around the year in the British Isles: Annual trends, seasonal structure of the year, singularities; *Quarterly Journal of the Royal Meteorological Society*, 76, 393–429.
LAMB, H. H., [1959], The southern westerlies: a preliminary survey; main characteristics and apparent associations; *Quarterly Journal of the Royal Meteorological Society*, 85, 1–23.
LA SEUR, N. E., [1954], On the asymmetry of the middle-latitude circumpolar current; *Journal of Meteorology*, 11, 43–57.
LA SEUR, N. E., [1960], Tropical synoptic models; in *Tropical meteorology in Africa*, (Ed., D. J. Bargman), Munitalp Foundation, (Nairobi), 47–54.
LEIGHLY, J., [1949], On continentality and glaciation; *Geografiska Annaler*, 31, 133–145.
LOCKWOOD, J. G., [1965], The Indian monsoon – a review; *Weather*, 20, 2–8.
LONG, R. R., [1952], The flow of a liquid past a barrier in a rotating spherical shell; *Journal of Meteorology*, 9, 187–199.
LUDLAM, F. H., [1961], The hailstorm; *Weather*, 16, 152–162.
LUND, I. A., [1963], Map-pattern classification by statistical methods; *Journal of Applied Meteorology*, 2, 50–65.
LYLE, H. H. and BRYSON, R. A., [1960], Harmonic analysis of the annual march of precipitation over the United States; *Annals of the Association of American Geographers*, 50, 157–171.
MALKUS, J. S., [1958], Tropical weather disturbances: why do so few become hurricanes? *Weather*, 13, 75–89.
MALKUS, J. S., [1962], Inter-change of properties between sea and air. Large-scale interactions; in *The Sea*, volume 1, (Ed., M. N. Hill), (Interscience Publishers, New York), 88–294.
MALKUS, J. S. and RIEHL, H., [1964], Cloud structure and distributions over the tropical Pacific Ocean; *Tellus*, 16, 275–287.
MANABE, S., SMAGORINSKY, J. and STRICKLER, R. F., [1965], Simulated climatology of a general circulation model with a hydrologic cycle; *Monthly Weather Review*, 93, 769–798.
MATHER, J. R., [1954], The present climatic fluctuation and its bearing on a reconstruction of Pleistocene climatic conditions; *Tellus*, 6, 287–301.
MERRITT, E. S., [1964], Easterly waves and perturbations, a reappraisal; *Journal of Applied Meteorology*, 3, 367–382.
MILES, M. K., [1962], Wind, temperature and humidity distribution at some cold fronts over S.E. England; *Quarterly Journal of the Royal Meteorological Society*, 88, 286–300.
MÖRTH, H. T., [1964], Primary factors governing tropospheric circulations in tropical and subtropical latitudes; in *Proceedings of the Symposium on Tropical Meteorology, Rotorua, New Zealand, 1963*, (Ed., J. W. Hutchings), (Wellington, New Zealand), 31–41.

MURRAY, R. and JOHNSON, D. H., [1952], Structure of the upper westerlies; a study of the wind field in the eastern Atlantic and western Europe in September 1950; *Quarterly Journal of the Meteorological Society*, 78, 186–199.

NAMIAS, J., [1950], The index cycle and its role in the general circulation; *Journal of Meteorology*, 17, 130–139.

NAMIAS, J., [1953], Thirty-day forecasting: A ten-year experiment; *Meteorological Monographs*, 2 (No. 6), 83 pp.

NAMIAS, J., [1963], Surface-atmosphere interactions as fundamental causes of drought and other climatic fluctuations; in *Changes of Climate*, Arid Zone Research XX, UNESCO, (Paris), 345–359.

NAMIAS, J. and CLAPP, P. F., [1951], Observational studies of general circulation patterns; in *Compendium of Meteorology*, (Ed., T. F. Malone), American Meteorological Society, (Boston, Mass.), 551–567.

NEWTON, C. W., [1959], Synoptic comparisons of jet stream and Gulf Stream systems; in *The atmosphere and the sea in motion*, (Ed., B. Bolin), (Rockefeller Institute Press, New York), 288–304.

NEWTON, C. W. and PERSSON, A. V., [1962], Structural characteristics of the sub-tropical jet stream and certain lower stratospheric wind systems; *Tellus*, 14, 221–241.

PALMÉN, E., [1951A], The aerology of extratropical disturbances; in *Compendium of Meteorology* (Ed., T. F. Malone), American Meteorological Society, (Boston, Mass.), 599–620.

PALMÉN, E., [1951B], The role of atmospheric disturbances in the general circulation; *Quarterly Journal of the Royal Meteorological Society*, 77, 337–354.

PALMÉN, E., [1954], General circulation of the tropics; in *Proceedings of the Symposium on Tropical Meteorology, Rotorua, New Zealand, 1963* (Ed., J. W. Hutchings), (Wellington, New Zealand), 3–30.

PALMÉN, E. and VUORELA, L. A., [1963], On the mean meridional circulation in the northern hemisphere during the winter season; *Quarterly Journal of the Royal Meteorological Society*, 89, 131–138.

PALMER, C. E., [1951], Tropical meterology; in *Compendium of Meteorology* (Ed., T. F. Malone), American Meteorological Society, (Boston, Mass.), 859–880.

PALMER, C. E., [1952], Tropical meterology; *Quarterly Journal of the Royal Meteorological Society*, 78, 126–164.

PÉDELABORDE, P., [1958], *The monsoon*; (translated 1963), (London), 196 pp.

PEDGLEY, D. E., [1962], *A course in elementary meteorology*; (H.M.S.O., London), 189 pp.

PENNER, C. M., [1955], A three-front model for synoptic analyses; *Quarterly Journal of the Royal Meteorological Society*, 81, 89–91.

PETTERSSEN, S., [1956], *Weather analysis and forecasting*, 2nd Edition, (New York), Vol. I 428 pp., Vol. II 266 pp.

PHILLIPS, N. A., [1956], The general circulation of the atmosphere: a numerical experiment; *Quarterly Journal of the Royal Meteorological Society*, 82, 123–164.

PISHAROTY, P. R. and ASNANI, G. C., [1960], Flow pattern over India and neighbourhood at 500 mb during the monsoon; in *Monsoons over the World*, India Meteorological Department, (Delhi), 112–117.

PRIESTLEY, C. H. B., [1949], Heat transport and zonal stress between latitudes; *Quarterly Journal of the Royal Meteorological Society*, 75, 28–40.

PRIESTLEY, C. H. B., [1951], Physical interactions between tropical and temperate latitudes; *Quarterly Journal of the Royal Meteorological Society*, 77, 200–214.

QUENEY, P., [1948], The problem of airflow over mountains: a summary of theoretical studies; *Bulletin of the American Meteorological Society*, 29, 16–26.

RAMAKRISHNAN, K. P., SREENIVASAIAH, B. N., and VENKITESHWARAN, S. P., [1958], Upper air climatology of India and neighbourhood in the monsoon seasons; in *Monsoons over the World*, India Meteorological Department, (Delhi), 3–34.

RAMANTHAN, K. R., [1958], Monsoons and the general circulation of the atmosphere – a review; in *Monsoons over the World*, India Meteorological Department, (Delhi), 53–64.

RAMASWAMY, C., [1962], Breaks in the Indian summer monsoon as a phenomenon of interaction between the easterly and the subtropical westerly jet streams; *Tellus*, 14, 337–349.

REITER, E. R., [1963], *Jet stream meteorology*; (Chicago), 515 pp.

REX, D. F., [1950], Blocking action in the middle troposphere and its effect on regional climate. Part II, The climatology of blocking action; *Tellus*, 2, 275–301.

RIEHL, H., [1951], Aerology of tropical storms; in *Compendium of Meteorology* (Ed., T. F. Malone), American Meteorological Society, (Boston, Mass.), 902–913.

RIEHL, H., [1954], *Tropical meteorology*, (New York), 392 pp.

RIEHL, H., [1962A], *Jet streams of the atmosphere*, Technical Paper No. 32, Department of Atmospheric Science, Colorado State University, (Fort Collins, Colorado), 117 pp.

RIEHL, H., [1962B], General atmospheric circulation of the tropics; *Science*, 135, 13–22.

RIEHL, H., and FULTZ, D., [1957], Jet stream and long waves in a steady rotating dishpan experiment: structure of the circulation; *Quarterly Journal of the Royal Meteorological Society*, 83, 215–231.

RIEHL, H., and FULTZ, D., [1958], The general circulation in a steady rotating dishpan experiment; *Quarterly Journal of the Royal Meteorological Society*, 84, 389–417.

RIEHL, H., YEH, T. C., MALKUS, J. C. and LA SEUR, N. E., [1951], The north-east trade of the Pacific Ocean; *Quarterly Journal of the Royal Meteorological Society*, 77, 598–626.

RIEHL, H. and Collaborators, [1952], Forecasting in middle latitudes; *Meteorological Monographs*, 1 (No. 5), 80 pp.

RICHARDSON, L. F., [1922], *Weather prediction by numerical process*, (Cambridge), 236 pp.

ROSSBY, C.-G., [1940], Planetary flow patterns in the atmosphere; *Quarterly Journal of the Royal Meteorological Society*, Supplement to 66, 68–87.

ROSSBY, C.-G., [1941], The scientific basis of modern meteorology; in *Climate and Man*, United States Department of Agriculture Yearbook, (Washington, D.C.), 599–655.

ROSSBY, C.-G., [1949], On the nature of the general circulation of the lower atmosphere; in *The atmospheres of the earth and planets* (Ed., G. P. Kuiper), (University of Chicago Press), 16–48.

ROSSBY, C.-G., and collaborators, [1939], Relations between variations in the intensity of the zonal circulation and displacements of the semi-permanent centres of action; *Journal of Marine Research*, 2, 38–55.
ROSSBY, C.-G. and WILLETT, H. C., [1948], The circulation of the upper troposphere and lower stratosphere; *Science*, 108, 643–652.
SABBAGH, M. E. and BRYSON, R. A., [1962], Aspects of the precipitation climatology of Canada investigated by the method of harmonic analysis; *Annals of the Association of American Geographers*, 52, 426–440.
SALTZMANN, B. and PEIXOTO, J. P., [1957], Harmonic analysis of the mean northern-hemisphere wind field for the year 1950; *Quarterly Journal of the Royal Meteorological Society*, 83, 360–364.
SANSOM, H. W., [1951], A study of cold fronts over the British Isles; *Quarterly Journal of the Royal Meteorological Society*, 77, 96–120.
SAWYER, J. S., [1958], Temperature, humidity and cloud near fronts in the middle and upper troposphere; *Quarterly Journal of the Royal Meteorological Society*, 375–388.
SAWYER, J. S., [1963], Notes on the response of the general circulation to changes in the solar constant; in *Changes of Climate*, Arid Zone Research XX, UNESCO, [Paris], 333–336.
SCORER, R. S., [1958], *Natural aerodynamics*, (Pergamon Press), 312 pp.
SCORER, R. S., [1966], Origin of cyclones; *Science Journal*, 2, 46–52.
SHAW, W. N. and LEMPFERT, R. K. G., [1906], *The life history of surface air currents*; Meteorological Committee, M.O. 174, (London), 107 pp.
SHEPPARD, P. A., [1958], The general circulation of the atmosphere; *Weather*, 13, 323–336.
SHEPPARD, P. A., [1962], Properties and processes at the earth's surface in relation to the general circulation; *Advances in Geophysics*, 9, 77–96.
SIMPSON, G. C., [1957], Further studies in world climate; *Quarterly Journal of the Royal Meteorological Society*, 83, 459–481.
SMAGORINSKY, J., [1953], The dynamic influence of large scale heat sources and sinks on the quasi-stationary mean motions of the atmosphere; *Quarterly Journal of the Royal Meteorological Society*, 79, 342–366.
SMAGORINSKY, J., [1963], General circulation experiments with the primitive equations; *Monthly Weather Review*, 91, 99–165.
SMAGORINSKY, J., [1964], Some aspects of the general circulation; *Quarterly Journal of the Royal Meteorological Society*, 90, 1–14.
STAFF MEMBERS, METEOROLOGY DEPARTMENT, UNIVERSITY OF CHICAGO; [1947], On the general circulation of the atmosphere in middle latitudes, *Bulletin of the American Meteorological Society*, 28, 255–280.
STARR, V. P., [1948], An essay on the general circulation of the earth's atmosphere; *Journal of Meteorology*, 5, 39–43.
STARR, V. P. and WHITE, R. M., [1951], A hemispherical study of the atmospheric angular-momentum balance; *Quarterly Journal of the Royal Meteorological Society*, 77, 215–225.
STARR, V. P. and WHITE, R. M., [1952], Schemes for the study of hemispheric exchange processes; *Quarterly Journal of the Royal Meteorological Society*, 78, 407–410.

STARR, V. P. and WHITE, R. M., [1954], *Balance requirements of the general circulation*; Geophysical Research Papers No. 35, Geophysics Research Directorate. (Cambridge, Mass.), 57 pp.

STEINER, D., [1965], A multivariate statistical approach to climatic regionalization and classification; *Tijdschrift van het Koninklij Nederlandsch Aardrijkskundig Genootschap.* 82 (4), 329–347.

SUTCLIFFE, R. C., [1947], A contribution to the problem of development; *Quarterly Journal of the Royal Meteorological Society*, 73, 370–383.

SUTCLIFFE, R. C., [1949], The general circulation – a problem in synoptic meteorology; *Quarterly Journal of the Royal Meteorological Society*, 75, 417–430.

SUTCLIFFE, R. C., [1950], Discussion on the general circulation; *Centenary Proceedings of the Royal Meteorological Society*, 180–183.

SUTCLIFFE, R. C., [1951], Mean upper contour patterns of the northern hemisphere – the thermal synoptic viewpoint; *Quarterly Journal of the Royal Meteorological Society*, 77, 435–440.

SUTCLIFFE, R. C., [1954], Cyclones and anticyclones – a comparative study; *Proceedings of the Toronto Meteorological Conference 1953*, 139–143.

SUTCLIFFE, R. C. and FORSDYKE, A. G., [1950], The theory and use of upper air thickness patterns in forecasting; *Quarterly Journal of the Royal Meteorological Society*, 76, 189–217.

SUTTON, O. G., [1960], *Understanding weather*, (Penguin Books), 215 pp.

SUTTON, O. G., [1962], *The challenge of the atmosphere*, (London), 227 pp.

TREWARTHA, G. T., [1961], *The earth's problem climates*, (University of Wisconsin Press, Madison), 334 pp.

TUCKER, G. B., [1959], Mean meridional circulation in the atmosphere; *Quarterly Journal of the Royal Meteorological Society*, 85, 209–224.

TUCKER, G. B., [1962], The general circulation of the atmosphere; *Weather*, 17, 320–340.

TUCKER, G. B., [1965], The equatorial tropospheric wind regime, *Quarterly Journal of the Royal Meteorological Society*, 91, 140–150.

VON HELMHOLTZ, H., [1888], Über atmosphärischer Bewegungen, trans. C. Abbé in The mechanics of the earth's atmosphere; *Smithsonian Institute Miscellaneous Collection*, 34 (843), 31–129.

WIIN-NIELSEN, A., BROWN, J. A. and DRAKE, M., [1964], Further studies of energy exchange between the zonal flow and the eddies; *Tellus*, 16, 168–180.

WILLETT, H. C., [1950], The general circulation at the last (Würm) glacial maximum; *Geografiska Annaler*, 32, 179–187.

WILLETT, H. C., [1951], The forecast problem; in *Compendium of Meteorology* (Ed., T. F. Malone), American Meteorological Society, (Boston, Mass.), 731–46.

WILLETT, H. C., [1961], The pattern of solar climatic relationships; *Annals of the New York Academy of Sciences*, 95, 89–106.

WILLETT, H. C. and SANDERS, F., [1959], *Descriptive meteorology*; 2nd Edn., (Academic Press, New York), 355 pp.

YANAI, M., [1964], Formation of tropical cyclones, *Reviews of Geophysics*, 2, 367–414.

YIN, M. T., [1949], A synoptic-aerological study of the onset of the summer monsoon over India and Burma; *Journal of Meteorology*, 6, 393–400.

CHAPTER FIVE

Hydrological Models and Geography

ROSEMARY J. MORE

INTRODUCTION

Hydrology is concerned with the science of water, its occurrence, circulation and distribution, its chemical and physical properties, its relation to the natural environment and its relation to living things, including man. The influence of water in moulding the landscape has been one of the principal concerns of physical geographers (e.g. Leopold, Wolman and Miller, 1964). The distribution of water has been important in the location of human settlements, and in the development of agriculture and industry. In Western Europe and North America the provision of adequate water supplies and the apportionment of water between competing demands are critical problems and ones which may set a limit to further economic developments.

Hydrological work has been traditionally dominated by engineers, whose scientific traditions have encouraged them to strive after generalizations and predictive schemes of a model character. Hence much of the present science of hydrology must fall naturally under the general heading of model building. One class of model associated with hydrology is that of hydraulic scale models of engineering projects, such as those built at the Wallingford Hydraulics Research Station or at the University of Delft's Hydraulics Laboratory. Although these hardware models are of immense practical value to the engineer (Allen, 1952), they are not the only models which have been used by hydrologists, for conceptual and mathematical models have also been applied to the appreciation and analysis of hydrological processes.

The basic conceptual model in hydrology is the idea of a cycle of water in its gaseous, liquid and solid form, and Figure 5.1 shows the principal components of this cycle. Water is evaporated from the oceans, stored as atmospheric moisture, and deposited as precipitation, which may be in the form of snow, sleet, hail, rain or dew. If the water resources of the earth are considered in terms of a budget, rainfall forms the largest item on the income side. However, much of the rain which falls does not immediately reach the river courses, but is lost by evaporation, transpiration, infiltration to soil moisture reserves, or deeper percolation to ground water in pervious rocks.

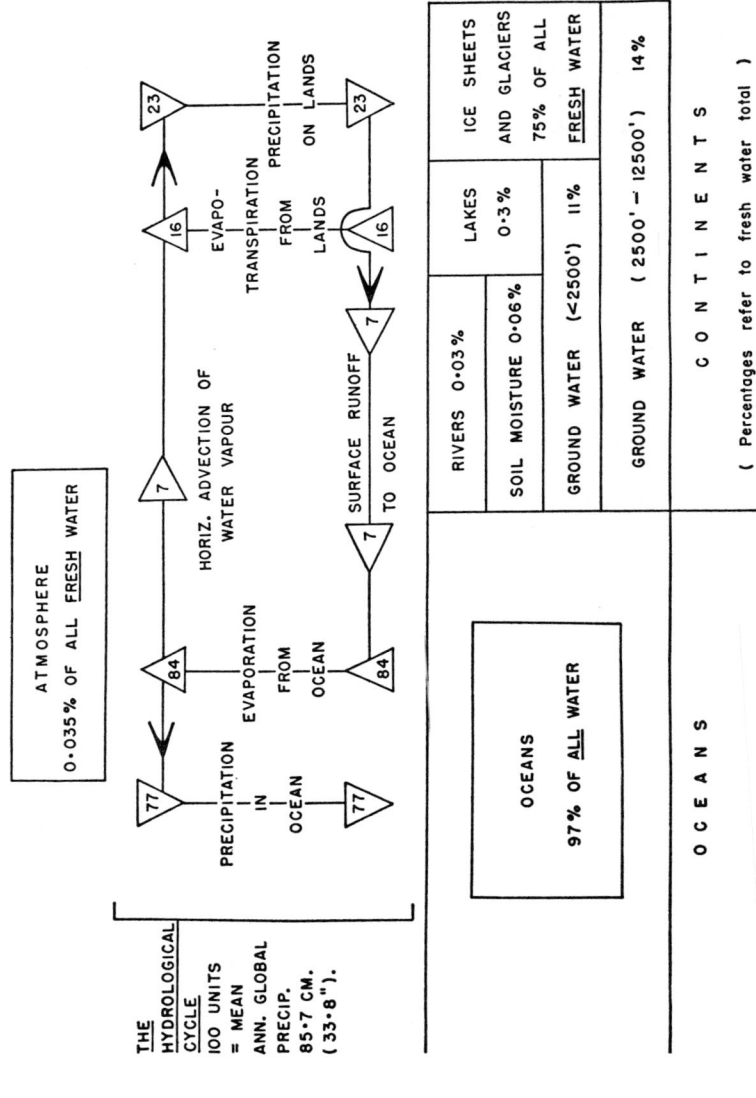

5.1 The hydrological cycle and terrestrial water storage. The oceanic percentage relates to *all* terrestrial water; the percentages for continental and atmospheric water to all *fresh* water. The units in the hydrological cycle are related to an assumed 100 units of mean global precipitation (77 oceanic and 23 continental).

The water surplus to these needs and processes flows as direct runoff into the streams, lakes and major watercourses. One of the chief problems in hydrology is to define the relationship between rainfall 'input' and that part of 'output' represented by direct runoff, but because of the number of intervening factors, their complexity and the problems involved in measuring them, the nature of this relationship is still only very partially understood.

Lack of meteorological and hydrological data has hindered a greater understanding of hydrological processes. Large areas of the globe, notably over the oceans, have no recording stations and many existing stations have only limited instrumentation. This deficiency is being remedied by work done during the International Hydrological Decade (1965–1975) and, increasingly, by information remotely sensed from spacecraft (Carter et al, 1966). The increased volume of data received from these new sources is creating problems both of data processing and analysis, and conventional methods of manual computation (see, for example, Beard, 1962) are being supplemented by computer techniques (Johnson, 1965; Johnson and Lang, 1965). These techniques have not only made routine calculations less laborious, but have enabled new techniques to evolve which seem likely to contribute to our understanding of fundamental hydrological principles (O'Donnell, 1966).

In relation to the limited knowledge which we have regarding the energy and water budgets of the earth, the hydrological cycle itself may be regarded as a large-scale conceptual model which attempts to simplify a complex reality. Within the 'great cycle' there are numerous smaller cycles. For example, water is evaporated from the surface of lakes and falls again as precipitation on the surrounding hills, returning by streams and rivers into the lakes from which it was evaporated. The smaller cycles are all related to the 'great cycle' system, and all its components can be thought of as interlocking subsystems, so that, if any one of them varies, others will change in response. The quantities of water in the various subsystems of the cycle are continually changing, so that mathematical models (i.e. those used to quantify hydrological processes) which are made to simulate the cycle numerically must be dynamic or self-adjusting (George, 1965). Furthermore, variations in the significance of the component parts of the cycle vary from one climatic region to another, such that, for example, areas with different climates experience differences in amounts, duration, intensity and occurrence of rainfall and runoff.

For practical convenience, a more local hydrological cycle can be identified in catchment area units, which correspond to river valleys bounded by watersheds. The water resources of catchment areas are assessed in terms of water balances, which are prepared in a similar fashion to a bank balance, and which show the 'income', 'expenditure' and 'credit balance' of water. These studies are most commonly based on the model used by Penman (1950) to assess the water balance of the River Stour in Essex. The mathematical model which

Penman used to express the water balance for an area over any period was:

Initial storage+Rainfall=Evaporation+Runoff+Final storage.

By using this formula Penman was able to estimate monthly changes in storage for the Stour catchment over the period 1933–1948. The catchment was generally assumed to be at field capacity (i.e. the level of moisture content at which the maximum amount of undrainable water is held) at the end of March. The calculation then proceeded by subtracting evaporation loss from rainfall and making estimates of the gain to storage under different conditions of vegetational root cover and its location over the catchment. Figure 5.2 shows that there was a very good agreement between Penman's calculations regarding total storage and movements of rest water levels in wells in the chalk.

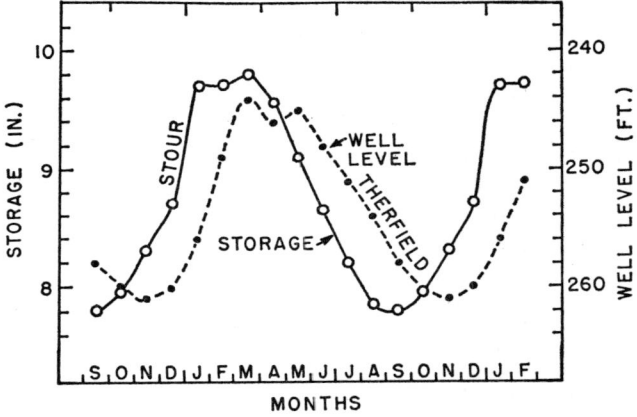

5.2 Estimated mean monthly changes in storage for the Stour catchment, Essex (1933–1948) compared with observed mean monthly well levels at Therfield on the same chalk ridge some 30 miles away. The discrepancies in the peak levels can be attributed to differences in precipitation between the two areas (i.e. the Stour had wetter winters and drier summers during the period), and the two-month phase difference to the time for percolating water to reach the deep water table. (*Source: Penman, 1950, p. 465*).

FUNCTION OF MODELS IN HYDROLOGY

Models are used for three main purposes in hydrology: (1) to simplify and generalize a complex reality, (2) to predict forthcoming hydrological events, and (3) to plan the future use of water resources.

Models for simplification and generalization

The design of hydrometric networks is based on model theory, in so far as, for example, the real pattern of rainfall distribution is generally so variable in time and in space that it would require a very large number of rain gauges to give a detailed picture of the distribution. However, the cost of rain gauges, the problems of maintenance and the lack of observers often make it impossible to have as close a grid as would be desirable, and make it necessary to sample the rainfall distribution with a small number of judicially-spaced gauges. (See Krumbein and Graybill, 1965, for general considerations of areal sampling.)

Two main approaches to estimating from point samples the average depth of rainfall over an area have been used (Chow, 1964, pp. 9-26 to 9-49). The Thiessen method (1911) assumes that the amount at any station has fallen over that part of the catchment nearest to the station. It is applied by constructing a Thiessen polygon network, the polygons being formed by the perpendicular bisectors of the lines joining nearby stations. The area of each polygon is determined and its proportion to the total catchment area is used to weight the rainfall amount of the station in the centre of the polygon when finding the catchment average fall. It is necessary to change the polygons each time a station is added to or taken away from the network.

The isohyetal method consists of drawing lines of equal rainfall amount, using observed amounts at stations and any additional factors available to adjust or interpolate between observed stations. The average depth is then determined by computing the incremental volume between each pair of isohyets, adding these incremental amounts and dividing by the total area.

A recent approach to the sampling and generalization of hydrological conditions has been suggested by means of the use of experimental and representative basins (International Association of Scientific Hydrology, 1965; UNESCO, 1965). The philosophy behind the use of these 'model' basins is that a relatively small area, in which accurate measurements of precipitation, runoff, evaporation, groundwater, sediment, etc., are made, may be used to generalize and predict conditions over larger areas. A small catchment, or sub-area of a larger catchment, can be held to be typical in two senses: (1) the data of the different elements may provide 'sample' statistics to supply information about the 'population' of the larger area, or (2) the relationships derived between some of the elements, e.g. rainfall-runoff, rainfall-evaporation, may be assumed equivalent to the relationships for the same elements in the larger areas. The sampled results would be particularly useful where larger areas have complications which are otherwise impossible to assess (e.g. 'foreign' water intrusion due to the presence of towns).

Experimental basins have been defined as basins which have been instrumented to measure most, or all, of the major components of the water balance, and where some land-use change is to be made during the period of investigation. Representative basins are similarly instrumented, but no land-use change is contemplated. The British International Hydrological Decade Committee list twenty experimental catchments which are to be studied during the decade period (1965–1975), including the Grendon Underwood experimental catchment on the River Ray in the Thames Valley, which is being studied by the Hydrological Research Unit at Wallingford, and three representative basins wherein the present land use and hydrology relationships are being investigated during the decade period.

An interesting variant of the concept of representative hydrological basins is that of the 'Vigil Network' (Leopold, 1962; Slaymaker and Chorley, 1964; Leopold and Emmett, 1965; Emmett, 1965), under which small basins in supposedly representative lithologic and climatic regions are instrumented in an attempted to obtain sediment budgets in a manner similar to hydrological budgets. Thus soil and talus creep, surface slope erosion, stream bank erosion, and channel and reservoir erosion and sedimentation are measured by a variety of instrumental procedures varying from soil strain gauges to the surveying of painted debris movements. There are three main objectives of this Vigil Network Project: (1) the estimation of regional rates of erosion; (2) the association of individual erosional and climatic events; and (3) the investigation of the areal patterns of erosion and sedimentation in an attempt to understand the nature of the geometrical transformations of the earth's surface in the most direct manner. It is hoped that erosional models of basins under differing lithologic and climatic conditions will ultimately emerge, similar to the type of hydrological basin models resulting from representative hydrological basin investigations.

The above models are merely examples of some of the simplest attempts at areal generalizations in hydrology. Much of the remainder of this chapter is concerned with other kinds of models designed to simplify and generalize a complex hydrological reality.

Models for hydrological prediction

Engineers responsible for the design of structures are naturally concerned with the magnitude and frequency of hydrological events experienced over long time periods. With the exception of records of discharges of the Nile and of some rivers in China, the majority of runoff records are for periods of less than a hundred years. This is a very short time span, and, within it, there is a great probability that extreme high and low values for water levels are not recorded. Yet, in the construction of flood-control dams, water-supply reservoirs and impounding works for irrigation schemes, it is important to know the 'recurrence intervals' for given magnitudes of floods and droughts. Thus there has arisen a practical need for predictive models. Wolf (1966A) has

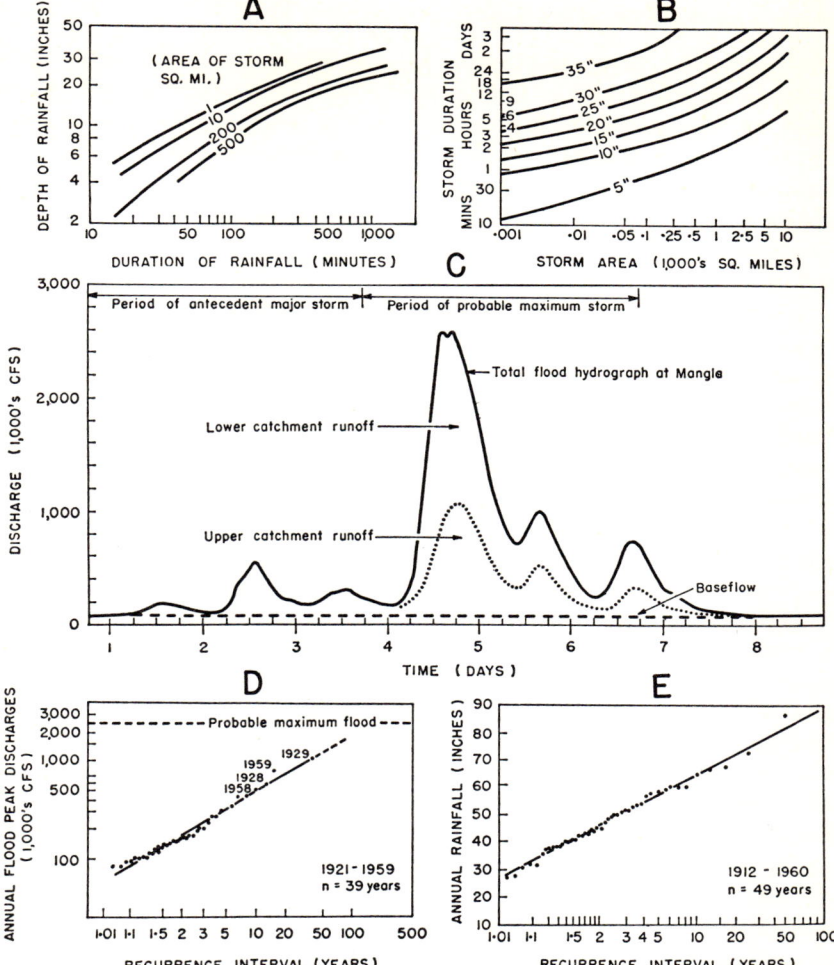

5.3 The prediction of hydrological events.

A Enveloping depth/duration curves of maximum thunderstorm rainfall for small areas in the United States (*Source: After Berry, Bollay and Beers. Redrawn from Wiesner, 1964, p. 162*).

B Enveloping isohyets of greatest observed depths of rainfall in the United States (*Source: After Berry, Bollay and Beers. Redrawn from Wiesner, 1964, p. 162*).

C Probable maximum flood hydrograph for the Jhelum River at Mangla, West Pakistan, calculated with respect to the worst characteristics of 14 major historic storms combined to give an estimate of the Probable Maximum Precipitation in a 72-hour storm. This is assumed to follow 72 hours after the start of an antecedent major storm for which the actual flood of 4–7 August, 1958, was used as a model. The abnormal form of the flood hydrograph results from the unusual drainage basin geometry (*Source: Binnie and Mansell-Moullin, 1966, Fig. 15*).

D Recurrence intervals of flood peaks at Mangla (1921–1959) on a logarithmic Gumbel plot. (*Source: Binnie and Mansell-Moullin, 1966, Fig. 16*).

E Recurrence intervals of annual rainfall amounts of different magnitude for Sydney, Australia (1912–1960) on an arithmetic Gumbel plot (*Source: Dury, G. H., 1964, Some Results of a Magnitude-Frequency analysis of Precipitation; Australian Geographical Studies, 2, p. 23*).

made a comparison of methods of flood estimation in which he describes the evolution of techniques from the use of empirical formulae and the 'rational method', to Sherman's development of the unit hydrograph theory and current attempts at non-linear analysis of catchment behaviour. Emphasis in predictive models for engineering purposes thus centres largely on the estimation of peak high values or extremely low values.

Two main approaches to prediction have been made. Wiesner (1964) advocates hydrometeorological analysis as a rational approach to flood prediction. He examines the meteorological elements and processes which produce a flood and estimates the highest value each can reasonably be expected to have. Based on this approach, meteorologists in the United States have calculated the maximum amounts of precipitation which can be expected for storms of different durations over areas of different size (see Fig. 5.3A and B). By this means probable maximum storms and their resultant floods can be estimated. Maximum possible flood estimates represent flood discharges expected from the most severe combination of critical meteorological and hydrological conditions reasonably possible in the region. They apply to projects where consideration is to be given to virtually complete security against risk to human life and are usually confined to the estimation of spillway requirements for high dams. The designs of many of the dams of the Australian Snowy Mountains scheme are based on probable maximum precipitation calculations. Binnie and Mansell-Moullin (1966) have described their use of this method to design the spillway capacity of the Mangla Dam in West Pakistan (Fig. 5.3C).

The other approach to prediction has been by statistical methods. E. J. Gumbel (1958A and 1958B) has been prominent in developing statistical distributions which describe the occurrence of extreme events. Gumbel suggested that the recurrence intervals of extreme events bear consistent relationships to their magnitudes (expressed in either arithmetic or logarithmic terms), and Figure 5.3D shows such a plot for floods at Mangla on the River Jhelum, and Figure 5.3E a similar analysis of rainfall records at Sidney, Australia (Dury, 1964). Gumbel proposed that the design of a structure should be based on the predicted flood discharge at a chosen recurrence interval (e.g. 50 years). While Gumbel's methods have great theoretical interest, they are not widely used at present for design purposes by practising engineers. Wiesner criticizes Gumbel's method in that he considers it statistically unjustifiable to extrapolate far beyond the period of record. He believes, for example, that the once-in-a-1,000-year flood, estimated from 50 years of stream-flow record, has confidence limits so wide that the results are of little value.

However, these two approaches to prediction are complementary rather than independent of one another in that statistical inferences are used to a considerable extent in hydrometeorology. Binnie and Mansell-Moullin

(1966), for example, use statistical frequency analysis to determine the recurrence intervals of flood peaks on the Jhelum (Fig. 5.3D). Both these approaches to flood and drought estimation are in their infancy and techniques are subject to constant revision, but this is a part of scientific hydrology in which the need to develop better techniques is critical and model theory plays a large part in their evolution.

Models for resource planning

The third function of models in hydrology – that of resource planning – is closely related to their predictive function. The management of water-supply reservoirs, both during a single year and over a longer period, requires the use of stochastic methods, such as queueing theory, so that it becomes possible to apportion a varying input to storage over specified demands, and within specified risks of failure. On a broader scale, models are used to achieve the optimum exploitation of resources in catchment areas which have already been partially developed, and to plan water-resource utilization in undeveloped catchments. Eckstein (1958), Krutilla and Eckstein (1958), McKean (1958), Maass et al (1962), Kuiper (1965), Smith and Castle (1965), Schwab et al (1966) and Wolf (1966B) have discussed the principles of water-resource management. Such multiple-purpose schemes are in various stages of planning and execution in the Rhône and Durance valleys in France (Giguet, 1957), in the Tennessee and Central Californian valleys of the United States and in the Mekong valley in South-East Asia (White et al., 1962). An opportunity has been given to River Authorities in England and Wales (under the terms of the Water Resources Act, 1963) to develop their rivers in the light of water-resource management techniques (Thorn, 1966). Further reference will be made to the resource-planning function of hydrological models in the section on hydro-economic models.

PHYSICAL AND SYSTEMS HYDROLOGY

Classification is always a difficult and arbitrary operation, and this is particularly true of hydrological models at this stage of their development. The use of computers has given model theory greater scope, and ideas as to the nature and function of models are developing rapidly (O'Donnell, 1966). Dawdy and O'Donnell (1965) and Amorocho and Hart (1964) agree, in broad terms, in their analyses of current methods of approach to hydrological research. They distinguish two main schools of thought in model building, one of which may be described as Physical Science Research, or Physical Hydrology, and the other as System Synthesis Investigations.

Physical hydrology involves the pursuit of scientific research into the basic operation of each component of the hydrological cycle in order to gain a full understanding of the mechanisms and interactions involved. Although the immediate motivation of an individual researcher may not go beyond a specific phenomenon, it is implicit that a full synthesis of the hydrological cycle *in toto* is the ultimate goal.

The other approach seeks to achieve the comprehensive simulation of catchment behaviour in the form of gross catchment models. In these models catchment components are treated in a lumped form and their behaviour is simulated in an approximate way by largely empirical relationships. The construction of the component parts of the model and the parameters of the relationships are adjusted until known responses, within an acceptable tolerance, are achieved from known inputs. Subjective decisions must be made in the choice of components and in the specification of behavioural relationships. The adjustments of the parameters has generally also been subjective, but objective techniques have been used. The construction and manipulation of such models rely to a great extent on certain methods of systems engineering, whereby the relationship between input and output is synthesized in a model of the prototype process.

These two approaches to model building in hydrology are in some respects complementary to one another. As the investigations of the physical hydrologists provide more information on the details of the component parts of the hydrological cycle, so the comprehensive simulation of catchment behaviour, by means of overall models, will show where further detailed specification is needed.

MODELS IN PHYSICAL HYDROLOGY

Because of the large number of interlocking component parts and the difficulties of scaling them down to manageable size, few attempts have yet been made to construct a complete working model of the physical hydrology of a catchment (e.g. Chery, 1966). However, much research has been carried out in an attempt to obtain a better understanding of each of the component parts with the long-term object of synthesizing this knowledge into a complete model. Figure 5.4A shows schematically the disposition of the main storm rainfall components (minus evaporation and transpiration) with time for an ideal drainage basin, assuming a constant rainfall intensity.

Precipitation simulation: A hardware scale model in hydrology

One of the basic requirements for many laboratory models is the ability to manufacture precipitation of the required intensity and duration, for, ideally,

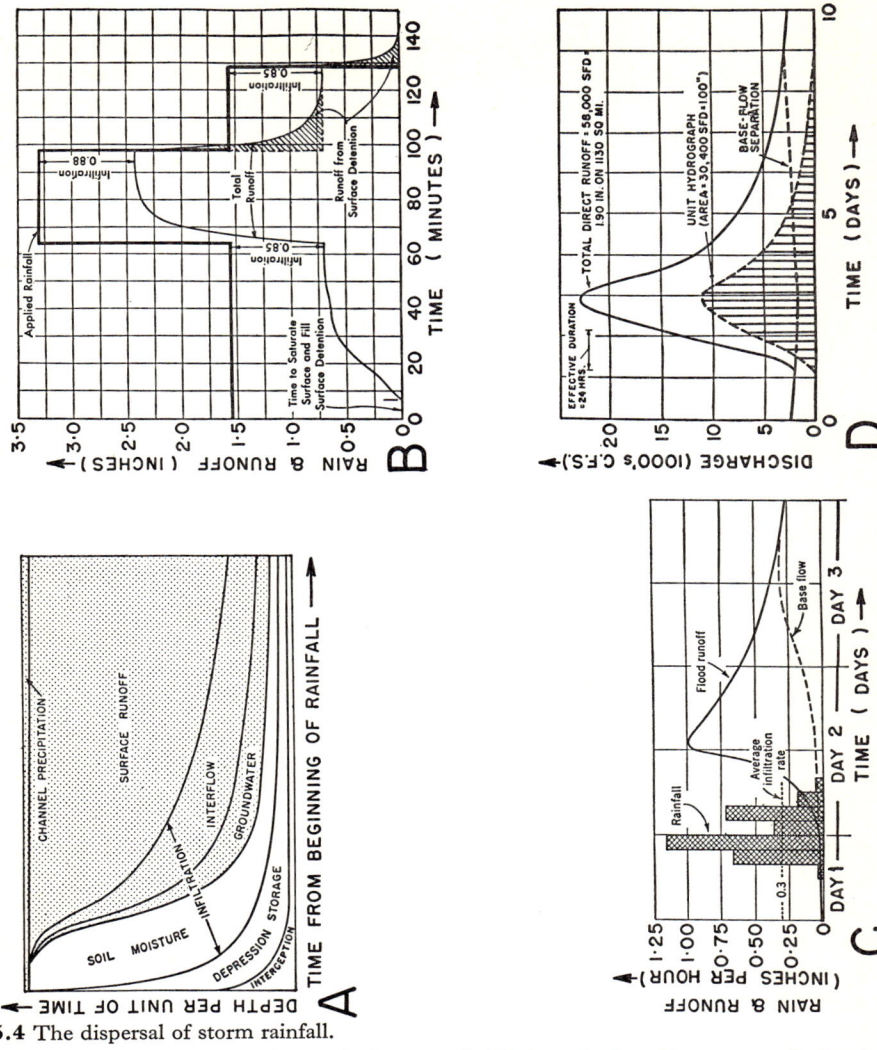

5.4 The dispersal of storm rainfall.

A Schematic diagram of the dispersal of storm rainfall through time. Evapo-transpiration is ignored, and rainfall is assumed to be constant and continuous. The dotted area represents that proportion which eventually becomes streamflow (*Source: Linsley, Kohler and Paulhus, 1949, p. 410*).
B Runoff on a plot resulting from artificially-applied precipitations of 1·55 and 3·30 inches per hour (*Source: After Sharp and Holton. From Foster,* Rainfall and Runoff, *1949, p. 309*).
C Hydrograph of Sugar Creek, Ohio (310 square miles), resulting from a storm of 6.3 inches lasting 14 hours. The immediate stream discharge was 3·0 inches – the rest evaporating, being taken up by vegetation, or going into soil or ground-water storage (*Source: After Hoyt and Langbein. From Strahler,* Introduction to Physical Geography, *1965, p. 274*).
D A simple derivation of a unit hydrograph for an isolated storm of 24-hours duration, yielding 1·9 inches over a basin of 1,130 square miles (*Source: Linsley, Kohler and Paulhus, 1949, p. 446*).

it should be ultimately possible to simulate actual storms moving across given catchment areas. Laboratory experiments with sprinkler equipment have been conducted widely, for example in the United Kingdom by Childs (1953). The purpose of these experiments has been: to generate controlled rainfall which can be equated with the resulting measured erosion and sediment transport; to produce rainfall for the simplification of a hydrological problem into a soluble hydro-mechanic state; or to apply rainfall to homogeneous porous media for the study of flow nets to drains and other infiltration phenomena (Luthin, 1957 and 1966). Figure 5.4B shows the infiltration/runoff relationships achieved for a controlled test plot under artificially-simulated precipitation.

Many of the major advances in physical hydrology in the last forty years have been made, however, not by the use of laboratory experiments, but by the application of mathematical models. The work of Penman (1948, 1956, 1963) on evaporation, Horton (1933) on infiltration and Sherman (1932) on the unit hydrograph theory of stream runoff has led to a quantitative evaluation of these components which, under the best conditions, give figures which are as accurate as the direct values obtainable in field measurements and do it very much more easily. Both Penman and Horton, and to a lesser extent Sherman, look at the component which they are studying as the focal point in the hydrological cycle. All three simplify the natural hydrological state by abstracting the major factors which are relevant to their hypothesis and ignoring the others.

Models for evaporation

The most direct approach to the calculation of evaporation is to employ the evaporation pan as a hardware scale model. This work has been based on the theory that there may be some determinable relationship between the rates of evaporation from the small water surface of an evaporation pan and from a large water surface in the same locality that has been subjected to the same natural influences. The analogy is difficult to check, however, for estimates of lake and reservoir evaporation losses require field measurements of all inflow of moisture from the atmosphere, and runoff and ground water measurements, such that the difference between measured liquid inflow and measured liquid outflow can be attributed to evaporation. This technique reflects the limited accuracy of hydraulic measurement and all observational errors show up as evaporation, so that the inherent error is probably not less than 20 per cent for large bodies of water. To add to the difficulties of this method a variety of evaporation pans, with different pan coefficients, are in common use, so that it is difficult to make wide-ranging comparisons between evaporation values obtained in this way. Another major difficulty is that most natural catchment evaporation losses take place from vegetated surfaces, rather than from standing water. It is in this latter respect that some of the following mathematical models are especially significant.

Penman has based such a mathematical model of evaporation on a combination of the energy balance theory and the aerodynamic approach. During summer the amount of energy used in converting water to vapour is the largest single term in the net exchange of energy between atmosphere and ground. Because of this, it is possible to draw up a balance sheet of energy income and expenditure in which evaporation is the only important unknown quantity. Penman formulated this in the equation:

$$H = E + K = R_C(1-r) - R_B$$

Where: H is the heat budget, or the essential equality between the energy used and the energy received.
E is the energy available for evaporation.
K is the energy used to heat the air.
R_C is the energy from the sun.
$R_C \times r$ is the energy reflected by the earth.
R_B is the energy radiated by the earth.

He then used the aerodynamic approach (i.e. that evaporation rates depend upon wind velocity and differences in vapour pressure of the air) to determine how the available energy is shared between E and K, the ratio between which can be used in conjunction with the energy balance equation. In this combination of the aerodynamic and energy balance methods of approach, some meteorological elements that are difficult to measure directly are eliminated from the equations to give an expression for evaporation in terms of those meteorological elements that can be measured; viz. mean air temperature, mean air humidity, duration of bright sunshine, and mean wind speed.

Since his first paper on evaporation (1948) Penman's formula has been modified by further research, notably by Monteith (1959) on values of the reflection coefficient. In his 1963 publication Penman published a generalized expression for evaporation as:

$$E = \left(\frac{\Delta}{\gamma}H + E_a\right) \bigg/ \left(\frac{\Delta}{\gamma} + x\right)$$

Where E_a is an expression for the 'drying power' of the air, involving wind speed and saturation deficit.
Δ is a temperature-dependent constant (= the slope of the saturation vapour-pressure curve at mean air temperature, which is obtainable from standard tables).
γ is the constant of the wet- and dry-bulb psychrometer equation.
H is the heat budget.

The ratio $\frac{\Delta}{\gamma}$ is dimensionless, and is effectively a weighting factor in assessing the relative effects of energy supply and ventilation on evaporation.

x is a factor dependent on stomatal geometry and day length.

Much practical value has been obtained from Penman's work on evaporation in that he has been able to show the appropriateness of his model in the estimation of water use by natural vegetation and in the calculation of irrigation need. Penman has shown that the water used by a plant is primarily dependent on atmospheric conditions, is largely independent of the soil or the type of crop, and, except for drainage (which virtually ceases when the soil moisture content falls below field capacity), water is only lost from the soil by evaporation from the surface and by transpiration from plants. The quantity of water removed from the soil for a given intake of solar energy can be thus stated simply in terms of air temperature, humidity and wind, such that the potential transpiration of a crop can be estimated from meteorological observations. Potential transpiration is defined as the amount of water transpired by a green crop of about the same colour as grass, which completely covers the ground, and which has an adequate supply of water. Potential transpiration is expressed as follows:

$$E_T = \left(\frac{\Delta}{\gamma} H_T + E_{aT}\right) \bigg/ \left(\frac{\Delta}{\gamma} + 1\right)$$

Where E_T is potential transpiration for a 30-day month.

H_T is the heat budget.

Δ is a temperature-dependent constant.

γ is the constant of the wet- and dry-bulb psychrometer equation.

E_{aT} is an expression for the drying power of the air.

Penman's formula is limited in its applicability to British and Western European climatic conditions. Formulae based on similar theoretical assumptions have been formulated by Blaney and Criddle (1950) and by Lowry and Johnson (1942) for use in the arid western part of the United States; by Thornthwaite (1948 and 1954) for use particularly in the eastern and central parts of the USA; and by Olivier (1963), who uses a correlated latitude/radiation factor in a simple formula by which plant water needs may be calculated for any part of the world.

A model of infiltration

R. E. Horton (1933) adopted a different approach to the relationship between rainfall, runoff and storage. He was able to express the characteristics of the infiltration curve (Fig. 5.5A) in terms of a mathematical model (Fig. 5.5B)

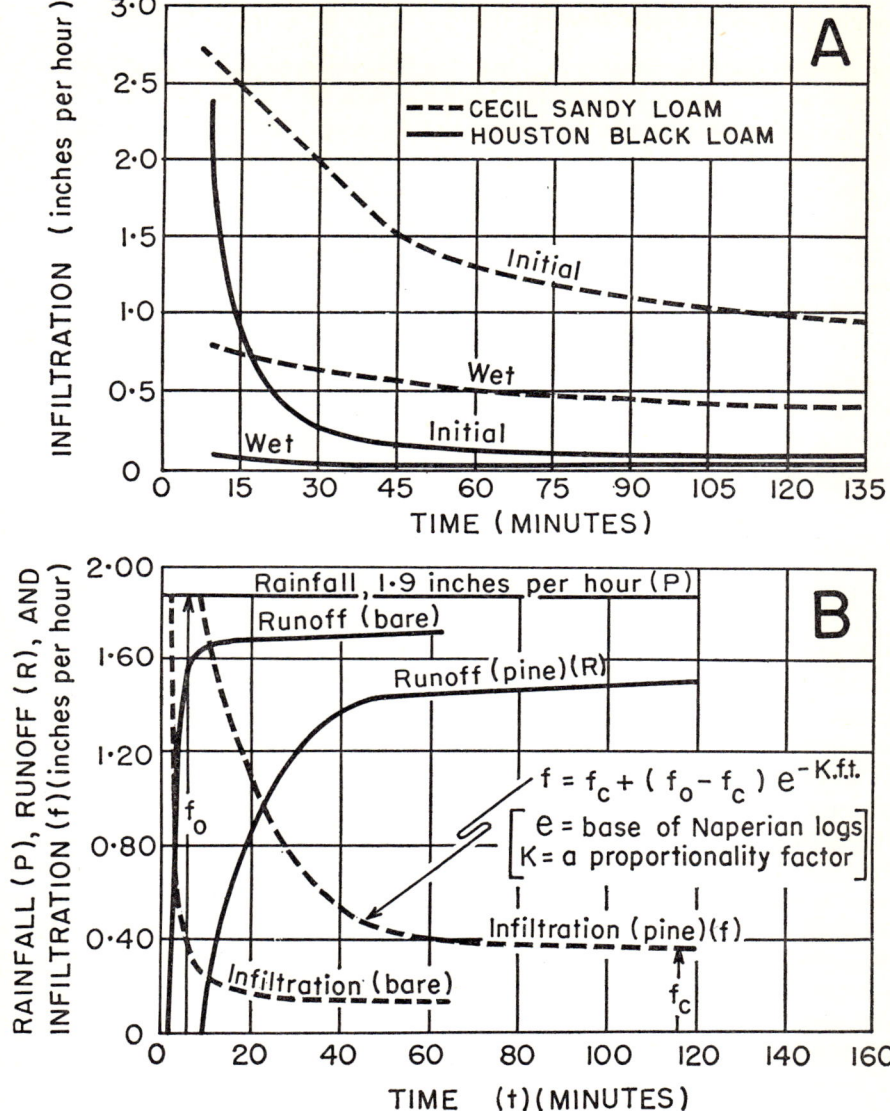

5.5 The effects of time, soil composition, initial soil moisture conditions and vegetational cover on surface infiltration.

A Comparative infiltration curves for the relatively permeable Cecil and the less permeable Houston loams. In both cases the initial (dry) tests yield higher infiltration rates than the (subsequent) wet tests, with a constant applied rainfall. The curve for the wet run is not simply the last portion of the curve for the dry run, and the initial soil moisture conditions seem to influence the character of the whole infiltration curve (*Source: After Free, Browning and Musgrave, From Linsley, Kohler and Paulhus, 1949, p. 313*).

B Infiltration and surface runoff from forested land (55-year-old pines) and bare abandoned land in the Tallahatchie River basin, Mississippi under a precipitation of 1·9 inches per hour. The infiltration curve equation is that suggested by Horton (*Source: After Musgrave. From Meinzer, O. E., Ed., Hydrology; 1942, p. 250*).

such that its main features under differing conditions of soil moisture, surface cover, length of storm, etc., could be predicted.

The reciprocal relationship between infiltration and runoff (Fig. 5.5B) led Horton to develop his ideas on infiltration capacity into a general theory for estimating runoff from rainfall. He defined rainfall excess as that part of the rainfall which falls at intensities exceeding the infiltration capacity and argued that, if the infiltration capacity of a drainage basin is known, it is possible, by an analysis of the rainfall excess in individual storms, to determine from rainfall data the surface runoff which will result. Later Horton (1945) extended his work to show the significance of infiltration capacity in an understanding of the erosional development of streams and their drainage basins, leading to the construction of morphometric models in geomorphology related to the process of surface runoff (Strahler, 1964) (See later section on Overall Catchment Models).

The unit hydrograph model of runoff

Probably the most important contribution to rainfall/runoff studies was made by L. K. Sherman (1932) who proposed the 'unit hydrograph' concept. Given storm inputs tend to result in recognizable patterns of stream runoff outputs for each basin (Fig. 5.4C) and Sherman postulated that the most important hydrological characteristics of each basin could be expressed in terms of the constant features of the direct runoff hydrograph resulting from an evenly-distributed unit rainstorm – i.e. 1 inch of rainfall over the whole basin occurring in 24 hours. The most simple method of unit-hydrograph construction is by taking the observed pattern of runoff resulting from a 24-hour storm, subtracting the base flow originating from ground water, and proportionately reducing the remaining hydrograph to the equivalent of a 1-inch storm (Fig. 5.4D). Thus, for any one basin, identical storms with the same antecedent conditions (one of the assumptions of this model) are expected to produce identical hydrographs, the geometry of which can be rationalized in terms of basin morphometry, infiltration characteristics, etc. Different physical characteristics can be expected to result in different unit hydrographs. For example, a topography with steep slopes and few pondage pockets gives a high, sharp peak and a short time period, whereas flat country with large pondage pockets gives a graph with a flat rounded peak and a long time period. Figure 5.6 shows how unit hydrographs for different watersheds reflect the variations in shape, size, morphometry, slope, pondage, etc.

The empirical unit hydrograph model has numerous advantages over other methods of determining runoff from given rainfalls, in that it is tailored to the characteristics of particular drainage basins such that patterns of runoff can, by adapting the basin unit hydrograph, be predicted for any rainfall input (O'Donnell, 1960; Dooge, 1959; Chow, 1964, pp. 14–13 to 14–35).

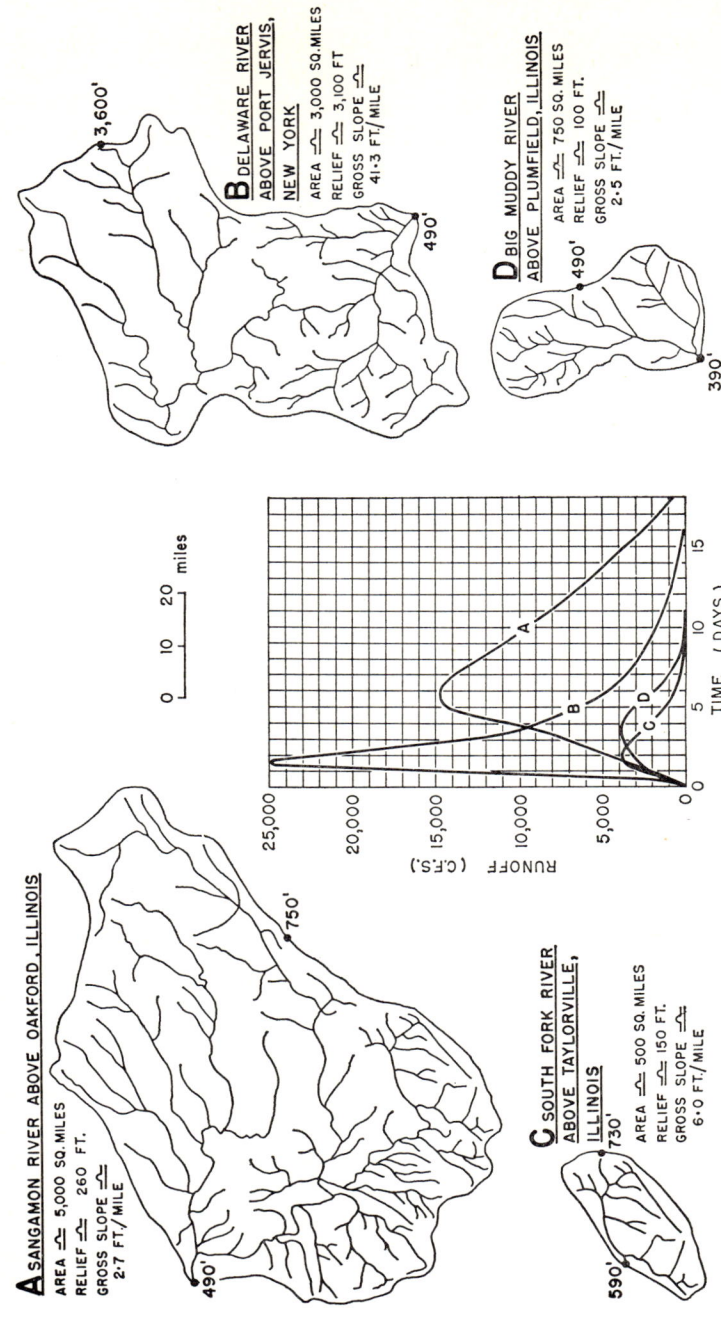

5.6 Unit hydrographs for (A) the Sangamon River, Illinois, (B) the Delaware River, New York, (C) the South Fork River, Illinois, and (D) the Big Muddy River, Illinois. These show clearly the effect of area on the magnitude of the unit hydrograph, and the effect of basin slope on its form (*Source: Partly from Sherman, 1932*).

Ground water models

The flow of ground water has been simulated by sand models, by viscous fluid models, by rubber membranes representing the water table around a well system, by resistance-capacity network analogues and by numerical analysis methods (Todd, 1959, pp. 307–325; De Wiest, 1965, pp. 318–348; Davis and De Wiest, 1966, pp. 241–256). Apart from the sand model, the other types of models are analogues of ground water flow, in that flow does not take place through permeable media. Viscous fluid models (also known as Hele-Shaw and parallel plate models) are based on the assumption that if a liquid such as oil or glycerin flows between two closely-spaced parallel plates its movement is analogous to that of ground water flow in a two-dimensional cross-section of an aquifer. The derivation follows from the generalized Navier-Stokes equation of motion. An important advantage of this model is that it can be used to simulate unsteady, as well as steady, flows in both confined and unconfined aquifers; permeability variations being introduced by attaching thin laminated sheets at appropriate positions between the plates. Viscous fluid models are also used to study sea water intrusion (as in the Netherlands), bank storage near flooding streams and the movement of ground water in earth dams.

Resistance capacity-network analogues are based on the relationship which can be postulated between Darcy's law for the flow of water through porous media, and Ohm's law for the flow of electrical current through a resistor (Fig. 5.7), such that the flow in a mesh of resistors can be held to represent the flow characteristics of an entire aquifer (Fig. 5.8). Voltages at mesh nodes can be arranged so as to be directly proportional to the observed head of water in the aquifer as determined by piezometer. This being so, the analogue model can be used to predict flow characteristics at all points in the aquifer and to estimate possible changes of ground water conditions attendant upon artificial abstraction or recharge of the ground water.

Electrical analogues can be used to model simple situations of two-dimensional flow, or more complex three-dimensional situations such as those investigated by the United States Geological Survey and described by Skibitzke (1963). Electrical analogue models have been made by the Thames Conservancy for the Lambourn Valley and the Kennet Valley in Berkshire, one of the functions of which is to study the possible effects of pumping wells at varying distances from streams. Figure 5.9B shows a theoretical cycle of recharge of the ground water, by pumping from streams with high discharges in winter, followed by pumping to lower the water table below its natural summer level to provide water for irrigation and other purposes. Thus the water in the aquifer is used to its full capacity in the same way as a surface reservoir might be managed for optimum supply.

Figure 5.9A shows a modification of this theoretical cycle which is to be

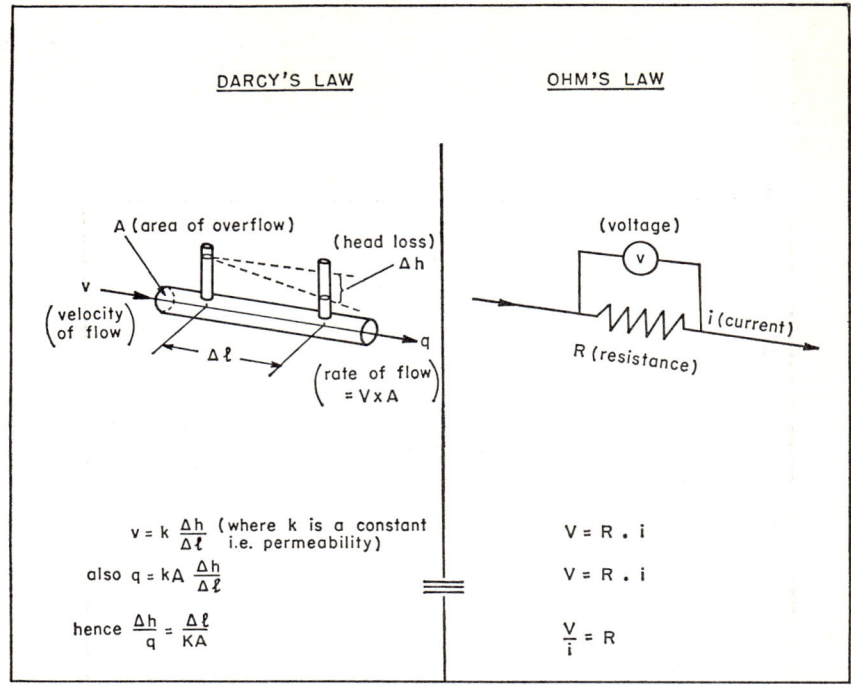

5.7 The analogy between Darcy's Law, for the flow of water through porous media, and Ohm's Law, for the flow of current through a resistor, which forms the basis of the ground water electrical analogue (*Source: Water Research Association, TH/H4*).

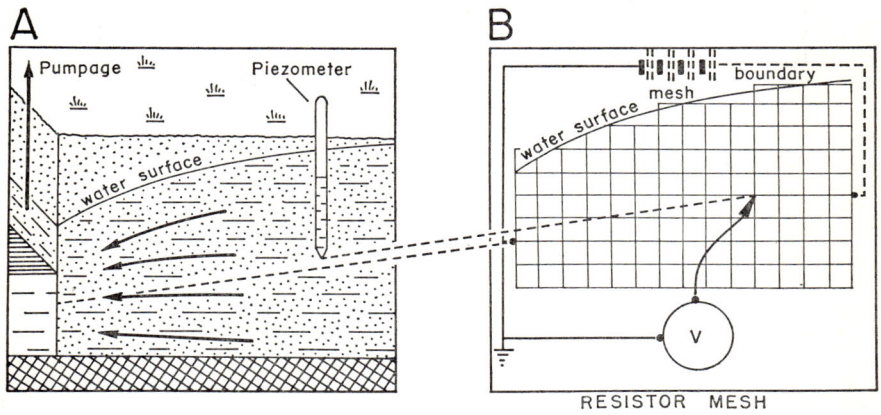

5.8 Electrical analogue for ground water flow. The flow of water in an underground aquifer (A) can be simulated by the flow of electricity through a network of electrical resistances (B). In this simple case of water pumpage from a ditch, the voltage at a mesh node is directly proportional to the water pressure in the aquifer, as determined by a piezometer (*Source: Water Research Association, Technical Paper TP 32*).

[164] MODELS IN GEOGRAPHY

developed by the Thames Conservancy (Thames Conservancy, 1965) for the Berkshire Downs, a large ground water reservoir in the Chalk. The scheme relies on the natural percolation from rainfall to the chalk aquifer. By pump-

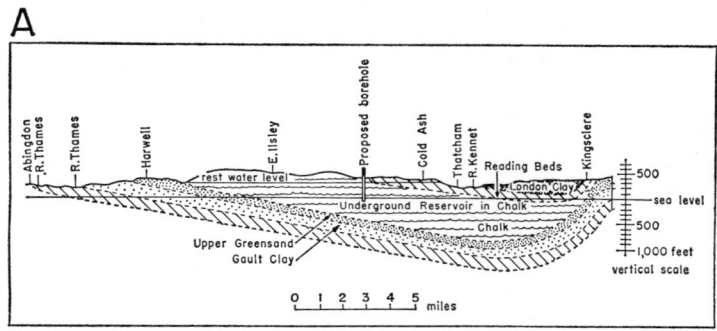

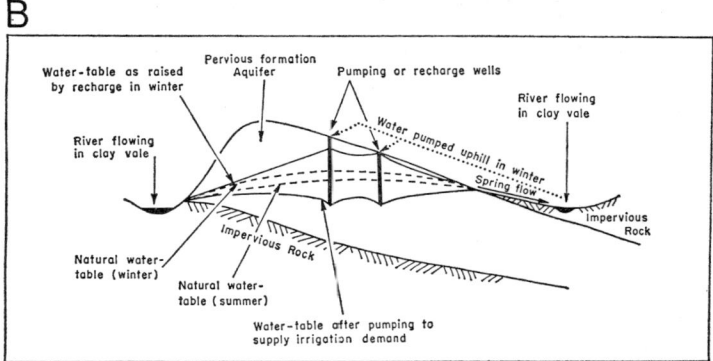

5.9 Increased utilization of the chalk aquifers of South-Eastern England.

A Proposed Lambourn Downs scheme of the Thames Conservancy to augment the flow of the Thames by pumping from boreholes into its tributaries (*Source: From* Thames Conservancy and the Water Crisis, *1965*).

B A more ambitious theoretical scheme to artificially recharge the chalk aquifers by pumping surface water *into* the boreholes in winter so that more water is available for pumping *out* during the water-deficient summer period (*Source: Suggested by P. O. Wolf. Illustrated in More, R. J., 1965*, A Geographical Analysis of the Distribution of Irrigated Land in South Eastern England; *unpublished Ph.D. thesis, Department of Geography, Liverpool University*).

ing from the underground reservoir, flows in the river may be augmented during dry periods in the summer. In the winter, the pumps are rested and the aquifer recharges by natural percolation from the winter rainfall. The Thames Conservancy contemplate the eventual construction of a network of boreholes, mainly in Berkshire and the Cotswolds, which may be capable of supplementing Thames flows by up to 270 million gallons per day, at an estimated cost of £8,000,000. It is hoped that the cost of pumping water from underground into the rivers will work out at about $3d.$ per 1,000 gallons, which compares very favourably with $3s.$ to $4s.$ per 1,000 gallons for many present-day reservoir schemes and much higher costs for desalinization. During a pilot scheme (1966–69) nine boreholes are to be sunk in the Lower Lambourn Valley in the area between Welford and Newbury, and the effects of pumping studied in detail. It is estimated that the output from the nine boreholes will be 16 million gallons per day, which will be pumped into the Lambourn, which joins the Kennet at Newbury, which in turn joins the Thames at Reading.

Recharging the chalk of the Berkshire Downs, by pumping from adjoining streams, may be a logical development at a much later date, if certain technical difficulties are overcome. Recharge of the chalk in the Lee Valley north of London has already been undertaken by the Metropolitan Water Board with encouraging results. (Boniface, 1959; Buchan, 1959 and 1963).

OVERALL CATCHMENT MODELS

General catchment models in hydrology can also be conveniently divided into the physical and the systems approaches (Amorocho and Hart, 1964); the former attempting to understand each mechanism and interaction of the hydrological cycle in the search for a complete, rational synthesis, the latter concerned with the establishment of workable relationships between subsystems of hydrological components such that hydrological events can be predicted from known climatic inputs. One of the most successful advocates of physical hydrology, Robert E. Horton, was much concerned with the interactions of hydrological events and drainage basin geometry ('morphometry') and, in an important paper (Horton, 1945), he showed both how a drainage system can be dissected into components of differing order having dynamic significance (later modified by Strahler, 1952 and 1964; see Fig. 5.10A) and how the characteristics of the network (e.g. the density of drainage lines) can be rationalized on the basis of infiltration runoff theory. It is interesting how, depending upon the time scale under consideration, drainage systems can be variously considered as both the effects and causes of runoff patterns. From the point of view of physical hydrology the latter case provides an interesting 'morphometric model' for hydrologists. Strahler (1964) has shown how the 'bifurcation ratio' $\left(R_b = \dfrac{N_u}{N_{u+1}}\right)$ may control the pattern of runoff from storms over individual basins (Fig. 5.10B); Leopold and Miller (1956) that there is a relationship between mean discharge and basin order (Fig. 5.10C); and Hack (1957) that there is a logarithmic relationship between

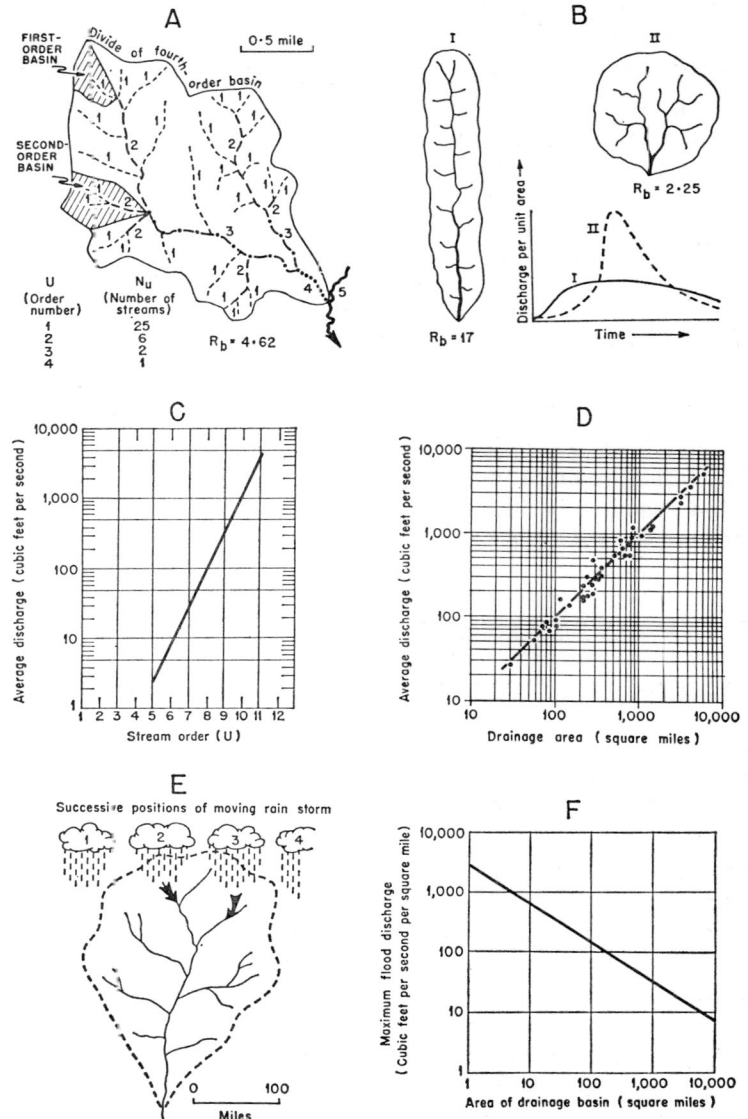

5.10 The influence of basin morphometry on hydrological characteristics.
A A channel ordering system, showing the derivation of the bifurcation ratio ($R_b = N_u/N_{u+1}$) (*Source: From Strahler, 1964, p. 4-44*).
B Schematic hydrographs for basins having high (I) and low (II) bifurcation ratios (*Source: From Strahler, 1964, p. 4-44*).
C A general relationship of discharge and stream order for arroyos in New Mexico (*Source: From Leopold and Miller, 1956, p. 23*).
D Relationship of average discharge to drainage area for all gauging stations on the Potomac River (*Source: After Hack, 1957. From Strahler, 1964, p. 4-50*).
E and F The small area covered by the most intense storms means that maximum flood discharges per unit area are recorded for small basins. (*Source: After Follansbee and Sawyer. From Leet and Judson, 1954, pp. 122-123*).

mean discharge and drainage area (Fig. 5.10D). Considerations of basin morphometry also help to rationalize extreme hydrological events and Follansbee and Sawyer (1948) have shown how the areal localization of most intense precipitation (Fig. 5.10E) provides for the maximum flood discharges (per unit area) to occur in the smallest basins (Fig. 5.10F). In Great Britain, Nash (1960) has attempted to correlate unit hydrograph parameters with the morphometric characteristics of drainage basins.

If a quantitative overall hydrological catchment model is to be acceptably accurate in all particulars it must inevitably be complex, yet it must be feasible to operate. These conflicting requirements were not met until high-speed computers came into use, but, even so, the evaluation of such models is still circumscribed by the limitations of current knowledge and understanding of the processes being simulated, and by the capabilities of the computers and the computing techniques available. Ideally the data for the construction of an overall model should include a specification of the properties of, and the processes that occur in, all the relevant components of a catchment. Specification should be given in terms of physical parameters and should involve behavioural relationships for and between the catchment components stated in terms of these parameters. At the present time hydrological knowledge and techniques do not permit more than a coarse approximation to this ideal. Overall models have been developed to the stage of being an effective and acceptable engineering tool (Crawford and Linsley, 1966), but Amorocho and Hart (1964) caution against an excessive reliance on synthetic models of catchment behaviour, naming as perhaps the most important causes of unreliability errors in the recorded data, effects due to lumping of components, imperfections of the structure of any synthetic model and non-uniqueness of the processes of synthesizing an unknown system.

O'Donnell (1966) emphasizes that digital and analogue computers differ in their structure and mechanical capabilities and that the advantages of each type of computer should be used in solving hydrological problems to which the machine is best adapted. Digital computers are particularly suited to calculations involving matrix procedures, such as the prediction of flood runoff using the unit hydrograph technique, and the Tennessee Valley Authority has published a number of papers describing the use of digital computers in hydrological analyses (TVA, 1961; 1963A; 1963B). The most advanced and comprehensive digital computer evaluation of catchment behaviour is the study which has been made by Crawford and Linsley (1962 and 1966). This model aims to represent the whole of the land phase of the hydrological cycle and has been tested and successfully adjusted to reproduce within acceptable limits the runoff behaviour of more than thirty catchments. The model is a general one and its operation is controlled by the characteristics of its component storage and routing elements. These characteristics are

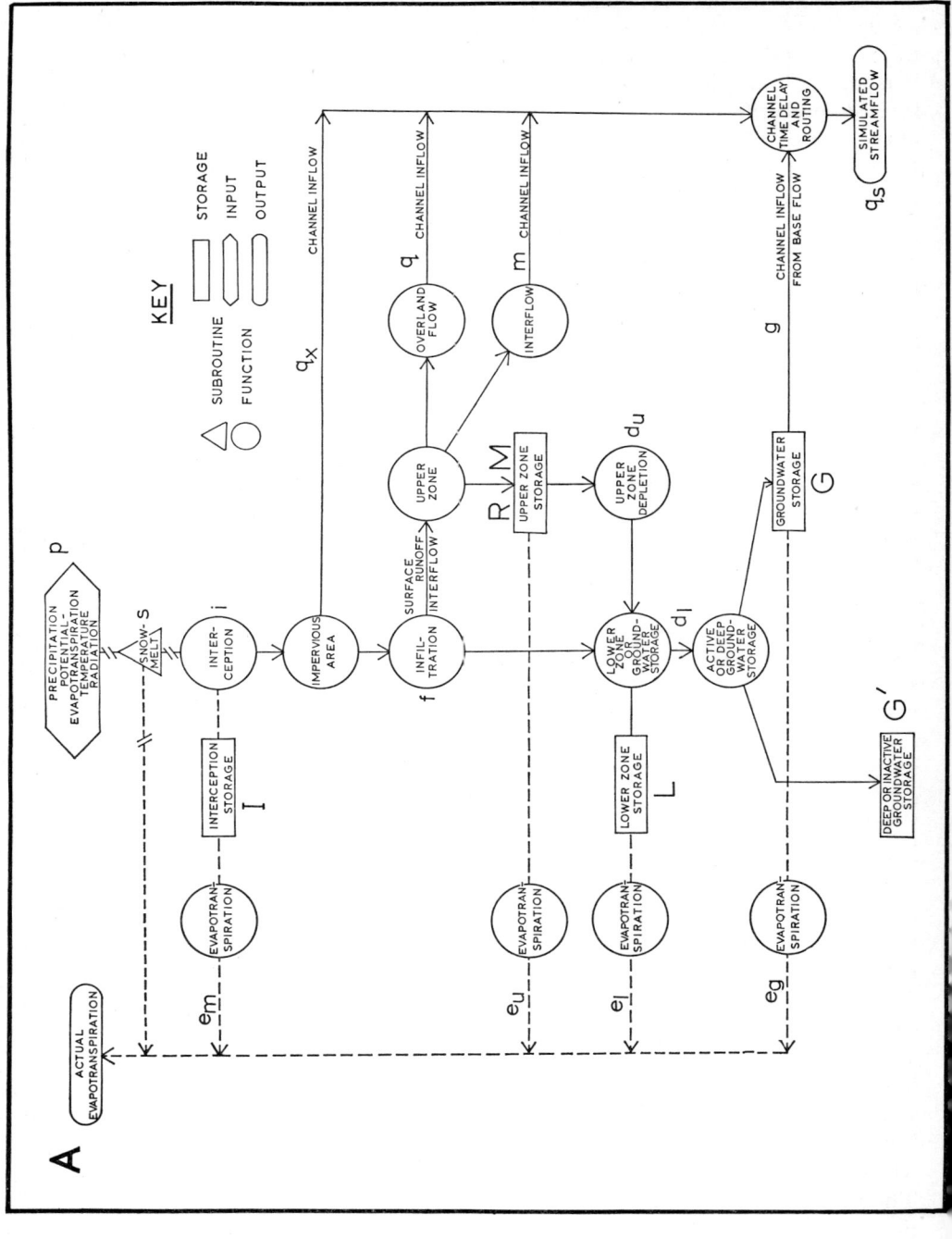

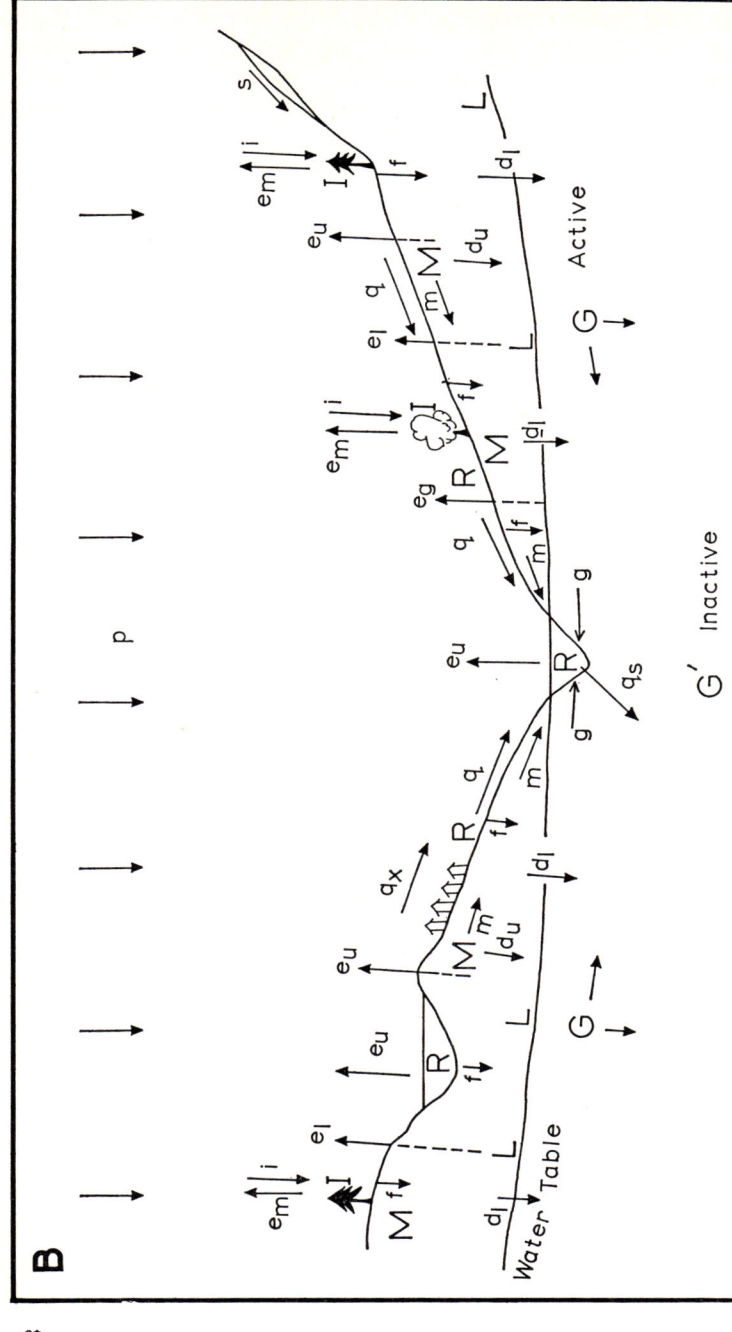

5.11 The Stanford Digital Computer Watershed Model IV.

A The flowchart for the model based on 5 storage elements (I, R and M, L¹, G, G') which are augmented or depleated by 13 transfers (*Source: From Crawford and Linsley, 1966*, Fig. 4.1).
B Schematic hydrologic cycle for a basin, employing the same nomenclature as A (*Source: Adapted from Linsley, Kohler and Paulhus, 1949*).

determined by relationships that are expressed in terms of certain parameters and that represent as rationally as possible the behaviour of various segments of the hydrological cycle. The model can be adapted for use in a particular basin by the adjustment of the parameters. The structure of the model is illustrated in Figure 5.11, and its operation is as follows. Estimates of initial storages are fed into the computer. Then hourly increments of rainfall enter the model. The incoming rain either becomes direct runoff, or is detailed into upper and lower soil moisture storage, the latter feeding a ground water storage. The upper zone storage absorbs a large part of the first few hours of rain in a storm; lower zone storage controls long-term infiltration and the ground water store influences the base flow. Evaporation takes place at the potential rate from upper zone storage and at less than the potential rate from the lower zone and from ground water. In applying this general model to a specific catchment (for example, one with short runoff records which are to be extended) the typical procedure is to select a five to six year portion of rainfall and runoff records for the catchment. This period is used to develop estimates of the model parameters that fit the general model to the given catchment. A second period of record is then used as a control to check the accuracy of the parameters obtained from the first period. The parameters are adjusted to the general model either by the operator or by the computer itself, using an internal looping routine of successive approximations. When the particular basin has been thus fitted by the general model, the working of the fitted model can be used to extend the period of record within certain confidence limits.

Dawdy and O'Donnell (1965) have been exploring objective methods of finding numerical values of the parameters of such synthetic hydrologic models using automatic optimization methods. They have used a simple mathematical model which appears in Figure 5.12. The overall model was deliberately kept simple so that emphasis could be given to the parameter sensitivity and optimization aspects of the work. The model is restricted to four storage elements – surface storage (R), channel storage (S), soil moisture storage (M) and ground water storage (G). The model has nine parameters:

R^* the threshold value of surface storage
f_o maximum infiltration rate
f_c minimum infiltration rate
k exponential die-away exponent
K_s storage constant of channel storage
M^* threshold value of soil moisture storage
G^* threshold value of ground water storage
c_{max} maximum rate of capilliary rise
K_G storage constant of ground water storage.

At the start of the first time interval during the running of the model trial volumes in each of the four storage elements and the potential infiltration rate have to be specified, and a set of trial values has to be given to the nine parameters. Thereafter the computations for the first interval, and for each suc-

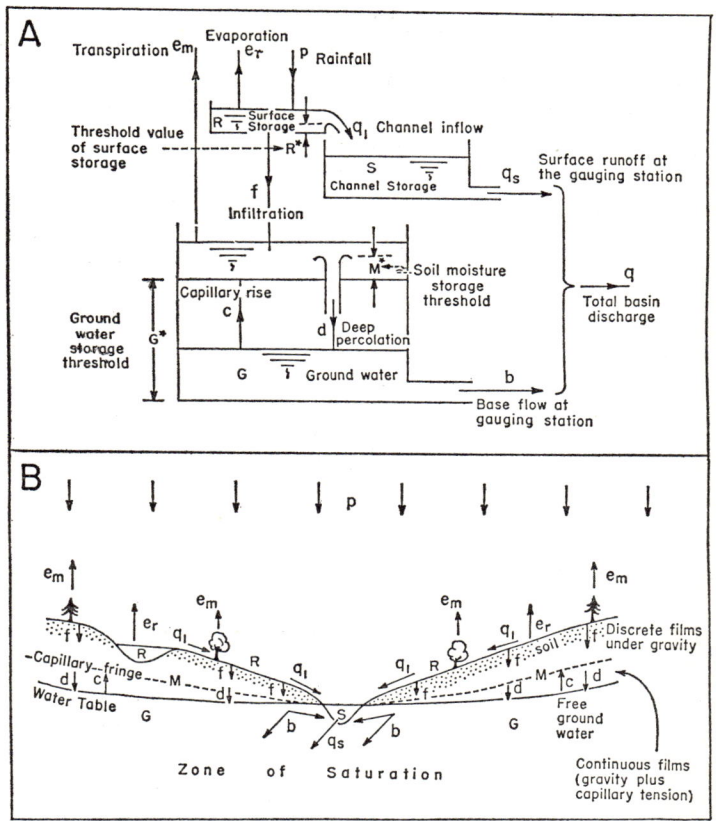

5.12 The basin hydrological cycle.
A A basin model based on four storage elements (R, S, M and G), three of which operate with respect to thresholds (R*, M* and G*), augmented and depleted by a number of transfers (p, e_m, e_r, q_1, f, c, d, q_s, b and q) (*Source: From Dawdy and O'Donnell, 1965*).
B Schematic basin hydrological cycle, employing the same nomenclature as A (*Source: Adapted from Linsley, Kohler and Paulhus, 1949*).

cessive interval, yield values for the four storage volumes and the potential infiltration rate for the start of the next interval. A completely general optimization would include the start-of-synthesis values of each of the four storage volumes and the potential infiltration rate, as well as the nine parameter values, but by postulating a long dry period before the start of a rainfall/

runoff synthesis it seems reasonably accurate to set all four initial storages to zero and to assume that the starting potential infiltration rate has recovered to a maximum value of f_o, so that, in practice, there are only nine input items. The input data to the model consist of precipitation, potential evaporation and runoff data for each of the intervals of a known record, and the initial trial values of the nine parameters. The model then works through the precipitation and evaporation data and calculates a runoff volume for each interval of the record, which in general will not agree with the known runoff values. The optimization technique sets out to adjust the initial parameter values so that the differences between the known and calculated values of runoff are eliminated to an acceptable tolerance.

The application of analogue computer techniques to hydrological problems may be accomplished by the direct simulation of a hydrological system (an analogue model), or by the solution of the mathematical relations describing such a system (an analogue analyser). The Water Resources Division of the United States Geological Survey have, for example, developed the analogy between the flow of water in a hydraulic system and the flow of electricity in an electrical circuit, and have applied it to flood routing, comprehensive modelling of catchment runoff behaviour, unsteady flow in open channels and the discharge computation for slope-rating stations (Shen, 1963A and 1963B).

STOCHASTIC MODELS

Another method of approach to model building in hydrology has been provided by the use of statistical models. It is assumed that each event involved in the generation of a series of hydrological data has a certain level of probability attached to its occurrence, i.e. the frequency of its occurrence is related to the magnitude of the event. If this assumption is valid, statistical probability distributions may be applied to lengths of historical hydrological records, and it may be possible to generate synthetic data from the statistical characteristics of the historical record. The major advantage of synthetic generation is to create records longer than the historical record and is particularly valuable in the study of reservoir operations and in the design of complex water-resource systems. (See earlier section on Models for Hydrological Prediction.)

Thus the hydrologic data series can be treated as a time series by considering it as a queue or waiting line (Cox and Smith, 1963; Duckworth, 1965). A queue involves arriving items or 'customers' that wait to be served at the facility which provides the service they seek. In a queueing system the customers arrive at the system, wait for service, receive service and then leave the system. The analogy between queueing theory and storage of runoff was first recognized by Moran and analysed theoretically by him and others to develop the theory of storage (Gani and Moran, 1955; Kendall, 1957; Moran, 1959; Harris, 1965).

Two principal attitudes towards synthetic data sequences have been adopted in the solution of hydrologic problems. Firstly, it may be assumed that hydrologic data are purely random. In this case Monte Carlo methods are used (Hammersley and Handscomb, 1964). The characteristics of the distribution of the recorded data are studied first. Then an appropriate simulation generation technique is employed (frequently involving the use of tables of random numbers) to produce artificial sequences of record having the same statistical structure as the measured historical record. By means of a table of random numbers Brittan (1961), for example, simulated stream flows in the Colorado River by selecting 100 random samples of 5 each corresponding to a 5-year runoff sequence. She was able to develop hydrologic records at Lees Ferry, Arizona, by determination of the probability distribution of mean flows in relation to the range.

The assumption of the randomness and independence of hydrologic events is not entirely realistic, however, particularly in regard to runoff where antecedent conditions often have a close connection with subsequent discharges. These 'carry-over' effects must be minimized by proper grouping of the data. For example, if 3-year carry-over effects are detectable in the historical record, the generation for Monte Carlo methods must be based on the total output for periods not shorter than 3 years, so that the period totals may approach statistical independence. However, the second statistical approach, the Markov chain, proceeds on the assumption that the outcome of any 'trial' depends on the knowledge of the state of the system at the immediately preceding time. Julian (1961) used a first-order Markov process for generating annual flows in the Colorado River at Lees Ferry whereby:

$$x_t = r x_{t-1} + \Sigma(y)_t$$

where x_t is the annual runoff at year t

x_{t-1} is the runoff at the preceding or the $(t-1)$st year

r is the first-order serial correlation coefficient for the runoff, or a Markov chain coefficient

$\Sigma(y)$ is a random uncorrelated component due to annual rainfall.

This equation indicates that the runoff at a given year is equal to a constant times the runoff of the preceding year plus a random component. Much more elaborate multivariate chains have been used by Brittan (1961) and Fiering (1962), but the underlying principle common to all these techniques is that the system may be characterized largely by the statistical structure of the historical record. The relative success or failure of any simulation process depends ultimately on the validity of this premise.

Ven Te Chow (1964), in reviewing the significance of stochastic hydrology, admits that this field of study requires further research, but feels that potentially it has a promising future, particularly in water-resources development and management. For example, Hurst, Black and Simaika (1965) have used statistical modelling techniques to investigate the volume of storage required to make the greatest possible use of the Nile waters. A recent work by Bowden (1965) applied a Monte Carlo innovation-type model based on an azimuthal grid to simulate the spread of irrigation in part of the high plains of Colorado up to 1962, and to predict the rate and areal disposition of the spread of irrigation in the future.

HYDRO-ECONOMIC MODELS

Model theory, as a whole, is becoming increasingly important in water-resource management, where the economic and social implications of projects have to be considered, as well as those of physical hydrology. Hydro-economic models attempt to simplify and simulate the combined hydrological and economic aspects of a project.

Cost-benefit analysis may be regarded as the classic procedure which has been used as a practical way of assessing the desirability of projects, where it is important to take a wide and long-term view (Grant and Ireson, 1964; Prest and Turvey, 1965). This technique involves the enumeration and evaluation of all the relevant costs and benefits, with the aim of maximizing the present value of all benefits, less that of all costs, subject to specified constraints. Although the technique has been discussed by economists for the last hundred years, its practical value has come to increasing prominence since Pigou's (1932) classic on welfare economics – the concept of social costs and benefits expressed in monetary terms. Cost-benefit analysis has been used by the United States Federal Government to evaluate the economic viability of a variety of public works engineering projects. 'The Green Book' published by the Inter-Agency River Basin Committee (1950), attempts to codify and agree the general principles of such evaluation. Another reason for the increasing interest by economists in cost-benefit analysis has been the rapid development in recent years of such techniques as operations research and systems analysis (McKean, 1958; Maass et al., 1962) which prompt linked consideration of assemblages of factors involved in a broad investigation.

Cost-benefit analysis has two important limitations. Basically it is only a technique for taking decisions within a framework which has to be decided upon in advance and which involves a wide range of considerations, many of them of a political or social character. Also, cost-benefit techniques, as so far developed, are least relevant and serviceable for large-size investment decisions, where finance and investment on a national scale are involved. There are also considerable difficulties in the enumeration of all the cost and benefit factors involved and in their evaluation, particularly where the benefits are hard to express in monetary terms (e.g. social benefits). Some attempts have recently been made to express the less tangible benefits (e.g. recreational benefits) in quantitative terms, as Foster and Beesley (1963) have done for benefits to travellers on the projected London Underground Victoria Line. The great advantage of cost-benefit studies is that they force those responsible for planning to quantify costs and benefits as far as possible rather than to be content with vague qualitative judgements or personal 'hunches'.

Gilbert White and his associates at the University of Chicago have given considerable attention to economic studies of flood-plain management (White, 1945; White *et al.*, 1958, 1961 and 1962). Burton (1962), for example, has described various types of agricultural occupance of flood plains in the United States, suggesting a classification of the main types of agricultural land use in relation to the physical geography of flood plains, and Kates (1962) has undertaken an investigation of the awareness of flood-plain inhabitants to the hazards of flooding. The latter made case studies of six towns in the United States which have experienced flooding and used questionnaires to assess the attitude of the people to the flood problem. His study of the psychological and sociological attitude to floods reveals that there is remarkably little conscious human adjustment to this danger and that, even in places where action is taken to reduce the flood hazard, it may be 'casual, improvized, ineffective and far from optimal'.

White's studies of applied hydrology have extended beyond flood plain management to integrated river-basin development, as in the economic and social study of the Lower Mekong development (1963) to which he was a contributor. The purpose of this study was to find out as much as possible about the physical, economic and social background into which engineering structures would have to be fitted. It was an attempt to understand a cultural scene, completely different to American or European experience, and to attempt to adjust Western technology to it.

Many river valley developments are now multi-purpose schemes, and in trying to design for optimum development it is necessary to look at a number of possible combinations of water use. Calculations become more complicated and sometimes exceed the limits of feasibility by conventional methods. Here a systems approach to the design of hydro-economic models may become increasingly useful. By means of high-speed digital computers it is possible to simulate by simplified models the behaviour of relatively complex water-resource systems for periods of any desired length. It is now possible to perform the numerous and repetitive computations needed for many combinations of the system variables, and to evolve an optimal or near optimal design of the system.

Dorfman (1965) and Hufschmidt (1965) distinguish two main types of model used in a systems approach to water-resources studies – the simulation model and the analytical model. In a simulation model the time structure of the real-life project is faithfully preserved. The core of a simulation model is a set of differential equations that express the relationships between the various magnitudes describing the state and operation of the system during each small segment of time in the course of its history. The use of simulation analysis was begun by the U.S. Army Corps of Engineers on the Missouri River in 1953 (DATAmatic Corporation, 1957). In this analysis the operation of six reservoirs on the Missouri River was simulated on the Univac I computer to maximize power generation subject to constraints for navigation, flood control and irrigation specifications. Subsequently several studies have been made using model theory in the optimization of the operation of hydroelectric schemes (King and Peel, 1960; Rockwood, 1961; Lewis and Shoemaker, 1962; Rockwood and Nelson, 1966). In 1955 Morrice and Allan simulated the Nile Valley plan on an IBM 650 computer to determine the particular combination of reservoirs, control works and operating procedures that would maximize the use of irrigation water. In the Missouri River simulation and in those of the Columbia and the Nile only the purely hydrological conditions were optimized. In the Harvard Water Programme simulation analysis was applied to the economic optimization of water-resources system designs. Simulation was made of a simplified river basin system consisting of four reservoirs having three purposes. This system was purely hypothetical, but it created an adequate number of relationships to typify the complexities of an actual system. By sampling techniques, this analysis was able to evolve a design that yielded net benefits of $811 million against $724 million for the best design by conventional methods. Because of the enormous number of possible combinations of the system variables, the point of greatest net benefit cannot be easily determined, even with the use of high-speed computers. The only time-saving practical method of locating the point of optimality is to sample the variables and eliminate undesirable combinations from the computation. Here analytical models are used to simplify the real situation as far as possible to enable the resultant equations to be manipulated. Thus analytical models and the simulation approach are used in tandem. Initially the problem is analysed into a set of manageable mathematical relationships that can be solved for an approximation to a good or optimal design, and then a range of plausible variation around that tentative solution can be explored by a sequence of simulations.

Fiering (1965) gives an example of the use of the systems approach in the control of salinity in West Pakistan. The accumulation of salts in the soil is a by-product of a method of crop irrigation which has been employed with increasing intensity over a period of about 3,000 years. During the early part of the twentieth century British hydraulic engineers initiated the barrage system of irrigation and began to divert large quantities from the River Indus and its five tributaries. The drainage provisions were inadequate and the ground water level rose, as shown in Figure 5.13A. As the ground water table came within 5 ft. of the surface, evaporation from the soil and transpiration from the crops caused salts to rise in increasing quantities to the surface, thereby putting large acreages of once-fertile agricultural land out of

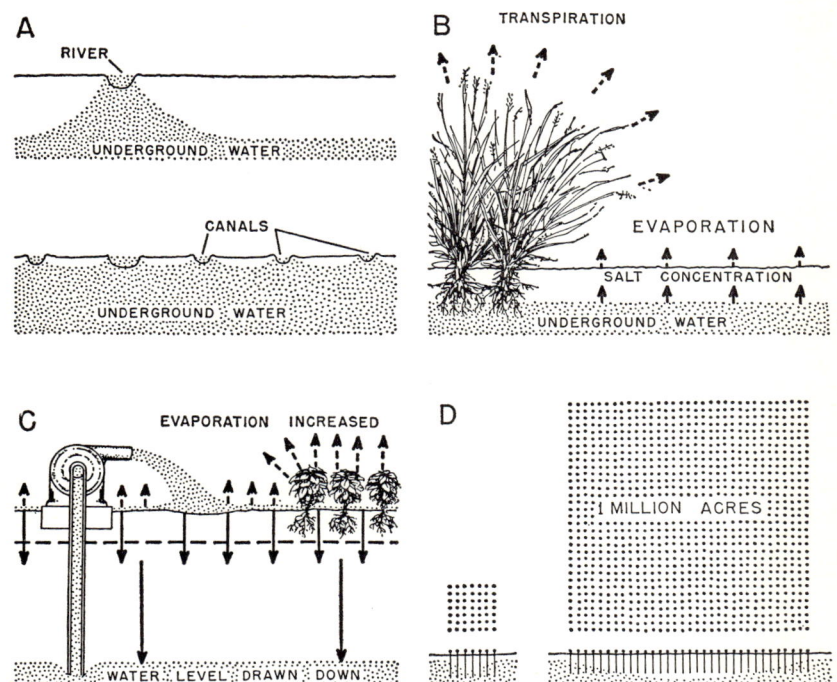

5.13 Ground water relationships in irrigated areas of West Pakistan.
A The construction of leaky irrigation canals has allowed a general rise of the water table.
B Salt accumulates through the evaporation of saline water which rises into the topsoil under capillary action from the high water table.
C Waterlogging and soil salinity can be ameliorated by controlled cased tubewell pumping, which lowers the water table and supplies enough water at the surface to leach down soil salts before evaporation occurs.
D Large-scale tubewell pumping is required to negate the effects of lateral ground water seepage and to cause an appreciable lowering of the water table (*Source: Redrawn from Revell, R., 1963, Water;* Scientific American, *209 (3), p. 100*).

cultivation (Fig. 5.13B). Several engineering consultants have studied this problem recently (White House – Interior Panel, 1964) and have suggested remedial action by means of tube wells. Figure 5.13C shows the action of one tube well in pumping out the ground water, so as to bring the water table to a deeper level. At the same time drainage should be greatly improved, so as to leach out and carry away the concentration of salts from the top layers of the soil. A certain amount of irrigation of salt-resistant crops could be carried on over the period of time needed to draw down the water table and leach the top soil. The well field would have to be dense and extensive to prevent saline ground water from seeping into the pumping area from surrounding areas (Fig. 5.13D). A simulation model of this ground water problem was con-

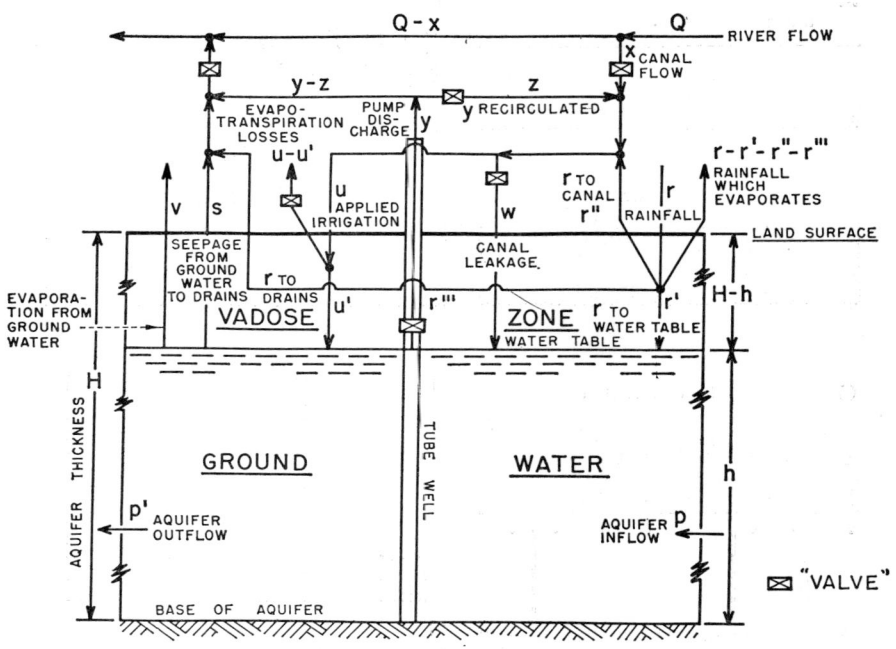

5.14 Schematic diagram of the basic hydrological model to simulate the action of a tubewell and appurtenant hydraulic control devices for combating waterlogging and salinity in the irrigated regions of West Pakistan. The 'valves' are economic decision points – e.g. the setting of the valve at W represents the efficacy of the canal lining. (*Source: From Fiering, 1965, p. 43*).

structed at the Harvard Water Resources Centre (Fig. 5.14). By use of this model attempts were made to formulate answers to the following questions:

1 What is the best well spacing and pump capacity?
2 What proportion of the effluent should be re-applied to the land?
3 How much lining should be installed in the canal system?
4 How fast will the ground water drop?
5 What is the best cropping pattern that can be devised?
6 How much surface drainage should be provided?

The 'valves' in Figure 5.14 represent critical points which can be controlled in the operation of the system. Statistical sampling was used to derive a workable number of combinations of these 'valve settings'. The various combinations were then compared to find an optimal solution in terms of cost, agricultural production and engineering feasibility.

This salinity model is the most ambitious systems analysis yet attempted, but both Dorfman (1965) and Fiering (1965) see this type of study as a beginning leading to a new range of analyses made possible by computing techniques.

HYDROLOGICAL MODELS AND GEOGRAPHY

Although there are large areas of overlap between hydrology and geography, especially in morphometry, hydraulic geometry and in the more economic aspects of hydrological planning, the two disciplines have developed quite separately. The work of Horton (1933 and 1945) in relating hydrology to geomorphology is a rare example of work which combines the two disciplines and shows how fruitful attempts to relate them can be.

Much of the research work in pure hydrology has been conducted by engineers, who have traditionally used model building in the solution of a variety of engineering problems. It is natural, therefore, that this approach should have extended into other types of hydrological investigations and that model building should be used as a tool throughout the whole field of hydrology. The generation of many of these models is a very specialist matter, but because the hydrologist deals so often with water in relation to man's activities, geographers should be extremely interested in them, in so far as the results of the hydrologists' work may very well be susceptible to remodelling into a geographically-orientated synthesis. The majority of the work described in this chapter is not geographical in its inception, and has not been done by geographers, but this does not mean that geographers can ignore the implications of this basic and fast-developing earth science, which possesses, as does geography, a strong bias towards the activity and welfare of man.

REFERENCES

ALLEN, J., [1952], *Scale Models in Hydraulic Engineering*, (London), 407 pp.
AMOROCHO, J., [1963], Measures of the linearity of hydrologic systems; *Journal of Geophysical Research*, 68, 2237–2249.
AMOROCHO, J. and HART, W. E., [1964], A critique of current methods in hydrologic systems investigation; *Transactions of the American Geophysical Union*, 45, 307–321.
AMOROCHO, J. and HART, W. E., [1965], The use of laboratory catchments in the study of hydrologic systems; *Journal of Hydrology*, 3, 106–123.
APPLEBY, F. V., [1954], Run-off dynamics; A heat conduction analogue of storage

flow in channel networks; *International Union of Geodesy and Geophysics*, (Rome).
BEARD, L. R., [1962], *Statistical Methods in Hydrology*, U.S. Army Engineer District, Corps of Engineers, Sacramento, California, Project CW-151, 62 pp.
BINNIE, G. M. and MANSELL-MOULLIN, M., [1966], The estimated probable maximum storm and flood on the Jhelum River – a tributary of the Indus; *The Institution of Civil Engineers, Proceedings of the Symposium on River Flood Hydrology*, 189–210.
BLANEY, H. F. and CRIDDLE, W. D., [1950], *Determining Water Requirements in Irrigated Areas from Climatological and Irrigation Data*; United States Department of Agriculture, Soil Conservation Service, 96 pp.
BONIFACE, E. S., [1959], Some experiments in artificial recharge in the Lower Lee Valley; *Proceedings of the Institution of Civil Engineers*, 14, 325–338.
BOWDEN, L. W., [1965], *Diffusion of the Decision to Irrigate*; The University of Chicago, Department of Geography, Research Paper No. 97, 146 pp.
BRITTAN, M. R., [1961], Probability analysis applied to the development of synthetic hydrology for Colorado River; part 4 of *Past and Probable Future Variations in Stream Flow in the Upper Colorado River*, University of Colorado, Bureau of Economic Research, (Boulder, Colorado).
BUCHAN, S., [1959], Artificial replenishment of aquifers; *Journal of the Institution of Water Engineers*, 9, 111–163.
BUCHAN, S., [1963], Conservation by integrated use of surface and ground water; *The Institution of Civil Engineers, Proceedings of the Symposium on Water Conservation*, 181–185.
BURTON, I., [1962], *Types of Agricultural Occupance of Flood Plains in the United States*; The University of Chicago, Department of Geography, Research Paper No. 75, 167 pp.
CARTER, D. B. et al., [1966], Energy and water budget, In *Spacecraft in Geographic Research*, National Academy of Sciences, National Research Council, Publication 1353, (Washington, D.C.), 23–47.
CHERY, D. L., [1966], Design and tests of a physical watershed model; *Journal of Hydrology*, 4, 224–235.
CHILDS, E. C., [1953], A new laboratory for the study of the flow of fluids in porous beds; *Proceedings of the Institution of Civil Engineers*, 2 (3), 134–141.
CHILDS, E. C., [1957], Physics of land drainage; In Luthin, J. N., (Ed.), *Drainage of Agricultural Lands*, (Madison, Wisconsin), 1–66.
CHOW, VEN TE. (Ed.), [1964], *Handbook of Applied Hydrology*, (New York).
CIRIACY-WANTRUP, S. V., [1963], *Resource Conservation, Economics and Policies*, (Berkeley, California), 395 pp.
COX, D. R. and SMITH, W. L., [1963], *Queues*, (London), 180 pp.
CRAWFORD, N. H. and LINSLEY, R. K., [1962], *Synthesis of continuous streamflow hydrographs on a digital computer*; Department of Civil Engineering, Stanford University, Technical Report No. 12.
CRAWFORD, N. H. and LINSLEY, R. K., [1966], *Digital simulation in hydrology: Stanford Watershed Model IV*; Department of Civil Engineering, Stanford University, Technical Report No. 39, 210 pp.

DATAMATIC CORPORATION AND RAYTHEON MANUFACTURING COMPANY for U.S. Army Corps of Engineers, [1957], *Report of Use of Electronic Computers for Integrating Reservoir Operations*.

DAVIS, S. N. and DE WEIST, R. J. M., [1966], *Hydrogeology*, (New York), 463 pp.

DAWDY, D. R. and O'DONNELL, T., [1965], Mathematical models of catchment behavior; *Proceedings of the American Society of Civil Engineers, Journal of the Hydraulics Division*, 91, No. HY4, Part 1, 123–137.

DE WEIST, R. J. M., [1965], *Geohydrology*, (New York), 366 pp.

DOOGE, J. C. I., [1959], A general theory of the unit hydrograph; *Journal of Geophysical Research*, 64, 241–256.

DORFMAN, R., [1965], Formal models in the design of water resource systems; *Water Resources Research*, 1, 329–336.

DUCKWORTH, W. E., [1965], *A Guide to Operational Research*, (London), 153 pp.

ECKSTEIN, O., [1958], *Water Resource Development: The economics of project evaluation*, (Cambridge, Mass.).

EMMETT, W. W., [1965], The Vigil Network: methods of measurement and a sampling of data collected; *International Association of Scientific Hydrology, Symposium of Budapest*, Publication No. 66, 89–106.

FIERING, M. B., [1962], Queueing theory and simulation in reservoir design; *Transactions of the American Society of Civil Engineers*, 127, 1114–1144.

FIERING, M. B., [1965], Revitalizing a fertile plain; *Water Resources Research*, 1, 41–61.

FOLLANSBEE, R. and SAWYER, L. R., [1948], Floods in Colorado; *United States Geological Survey, Water Supply Paper* 997.

FOSTER, C. D. and BEESLEY, M. E., [1963], Estimating the social benefit of constructing an underground railway in London; *Journal of the Royal Statistical Society*, Series A, 126, 46–78.

GANI, J. and MORAN, P. A. P., [1955], The solution of dam equations by Monte Carlo methods; *Australian Journal of Applied Science*, 6, 267–273.

GEORGE, F. H., [1965], *Cybernetics and Biology*, (Edinburgh), 138 pp.

GIGUET, R., [1957], A French dual-purpose scheme – the Durance project and the Serre Ponçon dam; *Proceedings of the Institution of Civil Engineers*, 6, 550–576.

GRANT, E. L. and IRESON, W. G., [1964], *Principles of Engineering Economy*, (New York), 574 pp.

GUMBEL, E. J., [1958A], *Statistics of Extremes*, (New York), 375 pp.

GUMBEL, E. J., [1958B], Statistical theory of floods and droughts; *Journal of the Institution of Water Engineers*, 12, 157–184.

HACK, J. T., [1957], Studies in longitudinal stream profiles in Virginia and Maryland; *United States Geological Survey, Professional Paper*, 294-B, 45–97.

HAMMERSLEY, J. M. and HANDSCOMB, D. C., [1964], *Monte Carlo Methods*, (London), 178 pp.

HARRIS, R. A., [1965], Probability of reservoir yield failure using Moran's Steady-State Probability Method and Gould's Probability Routing Method; *Journal of the Institution of Water Engineers*, 19, 302–328.

HORTON, R. E., [1933], The role of infiltration in the hydrologic cycle; *Transactions of the American Geophysical Union*, 14, 446–460.

HORTON, R. E., [1945], Erosional development of streams and their drainage basins:

Hydrophysical approach to quantitative morphology; *Bulletin of the Geological Society of America*, 56, 275–370.
HUFSCHMIDT, M. F., [1965], Field level planning of water resource systems; *Water Resources Research*, 1, 147–171.
HURST, H. E., BLACK, R. P. and SIMAIKA, Y. M., [1965], *Long-Term Storage*, (London), 145 pp.
INTER-AGENCY RIVER BASIN COMMITTEE (Sub-Committee on Costs and Budgets), [1950], *Proposed Practices for Economic Analysis of River Basin Projects, 'The Green Book'*, (Washington, D.C.).
INTERNATIONAL ASSOCIATION OF SCIENTIFIC HYDROLOGY, [1965], *Symposium of Budapest, Representative and Experimental Areas*; Publication No. 66.
JOHNSON, A. I., [1965], Computer processing of hydrologic and geologic data; *Ground Water*, 3 (3), 9 pp.
JOHNSON, A. I. and LANG, S. M., [1965], Automated processing of water information; *Proceedings of the First Annual Meeting, American Water Resources Association*, (Chicago), 324–350.
JULIAN, P. R., [1961], A study of the statistical predictability of stream-runoff in the Upper Colorado River basin; part 2 of *Past and Probable Future Variations in Stream Flow in the Upper Colorado River*, University of Colorado, Bureau of Economic Research, (Boulder, Colorado).
KATES, R. W., [1962], *Hazard and Choice Perception in Flood Plain Management*; The University of Chicago, Department of Geography, Research Paper No. 78, 157 pp.
KENDALL, D. G., [1957], Some problems in the theory of dams; *Journal of the Royal Statistical Society*, Series B, 19, 207–212.
KING, P. F. and PEEL, D. A., [1960], An analysis of a hydro-electric system; *The Computer Journal*, 3, 161–163.
KRUMBEIN, W. C. and GRAYBILL, F. A., [1965], *An Introduction to Statistical Models in Geology*, (New York), 475 pp.
KRUTILLA, J. V. and ECKSTEIN, O., [1958], *Multiple Purpose River Development*, (Baltimore), 301 pp.
KUIPER, E., [1965], *Water Resources Development, Planning, Engineering and Economics*, (London), 483 pp.
LEET, L. D. and JUDSON, S., [1954], *Physical Geology*, 1st Edn., (New York), 466 pp.
LEOPOLD, L. B., [1962], The Vigil Network; *Publication of the International Association of Scientific Hydrology*, Year 7 (2), 5–9.
LEOPOLD, L. B. and EMMETT, W. W., [1965], Vigil Network sites: A sample of data for permanent filing; *Publication of the International Association of Scientific Hydrology*, Year 10 (3), 12–21.
LEOPOLD, L. B. and MILLER, J. P., [1956], Ephemeral streams – Hydraulic factors and their relation to the drainage net; *United States Geological Survey, Professional Paper* 282-A, 1–37.
LEOPOLD, L. B., WOLMAN, M. G. and MILLER, J. P., [1964], *Fluvial Processes in Geomorphology*, (San Francisco), 522 pp.
LEWIS, D. J. and SHOEMAKER, L. A., [1962], Hydro system power analysis by

digital computer; *Journal of the Hydraulics Division, Proceedings of the American Society of Civil Engineers*, 88, 113–130.
LINSLEY, R. K., KOHLER, M. A. and PAULHUS, L. H., [1949], *Applied Hydrology*, (New York), 689 pp.
LOWRY, R. L. and JOHNSON, A. F., [1942], Consumptive use of water for agriculture; *Transactions of the American Society of Civil Engineers*, 107, 1243–1302.
LUTHIN, J. N. (Ed.), [1957], *Drainage of agricultural lands*, (Madison, Wisconsin), 620 pp.
LUTHIN, J. N., [1966], *Drainage Engineering*, (New York), 250 pp.
MAASS, A., HUFSCHMIDT, M. M., DORFMAN, R., THOMAS, H. A., MARGLIN, S. A. and FAIR, G. M., [1962], *Design of Water-Resource Systems*, (Cambridge, Mass.), 620 pp.
MCKEAN, R. N., [1958], *Efficiency in Government through Systems Analysis: With Emphasis on Water Resources Development*, (New York), 336 pp.
MONTEITH, J. L., [1959], The reflection of short-wave radiation by vegetation; *Quarterly Journal of the Royal Meteorological Society*, 85, 386–392.
MORAN, P. A. P., [1959], *The Theory of Storage*, (London), 110 pp.
MORRICE, H. A. W. and ALLAN, W. N., [1959], Planning for the ultimate hydraulic development of the Nile Valley; *Proceedings of the Institution of Civil Engineers*, 14, 101–156.
MURPHY, F. C., [1958], *Regulating Flood-Plain Development*; The University of Chicago, Department of Geography, Research Paper No. 56, 204 pp.
NASH, J. E., [1958], Determining run-off from rainfall; *Proceedings of the Institution of Civil Engineers*, 10, 163–184.
NASH, J. E., [1960], A unit hydrograph study, with particular reference to British catchments; *Proceedings of the Institution of Civil Engineers*, 17, 249–282.
O'DONNELL, T., [1960], Instantaneous unit hydrograph derivation by harmonic analysis; *Commission of Surface Waters, International Association of Scientific Hydrology*, Publication 51, 546–557.
O'DONNELL, T., [1966], Computer evaluation of catchment behaviour and parameters significant in flood hydrology; *The Institution of Civil Engineers, Proceedings of the Symposium on River Flood Hydrology*, 103–113.
O'KELLY, J. J., [1955], The employment of unit hydrographs to determine the flows of Irish arterial drainage channels; *Proceedings of the Institution of Civil Engineers*, 4, 365–445.
OLIVIER, H., [1961], *Irrigation and Climate*, (London), 250 pp.
PENMAN, H. L., [1948], Natural evaporation from open water, bare soil and grass; *Proceedings of the Royal Society, Series A*, 193, 120–145.
PENMAN, H. L., [1950], The water balance of the Stour catchment area; *Journal of the Institution of Water Engineers*, 4, 457–469.
PENMAN, H. L., [1956], Evaporation – an introductory survey; *Netherlands Journal of Agricultural Science*, 4, 9–29.
PENMAN, H. L., [1963], *Vegetation and Hydrology*; Commonwealth Bureau of Soils, Harpenden, Technical Communication No. 53, 124 pp.
PIGOU, A. C., [1932], *The Economics of Welfare*, (London), 837 pp.
PREST, A. R. and TURVEY, R., [1965], Cost-Benefit analysis: a survey; *The Economic Journal*, 75, 683–735.

ROCKWOOD, D. M., [1961], Columbia Basin streamflow routing by computer; *Transactions of the American Society of Civil Engineers*, 126, 32–56.

ROCKWOOD, D. M. and NELSON, M. L., [1966], Computer application to streamflow synthesis and reservoir regulation; *Proceedings of the Sixth Congress of the International Commission of Irrigation and Drainage*, (New Delhi, India).

RUTTAN, V. W., [1965], *The Economic Demand for Irrigated Acreage: New Methodology and Some Preliminary Projections 1954–1980*, (Baltimore), 154 pp.

SCHWAB, G. O., FREVERT, R. K., EDMINSTER, T. W. and BARNES, K. K., [1966], *Soil and Water Conservation Engineering*, 2nd Edn., (New York), 683 pp.

SHEN, J., [1963A], Use of hydrologic models in the analysis of flood runoff; *Administrative Report, United States Geological Survey (Water Resources Division)*.

SHEN, J., [1963B], The role of analogue in surface-water hydrological problems; *Administrative Report, United States Geological Survey (Water Resources Division)*.

SHERMAN, L. K., [1932], Streamflow from rainfall by unit-graph method; *Engineering News-Record*, 108, 501–505.

SKIBITZKE, H. E., [1963], The use of analogue computers for studies in groundwater hydrology; *Journal of the Institution of Water Engineers*, 17, 216–230.

SLAYMAKER, H. O. and CHORLEY, R. J., [1964], The Vigil Network System; *Journal of Hydrology*, 2, 19–24.

SMITH, S. C. and CASTLE, E. N., [1965], *Economics and Public Policy in Water Resource Development*, (Ames, Iowa), 463 pp.

STRAHLER, A. N., [1952], Hypsometric (area-altitude) analysis of erosional topography; *Bulletin of the Geological Society of America*, 63, 1117–1142.

STRAHLER, A. N., [1964], Quantitative geomorphology of drainage basins and channel networks; Section 4-II in Chow, Ven Te, (Ed.), *Handbook of Applied Hydrology*, 4–39 to 4–76.

TENNESSEE VALLEY AUTHORITY, Office of Tributary Area Development, [1961], *Matrix operations in hydrograph computations*; Research Paper No. 1, (The Authority, Knoxville, Tennessee).

TENNESSEE VALLEY AUTHORITY, Office of Tributary Area Development, [1963A], *A water yield model for analysis of monthly runoff data*; Research Paper No. 2, (The Authority, Knoxville, Tennessee).

TENNESSEE VALLEY AUTHORITY, Office of Tributary Area Development, [1963B], *T.V.A. computer programmes for hydrologic analyses*; Research Paper No. 3, (The Authority, Knoxville, Tennessee).

THAMES CONSERVANCY, [1965], *Thames Conservancy and the Water Crisis*, 11 pp.

THIESSEN, A. H., [1911], Precipitation for large areas; *Monthly Weather Review*, 39, 1082–1084.

THORN, R. B. (Ed.), [1966], *River Engineering and Water Conservation Works*, (London), 520 pp.

TODD, D. K., [1959], *Ground Water Hydrology*, (New York), 336 pp.

THORNTHWAITE, C. W., [1948], An approach towards a rational classification of climate; *Geographical Review*, 38, 85–94.

THORNTHWAITE, C. W., [1954], A re-examination of the concept and measurement of potential transpiration; In Mather, J. R. (Ed.), *The Measurement of Potential Evapo-transpiration*, Problems in Climatology, (Seabrook, N.J.), 200–209.

UNESCO INTERNATIONAL HYDROLOGICAL DECADE CO-ORDINATING COUNCIL, [1965], *Final Report*, (Paris).
VAJDA, S., [1961], *The Theory of Games and Linear Programming*, (London), 106 pp.
WHITE, G. F., [1945], *Human Adjustment to Floods*; The University of Chicago, Department of Geography, Research Paper No. 29, 225 pp.
WHITE, G. F., CALEF, W. C., HUDSON, J. W., MAYER, H. M., SHEAFFER, J. R. and VOLK, D. J., [1958], *Changes in Urban Occupance of Flood Plains in the United States*; The University of Chicago, Department of Geography, Research Paper No. 57, 235 pp.
WHITE, G. F. (Ed.), [1961], *Papers on Flood Problems*; The University of Chicago, Department of Geography, Research Paper No. 70, 228 pp.
WHITE, G. F., DE VRIES, E., DUNKERLEY, H. B. and KRUTILLA, J. V., [1962], *Economic and Social Aspects of Lower Mekong Development*; A Report to the Committee for Co-ordination of Investigations of the Lower Mekong Basin, 106 pp.
WHITE, G. F., [1963], The Mekong River Plan; *Scientific American*, 208 (4), 49–59.
WHITE HOUSE-INTERIOR PANEL, [1964], *Report on Land and Water Development in the Indus Plain*.
WIESNER, C. J., [1964], Hydrometeorology and river flood estimation; *Proceedings of the Institution of Civil Engineers*, 27, 153–167.
WOLF, P. O., [1966A], Comparison of Methods of Flood Estimation; *The Institution of Civil Engineers, Proceedings of the Symposium on River Flood Hydrology*, 1–23.
WOLF, P. O., [1966B], Notes on the management of water resource systems; *Journal of the Institution of Water Engineers*, 20 (2), 95–105.

PART III

Models of Socio-Economic Systems

CHAPTER SIX

Demographic Models and Geography
E. A. WRIGLEY

INTRODUCTION

Population has always figured prominently in geography. Consideration of the distribution and density of population has been a starting point in many geographical studies, a finishing point in others, and has occasionally played both roles when the purpose of the study was to bring a new understanding of some aspects of distribution patterns (Wrigley, 1965B). We are frequently referred to maps showing densities of population, to general models of population distribution in space in the manner first pioneered by Christaller and Lösch, to studies of the relative rates of growth of settlements of different sizes, to questions like rural depopulation, to ideas like the market orientation of industry (which usually means simply the location of plants at places where there are a lot of people to buy the products). But for these purposes often little more than the crude totals of population are used, though sometimes these may be broken down further by occupational divisions, by age-structure, or in some other way. Changes in population totals over time and migration have occasionally been analysed with some rigour, and the use of models is now widespread in the attempt to achieve a clearer understanding of migration (see, for example, Lund Studies, 1957). Little attention, however, has been given to the construction or use of demographic models: nor has their potential importance to the understanding of several issues which bulk large in human geography been widely discussed.

Demographic models *sensu stricto* deal with the interplay of fertility, mortality and nuptiality. If these can be measured accurately demographic models will determine what the stable age-structure of the population will ultimately be assuming existing characteristics are maintained, what its rate of growth or decline, and so on. Alternatively, if they cannot be measured accurately, it may be possible with imperfect data to establish the extreme limits within which the truth must lie. Each variable will interact with the other two and produce sympathetic changes in them in the manner characteristic of open systems, and the changes can be predicted accurately if sufficient information is available. Much work of fundamental importance was carried out between the wars by Lotka in this field (Lotka, 1934–1939).

In a looser sense demographic models may be taken to include models in which the parameters may extend to features of the economic life of the community in question (United Nations, 1953; Leibenstein, 1954) or to its social and political activity (Sauvy, 1952–54). For example, to take a single restricted issue, use may be made of demographic models to determine the maximum percentage of matrilateral cross-cousin marriages which can occur in populations with a range of different demographic characteristics (Levy and Westoff, 1965). Or, more generally, a relationship between a system of land tenure and marriage customs may be postulated and the economic and demographic consequences of such a web of relationships can be examined by constructing a model in which the relationships between the major variables are expressed.

It is no accident that until recently models of these types aroused little interest in geographers, and for two reasons. In the first place it was commonly assumed that whatever the theoretical range of possibilities of demographic behaviour, most pre-industrial societies behaved in a manner which made it possible to treat population as a given feature of the situation whose demographic characteristics were of minor interest. This was the product of what might be called a modified Malthusianism. Few later writers have been so firmly pessimistic as the early Malthus in their treatment of the press of population against resources, but most have nevertheless tended to assume that pre-industrial populations normally grow towards a ceiling imposed by the maximum flow of food and raw materials which is possible at any given level of material technology and usable resources. In these circumstances it is clearly more important to devote attention to the nature of the material technology and the resource base of the community rather than to its demography. Another common assumption of a more general nature which also tends to cause a neglect of demographic issues is that which sees the main demographic characteristics of a community as a function of its fundamental economic constitution. Once more this implies that there is little point in investigating the structure of fertility, mortality and nuptiality in the population. If the general shape of the economic environment is known the chief features of the demographic situation can be read off from it.

Secondly, even when the treatment of population as a dependent variable was viewed with suspicion there was until recently a dearth of well-documented case studies to illustrate effectively the complexity of the interplay between demographic, economic, sociological and geographical variables. There is now a comparative abundance of these studies (e.g. Goubert, 1960; Hatt, 1952; Lorimer, 1954; Glass and Eversley, 1965). In one recent work, indeed, a most interesting attempt has been made to invert completely the Malthusian view that population size is dependent upon the potential of a given agricultural technology. It has been argued instead that agricultural technology is normally a function of population density. The dependent and independent variables have exchanged places (Boserup, 1965).

The importance and independence of the demography of a population has been brought home with special force to a wide public in recent years by the population problems of the developing countries. That they illustrate the point very dramatically, however, should not be allowed to obscure the fact that the case is general, not special.

ANIMAL POPULATION BEHAVIOUR

It may prove helpful to preface the discussion of demographic models by remarking the great change in thought about the characteristics of animal populations which has occurred recently. It forms a fascinating chapter in intellectual history and makes a convenient starting point for the subsequent discussion of the behaviour of men. The two types of population study have, after all, been closely linked in the past and have profited much from each other.

The study of animal population characteristics has long held a prominent place in biological studies. Darwin, acknowledging most generously the extent of his debt to Malthus, made of this an engine to drive the mechanism of natural selection. He pointed out that the long-term trend in the numbers of most animal populations was neither up nor down whereas their powers of reproduction were in all cases great enough to secure a rapid increase in their numbers. Even the elephant, according to Darwin's calculations, in spite of a very long gestation period, was capable of increasing from one couple to 15,000,000 in a period of 500 years (Darwin, 1869). Most animals possessed much more spectacular powers of multiplication.[1] The difference between the inherent powers of reproduction possessed by the animal kingdom and the absence of long-term increase represented in Darwin's view the pressure of selection, the constant sifting out and elimination of those individuals whose constitution made them less well able than their fellows to survive and reproduce. Those which were better adapted were able to breed and so the next generation included a higher proportion of individuals with the advantageous traits and the less well adapted gradually died out.

It is natural in terms of a model of this type to expect populations to be at

[1] '—reflect on the enormous multiplying power *inherent and annually in action* in all animals; reflect on the countless seed scattered by a hundred ingenious contrivances, year after year, over the whole face of the land; and yet we have every reason to suppose that the average percentage of every one of the inhabitants of a country will *ordinarily remain constant*.' (Darwin, 1958, p. 118.)

or close to the maximum which their habitat and their relative success in competition with other species permits them to attain. In recent years, however, it has come to be realized that this attitude needs substantial qualification if it is to do justice to the patterns of animal behaviour which are observable. The classical mechanism of selection may always be at work but is mediated through elaborate social conventions, many of which have developed in order to relieve a group of animals of one species living in an established territory from the extreme rigours of selection implied by the full use of the fertility potential of the individuals present. This view involves not only the assertion that mortality levels are dependent upon the density of the population – a point always recognized; but also that fertility levels vary with population density, falling as numbers rise and vice versa. In some forms of life the fall in fertility may be a simple reaction to direct physical problems (for example, the absence of further suitable sites for oviposition – there is a general discussion of this and other related points in Watt, 1962). In others the relationship is indirect and involves elaborate social conventions. In the former case the restriction of fertility may be viewed as an extension of the classical view of density-dependent mortality levels. In the latter, however, the social conventions, though closely geared to population densities, are very flexible and introduce a new element into the analysis of animal population behaviour. Populations of this type may fluctuate round a figure far below the maximum.

The conventions may take many forms. For example in birds which nest in colonies, such as storks or rooks, the number of nesting sites in the colony may be restricted by convention. Only those birds can breed which establish tenure upon one of the nesting sites. To secure a good site is evidence of high rank in the hierarchical society. Those lowest in the hierarchy, though sexually mature, may be unable to gain a site and so be prevented from breeding. The number of breeding pairs thus remains constant from year to year and the pressure upon the food supplies of the territory belonging to members of a rookery does not rise as it might if there were no social controls upon breeding. The mechanism of control varies in different species. For example, in the tits, each male seeks to establish control over an area from which he can exclude his rivals and which enables him to secure a mate, nest, and feed his young (see Wynne-Edwards, 1962, esp. Chap. 9). In other bird and animal species the social conventions governing fertility may take still other forms, but they all tend to the same end – to prevent numbers rising to the point where the means of subsistence is endangered by excessive population pressure or where adult members of the population are unable to keep themselves in good health because of the intensity of the competition for food. Selection operates, in short, at a social as well as an individual level, and the success of the group may be enhanced by the development of social habits which restrict fertility severely, or (which amounts to much the same thing) cause an extremely high mortality among the newly-born (see, for example, Wynne-Edwards, 1962, p. 537).[1]

Not all animal societies have developed mechanisms of this sort. Indeed the process of selection has thrown up radically different patterns of population behaviour in some cases, especially where the species in question lives in a marginal and difficult environment. For example, the locust attempts to maintain and extend its range by building up numbers very rapidly from time to time and 'exploding' into neighbouring areas. During the peak of a population cycle of this sort the food base available to the locust is exhausted over large areas, a pattern of events very different from that described in the previous paragraph. It has been supposed that the logic of population behaviour of this type may lie in the greater danger that exceptionally severe seasons in marginal areas will wipe out the whole of a local population so that waves of 'peak' populations spreading out in 'explosion' periods may find suitable areas to re-colonize. In more stable environments this is much less likely to occur and the long-term advantages of population 'explosions' is correspondingly less (see Wynne-Edwards, 1962, Chap. 20). At all events it is clear that a wide variety of types of population behaviour can be observed in animal populations, that a great deal of the social life of animals is connected with the fine adjustments of numbers to opportunities, and that it is far from being universally the case that the full reproductive potential of an animal group is called into play in normal seasons.

A SIMPLE DEMOGRAPHIC MODEL

What is true of animal societies is also true of societies of men and women. Societies have always, or almost always, developed social conventions which inhibited fertility to some extent, and often very drastically. Moreover, customs which have the effect of limiting fertility may be supplemented by others which cause high mortality among the very young. These have much the same effect as checks upon fertility in limiting the reproductive effort of the society. Levels of fertility and mortality, and particularly the latter, may

---

[1] Darwin was puzzled by the question of the age incidence of the heavy mortality which must follow from unrestricted fertility, but inclined to the view that it must fall mainly on the very young. For example, 'Where man has introduced plants and animals into a new country favourable to them, there are many accounts in how surprisingly few years the whole country was fully stocked; and yet we have every reason to believe from what is known of wild animals that *all* would pair in the spring. In the majority of cases it is most difficult to imagine where the check falls, generally no doubt on the seeds, eggs, and young; but when we remember how impossible even in mankind (so much better known than any other animal) it is to infer from repeated casual observations what the average of life is ... we ought to feel no legitimate surprise at not seeing where the check falls in animals and plants'. (Darwin, 1958, pp. 117–118).

fluctuate greatly from year to year and produce marked changes in a population in the short term, but the long-term effect of any prevailing level of fertility and mortality may be illustrated quite simply.

It is convenient to make a number of simplifying assumptions initially. Let us suppose that the society in question is at a comparatively primitive level of material culture, and that the material culture is static so that it is not possible to secure a steady increase in the production of food in the community by taking advantage of technical advances in agriculture. In these circumstances a population cannot grow through the ceiling set by the food-producing potential of the area at the given level of material culture. It is mistaken, however, to suppose that population will necessarily rise to this maximum. An equilibrium level of population may establish itself at many different

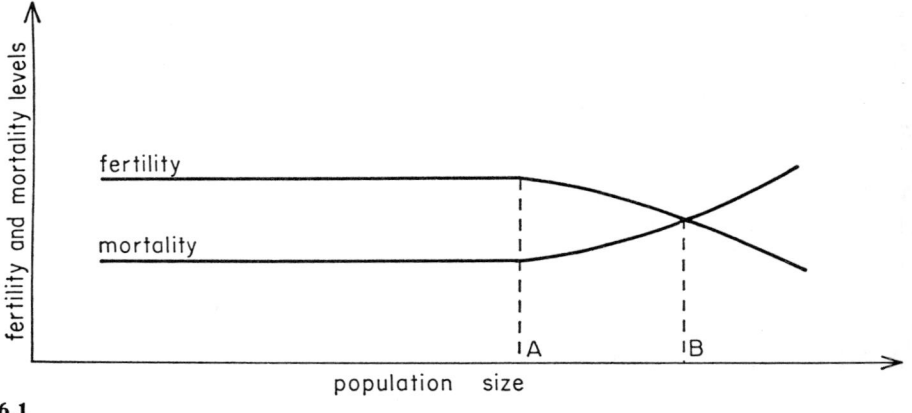

6.1

points. The level cannot, of course, be higher than the ceiling (except temporarily) and may well fall short of it by a substantial margin.

The combination of possibilities represented by the fertility and mortality schedules in Figure 6.1 describes a situation in which until population reaches A in total size the levels of fertility and mortality do not change and the population rises at a uniform speed. When population rises above A, however, both schedules are affected. Mortality rises and fertility falls as the density of population increases further until at B fertility and mortality are in balance and the population neither rises nor falls. There is, however, no *a priori* reason why fertility and mortality should both begin to be affected at the same density of population. In Figure 6.2 the pair of curves represents a situation closer perhaps to what Malthus had in mind in those moods when he doubted the efficacy of prudential checks (and which Darwin appears to have supposed typical of most animal populations). Here mortality follows the same path as in Figure 6.1, rising steadily when a certain critical density

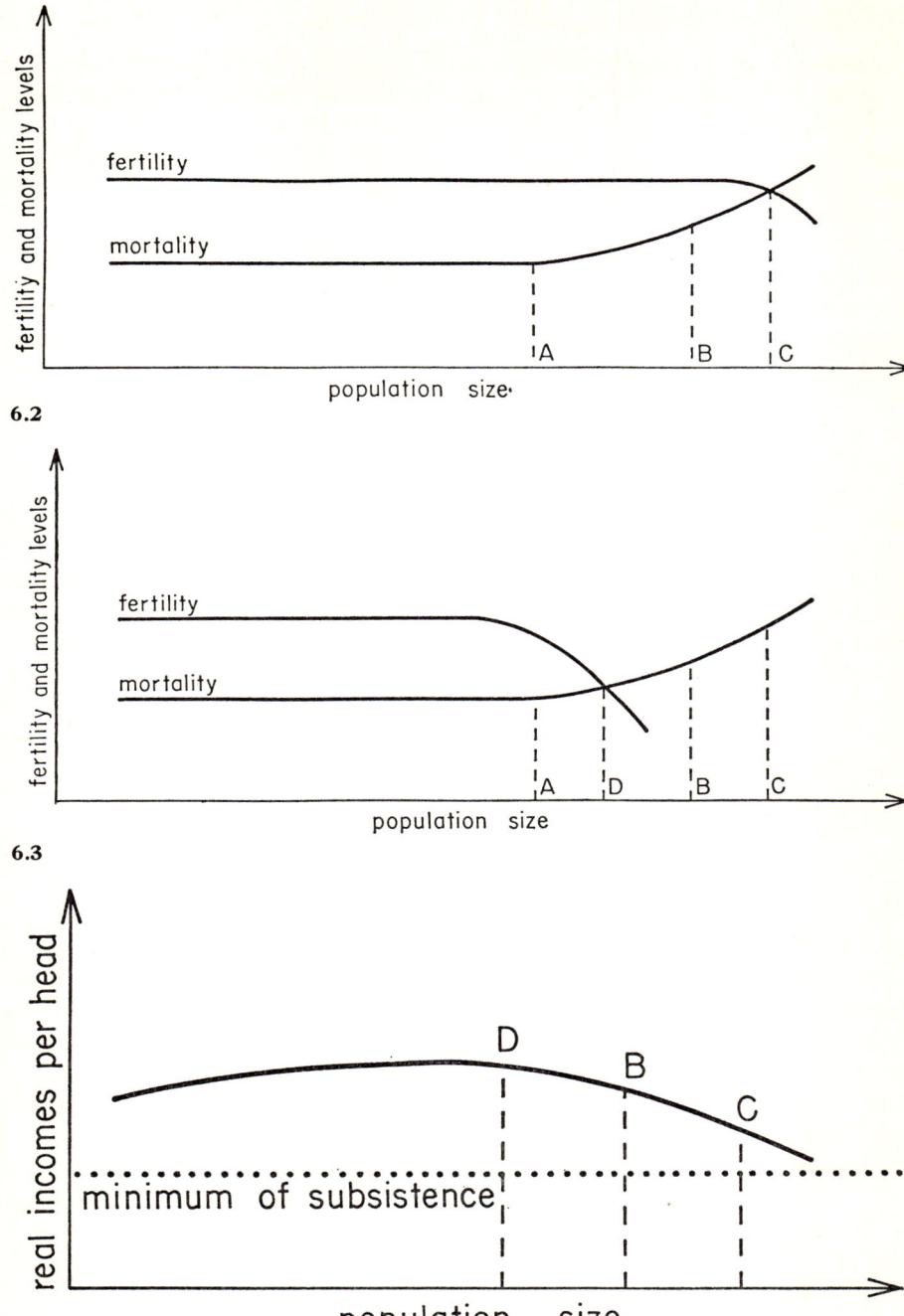

6.2

6.3

6.4

has been exceeded. But fertility is affected much later and less (indeed for the purpose of exposition, though not in reality, the fertility schedule might be a horizontal straight line). In these circumstances population will rise to C, a much higher level than B in Figure 6.1. Misery and vice, to use the language of Malthus, will be rife; or, in Darwinian terminology, the pressure of competitive selection will be very intense.

A third case may be imagined in which the mortality schedule is once more the same, but now fertility reacts very sensitively to any population pressure. The curve is deformed early and severely. This is shown in Figure 6.3. As long as this schedule describes the fertility characteristics of the population its total will never exceed D.

Figure 6.4 shows the implications of these three pairs of schedules in terms of real incomes per head. At any given level of material culture and with an inflexible resource base there may be expected to be some total of population at which real incomes per head will reach their maximum (see Sauvy, 1952–54 for a full discussion of this complex concept). It is reasonable to suppose that at some point after this optimum level has been passed mortality levels will begin to rise. If the fertility schedule intersects with the mortality schedule before the latter has begun to rise an optimum level of population may result enjoying the highest standard of living possible in the cultural and environmental context of the day and place. D, derived from Figure 6.3, represents a more advantageous position in Figure 6.4 than B which is the equilibrium level of total population from Figure 6.1, while C is the least happy position, reflecting the persistent high fertility found in Figure 6.2.

This account is both compressed and oversimplified. For example, it is obvious that mortality schedules may also vary considerably. In some societies infanticide was a common practice (Darwin thought Malthus had seriously underestimated its importance generally – Darwin, 1901, p. 69). It is conceivable that in such societies the mortality schedule might turn up as sharply as the fertility schedule turns down in Figure 6.3, with similar results as far as the equilibrium total of population is concerned.[1] Other qualifications and enlargements of the model will be made later in this essay, and many more would be called for in a larger treatment of the subject. But for the present purpose a sketch of this sort is sufficient if it makes it possible to discuss the importance to geography of the demographic chracteristics of populations. The elaboration of the model may now, therefore, be postponed in order to consider some of its implications in its present crude state.

[1] To calculate mortality from the moment of birth is, of course, a somewhat arbitrary proceeding. In all societies many lives are lost in the womb by miscarriage, abortion and stillbirth. In those societies in which infanticide is widely practised, it might be more useful for some purposes to measure mortality only from, say, the end of the first month of life, and to count death from infanticide, like stillbirth or abortion, as a feature which keeps fertility low rather than one which boosts mortality.

DEMOGRAPHIC CHARACTERISTICS AND GEOGRAPHICAL CONDITIONS

Many of the features of the human geography of an area which figure most prominently in orthodox discussions are in part a function of the demographic characteristics of the society in question. For example, it is evident that the subdivision of holdings may be carried much further in a given area when the population total is at, say, C in Figure 6.4 than would be the case if the total were at D. The pressure of high fertility cannot drive population increase up beyond a certain level because mortality will rise sufficiently to cause the increase to taper off, but it can push population totals up to a higher level than would be achieved at a lower level of fertility. This in turn may mean smaller holdings, more intensive use of the land (but at a lower output per worker), and the bringing into cultivation of land of poor quality (see Boserup, 1965, p. 118, for a lively exposition of the view that the ultimate result of an increasing density of population on the land may even be to better the prospects of sustained economic growth). Further examination of land use in the area may reveal other related effects. Where population pressure is great there may be a heavy premium placed on growing the crop which yields the largest return of calories to the acre – perhaps the potato; and the balance between livestock and crops may also be affected. Where it is necessary to maximize yield of food per acre, and where real incomes are low because the equilibrium level of total population is close to the Malthusian ceiling, livestock normally give way to crops and land that might otherwise produce good beef may be used instead for wheat (for a discussion of a very similar issue see Wrigley, 1962).

Through the same set of interrelated circumstances the local demography may have a great bearing on such matters as soil exhaustion, erosion, the upsetting of the hydrological balance, and so on. The progressive impoverishment of the soil leading ultimately to its destruction or to a permanent impairment of its fertility is not something which is solely or even primarily determined by the physical and chemical characteristics of the soil itself. It might indeed be said, if one wished to be paradoxical, that there is no marginal land *in se*. The manner in which the land is used is of paramount importance, and this in turn is in part a function of the pressure of population. Even the most unstable of soils and most delicately balanced ecological systems can be preserved if they are not subjected to undue pressure. Equally, even soils in areas not usually considered marginal may deteriorate if used without discretion. In late thirteenth-century England there are grounds for believing that population pressure at the contemporary level of agricultural technique was intense. There is indirect evidence that unsuitable land, lying on steep slopes or consisting of poor sands or chalk, was taken into cultivation for a short period but that the deterioration of the soil after a short period of use was so great that it dropped out of use and that as a result population fell. The Malthusian ceiling was probably closely approached with correspondingly severe effects (Postan and Titow, 1959: see also Herlihy, 1965, on the still more severe situation in northern Italy in the thirteenth century).

Or again it is easy to imagine an illuminating contrast in land use in a tropical area of slash-and-burn cultivation. In such areas it is impossible to keep the same piece of land in cultivation for more than a limited span of years because yields fall off sharply and the soils deteriorate fast under the strain of cultivation. Each small group of cultivators must therefore move on from time to time, and may need, say, twenty times as much land at its disposal as is in cultivation at any one time. Suppose that there are two groups possessing an identical material culture and living in the same environmental circumstances. They possess identical mortality schedules but different fertility schedules. It is quite possible to imagine that in one case the push of fertility may be strong enough to raise population to the point at which the group is cultivating at any one time not a twentieth of the total land in their possession but a fifteenth in order to meet their current food requirements. This will mean that each plot has insufficient time in which to recover from the last bout of cultivation. The land will yield progressively less and ultimately the population will be obliged to contract very severely and the land may be damaged permanently. The eroded land and exhausted soils which then occur in the area will have been damaged not for any reason intrinsic to their physical nature but because of the demographic characteristics of the group living there. The other group, with its lower fertility, may be able to continue to cultivate indefinitely and without danger land which was originally identical in nature.

FERTILITY LEVELS IN PRE-INDUSTRIAL SOCIETIES

The assumption that the mortality schedule is necessarily the same in societies living at similar levels of material culture and comparable physical environments is, of course, unrealistic (even if it were justified it would still leave room for many different levels of mortality depending upon the degree of population pressure). But at any time before the Industrial Revolution it is reasonable to assume that mortality was less under social control than fertility, except in the case of the new-born in societies practising infanticide.

It may therefore be helpful at this point to review the range of social customs which may serve to keep fertility below the full potential physiologically available.

In western European societies during the centuries immediately before the Industrial Revolution the most important institutional check upon fertility was probably the conventional age at first marriage for women. This was normally in the middle twenties, and might occasionally be substantially higher. In the parish of Colyton in Devon, for example in the later seventeenth century it was about 30 (see Wrigley, 1966). Similar results have been obtained for the parish of Sainghin-en-Mélantois in northern France at the same period (see Deniel and Henry, 1965). Perhaps the lowest age that is well attested by family reconstitution methods is 22 during the period 1700–79 in a part of western Flanders (see Deprez, 1965A, pp. 615–616). Lower ages have been found occasionally but for special groups within the population rather than for a whole community (19.7 years for the British peerage in 1575–99, for example: Hollingsworth, 1964, p. 25). Since fecundity declines rapidly in the later 30s and last children are seldom borne at a mean age much above 40, it is obvious that a late age at first marriage may drastically reduce the number of children a woman bears. An average age at first marriage of 25, other things being equal, might well reduce total fertility by a third when compared with an average age at first marriage of, say, 18 – by no means uncommon in other places and periods. The effect of a late age at first marriage for women, moreover, is frequently compounded by a comparatively high percentage of women never marrying. These two demographic characteristics appear to be in most cases positively correlated with each other (see Hajnal, 1965).

The age pattern for women at first marriage found in pre-industrial western Europe, however, is seldom if ever found in other pre-industrial societies. In most African and Asiatic pre-industrial societies, for example, a high proportion of all women may be married before they attain the age of 20 and very few women remain spinsters for life (in India in 1891 the mean age at marriage of women was 12.5 years; by the age of 15 the vast majority of girls were married or widowed; see Goode, 1963, pp. 232–235). But there may be many other customs practised which serve to reduce fertility considerably. For example, in a society where adult mortality is high a ban upon the remarriage of widows will operate as a brake upon fertility. The custom of suckling children for two, three or four years after birth (even higher ages have been recorded quite frequently) may reduce marital fertility considerably, especially if combined with a taboo upon intercourse for a period after the birth of a child. It is possible that polygyny may have a similar effect. There is some evidence that women who are partners to a polygynous marriage may normally have a lower fertility than those monogamously married (see, for example, Dorjahn, 1958).

Many other methods of escaping the burdens of a too great fertility may also be found in pre-industrial societies. The extent and efficacy of methods of controlling conception before the nineteenth century are still a matter for debate and their impact may have been relatively slight in most societies (but see Wrigley, 1966, Carr-Saunders, 1922, and Lorimer, 1954). Abortion and infanticide, however, were common in many societies in the past. The latter is especially effective, from the point of view of limiting population growth, if the majority of the infants who are exposed or killed in other ways are female, and this was sometimes the case (Westermarck, 1921, vol. 3, pp. 58, 162 and 166–169). Infanticide may take many forms from the almost casual over-laying of young babies to the killing of those unfortunate enough to be born with physical characteristics which the society deemed abhorrent. In some societies those who were born with teeth already cut or who cut teeth in the upper part of the jaw before those in the lower were put to death: in others all twins (in general roughly one child in every forty is a twin), or those born feet first.

Sometimes the functional necessity for measures of this sort seems clear. For example, amongst some Australian aboriginal tribes, for whom ease of movement was essential, it was courting disaster for any mother to be burdened with more than one child not capable of keeping up with the tribe without assistance (see Spencer and Gillen, 1927, vol. I, pp. 39–40, 221; Krzywicki, pp. 119–144). The too-fertile woman was a figure of fun and if a woman became pregnant again too quickly the new child might well be destroyed at birth or very soon after. This was not a symptom of exceptional lack of feeling on the part of the aborigines. They treated the children whom they dared to allow to live with as much affection as children are treated in other lands and times, but long experience threw up this as a satisfactory solution to a problem which would otherwise have led to constant difficulty and perhaps danger for the group as a whole. In other cases there is no clear functional relationship between the immediate problems facing a society and practices which tend to lower effective fertility, either in the minds of members of those societies or visible to the anthropologist observing them. Nevertheless, social habits of this sort may well have become established, and have proved their selectional worth, because of their effect in making it easier to maintain population well below the maximum possible, just as many of the social habits of animals may prove to be best explained in this way.

A MODIFIED DEMOGRAPHIC MODEL

The model of the inter-relationship between demography and matters of prime interest to geography developed so far is too simple to be of much value. It must now be made to resemble reality a little more closely by modifying it in some particulars. The most important difference between the population characteristics of men and animals can be put very simply. It is legitimate to treat the long-term trend of any animal population as in principle horizontal. If environmental and ecological circumstances do not change, however much the totals of a species may fluctuate from one year to the next, it is reasonable to treat its mean number as unchanging. With men this is not so. Even before the Industrial Revolution very important changes in the material culture of societies took place and radically altered population maxima and optima from time to time. Where men are pegged down at a particular level of material culture for very long periods it is no coincidence that their demographic characteristics show a great similarity to those of animals. Most of the examples quoted in recent paragraphs to illustrate methods of population control were drawn from societies of rather low and unchanging material culture. The tribes of Australian aborigines met and came to terms with these problems in a manner which can be assimilated to general models of animal population behaviour, but matters become more complex once changes in material culture become sufficiently frequent and important.

Suppose, by way of illustration, that an agricultural group is settled in an area of mixed soils and relief in which chalk uplands alternate with clay vales. For many centuries they have been able to make effective use of the light chalk soils for agricultural purposes, but have entered the wooded clay vales only for hunting purposes. Over a period of centuries their population has tended to fluctuate round a given level, the equilibrium level of the fertility and mortality schedules of that society. Then a method of clearing the forest from the clay vales and cultivating the soil in suitable areas is developed or acquired from contact with a more advanced material culture. The capacity of the land to support people is raised considerably by this change, say, to three times its former level. New land for settlement is readily available.

It is possible to imagine population numbers rising to meet the new opportunities made available in this way without any change in the general level and shape of the fertility and mortality schedules, which would simply be displaced to the right, as in Figure 6.5. This would happen because food supplies and other necessities were now more abundant per head at levels of total population which would previously have meant severe hardship and rising mortality. For example, at P before the change in material culture mortality is rising fast. After the change a population of this size creates no difficulties (similar observations hold good for fertility also, of course). Population would continue to rise decade by decade until the new opportunities had been exhausted. Then mortality and fertility would again reflect population pressure and a new equilibrium level of population would be established at Y, replacing the old equilibrium level at X. The general situation, however, would be similar in all respects to that which had obtained previously except that the absolute number of individuals in the population had tripled.

fertility and mortality levels
fertility
mortality
P
X
Y
population size

6.5

fertility and mortality levels
fertility (new hypothesis)
fertility (old hypothesis)
mortality
X
Y
Z
population size

6.6

real incomes per head
Y (and X)
Z
minimum of subsistence
population size

6.7

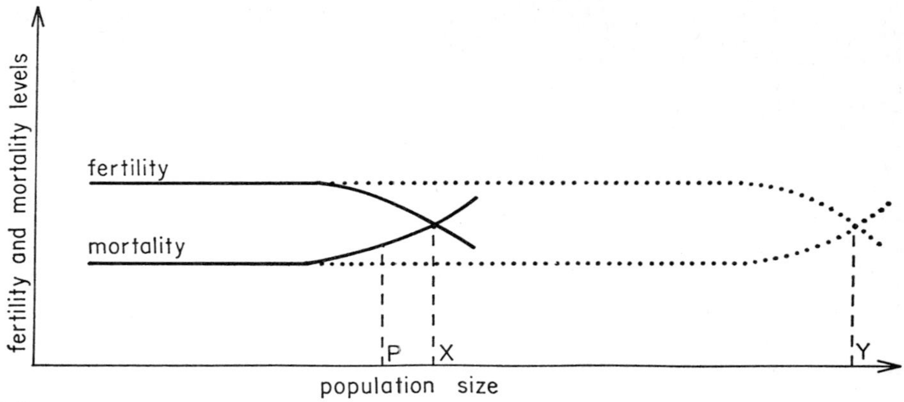

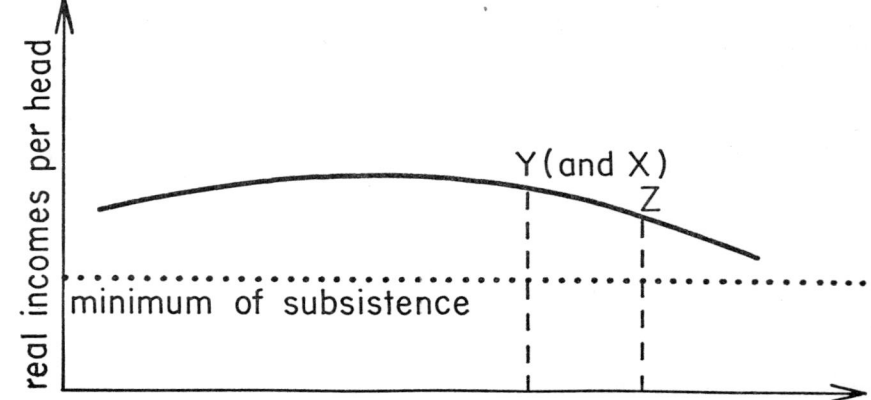

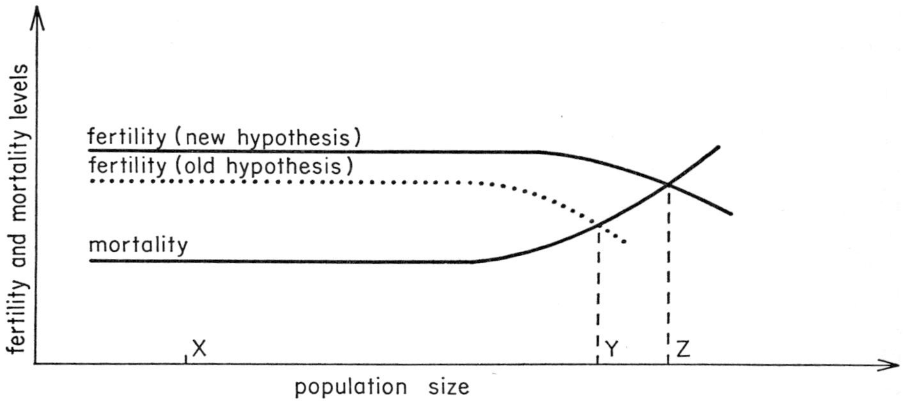

But in the new circumstances created by the change in material culture, things might fall out very differently. It might be, for example, that there would be changes in the level and shape of the fertility schedule in response to the new opportunities. These might occur because age at marriage for both sexes fell as a result of the abundance of good new land; or by the abandonment of practices tending to reduce fertility (perhaps a relaxation of attitude to the remarriage of widows; perhaps the abandonment of the procuring of abortions). Figure 6.6 shows a change of this sort. During the generations in which the clay vales were filling up, the new fertility schedule would carry with it no penalty in the form of lowered living standards such as would have been entailed if the change in fertility had occurred without any new economic opportunity. Nor would the soil itself be endangered by too fierce exploitation at this stage. But if, when the new population total implied by the new fertility schedule were approached at the end of the period of colonizing the clays, the new fertility characteristics had become firmly embedded in the social conventions of the community, a period of great difficulty might well ensue. The new equilibrium total of population at Z would be at a lower standard of living than the old because of the changed fertility schedule. Figure 6.7, taken in conjunction with Figure 6.6 shows how this could come about. The new total of population is much larger than the old, and although supported by a much larger agricultural output, is unable to sustain the living standards which once existed. There might well be danger that the more vulnerable soils (perhaps in the older chalk areas) would be used too intensively under the new pressure of population and become permanently impoverished. The society might then learn once more to adopt social conventions which would result in a lower fertility schedule; or, failing that, become inured to a lower long-term standard of living (the history of Irish population between, say, 1750 and 1900 is extremely instructive in this connection – see Connell, 1950 and 1962; and Drake, 1963).

This illustration might be much more extensively developed and modified but is meant only to establish one point – that a new uncertainty and dynamism is introduced into the situation once the idea of a flexible ceiling is introduced into the model. Indeed the ramifications of this may be very intricate since it is perfectly possible that, given changes of the sort just described, new advances in material culture would be thrown up in response to the growing adversity in which the society found itself at the end of a period of easy expansion. For example, the next important advance might be towards a more intensive agriculture rather than an extensive expansion of the sort experienced in the preceding phase.

The fact that since the Neolithic food revolution changes in the material culture of societies have been frequent has destroyed the simplicity of the model appropriate for a society of static material culture. Both in the study of the developing areas of the world today and in the conventional topics of historical geography the nature of the material culture has rightly come to occupy much attention (see Wrigley, 1965A). It is the dynamic factor in so many situations. Yet the importance of understanding the range of demographic possibilities before a community is not affected by this except in the sense that it is much less easy to deal with the matter briefly and categorically. It is still true that the intensity of land use, the extent to which unpromising land is used, the division of the land between plough and pasture, the degree of subdivision of the land, and so on, are all deeply affected by the interplay between the fertility, mortality and nuptiality of the community. The English peasants of the thirteenth century were pushed closer and closer to the edge of a Malthusian precipice by their demography. Those of the fifteenth century lived better, used the land in different ways (relatively much more pasture – see Postan, 1962), and held the land upon different conditions, in part at least because the demography of the countrymen of Henry V was quite different from that of their forebears 150 years earlier. This state of affairs was not the temporary aftermath of the Black Death but continued for more than a century thereafter. That it later gave way to renewed population pressure and another cycle of events reminiscent in some ways of the thirteenth century does not alter the fact that the demographic situation for a century or more was sufficiently different from what it had been before or was to become later to have a great influence on living standards and land use.

Similarly, it is possible that demography had something to do with the contrasts which existed between different parts of south-east Asia in the period before the recent drastic fall in mortality. There were substantial differences in production per head and in real incomes in this area in spite of the fact that in many instances agricultural techniques, food crops and even environmental conditions were broadly similar. If the populations had all been characterized by the same fertility and mortality schedules it would be reasonable to expect that real income, for example, would have been at much the same level in areas which had much in common in other respects (provided always that there had not been major recent changes in material technology). Indeed it would not be absurd to claim that given similar fertility and mortality schedules it would be largely a matter of indifference whether an area was very fertile or largely barren, whether the climate was favourable to the growth of crops or marginal, since in the course of time the relative pressure of population upon resources would become much the same. Where, however, fertility and mortality schedules vary considerably, there is much greater scope for regional diversity. This is one possible explanation of those situations where peasants living on the richest land prove less wealthy on an average than others living in more adverse conditions of soil or climate. If the demographic characteristics of the latter are such that the equilibrium population total in the poorer areas is close to the optimum attainable with existing techniques, whereas in the former the point of balance between births and deaths is nearer to the maximum, those living in the first area may be less well off than those in the second.[1]

In considering the possible effects of different combinations of fertility and mortality schedules upon the long-term tendency of a pre-industrial population to stabilize in numbers at a particular level two further points should be borne in mind. The first is that what has been written, in addition to having the drawbacks of any schematic model, is likely to prove true only in secular tendency. It was a very common, perhaps a universal, feature of pre-industrial societies that in the shorter term there were very large fluctuations in births and deaths and to a lesser extent in population totals as a result of the ravages of famine and disease. A roughly cyclical movement of population totals, often with a periodicity of about thirty years, has been found in many parts of pre-industrial Europe, for example. A typical sequence of events was a bad harvest, or still more two bad harvests on the run, leading to undernourishment and accompanied or followed by serious epidemic infections. Such a crisis often reduced populations by a fifth or a quarter (see especially Goubert 1960: also Drake, 1964). There might then follow a generation of recovery with a surplus of births over deaths in most years before the abrupt downswing in the cycle again took place. Cycles of this sort might occur whatever the relative pressure of population upon resources, but would be much more severe in areas where the long-run equilibrium total of population implied by the fertility and mortality schedules was close to the maximum.

Secondly, it is important to be aware of the secondary effects of the demographic characteristics of a given population. For example, the age composition of a population is determined by its fertility and mortality schedules (for the sake of simplicity of exposition, I ignore migration, though it is often, of course, of great importance). This may be a fact of some importance in helping to understand living standards and patterns of expenditure, which in turn have an influence on land use and the structure of demand for industrial products. The specimen stable population figures published by the United Nations (United Nations, 1956B, pp. 26–27) may be used to illustrate this. These show that in a population in which the gross reproduction rate is 2·5 and the expectation of life at birth is 30 years the percentage of the total population between 15 and 59 years will be 57·6 per cent. If fertility remains unchanged but expectation of life at birth improves to 60 years the percentage of people in this age range falls to 52·6 per cent. These figures are for stable populations, that is populations in which the implications of fertility and

---

[1] The opposite may also occur, of course. There are many examples of poor agricultural or pastoral areas which suffered from rural under-employment, developing rural industries to occupy idle hands, encouraging thereby a further growth of population, and ending with a large but miserably poor population (see, for example, Deprez, 1965A and 1965B; and also an interesting discussion of the same general issue in Thirsk, 1961).

mortality schedules have been fully worked out. Any change in mortality as drastic as that just mentioned would mean that several generations must elapse before the new stable age structure was established. Nevertheless it represents the type of change which has taken place recently in many developing countries. If their present high fertility levels do not fall and mortality continues to improve the dependent 'tail' in the population must grow. If the income of the breadwinners does not rise, they will be hard pressed to provide for their more numerous dependants. More and more of the total demand for goods will be devoted to the basic necessities of life, and above all food. If population pressure is in any case intense the situation will be aggravated by this unfavourable development in age structure. If, on the other hand, the fall in mortality is soon followed by a decline in fertility sufficient to reduce the gross reproduction rate from 2·5 to 1·5, then the percentage of population 15–59 in the stable population which would ultimately develop would be very close to the original figure, but now there would be a far higher percentage of people of 60 or more and a lower percentage under 15 (in the first case 36·9 per cent 0–14, 57·6 per cent 15–59, 5·5 per cent 60 plus: in the second case, 28·2, 58·7 and 13·1 per cent respectively).[1]

Again, within any pre-industrial population there will probably be important differences between one region and another, and between town and country. It is useful to bear it in mind, for example, that in addition to the economic problems of maintaining large cities in pre-industrial times, associated with small and variable food surpluses in rural areas and the difficulties of moving food from country to town, there were also important demographic checks upon unrestricted town growth. If, as was frequently the case, mortality in the towns was at such a high level that the urban populations were only maintained by a constant flow of immigrants from the countryside, it is clear that if urban populations had risen above a certain fraction of the whole their birth deficit would have outweighed any surpluses occurring elsewhere and would have caused total population to fall. Pre-industrial urban populations were probably seldom large enough to lend much substance to this point, though it may turn out to be of importance in considering the relationship between London and the rest of England in the late seventeenth and early eighteenth centuries, when London was growing quickly, but the country as a whole grew little if at all.

More generally it is useful to know something of the results of different combinations of fertility and mortality on rates of growth of total population. This is particularly important in any topic related to the contemporary growth of population in the developing countries, sometimes referred to as the 'population explosion'. However steep the fall in mortality in these countries

---

[1] It may be noted that where a population is both stable and unchanging in total number, the percentage of the population of working age, say 15–59, is remarkably constant, whatever the combination of fertility and mortality which brings this about.

and however slight the changes in fertility, rates of growth of population will almost certainly never exceed five per cent per annum, and are very unlikely to reach even four per cent. Indeed rates much in excess of three per cent are uncommon. These are still formidable figures (three per cent per annum means a doubling of numbers in 23 years), but it is a fortunate circumstance that human physiology, unlike that of most animal species, does not permit faster rates of growth. Otherwise the economic prospects of some parts of Asia, Africa and South America today might be poor indeed.

DEMOGRAPHIC MODELS AND POST-INDUSTRIAL SOCIETIES

I have now touched upon the implications of demographic models when the material culture may be regarded as unchanging, as with Australian aborigines or South African Bushmen; and have broadened the discussion to enable agricultural economies to be considered also, those economies in which an advance in the techniques of cultivation, or transport, or civil administration may change the level of the ceiling of population and so permit renewed growth, perhaps with a new set of fertility and mortality schedules. Further modifications to the simplicity of the original model are necessary to embrace post-industrial economies. Whereas in hunting and fishing economies one may posit an unchanging ceiling, and in agricultural societies a ceiling which may inch up gradually or even rise with a jerk from time to time but which is still a viable concept for certain analytic purposes (it was, after all, of such a society that Malthus himself wrote), the very concept of a ceiling to population growth is inappropriate to a country which has undergone a full Industrial Revolution. In industrialized countries today indeed it is very difficult to give much precision to the idea of an optimum or maximum population. Certainly it is true to say that the criteria which can best be used in the discussion of this issue are no longer closely tied to the food base and the land. Populations have almost everywhere risen immensely in the last two centuries but living standards have also risen fast. Mortality, especially in the younger age-groups, has fallen precipitately. Exogenous mortality[1] has ceased to affect general mortality significantly. Fertility has also fallen very substantially and family size has become a matter of conscious individual choice to a degree which has few if any parallels in pre-industrial societies.

[1] Exogenous mortality consists of deaths from infectious diseases such as tuberculosis, smallpox, dysentery, typhus, malaria and plague, which caused immense loss of life until recently. Endogenous mortality, on the other hand, is made up of deaths from circulatory defects, organic malfunction, and (perhaps) cancer, which are not produced by the entry into the body of harmful viruses and bacteria.

In industrialized countries the vast majority of people live in cities: only a tiny fraction still make a living from the land. In these circumstances the mortality and fertility schedules need not change significantly as population rises, nor if they change are the changes necessarily in conformity with what might be expected on the assumptions embodied in the earlier models. Increasing density of population need not mean a rise in mortality. Changes in fertility may occur in either direction. Fertility and mortality schedules in which fertility and mortality levels are plotted against population density might take the form of a horizontal straight line as plausibly as any other, at least within the range of densities known today. It no longer makes sense even as a theoretical exercise to construct fertility and mortality schedules of the type to be found in the figures earlier in this chapter in which the shape of the curves is directly related to population size. If plotted against time rather than population density the mortality schedule might show a small decline as the years pass, while the fertility schedule would be indeterminate on present evidence. Changes in material culture which in very primitive societies may be so slight over long periods as to be negligible, and which in pre-industrial societies which cultivated the land, though swifter, were still halting and slow, have now become so sweeping and frequent that models which depend upon the absence of insignificance of these changes are of no utility. Nor is the level of real income per head closely connected with population density in the manner familiar in pre-industrial societies. Demographic model building for the study of contemporary problems must therefore take a different form.

In industrialized countries very little life is lost before men and women have passed through the reproductive ages. No foreseeable change in mortality rates, therefore, can greatly change rates of population growth and attention is naturally concentrated on fertility, except in the study of certain special problems. If, for example, current fertility levels in England are maintained till the end of the century it is possible to predict within small margins of error (largely the product of uncertainty about migration) how large the population in the year 2000 will be, not only in total but by age-groups. Mortality changes will probably be very small and can be largely ignored, saving a nuclear disaster or other unpredictable catastrophe. In any pre-industrial society, harassed by disease and famine, to ignore possible changes in mortality in this way would have meant inviting large errors in forecasting. He would be a foolhardy man who placed much store upon the continuance of present fertility rates for a period of more than a generation, but by making a series of alternative assumptions about fertility (and migration and mortality, of course, though the second is of slight importance) a range of possible population sizes can be ascertained. This in turn can be coupled to other assumptions; for example about the ratio between school-teachers and children of school age. Estimates of future needs of this sort cannot be very exact because several assumptions are involved and errors may be compounded. The number of schoolteachers needed is a function not merely of the number of children to be taught but also of average class size, and so on. But population projections are of importance none the less. Reasoned estimates are much more valuable than intuitive hunches. The simple models of demographic change upon which estimates of this sort are based are of great importance to those geographers interested in the industrialized countries (see, for example, United Nations, 1956A). Studies of resource use, of changing land use, of water conservation, of urban growth, of new patterns of population distribution, of regional growth and decay, and of migration, all require a knowledge of the population movements of the immediate past, of the present demographic situation, and of the changes which are likely in the near future. If population in general can reasonably be thought of as central to the whole range of issues with which human geographers occupy themselves (see Wrigley, 1965B) then a knowledge of demography can hardly fail to be of importance also.

THE DEVELOPING COUNTRIES

The case for an understanding of demographic models needs least advocacy perhaps in the case of the countries of Asia, Africa and Latin America which are nowadays usually referred to as the developing countries. Many of them are placed in a position of grave difficulty by their present demographic constitution. The nature of the problem is well known. In these areas in the last generation or so mortality has fallen precipitately because of the use of modern public health measures and powerful drugs. Fertility has changed little. In these circumstances population increase at the rate of two or even three per cent per annum frequently occurs. If the economies of these countries had changed as completely as their mortality rates there would not necessarily be any cause for alarm in these developments. Unfortunately the changes have not always been accompanied by rapid economic growth. Whereas in Europe the mortality changes which occurred in the early decades of the Industrial Revolution down to the end of the nineteenth century were slow and were associated with economic growth rather than medical advance, in the developing countries today this is not the case. In them health and wealth have not always moved in step.

The danger is obvious. If a drastic fall in mortality occurs without any change in fertility and without revolutionary changes in the economy, the ultimate adjustment may be exceedingly severe. Mortality in these circumstances remains ultimately density-dependent in the classical manner and the mortality schedule must turn up very sharply if population continues to grow. The gloomiest fears have been entertained on this score. There are good reasons for supposing that the worst of these fears will prove groundless,[1] but in the meanwhile the life of these countries is dominated by the demographic situation. In Egypt, Indonesia and Ceylon today, for example, it must be the starting point for any adequate discussion of economy, society and human geography, and the implications of the current state of affairs can only be set out adequately in the light of a knowledge of the functioning of demographic models. For example, the population of Ceylon rose by 31·2 per cent in the years 1945–55; that of Japan by 23·3 per cent in the same period (United Nations, 1963, pp. 154–155). It might seem that in these two countries, both densely populated and both experiencing a rather rapid rise in population, the population problems were broadly similar. Yet demographically they had little in common apart from the rate of increase, for the rise in Japanese population in that period was largely the result of the age composition of the population and the repatriation of Japanese nationals after the war. The net reproduction rate was not very much above unity and was falling – that is to say that if the prevailing trends in fertility and mortality had continued population must soon have ceased to grow quickly. In Ceylon, on the other hand, the demographic situation was quite different. Any forecast of future population based on this knowledge was bound to include a considerable period of time during which the rate of growth would increase. Only a substantial and immediate fall in Ceylonese fertility rates could have falsified such a prediction.

The general possibilities of economic growth in countries like Ceylon is closely bound up with demographic developments. Each government in attempting to foster an economic 'take-off' is obliged to wrestle with the tendency for the rate of growth of population to accelerate. The Indian planners for more than a decade have been caught in the same predicament as the Red Queen. They must always run faster than they were running a little earlier simply in order to keep up. If they fail to keep the economy growing

---

[1] Reliable fertility statistics are few and far between for the developing countries. Undoubtedly in many of them fertility has not fallen at all as yet and may even in some cases be rising. But the crude birth rates for 1959–64 for Formosa, Hong Kong and Singapore do suggest the possibility of rapid change in favourable circumstances. Hong Kong and Singapore are, of course, special cases, but the Formosan figures, since they refer to a population with a large rural component, may be very significant.

Crude birth-rates per thousand total population

	1959	1960	1961	1962	1963	1964
Formosa	41·2	39·5	38·3	37·4	36·3	34·5
Hong Kong	35·2	36·0	34·2	32·8	32·1	29·4
Singapore	40·3	38·7	36·5	35·1	34·7	32·1

Source: *United Nations Monthly Bulletin of Statistics*, July 1965, p. 5.

more quickly than the population real incomes per head will decline, demand for the consumer goods produced by industry may be expected to fall, capital saving to become still more difficult, and so on. The chances of success grow slim. Coale and Hoover's analysis of India's economic prospects (Coale and Hoover, 1958: see also Schwartzberg, 1963) gains greatly in depth and subtlety because of their care to consider a number of possible courses of demographic change and to show how these might affect economic growth in India.

General economico-demographic models can generate a wide range of important results. For example, in order to relate future increases of population in India to the degree of pressure on the land, it is not enough simply to be aware of the probable increase in total population over the next decade or quarter century. What may be called the secondary as well as the primary material produced by models of this type is useful. It would be naïve to suppose that a population increase of 50 per cent over the next 25 years would necessarily involve an increase in rural population of the same order of magnitude. If real incomes per head did not increase this might prove a valid assumption, with all that it implies for further subdivision of holdings and rural underemployment. If, on the other hand, the model shows that this rise in population is consonant with rising real incomes, the percentage of people living on the land will fall. Indeed a sufficiently steep rise in real incomes might imply that the absolute numbers of the rural population would stabilize within this period, or even decline. It would probably emerge from the consideration of alternative possible rates of growth that the faster the rate of increase, the slower the decline of the percentage of the population directly dependent on the land for income. Since the highest percentages on the land might be associated with the largest absolute population totals, the range of possible rural population totals would be wide. This is an issue of great significance for many aspects of social and economic life in developing countries. The merits of the assumptions built into any one model used in this connection must be a subject for dispute. But the general merits of tackling the problem in this way seem clear. The equations expressing the relationship between rates of population growth, trends in income per head, the structure of demand, and the fraction of the labour force employed on the land, may be mistaken, but at least they make it possible to explore the effect of changes in one or more variables upon the other variables in the system in a coherent and logical fashion. No human geography of an area of this sort which seeks to analyse as well as describe can afford to neglect the opportunities offered by models either of the narrow or the broader type.

It may be appropriate to note further here that models expressing the relationship between the demography of a society and other economic, social and geographical variables are inevitably of importance in studies of Industrial Revolution and the 'take-off' generally. For example, if it is agreed that at least until the 1780's the growth of industrial production in England was sustained largely by the strength of home demand, then it follows that the demographic constitution of the country must have embodied some unexpected features. The growth of home demand was comparatively slow and related to slowly rising real incomes. If Malthus had been right in supposing that the mass of the people would marry earlier and bring up larger families when economic circumstances were favourable, it is hard to imagine real incomes rising steadily for several successive generations. If they had behaved as he supposed (and there is much evidence suggesting that his model holds good for, say, the sixteenth century and at other earlier periods), events could not have fallen out as they did (see Wrigley, 1966 for a fuller discussion of this issue). More generally, in as much as the level of real incomes per head is important in studies of this type, demography must also be important since population size must enter into any calculation of income per head as one half of the ratio.

CONCLUSION

In describing and discussing where people live, how they obtain a living, and in what numbers they are to be found, geographers have become more and more sophisticated as the years have passed. The attempt to deal with these issues by referring aspects of population density and distribution directly to features of the physical environment was seldom considered satisfactory even in the infancy of modern human geography. Models of economic activity and of general spatial organization have been called into play increasingly in recent years to supersede or supplement older ideas, especially when dealing with areas which have experienced an Industrial Revolution. Sociological models and models of urban function have made an appearance (some are dealt with in Chaps. 7 and 9). Models which are derived from the consideration of formal geometrical properties are invoked to foster the understanding of the development of route networks and other problems. And many others might be added to the list. The use of models drawn from so many different sources, while it greatly increases the range of conceptual tools available, brings in its train other difficulties. For example, it is desirable that all should be subsumed under some overarching general model so that what each implies for the others can be examined (hence the great influence which the writings of August Lösch have exerted on geographers of the last two decades: his was a brave attempt at an important general solution). An ideal general model should make it possible to examine the effect which change in any one aspect of the segment of reality subsumed within the model will have upon all others.

The justification for introducing demographic models into the normal ambience of analysis and thus complicating matters still further lies partly in the rapidly growing body of empirical evidence which shows the importance of demography to many points of interest to geographers, and partly in the rapid improvement in the demographic models themselves in recent years. The Population Branch of the United Nations Bureau of Social Affairs has played a notable part since the war in promoting work of both types. Equally notable has been the work on theoretical, historical and contemporary demography carried out at the *Institut National d'Etudes Démographiques* in Paris. As a result of work done under the aegis of these two bodies and by a host of individual scholars there is ample evidence available both from empirical studies and theoretical works that it is impossible to treat population as a simple, dependent variable. Populations have, as it were, a life of their own, in part a reflection of their social and physical environments, perhaps in an absolute sense wholly their product, but in any proximate analysis best considered as a partially independent variable which is capable of modifying profoundly many aspects of economy, society and geography. To neglect demography is to shut oneself off from one of the most important sources of insight into the question of where people live, in what numbers and by what means.

REFERENCES

Note: this bibliography is restricted largely to works referred to in the chapter but includes a few other works of general usefulness.

ACKERMAN, E. A., [1959], Geography and demography; In *The Study of Population* (Ed. Hauser, P. M. and Duncan, O. D.), (Chicago), 717–727.
BEAUJEU-GARNIER, J., [1956–58], *Géographie de la population*; 2 vols, (Paris).
BEAUJEU-GARNIER, J., [1966], *Geography of Population*, (London).
BOSERUP, E., [1965], *The Conditions of Agricultural Growth*.
CARR-SAUNDERS, A. M., [1922], *The Population Problem*, (Oxford).
COALE, A. J. and HOOVER, E. M., [1958], *Population Growth and Economic Development in Low-income Countries: a case study of India's prospects*, (Princeton).
CONNELL, K. H., [1950], *The Population of Ireland 1750–1845*, (Oxford).
CONNELL, K. H., [1962], Peasant marriage in Ireland: its structure and development since the famine; *Economic History Review*, 2nd ser., xiv, 502–523.
COX, P. R., [1959], *Demography*; 3rd ed., (Cambridge).
DARWIN, C., [1869], *On the Origin of Species*; 5th ed., (London).
DARWIN, C., [1901], *The Descent of Man*; new ed., (London).
DARWIN, C., [1958], *Essay of 1844*; In *Charles Darwin and Alfred Russel Wallace. Evolution by natural selection*, Pub. for XV International Congress of Zoology and the Linnean Society of London, (Cambridge).
DENIEL, R. and HENRY, L., [1965], La population d'un village du Nord de la France, Sainghin-en-Mélantois, de 1665 à 1851; *Population*, 563–602.

DEPREZ, P., [1965A], The demographic development of Flanders in the eighteenth century; In *Population in History* (Ed. Glass, D. V. and Eversley, D. E. C.), (London), 608–630.
DEPREZ, P., [1965B], *Evolution démographique et évolution économique en Flandre au 18e siecle*; mimeographed paper for Section VII, Third International Economic History Conference, (Münich), 1965.
DORJAHN, V. R., [1958], Fertility, polygyny and their interrelations in Temne society; *American Anthropologist*, 60, 838–860.
DRAKE, K. M., [1963], Marriage and population growth in Ireland, 1750–1845; *Economic History Review*, 2nd ser., xvi, 301–313.
DRAKE, K. M., [1964], *Marriage and Population Growth in Norway, 1735–1865*; unpub. Ph.D. thesis (Cambridge).
GLASS, D. V. and EVERSLEY, D. E. C., (Eds.), [1965], *Population in History*, (London).
GOODE, W. J., [1963], *World revolution and family patterns*, (New York).
GOUBERT, P., [1960], *Beauvais et le Beauvaisis de 1600 à 1730*; S.E.V.P.E.N.
HAJNAL, J., [1965], European marriage patterns in perspective; in *Population in history* (Ed. Glass, D.V. and Eversley, D. E. C.), (London), 101–143.
HATT, P. K., [1952], *Backgrounds of Human Fertility in Puerto Rico*, (Princeton).
HAUSER, P. M. and DUNCAN, O. D. (Eds.), [1959], *The Study of Population*, (Chicago).
HERLIHY, D., [1965], Population, plague and social change in rural Pistoia 1201–1430; *Economic History Review*, 2nd ser., xviii, 225–244.
HIMES, N. E.,]1936], *Medical History of Contraception*, (Baltimore).
HOLLINGSWORTH, T. H., [1964], The demography of the British peerage; supplement to *Population Studies*, xviii.
JAFFE, A. J., [1951], *Handbook of Statistical Methods for Demographers*; (Washington).
KRZYWICKI, L., [1934], *Primitive Society and its Vital Statistics*, (London).
LEIBENSTEIN, H., [1954], *A Theory of Economic-Demographic Development*, (Princeton).
LEVY, M. J. and WESTOFF, C. F., [1965], Simulation of kinship systems; *New Scientist*, xxvii, No. 459, 571–572.
LORIMER, F. and others, [1954], *Culture and Human Fertility*, (UNESCO).
LOTKA, A. J., [1934–39], *Théorie analytique des associations biologiques*; 2 parts, (Paris).
LUND STUDIES IN GEOGRAPHY, [1957], Ser. B, Human Geography, No. 13, *Migration in Sweden* (Lund).
POSTAN, M. M., [1950], Some economic evidence of declining population in the later Middle Ages; *Economic History Review*, 2nd ser., ii, 221–246.
POSTAN, M. M., [1962], Village livestock in the thirteenth century; *Economic History Review*, 2nd ser., xv, 219–249.
POSTAN, M. M. and TITOW, J., [1959], Heriots and prices on Winchester estates; *Economic History Review*, 2nd ser., xii, 392–417.
PRESSAT, R., [1961], *L'Analyse démographique*, (Paris).
SAUVY, A., [1952–54], *Théorie générale de la population*; 2 vols., (Paris).
SCHWARTZBERG, J. E., [1963], Agricultural labour in India: a regional analysis

with particular reference to population growth; *Economic Development and Cultural Change*, xi, 337–352.
SPENCER, B. and GILLEN, F. J., [1927], *The Arunta: a study of a Stone Age people*; 2 vols, (London).
THIRSK, J., [1961], Industries in the countryside; In *Essays in the Economic and Social History of Tudor and Stuart England* (Ed. Fisher, F. J.), (Cambridge), 70–88.
UNITED NATIONS, [1949], Population Branch of the Department of Social Affairs, Report No. 7, *Methods of using census statistics*.
UNITED NATIONS, [1953], Population Branch of the Department of Social Affairs, Report No. 17, *The determinants and consequences of population trends*.
UNITED NATIONS, [1956A], Population Branch of the Department of Social Affairs, Report No. 25, *Methods for population projections by sex and age*.
UNITED NATIONS, [1956B], Population Branch of the Department of Social Affairs, Report No. 26, *The aging of populations and its economic and social implications*.
UNITED NATIONS, [1963], *Demographic Yearbook*.
WATT, K. E. F., [1962], The effect of population density on fecundity in insects; *General Systems Yearbook*, 7, 231–244.
WESTERMARCK, E., [1921], *The History of Human Marriage*; 5th edn., 3 vols., (London).
WOLFENDEN, H. H., [1954], *Population Statistics and their Compilation*; 2nd edn., (Chicago).
WRIGLEY, E. A., [1962], The supply of raw materials in the Industrial Revolution; *Economic History Review*, 2nd ser., xv, 1–16.
WRIGLEY, E. A., [1965A], Changes in the philosophy of geography; In *Frontiers of Geographical Teaching*, (Ed. Chorley, R. J. and Haggett, P.), (London).
WRIGLEY, E. A., [1965B], Geography and population; In *Frontiers of Geographical Teaching*, (Ed. Chorley, R. J. and Haggett, P.), (London).
WRIGLEY, E. A., [1966], Family limitation in pre-industrial England; *Economic History Review*, 2nd ser., xviii, 82–109.
WYNNE-EDWARDS, V. C., [1962], *Animal Dispersion in Relation to Social Behaviour*, (Edinburgh and London).

CHAPTER SEVEN

Sociological Models in Geography[1]

R. E. PAHL

The Poet Wonders Whether the Course of Human History is a Progress, a Drama, a Retrogression, a Cycle, an Undulation, a Vortex, a Right- or Left-Handed Spiral, a Mere Continuum, or What Have You. Certain Evidence is Brought Forward, but of an Ambiguous and Inconclusive Nature. (Title to Ch. 18 of *The Sot Weed Factor* by JOHN BARTH)

The important question for the sociologist is not whether he should interpret observed human behaviour in terms of models, but what sort of model he should employ. (JOHN REX)

Our understanding of complex reality depends on the questions we ask: these in turn depend upon the culture into which we have been socialized and the viewpoint or discipline in which we have been trained. A fact is only a fact in terms of a theory and problems are defined as such within an implicit or explicit theoretical framework. That such conceptual schemes are more often only tacitly accepted in empirical investigation should be a matter of deep concern to social scientists. It seems clear that any attempt to explain the ordering and pattern of human activity is constricted if forced into a monocausal framework. Geographers in particular, who may attempt to 'explain' medieval field-systems or economic growth in modernizing countries by 'the facts' should see the need for a more rigorous theoretical framework. Seemingly every generation is obliged to rewrite the past in terms of its own values and ideologies. Yet however strongly we may react against a naïve historicism it cannot be denied that an understanding of the past is internalized in the individual – particularly in certain charismatic leaders – and may be partly responsible for spatial and temporal diversity of organized human activity.

[1] I am grateful to Geoffrey Hawthorn of the Department of Sociology at the University of Essex and my fellow sociologist Rex Taylor at the University of Kent at Canterbury for helpful comments on this essay. We were all trained as geographers.

INTERNALIZED MODELS

Everyday behaviour is based on a generalized and structured concept of reality. We react to situations and persons in accordance with our stored experience; only in unusual situations without precedent do we find ourselves at a loss, that is, when our mental models' approximation to reality is slight. A resistance to the acceptance of alternative models at the individual level is normal; at a more general level such resistance may take the form of ethnocentrism, which clearly demonstrates, on a comparative basis, the nature of a culturally bestowed model. It would be interesting to consider the degree of ethnocentricity as between nations and also between groups within nations. Certainly it is difficult to be objective about one's own cultural or sub-cultural values. Academics, who might be expected to be more sympathetic to the relativity of their socially bestowed attitudes, and who may discuss with considerable sophistication the cultural differences between, say, the French peasant and the English manual worker, may be surprised and somewhat shocked if a son of the latter has difficulty in accepting the different values of what might be a middle-class dominated residential university.

If we consider the internalized models we may have of the British nation state, it is possible that collectively the nation is suffering from the delay in rejecting an outdated model of Britain as an Imperial power. The newly independent states, once part of the British Empire, have in many cases rejected some of the distinctive cultural and political values of the administrative elite, often partly formed by the academics who trained them. This attack on a model, which many assumed did not need to be defended, is paralleled by the widespread acceptance of some form of socialism over most of the world for most people, leaving a much diluted form of capitalism to a geographically small but economically rich area in Europe and North America, with isolated outposts in Australasia and Japan. This double attack has had the effect of administering something of a national shock, which needs more than the growth of coffee drinking and central heating over the past fifteen years to cure – however much a hot drink and keeping warm may be the traditional cures for shock.

These attacks on our internalized models of the nature of our own nation in relation to the rest of the world, and the attempts to come to terms with the changed situation, have had repercussions on our whole intellectual climate and throughout our educational system. It is therefore not surprising that in this situation those responsible for teaching and research in geography should question some of the basic assumptions which had become fossilized in the conventional wisdom of the textbooks. Philip Abrams' strictures on the teaching of history are equally applicable to the teaching of geography – 'Sometimes one feels that . . . teachers and authors are incapable of recognizing a controversial subject when they see one – the number of fourth- and fifth-form histories that treat unions as things that simply "grew" or slumps as things that simply "set in" is truly remarkable.' Similarly in geography controversial aspects of economic development and regional planning are played down. Alternative models to explain, say, the local climate are not always matched by a similar sophistication when considering other aspects of economic development. One textbook will have to suffice as an example – *The Mediterranean Lands* by D. S. Walker, which just happened to be the first that came to hand.

In his discussion on the economic development of Spain, Walker states that 'The depression of 1929 and onwards coincided with a period of political experimentation which provoked such violent opposition that civil war broke out in 1936.' This led, we are told, to the near collapse of the Spanish economy. The Second World War prevented aid coming from abroad and 'Even after the end of hostilities she remained isolated economically and ostracized politically.' (The reasons for this are not explained.) Thus, Walker continues, 'In the light of these events the adoption by the present régime of an autarkic policy is understandable; they have made a virtue of necessity' (p. 110). And this is a sixth-form textbook! Turning to Portugal, where the main problem is said to be overpopulation, Walker notes: 'The traditional remedy is emigration and Portugal is fortunate in possessing outlets in Africa and cultural connexions with Brazil which make it possible for the emigrant to acclimatize himself rapidly' (p. 142). It appears that most emigrants prefer Brazil 'to the colonies'. (The reasons for this are not explained.) That the fortunate outlet could also be described as a repressively paternalistic forced labour state is not even hinted at, although James Duffy's *Portuguese Africa* was published in 1959, the year before Walker's book appeared.

THE MYTH OF A VALUE-FREE GEOGRAPHY

When geographers move away from the mapping of a static situation to consider aspects of change they inevitably become involved with values – both by the problems they choose to consider and the interpretation of the 'facts' they present. The interpretation of socialist economic development may vary according to the political sympathies of the geographer concerned: failure of the harvest in China might be held as an indictment of the political system, whereas a similar catastrophe in India may simply be seen as an Act of God. Clearly the issue is not as clear cut in actual practice, but there is need for considerable alertness in spotting the ideological bias. It is possible that the spatial element in a given situation may dictate action more forcibly in a capitalist society, whereas a socialist or centrally planned economy has a greater choice in the way resources are developed. Such greater choice would carry with it the possibility of greater error. When the state makes the wrong investment decision this is held to be an indictment of one system, yet when an individual firm goes bankrupt this is held to be a justification of the other. Certain geographers in France are held by some to be suspect because they are 'doctrinaire marxists'. This may be a justifiable accusation, but it would be nice to hear similar accusations of 'doctrinaire capitalists'. If we take as an example the problems of regional planning in Britain, it is clear that there can be strong divergence between those who project present trends, with some sort of historical inevitability, to predict the future, and those who instead propose alternative policies based on political interference with the so-called 'free play of market forces'. This distinction between 'drifters' and 'counter-drifters' is not always made explicit, although individual geographers have openly committed themselves by membership of technical advisory groups of all three political parties and there is by no means agreement by what criteria an abstract, 'ideal-type' geographer should advise the government of the day.

The geographer, as any other social scientist, cannot avoid being socialized into a specific culture – or, more accurately, sub-culture – at a particular period of time. Family, educational institutions, peer group, the mass media and so on have provided him with a certain model of the nature of society. By the material he chooses to teach from and the research problems he investigates, certain attitudes about the nature of society become incorporated into a pattern of thought, a system of values, which may or may not be made explicit. Whereas the relations between phenomena of the natural world can only be explained from the outside – and the geographical distribution of individuals and social groups can be observed in a sense as natural phenomena – relations between phenomena of the human world are relations of value and purpose.

SOCIOLOGICAL MODELS IN GEOGRAPHY

In the context of the other contributions to this symposium it is unnecessary to consider how models generally have been applied in sociology. That in itself is a fit subject for another symposium, aimed specifically at sociologists, who may be unfamiliar with the more refined mathematical techniques and their applicability to quantifiable data. Certainly the theory of games and other probabilistic models have considerable potential for developing certain branches of sociology, particularly those which view problems in a sub-social or materialistic way. It is perhaps significant that at the time of writing the most recent issue of the *European Journal of Sociology* is devoted to five articles on simulation models. Evaluation of the various elements in complex reality requires the use of the most rigorous and refined techniques available. The rather arid discussion about terminology – whether quantified theories should be called mathematical models (Brodbeck, 1959) and so on – need not detain us. The importance and relevance of such techniques in geography can be left to others in this book. Our concern here is with specifically *sociological* as opposed to mathematical or any other sort of models.

Although an interest in groups and 'societies' can, as with so many other interests, be traced back to the Greeks (Barnes and Becker, 1961), sociology is a very modern science. Social information does not constitute a science and sociology could not exist without someone first creating a model. 'The important question for the sociologist is not whether he should interpret observed human behaviour in terms of models, but what sort of model he should employ' (Rex, 1961, p. 60). However, it is one thing to discuss sociological models as such, about which sociologists may disagree amongst themselves, and it is quite another matter to relate such models to geography. Looking at the problems defined by one discipline with the conceptual tools of another is not easy within the existing intellectual climate, although this book is of itself a welcome sign of change. The social sciences have suffered individually from a parochial concern to defend the importance of the specific 'factors' with which they are most concerned. This 'fallacy of misplaced concreteness' has been attacked by Parsons – 'the effect of this tendency to "empirical closure" of a system is to make its application to any field, especially a new one, a rigidly simple question of whether it "applies" or not. Applications are interpreted in "all or none" terms – it is either a case or not' (1945, pp. 221–222). Factor theories are justifiable if they can demonstrate their effectiveness in solving empirical problems. Geographers would probably be the first to admit that in problems such as are involved in the economic development of some of the new states an understanding of the 'factors', such as the physical resource base or the spatial component, is of limited value. Similarly, economists accept the limitations of their more explicitly stated models. A crude reductionism heaps together perhaps the most significant elements in the situation in the geographer's 'other human factors' or the economist's escape clause 'other things being equal'. Both geographers and economists acknowledge the importance of the social factor in economic development but are not always clear how this can be understood in terms of some overall framework. Economists have different models which they apply in different situations – for example an economy of perfectly competitive enterprises (without monopoly), an economy of 'imperfect' competition – that is an economy containing some monopolistic power – or a socialist economy. They do not expect any one model to fit an actual society exactly. When we turn to sociology, as the author of a recent textbook puts it,

> Instead of making various simplifying assumptions about social institutions and the distribution of economic power and then asking how certain variables are interrelated within such a purposely simplified framework, the sociologist focuses attention on the framework itself and asks what the institutional patterns are

within which economic activity is carried on, in what ways they are alike in all societies, in what ways they differ from one society to another and how a society comes to change its institutional pattern. (Johnson, 1960, p. 210.)

Indeed the concept of the 'economy' can be analysed in terms of a specifically differentiated 'sub-system' of 'society' within the framework of a 'social system' (Parsons and Smelser, 1956). Social information thus gains considerably more significance in terms of such conceptual models as a social structure or a social system.

This should by no means be seen as an attack on geography as such, nor on any form of specialization, which, in view of the complexity of the subject matter of the social sciences, is both necessary and desirable. 'What is needed is a close collaboration between sociologists and other social scientists: and such collaboration implies both that the sociologist should have competence in one or other of the special social sciences, and that specialists should have some knowledge of general sociology' (Bottomore, 1962, p. 21). Clearly this is not the place to provide a geographer's introduction to sociology, although a book on this subject is certainly needed. Rather it is hoped that by discussing the sociologists' fundamental conceptual models, the geographer may gain a tool which may be of value in both formulating and clarifying his own empirical research, and which may also provide a framework into which other sociological work may be placed. It will be necessary to move to a fairly high level of abstraction in the discussion which follows. But it is perhaps useful at this point to pose the key question, which would presumably be of equal interest to both sociologists and geographers and on which we may test the conceptual models to be discussed: *Why are some societies different from others in the way they utilize their resources and distribute themselves in space, and, further, given a particular pattern of activities in relation to certain resources, what leads to change?*

As emerged above when discussing the role of values, it is the element of *change* in a situation which would appear to force the geographer away from a static, snap-shot approach. It is my view that mathematical and experimental models cannot be fully exploited in either geography or sociology until conceptual models of *changing* situations provide a better foundation from which to work. All social scientists talk about 'society' and yet its definition creates a host of problems, not to speak of the problem of what we mean when we talk of a society 'changing'. We turn now to a consideration of social structure and social system, outlining the way sociologists use the model of functional analysis, 'the most promising . . . of contemporary orientations to problems of sociological interpretations' (Merton, 1957, p. 19). In particular we consider the way this model stands up to the problem of change. Finally we consider the socio-ecological model in relation to the specific problem of evaluating the spatial component within a specific social structure.

THE NORMATIVE ORIENTATION OF ACTION SYSTEMS

Geographers, in so far as they take account of sociological theory, generally limit themselves to a demographic notion of social structure which of course lends itself more readily to the manipulation of data both cartographically and statistically. This is to impose an unnecessary restriction on the scope of the social sciences. Certainly it is possible to observe the external course of events with the accepted methods of natural science: elements of uniformity clearly do emerge in a study of social behaviour. But social scientists can, in addition, impute motives to men and 'interpret' their actions and words as expressions of their motives, for men act on the basis of *shared* values. A way of 'structuring social action' as a model, however rarified, is better than no model at all. The very important task of creating a specifically sociological frame of reference, not tied to environmentalism or psychological reductionism was greatly aided by Talcott Parsons in his somewhat involved but greatly influential work *The Structure of Social Action*, first published in 1937. The argument that the 'irrationality' of human action was simply due to the freedom of the will and therefore was not capable of scientific investigation (without calling on the laws of chance) was refuted by Max Weber. He pointed out that the sense of freedom would, if this were true, be associated primarily with irrational actions – that is those involving emotional outbreaks and so on – but in fact the reverse is more nearly true. It is when we act most rationally that we feel most free, not constrained by emotional elements. If we then accept that *rational action* is to a high degree both predictable and subject to analysis in terms of general concepts, it is of crucial importance to develop such concepts – which would clearly have relevance to all the social sciences.

'The normative orientation of action systems' can be a difficult concept to grasp because of the abstract nature of the concept and its somewhat tenuous connection with empirical investigations. Indeed, Parsons himself admits 'these concepts contain an element of "unreality" which is not involved in the physical sciences. Of course the only reason for admitting such concepts to a scientific theory is that they are in fact descriptive of an empirical phenomenon, namely the state of mind of the actor. They exist in this state of mind but not in the actor's external world' (Parsons, 1937, p. 295). In developing his voluntaristic theory of action Parsons refutes positivistic thought, caught in what he calls 'the utilitarian dilemma':

> That is, either the active agency of the actor in the choice of ends is an independent factor in action, and the end element must be random; or the objectionable implications of the randomness of ends is denied, but then this independence disappears and they are assimilated to the conditions of the situation, that is to elements

analysable in terms of nonsubjective categories, principally heredity and environment in the analytical sense of biological theory. (Parsons, 1937, p. 64.)

Individualistic positivism is seen, perhaps, in its most acute form in the 'Hobbesian Problem of Order'. To Hobbes man is guided by a plurality of passions, desires are random and, since the ultimate ends of action are diverse, conflict would seemingly be inevitable. The war of all against all is avoided, according to Hobbes, by means of the idea of a social contract, whereby the actors come to realize the situation as a whole, instead of pursuing their own ends in terms of their own situation. Nevertheless, he saw that it is a direct corollary of the postulate of rationality that all men should seek and desire power over one another. 'Thus', as Parsons significantly remarks, 'the concept of power comes to occupy a central position in the analysis of the problem of order' (p. 93). His aim was to deal with this problem without making use of 'such an objectionable metaphysical prop as the doctrine of the natural identity of interests' (p. 102).

Parsons, in a masterly analysis of the works of Marshall, Pareto, Durkheim and Max Weber demonstrates the thread of a normative theory of action running through the work of each, despite the very different backgrounds and assumptions from which they start. Marshall rejected the idea of the egoism of the traditional economic man: he did not accept that society may be completely understood in terms of utilitarian want satisfaction but felt that it also involves certain common values. As Parsons put it 'economic actions ... are also carried on for their own sake, they are modes of the immediate expression of ultimate value attitudes in action' (p. 167). Moving on to Pareto and Durkheim, Parsons argues that the former saw the individual integrated to some degree with others in a common value system, without seeing this 'social' element as a metaphysical entity in either a positivistic or an idealistic sense.

Durkheim went further, seeing the social factor as a system of ideas, which the actor passively contemplated. This led him on to identify the social factor with the *a priori* source of the categories, thus finally breaking the bond which had held it as part of empirical reality. Geographers interested in society could still read Durkheim with considerable profit – in particular his *Division of Labour in Society*. When discussing the incidence of suicide among Protestants and Catholics, Durkheim argues that the higher rate among the former is due to the different *content* of the different value systems: the individual responds to the norms and values of the group. It is easy to see from this sort of analysis how Durkheim came to see society 'out there' imposing itself on the individual. As Parsons puts it, in his evaluation of Durkheim's consideration of the social conditions of individual action:

> Among these he found a crucial role to be played by a body of rules, independent of the immediate ends of action. In the end these rules are seen to be capable of interpretation as manifestations of the common value system of the community;

it is because of this that they are able to exercise moral authority over the individual. In so far as the immediate ends of particular acts are removed from ultimate ends by many links of the means – end chain, even though these ultimate ends be in conformity with the common system of ultimate values, there is need for a regulatory system of rules, explicit or implicit, legal or customary, which keeps action . . . in conformity with that system. The breakdown of this control is *anomie* or the war of all against all.

This body of rules governing action in pursuit of immediate ends in so far as they exercise moral authority derivable from a common value system may be called social institutions. (Parsons, 1937, p. 407.)

On two occasions Durkheim defined sociology as the science of institutions. In *The Elementary Forms of the Religious Life* Durkheim was forced further from any lingering positivism as being man's sole significant cognitive relation to external reality. Religion may be seen as one mode of human orientation towards the non-empirical: religious ideas are an important, and perhaps essential, element in the normative order. Durkheim did not exploit the fundamental point that Parsons draws from his work, namely that 'the central importance of religion lies in relation to action not to thought' (p. 441). The work of Max Weber, as we shall see, made this point explicit.

Durkheim found the essential element of order in common values, manifested above all in institutional norms, but he was primarily concerned with a stable system, more useful in the definition of *categories* of sociological analysis than the functional interrelations between them. He provided a basis on which a theory of social change could be based: such a theory could not emerge until it was known *what* changes. It was primarily through Durkheim's work that ends and norms were seen to be no longer individual but also social. The functional relationships between different elements in the system was a subject of deep concern to social anthropologists: before considering this functional model we may turn to the work of Max Weber.

WEBER'S 'IDEAL TYPE' MODELS

The relationship between the Protestant ethic and the spirit of capitalism, as discussed by Weber, is the work for which to non-sociologists he is best known. His comparative analyses of Confucianism, Buddhism, Hinduism and Judaism are not so well known and yet it is on the basis of this larger work that his work on the *Protestant Ethic* gains its force. Certainly it is difficult to provide any simple answer to the question which asks why, given the coal resources of Britain, India and China, those in Britain should be developed first. One can hardly postulate the environment or heredity form of single factor analysis, but it is also clear that Weber's work on ideas as causes and consequences, however strongly based on the most thorough and wide ranging scholarship, cannot provide any sort of holistic model of the processes of economic development. Nevertheless the analysis of Weber's work by Parsons (1937, pp. 500–578) or Bendix (1962), together, of course, with some basic elements of the original work (as, for example, available in Gerth and Mills, 1948, pp. 302–359) is essential reading for the social geographer interested in economic development, as a brilliant example of sociological analysis. There is, of course, no quantitative assessment of religion as a causative element in modern capitalism: it was a necessary though not sufficient condition. Weber's work, as far as this present chapter is concerned, is mainly of importance in illustrating the use of the concept of an 'ideal type'.

Weber outlined his methodical position in Chapter I of his posthumously published '*Wirtschaft und Gesellschaft*'; his remarks on ideal types are of particular interest:

> the same historical phenomenon may be in one aspect 'feudal', in another 'patrimonial', in another 'bureaucratic', and in still another 'charismatic'. In order to give a precise meaning to these terms, it is necessary for the sociologist to formulate pure ideal types of the corresponding forms of action which in each case involve the highest possible degree of logical integration by virtue of their complete adequacy on the level of meaning. But precisely because this is true, it is probably seldom if ever that a real phenomenon can be found which corresponds exactly to one of those ideally constructed pure types. The case is similar to a physical reaction which has been calculated on the assumption of an absolute vacuum. Theoretical analysis in the field of sociology is possibly only in terms of such pure types. . . . The more sharply and precisely the ideal type has been constructed, thus the more abstract and unrealistic in this sense it is, the better it is able to perform its methodological functions in formulating the clarification of terminology, and in the formulation of classifications, and of hypotheses. (Weber, 1947, pp. 110–111.)

Weber's tortuous prose hardly gains in clarity on translation and the fact that his work has had such a remarkable impact on the development of American sociology owes much to the penetration of his commentators. *The Theory of Economic and Social Organization*, which first appeared in translation in 1947, has a valuable introduction by Parsons. However, returning to the discussion of an ideal-type as an important sociological model, David Lockwood has provided a succinct summary of the characteristics of Weberian ideal-type analysis (Lockwood in Gould and Kolb, 1964, pp. 312–313), based on the passage from which we have just quoted. He notes that Weber's discussion of 'bureaucracy' is in terms of a true ideal-type since this phenomenon recurs in a variety of historical contexts (cf. Eisenstadt, 1963) and so it is truly a generalizing concept. However 'the Protestant Ethic' although discussed by Weber in terms of an ideal type can hardly be considered as such, since it lacks the abstract general quality of the former example. Further discussion of Weber and ideal-types by Carl Hempel and H. Stuart Hughes provides deeper analysis, and geographers as well as other social scientists may find them useful and thought-provoking accounts.

FUNCTIONALISM AS A MODEL

During the inter-war period when Parsons was working on his interpretation of Weber and moving towards his conception of a normative system of action, social anthropologists, through their work on primitive societies, were much concerned with the role specific forms of social activity played in the working of the whole society. Radcliffe-Brown, in 1935, discussed the analogy between social life and organic life, pointing out that any such analogy should be used with care. Like the life of an organism the continuity of a social structure is preserved through its functional continuity. As he put it:

> – if we examine such a community as an African or Australian tribe we can recognize the existence of a social structure. Individual human beings, the essential units in this instance, are connected by a definite set of social relations into an integrated whole. The continuity of the social structure, like that of an organic structure, is not destroyed by changes in the units. Individuals may leave the society by death or otherwise: others may enter it. The continuity of structure is maintained by the process of social life, which consists of the activities and interactions of the human beings and of the organized groups into which they are united. The social life of the community is here defined as the *functioning* of the social structure. The function of any recurrent activity, such as the punishment of a crime, or a funeral ceremony, is the part it plays in the social life as a whole and therefore the contribution it makes to the maintenance of the social structure. (Radcliffe-Brown, 1952, p. 180.)

It is important to remember during the discussion which follows that Radcliffe-Brown made it quite clear that the idea of the functional unity of a social system was a *hypothesis*. He argued that in the same way that physics deals with the structure of atoms and colloidal chemistry with the structure of colloids, so also should there be a place for a branch of natural science dealing with the general characteristics of those social structures of which the component elements are human beings. The network of actually existing relations which makes up the social structure is, he argued, just as real as individual organisms.

The functionalism of social anthropologists has come under powerful attack, in particular from Rex and Dahrendorf. If we recall the question which we posed as basic to the argument of this chapter – the analysis of social change with a functionalist model – then we must consider what happens when a society falls into what Radcliffe-Brown calls 'a system of functional disunity or unconsistency'. The biological analogy suggest that ill-health or even death will follow in such a situation. It is easy to fall into the trap of regarding all changes as dysfunctional and of thinking that the old order represents a 'healthy' social organism which has to 'suffer' change. Certainly the analogy might be fruitfully pursued when considering recent events in the Congo, but it is difficult to apply it in the case of South Africa, where the social organism is seemingly maintained by the brute force of a minority. Clearly some things are more important than others – the body can still function without a limb but not without a heart or brain. As John Rex points out: 'The danger is that if the nature of the theoretical model which is unconsciously being used is not made explicit, the anthropologist is likely to interpret as 'functional' those activities which fit into his own scheme of goals and values, and what is more, represent them as essential to the survival of the society. It cannot be too strongly pointed out that explanations of social activities as performing functions in this sense carry no implication that without these activities the society would not survive. All that they imply is that without them certain goals would not be achieved' (Rex, 1961, p. 72).

The form of functionalism which has been developed by social anthropologists, based on a 'face-to-face' society where there are fairly clear-cut lines as between one society or tribe and another, is of less immediate relevance in the context of an advanced industrial society, where the important situations depend more on the relationships between institutions rather than between actors. Before making a more extended analysis of functionalism it will be useful to say something about *The Social System* by Talcott Parsons, itself a carefully worked out conceptual model.

Parsons aimed to create 'a body of logically interdependent generalized concepts of empirical reference' in such a state of logical purity that every logical implication of any combination of propositions in the system would be explicitly stated in some other proposition in the same system. He claimed that this would provide a 'genuinely technical analytical tool' which would avoid a situation in which vital elements were overlooked, and thus it would minimize 'the danger, so serious to common sense thinking, of filling gaps by resort to uncriticized residual categories'. (Parsons, 1954, p. 217.) This generalized theoretical conceptual framework he termed the structural-functional system (incidentally he also noted with approval the value of a physiological analogy).

Perhaps the most concise summary of Parsons' ideas appears in *Toward a General Theory of Action* (1951, esp. pp. 53–109) and those who wish to do full justice to them should read the original and not be misled by his 'translators' and commentators. 'Any behaviour of a living organism might be called action; but to be so called, it must be analysed in terms of the anticipated states of affairs toward which it is directed, the situation in which it occurs, the normative regulation (e.g. the intelligence) of the behaviour, and the expenditure of energy or 'motivation' involved. Behaviour which is reducible to these terms, then, is action' (p. 53). 'Actions are not empirically discrete but occur in constellations which we call systems. We are concerned with three systems, three modes of organization of the elements of action; those elements are organized as social systems, as personalities and as cultural systems' (p. 54). An analysis of cultural systems is essential to the theory of action 'because systems of value standards . . . and other patterns of culture, when *institutionalized* in social systems and *internalized* in personality systems, guide the actor with respect to both *the orientation to ends* and the *normative regulation* of means and of expressive activities' (p. 56).

The normative, ideal, aspects of the structure of systems of action, or the model of part of its culture are analysed by Parsons in terms of five pattern variables.[1] These are dichotomies 'one side of which must be chosen by an actor before the meaning of a situation is determinate for him, and thus before he can act with respect to that situation' (p. 77). The argument is that '*every* concrete need-disposition of personality, or every role-expectation of social structure involves a combination of values of the five pattern variables' (p. 93). Assuming that this list is exhaustive a cross-classification of each of the five against each of the others will provide a table of thirty-two cells and this can be seen as a first step towards the construction of a dynamic theory of systems of action. In fact, actuality is simpler than Parsons' theoretical complexities and there are empirical clusterings of the various structural components of society.

In so far as geographers are concerned with clarifying their conception of Parsons' notion of a social system there is much to be said for reading the original (Parsons, 1951, Chaps. 3, 4 and 5) although both Johnson (1961, Chap. 3) and Rex (1961, Chaps. 5 and 6) provide useful summaries. Parsons' social system is based on normative consensus: indeed he sees the integration of common value patterns with the internalized need – disposition structure of the actors involved as not only crucial for the stability of any social system but as the 'fundamental dynamic theorem of sociology' (1951, p. 42). This institutional integration may explain very little in detail but Parsons argues that his exposition, based on deductive reasoning, will enable empirical application of the conceptual scheme to follow. As he admits 'this is obviously a highly simplified model' (1951, p. 44) which hardly accommodates conflict and ambivalence in relation to the central value system of the society. 'Fulfilment of a given set of expectations will impose a greater 'strain' on one actor than another' (p. 45).

Unfortunately Parsons is unable to present a general theory of the processes of change of social systems, but he does concede the possibility of an 'increase of strains in one strategic area of the social structure which are

[1] The five pattern variables are: 1. Affectivity – Affective neutrality, 2. Self-Orientation – Collectivity orientation, 3. Universalism – Particularism, 4. Ascription – Achievement, 5. Specifity – Diffuseness.

finally resolved by a structural reorganization of the system' (p. 493). He feels that it is probably of more importance to trace the repercussions of change, once started, throughout the social system, including what he calls the 'backwash' of modification of the original direction of change (p. 494).

Following Johnson's admirable summary (1961, pp. 51–59) the main elements in the structure of a social system are:

1 Subgroups of various kinds, normatively related.
2 Roles both within the larger system and within subgroups – each role system being normatively related with each of the others.
3 Regulative norms governing subgroups and roles.
4 Cultural values.

Every social system must solve four functional problems:

1 Pattern maintenance and tension management.
2 Adaption
3 Goal attainment
4 Integration.

In very broad terms, the family, schools, religious groups and so on deal with the first problem, the economy deals with the adaptive subsystem of society, the polity or government deals with society's goals (although people may be more actively concerned with these goals at specific times) and the integrative subsystem is maintained by the legal profession, opinion formers of the mass media, religious leaders and so on. Each of these functional subsystems of a society can, in themselves, be analysed as a social system and the other systems are then seen as part of the 'environment' to which the subgroup must adapt if it is to survive and achieve its goals. Significantly Johnson comments 'It always happens that the cost of meeting societal problems is borne unequally by the subgroups of society. In our own society, for example, successful functioning of the economy requires that certain particular business firms must be allowed or 'forced' to fail; otherwise the economy as a whole would be less adaptive than it is' (p. 59).

FUNCTIONALISM AND THE PROBLEM OF CHANGE

There have been various hints in the preceding section suggesting some inadequacy in the functional model in coping with the problem of change. Much of the structural-functional model works simply by definition. For example, a given socio-technological system 'demands' an industrial organization of a certain degree of complexity within which, with the division of labour in society, actors perform the necessary roles to keep the enterprise functioning and others, in another sector of the system, are trained to fill the gaps at one end caused by death or retirement at the other.

It is important to see that 'society' is a mental construct or model of a web of interrelated social networks intermeshed in a highly complex manner and expanding over geographical and language barriers to cover the whole globe. These networks are 'bunched' into the subsystems of 'society' for purposes of analysis, and boundaries between 'total' social structures are delineated: political systems are perhaps the most fundamental differentiating variable between modern societies. Hence, even if an individual is socialized within British society and is trained to fill a role in the adaptive subsystem or economy, if he in fact emigrates and gives up his British political status to become a South African national, then it is likely he will acquire the political assumptions of that society. Many British nationals may have, for example, kinship links with South Africans without necessarily accepting their political goals or ideals. The importance of the polity can be challenged by the economy (as when it was said by some that 'the city' challenged the Wilson administration in the Autumn of 1964) or the economy can be fused with the polity (as in centrally planned economics), but rule is always rule of the few, whether or not it is accepted by the rest of society. The function of government is to maintain rules, with physical force as an absolute sanction. This does not mean that order implies consensus: certain aspects of the situation may be 'dysfunctional' and may conflict with the goal attainment patterns of some groups within the society but may be 'functional' for others. Similarly the *manifest* function, intended and recognized by the participants in the system, may have a *latent* function producing consequences neither intended nor recognized. As Merton points out in his classic discussion of the subject:

> Through the systematic application of the concept of latent function, therefore, *apparently* irrational behaviour may *at times* be found to be positively functional for the group. Operating with the concept of latent function, we are not too quick to conclude that if an activity of a group does not achieve its nominal purpose, then its persistence can be described only as an instance of 'inertia', 'survival' or 'manipulation by powerful sub-groups in society' (p. 65).... Findings concerning latent functions represent a greater increment in knowledge than findings concerning manifest functions. They represent, also, greater departures from 'commonsense' knowledge about social life. (Merton, 1957, p. 68.)

The comparative analysis of new states has helped to drive sociologists back to the larger problems which bothered a generation of scholars at the end of the nineteenth century. However revealing ad hoc explanations may be, they may have little potential for any logical extension and thus the growth of theoretical understanding. Shils sees comparative empirical analysis 'freed from its evolutionary encrustation, and brought into a dynamic conception of social systems' (Shils, 1963, p. 20) as being a more relevant and hopeful approach. He lays stress on the relationship between what he calls 'the centre' and 'the periphery' of society – 'consensus is the key phenomenon of macrosociology . . . How does this institution or practice or belief function in the articulation of the society, in attaching or detaching or fixing each sector in its relationship to the central institutional and value systems of the society?' (pp. 23–24). It is to this consideration of points of tenson and the introduction of change which we now turn.

Parsons has been criticized for being more concerned with what holds societies together than with what drives them on. David Lockwood (1956) has argued that the crucial issue of differential access to scarce resources has been neglected by Parsons and that by stressing the normative aspects he has diverted attention from the *substratum* of social action, especially as it conditions interests which are productive of social conflict and instability. This argument was picked up by Dahrendorf in 1958 who called for a model which would make understandable the structural origin of social conflict. Geographers in particular are likely to have much sympathy with Gouldner's highly apposite criticism of Parsons and Merton. He calls the exclusion of the physical environment by Parsons a form of 'academic monasticism' whereby men are cleansed of their basic passions for sex, food and material possessions by theoretical purification (Gouldner, 1959). The potency of ecological forces has been clearly demonstrated by anthropologists (e.g. Evans-Pritchard 1940) and by Wittfogel in his analysis of hydraulic society. Parsons does not own up to the likelihood of his model of a social system having an unequal capacity to account for variance in social behaviour in different situations. Gouldner goes on to make the crucial point that the consideration of systems in terms of interdependence and equilibrium obscures the fact that they can vary in degree and, further, they may vary differentially. Not all system parts have an equally deep involvement in the resolution of the tensions of the system, or in the mobilization of defences against these. He draws a distinction between functional reciprocity and functional autonomy: those parts with least functional autonomy may be unable to survive separation from a social system and hence may be more likely to be concerned with its conservation than those with greater functional autonomy. Thus it is that the polity may undergo frequent changes in certain Latin American republics without there being any real change in the basic social order, largely because of the polity's functional autonomy. (This example shows that functional autonomy may also militate against charge – a demonstration of Shils' point about central and peripheral subsystems).

Merton, curiously enough, made this point earlier when he said that 'It is not enough to refer to the 'institutions' as though they were all uniformly supported by all groups and strata in the society. Unless systematic consideration is given to the *degree* of support of particular 'institutions' by specific groups we shall overlook the important place of power in society' (Merton, 1957, p. 122). A similar sort of notion can be found in Parsons' work, as we have already noted, but it is not elaborated in any sort of satisfactory way. A valuable statement, which goes a long way to make the functionalist model more flexible, and hence able to accommodate social change, is provided by David Lockwood in his recent paper *'Social Integration and System Integration'* (1964). In this he defines social change as 'a transformation of the core institutional order of a society such that we can speak of a change in a type of society' (p. 244). The distinction is made between social integration of the actors as opposed to the integration of the parts of a social system. Normative functionalism provides a way of dealing with both aspects within the same conceptual framework. Lockwood powerfully refutes the conflict theorists, who wish to replace the functionalists' concepts of 'norm', 'consensus' and 'order' with the more fashionable concepts of 'power', 'alienation' and 'conflict', arguing that this is an unreal dichotomy. Conflict is the better grasped within a normative functionalist framework: there is no way of seeing how some conflict is associated with change and other not, except in relation to such a framework. Lockwood goes on to argue that one source of tension and possible change in a social system is 'lack of fit' between its core institutional order and its material substructure. As far as I understand it, he is doing much to integrate crude marxian theory with crude normative functionalism. The 'dominant' or 'core' institutional order may vary from one type of society to another and it is on this that analysis should be focused. Emphasis on the moral aspects of *social* integration has led to neglect of *system* integration by both functional and conflict theorists. We cannot accept that 'The integration of a set of common value patterns with the internalized need-disposition structure of the constituent personalities is the core phenomenon of the dynamics of social systems' (Parsons, 1951, p. 42). We must rather follow Gouldner, Lockwood, Merton and Shils to find a dynamic neo-functionalist model without the Parsonian sweep but with much greater analytical value.

Societies do not change simply through the impact of one factor – whether that factor be coal, geographical position, a decline in the death rate or the growth of an entrepreneurial class. Economic concepts have, perhaps, been accepted too readily by geographers, partly because economists have put forward *theories* of economic development, which are often highly persuasive to the non-economist, and partly because the economist uses quantified indices which are a great solace in a field lacking basic empirical date. The sociologist, whilst not wishing to substitute his 'factors' for those of sister disciplines, nevertheless 'sees' society in terms of a system in which the key points of strain can be isolated. Hence issues such as the flexibility of the system of stratification of that society or the social position of women, may be of crucial importance when considering the goals of a society in relation to its resources. It is most important to remember that 'the goal of the economy is not simply the production of income for the utility of an aggregate of individuals. It is the maximization of production relative to the whole complex of institutionalized value-systems and functions of the society and its subsystems' (Parsons and Smelser, 1956, p. 22). As the goal of the economy is defined by socially structured goals then it is inappropriate to refer to the measurability of utility among individuals. 'Utility, then, is the *economic value* of physical, social or cultural objects in accord with their *significance as facilities* for solving the adaptive problems of social systems' (loc cit.). *The 'spatial constraint' and the endowment of 'natural' resources must be evaluated in the context of a society's adaptive and goal attainment functions.*

Geographers, who have found the models of Myrdal or Rostow useful, should also find *Economy and Society* by Parsons and Smelser helpful in integrating a sociological model into their work. The publication of two admirably concise and reasonably-priced introductory outlines on *Social Change* and *The Sociology of Economic Life*, by Moore and Smelser respectively, should do much to spread these ideas into sixth-form teaching in schools. Indeed, if these books were widely used there would be less need for this chapter in its present form. The emphasis on Parsonian model-building here is felt to be justified because of its very great importance generally and (as far as I know) total neglect among geographers. 'Until Parsons, only economics among all the social sciences could be said to have a rational foundation for its theoretical formulations . . . Parsons has opened the way for other social disciplines to acquire distinctive rationalities of the same type' (Morse, 1961, p. 142).

There are of course a whole range of theories which purport to explain the sources and patterns of change. Herbert Spencer and Auguste Comte in the nineteenth century both believed in the unilinear development of society: in the twentieth century Spengler and Toynbee were less optimistic and saw civilizations rising and falling in waves or cycles. The Marxist theory of social change is perhaps the most powerful, if not the most accurate, since it has clearly been able to influence history, even though Marx might be surprised to find his prophecies having more application in, say, Cuba than in England. If Marx believed that revolutions are brought about by a gradual decline from bad to worse then this hardly squares with a comparative analysis of past revolutions. Davies has recently presented a 'Theory of Revolution' (Davies, 1962) in which he argues that 'Revolutions are most likely to occur when a prolonged period of objective economic and social development is followed by a short period of sharp reversal' (p. 6). He goes on to analyse the Dorr's Rebellion of 1842, the Russian Revolution of 1917 and the Egyptian Revolution of 1952. 'Far from making people into revolutionaries, enduring poverty makes for concern with one's solitary self or solitary family at best and resignation or mute despair at worst. When it is a choice between losing their chains or their lives, people will mostly choose to keep their chains, a fact which Marx seems to have overlooked' (p. 9). Davies' model can be diagrammatically represented in Figure 7.1.

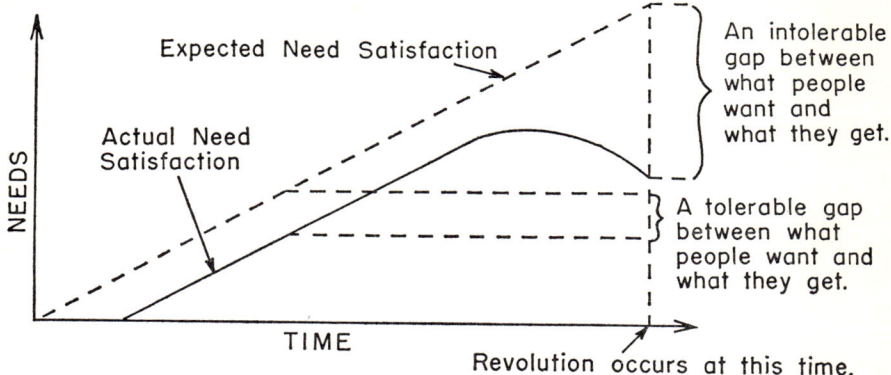

7.1 The relationship between relative deprivation and revolution (*After Davies, 1962*).

Moore makes a valuable distinction between 'mere *sequences of small actions*, that in sum essentially *comprise the pattern*, the system, and *changes in the system* itself, in the magnitude or the boundaries, in the prescriptions for action, in the relation of a particular system to its environment' (Moore, 1963, p. 6). Neither rebellions, nor indeed revolutions, necessarily lead to changes in the system: that is part of the tragedy of revolutions, but it also illustrates Gouldner's concept of functional autonomy. Moore underlines the point that we have already made – namely that an equilibrium model of society 'either forecloses questions about the sources of change, or if discordant internal elements are brought into the analysis, the theoretical model will predict one direction of change, and one only – change that restores the system to a steady state' (p. 10). If, on the other hand, society is seen as a tension-management system, where both order and change are problematical but also 'normal', then there is no need to expect that change to accommodate tension will necessarily remove tension, but will rather shift it to other points in the system. The crucial issue, which Moore very neatly poses, is thus to distinguish between changes in the *system* and changes which are inherent in the way the system operates. It is Moore's contention that one of the main change-producing strains which impinges on a tension-management social system is provided by the non-human environment.

It is somewhat ironic that sociologists should now be turning to the broad issues that geographers may feel they outgrew at the end of the nineteenth century. At the sixtieth annual meeting of the American Sociological Association, held in 1965, it is significant that the first plenary session was devoted to a re-evaluation of Karl Marx (in which Parsons argued that Marx was not sufficiently alert to other forms of conflict in society than that based on class) and the second session was devoted to 'Civilizations and their changes'. There seems no doubt that these macrosociological problems of social and cultural change, based on the models discussed in this chapter, will continue to exercise the minds of some of the most distinguished sociologists in the world. It would be unfortunate if geographers, preoccupied with models of man at a sub-social, materialist or mechanistic level, withdrew from their traditional interest in this field. So much of the discussion by geographers seems to take place in inaugural lectures, when the determinism – possibilism skeleton is taken out of the cupboard for its ritual dusting, that there is a danger that younger geographers will neglect the field for fear of getting tarred with the brush of their elders. Yet if one turns to the collection of readings on *Social Change* edited by the Etzionis, for example, the range and fertility of the *ideas* of contemporary sociologists is impressive. There is clearly a danger that over-concern with techniques may be at the expense of conceptual model building in geography. Certainly work by sociologists over the past ten years – particularly by Parsons, Smelser and Moore and the others mentioned above – is moving towards some general theory of change for whole societies. Smelser's recent essay '*Toward a Theory of Modernization*' is worth a long and careful look by geographers. He attempts to construct an ideal type, in the Weberian sense, to analyse the relationship between economic growth and social structure: in particular he builds up a *differentiation* model to clarify the way major social functions gain structural independence. For every social function there is a distinct set of structural conditions under which it is optimally served and this is the link between modernization and differentiation. Historical geographers would no doubt find Smelser's study of the Lancashire cotton industry which 'twisted, spiralled, reversed, and creaked as it accumulated the elements which carried it nearer to Weber's conditions of extensive role differentiation' (Smelser, 1959, p. 101) a particularly relevant introduction to his model.

In the light of Smelser's ideas it is useful to consider Epstein's study of *Economic Development and Social Change in South India*, based on the two villages of Wangala and Dalena. She shows convincingly that economic development does not necessarily lead to economic change. In the case of Wangala the introduction of irrigation changed the social system very slightly, as the traditional farming economy was simply strengthened. Only where the new economic system was incompatible with features of traditional economic organization did change in economic roles and relations occur. This was the case at Dalena, where the whole social system was changed when the village had to provide secondary and tertiary services for the neighbouring villages, where the land had been irrigated. The 'backwash' effects of the change in the economic system led to changes in political and ritual roles and relations and in the principles of social organization. Such empirical studies as this in different social and physical environments are urgently needed if the conceptual models of social change and modernization are to be refined and developed.

SOCIOLOGICAL MODELS AND URBAN GEOGRAPHY

It may be that geographers are more willing to accept quantitative techniques from elsewhere than conceptual models, and yet in certain areas the very lack of a conceptual base seriously impedes research. Much of the apparent precision of recent studies in locational analysis is based on inadequate operational definitions of populations. As Haggett (1965, p. 189) admits, 'The problem of standardizing definitions of cities has not been solved'. Furthermore, Berry, when discussing the disappointing results of research by geographers on the residential pattern in cities, concludes that 'geographers must take second place to urban sociologists in studies of the residential patterning of cities' (in Hauser and Schnore, 1965, p. 417).

Now why should this be so? Why must the geographer admit defeat both in spatial delimitation and in the understanding or explanation of a spatial pattern? The answer, surely, must be that he is working with a wrong model and is approaching the problem with the wrong assumptions. Norton Ginsburg is an urban geographer who has moved some way towards a more sociologically informed approach to urbanization, no doubt stimulated by the seminal atmosphere of the admirable multi-disciplinary tradition of the University of Chicago. Ginsburg outlines his problem thus: 'What kinds of cities can be expected to evolve in different societies as these societies make their decisions to select, adopt, and modify those elements that characterize Western city-building, functions, and structure?' (in Hauser and Schnore, 1965, p. 319) and he goes on to describe Japan as a highly urbanized country, not in demographic terms – the proportion of people living in sizable towns – but in more *sociological* terms of the impact of the city on the nation's life and 'the awareness among rural dwellers of the existence and nature of an urban way of life'.

Ginsburg appears to be right in making a distinction between ways of life, for how else is one to distinguish between, say, an Indian city, parts of which may be little more than villages joined together with a 'rural' pattern of life within them, and the situation in a Japanese city, the product of a different society with a strong tradition of urbanism? There is clearly no reason why geographers should not concern themselves with sociological problems – so long as they remember that differences which are internal to a social system cannot be explained only by reference to forces external to it. Those who see 'the necessities of economic expansion' as a sort of global 'prime mover' in stimulating 'urbanization' and an increase in the 'scale of society' have been criticized by Sjoberg for failing to explicate their theories (in Hauser and Schnore, 1965, pp. 168–177) Berry and Ginsburg have been more honest in rejecting the adequacy of non-sociologically-based frameworks.

Work by Duncan and Schnore (1959), influenced by Durkheim's *Division of Labour in Society* suggests a model with four basic components – environment, population, social organization and technology – which are functionally interrelated. Sjoberg accuses them of materialism and finds the concept of social organization 'particularly spongy'. Perhaps the most sensible statement on whether settlement size and other ecological concepts are useful for explaining ways of life has been provided in a succinct and satisfying essay by Gans (in Rose, 1962, pp. 625–648) and he deserves to be cited at length:

> Ecological explanations of social life are most applicable if the subjects under study lack the ability to *make choices*, be they plants, animals, or human beings. Thus, if there is a housing shortage, people will live almost anywhere, and under extreme conditions of no choice, as in a disaster, married and single, old and young, middle and working class, stable and transient will be found side by side in whatever accommodations are available. At that time, their ways of life represent an almost direct adaptation to the environment. If the supply of housing and of neighbourhoods is such that alternatives are available, however, people will make choices, and if the housing market is responsive, they can even make and satisfy explicit *demands*.
>
> Choices and demands do not develop independently or at random; they are functions of the roles people play in the social system. These can best be understood in terms of the *characteristics* of the people involved; that is characteristics can be used as indices to choices and demands made in the roles that constitute ways of life. Although many characteristics affect the choices and demands people make with respect to housing and neighbourhoods, the most important ones seem to be *class* – in all its economic, social, and cultural ramifications – and *life-cycle stage*. If people have an opportunity to choose, these two characteristics will go far in explaining the kinds of housing and neighbourhoods they will occupy and the ways of life they will try to establish within them (pp. 639–640).
>
> Characteristics do not explain the causes of behaviour; rather, they are clues to socially created and culturally defined roles, choices and demands. A causal analysis must trace them back to the larger social, economic, and political systems which determine the situations in which roles are played and the cultural content of choices and demands, as well as the opportunities for their achievement (p. 641).

Gans goes on to conclude that if ways of life do not coincide with settlement type but are rather functions of class and life-cycle stage, then a sociological definition of the city cannot be formulated.

With these ideas in mind it is possible to probe the implicit model in my study of the rural/urban fringe of the London Metropolitan Region (Pahl, 1965). This was an attempt to study the 'urban' influences on a 'rural' area and illustrates very well the need to abandon traditional geographical concepts for a sociological approach in such a problem area. Indeed I started with a naïve assumption that it would be possible to use a 'spatial' model so that 'accessibility' to urban functions could be taken as a key variable. It soon became clear that to group the whole population together, as so many social atoms showing certain mathematical regularities, was a meaningless exercise. The styles of life of two broad sub-groups of the population, empirically determined, were so very different that a non-spatial conceptual model was forced upon me. For members of the working class, who commute to work out of the area, space is a 'constraint' which has to be overcome and the cost of doing so is an added economic burden on the family. Not all the working class *choose* to live in a rural area but they are 'forced' to do so by a society which allocates council houses by area of residence and not place of present employment.

By contrast, for a section of the middle class, space, far from being a constraint, is valued as an amenity, which should be preserved. Space may become a symbol of a certain style of life to which such middle-class people aspire and the economic burden of crossing it, in order to reach urban employment and amenities, is an accepted concomitant of their way of life. Such people *choose* to live in an area where working class people may be *forced* to live and the residential pattern is a resolution of the two forces. Clearly this polar ideal type can be made more complex by adding further groups – for example, those who aspire to the middle-class life style but whose position in the larger socio-economic system makes it difficult for them to maintain it. A further group may be obliged to move into the rural-urban fringe area simply because the price of land nearer towns makes home ownership there impossible. Yet another group, at the top of the social hierarchy, may feel that the power and influence which they wish to achieve is more easily had by living in the centre of cities and then their country house becomes of secondary importance and is only used at week-ends or at certain times of the year.

I tried to show in my study that any attempt to understand the striking expansion of population in the Outer Metropolitan Region, outside the officially designated Urban Districts, cannot be achieved with simple materialistic models, be they called ecological, spatial, geographic, demographic or whatever. Yet this does not mean that a consideration of people's values leads to a confusion without order. People's choices reflect values that are shared. Other people find themselves in a similar position in the socio-technical system and, with more limited house-buying potential, are forced to certain areas. Yet other people occupy an even more restricted position in the social system so that they do not even have the choice of owning a home; and so on. The residential pattern is a reflection of the functioning of the social system. In Britain we have a conventional socio-economic stratification system related to the unequal distribution of social and economic resources, which is supported by the central value system of our society. In other societies with a different social system – say India – the residential pattern of the population will also be different. Yet in both societies the poorest will have no choice, so that their position can be explained with a materialistic model and they will be more amenable to study with quantitative techniques based on 'economic man'. At the other extreme the very rich have the greatest choice in where they shall live and what life-styles they will pursue (cf. Pahl, 1966).

If the geographer is not to admit complete defeat in accounting for the residential pattern he must try to understand the sociologists' model of social stratification for the appropriate society, together with a knowledge of the values, in particular the aspirations for certain life styles of major sub-groups of the population; this, as Gans pointed out in the quotation above, will lead the geographer to the larger social, economic and political systems. Certainly geographers must be sure to make a distinction in their quantitative analyses between least-cost locations, which involve no choice, and other locations based on other values. Failure to do this may lead the most exciting breakthrough of this century to lose its momentum in the materialistic determinism of a 'social physics', trapping the model makers in a strait-jacket of their own invention. It is vital that interdisciplinary co-operation and criticism continues: a fitting conclusion is provided by the following comment – the result of co-operation between an economist and a social anthropologist:

> Whether particular models and the development of particular lines of research are worthwhile or not must in the end be a matter of personal judgment and faith. All that we can plead for is that those who engage in this activity should not arrogantly assume that their system is necessarily the best or that their system will necessarily be relevant to the explanation of reality, or that a model which appears to explain development in one country will have universal application and validity. (Gluckman and Devons, 1964, p. 186.)

REFERENCES

ARON, R. and HOSELITZ, B. F., (Eds.), [1965], *Social Development*, (Paris).
ABRAMS, P., [1963], Notes on the Uses of Ignorance; *Twentieth Century*, 67–77.
BECKER, H. and BARNES, H. E., [1961], *Social Thought From Lore to Science*, (New York).
BENDIX, R., [1962], *Max Weber. An Intellectual Portrait*, (New York).
BESHERS, J. M., [1957], Models and Theory Construction; *American Sociological Review*, 22, 32–38.
BLACK, MAX. (Ed.), [1961], *The Social Theories of Talcott Parsons*, (Englewood Cliffs, N.J.).
BOSKOFF, A., [1964], Functional Analysis as a Source of a Theoretical Repertory and Research Tasks in the Study of Social Change; In Zollschan, G. K. and Hirsch, W., (Eds.), *Explorations in Social Change*, (London), pp. 213–243.
BOTTOMORE, T. B., [1962], *Sociology*, (London).

BRODBECK, M., [1959], *Models, Meaning and Theory*: In Gross, L., (Ed.), *Symposium on Sociological Theory*, 275–403.
COLEMAN, J. S., [1964], *Mathematical Sociology*, (London).
DAHRENDORF, R., [1957], *Class and Class Conflict in Industrial Society*, (London).
DAHRENDORF, R., [1958], Toward a Theory of Social Conflict, *Journal of Conflict Resolution*, 11, 170–183.
DAVIES, J. C., [1962], Toward a Theory of Revolution; *American Sociological Review*, 27, 5–19.
DUNCAN, O. D. and SCHNORE, L. F., [1959], Cultural, Behavioural and Ecological Perspectives in the Study of Social Organization; *American Journal of Sociology*, 65, 132–146.
DURKHEIM, E., [1947], *The Division of Labour in Society*, (trans. G. Simpson), (Glencoe, Illinois), (First published in 1893).
EISENSTADT, S. N., [1963], *The Political Systems of Empires: The Rise and Fall of Historical Bureaucratic Societies*, (London).
EPSTEIN, T. S., [1962], *Economic Development and Social Change in South India*, (Manchester).
ETZIONI, A. and E., (Eds.), [1964], *Social Change: Sources, Patterns and Consequences*, (London).
EVANS-PRITCHARD, E. E., [1940], *The Nuer*, (Oxford).
GERTH, H. H. and MILLS, C. W., [1948], *From Max Weber: Essays in Sociology*, (London).
GLUCKMAN, M. and DEVONS, E., (Eds.), [1964], *Closed Systems and Open Minds*, (Edinburgh).
GOULD, J. and KOLB, W. L., (Eds.), [1964], *A Dictionary of the Social Sciences*, (Paris).
GOULDNER, A., [1959], Reciprocity and Autonomy in Functional Theory; In Gross, L., (Ed.), *Symposium on Sociological Theory*, 241–270.
GOULDNER, A. W. and H. P., [1963], *Modern Sociology*, (London).
GROSS, L. (Ed.), [1959], *Symposium on Sociological Theory*, (Row, Peterson & Co.).
HAGEN, E. E., [1962], *On the Theory of Social Change*, (Homewood, Illinois).
HAGGETT, P. [1965], *Locational Analysis in Human Geography*, (London).
HAUSER, P. M. and SCHNORE, L. F., (Eds.), [1965], *The Study of Urbanization*, (London).
HEMPEL, CARL, [1952], Typological Methods in the Social Sciences; Reprinted in M. Natanson, (Ed.), *Philosophy of the Social Sciences: A Reader*; New York: Random House, 1963, pp. 210–230 (see section on Ideal Types and Theoretical Models, pp. 223–230).
HUGHES, H. S., [1959], *Consciousness and Society: The Reorientation of European Social Thought 1890–1930*, (London).
INKELES, A., [1964], *What is Sociology?*, (Englewood Cliffs, N.J.), (Ch. 3 Models of Society in Sociological Analysis).
JOHNSON, H. M., [1961], *Sociology*, (London).
LA PIERE, R. T., [1965], Social Change, (London), (Ch. 3 Models of Change and Stability).
LOCKWOOD, D., [1956], Some Remarks on 'The Social System'; *British Journal of Sociology* 7 (2).

LOCKWOOD, D., [1964], Social Integration and System Integration; In Zollschan, G. K. and Hirsch, W., (Eds.), *Explorations in Social Change*, (London), 244–257.
MEADOWS, P., [1957], Models, Systems and Science; *American Sociological Review*, 22, 3–9.
MERTON, R. K., [1957], *Social Theory and Social Structure*, (Glencoe, Ill.).
MILLS, C. W., [1959], *The Sociological Imagination*, (New York), (Ch. 2 Grand Theory).
MOORE, E., [1963], *Social Change*, (Englewood Cliffs, N.J.).
MOORE, W. E. and HOSELITZ, B. F., (Eds.), [1963], *Industrialization and Society*, (Paris and the Hague).
MORSE, C., [1961], The Functional Imperatives; In Black, M., (Ed.), *The Social Theories of Talcott Parsons*, (Englewood Cliffs, N.J.), 100–152.
PAHL, R. E., [1965], *Urbs in Rure*; London School of Economics Geographical Paper No. 2.
PAHL, R. E., [1966], The Rural-Urban Continuum; *Sociologia Ruralis*, 6 (3–4), pp. 299–326.
PARSONS, T., [1937], *The Structure of Social Action*, (New York).
PARSONS, T., [1951], *The Social System*, (London).
PARSONS, T. and SHILS, E. A., [1951], *Toward a General Theory of Action*, (Harper Torchbook Edn., New York 1962).
PARSONS, T., [1954], Present Position and Prospects of Systematic Theory in Sociology (1945); In *Essays in Sociological Theory*, (Glencoe, Ill.), 212–237.
PARSONS, T. and SMELSER, N. J., [1956], *Economy and Society*, (London).
PARSONS, T., [1960], The Principle Structures of Community; In *Structure and Process in Modern Societies*, (Glencoe, Ill.), 250–279.
PARSONS, T., [1961], Some Considerations on the Theory of Social Change; *Rural Sociology*, 3, 219–239.
RADCLIFFE-BROWN, A. R., [1952], *Structure and Function in Primitive Society*, (London).
PONSIOEN, J. A., [1962], *The Analysis of Social Change Reconsidered*, ('S – Gravenhage, Netherlands).
ROSE, A. M., (Ed.), (1962), *Human Behaviour and Social Processes*, (London), (Chs. 33 and 34).
REX, J., [1961], *Key Problems in Sociological Theory*, (London).
SHILS, E. A., [1963], On the Comparative Study of the New States; In Geertz, C., (Ed.), *Old Societies and New States*, (London).
SMELSER, N. J., [1959], *Social Change in the Industrial Revolution*, (London).
SMELSER, N. J., [1963], *The Sociology of Economic Life*, (Englewood Cliffs, N.J.).
SMELSER, N. J., [1964], Toward a Theory of Modernization; In Etzioni, A. and E., (Eds.), *Social Change: Sources, Patterns, and Consequences*, (London), 258–274.
STEIN, M. and VIDICH, A., (Eds.), [1963], *Sociology on Trial*, (Englewood Cliffs, N.J.), (esp. essays by Gouldner and Foss).
WEBER, M., [1947], *The Theory of Social and Economic Organisation*, (New York).
WITTFOGEL, K. A., [1957], *Oriental Despotism*, (New Haven and London).
ZOLLSCHAN, G. K. and HIRSCH, W., (Eds.), [1964], *Explorations in Social Change*, (London).

CHAPTER EIGHT

Models of Economic Development

D. E. KEEBLE

GEOGRAPHY, MODELS AND ECONOMIC DEVELOPMENT

Any examination of the professional geographical literature of recent years reveals an apparent and remarkable lack of interest among geographers in the study of the phenomenon of 'economic development' (Ginsburg, 1960, p. ix; Mountjoy, 1963, p. 13; Steel, 1964, p. 13; Lacoste, 1962, p. 248). For example, of the 251 major articles (excluding editorials and reviews) published between 1955 and 1964 inclusive in what is probably the most relevant professional geographical journal, *Economic Geography*, only ten were explicitly concerned in whole or part with problems of economic development. With the more general journal of the *Annals of the Association of American Geographers*, the percentage falls still further, to 2·5 per cent (i.e. six articles out of 242). This state of affairs is remarkable in view both of traditional geographical concern with countries now categorized as 'underdeveloped', and of the enormous surge of interest in problems of economic development which has occurred in other, often fairly closely allied, disciplines (e.g. history, sociology, politics and economics) since the Second World War (Goldsmith, 1959, p. 25; Meier and Baldwin, 1957, p. 1; Pen, 1965, p. 190; Gerschenkron, 1962, pp. 5–6; Meynaud, 1963, pp. 9–10).

The background to this apparent disinterest is undoubtedly complex. But in addition to 'the extreme separatism of geographers as a group' (Haggett, 1965A, p. 101; see also Chorley and Haggett, 1965, p. 375; Ackerman, 1963, pp. 431–432; Chisholm, 1966, pp. 1–2), geography's traditional preoccupation with the individuality and uniqueness of different countries and areas – i.e. the 'idiographic' approach – rather than with their general similarities – i.e. the 'nomothetic' approach (Hartshorne, 1939, pp. 378–384; Ackerman, 1958, pp. 13–16; Bunge, 1962, pp. 7–13) must surely have played some part. For decades, geographers have concerned themselves primarily with the description and analysis of those unique combinations of spatially-associated phenomena which are found in particular, individual, areas and countries (Hartshorne, 1959, pp. 146–149). The categorization in the post

Second World War period of many of these areas as 'underdeveloped' – i.e. part of a general group of areas, dominated by common features and problems – has failed to influence this attitude, a failure evident in the disinterest in economic development problems noted above.

But the effects of this idiographic approach go farther still. For even those geographical studies which focus explicitly on the problems and nature of economic development bear its imprint. Such studies may be divided into the four groups listed in Table 8.1, which also indicates the distribution among these groups of the 16 *Annals* and *Economic Geography* articles mentioned earlier.

TABLE 8.1

'Geographical' Articles on Economic Development – A Classification

Group	Main Focus of Article's Interest	Number of Articles
A	Relationship between Physical Environment (especially Natural Resources) and Economic Development	4
B	Classification of Areas in terms of Indexes of Economic Development	3
C	Unique Characteristics of an Individual Area, with peripheral reference to its Economic Development	6
D	Other	3
	Total	16

Although the sample presented above is perforce a small one, examination of other geographical studies relating explicitly to economic development strongly supports the fourfold grouping suggested. Into Group A, which already includes Ginsburg's *Natural Resources and Economic Development* (1957), and Tosi and Voertman's analysis (1964) of relationships between physical environment and economic development in the tropics, fall such studies as those of James (1951), Keller (1953), Stamp (1953 and 1963), Gribaudi (1965) and Fordham (1965). Group B, including Fryer's classificatory article (1958) in *Economic Geography*, is represented on the wider scale by the opening pages of his *World Economic Development* (1965, pp. 3–24), by Ginsburg's stimulating *Atlas of Economic Development* (1961), and by many of the contributions to the latter's pioneering *Essays on Geography and Economic Development* (1960), notably those by Hartshorne, Wagner, Guyol, Berry, Gosling and Rodgers. Amongst many further examples of Group C may be singled out the various case-studies included in

Hance's *African Economic Development* (1958), together with Ooi Jin-Bee's study of rural development in Malaya (1959), Dwyer's excellent analysis (1965) of Hong Kong, and Green and Fair's work (1962) on economic growth in Southern Africa. Only a very few geographical studies, such as Mountjoy's (1963), cannot be allocated to the first three groups, and must therefore be added to Group D.

Of the four groups, Group C, with the most articles in the two journals examined, is clearly closely identified with the idiographic approach. The accent throughout these articles is on the individual analysis of the particular areas chosen. The same approach is evident, however, in studies in Group A, since apart from Ginsburg's article (1957), most of these work explicitly or implicitly from the viewpoint that 'the specific variations imposed by the conditions of the total environment are unique', and that in analysing the effect of the physical environment upon economic development 'general concepts of wide applicability may be less important than the careful analysis of unique situations' (James, 1951, p. 230). Even studies in Group D appear to emphasize this idiographic approach at times. For example, Mountjoy devotes two chapters to exemplifying his comment that 'every nation runs an individual course in the cross-country development race' (Mountjoy, 1963, p. 157): while Orchard (1960) is undoubtedly as interested in East and South Asia as a unique area as in utilizing it as 'a laboratory for the economic geographer' (Orchard, 1960, p. 215). Only in Group B are the majority of studies concerned explicitly with a search for underlying similarities in patterns of economic development. The nomothetic approach of such classificatory work is exemplified by Ginsburg's study (1960), which concludes that geography's contribution to the study of economic development 'will be greatest through the careful examination and analysis of reality so as to test, appraise, and modify generalizations, rather than through the idiographic study of presumably isolated events' (Ginsburg, 1960, p. xx).

The predominantly idiographic approach of what little geographical analysis of economic development has been undertaken stands in marked contrast to the approach adopted in other disciplines. Only amongst historians is the study of individual, unique, cases of economic development of major disciplinary importance (Hicks, 1953; Hoselitz, 1959); and even here, the search for more general similarities and concepts appears to have intensified in recent years (Hoselitz, 1955A and 1959; Supple, 1963, pp. 7–8; Conrad and Meyer, 1965, pp. 3–28; Rostow, 1960, p. 1). In other disciplines, particularly the most significant one in this context, economics, the nomothetic approach predominates. Indeed, Goldsmith (1959, p. 27) for one claims that idiographic attitudes have no place whatever in economic analysis as applied to problems of economic growth. This major contrast between geography and economics clearly reflects the different historical origins of the two disciplines (McNee, 1959; Chisholm, 1966, pp. 4–25), the former's largely 'empirical and descriptive' character stemming from nineteenth-century Darwinian thinking, the latter's 'more abstract' character as 'a deductive system of logic' being derived chiefly from eighteenth-century rational thinking (McNee, 1959, p. 191).

The nomothetic approach of economics finds its greatest expression in the construction of economic models. An economic model may be defined as 'an organized set of relationships that describe the functioning of an economic entity . . . under a set of simplifying assumptions' (United Nations, 1961, p. 7). By selecting those aspects of economic reality which are deemed particularly significant, and concentrating on the relationships between these few aspects, economists have developed models of economic activity which are 'of great value in interpreting, if not predicting, economic behaviour in the real world' (McNee, 1959, p. 191). Model-building of this kind has in recent years become a key element in economic analysis (Pen, 1965, p. 65; Orcutt, 1960, p. 897). In particular, it has been very rapidly applied to the study of economic development, both conceptually and for planning purposes. This is not to say that economists have not followed other approaches. Even studies of the influence of natural resources upon development (Spengler, 1961; Clawson, 1964) and of methods of classifying countries according to level of development (El-Kammash, 1963) have been included in the vast flood of economic literature on economic development in the last decade. But the building, testing and application of growth models has undoubtedly formed a very important, if not dominant, part of economics' contribution to the study of economic development.

Until recently, one major defect of this model-building activity was its lack of concern with the spatial changes inherent in economic growth. Since the Second World War, however, economics has at last come to realize the significance of this omission, and begun 'balancing its spaceless models with others including the spatial variable' (McNee, 1959, p. 198; see also Chisholm, 1966, p. 2). Despite certain problems (Meyer, 1963, p. 41; Paauw, 1961, p. 180), this process has proceeded so rapidly that a general survey of model-building attempts to analyse the spatial aspects of economic growth is fully justified. Such a survey is perforce concerned primarily with the work of economists who, unfettered by an idiographic tradition, have at last moved to fill the wide intellectual void left open by geographers.

Figure 8.1 portrays diagrammatically a simple typology of economic growth models, organized on the basis of both spatial content and scale coverage. The breakdown of the latter into Supra-National, National and Sub-National units, quite apart from its similarity to scale classifications adopted by other analyses of similar topics (Isard and Reiner, 1961, p. 19; Isard and Smolensky, 1963, p. 105; Friedmann, 1963, pp. 43–44), may be justified in two ways. Firstly, on practical grounds, most models have been developed specifically in terms of one or other of these scale groups. Secondly, on conceptual grounds, both the degree of internal homogeneity of those factors influencing economic development – government economic policy, laws, currency, language, financial institutions, communication systems, etc. – if not of the level of economic development itself, and the degree of 'openness' of the economy concerned to external economic stimuli, vary fairly sharply between these groups.

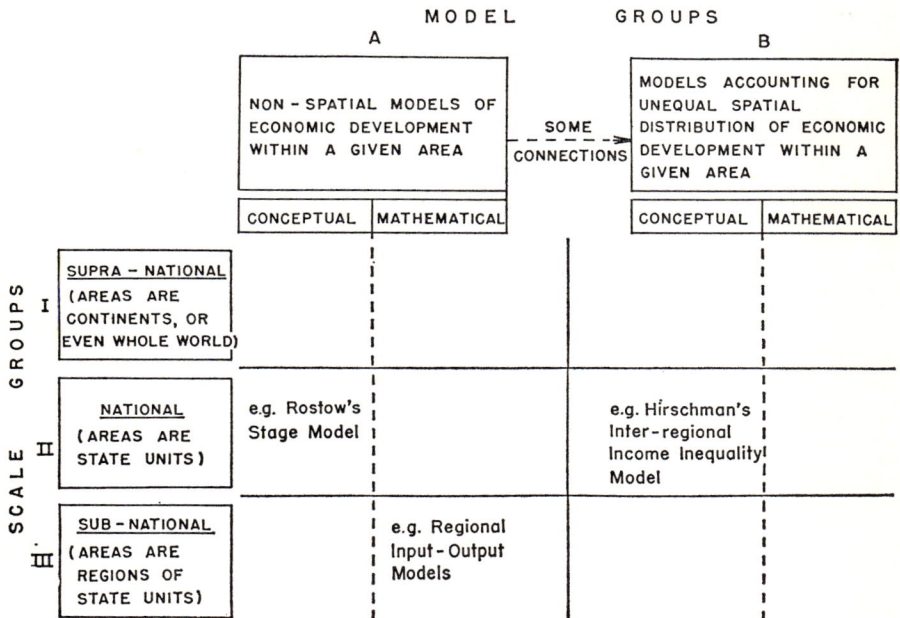

8.1 A Typology of Economic Growth Models.

The key scale group here is undoubtedly the national one. Most economists (e.g. Kuznets, 1951; Robinson, 1960, pp. xiv–xv) appear to agree that 'in analysing economic growth and structure . . . national states are the natural units of comparison' (Goldsmith, 1959, p. 23). State units generally exhibit not only a high level of internal economic homogeneity but also a low degree of openness, in that they are surrounded by considerable barriers to the free flow of commodities — labour, capital, goods, ideas – vital to the process of economic growth (Kuznets, 1951, pp. 29–33; Robinson, 1960, p. xiv). Interestingly enough, both these criteria appear to vary to some extent with the overall level of development of the state unit concerned (Williamson, 1965; Berry, 1960), although this cannot be pursued here. At scales greater than the national unit, however – i.e. continental and world scales, both of which fall into the Supra-National group – internal homogeneity and degree of openness become far less. Indeed, at the world scale, the economic system under consideration is of course completely closed (Tiebout, 1956, p. 161). Conversely, at the Sub-National scale, the economic system becomes much more open, often with virtually complete mobility of commodities across its borders (Harris, 1954, p. 369; Sickle, 1954, p. 382) – while the level of internal homogeneity may also rise. The last point is of course particularly true of those regions which have been delimited in terms of homogeneous developmental characteristics – an exercise in which geographers have hitherto played very little part, despite the discipline's traditional preoccupation with 'regional geography' (see McLoughlin, 1966). For the above reasons, therefore, it appears more realistic to organize the scale breakdown in terms of national units (and subdivisions, or aggregations, of these) than in terms of some absolute scale index, such as that proposed by Haggett, Chorley and Stoddart (1965). The variability in absolute size of state units, and its influence upon levels of economic development, will however be referred to later (see p. 254).

Although it is clear that geographers will primarily be interested in Group B models, certain aspects of those in Group A may also be of value in geographical study. These latter models will be discussed first.

NON-SPATIAL MODELS OF ECONOMIC DEVELOPMENT

The great majority of the interpretations and models of economic growth produced by economists have been non-spatial, a situation arising from the early 'elimination of the spatial variable from theoretical economics' (McNee, 1959, p. 192). None the less, it seems useful to consider some of them in order to illustrate the ways in which economic models have been constructed, to indicate techniques which could be adapted to the building of spatial models, and to demonstrate the possible use of growth models for classifying and comparing different economies. In that an economic model is basically only 'a simplified description of reality' which stresses 'crucial variables' at the expense of the 'myriad of variables that are of secondary importance' (Borts and Stein, 1964, p. 48), apparently dissimilar intellectual constructs may none the less rightly be termed models. For example, a verbal analysis of the generalized pattern of economic change which is thought to have occurred in advanced economies is as much a model as any set of algebraic formulae used under mathematically-specified assumptions to calculate changing values of economic parameters (Enke, 1964, p. 189; Pen, 1965, p. 65). The former type of construct will be referred to here as a conceptual/historical model, the latter as a mathematical model.

National-scale non-spatial models

(1) *Conceptual/historical models.* Although acknowledging the difficulty of reducing to some common denominator the economic history of such widely differing nations as Britain, the United States and Denmark, Hoselitz (1959) has stressed the need for economic growth models to be based on the history of already developed countries. Such models should single out 'from the unique economic history of each country ... some of the variables that seem to have a crucial impact' (Hoselitz, 1959, p. 146), so that the process of growth in these different economies may be more easily compared. At the same time, however, Paauw (1961) has emphasized that the building of historical growth models on the basis of 'armchair theorizing or almost intuitive observation' has 'tended to outstrip empirical verification'. Such models must therefore also be 'formulated as meaningful propositions (i.e. propositions which are verifiable, or may be refuted, if only under ideal conditions)' (Paauw, 1961, p. 180). Finally, Enke (1964, p. 189) has pointed out that any model based on historical data 'must be much more than a description' of growth regularities. It should rather develop 'its own inner logic' (Enke, 1964, p. 190), by showing how changes during one period of growth are related to those occurring during preceding and subsequent ones.

These criteria are very useful in assessing proposed conceptual/historical models. The earliest such were the 'stage theories' of certain late nineteenth-century German economists, such as List, Hildebrand, Bucher and Smoller. These models organized the economic history of the then advanced countries into stages distinguished from each other by a variety of criteria – by character of exchange system in Hildebrand's model (barter; money; credit), by dominant occupations in List's (savage; pastoral; agricultural; agricultural and manufacturing; agricultural, manufacturing and commercial) (Enke, 1964, pp. 191–194). However, these early stage models were in fact little more than descriptive classifications of different types of economic organization supposed to follow each other in time (Goldsmith, 1959, p. 25): and when tested against the real world, their extremely low level of applicability rendered them virtually useless as analytical tools (Hoselitz, 1960, pp. 193–238; Meier and Baldwin, 1957, pp. 143–147; Gras, 1930).

The attempt to develop a more sophisticated and meaningful stage model of economic growth has only been made in quite recent years. Since its initial formulation (1955, Chap. 7, and 1956), extension (1959) and final elaboration (1960), Rostow's generalization of 'the sweep of modern history' as 'a set of stages-of-growth' (Rostow, 1960, p. 1) has attracted remarkable attention. Accepted with alacrity both by economists concerned with underdeveloped countries (Haq, 1963, p. 13; Das-Gupta, 1965, p. 56; Enke, 1964, p. 201; Meier, 1964, p. 3; Brandenburg, 1964, p. 96) and by the intelligent public (Ohlin, 1961, p. 648; Magee, 1965, p. 76), it has none the less provoked strong criticism from economists in developed countries. Its five stages are portrayed diagrammatically in Figure 8.2. Rostow's own chart applying the model to particular countries is reproduced in Figure 8.3.

The model's base-level, as it were, is the traditional society, characterized by limited technology, pre-Newtonian attitudes to science and the physical

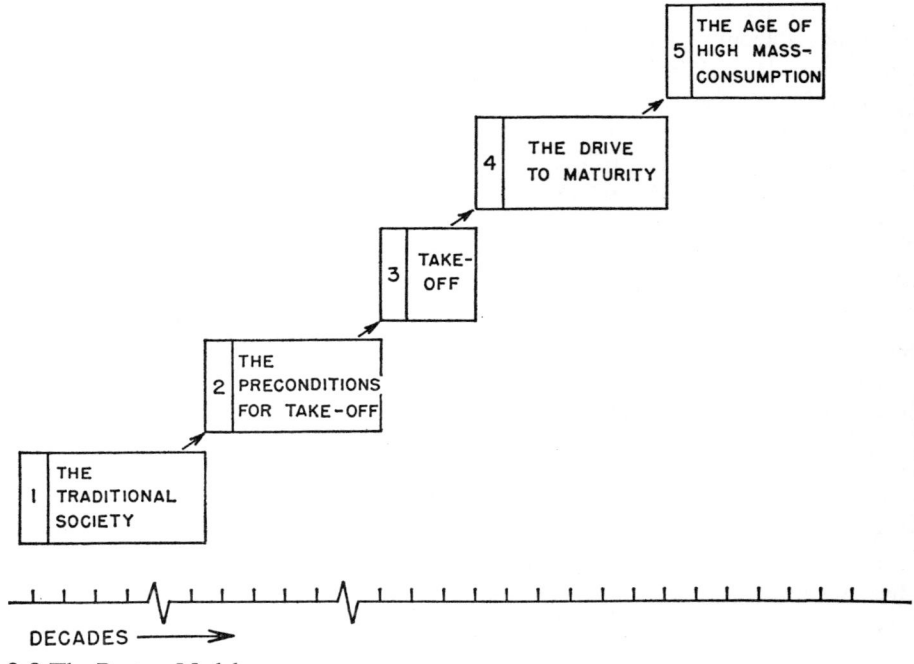

8.2 The Rostow Model.

world, and a static and hierarchical social structure. As primarily exogenous influences stimulate the beginnings of a rise in the rate of productive investment, the installation of 'social overhead capital' (roads, railways, etc.), and the evolution of a new social/political elite, the preconditions for take-off stage develops, with agriculture and extractive industry playing a key role. The crucial stage, however, is take-off, the 'decade or two' when economy and society 'are transformed in such a way that a steady rate of growth can be, thereafter, regularly sustained' (Rostow, 1960, pp. 8-9). In practical terms, take-off is launched by some initial stimulus, and characterized by 'a rise in the rate of productive investment . . . to over 10 per cent of national income', the 'development of one or more substantial manufacturing sectors, with a high rate of growth', and the emergence of 'a political, social and institutional framework' which encourages growth (Rostow, 1960, p. 39). After take-off follows the drive to maturity, during which the impact of growth is trans-

MODELS OF ECONOMIC DEVELOPMENT [251]

mitted to all parts of the economy. Finally, with the shift in sectoral leadership to industries such as durable consumer goods, ensues the age of high mass-consumption – although other alternatives, such as the pursuit of international power, or of the welfare state, may be chosen by particular societies instead.

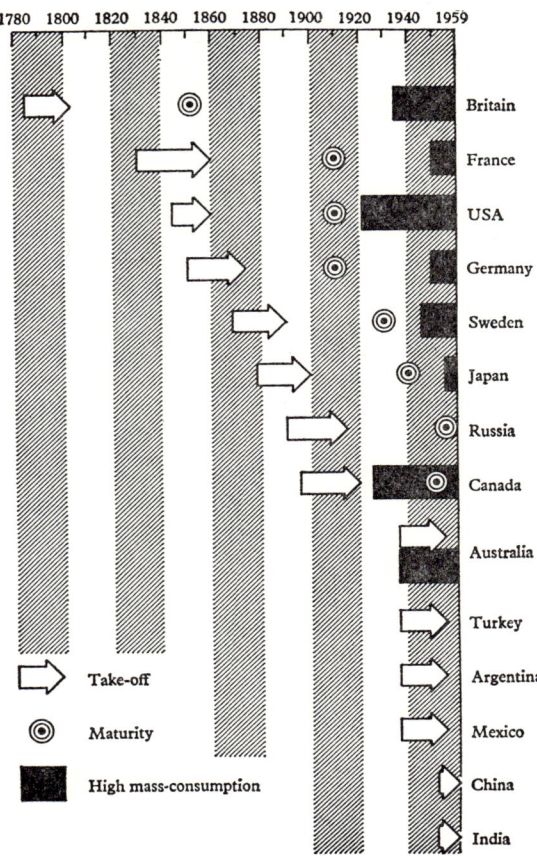

8.3 The Rostow Model Applied to Selected Countries.
(*Source: Rostow, 1960, p. xii*. The chart was originally published in *The Economist*, August 15, 1959, p. 413).

The model's 'provocative suggestions' (Hagen, 1962, p. 522) represent an attempt to 'isolate the strategic factors' in economic growth which is undoubtedly 'more substantial ... more analytical and related to a wider range of issues' than that of any previous stage model-builder (Meier, 1964, p. 25). Despite this, however, numerous criticisms have been levelled against it – criticisms which appear to operate on three distinct levels. On the first level,

Rostow is criticized for having tried to build a model at all. For example, Cairncross (1961, p. 451) claims that any attempt to explain economic growth in terms of 'one or two all-embracing variables' is impossible: growth patterns vary too greatly between different countries. This criticism is to a large extent refuted by Hagen (1962) and Higgins (1964), who agree that the search for regularities in growth patterns is a very necessary task; and that what really matters is not whether Rostow's model and its stages 'ignore some complexities of reality, but whether they are congruent with reality in respects which make them useful for its analysis' (Hagen, 1962, p. 514). This point will be discussed further below.

The second level of criticism is that which accepts the value of model-building, but criticizes Rostow's model as being a poor example of the species. In other words, it attacks the model *as a model*. At this level, the three criteria listed earlier provide valuable yardsticks for assessment. Does Rostow's model single out crucial variables, define its concepts precisely enough to permit comparison with the real world, and possess an analytical rather than descriptive framework? On the first score, the model comes out well. In stressing, for example, the role of 'leading sectors' (i.e. economic activities which exhibit high growth rates and from which growth is transmitted to other parts of the economy), of agriculture and the provision of social overhead capital during early stages in growth, and of a growth-orientated political and social framework, Rostow does appear to have focused attention on variables judged very important by other economists (Hagen, 1959, p. 135; Hoselitz, 1959, pp. 154-155; Johnston and Mellor, 1961; Meier, 1964, pp. 266-272; Meynaud, 1963). Over the second criterion, however, commentators appear divided. Hagen (1962, p. 515), for example, applauds Rostow for identifying 'specific conditions whose presence or absence can be tested'. Certainly this appears true of one of the most crucial variables in his model, the concept of a sudden sharp rise in rate of productive investment at take-off. However, other observers (Enke, 1964, p. 201; Cairncross, 1961, p. 451; Drummond, 1961, p. 113) strongly disagree. In Cairncross' words, 'there are no definitions of the successive stages that admit of their identification by reference to verifiable criteria': and the balance of opinion appears to support a conclusion of this kind.

Criticism over the third criterion is even stronger. Although Rostow in fact claims that his stages 'are not merely descriptive' but 'have an inner logic and . . . analytic bone-structure' (Rostow, 1960, pp. 12-13), nearly all reviewers (Enke, 1964, p. 201; Drummond, 1961, p. 113; Cairncross, 1961, p. 451) agree that in fact his model 'simply fails to specify any mechanism which links the different stages' (Baran and Hobsbawm, 1961, p. 236) and 'is essentially an essay in classification' (Habakkuk, 1961, p. 601). Rostow himself places great stress on the concept of leading sectors, whose 'changing sequence' provides an essential raison d'être for his stage breakdown (Rostow,

1960, p. 14). But as Ohlin (1961, p. 649) points out, leading sectors are not even mentioned in detailed analysis of the first two stages; and discussion of their significance during take-off and post-take-off stages is insufficient to provide the model with the 'analytic bone structure' which Rostow claims for it.

The third level of criticism concerns the testing of the model against the real world. As Chorley (1964, p. 136) has pointed out, such testing is crucial to any final assessment of a particular model's value. Unfortunately, it is here that Rostow's model appears most open to criticism. Firstly, most economic historians (Ohlin, 1961, pp. 649-650; Hagen, 1959, p. 132; 1962, pp. 519-520; North, 1958, p. 75; Cairncross, 1961, pp. 454-456) deny that the histories of present-day advanced countries reveal any signs of a twenty-to-thirty year period in which investment rates suddenly rose sharply, denoting take-off. In Kuznets' words (1963, p. 35), 'the available evidence lends no support to Professor Rostow's suggestions' on this crucial point. Secondly, empirical observations have thrown serious doubt on the very separateness of Rostow's different stages. He himself admits that, for example, the age of high mass consumption can be coincident with the drive to maturity (see Fig. 8.3). However, most observers are agreed that historically, the preconditions and take-off stages are also often indistinguishable (Hagen, 1962, pp. 517-519; Kuznets, 1963, p. 37; Cairncross, 1961, p. 456; Habakkuk, 1961, p. 602). If this is so, it surely casts considerable doubt on the value of Rostow's model as an analytical and predictive tool.

Whatever the final verdict on the model, however, it must be admitted that it has stimulated an enormous amount of research into regularities in economic growth, ranging from empirical testing (Rostow, 1963) to the construction of mathematical models based explicitly on Rostow's ideas (Ranis and Fei, 1961). In addition, it has recently been used, apparently successfully, for setting economic growth in a particular country in a general context (Houghton, 1964). Such use suggests that it might well be valuable and illuminative as a teaching device, for comparing and classifying apparently different economies.

Rostow's is the only major example of a conceptual model which attempts to fit all countries. However, as Enke (1964, p. 204) has pointed out, one way of increasing its 'degree of fit' would have been to have framed it in terms of a smaller and more homogeneous group of state-units. A more limited approach of this kind is favoured by several economists (e.g. North, 1958, p. 75; Ruttan, 1959). The criterion of homogeneity envisaged, however, varies considerably. Enke himself suggests culture. Myint (1964, p. 36) on the other hand regards degree of overpopulation as the crucial factor, at least as far as the construction of models relating to underdeveloped countries is concerned. Hoselitz (1955A, pp. 417-418) favours a division between countries in which growth has occurred by an 'expansionist' process and those in which it has taken place by an 'intrinsic' one. Still other economists, notably Kuznets

(1960, p. 15), stress the importance of size, and the fact that partly because of economies of scale (Ewing, 1964, pp. 356–358; Chenery, 1960, pp. 645 and 651), partly because of different resource potential (Hicks, 1959, pp. 182–183), the nature and problems of economic development in a large country are often significantly different from those of a small one (Kuznets, 1951, pp. 29–31; 1953–54, pp. 14–16; and 1958; Deane, 1961, p. 18; Hoselitz, 1959, p. 145; Robinson, 1960). However, the only commentators who support their particular criterion by actual model construction are Fairbank, Eckstein and Yang (1960). They suggest a simple division of the world into more and less developed economies, and construct a five-phase historical model designed to fit most of the latter, including in particular India, China and other Far Eastern areas. Conceptually, the model clearly owes a vast debt to Rostow, being framed in terms of 'five phases characterized by (1) traditional equilibrium, (2) the rise of disequilibrating forces, (3) gestation, (4) breakthrough or as some prefer to call it, take-off, and (5) self-sustaining growth' (Fairbank, Eckstein and Yang, 1960, p. 1). However, in that its ideas apply specifically to one group of economies, the model's 'degree of fit' to individual cases such as China is clearly greater than that of Rostow's own general model. The development of more limited models of this kind referring to particular groups of countries might well prove a fruitful avenue for future research.

(2) *Mathematical models*. The construction of mathematical models of economic growth within national economies has become one of economics' fastest growth points in recent years. The reason for this lies not only in the growing realization of the value of such models for government planning purposes (United Nations, 1961; Hart, Mills and Whitaker, 1964), but also in the development of new statistical techniques and of computers capable of handling vast quantities of data and calculations (Orcutt, 1960; Cohen and Cyert, 1961; Mills, 1964; Stone, 1964A). Mathematical growth models contain at least three types of elements; components, which are the units for which data are collected (e.g. major industries, or individual households); variables, which describe some aspect of the components (e.g. labour force, or yearly expenditure); and relationships, which 'specify how the values of different variables in the model are related to each other' (Orcutt, 1960, p. 899; see also Tinbergen and Bos, 1962, p. 6). Once the elements of a particular model have been mathematically specified, usually in the form of a series of equations (Tinbergen, 1959; Enke, 1964, Chap. 9; Goldsmith, 1959, pp. 75–78), the implications of changes in any of them produced by growth can be calculated.

A simplified classification of mathematical growth models involves a division into 'aggregate' and 'inter-industry' models (United Nations, 1961). The former usually consist of a series of formulae, specifying relationships between variables such as production, consumption, investment, etc., within an entire economy. Good examples are the models developed by Klein (1961) for Japan, Mahalanobis (1963) for India, and Valavanis-Vail (1955) for the United States. Inter-industry models, on the other hand, are 'concerned with the quantitative analysis of the interdependence of producing and consuming units in a modern economy', and focus particular attention on the interrelationships among different producers 'as buyers of each others' outputs, as users of scarce resources, and as sellers to final consumers' (Chenery and Clark, 1959, p. 1). The simplest and commonest type is the 'input-output' model, developed by Leontief (1936; 1951A; 1951B and 1953). Input-output models group production activities in an economy into a number of sectors – e.g. in simplest form, agriculture, extractive industry, manufacturing industry, services, etc. (Chenery and Clark, 1959, pp. 13–15). These sectors are then listed as headings in a table, or 'input-output matrix', both down the left-hand side – where they are thought of as producing units – and along the top – where they are thought of as consumers. The spaces in each horizontal row of the matrix are filled in with the value of that sector's production (output) which is sold to each of the other sectors (as their inputs) during the period under study (Fisher, 1964, p. 20). An extra vertical column is usually included to contain values of output sold directly to consumers who are not themselves producers – i.e. to final consumers. From the values thus plotted, coefficients relating input from one sector to total output of another sector can be calculated (Leontief, 1965A, p. 33); and these, under certain assumptions (United Nations, 1961, p. 13), can be used to analyse the effects which growth in one sector – e.g. in total agricultural output – will have upon each of the others, and therefore on the whole economy. In other words, input-output models explicitly recognize that changes in production in one sector inevitably affect production in many other sectors as well – by means of altered demands for outputs from these other sectors as inputs to the original sector. Calculation of the full effects on other sectors is carried out by the 'iterative' method (Isard, 1960, p. 331), or 'round-by-round' computation of input requirements. Input-output models have proved very valuable for assessing the impact of changes of this kind, and for planning economic development generally in a number of countries (Barna, 1963, Part I; United Nations, 1961; Leontief, 1965B).

Another and similar type of model (Stone, 1956) is the 'social accounting' model. This also organizes its information in terms of a matrix, but interest is focused here on all the different kinds of monetary transactions (e.g. government payments, property income) which occur within an economy (Stone, 1956, p. 156). A social accounting model is at present being used by the Cambridge University Department of Applied Economics to analyse the sectoral implications of future economic growth in the United Kingdom (Department of Applied Economics, 1962; Stone, 1961; 1964B and 1965;

Hart, Mills and Whitaker, 1964). Finally, 'linear programming' models, representing an elaboration of input-output ones, are also now being used to analyse economic growth, and to determine 'the most economical way of achieving a given set of objectives' in economic development (United Nations, 1961, p. 14; see also Chenery, 1961 and 1963; Chenery and Clark, 1959, Chap. 4; Sandee, 1959).

Although non-spatial, most of the mathematical models discussed above can be used to throw light on the differences between countries and regions in nature and level of economic development, and it is because of this that they have been discussed here.

Sub-national-scale non-spatial models

Although most economists have studied non-spatial aspects of economic growth in terms of national units, a few (Gras, 1922; Robock, 1956; Isard and Smolensky, 1963; Fisher, 1955) have stressed the need for regional development studies and models, particularly in underdeveloped and very large countries. Hoover's conceptual model of regional growth (Hoover, 1937, pp. 284–285 and 1948, pp. 187–196), involving a shift in regional economic activity from self-sufficient subsistence agriculture, through agricultural- and mineral-based industries, to tertiary activity oriented to export markets, is very similar to that suggested by Fisher (1955, pp. 3–14). The key to both is regional industrialization, and the focus on the changing economic structure of the region is largely non-spatial. So too is Rostow's model which, as he points out (Rostow, 1960, p. 1), is capable of application at the regional level. Later conceptual regional models, however, have stressed far more the spatial relationships of developing regions with other areas in the same country, and will therefore be discussed in the context of Model Group B.

The development of non-spatial mathematical models of regional growth has proceeded very rapidly in the last few years (Meyer, 1963). Nearly all of these however represent only the application at the regional level of models, particularly inter-industry models, developed initially at national scales. Input-output growth models, for example, have in recent years been increasingly applied to regional as well as national economies (Barna, 1963, Part II; Isard, 1951 and 1960, Chap. 8; Isard and Cumberland, 1961; Maki and Yien-I Tu, 1962; Chatterji, 1964). Indeed the size of region involved has occasionally been as small as an urban area (Hirsch, 1959 and 1963). More general accounting models, too, have been applied at the regional scale (Jouandet-Bernadat, 1964; Hirsch, 1962; Hochwald, 1961): and simple forms of aggregate models have been used for planning regional economic growth, as in Pakistan (Haq, 1963). These regional models generally and necessarily take more account of external relationships than their counterparts on the national scale; but they are still largely non-spatial in character.

MODELS OF SPATIAL DISTRIBUTION OF ECONOMIC DEVELOPMENT

As pointed out earlier, only in recent years have economists turned from their preoccupation with temporal variations in economic growth to consider in addition spatial variations in development (Hirschman, 1959, p. 144). Since then, however, the construction and testing of models of the spatial distribution of economic development has proceeded very rapidly. Indeed, even Rostow (1964, pp. 103 and 122–131) now stresses the major importance of such variation within countries, despite the fact that his earlier model almost entirely ignored it (Kindleberger, 1964, pp. 248–249). Since most work on spatial variation in development has been focused on the national scale (see Figure 8.1), this will be considered first.

National-scale spatial models

(1) *Regional income inequality models.* The existence of regional inequalities in income – in both absolute and per capita terms – within virtually all countries is by now well attested. Put another way, economic development is scarcely ever spread evenly over the whole area of a given state unit, but rather concentrated at certain points, producing a mosaic of regions at different levels of economic prosperity (Vinski, 1962; Hemming, 1963; Hirschman, 1958, p. 183; Paauw, 1961, p. 186; Milhau, 1956; Gannagé, 1962, pp. 62–65). It is true, as Klaassen, Kroft and Voskuil (1963, p. 77) have pointed out, that the range of spatial variation in level of development within a country will in large part depend upon the scale of regional sub-division adopted (see also Haggett, 1965B). But since spatial variation in development exists in most countries even on the simplest scale of sub-division – i.e. a 'north–south' division (see Eckaus, 1961; Paish, 1964) – this point does not affect the general statement above.

Once recognized, this kind of spatial variation has proved difficult to explain in terms of traditional models of interregional economic relationships, particularly trade models. Such models, based on the concept of static equilibrium, assume that given relatively free mobility of the factors of production – a situation which to some extent obtains within most countries (Harris, 1954, p. 369) – 'factor movements tend to bring about an equalization of income among regions' (Harris, 1957, p. 191). Any differences in income levels between regions must therefore be viewed as only temporary, due to some slight lag in adjustment. Unfortunately, however, as other authorities have pointed out (Borts and Stein, 1964, pp. 49–55; Borts, 1960; Myrdal, 1957B, p. 13; Williamson, 1965, p. 5; Bachmura, 1959, p. 1012; Sisler, 1959, p. 1100; Schuh and Leeds, 1963, p. 296), equalization models are of little use

[258] MODELS IN GEOGRAPHY

in illuminating the development of spatial variation in the real world, since such variation is not only remarkably persistent, but apparently increasing in many countries (Williamson, 1965, pp. 16–17; Economic Commission for Europe, 1955, pp. 143–144; O'Connor, 1963).

The inadequacy of equalization theories has prompted economists in recent years to put forward new conceptual models of the development of spatial variation in economic prosperity. Probably the most important of these is the model elaborated by Myrdal during the mid-1950's (Myrdal, 1956; 1957A and 1957B). Framed implicitly in terms of countries already populated but exhibiting a low level of economic development, Myrdal's

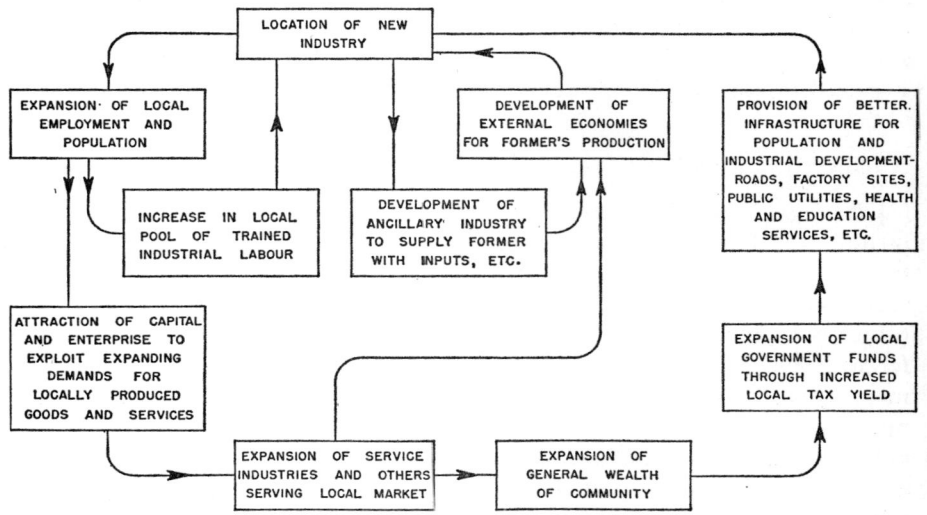

8.4 Myrdal's Process of Cumulative Causation – A Simple Illustration (the two subcircles are included to illustrate the ramifications of the cumulative causation process).

model is based on the contention that in a free economy, particular changes do not, as equalization models contend, 'call forth countervailing changes but, instead, supporting changes, which move the system in the same direction as the first change but much further' (Myrdal, 1957B, p. 13). Applying this idea of 'cumulative causation' to the problem of economic development within countries, Myrdal concludes that 'the play of the forces in the market normally tends to increase, rather than to decrease, the inequalities between regions' (Myrdal, 1957B, p. 26): that is, that once particular regions have by virtue of some initial advantage moved ahead of others (see also Clark, 1966, p. 6), new increments of activity and growth will tend to be concentrated in the already-expanding regions *because of their derived advantages* rather than in the remaining areas of the country. The flow diagram of Figure 8.4 illustrates one possible example of this cumulative process (see also Pred, 1965, p. 165).

The concept of cumulative concentration is not however the only important feature of Myrdal's model. Closely involved with it in his explanation of differential regional growth is spatial interaction between the growing and stagnating regions. Once growth has begun in the former, Myrdal claims, spatial flows of labour, capital and commodities develop spontaneously to support it. Such flows operate, however, as 'backwash effects' upon the remaining regions of the country, since faced with the higher returns obtainable in the growth regions, these other regions tend to lose not only their more skilled and enterprising workers, but also much of their locally-generated capital. At the same time, goods and services originating in the expanding regions flood the markets of the remaining regions, putting out of business what little local secondary and tertiary industry may already have developed there. A further type of backwash effect operates through non-economic factors, such as the provision of poorer health and education services in the stagnating as compared with the expanding regions. In all these ways, backwash effects, particularly those working through spatial interaction, come into operation to frustrate growth in the former and sustain it in the latter (Myrdal, 1957B, pp. 27-31).

However, these backwash effects are not the only interregional relationships which the model postulates as developing within a growing economy. Also of significance are 'certain centrifugal "spread effects" of expansionary momentum from the centres of economic expansion to other regions' (Myrdal, 1957B, p. 31). By stimulating demand (e.g. for agricultural and mineral products) in other, particularly neighbouring, regions, expansion in the growing areas may initiate economic growth elsewhere. If the impact is strong enough to overcome local backwash effects, a process of cumulative causation may well begin, leading to the development of new centres of self-sustained economic growth. Such spread effects are however strongest in economies which have already achieved a fairly high level of economic development, since this 'is accompanied by improved transportation and communications, higher levels of education and a more dynamic communion of ideas and values – all of which tends to strengthen the forces for the centrifugal spread of economic expansion or to remove the obstacles for its operation' (Myrdal, 1957B, p. 34). At the same time, stronger spread effects will boost the economic growth of the country as a whole, by utilizing properly the resources of the formerly stagnant regions.

A basic assumption of Myrdal's model is government non-intervention in economic development. However, he does suggest that in advanced economies, stronger spread effects are aided by government policies aimed at fostering growth in backward regions, and that such action can be interpreted as only another aspect of cumulative causation in development. Certainly government concern with regional inequalities does appear to be greater in advanced countries such as the United Kingdom (National

Economic Development Council, 1963, pp. 14–29; Needleman and Scott, 1964; Humphrys, 1962 and 1963; Lonsdale, 1965; Manners, 1962; Fyot and Calvez, 1956), those of Western Europe (European Coal and Steel Community, 1961; Political and Economic Planning, 1962; European Free Trade Association, 1965; Barzanti, 1965; Romus, 1958; International Information Centre for Local Credit, 1964; Meyers, 1965; Ginsburg, 1957), and those of North America (Committee for Economic Development, 1958; Gilmore, 1960; Wood and Thoman, 1965; Wilson, 1964), than in the underdeveloped countries of the world, where government-inspired studies such as those of Yates (1961) and Furtado (1964) are few and far between.

Myrdal's interregional income inequality model, the first major example of its kind, was quickly followed by others. Hirschman, in his *Strategy of Economic Development* (1958, pp. 183–194) develops a model which although arrived at independently of Myrdal's is in many ways remarkably similar to it. The key role in differential growth is here accorded to spatial interaction between growing 'Northern' and lagging 'Southern' regions, in the form of 'trickling-down' and 'polarization' effects. As Hirschman agrees, these are the exact counterparts of Myrdal's spread and backwash effects (Hirschman, 1958, p. 187), involving just those movements of capital, labour and commodities which have been previously described. However, the model, far from assuming a cumulative causation mechanism, implies that if an imbalance between regions resulting from the dominance of polarization effects develops during earlier stages of growth, counter-balancing forces will in time come into operation to restore the situation to an equilibrium position. Such forces, chief of which is government economic policy, are not to be thought of as intensified trickling-down effects, but as a new element in the model, arising only at a late stage in development. Their inclusion, together with the exclusion of any cumulative mechanism (Hirschman, 1958, p. 187), represent the model's chief structural differences from that of Myrdal.

Yet another similar interregional growth model has been suggested by Hicks (1959, pp. 162–166). After stressing the significance of initial advantages in determining the exact location of growth regions, Hick's model follows Myrdal's in emphasizing the importance of derived, cumulative, advantages in their subsequent development. It also focuses attention on spatial interaction, and flows of goods, labour and capital between growing and lagging regions. However, these flows are interpreted only as representing the 'tendency for wealth to spill over' from growth centres (Hicks, 1959, p. 163), rather than as a hindrance to growth in lagging regions. The movement of labour, for example, is seen as aiding the latter, not hindering their expansion: while only capital transfers from growing regions to other parts of the country are examined – a direction of flow opposite to the main flows suggested in both Myrdal's and Hirschman's models.

Least elaborated of this group of interregional growth models is Ullman's

(1958 and 1960). In his formulation, the crucial variable is the 'selfgenerating momentum' (Ullman, 1958, p. 180) experienced by growth regions – a momentum which reflects the development in them of 'notable external economies of scale and the largest market in the country' (Ullman, 1958, p. 184). Spatial interaction, though implicit in his analysis, is not really examined, probably because he is far more concerned with sheer concentration of new development than with the creation of differences in per capita regional income.

The two most important of these interregional income inequality models – Myrdal's and Hirschman's – have attracted a great deal of attention since their formulation; and although much remains to be done, preliminary testing of these, both as models and against the real world, has proceeded far enough to justify certain interim conclusions. Firstly, the models, especially Myrdal's, do appear to possess a well-defined structure, in which important variables are clearly related to each other. Indeed, Myrdal's claim that his cumulative causation hypothesis could in principle be stated 'in the form of an inter-connected set of quantitative equations' (Myrdal, 1957B, p. 19) has partly at least been justified by an attempt at mathematical formulation (Singer, 1961). More important, Borts and Stein (1964, pp. 4–7) have shown by a simple mathematical model that 'Myrdal's views are logical; i.e. they contain no internal contradictions' (Borts and Stein, 1964, p. 4): while despite Paauw's criticism (1961, p. 186) of backwash/polarization effects as 'hardly a satisfactory explanation' of regional backwardness, most commentators (e.g. Friedmann, 1959, p. 174; Lasuen, 1962; Hughes, 1961) seem to agree that these do represent important variables in explaining differential development.

Testing and criticism of the models as they relate to the real world has concentrated on four main topics. The first of these is Myrdal's hypothesis of cumulative causation, a concept stressed in slightly different form by Hicks and Ullman. Many authorities (e.g. Economic Commission for Europe, 1955, p. 142; Chisholm, 1962, p. 159; Caesar, 1964, p. 238; Nicholson, 1965, pp. 164–169) appear to agree with O'Connor's comment (1963, p. 42) that in most countries 'economic development tends to be concentrated in the areas where most has already taken place'. From this proposition, it is only a very short step to Myrdal's, and the stressing of the 'cumulative advantage' of growth regions (Perloff and Wingo, 1961, p. 106), or of 'the cumulative process of growth in the concentration areas' (Economic Commission for Europe, 1955, p. 154). As a result, Myrdal's concept receives considerable support, both implicitly in studies which actually precede his (Royal Commission, 1939, pp. 29, 49 and 170; Sturmthal, 1955, p. 200), and explicitly in subsequent work (Spencer, 1960, p. 46; Wonnacott, 1964, p. 418; Cairncross, 1959, p. 109; Baer, 1964, p. 269; Lasuen, 1962, p. 188; Pred, 1965, pp. 160–166; Brandenburg, 1964, pp. 208–209). In most cases, commentators are concerned primarily with growth regions, and tend to stress as Myrdal does (1957B, p. 27) the role of external economies (Lasuen, 1962, p. 180), together with growth above certain 'thresholds' of economic activity (Pred, 1965, p. 165), and higher rates of technological innovation (Perloff and Wingo, 1961), in their cumulative expansion. The prospect of cumulative growth has even stimulated policy recommendations regarding concentration of government investment in underdeveloped countries (Mabogunje, 1965, pp. 436–438; Friedmann, 1963, p. 53; Lutz, 1960, p. 45).

The cumulative causation hypothesis appears therefore to have been accepted by most authorities as a relevant and useful concept in analysing spatial concentration of economic growth, at least during early stages of development. So, too, with perhaps one major exception, has the significance of backwash/polarization effects. The most important of these effects, specified in both Myrdal's and Hirschman's models, are interregional flows of capital, commodities and labour; and tentative conclusions about the existence and nature of these flows can be drawn from recent empirical studies. For example, it seems generally agreed that in many underdeveloped countries there does indeed occur a net capital transfer from lagging to growing regions. This pattern has been reported for Indonesia (Williamson, 1965, p. 7), with transfers from the outer to the central islands; Pakistan, with flows from East to West (Haq, 1963); Brazil, with private capital movements from the north to south (Robock, 1963, p. 108; Baer, 1964); Nigeria, with currency flows from 'the West, Mid-West, and North towards the East and Lagos' (Hay and Smith, 1966, p. 23); and Spain (Lasuen, 1962). Most of these observers (e.g. Haq, 1963, pp. 103–104) consider the pattern an important factor behind differential regional growth. The models' forecasts on commodity flows also tend to be supported by what little evidence exists. In Belgium (Verburg, 1964A, p. 144), Italy (Clough and Livi, 1956, p. 336; Eckaus, 1961, p. 314) and to some extent Britain (Smith, 1953, pp. 93–121), goods manufactured in the growth regions of the country during the eighteenth and nineteenth centuries helped to depress what industry already existed in the more backward areas. The same process may have taken place in Brazil in the nineteenth and twentieth centuries (Furtado, 1963, pp. 264–265).

In so far as the sheer existence of net population migration from lagging to growing regions is concerned, the Myrdal/Hirschman analysis of labour flows has been fully accepted. However, their views concerning the selectivity – in terms of age, ability or skills – of such migration, and its effect upon the lagging and growing regions, appear to be far more open to question. Some commentators (Hughes, 1961; Lasuen, 1962; Williamson, 1965, p. 6; Friedmann, 1959, p. 174; Parr, 1966, pp. 152–154) explicitly support them, as do certain other independent studies (Eckaus, 1961, p. 317; Hathaway, 1960; Randall, 1962, p. 78). However, Okun and Richardson (1961), in an important contribution, claim that the Myrdal/Hirschman argument on selectivity 'does not adequately deal with the complexities inherent in the relationships between migration and inequality of per capita income' (Okun and Richardson, 1961, p. 132). To demonstrate this, they develop a simple conceptual model which relates interregional migration within a closed economy to level and rate of economic growth in different regions. When applied to migration within the United States, this model does appear to show that the influence of migration – even age-selective migration – upon regional per capita income differences varies considerably according to the circumstances, particularly the direction and duration of movement. Okun and Richardson's conclusions therefore throw considerable doubt on the view of selective migration as a normal feature of backwash/polarization effects.

Empirical evidence concerning the third important aspect of these models – spread or trickling-down effects – is much scantier than that for backwash effects. Lasuen, for example, sees little sign of their operation in Spain, except in the immediate vicinity of Barcelona and Bilbao (Lasuen, 1962, p. 177): while Chenery's analysis (1962) of economic development in southern Italy shows that although the area's 'economic structure is influenced by being part of a more advanced economy' (Chenery, 1962, p. 517), this influence has operated far more through massive government income transfers to the South than through any natural spread of expansionary momentum from the North. This apparent absence of spread effects in many, particularly underdeveloped, countries is possibly partly due, as Hicks (1961, p. 77) has pointed out, to the considerable cultural, social and economic differences between regions. However, in more advanced countries where such differences are less, the geographical spread of economic development appears to have occurred partly by a mechanism ignored in the Myrdal/Hirschman model – that of industrial dispersal in search of labour, raw materials, or markets (McLaughlin and Robock, 1949; Sickle, 1951, p. 387; McGovern, 1965; Keeble, 1965; Cameron and Clark, 1966). In at least one case, this has been a major factor stimulating regional economic growth (Manners, 1964, p. 51). By contrast, governments in underdeveloped countries have generally found it very difficult to promote industrial dispersal (see Coutsoumaris, 1964, pp. 85–86). It can be suggested therefore, that although no general conclusion concerning the existence of spread effects is as yet possible, at least one further mechanism for the interregional spread of economic growth in advanced economies needs to be included in any analysis of them.

The final aspect of these models to have provoked considerable empirical testing is the question of divergence or convergence of regional per capita incomes. On its own, Myrdal's cumulative causation concept suggests continuing divergence in such incomes as a typical feature of developing countries. Hirschman's model, on the other hand, provides a theoretical background for regarding convergence as the norm. Unfortunately, however, empirical studies at first sight appear to arrive at conflicting conclusions.

Indeed, in the case of Brazil authorities are divided even as regards the same country, Baer (1964) and Furtado (Robock, 1963, pp. 107–110) claiming that the gap between regional incomes per capita has been widening, Robock (1963, p. 46) that it has been converging. Elsewhere the situation appears only slightly less confused. Disparity in regional incomes appears to have been increasing in Italy (Clough and Livi, 1956, p. 335), Mexico (Sturmthal, 1955, p. 201) and possibly Uganda (Elkan, 1959, p. 137), while convergence, of wage levels as well as regional incomes, seems to have characterized the recent history of the best-attested case, the United States (Smolensky, 1961, p. 68; Easterlin, 1958 and 1960; Fulmer, 1950, p. 273; Sickle, 1951, p. 389; Wonnacott, 1964).

This conflict is however more apparent than real; for both models in fact recognize the greater likelihood of divergence during earlier periods of growth, and of convergence during much later periods. In Myrdal's case, cumulative growth in lagging regions may well eventually be stimulated by a combination of spread effects and a stimulus external to his model, government intervention. In Hirschman's, divergence during an early stage is in fact anticipated, convergence only occurring later. This idea of a shift from divergence to convergence with increasing economic development – an idea referred to by other authorities (e.g. Balassa, 1961, p. 201) and paralleled by trends in personal income inequalities (Kuznets, 1955, p. 18) – has been taken up by Williamson (1965) in a major empirical investigation. His work shows that what little statistical information on regional income inequalities in different countries is available does support the view that poor but developing countries are characterized by increasing regional disparities, while more developed ones exhibit decreasing disparities. A statistical index of regional income disparities in any one country might thus trace out a path on a graph similar to one of those shown in Figure 8.5. An interesting point brought out in Williamson's study is the apparent secondary influence of geographical size upon degree of regional inequality. His analysis (Williamson, 1965, p. 15) lends some support to Kuznets' view (1960, p. 30) that 'the developed small states seem to have succeeded in spreading the fruits of economic growth more widely among their populations than the larger states at comparable levels of income per capita'.

Initial testing of the Myrdal/Hirschman models on the whole therefore suggests that they are congruent with reality – particularly the reality of present-day underdeveloped economies – in respects which make them valuable in its interpretation. However, this can only be an interim conclusion, and considerable further empirical work needs to be done before a full assessment can be made.

Although the interregional income inequality models just discussed have attracted most attention, a few other conceptual models may perhaps also be included under this general heading. For example, Keirstead (1948, pp. 265–

313) has put forward a model, based clearly on Canadian experience, of the process of regional depression and growth in an advanced country. Framed in terms of a two-region ('Eastland' and 'Westland') closed economy, this model suggests that the volume and nature of industry in these regions is normally in some kind of 'location equilibrium'; and that if technological changes favouring concentration of industry in the region (Westland) with the larger market upset this equilibrium, shifts in industrial location will occur 'until a new equilibrium in location, satisfying the new conditions, is achieved' (Keirstead, 1948, p. 269). The creation of a labour cost differential favouring the depressed region plays a major part in this adjustment process.

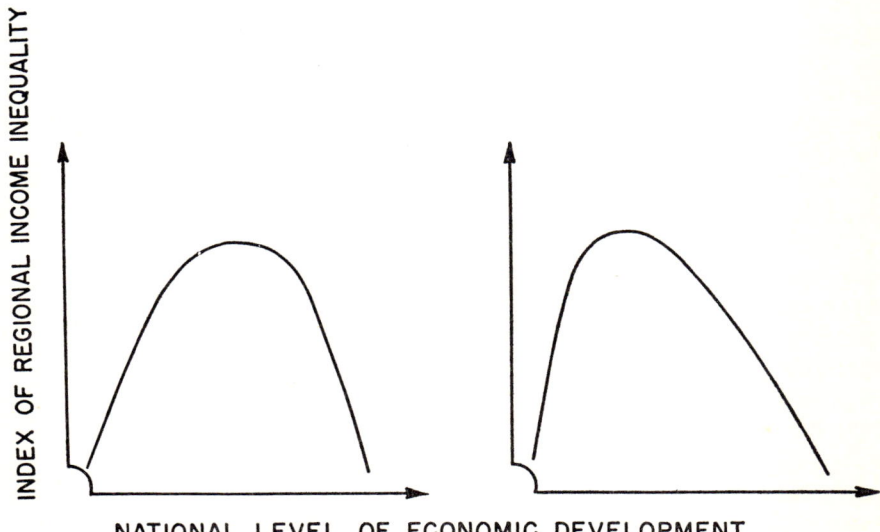

NATIONAL LEVEL OF ECONOMIC DEVELOPMENT

8.5 Possible Relationships between Indexes of Regional Income Inequality and National Development (*After Williamson, 1965, pp. 9–10*).

Two further conceptual models have been suggested by Friedmann. The earlier (Friedmann, 1956) represents an attempt to relate certain insights of location theory, such as the concept of functional hierarchies of cities and city regions, and of agglomeration economies, to the question of spatial changes associated with economic growth. As economic development proceeds, Friedmann claims, the degree of areal specialization, functional differentiation and spatial interaction between different regions increases. As a result, the economy's spatial structure develops from one characterized by 'small, isolated, and functionally undifferentiated communities', into a hierarchy of interdependent regions, this giving way at the highest stages of development to 'more or less "autonomous" linear cities and conurbations of very large

9*

proportions' (Friedmann, 1956, p. 226). His later model (1963) relates specifically to Latin American countries, and is couched in terms of four phases of development – initial coastal settlement; the phase of semi-autonomous, externally-orientated regions; industrialization and development of a centre-periphery structure; and final integration of the national space economy through the spread of metropolitan regions. Although bringing together many useful ideas, neither of these two models throws much new light on spatial inequalities in development, or has been followed up by empirical work.

(2) *Export base models*. Regional income inequality models are not, however, the only intellectual constructs to have been developed for studying variations in regional economic growth. A different approach has been followed by certain economists concerned with regional economic history and growth in the United States. Impressed by historical evidence, these have stressed the key role of a region's 'export base' – that is, 'collectively the exportable commodities (or services) of a region' (North, 1955, p. 248) – in determining the rate of its economic growth. Indeed, Perloff and Wingo (1961, p. 200) claim that in the United States 'regional growth typically has been promoted by the ability of a region to produce goods or services demanded by the national economy and to export them at a competitive advantage with respect to other regions'. This view has led such writers (e.g. North, 1955; Thomas, 1963) to abandon earlier regional growth models (see page 256) and to develop ones stressing the role played by the export base. North (1955), for example, has put forward a five-stage regional export base model which he suggests can be applied not only to the United States, but also to other capitalist countries in which population pressure has been slight. After a very brief subsistence stage, he envisages the rapid development of exporting of staple commodities to more advanced regions as the basis of the regional economy. With the growth of external economies, inflow of capital, and provision of an export-orientated infrastructure, a further stage of export intensification and regional development ensues, leading in time to the development of 'residentiary' industry, serving local markets. Finally, the expansion of residentiary industry, together with 'footloose' industries located more or less by chance in the area, may reach a point at which they too enter export markets, thereby diversifying the region's export base. This model is closely similar to ideas put forward by Perloff and Wingo (1961, pp. 200–201), and shares a common frame of reference with Baldwin's analysis (1956) of the role of the production function of the main export commodity in influencing the development of other sectors of economic activity in a newly-settled region.

North's model focuses attention on the mechanism of economic growth of one particular region. Duesenberry's export base model (Duesenberry, 1950, pp. 96–102), however, is more concerned with the impact of economic growth in a newly-settled, exporting region, upon that of an older, already developed region. Clearly based upon the historical experience of the East and West North Central regions of the United States, this model suggests that in addition to 'western' economic growth consequent upon the development of exports of farm products, 'eastern' economic growth will also be stimulated by increased demand from the former region for more sophisticated products (e.g. manufactured goods) which are not as yet produced in the west. Given a sufficient rate of western agricultural expansion, growth induced by this demand will be more than enough to offset local agricultural decline due to increased western competition. Finally, rise in income in both regions will, as in North's model, generate expansion in local industries and services to serve expanding local markets.

Duesenberry's two-region export base model has attracted little comment, although later studies of the United States (Conrad and Meyer, 1965, pp. 225–226), Canada (Meier, 1953, p. 5) and Mexico (Glade and Anderson, 1963, p. 77) lend some support to its analysis of growth relationships between older, developed regions, and newly-settled, 'frontier' areas. The more general export base model exemplified by North's analysis, however, has stimulated considerable controversy concerning both its conceptual framework and empirical validity. Tiebout's criticisms (1956) are largely in terms of the former. He claims that variables other than exports – such as 'business investment, government expenditures, and the volume of residential construction' (Tiebout, 1956, p. 161) – affect regional income, and that the development of residentiary industry is of great importance to a regional economy in lowering costs in export industries. Tiebout (1956, p. 161) also stresses that 'the quantitative importance of exports as an explanatory factor in regional income determination depends, in part, on the size of the region under study' – a point on which Harris (1957, p. 169) is in agreement. North's immediate reply (1956) concedes the point about residentiary industry, while his subsequent article (North, 1959) admits that the diversion of income received from exports into other activities within the region plays a more important role in sustained regional growth than he formerly allotted to it. This slightly amended model provides the conceptual framework for his major analysis (North, 1961) of nineteenth-century economic growth in the United States.

Although Wilhelm's earlier discussion (1950) of regional economic growth in Soviet Central Asia provides considerable support for the export base model, explicit testing has been confined to examination of United States' experience. An important contribution here is that of Borts (1960), whose analysis of factors influencing changing regional wage levels in the United States 'indicates strong support for a model of regional growth based on the demand for a region's exports' (Borts, 1960, pp. 342–343). Detailed study of one particular region, the Pacific Northwest, confirms this view for that area (Tattersall, 1962), while other studies (Perloff, 1960; Perloff and Wingo, 1961; North, 1961) do appear to demonstrate the relevance of the model to the economic history of other parts of the United States. The most recent of these, and one of particular interest, is Borts and Stein's sophisticated analysis (1964). These authors use mathematical growth models to explore changing economic relationships during regional economic development, and accord export activities, and changing demand for regional exports, a major role in explaining regional growth differentials in the United States (Borts and Stein, 1964, pp. 121 and 132). However, their analysis also places considerable weight upon another variable in regional economic growth – the role of regional differentials in wage levels, and related changes in labour supply. Both wage levels and exports are held to be of equal importance since either or both of them may stimulate 'an expansion of investment in the region' (Borts and Stein, 1964, p. 121) and regional economic growth. This conclusion is reinforced by careful statistical comparison with United States experience, and suggests that although 'supported by considerable evidence' (Friedmann and Alonso, 1964, p. 210), the simple export base model may well now need refinement if it is to remain a useful tool in the comprehension of regional economic growth.

(3) *Mathematical models.* The development of mathematical models focusing explicitly on variations in the spatial distribution of economic development within a given state-unit has lagged far behind that of the non-spatial models discussed earlier. Perhaps the simplest work along these lines is represented by the use of regression analysis to determine the degree of statistical correlation between level of economic development and other variables presumed to influence this, in different regions of a given country. Fulmer's study (1950) is an early example of this, while Thompson and Mattila (1959) use more sophisticated multiple correlation techniques in their study of the factors associated with rapid employment growth in different states of the United States. In one sense, these studies do represent attempts at model-building, since the end-product is the specification of a generalized relationship between certain, possible causative, variables, and level of economic development. Indeed, in Klaassen, Kroft and Voskuil's study (1963), the use of regression analysis is closely integrated with the development of an actual mathematical model, constructed to throw light on the factors behind differences in regional per capita income in Holland.

This particular model developed by Klaassen, Kroft and Voskuil is an aggregate model, relating such variables as size and rate of natural increase of the working population, economic structure, etc., to regional economic development. As such, it illustrates the way in which aggregate models, although originally non-spatial (see page 254), are now also being developed explicitly for the purpose of illuminating regional differences in economic growth within a given state-unit. Other examples of this trend are the aggregate models developed by Borts and Stein (1964) to explore variations in regional development in the United States, and that put forward by Klaassen in his recent study (1965, pp. 43–49) on area economic development. This latter model takes the form of seven equations, relating such variables as demand for labour in 'basic' and 'non-basic' industries (see Alexander, 1954), net migration, natural growth of the working population, and the wage or income differential between the area concerned and the outside world. When certain of its predictions are tested against actual per capita income data for the regions of Belgium, a reasonably close fit is obtained, justifying Klaassen's claim that in spite of their simplicity (Klaassen, 1965, p. 38), such aggregate models 'can help considerably to understand why and how differences in income level arise between areas and regions' (Klaassen, 1965, p. 43).

Though they do not fit neatly within the model classification adopted here, Markov chain models have also been used to illuminate the question of differential regional economic growth. In his pioneering study, Smith (1961) examines the implications of increasing the trade of a depressed region under equilibrium trading conditions, in terms of a stochastic or Markov matrix of transition probabilities. Given certain limiting theorems of Markov chain theory, Smith's analysis clearly indicates that a permanent increase in the depressed area's income relative to other regions can only be achieved under equilibrium trading conditions 'by continuous injections' of income. The application of such Markov chain models to the problems of regional economic development seems bound to increase.

The most developed form of mathematical model relating to differential regional economic growth, however, is the inter-industry model. Such models, particularly input-output ones, are easily adaptable to interregional analysis, as Isard (1951 and 1960, Chap. 8), Leontief (1953, Chaps. 4 and 5) and Chenery (1956) have shown. For 'n' regions, the full list of 'x' sectors into which each regional economy is divided for input-output purposes is repeated 'n' times, both down the left-hand edge and along the top of the matrix. This enables inter-sectoral flows to be recorded not only within each region, but between different sectors in different regions (Isard, 1960, p. 318). Alternatively, separate input-output matrixes may be drawn up for each region, flows from different sectors in other regions being recorded in the form of supply coefficients in columns in the right half of the matrix (Chenery, 1956, pp. 345–347). As with non-spatial input-output matrixes, coefficients relating flow of inputs from any given sector of any region to another sector of any region are derived from recorded transactions.

The usefulness of input-output models for analysing the locational implications of economic development is considerable. For example, they are particularly valuable for determining the impact of increased investment or other expenditure in one region upon economic activity in others. Chenery (1956, p. 353) and others (Chenery, Clark and Cao-Pinna, 1953) have applied interregional input-output models to the Italian economy for just such a purpose. Again, they may be used to discover the optimum locational pattern of new investment for the achievement of 'balanced interregional growth' (Tinbergen, 1960, pp. 8 and 12): or to discover the minimum investment outlay required in each region to achieve given regional income targets (Tinbergen, 1964, pp. 2–7). In fact, as these examples suggest, interregional input-output models have been developed primarily as analytical and predictive tools for government planning of economic development – an association clearly illustrated in two important recent publications (Barna, 1963, Part II; Isard and Cumberland, 1961, pp. 287–338) referring to them. The use of these models in government planning has, it is true, thrown up certain problems. One is the enormous difficulty of obtaining the detailed regional data needed for input-output analysis from any but the most advanced countries. Another, perhaps more important, is the failure of models framed in terms of constant interareal input coefficients (Isard, 1960, p. 333) to account for the effects of scale and external economies available in particular regions (Isard and Smolensky, 1963, p. 109; Isard, 1960, pp. 338–343). However, even with these problems, inter-industry models have proved very useful in assessing the locational implications of government plans for economic growth; and with the refinement of such models, as for example in the direction of interregional linear programming techniques (Isard, 1958 and 1960, Chap. 10; Chenery, 1963; Stevens, 1958; Berman, 1959), their application for policy purposes to the analysis of interregional differentials in economic growth seems bound to increase.

Supra-national-scale models

The remarkable postwar growth of interest in the mechanism and characteristics of economic development referred to earlier has partly been stimulated by the apparent steady increase in recent years of income disparities between state-units. Many authorities (Andic and Peacock, 1961; Kuznets, 1953–54 and 1956; Briggs, 1965; Myrdal, 1957B, p. 6) support Deane's contention (1961, p. 16) that as far as income per capita is concerned, 'international inequality is increasing and that it is appreciably greater than the intranational inequalities'. However, although at first sight well documented statistically, this conclusion needs to be qualified in at least two ways. Firstly, as a number of writers have pointed out (Usher, 1965; Rao, 1964, pp. 66–104; Kindleberger, 1958, p. 3; Hagen, 1960, pp. 63–64), apparent income differences between different countries may well be 'equalizing' – i.e., the real value of a given amount of income in an apparently poorer country may be much greater (because of lower prices for goods and services) than that of the same amount in an apparently richer country. This tendency must reduce the apparent income gap in many cases, particularly as between industrial and underdeveloped economies (Rao, 1964, p. 103). Secondly, however, a widening gap in per capita incomes between countries may occur statistically even when the poorer country's economy is expanding faster than the richer one's. This is because, as Myint (1964, p. 18) points out, 'arithmetically, the widening gap depends not only on the differences in rates of growth but also on the initial width of the gap'. Again, this suggests that a widening gap should not cause too great a concern, since considerable economic growth may none the less be occurring in the poorer country concerned.

These two qualifications are important. However, they are clearly insufficient to justify ignoring the observed growth in international income disparities over recent years. This growth runs counter to the theoretical predictions of static equilibrium analysis, which suggests that trade stimulates economic development in all participating countries, and in the long run helps to bring about equalization of per capita incomes (Haberler, 1959; Cairncross, 1962, Chap. 13; Neumark, 1964; Caves, 1960, p. 259). This discrepancy between traditional assumptions and observed reality has led certain economists in underdeveloped countries to discard static equilibrium models, despite their internal consistency (Das Gupta, 1965, p. 123), and to put forward other ideas about the nature of spatial interaction between poor and rich countries. Probably the most influential of these economists is Prebisch (1950 and 1959), who has developed what might be called a 'centre/periphery' model of international economic development. Although on the world scale the 'contour lines of international economic inequality' (Briggs, 1965, p. 15) clearly identify the centre as the zone of highly developed economies stretching from European Russia to the United States and Canada, the terms 'centre' and 'periphery' are, as Friedmann and Alonso point out (1964, p. 211) 'more than a description of geographic position'. They further connote 'a set of structural relations that hold the periphery in nearly permanent subordination to the urban-industrial heartland' (Friedmann and Alonso 1964, p. 211). Chief amongst these relations are the enforced exploitation of natural resources in peripheral countries, at least during earlier colonial periods; the flooding of peripheral markets with manufactured goods produced in the countries of the industrial centre; and the unfavourable secular trend in international terms of trade as far as the primary, chiefly agricultural, goods which are produced at the periphery are concerned (Prebisch, 1950; Friedmann, 1963, pp. 44-45; Meier and Baldwin, 1957, p. 147). That such production characterizes the periphery, and is a response to the demands of the centre, is verified by Melamid (1955), in terms of Von Thünen's model of agricultural location.

This centre/periphery model, as Friedmann (1963) has pointed out, is

[272] MODELS IN GEOGRAPHY

applicable not only on the world but on the continental scale. For example, United Nations experts (Economic Commission for Europe, 1955) have drawn attention to just such a spatial pattern of per capita income inequality within Europe itself (see Figure 8.6). They find not only that 'the countries situated near the economic centre of Europe are, in general, richer and more developed than those at the periphery', but also that within European coun-

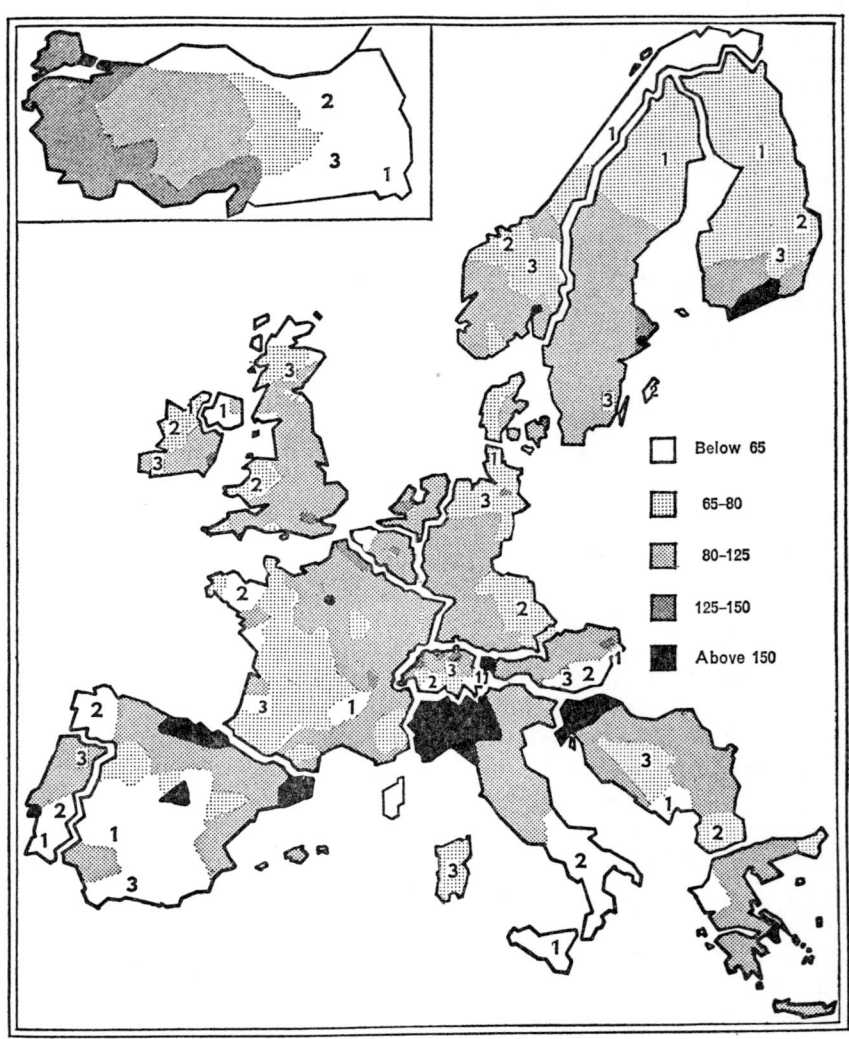

8.6 Regional Income Disparities within West European Countries (shading indicates estimated per capita income in each region expressed as a percentage of national income in each country: *Source: Map 2, Economic Commission for Europe, 1955*).

tries 'the levels of economic development tend to be lowest in the regions furthest removed from the relatively small area which developed as the main European centre of industrial activity, embracing England and the valley and outlet of the Rhine' (Economic Commission for Europe, 1955, p. 138). Of course, the restrictions on regional economic growth imposed by locations which are peripheral within particular European countries have by no means yet fully disappeared (see Verburg, 1964B). But on the whole, the supra-national pattern is supported by national-scale studies, too, such as that by Vinski (1962, p. 138) on Yugoslavia. Again, this pattern seems to reflect a set of economic relationships which restrict economic development at the periphery, by, for example, 'preventing the spreading of industry to the poor regions' (Economic Commission for Europe, 1955, p. 154), in the kind of way suggested by Prebisch.

Prebisch's model, and the ideas put forward in the Economic Commission for Europe's survey, can however be viewed as only a special case of Myrdal's or even Hirschman's income inequality model; that is, the case in which backwash/polarization effects are dominant. The authors of these two latter models in fact explicitly extend them to the supra-national scale, where Myrdal's thinking in particular closely parallels that of Prebisch – 'internationally, however, the backwash effects of trade and capital movements would dominate the outcome much more, as the countervailing spread effects of expansionary momentum are so very much weaker' (Myrdal, 1957B, p. 54). However, the inclusion in Myrdal's analysis of the cumulative causation mechanism (e.g. Myrdal, 1957B, p. 19: see also Das Gupta, 1965, pp. 119–120), together with an acknowledgement of the operation of some spread effects, though on a very limited scale, does provide his model with a more flexible and meaningful structure than is possessed by the earlier centre/periphery analysis.

This latter point on spread effects is important in that it allows consideration of the significance for economic development in a poor country of spatial proximity to a richer one – development which empirical evidence suggests is sometimes aided by certain spread effects from the latter. An inference of this kind could be drawn merely from the visual evidence of Figure 8.6. Firmer support, however, is available from empirical studies of particular countries. Glade (Glade and Anderson, 1963, p. 15), for example, has pointed out that as far as Mexico is concerned, 'without doubt, too, the geographical proximity of the United States, combined with tourism and the bracero movement, has facilitated the cross-cultural transmission of new values, tastes and attitudes' necessary for economic development; while at the same time, 'geographical proximity to the large and growing United States markets' has greatly aided the expansion of Mexican exports of primary products (Glade and Anderson, 1963, p. 16). Glade might also have pointed out that the massive annual inflow of foreign currency from American tourists, and the considerable remittances from temporary migrant workers – braceros – in the United States, are themselves largely a function of the geographical proximity of the two countries, and represent a spilling-over of economic prosperity from the United States to Mexico of great value for the latter's economic development. A second area benefiting in this way is Bechuanaland. In his recent study, Munger (1965, pp. 38–39) stresses the considerable advantages enjoyed by the Protectorate because of its 'physical proximity to South Africa, and especially the booming Witwatersrand'. For example, 'it is hard to realize the savings in money, time and general efficiency of having nearby specialists and research stations'; while proximity permits temporary migration of many Bechuanas to South Africa, and the remitting of wages back to the Protectorate (Munger, 1965, pp. 86–87). This latter phenomenon is not, it is true, specified as a spread effect in Myrdal's original model, but may surely here be considered as one, particularly in view of Okun and Richardson's conclusions discussed earlier.

The above evidence supports the contention that some spread effects do operate today on the supra-national scale, even if only between neighbouring countries. To that extent, Myrdal's model appears more useful in analysis of spatial interaction in economic development between states than does the centre/periphery one. This is probably even more true if international differences in economic growth during earlier centuries – particularly the nineteenth – are the subject of study. Reversing the centre/periphery model, in fact, Nurkse (1961, p. 14) has claimed that during the nineteenth century 'a vigorous process of economic growth came to be transmitted from the center to the outlying areas of the world' by the mechanism of trade in primary products – that is, exactly one kind of mechanism specified by Myrdal (1957B, p. 31) as a possible spread effect. Again, Kuznets seems to be suggesting in one or two publications (e.g. 1953–54 and 1959) that international differences in economic development in past centuries must be viewed within the framework of the 'gradual . . . and uneven spread to other countries' (Kuznets, 1953–54, p. 18) of the 'industrial system', and the economic prosperity which it brought. This rather special kind of spread effect did not, of course, necessarily influence all countries located close to the birthplace of the system, Great Britain. More important for successful transmission were 'similar material conditions' in the poorer country, or close association 'by social and cultural antecedents' (Kuznets, 1953–54, p. 22) – conditions met, for example, in New England, which early adopted factory production of textiles on the basis of skills and techniques imported directly from Great Britain, rather than in Portugal, where little penetration of the industrial system occurred. Whatever the direction of transmission, however, the stimulus which this spread of industrial technology and attitudes gave to sustained economic growth in some countries strongly supports its inclusion in any discussion of international differences in economic growth during the nineteenth century; and this in turn lends added weight to the conclusion regarding the usefulness of Myrdal's model reached above.

The use of regression analysis and other statistical techniques to determine the degree of statistical association on the supra-national scale between variations in level of economic development (usually measured in per capita income terms), on the one hand, and variations in such factors as total output of manufacturing industry (Chenery, 1960), size of state-unit (Kuznets, 1953-54), city size distributions (Berry, 1961A), energy consumption (Schurr, 1965), and latitude, climate, political status and economic structure (Berry, 1960 and 1961B), on the other, has expanded rapidly in recent years. The role that geographers, especially Berry, have played in this is extremely important, and the work done of great interest. However, actual mathematical models of the growth of differences in economic development between state-units have not yet been developed to any great extent. Leontief's recent use (1965B) of input-output models to compare the economic structure of countries at different levels of development perhaps comes close to it, particularly since he shows how to develop a hypothetical input-output model of one underdeveloped economy (Israel) 'as it would appear if it enjoyed self-sufficiency' (Leontief, 1965B, p. 140), and how to use such a model in assessing the changes needed to achieve higher levels of economic development in different countries. Obviously, too, interregional input-output models can also be used on this scale, to investigate the impact of changes in final demand in one country upon the economy of another. Generally, however, mathematical models have been developed for analysis of economic development at national and sub-national scales, rather than to account for spatial interaction and areal differentiation in development between state-units.

Sub-national-scale models

(1) *The regional multiplier concept*. In analysing the development of spatial variation in economic prosperity within regions of countries, the regional multiplier concept provides an important starting point. This concept has already been briefly introduced in discussion of the export base model of regional development (see page 266). Indeed, that model is in one sense incomplete without it, since the operation of a multiplier process following the inflow of export earnings to the region is essential to overall regional growth. However, in that the focus of attention in multiplier analysis is internal change within the region, discussion of the concept, though intimately related to the export base model, appears to fit more logically within this sub-national scale group.

Broadly speaking, the regional multiplier concept concerns the way in which a rise in income, production or employment in one group of economic activities in a region stimulates the expansion of other groups, through an increased demand from the former group and its workers for the goods and services produced by the latter. This rise is typically induced by changes external to the region. The inter-industry stimulus may take the form of an expansion of demand for actual production inputs needed by the original group; or it may operate indirectly through growth in demand for consumer-oriented goods and services from the better-paid and/or increased number of workers employed by that group. The activities which benefit in the latter case are of course the 'residentiary' industries, primarily engaged in serving local markets, referred to earlier (see page 266) – and these will also gain from the expansion induced in industries producing inputs for the original group, since their workers will also presumably generate a larger demand for local goods and services.

One of the first economists to recognize the importance of the multiplier process in local economic growth was Barfod, in his study (1938) of the economic impact upon the city of Aarhus, Denmark, of a large local oil factory, Aarhus Oliefabrik, A/S. Starting with local payments, in the form of wages, salaries and payments to local input suppliers, Barfod demonstrated that what he termed the 'primary elementary income' derived by the area from the factory was somewhat less than total local payments, owing to income leakages from the local economy. This kind of effect is illustrated by Figure 8.7 for the case of payments to local suppliers. However, he then showed that such 'primary investment or expenditure calls forth a series of secondary effects as links in a geometrical progression' (Barfod, 1938, p. 39), since part of this primary income would again be spent locally, thus providing further local wages and income, part of which would again be spent locally, and so on, until these secondary effects were exhausted, As a result, the total local income derived from total payments made locally by the Aarhus factory (based upon figures for 1937) was in the final result 27 per cent greater than the original sums paid (Barfod, 1938, p. 55).

This pioneering study was primarily concerned with elucidating the existing structure of income flows between one economic activity and the rest of the economy within which it was located. Daly, however, in his slightly later analysis (1940) of interwar employment growth in southern and midland England, not only adopted the term 'geographical multiplier', but stressed the now-accepted view that this should be 'a dynamic concept', concerned not with 'a static relationship between the unimpeded and localized industries' but with 'the effect of a change in the numbers employed in one type of industry upon the magnitude of the numbers employed in the other' (Daly, 1940, p. 250). Daly's 'unimpeded' industries were those which developed in an area for reasons unconnected with serving local markets – i.e., logically, export-oriented industries; while his 'localized' industries were those serving local needs, such as building and service industries – i.e. virtually the same as North's residentiary industries. Assessment of the multiplier effect therefore involved defining these two kinds of activity in some way and measuring over some period of time the ratio of employment growth in the former to that in the latter. For Britain over the 1921–31 period, the ratio was 1 : 1·042.

Probably because of its predictive possibilities, Daly's approach rather than Barfod's was the one adopted by subsequent multiplier studies. For example, Vining (1946, p. 203) also divided regional economic activity into two groups – 'carrier' industries and 'passive' industries – representing export and resi-

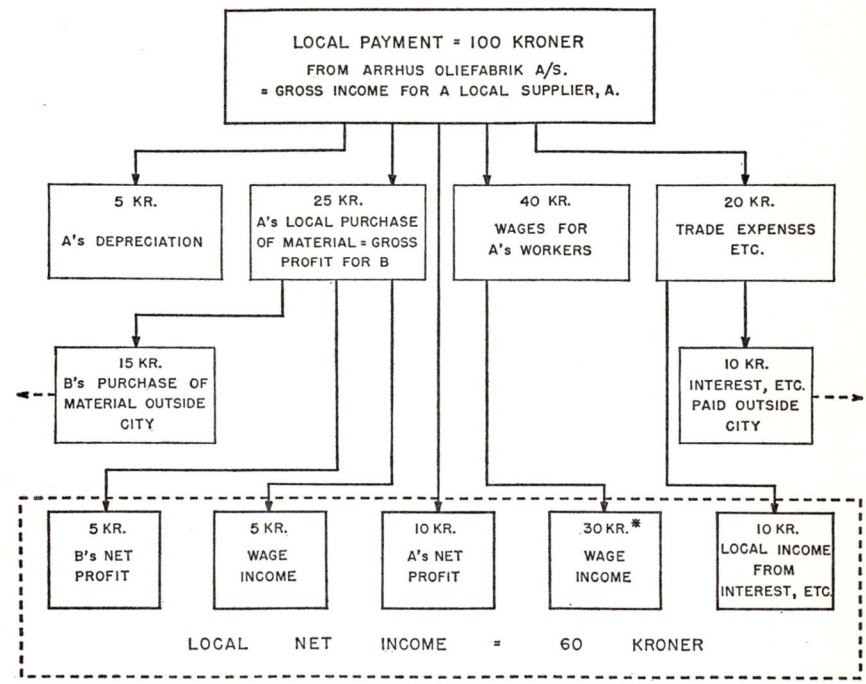

8.7 Hypothetical Income Flows within the Aarhus Economy (*After Barfod, 1938, p. 31*). Barfod reduces the contribution of wage income to local net income (*) by a factor of ·75, to allow for loss of the former, presumably outside the town.

dentiary industries respectively; and he too derived a multiplier ratio – of about 1 : 2 – for employment growth in these industries in one particular area, the Pine Bluff trading region of Arkansas. Hildebrand and Mace (1950), too, in their multiplier analysis of industrial growth in Los Angeles, built explicitly on Daly's work. Concerned with predicting employment growth in this rapidly expanding area, they classified industries as 'non-localised' or 'localised' according to their 'location quotient', an index of the degree of spatial concentration in a given area of particular industries (see Florence, 1948; Isard, 1960, pp. 123–126). From employment data, they then calculated a multiplier value of 1:1·248 for the period from just before to just after the Second World War. Their stress on the short-term applicability of this value was fully justified by their own calculations, which revealed a massive difference between the early (1 : 0·929) and later (1 : 2·785) parts of this period, if separate multiplier values are obtained for them.

These early multiplier studies, and associated 'economic base' analyses (see Andrews, 1953–56; Lane, 1966), were valuable in focusing attention on the importance of internal relationships in overall regional growth. However, as Isard (1960, p. 204) has pointed out, 'a regional multiplier derived from a basic-service ratio . . . has a strictly limited degree of usefulness and validity'. A full analysis of its shortcomings has been made elsewhere (Isard, 1960, pp. 194–205), but it is worth stressing two important ones. The first is the practical difficulty of separating export (basic) from residentiary (service) industries. Whatever method is used for such classification, no fully satisfactory result seems possible, since in any region some if not many industries ship large quantities of their products to both extra- and intra-regional consumers. The use of the location quotient for classification is open to serious criticisms (Isard, 1960, pp. 125–126). Secondly, multiplier analysis of this kind, particularly that based on a division by means of the location quotient, tends to ignore the direct kind of stimulus to other industries which increased demand for inputs from a growing export industry occasions (see above). This is because many of the former industries will, if already geared to the latter, probably possess a high location quotient, and therefore will have been classified in the export rather than residentiary sector. The full effect on regional economic growth of expansion in one particular export activity is therefore very difficult, if not impossible, to assess by this means (Isard, 1960, pp. 201–203).

For these and other reasons, later writers have on the whole adopted a different approach to multiplier analysis. This involves the use of input-output models. Such models can easily be adapted to focus on the changes which growth in one industry will stimulate in all the other industries of a given region. Probably the earliest study to attempt this was Isard and Kuenne's analysis (1953) of the direct and indirect repercussions upon local production and employment of the location of a new integrated iron and steel plant in the Greater New York – Philadelphia industrial region. Beginning with an estimate of production and employment in the new steel plant, the authors then calculated the probable growth of directly related steel-fabricating industries – this on the basis of historical trends and empirically derived indexes of association. The next step, involving the use of an existing matrix of input coefficients, was the determination (in value terms) of the first round of inputs from all other sectors needed for full-scale production of both steel and the products of the expanded steel-fabricating industries. The impact upon the household sector was included here. After reduction to allow for inputs produced outside the region, the second round of inputs needed to produce this first round of output expansions was calculated – and so on, until the increase represented by the 'n'th round was negligible. The total value of production, and corresponding employment, likely to be stimulated in the Greater New York—Philadelphia industrial region by the new steel plant could then be determined by summing the totals of each round. In this case, an extraordinarily high multiplier value – 1:19·66 – was obtained, due largely to an anticipated clustering of new or expanded steel-fabricating plants in the area, and their additional multiplier effects. Subsequent discussion of this result and method (Moses, 1955; Kuenne, 1955) added little of value, although Isard's later comments (1960, pp. 357–358) do point out certain, probably compensating, shortcomings.

Isard and Kuenne's study showed the value of input-output models for multiplier analysis, particularly when the impact of growth in one particular industry needs to be examined. Later studies have also utilized such models for this purpose. In their work on Utah, Moore and Petersen (1955) used an input-output model to assess the impact of variations in external demand for the products of Utah's main export industries upon the state's economy. Not only did this involve estimation of multiplier effects, but multiplier values were used as one index of the actual importance to the regional economy of different export industries. Moore's further work (1955) on Utah and California also assessed multiplier effects by an input-output model, and is interesting because it acknowledges that such effects 'may arise either internally or externally' (Moore, 1955, p. 136). That is, that both growth in an export industry, and, for example, internal population expansion, may produce a multiplier effect on regional income and employment. That the latter kind of change is also important in this respect is evident from empirical study of an area such as California. As a final example, Hirsch's work (1963) on St. Louis has demonstrated the value of input-output models for determining multiplier effects even within very small regions – in this case, urban regions. Such studies, potentially at least, are of great relevance for forward planning by local government authorities, and the use of input-output models in this way seems bound to increase.

Although the main approaches to regional multiplier analysis have now been outlined, a final comment concerning a topic often associated with it seems necessary. This topic is the so-called 'threshold' concept, which takes as its starting point the view that 'the size of a region's markets has an important bearing on the outcome of its economic development' (Pfister, 1963, p. 151). Clearly, this is true in that the larger the regional market, the greater the volume of local residentiary industry that can be supported. However, it is even more true in that for many manufacturing industries, economies of scale dictate a minimum size of plant (Pratten and Dean, 1965) and therefore a minimum size of regional market capable of supporting the particular industry concerned (Guthrie, 1955). This minimum market size is termed a 'threshold' (Pred, 1965, p. 168), and protagonists of the concept hold that attainment of a certain market 'threshold' is essential if regional growth is to become self-generating (Tattersall, 1962, p. 229). Clearly, for individual industries, the idea of a particular threshold is often justified. For example, the building of petroleum refineries at Seattle almost certainly reflects the growth of the Pacific Northwest regional market to a size which can just support such activity (Pfister, 1963, p. 151). However, as Tattersall (1962, p. 229) points out, the threshold concept is far less valid if applied to a region as a whole, in the sense that attainment of one particular, crucial, level of regional demand suddenly initiates a great leap forward in regionally-oriented industrial activity. This view, implicit for example in Pfister's comments (1963, p. 151) on the Pacific Northwest, and Perloff and Wingo's argument (1961, p. 216) on regional growth in the United States generally, is rightly criticized by Tattersall (1962, p. 229) on the grounds that 'accretion of this type of industry is much more gradual than the "threshold" concept suggests', and that there 'seems no reason why a "threshold" should be achieved in all, or even many, industries at about the same time'. Only when applied to individual industries can the threshold concept really be justified. However, the existence of a threshold for different industries ought certainly to be recognized when regional input-output multiplier analysis is being carried out, since, for example, the attainment of such a threshold as a result of multiplier effects may well alter the assumptions on which the input-output analysis is based – e.g. as regards proportion of regional demand satisfied by internal as compared to external producers. Indeed, input-output multiplier analysis might even be used to assess the scale of expansion in, say, an export industry, which is needed to stimulate growth to the threshold level of regional demand for the products of another, potentially residentiary, activity.

At the beginning of this section, it was suggested that regional multiplier analysis provides a valuable starting point for considering models of spatial variation in economic development within regions. Logically, such analysis has major locational implications, in that, for example, a growth impetus transmitted between linked manufacturing industries must involve interaction in space, even if only over a short distance. It is therefore remarkable that, to the best of the writer's knowledge, very few attempts indeed have been made to analyse these implications, and in particular to formulate any kind of model of the intra-regional locational impact of the multiplier process. As a result, the simple locational model put forward by Tattersall (1962), extending some of North's ideas (1955, pp. 250–251), appears as a pioneering study in a field in which much more work needs to be done. This model, aimed specifically at regions whose early economic growth has been geared to exporting, usually of primary products, suggests that although the actual production of such commodities is usually dispersed throughout different parts or the whole of the region, shipment and organization of exports soon become concentrated in certain urban centres. These centres usually grow up 'at points (frequently at natural breaks in transport routes, e.g. ports and river junctions) where transportation costs for exports and imports are minimized' (Tattersall, 1962, p. 217), and naturally handle both collection of exports and distribution of imports. As export activity expands, the operation of the multiplier process naturally stimulates the development of locally-orientated residentiary industry (interpreted here in its widest sense as referring to the production of both goods and services) within the region. The locational impact of this development is, however, also concentrated in these urban centres, partly because of the external economies available in them, partly because of 'their strategic role as distribution centres, plus their growing importance as markets'. They therefore attract 'a very large proportion of the region's domestically oriented manufacturing, and . . . a large proportion of regional trade and service employment', and grow even more rapidly (Tattersall, 1962, p. 226). Tattersall's model, although extremely simple, is interesting both for its identification of the spatial consequence of multiplier growth, and for the way in which it dovetails into North's export base model – a feature which clearly reflects the interest of both authors in the same region, the Pacific Northwest. However, much more work on this subject, aiming in particular at the integration of multiplier analysis with location theory, does appear to be needed.

(2) *The growth pole model.* Although perhaps less clearly a model than concepts already discussed, the development in recent years of a body of ideas about spatial variation in economic prosperity within regions demands attention here. Many of these ideas, explicitly or implicitly, centre around the concept of a 'growth pole', or 'growth centre'. At its simplest level, this concept refers to the fact, pointed out by many, particularly French, writers (Milhau, 1956; Gannagé, 1962, pp. 62–65; Perroux, 1961, p. 167; Beguin, 1963; Hicks, 1961, p. 77; Nicholls, 1961) and already demonstrated on other scales, that economic development never occurs uniformly over a particular region. Rather, it tends to be concentrated in certain parts, which thereby develop as 'growth poles', expanding much more rapidly than surrounding areas. These growth poles are almost invariably regarded as urban-industrial in character (e.g. Rostow, 1964, pp. 123–124; Ruttan, 1955, p. 56), while some writers have implied that they are also usually to be found close to the centre of a particular region. This latter view is represented by Schultz's claim that 'the existing economic organization works best at or near the the center of a particular matrix of economic development' (Schultz, 1953, p. 147; see also Schultz, 1950), 'matrix' here referring to a region in which economic growth is taking place, as well as by comments in two of

Friedmann's articles (1956, p. 215 and 1963, p. 55). The map in the former of these clearly indicates a central position for a growth pole, while the latter specifically outlines an 'illustrative model' in which 'metropolitan development areas' are located 'as the principal growth poles at the core of the spatial system', surrounded by concentric zones possessing other economic characteristics (Friedmann, 1963, p. 55).

At this level, of course, the growth pole concept is little more than a descriptive device. However, other writers have extended the meaning of the term to include by implication both the centre's internal mechanism of growth, and its relationships with the surrounding region. At these levels, the concept may justifiably be termed a model, in that it singles out crucial variables in the development of spatial variation in economic prosperity within a region, and specifies how they operate. The analysis of the internal expansion mechanism of a growth pole is closely associated with the work of Perroux (1955 and 1961, Part II) and other French economists. To Perroux, the growth pole ('pôle de croissance') owes its existence to the location within it of one main 'growth industry', or as he terms it 'une industrie motrice'. The growth of this industry attracts other, linked, industries (i.e. those which provide it with inputs, or derive their inputs from it) by virtue of the external economies created in the locality; and as these industries grow under stimulus from the 'industrie motrice', the growth pole as a whole expands still further. Other agglomeration economies (Friedrich, 1929, pp. 124-131; Losch, 1954, pp. 68-78) come into play to encourage further growth, a high rate of technological change is engendered by spatial proximity and ease of communication between these industries, and psychological factors, such as the development of a 'growth mentality' amongst businessmen of the district, aid further expansion (Perroux, 1955; Balassa, 1961, pp. 151 and 197).

That this kind of mechanism does work in many growth centres is attested by various studies, as for example those by Boudeville (1957) on Brazil and Lacroix (1964) on Congo. More specifically, the role of a key industry in creating external economies for supporting activities has already been noted (North, 1955, p. 252; Tattersall, 1962, p. 217), while further evidence on this is provided by Chinitz (1961) in his study of industrial agglomeration in Pittsburgh, Harris (1959), dealing with urbanization in India, and Pred (1965), analysing the relationship between industrialization and urban growth in the United States. The psychological advantages of a growth centre are stressed by Hirschman (1958, pp. 185-186). However, although the development of agglomeration economies does provide a powerful mechanism for continued industrial and urban growth, Perroux's insistence on the role of an 'industrie motrice' cannot really be accepted, for two reasons. Firstly, as Chinitz (1961) has shown, the very presence of just such a growth industry in an urban centre may in the long run inhibit industrial and economic development, by preventing the ingress of firms in other industries. In Pittsburgh, for example, the dominance of the primary metals industry over local capital and labour markets acts as a barrier to the development of such new firms; and Pittsburgh's recent overall economic growth has almost certainly suffered because of this. Secondly, and more important, many obvious growth poles exist which are not dominated by an 'industrie motrice'. For instance, Burley (1962, p. 184) rightly describes Mexico City as 'an outstanding example' of recent concentration of industrial and economic development (see also Bird, 1963); yet no one growth industry can be identified as stimulating this process, a very wide range of unconnected activities having grown up here over the last forty years. The same is true of other growth centres, such as those of Jinja, Uganda (Hoyle, 1963 and 1964) and Nairobi, Kenya (Pollock, 1960, p. 352), or the booming metropolis of São Paulo, Brazil (James, 1959, p. 500). For these reasons, Perroux's analysis of the role of an 'industrie motrice' in growth pole development, though valid in some cases, and probably of value for planning government-sponsored growth centres, cannot be accepted as the typical mechanism of internal expansion of a growth pole.

The third important feature of the growth pole model is its analysis of relationships between the growth centre and surrounding region – an analysis which, together with that of agglomeration economies, largely explains the recent acceptance of the model by several governments as a key concept in regional economic planning (Harris, 1966, p. 577). Probably the earliest example of this acceptance is represented by the Brazilian government's decision to locate the country's new capital, Brasilia, in the undeveloped interior of Brazil, 'the most often heard and perhaps the most persuasive single argument' for this being 'the profound reciprocal effect' which the city was intended to exercise on the economy of its surrounding region (Snyder, 1964, p. 35). However, other countries, notably France (Fourastie and Courthéoux, 1963, p. 134), Italy (Coquery, 1964) and the United Kingdom (National Economic Development Council, 1963, pp. 14–29; Board of Trade, 1963, p. 6; Scottish Development Department, 1963, pp. 27–30) have in recent years also apparently come to regard the model as providing a basis for the practical planning of regional development (Parr, 1965, pp. 1, 5). Indeed, in the United Kingdom if not elsewhere, certain new towns are now being planned deliberately as growth centres (Diamond, 1965, p. 183), in contrast to earlier new town policy (Rodwin, 1955, pp. A2–A3). This general acceptance partly reflects the belief, derived by logical extension from the model's analysis of agglomeration economies, that concentration of government investment and industrial activity within a growth pole will, in the long run, stimulate a higher level of industrial and economic development than if it had been spread over a wider region (see, for example, Scottish Development Department, 1963, p. 27). However, it also clearly reflects the idea that the 'benefit of new growth in any part would repercuss fairly quickly throughout the region' (National Economic Development Council, 1963, p. 26), or, in Wright's phrase (1965, p. 150), 'the hope that prosperity will spread outwards from the chosen points in concentric ripples'. If this prospect were not in fact suggested by the model, governments would undoubtedly have been far more chary of basing their regional development programmes on it.

This idea of centrifugal spread of economic prosperity from a growth pole probably owes something to earlier views on the functional relationship between a large city and its surrounding hinterland, such as those of Gras (1922, p. 700). However, the first explicit discussion of the relationship between urban economic growth and that of the surrounding region came in a group of articles published in the early 1950's. In the earliest of these, Hoselitz (1953) examined the role of medieval European cities in regional economic growth, and concluded that the 'increase in average real income' in these cities produced by industrialization and the development of efficient governmental services 'also strongly affected the non-urban regions located near the centers of development' (Hoselitz, 1953, p. 203). One important way in which this occurred was through a growth in urban demand for labour, and the development of commuting from villages close to the towns. This analysis was extended in his later study (Hoselitz, 1955B), which classified cities as 'generative' or 'parasitic' on the basis of whether or not they stimulated the economic growth of the wider region in which they were located. Many of the early colonial settlements in the New World and South Africa, Hoselitz claimed, were parasitic, enjoying a certain degree of economic growth 'within the city itself and its surrounding environs' only at the expense of the rest of the region, which was ruthlessly exploited for its natural and agricultural resources (Hoselitz, 1955B, p. 280). Overall regional growth was thereby retarded. However, Hoselitz does not appear to regard this as the normal regional impact of urban development in underdeveloped countries, since he points out that even these parasitic colonial settlements eventually developed into generative centres. Such centres transmit a growth stimulus to the surrounding area in various ways. In addition to the labour aspect already mentioned, economic and industrial development in generative cities (a) creates a new demand for industrial raw materials from the surrounding region, and (b) attracts new population to the cities, thereby increasing the demand for food from the countryside. The net effect of these forces is a 'widening of economic development over an increasing area affecting a growing proportion of the population outside the city' (Hoselitz, 1955B, p. 282).

Other writers appear to support Hoselitz's conclusions. For example, both Lampard (1954–55) and Stolper (1954–55) agree that large cities in underdeveloped countries have sometimes been parasitic, acting as 'a curb rather than a stimulus to wider economic growth' (Lampard, 1954–55, p. 131). Both writers suggest that a key factor in this is the dissipation of wealth derived from the surrounding region in non-productive urban consumption. Rather than investment in industry, for example, such wealth is used for 'grandiose urban construction' (Lampard, 1954–55, p. 131), which stimulates little or no economic development in the wider region (see also Tangri, 1962, p. 209). Cultural and racial factors may also act as major barriers to growth transmission (Stolper, 1954–55; Deane, 1961, p. 19; Hicks, 1961, p. 77). On the other hand, where these barriers are weaker, and urban investment is channeled into productive enterprises, urban economic development in an underdeveloped country does apparently stimulate growth in the surrounding region. Such at least is claimed by Myrdal (1957B, p. 31) and Friedmann (1961, pp. 95–96), and seems to have occurred, for example, in Uganda where 'the process of economic development which changed all of southern Uganda' during the colonial period 'radiated from the Protectorate center of commerce and colonial policy, Kampala in the Kingdom of Buganda' (Larimore, 1960, p. 120).

Hoselitz's ideas on this topic are more developed than those of other writers such as Myrdal (1957B, p. 31). However, they are clearly aimed primarily at underdeveloped countries, and, as has been pointed out (Lombardini, 1964; Kindleberger, 1964, p. 263), regional development in an underdeveloped area is a different matter from that in the more advanced economies. In the latter, whether the development problem is posed by a distressed industrial region (Estall, 1964) or a growing metropolitan region (Wright, 1965, pp. 148–149), economic and social conditions (such as much higher levels of car ownership) would appear to be even more conducive to the transmission of economic prosperity from a growth pole to its surrounding region.

This has been explicitly recognized by several commentators. Gerschenkron (1963), for example, has pointed out that in the United States 'the economic difference between the city and the countryside is no longer what it used to be' (Gerschenkron, 1963, p. 59) – i.e. economic prosperity has spread from the towns to surrounding areas. Key factors in this spread have been growing decentralization of industry, housing and even shops from the towns to surrounding areas, together with increasing car and refrigerator ownership, the latter permitting a considerably enlarged spatial scale of journeys-to-work and journeys-to-shop. Though Gerschenkron himself does not mention this, increased recreational use of surrounding rural areas by urban dwellers represents another mechanism by which prosperity has been spread within regions of the United States. Interestingly enough, the value of a simple growth pole model in planning resultant recreational development has recently been stressed by Harper, Schmudde and Thomas (1966). Thus intra-regional spread of prosperity is a direct function of the already high level of economic development in the United States, or, as Gerschenkron (1963, p. 61) puts it, of 'the rise in incomes and technological progress'. Similarly,

Wright (1965, p. 161) has stressed that in Britain 'the best position now for the go-ahead family and firm is a position near but not in a big city', a point clearly illustrated by both recent industrial and residential decentralization from major growth poles such as Greater London (Keeble, 1965, pp. 24-28; Pahl, 1965). Again, an intra-regional spread of economic prosperity is the result. However, the only commentator to develop these ideas into a growth pole model is Martin (1957). Adapting Schultz's views, Martin (1957, p. 173) describes 'a model of a dynamic, industrializing society in which economic development occurs primarily in the urban type of locational matrix'. This development in time produces a spread of prosperity to surrounding areas (a) by tapping them for industrial and urban raw materials and customers, (b) by stimulating food production for urban markets, and the introduction to the countryside of industrial-type farming techniques, (c) by encouraging migration of surplus rural population to the town, and (d) by deconcentration of urban population, industry and other institutions (Martin, 1957, pp. 173-174). Like Gerschenkron, Martin stresses that the fourth kind of impact is dependent upon a very high existing level of economic development.

Very little empirical testing of this kind of growth model has as yet been carried out. However, two studies of the changing pattern of agricultural development and farm income in the Tennessee Valley do throw limited light on its relevance to that area. In the earlier, Ruttan (1955) demonstrated by regression analysis that median farm family incomes in different parts of the Valley were closely correlated with the level of urban-industrial development in the same areas. In other words, proximity to urban-industrial centres had stimulated agricultural prosperity in some areas, chiefly through provision of increased off-farm jobs for members of farm families, but also through greater availability of local capital and markets. On the whole, this finding supports Martin's model. Nicholl's later study (1961) also found a close connection between proximity to urban centres and agricultural prosperity: but in stressing the increasing divergence over the 1900-50 period of intra-regional indexes of farm prosperity, his analysis implies that spread from urban growth poles has only influenced areas within a relatively short distance of these centres. Clearly, scale of regional size is very important when assessing the effectiveness of intra-regional spread of economic development from a growth pole.

The limitations of these two studies clearly illuminate the need for much more empirical testing and refining of the growth pole model – whether applied to underdeveloped or developed regions. Indeed, in view of its acceptance by governments in advanced economies as an important concept in regional planning, it is remarkable that so little work has been done on it, either in terms of spatial interaction between pole and region, or of internal growth mechanisms. What limited analysis has been carried out, as for example by Humphrys (1965) on the role of service industry in the expansion of growth centres, tends to throw up as many questions as it answers. Here surely is an important field for future geographical work.

CONCLUSION

Particularly at national and sub-national scales, then, the building of models of economic development is now proceeding rapidly, stimulated by a growing realization on the part of economists and government planners of the value of this approach for practical and theoretical purposes. Geographers have so far played very little part in this, despite their genuine interest in the variation over the earth's surface of many of the phenomena which contribute to spatial variations in economic development. In the light of a changing internal and external intellectual environment (Wrigley, 1965, pp. 13-19), however, geography surely now needs to accept model-building as an important avenue of study, and, in particular, to formulate and refine models of spatial variation and interaction in economic development.

REFERENCES

ACKERMAN, E. A., [1958], Geography as a Fundamental Research Discipline; *University of Chicago, Department of Geography Research Paper*, 53, 37 pp.
ACKERMAN, E. A., [1963], Where is a Research Frontier?; *Annals of the Association of American Geographers*, 53 (4), 429-440.
ALEXANDER, J. W., [1954], The Basic-Nonbasic Concept of Urban Economic Functions; *Economic Geography*, 30 (3), 246-261.
ANDIC, S. and PEACOCK, A. T., [1961], The International Distribution of Income, 1949 and 1957; *Journal of the Royal Statistical Society, Series A*, 124 (2), 206-218.
ANDREWS, R. B., [1953-56], Mechanics of the Urban Economic Base; *Land Economics*, 29-31.
BACHMURA, F. T., [1959], Man-Land Equalization through Migration; *American Economic Review*, 49 (5), 1004-1017.
BAER, W., [1964], Regional Inequality and Economic Growth in Brazil; *Economic Development and Cultural Change*, 12 (3), 268-285.
BALASSA, B., [1961], *The Theory of Economic Integration*, (Homewood, Ill.), 304 pp.
BALDWIN, R. E., [1956], Patterns of Development in Newly Settled Regions; *Manchester School of Economic and Social Studies*, 24 (2), 161-179.
BARAN, P. A. and HOBSBAWM, E. J., [1961], The Stages of Economic Growth; *Kyklos*, 14, 324-342.
BARFOD, B., [1938], *Local Economic Effects of A Large-scale Industrial Undertaking*, (Copenhagen), 74 pp.
BARNA, T., (Ed.), [1963], *Structural Interdependence and Economic Development*, (London), 365 pp.

BARZANTI, S., [1965], *Underdeveloped Areas within the Common Market*, (Princeton), 456 pp.
BEGUIN, H., [1963], Aspects Géographiques de la Polarisation; *Tiers-Monde*, 4, 16, 559–608.
BERMAN, E. B., [1959], A Spatial and Dynamic Growth Model; *Papers & Proceedings of the Regional Science Association*, 5, 143–150.
BERRY, B. J. L., [1960], An Inductive Approach to the Regionalization of Economic Development; Ch. 6, in Ginsburg, N., (Ed.), Essays on Geography and Economic Development; *University of Chicago, Department of Geography Research Paper*, 62, 173 pp.
BERRY, B. J. L., [1961A], City Size Distributions and Economic Development; *Economic Development and Cultural Change*, 9 (4), Part I, 573–587.
BERRY, B. J. L., [1961B], Basic Patterns of Economic Development; Part VIII, In Ginsburg, N., *Atlas of Economic Development*, (Chicago), 110–119.
BIRD, R., [1963], The Economy of the Mexican Federal District; *Inter-American Economic Affairs*, 17 (2), 19–51.
BOARD OF TRADE, [1963], *The North East. A Programme for Regional Development and Growth*, (London), 48 pp.
BORTS, G. H., [1960], The Equalization of Returns and Regional Economic Growth; *American Economic Review*, 50 (3), 319–347.
BORTS, G. H. and STEIN, J. L., [1964], *Economic Growth in a Free Market*, (New York), 235 pp.
BOUDEVILLE, J. R., [1957], Contribution a l'étude des pôles de croissance brésiliens; *Cahiers de l'Institute de Science Economique Appliquée, Cahiers disponibles, Série F*, 10, 71 pp.
BRANDENBURG, F. R., [1964], *The Making of Modern Mexico*, (Englewood Cliffs), 379 pp.
BRIGGS, A., [1965], Technology and Economic Development; In *Technology and Economic Development*, (Harmondsworth, Middlesex), 15–32.
BUNGE, W., [1962], Theoretical Geography; *Lund Studies in Geography, Series C, General and Mathematical Geography*, 1, 201 pp.
BURLEY, T. M., [1962], Industrial Expansion in the Federal District, Mexico; *Geography*, 47 (2), 184–185.
CAESAR, A. A. L., [1964], Planning and the Geography of Great Britain; *Advancement of Science*, 21, 91, 230–240.
CAIRNCROSS, A. K., [1959], Research on Comparative Economic Growth; In National Bureau of Economic Research, *The Comparative Study of Economic Growth and Structure*, (New York), 106–109.
CAIRNCROSS, A. K., [1961], Essays in Bibliography and Criticism, XLV, The Stages of Economic Growth; *Economic History Review, Second Series*, 13 (3), 450–458.
CAIRNCROSS, A. K., [1962], *Factors in Economic Development*, (London), 346 pp.
CAMERON, G. C. and CLARK, B. D., [1966], Industrial Movement and the Regional Problem; *University of Glasgow Social & Economic Studies, Occasional Papers*, 5, 220 pp.
CAVES, R. E., [1960], *Trade and Economic Structure: Models and Methods*, (Cambridge, Mass.), 317 pp.

CHATTERJI, M. K., [1964], An Input-Output Study of the Calcutta Industrial Region; *Papers of the Regional Science Assocation*, 13, 93–102.
CHENERY, H. B., [1956], Inter-Regional and International Input-Output Analysis; In Barna, T., (Ed.), *The Structural Interdependence of the Economy*, (New York), 339–356.
CHENERY, H. B., [1960], Patterns of Industrial Growth; *American Economic Review*, 50 (4), 624–654.
CHENERY, H. B., [1961], Comparative Advantage and Development Policy; *American Economic Review*, 51 (1), 18–51.
CHENERY, H. B., [1962], Development Policies for Southern Italy; *Quarterly Journal of Economics*, 76 (4), 515–547.
CHENERY, H. B., [1963], The Use of Interindustry Analysis in Development Programming; In Barna, T., (Ed.), *Structural Interdependence and Economic Development*, (London), Ch. 1.
CHENERY, H. B., CLARK, P. G. and CAO-PINNA, V., [1953], *The Structure and Growth of the Italian Economy*, (Rome), 165 pp.
CHENERY, H. B. and CLARK, P. G., [1959], *Interindustry Economics*, (New York), 345 p.
CHINITZ, B., [1961], Contrasts in Agglomeration: New York and Pittsburgh; *American Economic Review*, 51 (2), 279–289.
CHISHOLM, M., [1962], Tendencies in Agricultural Specialization and Regional Concentration of Industry; *Papers of the Regional Science Association, European Congress*, 10, 157–162.
CHISHOLM, M., [1966], *Geography and Economics*, (London), 230 pp.
CHORLEY, R. J., [1964], Geography and Analogue Theory; *Annals of the Association of American Geographers*, 54 (1), 127–137.
CHORLEY, R. J. and HAGGETT, P., (Eds.), [1965], *Frontiers in Geographical Teaching*, (London), 378 pp.
CLARK, C., [1966], Industrial Location and Economic Potential; *Lloyds Bank Review*, 82, 1–17.
CLAWSON, M. (Ed.), [1964], *Natural Resources and International Development*, (Baltimore), 462 pp.
CLOUGH, S. B. and LIVI, C., [1956], Economic Growth in Italy: An Analysis of the Uneven Development of North and South; *Journal of Economic History*, 16 (3), 334–349.
COHEN, K. J. and CYERT, R. M., [1961], Computer Models in Dynamic Economics; *Quarterly Journal of Economics*, 75 (1), 112–127.
COMMITTEE FOR ECONOMIC DEVELOPMENT, [1958], *The 'Little Economies': Problems of United States Area Development*, (New York), 60 pp.
CONRAD, A. H. and MEYER, J. R., [1965], *Studies in Economic History*, (London), 241 pp.
COQUERY, M., [1964], Problèmes de Développement et Aspects Nouveaux de L'Industrialisation en Italie Méridionale; *Bulletin de la Section de Géographie, Ministère de L'Education Nationale, Comité de Travaux Historiques et Scientifiques*, 76, Etudes Méditerranéennes, 393—498.
COUTSOUMARIS, G., [1964], Regional Activity Relocation Problems in a Developing

Economy; *Papers of the Regional Science Association, European Congress*, 12, 79–86.
DALY, M. C., [1940], An Approximation to a Geographical Multiplier; *Economic Journal*, 50, 248–258.
DAS-GUPTA, A. K., [1965], *Planning and Economic Growth*, (London), 185 pp.
DEANE, P., [1961], The Long Term Trends in World Economic Growth; *Malayan Economic Review*, 6 (2), 14–26.
DEPARTMENT OF APPLIED ECONOMICS, University of Cambridge, [1962], A Computable Model of Economic Growth; *A Programme for Growth*, 1, (London), 91 pp.
DIAMOND, D. R., [1965], Regional Planning: The Scottish Approach; pp. 183–184, in Caesar, A. A. L. and Keeble, D. E., (Eds.), Regional Planning Problems in Great Britain; *Advancement of Science*, 22, 97, 177–185.
DRUMMOND, I., [1961], Review of 'The Stages of Economic Growth' by W. W. Rostow; *Canadian Journal of Economics and Political Science*, 27 (1), 112–113.
DUESENBERRY, J. S., [1950], Some Aspects of the Theory of Economic Development; *Explorations in Entrepreneurial History*, 3 (2), 63–102.
DWYER, D. J., [1965], Size as a Factor in Economic Growth: Some Reflections on the Case of Hong Kong; *Tijdschrift voor Economische en Sociale Geografie*, 56 (5), 186–192.
EASTERLIN, R. A., [1958], Long Term Regional Income Changes: Some Suggested Factors; *Papers & Proceedings of the Regional Science Association*, 4, 313–325.
EASTERLIN, R. A., [1960], Interregional Differences in Per Capita Income, Population, and Total Income, 1840–1950; In National Bureau of Economic Research, Conference on Research in Income and Wealth, *Trends in the American Economy in the Nineteenth Century*, (Princeton), 73–140.
ECKAUS, R. S., [1961], The North–South Differential in Italian Economic Development; *Journal of Economic History*, 21 (3), 285–317.
ECONOMIC COMMISSION FOR EUROPE, Research and Planning Division, [1955], Problems of Regional Development and Industrial Location in Europe; In *Economic Survey of Europe in 1954*, (Geneva), 136–171.
EL-KAMMASH, M. M., [1963], On the Measurement of Economic Development using Scalogram Analysis; *Papers & Proceedings of the Regional Science Association*, 11, 309–334.
ELKAN, W., [1959], Regional Disparities in the Incidence of Taxation in Uganda; *Review of Economic Studies*, 26, 70, 135–143.
ENKE, S., [1964], *Economics for Development*, (London), 616 pp.
ESTALL, R. C., [1964], Planning for Industry in the Distressed Areas of the U.S.; *Journal of the Town Planning Institute*, 50 (9), 390–396.
EUROPEAN COAL AND STEEL COMMUNITY, HIGH AUTHORITY, [1961], *Les Politiques Nationales de Développement Régional et Conversion*, (Brussels), 196 pp.
EUROPEAN FREE TRADE ASSOCIATION, [1965], *Regional Development Policies in EFTA*, (Geneva), 78 pp.
EWING, A. F., [1964], Industrialisation and the U.N. Economic Commission for Africa; *Journal of Modern African Studies*, 2 (3), 351–363.
FAIRBANK, J. K., ECKSTEIN, A. and YANG, L. S., [1960], Economic Change in

Early Modern China: An Analytic Framework; *Economic Development and Cultural Change*, 9 (1), 1–26.
FISHER, J. L., [1955], Concepts in Regional Economic Development; *Papers & Proceedings of the Regional Science Association*, 1, W1–W20.
FISHER, M. R., [1964], Macro-Economic Models: Nature, Purpose and Limitations; *Eaton Papers*, 2, 40 pp.
FLORENCE, P. S., [1948], Investment, Location, and Size of Plant; *National Institute of Economic and Social Research, Economic and Social Studies*, 7, 211 pp.
FORDHAM, P., [1965], Natural Resources and Economic Development; In *The Geography of African Affairs*, (Harmondsworth), Ch. 4.
FOURASTIE, J. and COURTHÉOUX, J-P., [1963], *La Planification Economique en France*, (Paris), 208 pp.
FRIEDMANN, J. R. P., [1956], Locational Aspects of Economic Development; *Land Economics*, 32 (3), 213–227.
FRIEDMANN, J. R. P., [1959], Regional Planning: A Problem in Spatial Integration; *Papers & Proceedings of the Regional Science Association*, 5, 167–179.
FRIEDMANN, J. R. P., [1961], Integration of the Social System: An Approach to the Study of Economic Growth; *Diogenes*, 33, 75–97.
FRIEDMANN, J. R. P., [1963], Regional Economic Policy for Developing Areas; *Papers & Proceedings of the Regional Science Association*, 11, 41–61.
FRIEDMANN, J. R. P. and ALONSO, W. (Eds.), [1964], *Regional Development and Planning: A Reader*, (Cambridge, Mass.), 722 pp.
FRIEDRICH, C. J., [1929], *Alfred Weber's Theory of the Location of Industries*, (Chicago), 256 pp.
FRYER, D. W., [1958], World Income and Types of Economies: The Pattern of World Economic Development; *Economic Geography*, 34 (4), 283–303.
FRYER, D. W., [1965], *World Economic Development*, (New York), 627 pp.
FULMER, J. L., [1950], Factors Influencing State Per Capita Income Differentials; *Southern Economic Journal*, 16 (3), 259–278.
FURTADO, C., [1963], *The Economic Growth of Brazil. A Survey from Colonial to Modern Times*, (Berkeley), 285 pp.
FURTADO, C., [1964], *Regional Development in Brazil*, (Berkeley).
FYOT, J-L. and CALVEZ, J-Y., [1956], *Politique Economique Régionale en Grande-Bretagne*, (Paris), 312 pp.
GANNAGÉ, E., [1962], *Economie du Développement*, (Paris), 356 pp.
GERSCHENKRON, A., [1962], *Economic Backwardness in Historical Perspective*, (Cambridge, Mass.), 456 pp.
GERSCHENKRON, A., [1963], City Economies – Then and Now; In Handlin O. and Burchard, J., (Eds.), *The Historian and the City*, (Cambridge, Mass.), 56–62.
GILMORE, D. R., [1960], *Developing the Little Economies; A Survey of Area Development Programs in the United States*, (New York), 200 pp.
GINSBURG, L. B., [1957], Regional Planning in Europe; *Journal of the Town Planning Institute*, 43 (6), 142–147.
GINSBURG, N., [1957], Natural Resources and Economic Development; *Annals of the Association of American Geographers*, 47 (3), 197–212.
GINSBURG, N. (Ed.), [1960], Essays on Geography and Economic Development; *University of Chicago, Department of Geography Research Paper 62*, 173 pp.

GINSBURG, N., [1961], *Atlas of Economic Development*, (Chicago), 119 pp.
GLADE, W. P. and ANDERSON, C. W., [1963], *The Political Economy of Mexico*, (Madison, Wisconsin), 242 pp.
GOLDSMITH, R. W., [1959], Explanatory Report; In National Bureau of Economic Research, *The Comparative Study of Economic Growth and Structure*, (New York), Part I, 3–100.
GRAS, N. S. B., [1922], The Development of Metropolitan Economy in Europe and America; *American Historical Review*, 17 (4), 695–708.
GRAS, N. S. B., [1930], Stages in Economic History; *Journal of Economic and Business History*, 2, 397.
GREEN, L. P. and FAIR, T. J. D., [1962], *Development in Africa*, (Johannesburg), 203 pp.
GRIBAUDI, F., [1965], Some Geographic Aspects of Economic Development; *Tijdschrift voor Economische en Sociale Geografie*, 56 (2), 69–72.
GUTHRIE, J. A., [1955], Economies of Scale and Regional Development; *Papers & Proceedings of the Regional Science Association*, 1, J1–J10.
HABAKKUK, J. H., [1961], Review of The Stages of Economic Growth by W. W. Rostow; *Economic Journal*, 71, 283, 601–604.
HABERLER, G., [1959], *International Trade and Economic Development*, (Cairo), 36 pp.
HAGEN, E. E., [1959], Economic Structure and Economic Growth: A Survey of Areas in which Research is Needed; In National Bureau of Economic Research, *The Comparative Study of Economic Growth and Structure*, (New York), 124–141.
HAGEN, E. E., [1960], Some Facts about Income Levels and Economic Growth; *Review of Economics and Statistics*, 42 (1), 62–67.
HAGEN, E. E., [1962], *On the Theory of Social Change: How Economic Growth Begins*, (Cambridge, Mass.), 557 pp.
HAGGETT, P., [1965A], Changing Concepts in Economic Geography; Ch. 6 in Chorley, R. J. and Haggett, P. (Eds.), *Frontiers in Geographical Teaching*, (London), 378 pp.
HAGGETT, P., [1965B], Scale Components in Geographical Problems; Ch. 9 in Chorley, R. J. and Haggett, P. (Eds.), *Frontiers in Geographical Teaching*, (London), 378 pp.
HAGGETT, P., CHORLEY, R. J. and STODDART, D. R., [1965], Scale Standards in Geographical Research: A New Measure of Areal Magnitude; *Nature*, 205 4974, 844–847.
HANCE, W. A., [1958], *African Economic Development*, (London), 307 pp.
HAQ, M. U., [1963], *The Strategy of Economic Planning*, (Karachi), 266 pp.
HARPER, R. A., SCHMUDDE, T. H. and THOMAS, F. H., [1966], Recreation Based Economic Development and the Growth-Point Concept; *Land Economics*, 42 (1), 95–101.
HARRIS, B., [1959], Urbanisation Policy in India; *Papers & Proceedings of the Regional Science Association*, 5, 181–203.
HARRIS, D., [1966], The Idea of the Growth Area; *Official Architecture and Planning*, 29 (4), 577–581.
HARRIS, S. E., [1954], Interregional Competition: With Particular Reference to North–South Competition; *American Economic Review*, 44 (2), 367–380.
HARRIS, S. E., [1957], *International and Interregional Economics*, (New York), 564 pp.

HART, P. E., MILLS, G. and WHITAKER, J. K. (Eds.), [1964], *Econometric Analysis for National Economic Planning*, (London), 320 pp.
HARTSHORNE, R., [1939], *The Nature of Geography*, (Lancaster, Penn.), 482 pp.
HARTSHORNE, R., [1959], *Perspective on the Nature of Geography*, (Chicago), 201 pp.
HATHAWAY, D. E., [1960], Migration from Agriculture: The Historical Record and Its Meaning; *American Economic Review*, 50 (2), 379–391.
HAY, A. M. and SMITH, R. H. T., [1966], Preliminary Estimates of Nigeria's Inter-regional Trade and Associated Money Flows; *Nigerian Journal of Economic and Social Studies*, 8 (1), 9–35.
HEMMING, M. F. W., [1963], The Regional Problem; *National Institute Economic Review*, 25, 40–57.
HICKS, J. R., [1953], Review of The Process of Economic Growth by W. W. Rostow; *Journal of Political Economy*, 61 (2), 173–174.
HICKS, J. R., [1959], *Essays in World Economics*, (Oxford), 274 pp.
HICKS, U. K., CARNELL, F. G., NEWLYN, W. T., HICKS, J. R. and BIRCH, A. H., [1961], *Federalism and Economic Growth in Underdeveloped Countries*, (London), 185 pp.
HIGGINS, B., [1964], Review of On the Theory of Social Change: How Economic Growth Begins, by E. E. Hagen; *Journal of Political Economy*, 72 (6), 627–630.
HILDEBRAND, G. H. and MACE, A., [1950], The Employment Multiplier in an Expanding Industrial Market: Los Angeles County, 1940–47; *Review of Economics and Statistics*, 32 (3), 241–249.
HIRSCH, W. Z., [1959], Interindustry Relations of a Metropolitan Area; *Review of Economics and Statistics*, 41 (4), 360–369.
HIRSCH, W. Z., [1962], Design and Use of Regional Accounts; *American Economic Review*, 52 (2), 365–373.
HIRSCH, W. Z., [1963], Application of Input-Output Techniques to Urban Areas; Ch. 8 in Barna, T. (Ed.), *Structural Interdependence and Economic Development*, (Geneva), 365 pp.
HIRSCHMAN, A. O., [1958], *The Strategy of Economic Development*, (New Haven, Connecticut), 217 pp.
HIRSCHMAN, A. O., [1959], Some Suggestions for Research on Comparative Development; In National Bureau of Economic Research, *The Comparative Study of Economic Growth and Structure*, (New York), 142–144.
HOCHWALD, W., (Ed.), [1961], *The Design of Regional Accounts*, (Baltimore), 281 pp.
HOOVER, E. M., [1937], *Location Theory and the Shoe and Leather Industries*, (Cambridge, Mass.), 323 pp.
HOOVER, E. M., [1948], *The Location of Economic Activity*, (New York), 310 pp.
HOSELITZ, B. F., [1953], The Role of Cities in the Economic Growth of Underdeveloped Countries; *Journal of Political Economy*, 61 (3), 195–208.
HOSELITZ, B. F., [1955A], Patterns of Economic Growth; *Canadian Journal of Economics & Political Science*, 21 (4), 416–431.
HOSELITZ, B. F., [1955B], Generative and Parasitic Cities; *Economic Development and Cultural Change*, 3, 278–294.
HOSELITZ, B. F., [1959], On Historical Comparisons in the Study of Economic Growth; In National Bureau of Economic Research, *The Comparative Study of Economic Growth and Structure*, (New York), 145–161.

HOSELITZ, B. F., [1960], *Theories of Economic Growth*, (New York), 344 pp.
HOUGHTON, D. H., [1964], *The South African Economy*, (Cape Town), 261 pp.
HOYLE, B. S., [1963], The Economic Expansion of Jinja, Uganda; *Geographical Review*, 53 (3), 377–388.
HOYLE, B. S., [1964], Further Industrial Growth at Jinja; *East African Geographical Review*, 2, 44–45.
HUGHES, R. B., [1961], Interregional Income Differences; Self-Perpetuation; *Southern Economics Journal*, 28 (1), 41–45.
HUMPHRYS, G., [1962], Growth Industries and the Regional Economies of Britain; *District Bank Review*, 144, 35–56.
HUMPHRYS, G., [1963], Governmental Policy and the Growth Industries; *Professional Geographer*, 15 (4), 13–16.
HUMPHRYS, G., [1965], Services in Growth Centres, and their Implications for Regional Planning, with Special Reference to South Wales; In Caesar, A. A. L. and Keeble, D. E., (Eds.), Regional Planning Problems in Great Britain; *Advancement of Science*, 22, 181–182.
INTERNATIONAL INFORMATION CENTRE FOR LOCAL CREDIT, [1964], *Government Measures for the Promotion of Regional Economic Development*, (The Hague), 159 pp.
ISARD, W., [1951], Interregional and Regional Input-Output Analysis: A Model of a Space-Economy; *Review of Economics and Statistics*, 33 (4), 318–328.
ISARD, W., [1958], Interregional Linear Programming: An Elementary Presentation and a General Model; *Journal of Regional Science*, 1 (1), 1–59.
ISARD, W., [1960], *Methods of Regional Analysis: an Introduction to Regional Science*, (New York), 784 pp.
ISARD, W. and CUMBERLAND, J. H., (Eds.), [1961], *Regional Economic Planning, Techniques of Analysis for Less Developed Areas*, (Paris), 450 pp.
ISARD, W. and KUENNE, R., [1953], The Impact of Steel upon the Greater New York-Philadelphia Industrial Region: A Study in Agglomeration Projection; *Review of Economics and Statistics*, 35 (4), 289–301.
ISARD, W. and REINER, T., [1961], Regional and National Economic Planning and Analytic Techniques for Implementation; In Isard, W. and Cumberland, J. H., (Eds.), *Regional Economic Planning, Techniques of Analysis for Less Developed Areas*, (Paris), 19–38.
ISARD, W. and SMOLENSKY, E., [1963], Application of Input-Output Techniques to Regional Science; Ch. 6, in Barna, T. (Ed.), *Structural Interdependence and Economic Development*, (London), 365 pp.
JAMES, P. E., [1951], An Assessment of the Role of the Habitat as a Factor in Differential Economic Development; *American Economic Review*, 41 (2), 229–238.
JAMES, P. E., [1959], *Latin America*, (London), 942 pp.
JOHNSTON, B. F. and MELLOR, J. W., [1961], The Role of Agriculture in Economic Development; *American Economic Review*, 51 (4), 566–593.
JOUANDET-BERNADAT, R., [1964], Les comptabilités économiques régionales; *Revue D'Economie Politique*, 74 (1), 136–168.
KEEBLE, D. E., [1965], Industrial Migration from North-West London, 1940–1964; *Urban Studies*, 2 (1), 15–32.
KEIRSTEAD, B. S., [1948], *The Theory of Economic Change*, (Toronto), 386 pp.

KELLER, F., [1953], Resources Inventory – A Basic Step in Economic Development; *Economic Geography*, 29 (1), 39–47.
KINDLEBERGER, C. P., [1958], *Economic Development*, (New York), 325 pp.
KINDLEBERGER, C. P., [1964], *Economic Growth in France and Britain, 1851–1950*, (Cambridge, Mass.), 378 pp.
KLAASSEN, L. H., [1965], *Area Economic and Social Redevelopment*, (Paris), 113 pp.
KLAASSEN, L. H., KROFT, W. C. and VOSKUIL, R., [1963], Regional Income Differences in Holland; *Papers of the Regional Science Association, European Congress*, 10, 77–81.
KLEIN, L. R., [1961], A Model of Japanese Economic Growth; *Econometrica*, 29 (3), 277–292.
KUENNE, R. E., [1955], A Rejoinder; *Review of Economics and Statistics*, 37 (3), 312–314.
KUZNETS, S., [1951], The State as a Unit in Study of Economic Growth; *Journal of Economic History*, 11 (1), 25–41.
KUZNETS, S., [1953–54], International Differences in Income Levels: Reflections on their Causes; *Economic Development and Cultural Change*, 2, 3–26.
KUZNETS, S., [1955], Economic Growth and Income Inequality; *American Economic Review*, 45 (1), 1–28.
KUZNETS, S., [1956], Quantitative Aspects of the Economic Growth of Nations, 1 – Levels and Variability of Rates of Growth; *Economic Development and Cultural Change*, 5, 5–94.
KUZNETS, S., [1958], Economic Growth of Small Nations; In Bonne, A., (Ed.), *The Challenge of Development*, (Jerusalem), 9–25.
KUZNETS, S., [1959], On Comparative Study of Economic Structure and Growth of Nations; In National Bureau of Economic Research, *The Comparative Study of Economic Growth and Structure*, (New York), 162–176.
KUZNETS S., [1960], Economic Growth of Small Nations; Ch. 2 in Robinson, E. A. G., (Ed.), *Economic Consequences of the Size of Nations*, (London), 447 pp.
KUZNETS, S., [1963], Notes on the Take-Off; Ch. 2 in Rostow, W. W., (Ed.), *The Economics of Take-Off into Sustained Growth*, (London), 482 pp.
LACOSTE, Y., [1962], Le sous-développement: quelques ouvrages significatifs parus depuis dix ans; *Annales de Geographie*, 71, 247–278 and 387–414.
LACROIX, J. L., [1964], Les Poles de Développement Industrial en Congo; *Cahiers Economiques et Sociaux, Institute de Recherches Economiques et Sociales, Université Lovanium*, 11 (1), 146–191.
LAMPARD, E. E., [1954–55], The History of Cities in the Economically Advanced Areas; *Economic Development and Cultural Change*, 3, 81–136.
LANE, T., [1966], The Urban Base Multiplier: An Evaluation of the State of the Art; *Land Economics*, 42 (3), 339–347.
LARIMORE, A., [1960], A Measure of Economic Change: Sequent Development of Occupance in Busoga District, Uganda; In Ginsburg, N., (Ed.), Essays on Geography and Economic Development; *University of Chicago, Department of Geography Research Paper* 62, 111–123.
LASUEN, J. R., [1962], Regional Income Inequalities and the Problems of Growth in Spain; *Papers of the Regional Science Association, European Congress*, 8, 169–191.

LEONTIEF, W. W., [1936], Quantitative Input-Output Relations in the Economic System of the United States; *Review of Economics and Statistics*, 18 (3), 103-125.
LEONTIEF, W. W., [1951A], Input-Output Economics; *Scientific American*, 185, 4.
LEONTIEF, W. W., [1951B], *The Structure of the American Economy, 1919-1939*, (New York), 264 pp.
LEONTIEF, W. W. et al, [1953], *Studies in the Structure of the American Economy*, (New York), 561 pp.
LEONTIEF, W. W., [1965A], The Structure of the U.S. Economy; *Scientific American*, 212 (4), 25-35.
LEONTIEF, W. W., [1965B], The Structure of Development; In *Technology and Economic Development*, (Harmondsworth, Middlesex), 129-148.
LOMBARDINI, S., [1964], Les Analyses Economiques pour la Préparation d'un Plan Régionale; *Revue D'Economie Politique*, 74 (1), 45-64.
LONSDALE, C., [1965], Planning Britain's Regions; *Town and Country Planning*, 33 (2), 83-90.
LÖSCH, A., [1954], *The Economics of Location*, (New Haven), 520 pp.
LUTZ, V., [1960], Italy as a Study in Development; *Lloyds Bank Review*, 58, 31-45.
MABOGUNJE, A. L., [1965], Urbanization in Nigeria – A Constraint on Economic Development; *Economic Development and Cultural Change*, 8 (4), Pt. 1, 413-438.
MAGEE, B., [1965], *Towards 2000*, (London), 156 pp.
MAHALANOBIS, P. C., [1963], *The Approach of Operational Research to Planning in India* (Bombay), 168 pp.
MAKI, W. R. and TU, YIEN-I, [1962], Regional Growth Models for Rural Areas Development; *Papers and Proceedings of the Regional Science Association*, 9, 235-244.
MANNERS, G., [1962], Regional Protection: a factor in Economic Geography; *Economic Geography*, 38 (2), 122-129.
MANNERS, G., (Ed.), [1964], *South Wales in the Sixties*, (Oxford), 265 pp.
MARTIN, W., [1957], Ecological Change in Satellite Rural Areas; *American Sociological Review*, 22 (2), 173-183.
MCGOVERN, P. D., [1965], Industrial Dispersal; *Planning*, 31, 485, 39 pp.
MCLAUGHLIN, G. E. and ROBOCK, S., [1949], Why Industry Moves South; *National Planning Association, Committee of the South*, Report No. 3, 148 pp.
MCLOUGHLIN, P. F. M., [1966], Development Policy-Making and the Geographer's Regions: Comments by an Economist; *Land Economics*, 42 (1), 75-84.
MCNEE, R. B., [1959], The Changing Relationships of Economics and Economic Geography; *Economic Geography*, 35 (3), 189-198.
MEIER, G. M., [1953], Economic Development and the Transfer Mechanism; *Canadian Journal of Economics and Political Science*, 19 (1), 1-19.
MEIER, G. M., [1964], *Leading issues in development economics*, (New York), 572 pp.
MEIER, G. M. and BALDWIN, R. E., [1957], *Economic Development. Theory, History, Policy*, (New York), 588 pp.
MELAMID, A., [1955], Some Applications of Thunen's Model in Regional Analysis of Economic Growth; *Papers and Proceedings of the Regional Science Association*, 1, L1-L5.
MEYER, J. R., [1963], Regional Economics: A Survey; *American Economic Review*, 53 (1), 19-54.

MEYERS, F., [1965], *Area Redevelopment Policies in Britain and the Countries of the Common Market*, (Los Angeles).
MEYNAUD, J. (Ed.), [1963], *Social Change and Economic Development*, (Paris), 210 pp.
MILHAU, T., [1956], La Théorie de la Croissance et L'Expansion Régionale; *Economie Appliquée*, 9 (3), 349–366.
MILLS, G., [1964], Economics and Computers; *The Technologist*, 1 (4), 15–22.
MOORE, F. T. and PETERSEN, J. W., [1955], Regional Analysis: An Interindustry Model of Utah; *Review of Economics and Statistics*, 37 (4), 368–383.
MOORE, F. T., [1955], Regional Economic Reaction Paths; *American Economic Review*, 45 (2), 133–148.
MOSES, L. N., [1955], Location Theory, Input-Output, and Economic Development: An Appraisal; *Review of Economics and Statistics*, 37 (3), 308–312.
MOUNTJOY, A. B., [1963], *Industrialization and Under-Developed Countries*, (London), 223 pp.
MUNGER, E. S., [1965], *Bechuanaland*, (London), 114 pp.
MYINT, H., [1964], *The Economics of Developing Countries*, (London), 192 pp.
MYRDAL, G. M., [1956], *Development and Underdevelopment*, (Cairo), 88 pp.
MYRDAL, G. M., [1957A], *Rich Lands and Poor: the road to world prosperity*, (New York), 168 pp.
MYRDAL, G. M., [1957B], *Economic Theory and Under-Developed Regions*, (London), 168 pp.
NATIONAL ECONOMIC DEVELOPMENT COUNCIL, [1963], *Conditions Favourable to Faster Growth*, (London), 54 pp.
NEEDLEMAN, L. and SCOTT, B., [1964], Regional Problems and Location of Industry Policy in Britain; *Urban Studies*, 1 (2), 153–173.
NEUMARK, D. S., [1964], Foreign Trade and Economic Development in Africa; *Stanford Food Research Institute, Miscellaneous Publications* 15, 222 pp.
NICHOLLS, W. H., [1961], Industrialization, Factor Markets, and Agricultural Development; *Journal of Political Economy*, 69 (4), 319–340.
NICHOLSON, I., [1965], *The X in Mexico: Growth within Tradition*, (London), 319 pp.
NORTH, D. C., [1955], Location Theory and Regional Economic Growth; *Journal of Political Economy*, 63 (3), 243–258.
NORTH, D. C., [1956], A Reply; *Journal of Political Economy*, 64 (2), 165–168.
NORTH, D. C., [1958], A Note on Professor Rostow's 'Take-Off' Into Self-Sustained Economic Growth; *Manchester School of Economic and Social Studies*, 26 (1), 68–75.
NORTH, D. C., [1959], Agriculture in Regional Economic Growth; *Journal of Farm Economics*, 41 (5), 943–951.
NORTH, D. C., [1961], *The Economic Growth of the United States, 1790–1860*, (Englewood Cliffs), 304 pp.
NURKSE, R., [1961], *Patterns of Trade and Development*, (Oxford), 62 pp.
O'CONNOR, A. M., [1963], Regional Contrasts in Economic Development in Uganda; *East African Geographical Review*, 1, 33–43.
OHLIN, G., [1961], Reflections on the Rostow Doctrine; *Economic Development and Cultural Change*, 9 (4), Part I, 648–655.

OKUN, B. and RICHARDSON, R. W., [1961], Regional Income Inequality and Internal Population Migration; *Economic Development and Cultural Change*, 9 (2), 128–143.
OOI JIN-BEE, [1959], Rural Development in Tropical Areas, with Special Reference to Malaya; *Journal of Tropical Geography*, 12, 1–222.
ORCHARD, J. E., [1960], Industrialization in Japan, China Mainland, and India – Some World Implications; *Annals of the Association of American Geographers*, 50 (3), 193–215.
ORCUTT, G. H., [1960], Simulation of Economic Systems; *American Economic Review*, 50 (5), 893–907.
PAAUW, D. S., [1961], Some Frontiers of Empirical Research in Economic Development; *Economic Development and Cultural Change*, 9 (2), 180–199.
PAHL, R. E., [1965], Urbs in Rure; *London School of Economics and Political Science, Geographical Papers* 2, 83 pp.
PAISH, F. W., [1964], The Two Britains; *The Banker*, 114, 456, 88–98.
PARR, J. B., [1965], *The Nature and Function of Growth Poles in Economic Development*, (Seattle), mimeo., 13 pp.
PARR, J. B., [1966], Outmigration and the Depressed Area Problem; *Land Economics*, 42 (2), 149–159.
PEN, J., [1965], *Modern Economics*, (Harmondsworth), 266 pp.
PERLOFF, H. S., et al., [1960], *Regions, Resources and Economic Growth*, (Baltimore), 716 pp.
PERLOFF, H. S., and WINGO, L., [1961], Natural Resource Endowment and Regional Economic Growth; In Spengler, J. J. (Ed.), *Natural Resources and Economic Growth*, (Washington, D.C.), 191–212.
PERROUX, F., [1955], Note sur la notion de 'pôle de croissance'; *Economie Appliquée*, 8 (1–2), 307–320.
PERROUX, F., [1961], *L'économie du XXème siècle*, (Paris), 598 pp.
PFISTER, R. L., [1963], External Trade and Regional Growth: A Case Study of the Pacific Northwest; *Economic Development and Cultural Change*, 11 (2), Part I, 134–151.
POLITICAL AND ECONOMIC PLANNING, [1962], *Regional Development in the European Economic Community*, (London), 95 pp.
POLLOCK, N. C., [1960], Industrial Development in East Africa; *Economic Geography*, 36 (4), 344–354.
PRATTEN, C. and DEAN, R. M., [1965], The Economies of Large Scale Production in British Industry; *University of Cambridge, Department of Applied Economics, Occasional Paper*, 3, 105 pp.
PREBISCH, R., [1950], *The Economic Development of Latin America and its Principal Problems*, (Lake Success), 59 pp.
PREBISCH, R., [1959], Commercial Policy in the Underdeveloped Countries; *American Economic Review*, 49 (2), 251–273.
PRED, A., [1965], Industrialization, Initial Advantage, and American Metropolitan Growth; *Geographical Review*, 55 (2), 158–185.
RANDALL, L., [1962], Labour Migration and Mexican Economic Development; *Social and Economic Studies*, 11 (1), 73–81.

RANIS, G. and FEI, J. C. H., [1961], A Theory of Economic Development; *American Economic Review*, 51 (4), 533–565.
RAO, V. K. R. V., [1964], *Essays in Economic Development*, (London), 333 pp.
ROBINSON, E. A. G. (Ed.), [1960], *Economic Consequences of the Size of Nations*, (London), 447 pp.
ROBOCK, S., [1956], Regional Aspects of Economic Development, with special reference to recent experience in Northeast Brazil; *Papers and Proceedings of the Regional Science Association*, 2, 51–69.
ROBOCK, S. H., [1963], *Brazil's Developing Northeast: A Study of Regional Planning and Foreign Aid*, (Washington, D.C.), 213 pp.
RODWIN, L., [1955], Planned Decentralisation and Regional Development with Special Reference to the British New Towns; *Papers & Proceedings of the Regional Science Association*, 1, A1–A8.
ROMUS, P., [1958], *Expansion économique régionale et Communauté Européene*, (Leyden), 376 pp.
ROSTOW, W. W., [1955], *An American Policy in Asia*, (New York), 59 pp.
ROSTOW, W. W., [1956], The Take-Off into Self-Sustained Growth; *Economic Journal*, 66, 25–48.
ROSTOW, W. W., [1959], The Stages of Economic Growth; *Economic History Review*, Second Series, 12 (1), 1–16.
ROSTOW, W. W., [1960], *The Stages of Economic Growth: A Non-Communist Manifesto*, (Cambridge), 179 pp.
ROSTOW, W. W., (Ed.), [1963], *The Economics of Take-Off into Sustained Growth*, (London), 482 pp.
ROSTOW, W. W., [1964], *View from the Seventh Floor*, (New York), 178 pp.
ROYAL COMMISSION ON THE DISTRIBUTION OF THE INDUSTRIAL POPULATION, [1940], *Report*, (London), Cmd 6153, 320 pp.
RUTTAN, V. W., [1955], The Impact of Urban-Industrial Development on Agriculture in the Tennessee Valley and the Southeast; *Journal of Farm Economics*, 37 (1), 38–56.
RUTTAN, V. W., [1959], Discussion: the Location of Economic Activity; *Journal of Farm Economics*, 41 (5), 952–954.
SANDEE, J., [1959], *A Long-Term Planning Model for India*, (New York), 39 pp.
SCHUH, G. E. and LEEDS, J. R., [1963], A Regional Analysis of the Demand for Hired Agricultural Labor; *Papers & Proceedings of the Regional Science Association*, 11, 295–308.
SCHULTZ, T. W., [1950], Reflections on Poverty within Agriculture; *Journal of Political Economy*, 58 (1), 1–15.
SCHULTZ, T. W., [1953], *The Economic Organisation of Agriculture*, (New York), 374 pp.
SCHURR, S. H., [1965], Energy; In *Technology and Economic Development*, (Harmondsworth, Middlesex), 91–106.
SCOTTISH DEVELOPMENT DEPARTMENT, [1963], *Central Scotland. A Programme for Development and Growth*, (Edinburgh), 47 pp.
SICKLE, J. V. VAN, [1951], The Southeast: A Case Study in Delayed Industrialisation; *American Economic Review*, 41 (2), 384–393.

SICKLE, J. V. VAN, [1954], Regional Economic Adjustments: the Role of Geographical Wage Differentials; *American Economic Review*, 44 (2), 381-392.
SINGER, M., [1961], Cumulative Causation and Growth Economics; *Kyklos*, 14, 533-545.
SISLER, D. G., [1959], Regional Differences in the Impact of Urban-Industrial Development on Farm and Nonfarm Income; *Journal of Farm Economics*, 41 (5), 1100-1112.
SMITH, P. E., [1961], Markov Chains, Exchange Matrices, and Regional Development: *Journal of Regional Science*, 3 (1), 27-36.
SMITH, W., [1953], *An Economic Geography of Great Britain*, (London), 756 pp.
SMOLENSKY, E., [1961], Industrialisation and Income Inequality – Recent United States Experience; *Papers & Proceedings of the Regional Science Association*, 7, 67-88.
SNYDER, D. E., [1964], Alternative Perspectives on Brasilia; *Economic Geography*, 40 (1), 34-45.
SPENCER, J. E., [1960], The Cultural Factor in 'Underdevelopment': The Case of Malaya; Ch. 3 in Ginsburg, N. (Ed.), Essays on Geography and Economic Development, *University of Chicago, Department of Geography Research Paper*, 62, 173 pp.
SPENGLER, J. J., (Ed.), [1961], *Natural Resources and Economic Growth*, (Washington, D.C.), 306 pp.
STAMP, L. D., [1953], *Our Undeveloped World*, (London), 187 pp.
STAMP, L. D., [1963], *Our Developing World*, (London), 195 pp.
STEEL, R. W., [1964], Geographers and the Tropics; In Steel, R. W. and Protheroe, R. M. (Eds.), *Geographers and the Tropics: Liverpool Essays*, (London), 1-29.
STEVENS, B., [1958], Interregional Linear Programming; *Journal of Regional Science*, 1 (1), 60-98.
STOLPER, W., [1954-55], Spatial Order and the Economic Growth of Cities; *Economic Development and Cultural Change*, 3, 137-146.
STONE, R., [1956], Input-Output and the Social Accounts; Ch. 6 in Barna, T. (ed.), *The Structural Interdependence of the Economy*, (New York), 429 pp.
STONE, R., [1961], An Econometric Model of Growth: The British Economy in Ten Years Time; *Discovery*, 22 (5), 216-219.
STONE, R., [1964A], Computer Models of the Economy; *New Scientist*, 21, 381, 604-605.
STONE, R., [1964B], Mathematics in the Social Sciences; *Scientific American*, 211 (3), 168-182.
STONE, R., [1965], Social Accounting Matrix Models – a Framework for Economic Decisions; In Berners-Lee, C. M. (Ed.), *Models for Decision*, (London), 136-149.
STURMTHAL, A., [1955], Economic Development, Income Distribution, and Capital Formation in Mexico; *Journal of Political Economy*, 63 (3), 183-201.
SUPPLE, B. E., (Ed.), [1963], *The Experience of Economic Growth. Case Studies in Economic History*, (New York), 458 pp.
TANGRI, S., [1962], Urbanization, Political Stability, and Economic Growth; In Turner, R. (Ed.), *India's Urban Future*, (Berkeley), 192-212.
TATTERSALL, J. N., [1962], Exports and Economic Growth: The Pacific Northwest

1880 to 1960; *Papers & Proceedings of the Regional Science Association,* 9, 215–234.
THOMAS, M. D., [1963], Regional Economic Growth and Industrial Development; *Papers of the Regional Science Association, European Congress 1962,* 10, 61–75.
THOMPSON, W. R. and MATTILA, J. M., [1959], *An Econometric Model of Postwar State Industrial Development,* (Detroit), 116 pp.
TIEBOUT, C. M., [1956], Exports and Regional Economic Growth; *Journal of Political Economy,* 64 (2), 160–164 and 169.
TINBERGEN, J., [1959], Comparative Studies of Economic Growth; In National Bureau of Economic Research, *The Comparative Study of Economic Growth and Structure,* (New York), 193–200.
TINBERGEN, J., [1960], Regional Planning: Some Principles; *Netherlands Economic Institute, Division of Balanced International Growth, Publication* 21/60, 13 pp.
TINBERGEN, J., [1964], *Regional Planning,* (Rotterdam), 11 pp.
TINBERGEN, J. and BOS, H. C., [1962], *Mathematical Models of Economic Growth,* (New York), 131 pp.
TOSI, J. A. and VOERTMAN, R. F., [1964], Some Environmental Factors in the Economic Development of the Tropics; *Economic Geography,* 40 (3), 189–205.
ULLMAN, E. L., [1958], Regional Development and the Geography of Concentration; *Papers & Proceedings of the Regional Science Association,* 4, 179–198.
ULLMAN, E. L., [1960], Geographic Theory and Underdeveloped Areas; Ch. 2 in Ginsburg, N. (Ed.), *Essays on Geography and Economic Development; University of Chicago, Department of Geography Research Paper,* 62, 173 pp.
UNITED NATIONS, Department of Economic and Social Affairs, [1961], Use of Models in Programming; *Industrialization and Productivity,* 4, 7–17.
USHER, D., [1965], 'Equalizing Differences' in Income and the Interpretation of National Income Statistics; *Economica, New Series,* 32, 127, 253–268.
VALAVANIS-VAIL, S., [1955], An Econometric Model of Growth: U.S., 1869–1953; *American Economic Review,* 45 (2), 208–221.
VERBURG, M. C., [1964A], The Gent-Terneuzen Developmental Axis in the Perspective of the European Economic Community; *Tijdschrift voor Economische en Sociale Geografie,* 55 (6/7), 143–150.
VERBURG, M. C., [1964B], Location Analysis of the Common Frontier Zones in the European Economic Community; *Papers of the Regional Science Association, European Congress,* 12, 61–78.
VINING, R., [1946], The Region as a Concept in Business-Cycle Analysis; *Econometrica,* 14 (3), 201–218.
VINSKI, I., [1962], Regional Distribution of National Wealth in Yugoslavia; *Papers of the Regional Science Association, European Congress,* 3, 127–168.
WILHELM, W., [1950], Soviet Central Asia: Development of a Backward Area; *Foreign Policy Reports,* 217–228.
WILLIAMSON, J. G., [1965], Regional Inequality and the Process of National Development: A Description of the Patterns; *Economic Development and Cultural Change,* 13 (4), Part II, 84 pp.
WILSON, T., [1964], Policies for Regional Development; *University of Glasgow Social & Economic Studies, Occasional Papers,* 3, 93 pp.
WONNACOTT, R. J., [1964], Wage Levels and Employment Structure in United

States Regions: A Free Trade Precedent; *Journal of Political Economy*, 72 (4), 414–419.
WOOD, W. D. and THOMAN, R. S., (Eds.), [1965], *Areas of Economic Stress in Canada*, (Kingston, Ontario), 221 pp.
WRIGHT, M., [1965], Regional Development: Problems and Lines of Advance in Europe; *Town Planning Review*, 36 (3), 147–164.
WRIGLEY, E. A., [1965], Changes in the Philosophy of Geography; Ch. 1 in Chorley, R. J. and Haggett, P., (Eds.), *Frontiers in Geographical Teaching*, (London), 378 pp.
YATES, P. L., [1961], *El Desarollo Regional de Mexico*, (Mexico City), 405 pp.

CHAPTER NINE

Models of Urban Geography and Settlement Location

B. J. GARNER

INTRODUCTION

Settlement studies form a traditional part of Human geography. They have held a dominant place in early statements, like that of Brunhes (1925), and in contemporary reviews like that by Jones (1964). Forming such distinctive features in the landscape, they were viewed as a fundamental expression of 'Man-Land' relationships. Perhaps not surprisingly, studies have traditionally emphasized strong links between the physical environment and various aspects of human occupance of regions; patterns of settlement distribution and morphology were all too often 'accounted' for by physical features. With the realization that urban areas themselves are regions full of interest for study it was merely a matter of applying the traditional 'Man-Land' concepts to the untapped sources for geographical investigation. The 'Townscape' became the urban equivalent of landscape and attention was drawn immediately to microscopic differences in the 'feel' or character of the various parts of urban areas. Emphasis was placed on form and patterns were essentially viewed as a reflection of physical controls supplemented heavily by historical influences. Although there is no question of the usefulness of these kinds of study, new frameworks are needed if a deeper understanding of the spatial organization of this particular aspect of human activity is to be obtained.

Settlements are considered as comprising a complex set of 'Man-Man' relationships here and this concept underlies the models discussed in this chapter. The importance of physical agents is overshadowed by the emphasis placed on various economic and social factors. But the impressive literature dealing with the many aspects of settlements is too large to be included in its entirety; we must be selective. Emphasis is given to static rather than dynamic or predictive models and little explicit attention is paid to intra-urban transport and movement models. The chapter is divided into two parts: (1) a review of the major models and empirical studies dealing with the location and arrangement of settlements, and (2) a discussion of some of the models pertaining to the internal structure of urban areas.

Some underlying regularities in models

Virtually all models of settlement location and urban structure have one thing in common; they assume a measurable degree of order in spatial behaviour. This seems to be founded on the following six premises which form the basis of, or are implied in, most models.

1. *The spatial distribution of human activity reflects an ordered adjustment to the factor of distance.* Distance is basic to geography. Watson (1955) even goes as far as to state that geography itself is a 'discipline in distance'. This is not hard to see, for if all things were concentrated at a given place at a given time, there would be no patterns, no spatial variation or areal differentiation, in short there would be no geography. However, regularities with distance may not be immediately apparent. Writing on *Locational Analysis in Human Geography*, Haggett (1965, p. 2) quotes Sigwart's words, '... that there is more order in the world than appears at first sight is not discovered till order is looked for'.

The search for order in spatial behaviour must be accompanied by greater flexibility in thinking about distance. Different distance measures can be justified in model building by the fact that different things are more or less relevant in different types of studies. For example, in movement models, travel time (Voorhees, 1955), transport costs (Harris, 1954) or road distances weighted according to different kinds of road surface (Garrison, 1956) have proved more important than linear distance. Non-linear distance measures are equally important in locational models; Olsson and Persson (1964) have used density-distance, Getis (1963) income-distance and Garner (1966) land value-distance to reveal order in the distribution of urban functions.

2. *Locational decisions are taken, in general, so as to minimize the frictional effects of distance.* This concept, generally known as the 'law of minimum effort' (Lösch, 1954, p. 184) or the 'principle of least effort' (Zipf, 1949), suggests that events reach their goal by the shortest route. In settlement models in particular, movement-minimization is fundamental to an understanding of the geometry of settlement patterns and intra-urban location. Its importance is reflected in the place accorded the circle as a theoretical trade area and city shape (Haggett, 1965, p. 48).

3. *All locations are endowed with a degree of accessibility but some locations are more accessible than others.* Accessibility is difficult to define explicitly, but the

term generally implies 'ease of getting to a place' (Forbes, 1964). As such it is a variable quality of location. In a technical sense, accessibility is a relative quality accruing to a piece of land by virtue of its relationship to a system of transport (Wingo, 1961, p. 26). In an operational sense, it is the variable quality of centrality or nearness to other functions and locations. Clearly the notion of accessibility is closely related to the concept of movement-minimization, especially when this is measured by the costs involved in overcoming distance. In this context, it is also generally accepted as the basis of the rent paid for and the value attached to sites in urban land use models.

4. *There is a tendency for human activities to agglomerate to take advantage of scale economies.* Scale economies mean the savings in costs of operation made possible by concentrating activities at common locations. In the organization of an industrial firm it is exemplified in mass production techniques. Proximal concentration of a number of firms also make savings possible. Agglomerations can thus be viewed as nodes in the economic landscape arising from centripetal forces in spatial organization. The concentration of activities to form settlements themselves can be viewed as a reflection of scale economies but more important perhaps are the various agglomerations within urban areas such as shopping centres and industrial districts. Residential zones can also be viewed as agglomerations for scale economies in household costs of utilities and public services.

5. *The organization of human activity is essentially hierarchical in character.* This is true of both spatial and non-spatial aspects of human activity. For example, it is true of political organization although it may not be explicitly expressed spatially. However, the latter is a fundamental aspect of the spatial structure of settlement patterns and appears to result from interrelationships between agglomeration tendencies and accessibility. More accessible locations appear to be the sites of larger agglomerations. One of the implications from this is that there exists in an area a hierarchy of locations in terms of accessibility.

6. *Human occupance is focal in character.* This notion underlies the concept of the nodal or functional region and is basic to movement models and the spacing of certain activities in areas. The nodes about which human activity is organized are agglomerations of varying size. Since these are hierarchically arranged it follows that there is a hierarchy of different sized focal regions. Philbrick (1957) argues that the areal structure of the occupance of the earth's surface is composed of a number of hierarchically 'nested' orders of spatial functional organization. In this way, movement-minimization, accessibility, agglomerations and hierarchies are linked together to form a system of human organization in space.

SETTLEMENT PATTERNS

The existence of varying sized population clusters in the landscape is an inevitable feature in the spatial organization of human activity. Settlements exist because certain activities can be carried on most efficiently if they are clustered together rather than dispersed. No matter what the particular activities are, they can be generally viewed as services which are provided not just for settlements themselves, but for people living in surrounding tributary areas. Since settlements are spatially separated one from another, linkages between them are essential, and one framework for study is to view them as nodes or focal points in a transport network. We are concerned in this section with models of the arrangement of these varying sized nodes. For convenience our analysis will be divided into two components: (1) a 'horizontal' component in which the spatial parameter is explicit. Focus here is on maps and concerns regularities in the size and spacing of settlements; (2) a 'vertical' or organizational component, in which the spatial parameter need *not* be explicit. Focus here is on regularities which appear on graphs. However, it must be realized that the two components are mutually interrelated in reality. For example, the size and spacing of settlements may affect, and give rise to, regularity in vertical organization and *vice versa* – our division is consequently merely one of expediency.

Simple horizontal arrangements

A map showing the pattern of settlement in a region can be broken down into three basic parts (Harris and Ullman, 1945); (1) a linear pattern consisting of transport centres performing break-of-bulk and allied services and for which location is related to the disposition of transport routes; (2) a cluster pattern, consisting of places performing specialized services such as manufacturing, mining or recreation, and for which location is related to the localization of resources, and (3) a uniform pattern, consisting of places whose prime function is the provision of a wide range of tertiary goods and services and for which location is related to a dispersed population.

So far no model has satisfactorily taken into account all three aspects of urban causation and location, although steps have been taken in this direction by Lösch (1954), Isard (1956) and Bunge (1962, p. 59). Much recent geographical research has used models developed to explain the distribution and arrangement of the third category of centres, commonly referred to as central places, as a framework for study, and it is with these that we will concern ourselves here.

A. *The regular lattice model.* This is the basis of the central place model of

Christaller (1933; 1966), which he viewed as a 'general deductive theory' to explain the 'size, number and distribution of towns' in the belief that 'there is some ordering principle governing the distribution' (Berry and Pred, 1961, p. 15). Although this claim may be too great, there is no doubt that the model has had a profound impact on geographical research in the last decade and Bunge (1962) at least views it as crucial for the existence of a theoretical geography. In view of the importance of Christaller's work, it is surprising that there was not a satisfactory translation from the original German until that by Baskin in 1966. Perhaps even more surprising is that there is no complete review of the ideas contained in the model although discussion of its spatial aspects can be found in Berry and Pred (1961, pp. 3–18), Ullman (1941), Berry and Garrison (1958A, B and C) and Bunge (1962), and a critique of its economic assumptions is given by Baskin (1957).

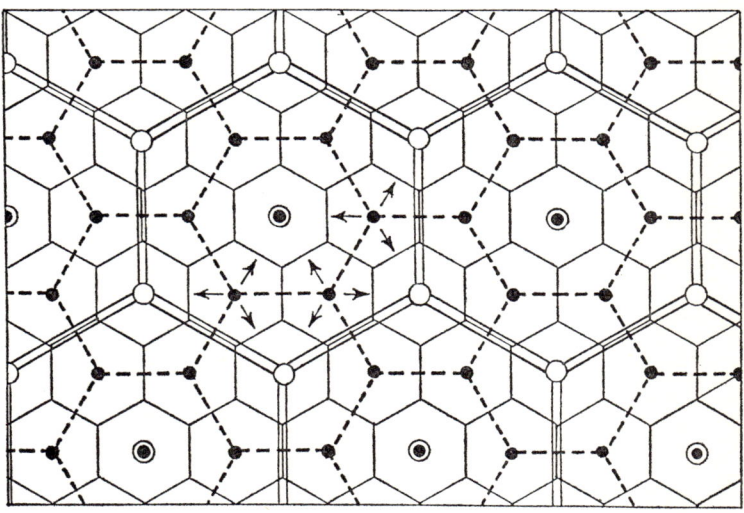

9.1 K=3 settlement pattern according to Christaller's marketing principle.

The basic elements of the regular lattice model, shown in Figure 9.1, are developed under assumptions of an isotropic surface – that is under conditions of a uniform distribution of population and purchasing power, uniform terrain and resource localization, and equal transport facility in all directions. In this ideal situation, the essential features of the 'Horizontal' arrangement of settlements are: (H-1) they are regularly spaced to form a triangular lattice, and (H-2) they are centrally located within hexagonal shaped trade areas. 'Vertical' organization hinges on the assumption that a hierarchy of discrete groups or orders of settlement exists in which (V-1) higher order places supply all the goods of lower order places plus a number of higher order goods and services that differentiates them from, and at the same time sets them above, central places of lower order, and (V-2) higher order places offer a greater range of goods and services, have more establishments, larger populations, trade areas and trade area populations and do greater volumes of business than lower order settlements. This 'vertical' organization has 'horizontal' expression in the following ways: (C-1) higher order central places are more widely spaced than lower order places, and (C-2) lower order central places, to be provided with higher order goods and services, are contained or 'nest' within the trade areas of higher order places according to a definite rule.

As shown in Figure 9.5, a hierarchy of settlements can be organized in various ways, each with its own geometrical arrangement of central places and trade area boundaries. In Christaller's basic model, organized on what he calls the marketing principle, the hierarchy and nesting pattern results in the maximum number of central places – a necessary condition if the supply of goods from central places is to be as near as possible to the consumer in accordance with the notion of movement-minimization. This particular system, known as a K-3 network, is the one shown in Figure 9.1. Three orders in the hierarchy are shown in the following way; (1) the lowest order places (e.g. hamlets) by filled circles, (2) intermediate order places (e.g. villages) by open circles, and (3) high order places (e.g. towns) by double circles. Trade area boundaries of the three orders of settlements are indicated by solid lines, dashed lines and double lines respectively.

The K-value, here three, refers to the number of settlements at a given level in the hierarchy served by a central place at the next highest order in the system. For example, in Figure 9.1, each village serves the equivalent of three hamlets. This number is made up of the hamlet part of the functional structure of the village itself (from V-1 above), plus a one-third share of the six border hamlets since each of them is shared between three villages as indicated by the arrows. Similarly, towns will provide town-level goods to three villages and it follows from the geometry of the hexagonal trade areas, they will serve nine hamlets. This regular progression, which can be extended upward from the town level, exists because Christaller assumed that once the K-value was adopted in any region, it would be fixed. Consequently, it would apply equally to the relationship between hamlets and villages, villages and towns, and so on up through the central place hierarchy. Because of this the total number of settlements in any area should follow a regular progression. In the case when the K-value equals three, this would be one, three, nine, twenty-seven, eighty-one . . . , starting with the highest order place in the region.

Although Christaller's K-3 model has received most attention in empirical studies, he did postulate two other forms of hierarchical arrangements to take account of deviations from the marketing principle. Firstly, a K-4 network (see Fig. 9.5) organized according to the traffic principle was proposed to account for situations in which costs of transport were significant. This arrangement enables '... as many important places as possible to lie on one traffic route between larger towns, the route being established as cheaply as possible' (Berry and Pred, 1961, p. 16), and gives rise to linear patterns and distortions of trade areas at right angles to traffic routes. 'Nesting' is according to the rule of fours because connections will be made with the equivalent of three of the six nearest dependent places. Secondly, a K-7 network was proposed to take account of the administrative principle or principle of 'separation', in which connections are made between a given order of central place and all six of the nearest immediately lower order places (see Fig. 9.5). Nesting is therefore according to the rule of sevens.

The literature dealing with various aspects of Christaller's models is both voluminous and contradictory. The reader is referred to the excellent annotated bibliography of central place studies by Berry and Pred (1961), to Haggett (1965), and to Berry (1967) for a summary of findings. Here we will look at the evidence for a regular lattice distribution of settlement in reality.

Hexagons: One of the commonest criticisms of the regular lattice and the hexagonal framework is that it is far too rigid and abstract. Although this is certainly true up to a point since the hexagon is a pure concept much as perfect competition is a pure concept to the economist, there are forceful theoretical arguments (see Haggett, 1965, p. 49, and Lösch, 1954, p. 105) for thinking in this framework if movement-minimization is at all relevant in spatial behaviour. It is not surprising to find, therefore, that very few studies have actually tested whether hexagonal arrangements do in fact exist in reality, and more are needed. In part this has perhaps been due to the subjective delimitation of trade areas (compare the methods used by Smailes (1947) with those by Green (1950) for example) and in part because until recent suggestions by Bunge (1962) and Boyce and Clark (1964) the concept of shape was difficult to quantify.

One study which provides evidence of hexagons in territorial organization is that by Haggett (1965, p. 50), who studied shape patterns of political division (*municipios*) in Brazil. Although the pattern was dominated by elongated shapes, the number of contacts between one territory and adjacent territories showed evidence of hexagonal arrangement. Nearly one out of three *municipios* bordered on exactly six neighbours while the mean number of contacting sides was 6·21 compared to six in the perfectly hexagonal system shown in Figure 9.1. Although this kind of investigation must be extended to less easily defined trade area boundaries before we can definitely say that the hexagon is a basic part of reality, the rather striking approximations of Haggett's findings to the model suggests that '... criticism of the hexagonal system as being over-theoretical may have been too hasty' (Haggett, 1965, p. 51).

Regular lattices: Until the recent introduction of a statistical definition of spatial uniformity based on nearest neighbour analysis (Clark and Evans, 1954; Dacey, 1963), it was difficult rigorously to measure dot-patterns. The more traditional 'eye-ball' methods are not really satisfactory. Nearest neighbour analysis enables patterns to be measured by an index, *Rn*, which ranges between zero for a perfectly clustered pattern and a maximum of 2·15 for a perfectly uniform pattern. A value of one is associated with random patterns. These three patterns are illustrated in Figure 9.2. Applications of this statistical measure to settlement patterns have shown them to tend toward randomness rather than the uniform distributions postulated in the model.

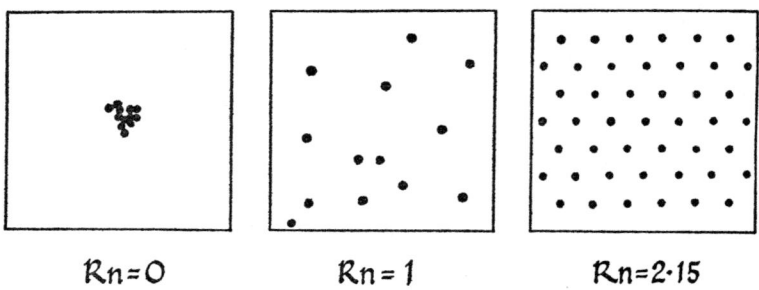

9.2 Clustered, random and perfectly uniform distributions.

Typical of studies using the statistical approach is that by King (1962) who analysed the settlement patterns in twenty sample areas in the United States (Fig. 9.3A). Comparison of the observed linear distances between a place and its nearest neighbour, regardless of size, with the spacing expected in a random distribution yielded a range in *Rn* values. As Figure 9.3B illustrates, these show a small range from 0·7 for Utah with a relatively clustered pattern to 1·38 for Missouri which tends towards a uniform spacing (Fig. 9.3C). The concentration of *Rn* values for the remaining areas between these limits indicates that the settlement pattern of the United States approximates a random more than the hypothesized uniform distribution. This finding concurs with that found for part of Wisconsin by Dacey (1962). Although size of centre was explicitly considered in this study, at no level in the hierarchy was a uniform distribution indicated.

The failure of the regular lattice model of settlement distribution is hardly surprising in view of the idealized conditions under which it was hypothesized. Haggett (1965) illustrates the distorting effects of agglomeration, resource localization and of time lags and it is not difficult to think of others. But it might be that the lack of evidence for the regular lattice in empirical work is due to shortcomings in study design and measurements used. Thus

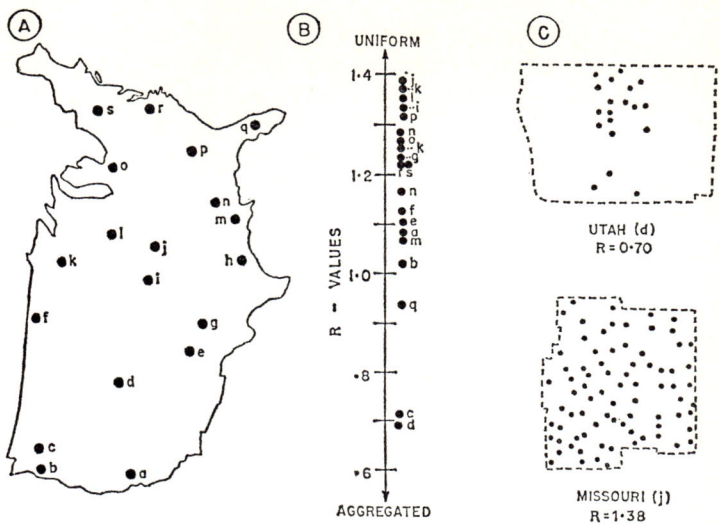

9.3 Sample study areas within the United States (A), scale of Rn-values (B), and clustered settlement patterns for Utah contrasted to uniform settlement pattern for Missouri (C) (*Source: Haggett, 1965, p. 91. After King, 1962, pp. 3-4*).

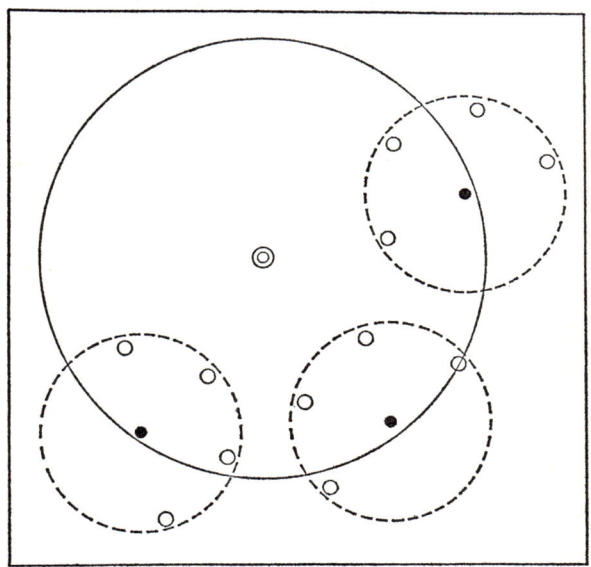

9.4 Regular cluster settlement pattern (*Source: Brush, 1953, p. 391*).

Dacey (1962) suggests that the notion of a lattice could be accepted for south-western Wisconsin but that the central place hierarchy on which he based his analysis had been incorrectly identified by Brush (1953) in the original study of the area. If this is so, then the definition of nearest neighbour is crucial in study design. Normally, measurements are made to the nearest neighbour of the *same* size. The problem is in deciding what constitutes a place of the same size, for apart from chance circumstances, two places are very unlikely to be exactly the same size, however measured. Rather, as pointed out by Thomas (1961), it is better to consider the nearest place of *approximately* the same size. To sharpen the definition of 'approximately', he introduces a statistical definition based on probability notions but the refinement has not so far been used to test for the regular lattice.

A more serious shortcoming concerns the measure of distance used. It can be argued that regularities are not found because the models have been tested under non-isotropic conditions for which they were never designed, and that to compensate for this disparity between 'model world' and real world, more functional measures of distance are wanted. Evidence that this might be worthwhile investigating is provided by Olsson and Persson (1964) who are able to explain a higher amount of variation in the spacing of settlements in Sweden by measuring distance as the number of people living between two places. This idea is consistent with a model of consumer movement proposed by Huff (1961) which postulates that shopping trips are restricted not only by physical distance but also by intervening population density. Although Huff's model was developed for intra-city movements there is no logical reason why similar arguments cannot apply to movements between cities and to their spacing.

Transformations of distance in this and other ways would be consistent with Christaller's original formulation which emphasized time-cost distance in the spacing of settlements. However, as Blome (1963) shows, there are shortcomings in the use of this distance transformation. A further possibility of eliminating the problem of non-isotropism of surface might be to abandon classical concepts of Euclidean geometry. Tobler (1963) has drawn attention to the possibilities for this and encouraging evidence is presented in a study by Getis (1963) which shows that when variations in income are smoothed out, uniform patterns are identified in the distribution of grocery shops in Tacoma city, USA.

B. *Regular cluster models*. An alternative hypothesis to the regular lattice model is that settlements form regular clusters. This is the basis of a model proposed by Kolb and Brunner (1946). In Figure 9.4, towns are shown by double circles, villages by filled circles and hamlets by open circles; solid lines indicate the trade area boundary of towns, and dashed lines the boundary of the influence of villages. The pattern is one in which (1) villages are located near the boundary of the influence of towns, (2) hamlets are clustered around villages at the edge of village trade areas, and (3) towns are centrally located.

The spacing of settlements is consequently one of clusters resulting from the influence of size on location. Although elements of the latter are included in the regular lattice model (e.g. C-1 above), the cluster model is less rigorous in that neither distances nor direction are specified and because of this it is perhaps more in accord with reality. Moreover, the model is in agreement with other models of spatial interaction including the 'Law of Retail Gravitation' (Reilly, 1931) and the 'Proportional Range of Influence' (Tuominen, 1949) which specify that the 'pull' exerted by a place varies directly with its size and decreases outwards with distance. Consequently smaller places are not likely to develop as close to large clusters as they are to one another.

Evidence from empirical studies of clustering is not conclusive and if anything is contradictory. Brush (1953) found hamlets crowding together in areas farthest from the larger towns in Wisconsin, USA, and Dacey (1962) confirmed this observation using nearest neighbour analysis. In other studies, for example that by Berry, Barnum and Tennant (1962) of central places in Iowa, USA, there is less evidence of clustering although it is shown that villages are functionally linked with cities rather than with towns, as envisaged in the model. Further evidence that smaller places grow up at the boundaries of trade areas between larger settlements is given for Sweden by Godlund (1956).

On the other hand, there is abundant evidence illustrating that the spacing of settlement is governed by the size of settlement. Brush and Bracey (1955) found from a comparative study of settlement patterns in south-western Wisconsin, USA and southern England that towns were spaced at an average distance of twenty-one miles apart and villages between eight and ten miles apart. Similar evidence of the wider spacing of larger settlements is given by Christaller (1933) for southern Germany and by Lösch (1954) for Iowa, USA. The implication from these studies is that regularities of this sort may well exist in many areas despite differences in population density, regional economy and social and economic history.

The general evidence of these earlier studies has been specified more exactly by Thomas (1961) for Iowa, USA, and Olsson and Persson (1964) for Sweden using regression analysis. These studies both show rather low, but statistically significant, correlations between size of centre and spacing. Only about a third of the variation in spacing could be 'explained' ($R^2=0.35$) in Iowa, and this seemed to be fairly stable over time (Thomas, 1962). In Sweden, a similar low 'explanation' ($R^2=0.31$) was found although this improved slightly ($R^2=0.46$) when the number of people living in theoretically delimited trade areas surrounding central places was used as a modification of distance, implying that the simple regularity is distorted by population densities not considered in the model. Although we can conclude from these studies that there is a significant relationship between size of city and spacing, it is obviously not as simple as postulated in the regular cluster model. The spacing of settlement must be considered in a wider context than size alone.

One of the few studies in which this has been attempted is that by King (1961) who used multiple regression analysis on a sample of 200 towns in the USA. His model specified that spacing was a complex function of size of centre, its occupational structure and the character of the region in which it is located. Table 9.1 shows that only two per cent of variation could be explained although, as to be expected, the spacing was more predictable for central places than for non-central places. The difficulties of generalizing over large

TABLE 9.1

Relationship between settlement spacing and other variables in the United States, 1950

Hypotheses:	Single hypothesis (*Size of settlement*)	Multiple hypothesis (*Six-factors*)
Coefficients of determination (R^2):		
National results	0·02*	0·25*
Centre classification:		
Central places	0·09*	0·26*
Non-central places	0·01	0·42*
Regional agricultural classification		
Grazing and wheat zone	0·42*	0·67*
Specialized farming zone	0·01	0·20*
General farming zone	0·07	0·67*
Feed grain and livestock zone	0·22*	0·34*
Dairying zone	0·04*	0·36*

* Significant at the 95 per cent confidence level

Source: King, *1961, pp. 227–31.*

areas such as the entire USA is also clearly revealed by the marked regional differences in spacing. In the Great Plains where physical conditions most closely approximate those assumed by Christaller, over forty per cent of variation in spacing could be explained compared to only four per cent in the more undulating country of the Dairy Region in the Middle West.

The low correlation between size of centre and spacing prompted King to investigate the effects of other factors on spacing. He argued that settlements of a given size were likely to be more widely spaced in areas of extensive farming, and where rural population densities, overall population densities, agricultural production and the proportion of workers in manufacturing were low. Regression analysis showed that all five variables were only slightly more valuable than town size in predicting spacing. Only overall population density explained as much as 10 per cent of the variation in spacing. Even when spacing was considered a function of all six variables taken together

MODELS OF URBAN GEOGRAPHY AND SETTLEMENT LOCATION

only one quarter of the variation in spacing could be accounted for, although the model gave a better fit when regional differences were considered.

Complex horizontal arrangements

Since Christaller's contribution, the hexagonal based model has been elaborated and extended by Lösch (1954) into a more complex model which approximates more nearly the patterns in the real world. The more flexible model is made possible by regarding the K-value as being free to vary. Thus

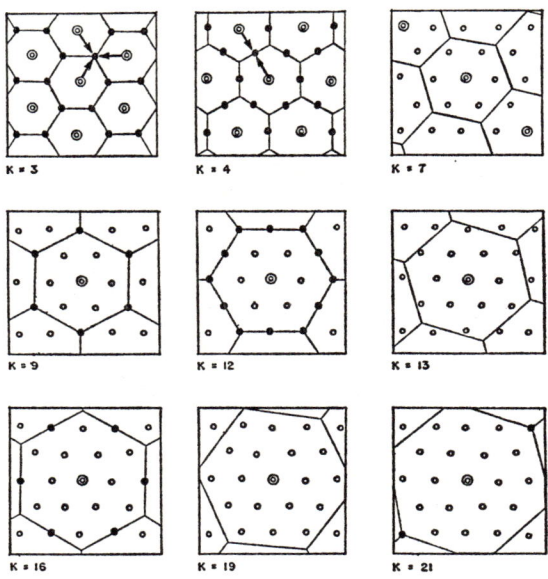

9.5 Nine smallest hexagonal trade areas in a Löschian landscape (*Source: Haggett, 1965, p. 119. After Lösch, 1954, p. 118*).

the fixed-K assumptions of the marketing, traffic and 'separation' principles of the Christaller model are considered to be special cases of a larger number of possible hexagonal systems and settlement distributions. The nine smallest feasible hexagonal arrangements associated with the various K-values are illustrated in Figure 9.5. These are obtained by varying the orientation and size of the hexagon. In the figure, high order places are shown by a double circle, the dependent lower order places by open circles if they lie completely within the trade area of the high order centre and by filled circles if they lie on the perimeter of it.

The nine arrangements shown in Figure 9.5 and further nets up to K-25 derived in a similar way are superimposed on one another in Figure 9.6 so that they all have the same central point. This point, Lösch argues, will be the site of a metropolis – the largest order of central place in his system. By rotating the nets about this point it is possible to derive a pattern which has six sectors with many, and six sectors with few, settlements. The sectoral

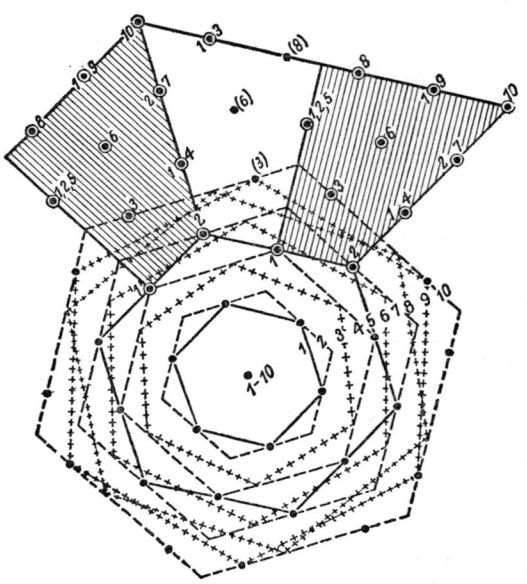

9.6 Ten smallest hexagons superimposed over a common central point. The shaded sectors contain many centres. Simple points represent original settlements; numbers beside encircled points indicate the size of trade areas (*Source: Lösch, 1954, p. 118*).

pattern, called an 'economic landscape', is shown in Figure 9.7. In this arrangement, Lösch claims that the greatest number of locations coincide, the aggregate distance between all settlements is minimized, and the maximum number of goods can be locally supplied.

The basic features of the 'horizontal' arrangement of settlement in the Lösch model are; (1) concentration of settlement into sectors separated by interstitial areas in which settlement is less dense, as shown in Figure 9.8A. Within the 'city rich' sectors, (2) settlement increases in size with distance from the central metropolis, as shown in Figure 9.8B, and (3) small settlements are located 'about half way' between two larger ones. This is shown in Figure 9.8C by the two larger classes of dots, representing places which are the centres of over four and over eight coincident hexagonal nets respectively.

Lösch further asserts, but does not demonstrate, that settlements comprise

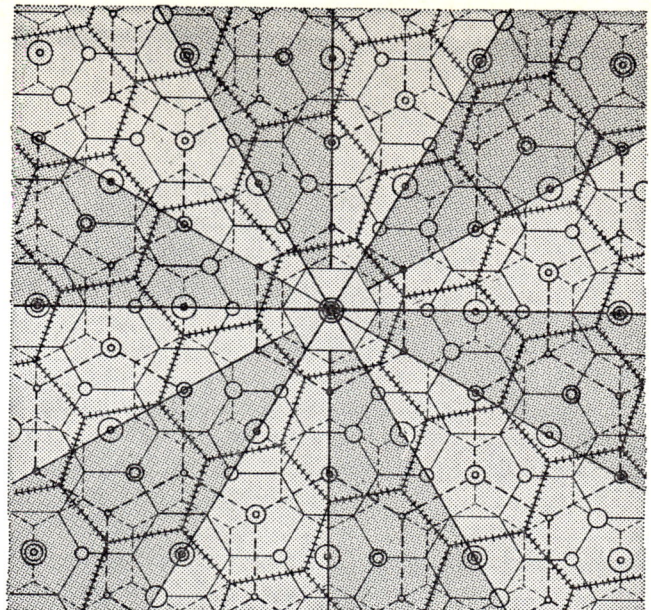

9.7 Simplified Löschian landscape and system of hexagonal trade areas (*Source: Haggett, 1965, p. 123. After Isard, 1956, p. 270*).

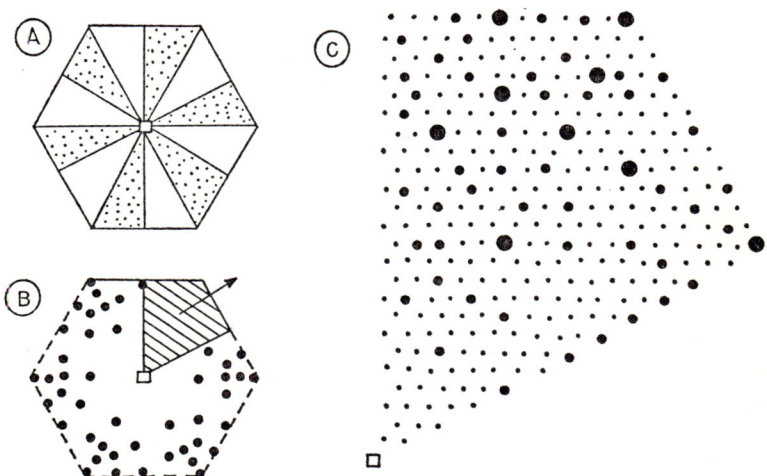

9.8 Alternating city-rich and city-poor sectors (A), distribution of large cities (B), and pattern of settlement within one sector (C) (*Source: Haggett, 1965, p. 123. After Lösch, 1954, p. 127*).

a hierarchy. Since this is based on a variable-K rather than the fixed-K of Christaller, 'vertical' organization is far less rigid. Its basic features are; (1) a continuous array of centres in which, (2) higher order places do not necessarily provide all the functions typical of lower order places, and (3) settlements performing the same *number* of functions do not necessarily provide the same *kinds* of functions. For example, because of the rotation of nets, a settlement serving seven smaller places may be either a K-7 settlement or merely the coincident centre for both a K-3 and a K-4 size market area.

Partial evidence in support of Lösch's 'city rich' and 'city poor' landscape is presented by Bogue (1949) from a study of population densities around the sixty-seven largest metropolitan centres in the USA. Bogue divided the hinterlands of each metropolitan area into three types of sector; (1) *route* sectors containing major highways leading from the central metropolis to other metropolitan centres, (2) *subdominant* sectors, which contained at least one city of more than 25,000 inhabitants, and (3) *local* sectors, which contained neither large cities nor inter-metropolitan routes. Figure 9.9E shows that urban densities are greatest in the subdominant sectors. Densities are lower than might be expected from the model in the route sectors (Fig. 9.9D), while local sectors were well below the level of the other two (Fig. 9.9F).

Bogue's study clearly revealed a logarithmic decline in population densities with distance away from the central metropolis. In Figure 9.9 this is shown by the broken line in all the graphs. At 25 miles from the city densities are over 200 per square mile whereas at 250 miles from it, densities have fallen to only about 4 persons per square mile. In addition, the findings suggest that densities are greater around larger cities than smaller ones, presumably a reflection of the differential forces of agglomeration, and that density patterns are markedly different from one region to another in the United States, (Fig. 9.9A, B and C).

Although partly confirming the Lösch model, Bogue's study points up one of its major deficiencies, namely the disparity between the postulated uniform size of hexagonal market areas and the implied decline in density of population with distance from the central metropolis. Under the latter conditions, market areas should be smaller close to the metropolis and should increase in size with distance from it. It follows from this that settlements should be more widely spaced the further they are away from the metropolis. However, this part of Lösch's work is not at all clear and more detailed information needs to be supplied, although Isard (1956, p. 272) has attempted a graphic modification, as shown in Figure 9.10, to take into account density differences. Considerable difficulty appears to have been encountered in trying to adhere to the hexagonal shape of market area and in view of his criticisms of the Lösch model (Isard, 1956, p. 271) it is not clear why this basic shape was retained.

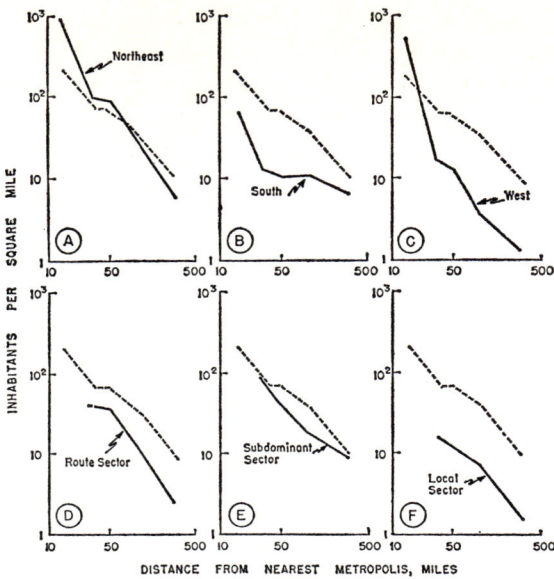

9.9 Variations in urban population densities by region (A, B, C) and by sector (D, E, F) in the United States. Both axes are transformed to logarithmic scales (*Source: Haggett, 1965, p. 93. After Bogue, 1949, pp. 47 and 58*).

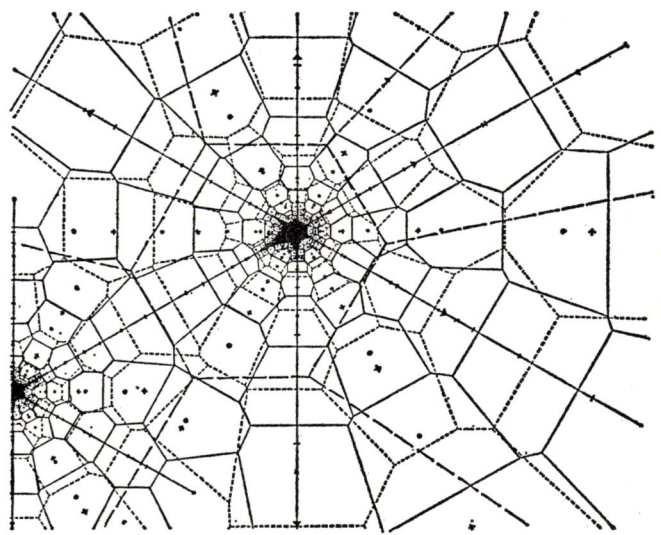

9.10 Löschian system of hexagonal trade areas modified by agglomeration (*Source: Isard, 1956, p. 272*).

Periodic central places

In many parts of the world, especially the lesser developed countries, consumers are not normally supplied with goods and services from permanent central places. Rather, central functions are performed by mobile agents and their physical facilities moving from place to place during the short run of time. Such spatial behaviour of individual producers gives rise, in the aggregate, to the 'fair' or periodic market. Two or more merchants meeting at the same place at the same time are apt to sell more than if the time of their visits to a place is not co-ordinated. Thus, a regular schedule of markets emerges as the important feature of the central place system.

Even where permanent central places exist in the landscape, the number and kinds of functions they perform pulsates periodically when they become the site of markets. This is true, for example, of many small rural settlements in England today which still hold regular weekly markets. On such days, the normal hierarchy of central places is temporarily disturbed; low order places, enlarged by the addition of other functions comprising the market, assume higher order status. Concomitant with their increased importance, they serve larger trade areas than normal. In England this is reflected in the special bus services provided for the people living in surrounding rural areas on market days. In other areas of the world, it is reflected in the convergence of people from rural areas on the market centre. In both cases consumers, by submitting themselves to the discipline of time in trip-making, are able to free themselves from the discipline imposed by distance in order to purchase the goods they need.

An interesting model which accounts for the periodicity of central places and the behaviour of mobile merchants has been proposed by Stine (1962). The model is founded on the notions of the range of a good and the threshold. The maximum range of a good may be regarded diagrammatically as the radius generating a circle within which all consumers are willing to purchase at least some of a particular commodity. The threshold, or minimum range of a good, can be similarly viewed as the radius of a circle, but one containing the minimum demand necessary for the firm to earn normal profits, to stay in business and therefore become a viable entity in the economic landscape. Under normal conditions, these ranges are spatially related as shown in Figure 9.11A. The outer range, shown by the solid line, lies beyond the inner range. The establishment providing the particular good is permanent and has a fixed location. However, in areas characterized by a relatively primitive state of transport, consumers are unable to travel very far to buy goods because of the time needed to overcome the 'friction of distance'. Under these conditions, the position of the two ranges is likely to be reversed as shown in Figure 9.11B. When this situation pertains, there are two alternatives: (1) the firm cannot survive at a fixed location and so disappears from the land-

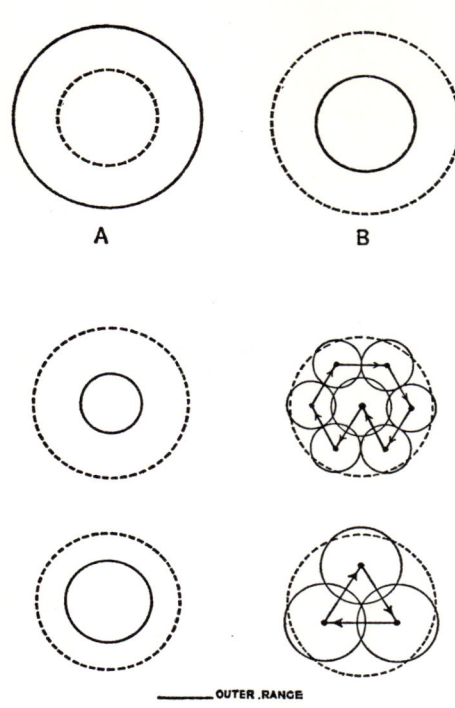

9.11 Normal (A) and inverted relationships (B) of the outer and inner ranges of a good. The number and length of paths for a mobile firm is a function of the radii of the two ranges of a good (C) (*Source: Stine, 1962, p. 22*).

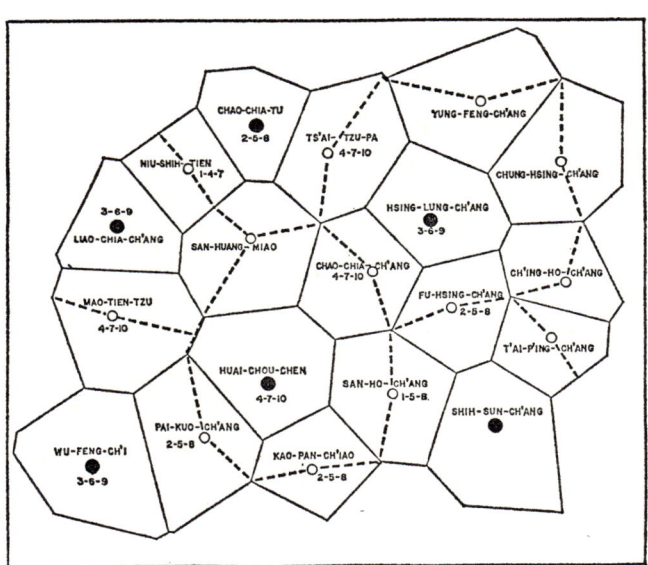

9.12 The market circuit in part of Szechwan (*Source: Skinner, 1964, p. 22*).

scape, or (2) the firm must become mobile and move from place to place. When the latter course is decided upon, the number and length of the moves that the firm must make in order to stay in business would seem to be, as shown in Figure 9.11C, a function of the difference between the length of the radii for the maximum and minimum ranges for the particular good in question.

A good illustration of the periodic nature of central places has been provided recently by Skinner (1964) from a study of marketing in rural China. It appears that both the movement paths of travelling merchants and the spatial pattern of market locations support many of the basic elements of Christaller's central place model. For example, Figure 9.12 shows the location of six 'intermediate level' permanent central places (shaded circles) in the hierarchy of settlement in part of Szechwan. The places at which periodic markets are held according to a regular calendar are shown by open circles. The links between the larger settlements and the location of the periodic markets are shown by the dotted lines. Merchants travel from their base at an intermediate centre to six of the nearest surrounding smaller places. Since each of these is shared between two intermediate centres, a K-4 system seems to be typical of this area, but Skinner also shows that a K-3 system is typical of other parts of China.

These findings are interesting since they suggest that the basic notions of central place theory may have much wider application than was originally thought by Christaller, or has hitherto been imagined by recent researchers. The ideas would seem to provide a particularly good framework for the study of the weekly markets typical of English rural settlement.

Vertical arrangements

Hierarchies. It is a common observation that there are fewer larger places than smaller ones in a region and that the larger centres provide a greater number and variety of goods than the small places do. As we have seen this hierarchical organization is fundamental to the central place models of Christaller (1933) and Lösch (1954). More recently, Berry and Garrison (1958B) have demonstrated by using the notion of threshold how hierarchies arise in the landscape. Their model is more flexible since it does not assume a uniform distribution of population or a hexagonal system of trade areas. Thus it can be applied more widely and is especially important because it provides a logical explanation of a hierarchy of shopping centres, the equivalent of central places, within urban areas.

The relevant 'vertical' aspects of the model have been simplified and diagrammatically represented in Figure 9.13. The argument can be simply stated as follows. Figure 9.13A shows five goods with their respective threshold sizes; function A requires only 10 people to support it whereas function

E requires 160 people. Let us assume that these goods are to be supplied for an area with only 160 inhabitants. Once the population to be served and threshold sizes for each function are specified, the number of each function appearing in the landscape is determined. Figure 9.13B shows the numbers of each function in the example. Since the threshold of function E is equal to the total population to be served, only one of it can be provided. Conversely, the area can support 16 establishments of function A with its lower threshold

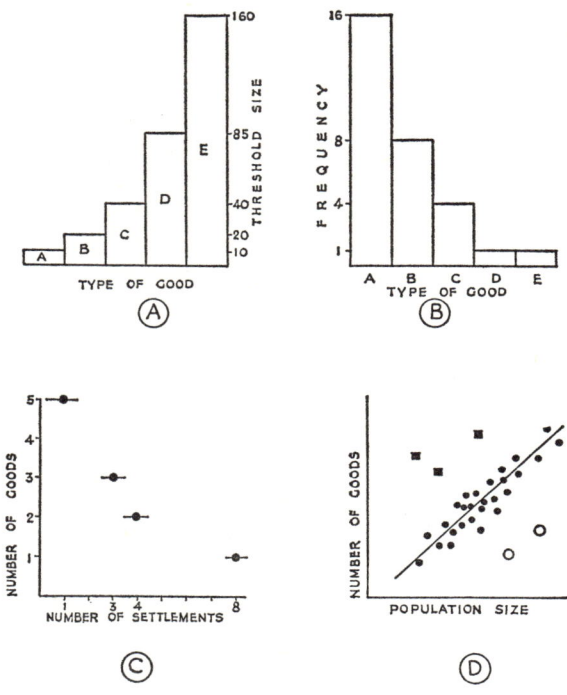

9.13 Threshold sizes for five functions (A), number of each type of function (B), size and number of settlements (C) and the simplified relationship of number of functions to size of settlement (D).

of 10 people. The numbers of each other good can be calculated in the same way. As a result some functions will be provided more frequently in the landscape than others. These are goods that must be provided conveniently (e.g. grocers). Other goods are supplied less frequently in the landscape, and are typical of goods for which the demand is more periodic (e.g. men's clothing shops). If a high order place supplies all lower threshold functions, then the number and size of settlements can be calculated. In this example, as Figure 9.13C shows, there will be a total of sixteen places; the largest provides all five functions, three supply goods A, B and C, four places supply the two lowest threshold functions, and eight places provide only function A.

A stepped distribution of different size settlements results. From $(V\text{-}2)$ above, it was stated that higher order centres have larger populations. The regularity between functional size and population size, and two kinds of deviant cases are shown in Figure 9.13D. Some places may have fewer functions than expected from their population size (circles in Figure 9.13D) and thus occur below the line of the general relationship. Berry (1960) provides evidence that this is typical in hierarchies around large cities where effects of suburbanization give rise to 'dormitory' settlements lacking a complete range of functions. Alternatively, some places may have more functions than expected from their population size, and thus fall above the line of general relationship. These are shown in Figure 9.13D as squares. Thomas (1961) has identified deviants of this sort in a recreational area of northwest Iowa where central places, in catering for tourists, have a range of additional functions which are not characteristic of other Iowan cities.

Although hierarchies have been identified in many areas, for example by Brush (1953) in south-western Wisconsin, USA, by Bracy (1962) in southern England, by Berry, Barnum and Tennant (1962) in part of Iowa, and by Mayfield (1962) in India, there is considerable controversy as to whether in fact such hierarchies may be identified empirically or whether instead only a continuous functional relationship exists. Like all classificatory systems, it could be argued that the levels they postulated are merely ones of convenience for handling empirical data. Vining (1955, p. 169) summarizes this when saying, 'Like pool, pond and lake, the terms hamlet, village and town are convenient modes of expression, but they do not refer to structurally distinct entities'.

Perfect uniformity is rarely obtained in the real world. The relationship between number of functions and population size varies. Berry and Mayer (1962) argue that for large areas such variations are enough to provide the appearance of a continuous linear relationship even though the underlying spatial pattern may be that of a hierarchy, while Beckman (1958) has shown how, with the addition of random elements, the discrete steps of Christaller's hierarchy can be blurred into a continuous relationship. Moreover, it would appear that aggregate analysis inevitably emphasizes the importance of continuous arrangements, whereas elemental investigations usually identify a hierarchy.

Evidence of continuous relationships between the population and functional size of settlement in agreement with Lösch's model provided by Stafford (1963) for southern Illinois and Gunawardena (1964) for the southern part of Ceylon is shown in Figure 9.14. In both areas high positive correlations are found; $(R^2 = 0.79)$ in the former area and $(R^2 = 0.83)$ in the latter. A slightly lower correlation was found by Berry and Garrison (1958A) for Snohomish

MODELS OF URBAN GEOGRAPHY AND SETTLEMENT LOCATION [325]

County, Washington, USA ($R^2 = 0.55$). These findings confirm the hypothesis that larger places provide a greater range of functions than smaller places and, in addition, it appears from the curvilinear form of the relationship that as settlements become larger they add fewer functions for each additional increment in population size.

Evidence of breaks in the overall continuous relationship more in line with the Christaller hypothesis is provided by Berry and Garrison (1958A) for

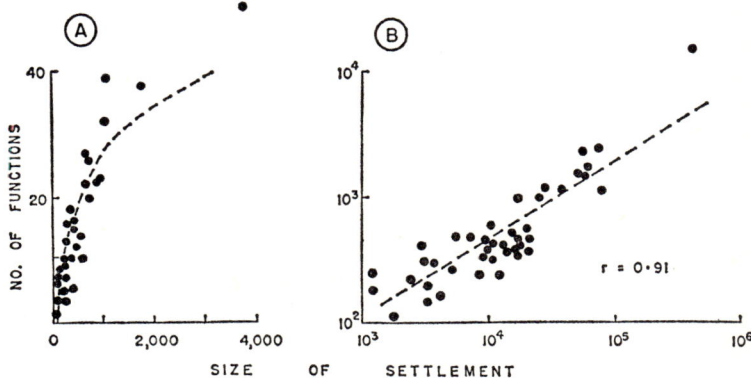

9.14 Relationships of functional size to settlement size in southern Illinois, USA (A), and southern Ceylon (B). The y-axis is transformed to a logarithmic scale in the second graph (*Sources: Haggett, 1965, p. 115. After Stafford, 1963, p. 170; Gunawardena, 1964*).

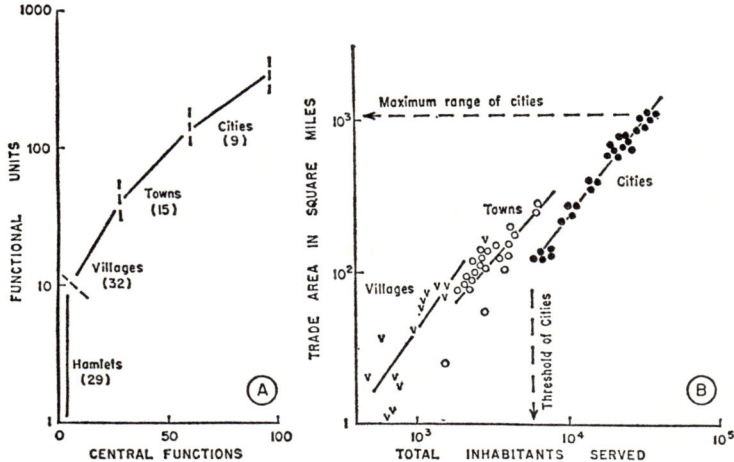

9.15 Four orders of central places in the settlement hierarchy, southwestern Iowa, USA (*Source: Haggett, 1965, p. 118. After Berry, Barnum and Tennant, 1962, pp. 79 and 80*).

Snohomish, County, Washington and Berry, Barnum and Tenant (1962) for south-western Iowa. In the latter study, factor analysis was used to group settlements on the basis of numbers and kinds of functions, and numbers of outlets (functional units). As Figure 9.15A shows, three distinct classes were recognized: (1) *cities* with more than 55 functions, (2) *towns* with from 28–50 functions, and (3) *villages* with between 10 and 25 functions. The lowest level of places, the *hamlets* were not included in the factor analysis. The horizontal expression of this vertical organization can be inferred from Figure 9.15B where each order of central place is shown on a logarithmic graph in relation to the size of trade area and total population served. In spite of overlap in the graph there is general agreement with the Christaller model in that higher order places serve larger tributary areas and populations than places at lower levels in the hierarchy.

Rank-size regularities. During the past fifty years interest has centred on another regularity in the size distribution of cities. This is the graphical relationship between the larger number of smaller places and fewer number of larger places in a region. When all cities in an area are ranked in decreasing order of population size, the size of a settlement of a given rank appears to be related to the size of the largest, or primate city, in the region. This regularity, generally known as the 'rank-size' rule, can be expressed by the formula,

$$Pr = Pl/r^q$$

where Pr is the population size of the city of rank r in the descending array of towns, Pl is the population of the largest city. This formula simply states that we should expect the population of the largest town to be four times the size of the fourth largest town in the region. Moreover, when the relationship is expressed in its logarithmic form

$$\log Pr = \log Pl - q \log r$$

it can be considered as lognormally distributed and a plot of rank on the x-axis against city size on the y-axis on double logarithmic graph paper should give a straight line with a slope $-q$ as shown by the upper line in Figure 9.16A.

Some of the alternative explanations of this empirical regularity are reviewed by Berry and Garrison (1958D). The most acceptable, though by no means easily grasped, of these seems to be that offered in terms of general systems theory by Simon (1955). Noticing that the form of the rank-size frequency distribution is identical to many probability distributions, among them the Yule and Lognormal, Simon argues that the regularity of city sizes is generated by some stochastic process. The close agreement between observed and expected frequency distributions in a recent study by Berry and Garrison (1958D) tend to confirm the Simon model as a plausible explanation of rank-size regularities.

This city relationship has been noted for many areas by as many writers. The formidable empirical evidence presented has been reviewed by Vining (1955) and Isard (1956, pp. 55–60); and Haggett (1965, p. 101), in summarizing the general evidence, concludes that although it may not be borne out exactly in reality, the rank-size regularity provides a useful framework within which generalizations about the population distribution of a region can be made.

The work of Stewart (1958) is perhaps typical of the application of the rank-size idea to large cities. From comparisons of the populations of the

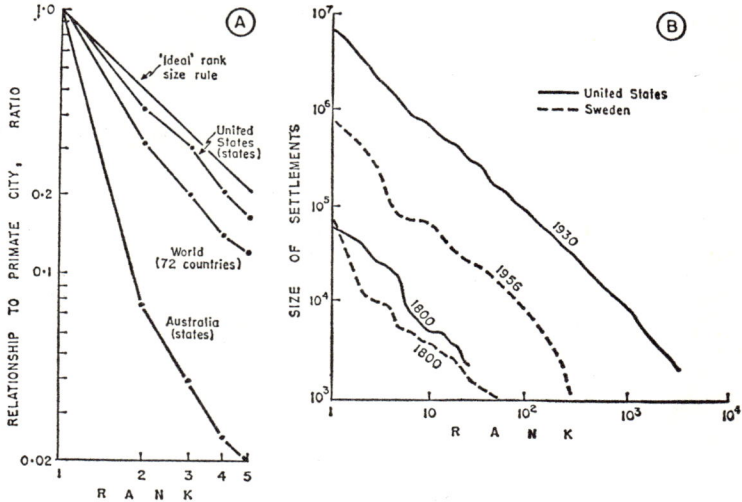

9.16 Median size of cities as a ratio of the largest city (A) and changes in the size-distribution of cities in the USA and Sweden (B) (*Source: Haggett, 1965, p. 102. After Stewart, 1958, pp. 228 and 231; Zipf, 1949*).

largest and second largest cities in seventy two different countries he concluded that the relationship was more varied than expected from the formula. For all countries sampled, the largest city was characteristically three and a quarter times, rather than twice, the size of the second largest city. Ratios varied from as high as 17·0 for Uruguay to 1·5 for Canada. The only regularity that Stewart could find was that larger countries had lower ratios.

Variation in rank-size relationships also appears to be characteristic of places lower down the urban spectrum. Figure 9.16A shows the reasonably close correspondence with the expected sequence for the five largest cities in the United States, and the strong divergence in Australia. Even when the rank-size relationship is extended over the full range of towns, straight line relationships are not always found. The curves shown in Figure 9.16B show two contrasted cases; (1) the linear form for the United States which generally conforms to the rank-size rule and (2) the irregular curve for Sweden. Comparison over time shows increasing linearity for the United States but increasing irregularity for Sweden.

Divergence from the expected straight line relationship seems to be due to a wide number of factors. Zipf (1949) suggests that the regularity is only characteristic of complete regions; that is, areas which are self-contained and not part of a larger region. Thus it could not be expected to hold for an English county but might for the entire British Isles. Stewart (1958) suggests that divergence from the rule is greater for homogeneous fairly well populated, mainly agricultural countries in which there are many smaller cities. Conversely, the rule holds well for industrialized countries principally because of the large size of some industrial centres, for areas with high rural population densities and for areas where population is well distributed spatially.

A recent study by Berry (1961) suggests, however, that it is not quite so

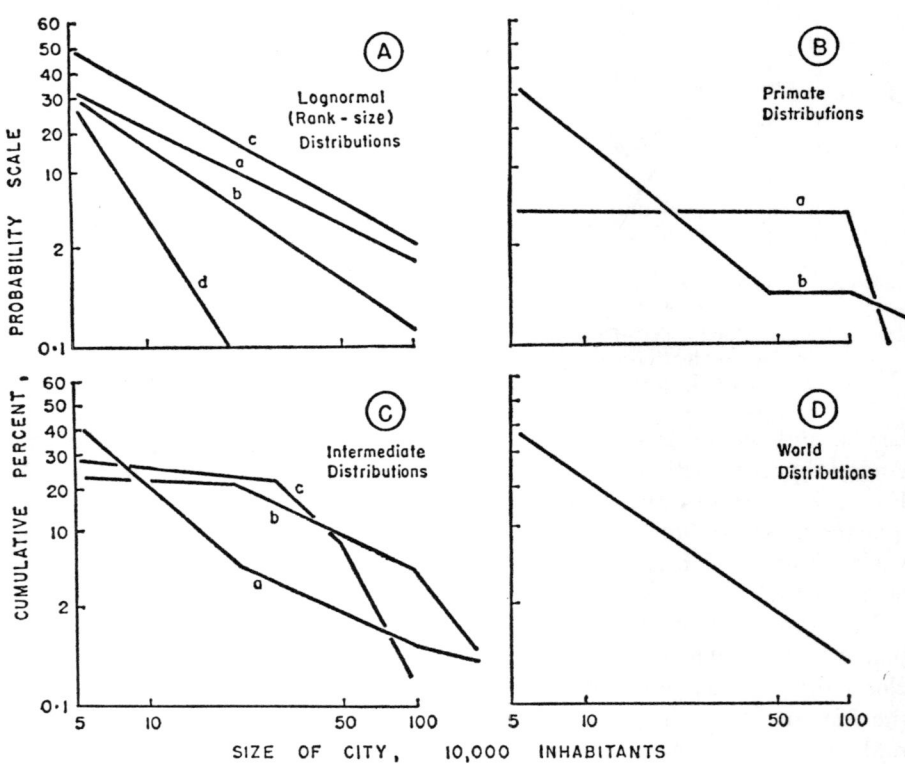

9.17 Alternative forms of city size distributions (*Source: Haggett, 1965, p. 104. After Berry, 1961, pp. 575–578*).

easy to relate the existence of rank-size relationships with general characteristics of a country. Berry plotted for each of thirty-eight countries the number of centres of over 20,000 inhabitants as a cumulative percentage on a normal probability scale on the y-axis and the size of centres on a logarithmic scale on the x-axis of a graph. Countries could be classified into three types according to their city size distributions. Firstly, 13 with *lognormal* (rank-size) distributions which appear as straight lines on this type of graph (Fig. 9.17A). This group included both highly developed countries like the United States (a) and underdeveloped countries like Korea (b); large countries like China (c) and small ones like El Salvador (d). Secondly, 15 countries with *primate* distributions in which a stratum of small towns and cities is dominated by one or more very large ones but in which there are very few centres of intermediate size. On the graph these appear as curved lines as shown in Figure 9.17B. Although all countries in this group are small, they show marked differences in types of curve. For example Thailand (a) lacks any signs of the lognormal curve, whilst Denmark (b) shows evidence of the lognormal distribution for its smaller cities. Thirdly, 9 countries showed *intermediate* distributions (Fig. 9.17C) including England and Wales (a), Australia (b) and Portugal (c). Figure 9.17D shows the tendency for the lognormal distribution to hold for all cities in the world.

As far as is possible to generalize from these graphs, lognormal (rank-size) distributions appear to be typical of larger countries which have long traditions of urbanization and which are politically and economically complex, whereas the exact converse holds for countries with primate distributions. They are generally smaller, have shorter traditions of urbanization and are economically and politically simple. More importantly perhaps, lognormal distributions are not only characteristic of technically advanced countries neither are primate distributions related in any significant way with underdeveloped countries as might intuitively be expected.

City classifications. When we refer to Sheffield as a 'steel' town or to Stoke-on-Trent as a 'pottery' town, we are recognizing a classification of towns by the economic functions they perform. Although all towns may be considered in part as central places for the provision of goods and services to local surrounding areas, for many this is subordinate to other specialized functions they perform for the wider regional and national markets. Groups of towns with similar functional specialization have been most frequently identified from the analysis of employment or occupation data. Specialization is said to exist when employment in a given industry category exceeds some 'normal' level. It is only when an abnormally large proportion of the labour force is employed in a particular activity that it becomes a distinguishing feature in differentiating that town from others.

Vital to the classification of towns is the problem of defining the 'normal' level of employment for a given industry group. In a number of studies, typified by the classification of American cities by Harris (1943), the breakpoint has been identified intuitively from observations of the employment-occupation structure of towns of well-defined types. The criteria used by Harris for identifying towns specializing in manufacturing, graphically presented in Figure 9.18A, are: (1) employment in manufacturing must equal at least 74 per cent of the total employment in manufacturing, retailing and wholesaling (*y* axis), and (2) manufacturing and mechanical industries must contain at least 45 per cent of the total gainfully employed labour force (*x* axis). Towns for which employment-occupation figures satisfy both of these criteria fall in the shaded upper right quadrant of the graph. Similar procedures using arbitrarily selected criteria were used to identify eight other types of specialized cities.

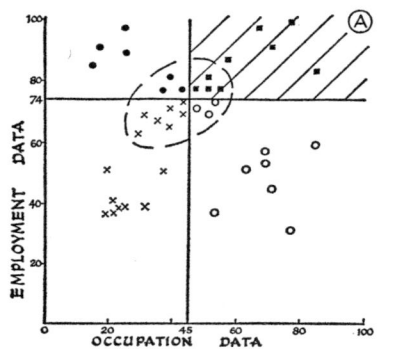
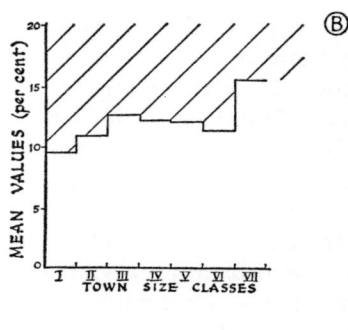

9.18 Methods of city classification used by Harris (A) and Pownall (B) (*Source: Harris, 1943, p. 87; Pownall, 1953, p. 333*).

A second group of studies has used average employment figures in a given industry calculated from the data for all towns as the break-point. On this basis, a town is considered as specializing in a given activity when its employment exceeds the national average. The classification of towns in the Netherlands by Steigenga (1955) is typical of studies using this approach. However, the use of national averages implies that the economy of every town is a miniature replica of the national economy. Most studies of cities have shown that the proportion of the labour force employed in a given activity varies directly with city size. An alternative, and perhaps more meaningful approach, is to use averages calculated for towns of different size classes. This method was adopted by Pownall (1953) in the classification of New Zealand towns. Figure 9.18B shows the variation in mean per cent values of total employment in the manufacturing, building and construction category (*y* axis) for towns of different sizes (*x* axis) in New Zealand. Towns for which per cent employment in this category exceed the average for its size class, and thus fall in the shaded area in Figure 9.18B, were classified as specialized manufacturing towns.

When towns are classified in this way, no account is taken of the magnitude of deviation above the average. Consequently, towns for which employment is just above the average are grouped in the same category as those with extremely large deviations. To differentiate towns taking this into account, Nelson (1955) has suggested the use of the standard deviation. In his classification, three classes of specialization can be recognized for any given activity depending on whether deviation is in excess of one, two or three standard deviations above the mean respectively.

Smith (1965) points out that if the immediate purpose of classification is to identify classes in which towns of similar functional composition are grouped so that its members are more like each other than they are other towns not included in the group, this will not necessarily be achieved by the procedures used in the above-mentioned simple classifications. This is very clearly illustrated in Figure 9.18A, where the cluster of towns enclosed by the broken circle around the intersection of the two break-points would be split among four groups despite the fact that each town is closer to, and is therefore functionally more similar to, other towns in the cluster than to other towns in the graph. At best the simple methods of classification yield functionally heterogenous classes in which significant functional combinations may be concealed. More sophisticated taxonomic methods are needed if classifications are to be at all meaningful. Some of the available techniques have been reviewed by Berry (1958). Perhaps the most important attribute of these techniques is that they permit classifications based on multiple rather than single variables.

Typical of multi-variable classifications is that by Smith (1964) of Australian towns. Using correlation techniques on employment data for 422 towns, he identified 91 initial groups comprising towns with similar functional structure. Cluster analysis of the resulting correlation matrix yielded a final classification of 17 groups. When many characteristics of cities are used in this way as a basis for classification, it may well turn out that some of the measures vary in similar ways from city to city and thus tell roughly the same story about a place. When this is the case, one might suspect that a more basic pattern underlies the variation of the different characteristics. Approximations to the basic underlying patterns may be revealed by the use of factor analysis. This method was used in the classification of British towns by Moser and Scott (1961). Fifty-eight measures covering eight aspects of social, economic and demographic characteristics were included for each of the 157 towns of over 50,000 inhabitants in 1951. In the resulting classification, towns are grouped on the basis of overall character rather than economic specialization alone.

Variation in the original characteristics was summarized by four underlying principal components. These were identified with (1) social class structure, (2) population growth, (3) a mixture of developments after 1951 and employment structure in 1951, and (4) housing conditions. The four components accounted for about 60 per cent of the basic differences between the 157 towns. Allocation of each town to a group on the basis of 'scores' on these components yielded a classification into 14 classes each containing about 10 towns. Three main urban types emerged: (1) the resorts, administrative and commercial centres, (2) industrial towns, and (3) suburbs and 'residential' centres.

The economic base. It has been implied so far that towns cannot exist by taking in their own washing; their existence and growth depend in large measure on their ties with other areas. A town flourishes because a proportion of the goods it produces is sold beyond its borders. It follows from this that a proportion of the total labour force in any town is directly concerned with the production of goods for 'export'. These are called *basic* or city forming workers because their efforts bring money into the town, thereby enabling the purchase of raw materials, food and manufacturing goods which the town cannot produce for itself. The remaining workers can be considered as *non-basic* or city serving since their primary role in the city's economy is to service the basic sector. Consideration of this dichotomy in the labour force has long been recognized as a useful concept in the economic analysis of urban areas and, since towns can be differentiated on the basis of basic employment, as a more satisfying means of city classification. The difference in the relative importance of various industry categories when basic rather than total employment is considered is nicely revealed in Figure 9.19. In both of these towns in Wisconsin, USA, the service sector becomes less important as expected while the dominance of manufacturing in Oshkosh and government in Madison, the State capital, are clearly revealed as the major basic functions.

Crucial for the application of the economic base concept is a satisfactory method of calculating the size of basic employment. Many methods have been suggested. Alexander (1954) used the 'firm by firm' approach in which individual businesses provide information about total employment and per cent of sales to local and regional markets. Using this information, employment can be pro-rated into its basic and non-basic components. For example, a firm employing 100 workers and selling 60 per cent of its total finished product in the regional market would be assumed to have 60 basic and 40 non-basic workers. Total basic workers in a city can be obtained by summing the figures over all firms. The obvious disadvantage to the use of this method is that although it may be alright in small towns like Madison and Oshkosh shown in Figure 9.19, it becomes virtually impossible to apply to large urban areas.

A more general approach was used by Alexandersson (1956) in a study of the industrial structure of 864 cities with more than 10,000 inhabitants in the United States. The per cent of the city's total employment was calculated for each of 36 industrial categories. For each category, towns were arrayed from smallest to largest by per cent employment in a cumulative distribution. Then, for a given industry, the value of the fifth percentile which corresponded to the value for the 43rd town in the array, was arbitrarily selected as the minimum per cent employment needed to serve the needs of the city's own population. This was designated the K-value for that industry. The sum of

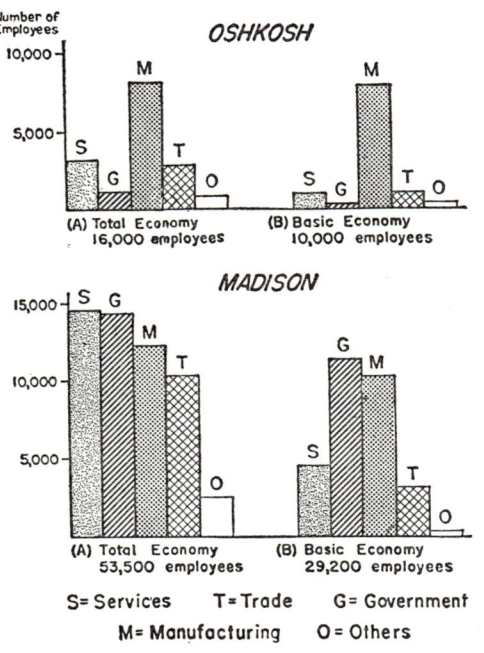

9.19 Differences in total and basic employment (*Source: Alexander, 1954, p. 250*).

all K-values for the 36 industry groups yielded the total non-basic employment in any city which was subtracted from the total employed labour force to give the number of basic workers. However, Alexandersson's interpretation of the K-values is questionable because of the weight given to small cities in the analysis; over three quarters of the towns included were smaller than 50,000 inhabitants. Morrissett (1958) in particular has provided insights into the variation of the K-values with city size and also shows that there are marked differences between regions within the United States.

A reliable short-cut method for estimating the non-basic employment in a city, which partly overcomes the objections to Alexandersson's technique, has been proposed by Ullman and Dacey (1960). Cities in the United States were divided on the basis of population size into six classes. A random sample of 38 cities was taken from each size class except from the largest class comprising places with more than one million inhabitants, from which only 14 cities were selected. For each city, the per cent of the labour force employed in each of 14 census categories was calculated. Then for each size class the figure for the city with the *minimum* per cent employed in each of the 14 categories was entered in a table. The sum of the minima for towns in a given size class was interpreted as the non-basic employment needed for cities of that size. As might be expected, non-basic employment in a given category

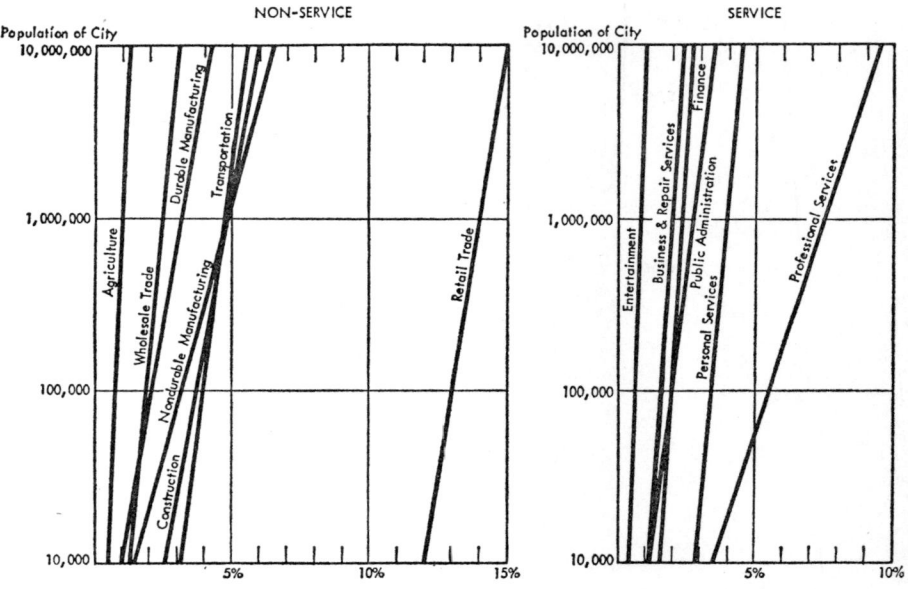

9.20 Regression lines for minimum-requirements by city size (*Source: Ullman and Dacey, 1962, p. 129*).

is directly related to the size of the city. Moreover, as Figure 9.20 shows, when the minima are plotted on a semi-logarithmic graph, straight line relationships are obtained. Minimum per cent employment figures for a town of any population size can be read off the graphs for each of the 14 categories, and summed to give the total non-basic employment. Basic employment can then be determined by subtracting this percentage from the total employment for the city.

The methods used to identify the basic employment of cities should be used with caution. Isard (1960) has listed some of the difficulties encountered in the use of employment data for this purpose and Roterus and Calef (1955) have stressed the effect that the size of area used as a base for measuring basic employment has on results. The strength of economic base models is their simplicity; it might however appear that simplicity is an attribute of declining importance in geographical studies. If the methods of determining basic employment become too complex, for example see recent methods proposed by Hoyt (1961), the appeal of economic base studies is largely lost and the concept must compete on unfavourable terms with other analytical tools like inter-industry linkage analysis (Isard and Kavesh, 1954) which are capable of providing more elegant results.

THE INTERNAL STRUCTURE OF CITIES

Models of urban structure are basically of two kinds: (1) *partial*, which deal with the location of a specified set of activities (e.g. residential land use) based on assumptions about the locational characteristics of all other activities in the urban area, and (2) *comprehensive*, which deal with the location of all activities together within the urban area. In both types of models the importance of transport in determining land use patterns is stressed, either explicitly in terms of the substitution of rents for transport costs, or implicitly in terms of land value-accessibility relationships which Hoyt (1939A) thought of as a common denominator of all land uses.

The urban land market

Land use patterns result from a multitude of decisions made by individuals about location. It is not at all clear how these decisions are reached although recent studies using gaming-simulation techniques are beginning to shed some light on the underlying mechanisms involved (Duke, 1964). But no matter what the underlying considerations are, it appears that decisions are regulated in varying ways by the economic processes operating in society. Ratcliff (1949), elaborating the earlier ideas of Hurd (1911) summarises this when he says that '. . . the locational patterns of land use in urban areas result from basic economic forces, and the arrangements of activities at strategic points on the web of transportation is a part of the economic mechanism of society'.

The pertinent aspects of this mechanism as it relates to the generation of land use patterns can be briefly summarized as follows. Each activity has an ability to derive utility from every site in the urban area; the utility of a site is measured by the rent the activity is willing to pay for the use of the site. The greater the derivable utility, the greater the rent an activity is willing to pay. In the long run, competition in the urban land market for the use of available sites results in the occupation of each site by the 'highest and best' use, which is the use able to derive the greatest utility from the site and which is, therefore, willing to pay most to occupy it. As an outgrowth of the occupation of sites by 'highest and best' uses, an orderly pattern of land uses results in which rents throughout the systems are maximized and all activities are optimally located. This process is, of course, identical with the original formulation by von Thünen (1826), and later modifications by Dunn (1955), of the ordering of land uses in agricultural regions in relation to market centres. However, as Alonso (1960) and Wingo (1961) have pointed out, the operation of the urban land market is rather more complex than its agricultural counterpart.

The rent paid for the use of a site is affected by many factors, but most importantly by the location of the site relative to other uses. The logic of this relationship is founded on the assumption that site rents represent a saving in transport costs in overcoming the 'friction' of distance. From this it is argued that competition for the use of land results in the minimization of the 'friction' of distance in the entire urban area and, since accessibility increases inversely with distance, the resulting pattern of urban rents is essentially a function of transport. Savings in transport costs can be traded off for extra rent payments to ensure the use of a particular site. Therefore those activities which enjoy the greatest benefits from occupying accessible locations will have greater surpluses available with which to bid for land. Consequently, sites in the urban area are not merely occupied by activities which can pay most for their use, but more specifically, by those activities which are able to derive the greatest positive transport advantages from the use of a given piece of land. When rents are viewed in this framework they will be represented by land values, which in turn can be considered as a direct reflection of differences in intra-urban accessibility. Thus, high land values will be associated with highly accessible locations and *vice versa*.

The land value surface

If land values are the common denominator of land uses, an understanding of the pattern of land values would help in understanding the internal structure of cities. It is perhaps somewhat surprising therefore that until recently little attention has been paid to this matter by geographers. The specific pattern of land values, just like its equivalent pattern of land uses, obviously varies from city to city depending upon local circumstances. At least three elements are present in the pattern in all cities, namely (1) land values reach a grand peak in the centre of the city and decrease by varying amounts outward toward the periphery of the urban area, (2) land values are higher along the major traffic arteries than in the areas away from them, and (3) local peaks of higher value than the general level at a given distance from the city centre occur at the intersection of major traffic arteries. Superimposition of these three components results in a general surface of land values similar to that shown

MODELS OF URBAN GEOGRAPHY AND SETTLEMENT LOCATION [337]

diagrammatically in Figure 9.21. This surface can be likened to a conical hill, the smooth surface of which is disturbed by ridges, depressions and minor peaks.

Although many factors affect the value of a given piece of land, for example elevation has a noticeable effect on residential values (Brigham, 1964), we argue here that the land value surface is essentially a direct reflection of accessibility within the urban area. Accessibility is highest at the city centre because it has developed through time as the major focus of routes and consequently is the most easily reached part of the entire urban area. Competition

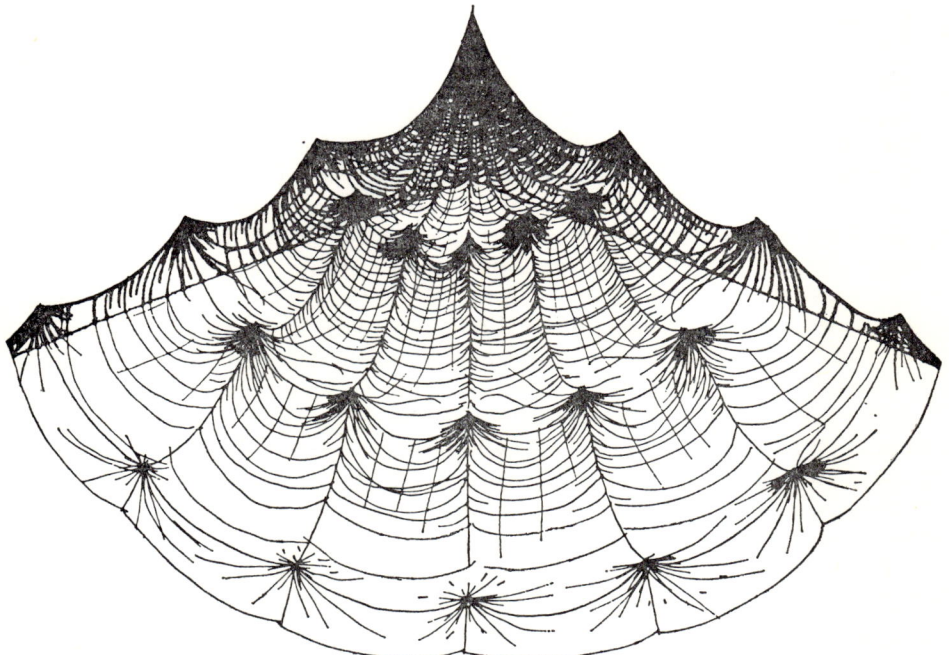

9.21 Generalized land value surface within a city (*Source: Berry, Tennant, Garner and Simmons, 1963, p. 14*).

for the use of land is most intense here. In a similar way, accessibility is greater at sites located along the radial and circumferential routes and at the intersections of these than it is away from them. Moreover, since some parts of the urban area are better served with transport than others, the value surface will not be characterized by uniform slope in every direction but will decline more rapidly with distance from the central peak in some directions than in others. The result is a marked sectoral variation in the general level of the value surface.

Accessibility means different things for different activities. For commercial functions, proximity to sales potential would appear to be crucial for retail activities while proximity to complementary uses is critical for offices requiring face-to-face contacts. For the variety of industrial uses, accessibility is in terms of availability of suitable land, transport facilities, public utilities and, in the case of agglomerations, proximity to other industries with which there are technological linkages. In all these instances, accessibility relates directly to costs of operation and profit levels. This is not so in the case of the third general land use type, residential activity. Although proximity to workplaces, shopping centres and other amenities is important, decisions about residential location are made within the framework of the more intangible environmental qualities that Alonso (1960) argues underlie 'satisfaction' for the household.

Following the classic work of Hoyt (1939A) in Chicago, the findings of other empirical studies support the general nature of the land value surface. Typical of these studies is that by Seyfried (1963) in which directional differences in the exponential slope of land value profiles were clearly identified in Seattle, USA. Significant correlations ($R^2=0.64$) were obtained between land values and distance away from the centre of the city but comparison of the slopes of regression lines for sectors going north, south, east and west from the city centre indicated the western profile was nearly three times as steep as that for the northern sector, while the profiles for the other two sectors were intermediate in slope. Similar evidence is given by Knoss (1962) for Topeka, Kansas, USA, where values were found to decline inversely with the reciprocal of distance from the city centre and from major radial routes, while they appeared to vary directly with the direction of growth within the city. Sectoral variation in land values has been identified in Chicago by Hayes (1957) and Mayer (1942), and more recently Berry, Tennant, Garner and Simmons (1963) have indicated the relationship between local peakings on the value surface and the relative importance of highway intersections.

The most comprehensive study to date is that by Yeates (1965) who hypothesized six variables in a multiple regression model to explain variation in the value surface in the city of Chicago. These were namely, (a) distance from the central business district, (b) distance from the nearest regional level shopping centre, (c) distance from Lake Michigan, (d) distance from the nearest elevated-subway line, (e) per cent non-white population of the block in which a particular site is located, and (f) population density. The model provided a low degree of explanation ($R^2=0.18$) of variation in the surface when applied to data for 1960. However, when a sectoral component was added, about half of the total variation ($R^2=0.51$) was accounted for, which attests the importance of radial routes in determining land values.

Land use models

The internal structure of any city is unique in its particular combination of

MODELS OF URBAN GEOGRAPHY AND SETTLEMENT LOCATION [339]

detail. In spite of this, it appears that in general there is a degree of order underlying the land use patterns of individual cities. Since there is as yet little agreement on the specific nature of this order, a variety of models have been proposed. In line with the three basic components identified in the land value surface, they will be discussed here under concentric, sector and nucleii headings.

Concentric models. Models of this kind are developed on the assumption that land values, and by implication accessibility, decline with equal regularity in all directions from a common central point in the city. Since no account is taken of distortions caused by differential accessibility, land use patterns are assumed to be arranged in regular concentric zones about the city centre.

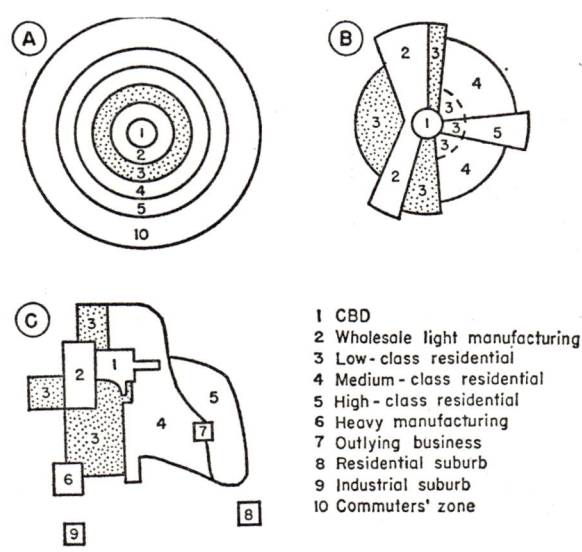

1 CBD
2 Wholesale light manufacturing
3 Low-class residential
4 Medium-class residential
5 High-class residential
6 Heavy manufacturing
7 Outlying business
8 Residential suburb
9 Industrial suburb
10 Commuters' zone

9.22 Concentric (A) sector (B), and multiple nucleii (C) models of urban structure (*Source: Haggett, 1965, p. 178*)

The classic concentric zone model (Fig. 9.22A) was proposed by Burgess (1925) based largely on his studies of the Chicago region. The model states that at any given moment in time land uses within the city are organized into zones differing in age and character and located in a definite order from the city centre. For Chicago, five zones were identified; these were in order from the centre outwards: (1) an inner central zone which formed the 'heart' of the city's commercial, social, cultural and industrial life and which was the focus of urban transport, surrounded by (2) a transition zone of mixed land uses in which deteriorating residential property predominates. Its unattractiveness is emphasized by the blighted conditions and slums intermixed with light industries and other business uses which have spilled over from the inner core; (3) a working class residential zone in which second generation immigrants form an important element in the population structure; (4) a zone of better housing characterized by single-family dwellings interspersed with pockets of exclusive residences and high class apartment buildings; and (5) a fringe zone of suburban and satellite communities forming dormitory suburbs for people working in the central city.

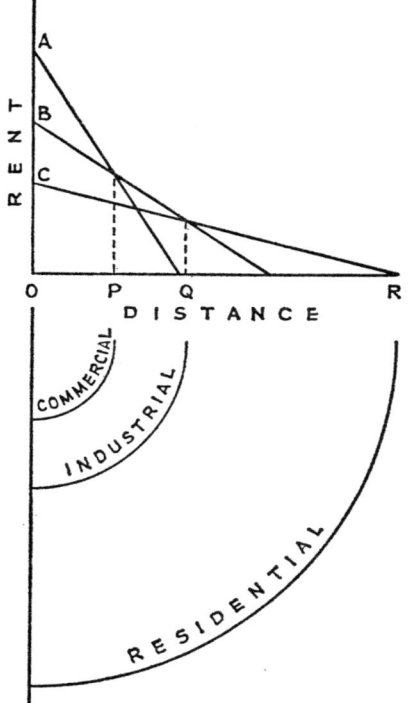

9.23 Hypothetical rent-distance relationships within a city (*Source: Garrison*, et al., *1959, p. 64*).

Although no explanations were offered by Burgess for this particular arrangement of land uses, Berry (in Garrison *et al* 1959) and Isard (1956, p. 200) provide a partial understanding in terms of the substitution of rents for transport costs. Accessibility is assumed to decrease uniformly in all directions from the city centre, and by implication the rent that an activity is prepared to pay for site use decreases from the central point. The resulting rent curves will slope downward to the right, as shown for example, for activity A in Figure 9.23. Not all activities are equally susceptible to differences in accessibility and their rent curves will vary in steepness accordingly, as shown for activities B and C in Figure 9.23. Since the rent curves for activities

A, B and C are ranked in steepness, they are also ranked in order of accessibility from the city centre. In competition for the use of sites, activity A outbids all other uses and occupies locations from the centre of the city (O) to distance (P); activity B is able to pay the highest rent and occupies sites between (P) and (Q); while activity C is only successful in bidding for sites between (Q) and the edge of the city (R). When this arrangement is generalized from one to two dimensions by rotation of (OR) about (O), a simple concentric pattern of land uses is obtained and if activities A, B, and C are imagined as commercial, industrial, and residential uses respectively, the result is a simplified version of the concentric zone model.

One of the interesting features of the concentric model is the anomaly of poor people living close to the city centre on high value land and the rich living at the periphery where land is cheaper. Both Alonso (1960) and Wingo (1961) have partially explained this in models of transport and urban land use. The specific details of the economic arguments proposed are too complex to be elaborated here although the general argument is simplified as follows. Each household has a limited budget with which it must satisfy all its needs, viewed essentially as (a) basic costs of living, (b) costs of housing, and (c) costs of getting to and from the centre of the city where it is assumed all workplaces are concentrated. Once the basic living costs of the household are established, the remaining funds can be allocated in varying proportions between housing and commuting costs. Poorer families, with less money available for commuting live close to workplaces but can only afford to pay for small amounts of the high value land near the city centre. Conversely, richer households can afford to live more luxuriously at the edge of the city where they consume larger amounts of lower value land at the expense of paying higher commuting costs to the city centre.

Although Blumenfeld (1949) claims that the concentric pattern is found in Philadelphia and Smith (1962) has recognized some of the zones in Calgary, Canada, distortions will inevitably be introduced by natural barriers and the pattern of transport routes. Hartman (1950) shows that the circular shaped city form necessary for the development of pure concentric land use patterns will only result if there is a strong radial component in the transport system, and then only if the number of radial routes is quite large and they are closely spaced. Fewer radial routes spaced more widely apart give rise to marked differences in intra-urban accessibility to result in a 'star' shaped form in which concentric arrangements of land use are distorted and even destroyed.

Sector Models. Models of this type are developed on the assumption that the internal structure of the city is conditioned by the disposition of routes radiating outwards from the city centre. Differences in accessibility between radials causes marked sectoral variation in the land value surface and correspondingly an arrangement of land uses in sectors. This is the basis of the model proposed by Hoyt (1939B) who hypothesizes that similar land uses concentrate along a particular radial route from the city centre to form sectors as shown in Figure 9.22B. Thus a high rent residential district in one sector of the city would migrate outwards in that direction by the addition of new growth on its outer arc. Similarly, low rent districts in another part of the city would develop by the same process in that direction. The sectors of high-class residential areas would seem to be particularly pronounced in the direction of high ground and open spaces.

Although the model is descriptive, it is clearly an improvement on the earlier concentric zone idea since both distance and direction from the city centre are taken into consideration. Perhaps because of this, sectoral arrangements have been identified in many cities. There is evidence of a sector of high-class residential land use with high land values along the northern shore of Lake Michigan and sectors of low-class residences in the south and industry in the west of Chicago (Yeates, 1965). For Belfast, Jones (1960) found that the pattern of high-class residential areas was consistent with the sector model, while Smith (1962) claims that sectors rather than concentric zones seem to be most meaningful in Calgary, Canada. Although approximations to an inner core area surrounded by a transition zone and an outer commuting zone are found in this city, the two middle residential zones postulated in the concentric model are not apparent in any form. Most of the major land uses however, formed something like sectors.

Nucleii models. Land use patterns in most cities are not built around the single centre postulated in the above models; rather they are developed around several discrete centres within the urban area. This is the basis of the multiple nucleii model of Harris and Ullman (1945) shown in Figure 9.22C. The number and location of the nucleii within the urban area depends on the size of the city, its overall structure and historical development. The larger cities have a greater number and more specialized nucleii than smaller places. For American cities, five districts were identified: (1) the central business district, (2) a wholesaling and light-manufacturing area near the focus of extra-city transport facilities, (3) a heavy industrial district near the present or former outer edge of the city, (4) various different residential districts and (5) peripheral dormitory suburbs. The reasons given for the existence of separate nucleii and differentiated districts were combinations of (1) the specialized requirements of certain activities, (2) the tendency for activities to agglomerate, (3) the repulsion of some activities by others which is linked to (4) the differences in rent paying ability which force activities to cluster in separate districts within the city.

One of the many implications of this model is that marked differences in type of residential land use should be noted around the various business nucleii in the city. Marble (in Garrison *et al.* 1959) has proposed a series of models in which residential site selection, measured by land values, is hypothesized as a function of location within the city in respect to its centre and other major business nucleii. Application of the models to randomly selected city blocks in Spokane (Washington) and Cedar Rapids (Iowa), USA, yielded non-significant results. It was concluded that the distribution of residential land values, and by implication the type of residential property, provides no support for the concept of multiple nucleii or even concentricity in residential pattern.

The three models are not mutually exclusive and elements of all three patterns might be expected in cities, especially where they have fused with one another to form the large conurbation areas or Megalopolis (Gottmann, 1961), the urbanized north-eastern part of the United States, with more complex internal land use patterns. Marble (in Garrison *et al.* 1959) has suggested a fused model in which growth proceeds radially from the city centre and from various other nucleii but is intercepted by axial growth pushing outwards along the lines of least resistance from the main centre to result in a star shaped city in which distinct social, economic and technical zones are developed. Haggett (1965, p. 180) offers evidence from an area south of Cambridge, England, suggesting that Marble's views may be more appropriate. In this area, the spread of housing since World War II shows distinct gradients close to the city, along major radials and in the outlying villages which were identified with the concentric, sector and nucleii models respectively. Clearly, all three models are useful in generalizing about the internal structure of cities and it might well be that judicious application of analytical tools like factor analysis will enable a more detailed description of the underlying components of city structure in terms of these simple models and enable the formulation of even more general and meaningful models to supplement them.

Urban population densities

Regardless of the particular land use arrangement within cities, the distribution of population densities appears to be the same in all cities. Densities decline as a negative exponential function of distance from the city centre. This can be generalized as

$$P_d = P_o e^{-gd}$$

in which P_d is the population density at a given distance (d) from a point in the central business district; ($-g$) is the slope of the density decline curve, and P_o is the density of the central area (extrapolated from the slope for outer areas). The regularity, shown for Hyderabad in India and Chicago, USA, in Figure 9.24, was originally identified by Clark (1951) from a study of 36 cities in various parts of the world, and many of the subsequent studies of urban densities are summarized by Berry, Simmons and Tennant (1963).

Although to date no city has been studied for which the regularity does not hold, marked variation in the slope of the gradient is characteristic. Muth (1962), for example, has shown that the g-values for 36 American cities ranged between 0·7 and 1·2, although the majority of values clustered in the range 0·2 to 0·5. Attempts to explain these variations led Muth to carry out a regression study of density gradients against twelve controlling variables of which only four (car registration per capita in the urban area, the proportion of manufacturing employment and of substandard dwellings in the central area, and the size of the urban area) were significant in accounting for differences in slope of the curves. This led to the conclusion that larger cities with low transport costs, dilapidated central parts and dispersed employment centres were more compact than other cities. However, a study by Berry

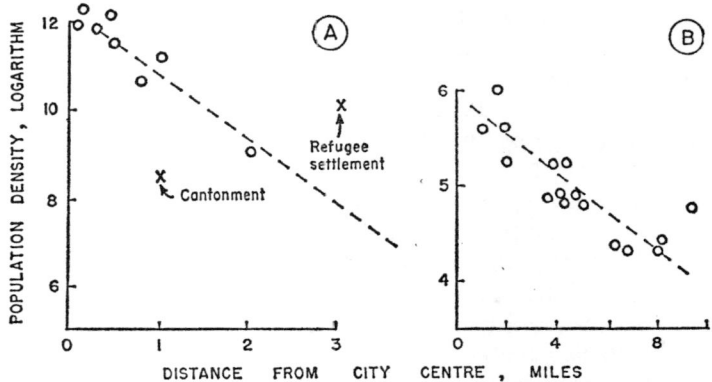

9.24 Population density-distance relationships for Hyderabad, India (A) and Chicago, USA (B) (*Source: Haggett, 1965, p. 157. After Berry, Simmons and Tennant, 1963, pp. 392 and 394*).

(Berry *et al.*, 1963) of 46 American cities suggests that the spatial pattern of manufacturing activity is not one of the factors associated with variation in the slope of density gradients. Variation also typifies central densities, which Winsborough (1962) has shown to be a function of overall population density within the city, which in turn is directly related to population size of the town, its importance as a manufacturing centre and the proportion of old dwellings in the city. Added evidence of the relationship between population densities and age of cities is given by Berry (Berry *et al.*, 1963) who found that a regression equation with age of city and the slope of the density gradient as independent variables could account for approximately sixty per cent of the variation in the size of central densities in his sample cities. The implication from these studies is that larger, older industrial cities tend to have higher overall population densities than other kinds of cities.

Although there is no complete explanation for the empirical regularity, recent works by Alonso (1964) and Muth (1962) suggest that it results from substitution on the part of households of rents for transport costs as discussed above, and confirm Clark's (1951) original speculation that the observed regularity had something to do with transport costs. If the rent curves for lower income groups is steeper for any pair of households with identical tastes, the poor will live at high densities near the city centre while the rich will live at lower densities near the periphery. The reason for the negative exponential nature of the density-decline function is claimed by Muth (1962) to stem from the similar form of the function for housing production. Both these sets of findings suggest that the negative exponential equation is a general one which can be derived as a logical extension of the theory of the urban land market. Moreover, when placed in the framework of general systems theory, Berry (1964) shows that the pattern of population densities within cities can be fairly accurately predicted if the population size and age of city are known, for the density gradient (g) and central densities (P_0) are related to the position of the city in, and the slope of, the rank-size curve. It seems that what were considered hitherto as disparate empirical regularities may be synthesized as part of a wider model of city structure.

Commercial structure. Apart from references to the central business district, the general models of internal structure do not say much about the nature of the urban business complex. A descriptive model of this complex is given by Berry (1959) for American cities, and the major components recognized are shown in Figure 9.25. This description is the result of analysis of locational

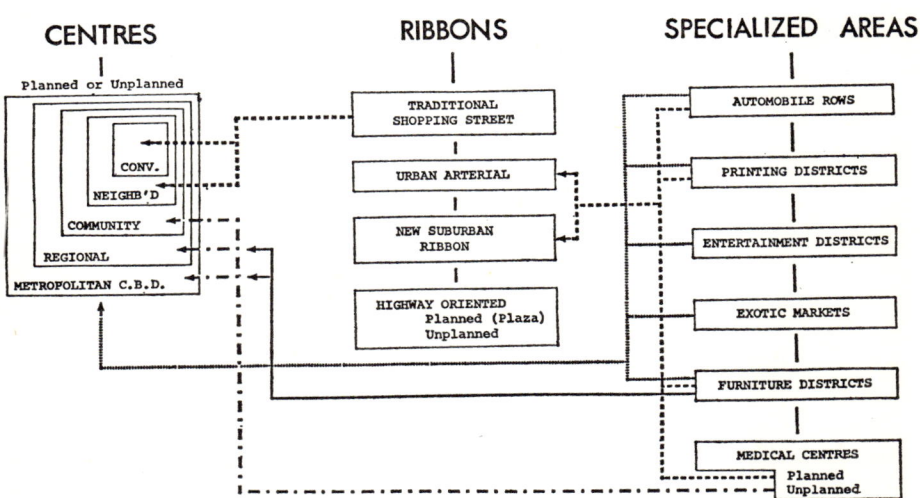

9.25 Elements of the commercial structure of American cities (*Source: Berry, Tennant, Garner and Simmons, 1963, p. 20*).

requirements and spatial associations of different business functions and is unlike earlier studies of commercial structure, for example the works of Mayer (1942) or Kelly (1956), which stressed differences in morphological structure rather than the functions typical of the different conformations.

Three basic conformations are recognized: (1) older unplanned and newer planned shopping centres, which provide a variety of convenience and shopping goods from functionally linked establishments for essentially home-based shopping trips of various frequency and duration; (2) a variety of ribbon developments which, with the exception of the traditional shopping streets, consist of a range of space consuming services catering primarily to demands originating on highways, and (3) specialized areas catering to special consumer demands. Differences within each of the major types are related to the numbers and kinds of functions performed and a detailed discussion of their functional structure is provided in Berry, Tennant, Garner, and Simmons (1963).

Although there are studies of various aspects of ribbons, for example Faithfull (1959) in Australia, Foster and Nelson (1958) in Los Angeles, Berry (1959) in part of Washington State, USA, and Boal and Johnson

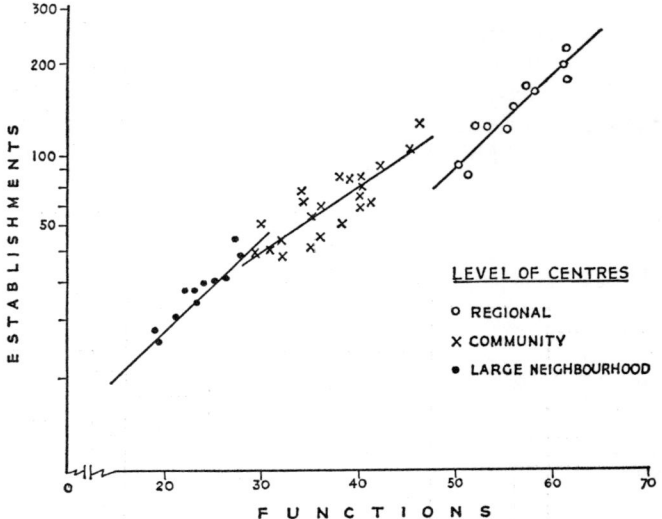

9.26 Three orders in the hierarchy of shopping centres in the city of Chicago, USA (*Source: Garner, 1966*).

(1965) in Canada, most emphasis has been given to the study of nucleated centres, and especially the central business district in studies of commercial structure. Shopping centres can be viewed as the urban equivalent of rural central places and the basic notions of central place theory can be applied to their study. Analysis of functional structure, size and trade area characteristics shows that, like central places in rural areas, there is a hierarchy of business centres within the city. This has been recognized by many researchers, for example by Smailes and Hartley (1961) and Carrothers (1962) in Greater London, and Thorpe and Rhodes (1966) in the Tyneside urban region, in

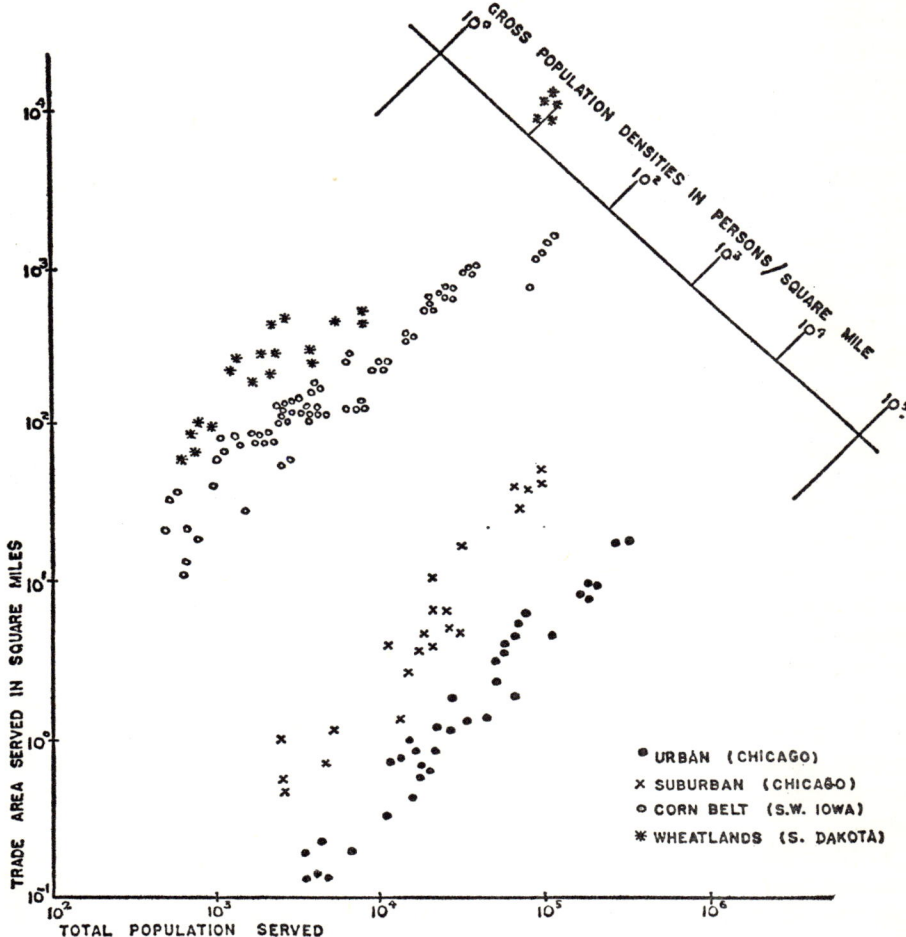

9.27 Hierarchies under different density conditions in the United States (*Source: Berry, and Barnum, 1962, p. 40*).

which subjectively selected criteria were used in identification and delimitation. Other studies have used more rigorous analytical techniques, for example Berry, Tennant, Garner, and Simmons (1963) used factor analysis in a study of Chicago, and Garner (1966) used regression and covariance analysis to identify the hierarchy in the city of Chicago. For this city these studies have identified three levels below the central business district. In ascending order, these are: (1) *neighbourhood* centres offering convenience goods for people living locally, (2) *community* centres providing infrequently demanded goods for several neighbourhoods, and (3) *regional* centres supplying specialized goods for people living in a major portion of the urban area. The number of levels in the hierarchy depends very much on the size of the central business district, which in small towns may be the equivalent of a regional centre for a large metropolitan area like Chicago. A hierarchy for the city of Chicago is illustrated in Figure 9.26 from the relationship between number of functions (x axis) and number of shops (y axis) in 68 shopping centres. The average regional, community and neighbourhood centre provide 56, 37 and 24 functions respectively. Comparison of these sizes with those shown in Figure 9.15A for cities, towns and villages in rural Iowa, USA, show a surprising degree of consistency in functional size, and suggests that intra-urban hierarchies may be viewed as a logical part of a series of hierarchies developed under different population density conditions (Berry and Barnum, 1962) as shown in Figure 9.27.

More detailed analysis of the hierarchy of shopping centres in Chicago by Berry (Berry et al, 1963) suggests the existence of more than one hierarchy within cities. The overall relationship between number of functions and shops is shown in Figure 9.28 to mask noticeable differences in the relationship for various parts of the city on the one hand and between the older and newer centres on the other. The regime in the poorer part of the city is of a different order to that in the high income area, and the lower correlation between numbers of functions and shops ($R^2 = 0.69$) suggests greater duplication of functional types in areas where demand is lower and of a different nature. Both Pred (1963) and Garner and Harvey (1965) have provided evidence of the different functional composition and quality of establishments in shopping centres serving lower income groups in the city. Planned centres appear to provide fewer functions less frequently than their unplanned counterparts owing to the greater selectivity in the design of centre size and form.

Central place theory does however have limitations for studying internal business structure; without considerable mental gymnastics it is difficult to apply to the study of urban business ribbons. This shortcoming has prompted Curry (1962) to suggest that a completely new approach to central place studies of the internal business structure of cities is required and he attempts to provide the framework of an operational approach in a probabilistic formulation of consumer behaviour using the poisson series.

A definite relationship appears to exist between the location of the various components of urban business structure and land values. Ribbons are associated with the ridges of higher land values along major traffic arteries while centres are related to the local peakings in value at highway intersections.

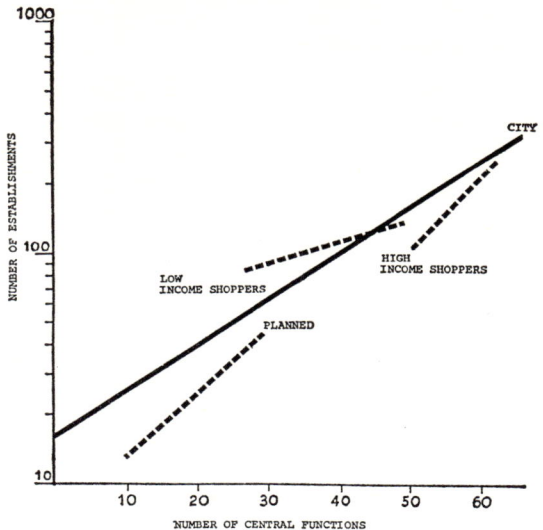

9.28 Different relationships of shops (y-axis) to number of functions (x-axis) for Chicago, USA (*Source: Berry, Tennant, Garner and Simmons, 1963, p. 134*).

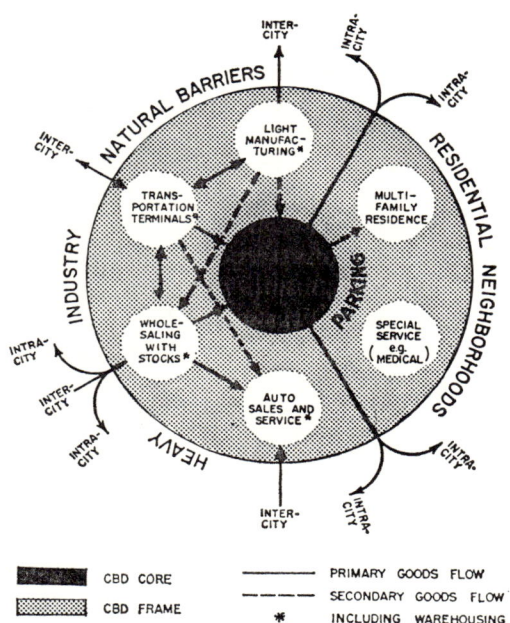

9.29 Diagrammatical structure of the central business district (*Source: Horwood and Boyce, 1959, p. 21*).

Stratification in value peaking for the latter might be expected to correspond to stratification of centres within the hierarchy and there is some general evidence for this in Chicago (Berry et al., 1963). Here the smaller neighbourhood centres are generally found at places with values about fifty dollars above the surrounding general level, while regional centres, needing more accessible locations to serve larger trade areas, are found where peaks rise about 4,000 dollars above the general level. However, such factors as stage of growth of centres, distance from the central business district, quality of market area served and lag in values accompanying decline distort the regularity and prevent the use of peak values as a proxy variable in the identification of centres (Garner, 1966).

The relationship between land values and location of commercial activities is more clearly apparent in the arrangement of functions *within* shopping centres. Many studies have shown this for the central business district, for example in Glasgow's central area studied by Diamond (1962), and it can be viewed as one of the explanatory variables underlying the 'core-frame' model of central business district structure proposed by Horwood and Boyce (1959). They argue that the commercial centre comprises two distinct parts, as shown in Figure 9.29: (1) an inner core of intensive use of high value land resulting in marked vertical expansion, where strong functional links between shops and between various offices is reflected in clusters of functions forming distinct micro-land use zones, and (2) a less intensively developed frame where values are relatively lower and functions have very little in common except location.

The relationship between land uses and functional arrangement is more clearly specified for centres below the central business district in the hierarchy in a model proposed by Garner (1966). He suggests that competition between functions of different threshold sizes for the use of sites leads to an ordered arrangement of land use within shopping centres, the salient features of which are shown for three levels of centre in Figure 9.30. For a given level of centre, the inner core area of high value land is occupied by functions which set the centre above the level of others in the hierarchy and is surrounded by functions typical of each lower level of centre on successively lower value land. Thus, in a regional centre (Fig. 9.30A) the core comprises high threshold, and as it turns out, high rent paying regional level functions surrounded in turn by community level and neighbourhood level functions. At the community level (Fig. 9.30B) regional types are excluded by definition and the core area is pre-empted by community functions which are surrounded on lower value land by neighbourhood types. Neighbourhood centres (Fig. 9.30C) have a simple structure since only one level of function is provided. Tests of the model in the city of Chicago show that the hypothesized patterns agree with the general structure of centres but that at the regional level, the model tends to oversimplify the real world conditions.

MODELS OF URBAN GEOGRAPHY AND SETTLEMENT LOCATION [351]

Before closer agreement between model and reality is to be obtained, refinements are needed in classification of business types and a more thorough appraisal of the threshold concept and product differentiation between establishments must be developed.

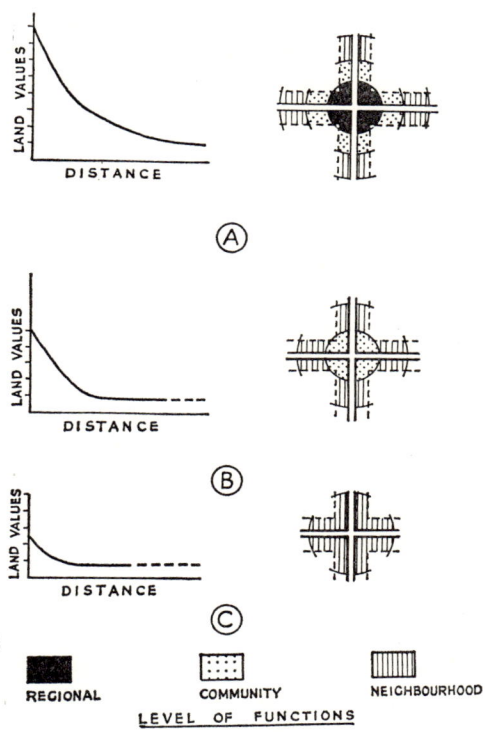

9.30 Internal structure of different order shopping centres in cities (*Source: Garner, 1966*).

Growth and change in internal structure. The models discussed so far represent cross-sections of the urban scene at a given moment of time. But the city is dynamic and is constantly in a state of flux accompanying growth of its constituent parts, either by extension at the periphery of the urban area or by internal re-arrangement of existing land uses. Both result in changes in intra-urban accessibility and attendant changes in the pattern of land values, which in turn is reflected in the changing pattern of locations. Yeates (1965) has amply demonstrated this in Chicago where his model of land values (discussed above) accounted for only 18 per cent of variation in the surface for 1960 but accounted for over 75 per cent of the total variation in 1910. Changes in mobility associated with increasing use of the automobile and continued decentralization of population has caused the effects of distance from the central business district and the influence of rapid transit routes to be less important and an increase in values along new radial and circumferential routes.

Growth in the normative models discussed above is accounted for by the ecological concept of invasion-succession. This states that occupancy of land by new uses tends to make it unsuitable for further occupancy by the original tenants who move out to new locations. In the concentric zone model new locations for functions of one zone are found in the next outer adjacent zone, so that growth consists essentially of colonization outwards on a broad front. In this way, the innermost zones are subject to continued decay and outer zones by gradual modification as lower status occupants move in. In a similar way, the innermost part of a given sector of land use changes in character as new uses take over and the sector pushes out by accretion on its outer arc.

Empirical studies of invasion-succession are abundant in the literature of Human Ecology, for example see Hawley (1950). Recently, Morrill (1965) has provided a more detailed understanding of the processes as they pertain to the extension of negro areas in American cities using Monte Carlo simulation techniques, and Hoyt (1964) has specified the factors responsible for distorting the classical models of urban structure. According to Smith (1962) it appears that these result in a progressive breakdown of the rigid concentric and sectoral zonations through time causing patterns to conform increasingly to the more general arrangement hypothesized in the multiple nucleii model.

Concomitant with reorganization of the internal land use patterns, growth by extension of the built-up area takes place at the periphery. Chapin and Weiss (1962) provide some insight into the factors influencing the development of urban land using multiple regression models in a study of growth in a cluster of towns in North Carolina, USA. They conclude that development is discouraged by poor drainage and proximity to blighted and non-white areas. Conversely, growth is intensified in areas with good transport facilities, near to large employment centres and in fringe settlements provided with a wide range of community services.

An attempt to specify the relationship between population growth and the characteristics of suburbs has been made by Thomas (1960) for the urbanized area of Chicago in the decade 1940–50. He hypothesized that population growth in suburbs at the periphery of the built-up area is a function of population size and density (a proxy variable for vacant land) of the place, age of suburb, cost of housing, quality of schools, birth-death differentials, degree of industrialization and accessibility to the central city. However, these only accounted for about a third ($R^2=0\cdot 36$) of variation in population growth in the entire study area, and birth-death differential and population density alone were significant. The model was more successful when fitted separately to three sectors within the city and presumably illustrates differences in residential desirability within the urban area.

MODELS OF URBAN GEOGRAPHY AND SETTLEMENT LOCATION [353]

Other studies have investigated the way in which growth takes place at the edge of the city. Blumenfeld (1959) has formulated a model in which growth takes place in the form of waves at the edge of the city although studies by Garner (1960) tend to indicate that growth 'leapfrogs' backwards and forwards between suburbs with no discernible regularity. Something similar to a wavelike expansion exists in relation to the disposition of transport routes. He suggests a model in which growth takes place first at locations on radial

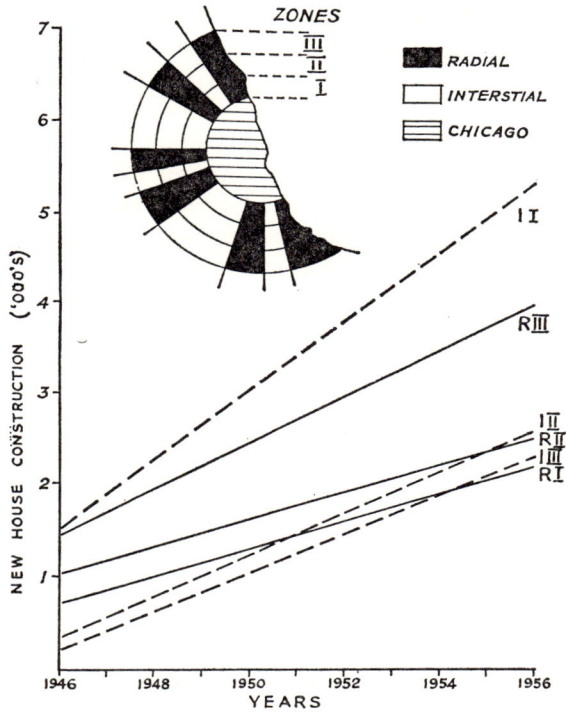

9.31 Regression lines for residential expansion in the Chicago urbanized area, USA (*Source: Garner, 1960, p. 30*).

routes and occurs at a later period in the interstitial areas between them. Evidence in support of this hypothesis is presented in Figure 9.31 which shows trends in new house construction at radial and interstitial suburbs at various distances from the central city during the period 1946–56, in the Chicago urbanized area. The number of new houses constructed in zone one at the edge of the city (10–15 miles from the centre) is much greater at interstitial suburbs (I_1) than at places on radial routes (R_1) for that distance. In zone II, building continued to push out along radial routes (R_2) but by 1956

12

this growth had been surpassed by residential growth at interstitial suburbs (I_2). In the outermost zone III, growth is predominantly at radial sites (R_3) with relatively little new house construction at corresponding interstitial locations (I_3). The implications from this are: (1) growth occurs at different rates in sectors which have good accessibility with the central city, and (2) expansion does not take place along a broad radial front as posited in the concentric zone model but is more in line with the general ideas of the sector model although the lag in development of interstitial areas is noticeable.

The continued outward expansion of the urban area and redistribution of population seems to be reflected in changes in the population density gradients. Berry, Barnum, and Tennant (1963) have shown that the gradient tends to flatten out accompanying low density extension of the periphery and a lowering of the high central densities as people spread out into adjoining residential zones. These changes are in turn related to modification of the commercial pattern. As areas are invaded by lower income groups, changes in the nature of demands result in the reduction of numbers of shops and functions provided, and perhaps more importantly, in the kinds of functions offered as blight conditions set in. At the same time, new shopping centres grow up to serve the increasing demands in outlying districts, shopping habits change and older centres decline (Simmons, 1964). The central business district becomes less important and sales decline in face of competition from outlying shopping centres (Boyce and Clarke, 1963). This in turn is associated with a shrinkage of the core area and the gradual encroachment of the frame uses to give rise to a 'zone of discard' which Murphy and Vance (1955) have shown to be characterized by the invasion of low grade establishments.

CONCLUSION

We have illustrated some of the ways in which models are being used in the study of settlement location and urban geography. The coverage is by no means complete; the various alternative approaches proposed by sociologists, econometricians, and the recent wealth of studies of urban structure provided by traffic engineers and planners have been passed over. Three remarks may be in order by way of summary: (1) Urban research is still in a relatively primitive stage of development and in spite of the voluminous literature much more work is needed to provide a comprehensive understanding of the complex integrated growth and structure of urban areas. (2) The study of settlement by geographers must progress hand in hand with workers in allied fields if spatial models are to be at all meaningful. (3) The paucity of references in the chapter to the use of models in British studies speaks for itself; there is a very obvious need for the development of models to be used in research and teaching if geographers are to contribute to the study and understanding of the spatial structure of urban areas.

REFERENCES

ALEXANDER, J. W., [1954], The basic-nonbasic concept of urban economic functions; *Economic Geography*, 3c, 246–261.

ALEXANDERSSON, G., [1956], *The Industrial structure of American cities*, (Lincoln, Nebraska).

ALONSO, W., [1960], A theory of the urban land market; *Regional Science Association, Papers and Proceedings*, 6, 149–157.

ALONSO, W., [1964], *Location and Land use*, (Cambridge, USA).

BASKIN, C. W., [1957], *A Critique and Translation of W. Christaller's 'Die zentralen Orte in Süddeutschland'*, (University of Virginia, Ph. D. Thesis).

BECKMAN, M. J., [1958], City hierarchies and the distribution of city size; *Economic Development and Cultural Change*, 6, 243–248.

BERRY, B. J. L., [1958], A note concerning methods of classification; *Annals of the Association of American Geographers*, 48, 300–303.

BERRY, B. J. L., [1959], Ribbon developments in the urban business pattern; *Annals of the Association of American Geographers*, 49, 145–155.

BERRY, B. J. L., [1960], The impact of expanding metropolitan communities upon the central place hierarchy; *Annals of the Association of American Geographers*, 50, 112–116.

BERRY, B. J. L., [1961], City size distributions and economic development; *Economic Development and Cultural Change*, 9, 573–588.

BERRY, B. J. L., [1964], Cities as systems within systems of cities; *Regional Science Association Papers and Proceedings*, 13, 147–163.

BERRY, B. J. L., [1967], *Geography of Market Centers and Retail Distribution*, (Englewood Cliffs, N. J.).

BERRY, B. J. L. and BARNUM, H. G., [1962], Aggregate relations and elemental components of central place systems; *Journal of Regional Science*, 4, 35–68.

BERRY, B. J. L., BARNUM, H. G. and TENNANT, R. J., [1962], Retail location and consumer behaviour; *Regional Science Association, Papers and Proceedings*, 9, 65–106.

BERRY, B. J. L. and GARRISON, W., [1958A], Functional bases of the central place hierarchy; *Economic Geography*, 34, 145–154.

BERRY, B. J. L. and GARRISON, W., [1958B], Recent developments of central place theory; *Regional Science Association, Papers and Proceedings*, 4, 107–120.

BERRY, B. J. L. and GARRISON, W., [1958C], A note on central place theory and the range of a good; *Economic Geography*, 34, 304–311.

BERRY, B. J. L. and GARRISON, W., [1958D], Alternate explanations of urban rank-size relationships; *Annals of the Association of American Geographers*, 48, 83–91.

BERRY, B. J. L. and MAYER, H. M., [1962], Comparative studies of central place systems; *Office of Naval Research, Contract* NONR 2121-18.

BERRY, B. J. L. and PRED, A., [1961]. Central place studies: a bibliography of theory and applications; *Regional Science Research Institute, Bibliography Series*, 1.

BERRY, B. J. L., SIMMONS, J. W. and TENNANT, R. J., [1963], Urban population densities: structure and change; *Geographical Review*, 53, 389–405.

BERRY, B. J. L., TENNANT, R. J., GARNER, B. J. and SIMMONS, J. W., [1963], Commercial structure and commercial blight; *University of Chicago, Department of Geography, Research Paper*, 85.

BLOME, D. A., [1963], A map transformation of the time-distance relationships in the Lansing Tri-County area; *Michigan State University, Institute for Community Development and Services*, (East Lansing, Michigan).

BLUMENFELD, H., [1949], On the concentric circle theory of urban growth; *Land Economics*, 25, 209–212.

BLUMENFELD, H., [1959], The tidal wave of metropolitan expansion; *Journal of the American Institute of Planners*, 25, 3–14.

BOAL, F. W. and JOHNSON, D. B., [1965], The functions of retail and service establishments; *Canadian Geographer*, 9, 154–169.

BOGUE, D. J., [1949], *The structure of the metropolitan community: a study of dominance and subdominance*, (Ann Arbor, Mich.).

BOYCE, R. and CLARK, W. A. V., [1963], Selected spatial variables and central business district retail sales; *Regional Science Association, Papers and Proceedings*, 11, 167–193.

BOYCE, R. and CLARK, W. A. V., [1964], The concept of shape in geography; *Geographical Review*, 54, 561–572.

BRACEY, H. E., [1962], English central villages: identification, distribution and functions; *Lund Studies in Geography, Series B, Human Geography*, 24, 169–190.

BRIGHAM, E. F., [1964], A model of residential land values; *RAND Corporation*, (Santa Monica, California).

BRUNHES, J., [1925], *La Géographie Humaine*, (Paris).

BRUSH, J. E., [1953], The hierarchy of central places in southwestern Wisconsin; *Geographical Review*, 43, 380–402.

BRUSH, J. E. and BRACEY, H. E., [1955], Rural service centres in southwestern Wisconsin and southern England; *Geographical Review*, 45, 559–569.

BUNGE, W., [1962], Theoretical Geography; *Lund Studies in Geography, Series C, General and Mathematical Geography*, 1.

BURGESS, E. W., [1925], The growth of the city; *The City*, 47–62, (Chicago).

CARROTHERS, W. J., [1962], Service centres in Greater London; *The Town Planning Review*, 33, 5–31.

CHAPIN, F. S. JR., and WEISS, S. F. (Editors) [1962], *Urban Growth Dynamics*, (New York).

CHRISTALLER, W., [1933], *Die zentralen Orte in Süddeutschland: Eine ökonomisch-geographische Untersuchung über die Gesetzmässigkeit der Verbreitung und Entwicklung der Siedlungen mit städtischen Funktionen*, (Jena).

CHRISTALLER, W., [1966], *Central Places in Southern Germany*, (Trans. C .W. Baskin), (Englewood Cliffs, New Jersey).

CLARK, C., [1951], Urban population densities; *Journal of the Royal Statistical Society, Series A*, 114, 490–496.

CLARK, P. J. and EVANS, F. C., [1954], Distance to nearest neighbour as a measure of spatial relationships in populations; *Ecology*, 35, 445-453.
CURRY, L., [1962], The geography of service centres within towns: the elements of an operational approach; *Lund Studies in Geography, Series B, Human Geography*, 24, 31-53.
DACEY, M., [1962], Analysis of central place and point patterns by a nearest neighbour method; *Lund Studies in Geography, Series B, Human Geography*, 24, 55-75.
DACEY, M., [1963], Order neighbour statistics for a class of random patterns in multidimensional space; *Annals of the Association of American Geographers*, 53, 505-515.
DIAMOND, D. R., [1962], The central business district of Glasgow; *Lund Studies in Geography, Series B, Human Geography*, 24, 525-534.
DUKE, R. L., [1964], Gaming-simulation and urban research; *Michigan State University, Institute for Community Development and Service*, (East Lansing, Michigan).
DUNN, E. S., [1954], *The location of agricultural production*, (Gainesville, Florida).
FAITHFULL, W. G., [1959], Ribbon development in Australia; *Traffic Quarterly*, 13, 34-54.
FORBES, J., [1964], Mapping accessibility; *Scottish Geographical Magazine*, 80, 12-21.
FOSTER, G. J. and NELSON, H. J., [1958], *Ventura boulevard: a string-type shopping street*, (Los Angeles).
GARNER, B. J., [1960], *Differential residential growth of incorporated municipalities in the Chicago suburban region*, (Mimeographed).
GARNER, B. J., [1966], The Internal Structure of Shopping Centres; *Northwestern University, Studies in Geography*, 12.
GARNER, B. J. and HARVEY, D., [1965], work in progress.
GARRISON, W., [1956], Allocation of Road and street costs; *Washington State Council for Highway Research, The Benefits of Rural Roads to Rural Property*, IV, (Seattle).
GARRISON, W., BERRY, B. J. L., MARBLE, D. F., NYSTUEN, J. D. and MORRILL, R. L., [1959], *Studies of highway development and geographic change*, (Seattle).
GETIS, A., [1963], The determination of the location of retail activities with the use of a map transformation; *Economic Geography*, 39, 1-22.
GODLUND, S., [1956], Bus service in Sweden; *Lund Studies in Geography, Series B, Human Geography*, 17.
GOTTMANN, J., [1961], *Megalopolis: the urbanised northeastern seaboard of the United States*, (New York).
GREEN, F. H. W., [1950], Urban hinterlands in England and Wales: an analysis of bus services; *Geographical Journal*, 116, 65-88.
GUNAWARDENA, K. A., [1964], Service centres in southern Ceylon; *University of Cambridge, Ph.D. Thesis*.
HAGGETT, P., [1965], *Locational analysis in human geography*, (London).
HARRIS, C. D., [1943], A functional classification of cities in the United States; *Geographical Review*, 33, 86-99.

HARRIS, C. D., [1954], The market as a factor in the localisation of industry in the United States; *Annals of the Association of American Geographers*, 44, 315–348.
HARRIS, C. D. and ULLMANN, E. L., [1945], The nature of cities; *Annals of the American Academy of Political and Social Science*, 242, 7–17.
HARTMAN, G. W., [1950], The central business district: a study in urban geography; *Economic Geography*, 26, 237–244.
HAWLEY, A. H., [1950], *Human Ecology*, (New York).
HAYES, C. R., [1957], Suburban residential land values along the C.B. & Q. railroad; *Land Economics*, 33, 177–181.
HORWOOD, E. and BOYCE, R., [1959], *Studies of the central business district and urban freeway development*, (Seattle).
HOYT, H., [1939A], *One hundred years of land values in Chicago*, (Chicago).
HOYT, H., [1939B], *The structure and growth of residential neighbourhoods in American cities*, (Washington).
HOYT, H., [1961], A method for measuring the value of imports into an urban community; *Land Economics*, 37, 150–161.
HOYT, H., [1964], Recent distortions of the classical models of urban structure; *Land Economics*, 40, 199–212.
HUFF, D. L., [1961], Ecological characteristics of consumer behaviour; *Regional Science Association, Papers and Proceedings*, 7, 19–28.
HURD, R. M., [1911], *Principles of city land values*, (New York).
ISARD, W., [1956], *Location and Space-economy*, (New York).
ISARD, W., [1960], *Methods of regional analysis: an introduction to Regional Science*, (New York).
ISARD, W. and KAVESH, R., [1954], Economic structural interrelations of metropolitan regions; *American Journal of Sociology*, 60, 152–162.
JONES, E., [1960], *A social geography of Belfast*, (London).
JONES, E., [1964], *Human geography*, (London).
KELLEY, E. J., [1956], *Shopping centres*, (Saugatuck, Connecticut).
KING, L. J., [1961], A multivariate analysis of the spacing of urban settlements in the United States; *Annals of the Association of American Geographers*, 51, 222–233.
KING, L. J., [1962], A quantitative expression of the pattern of urban settlements in selected areas of the United States; *Tijdschrift voor Economische en Sociale Geografie*, 53, 1–7.
KNOSS, D., [1962], *Distribution of land values in Topeka, Kansas*, (Lawrence, Kansas).
KOLB, J. H. and BRUNNER, E. DE S., [1946], *A study of human society*, (Boston).
LÖSCH, A., [1954], *The economics of location*, (New Haven).
MAYER, H., [1942], Patterns and recent trends of Chicago's outlying business centres; *Journal of Land and Public Utility Economics*, 18, 4–16.
MAYFIELD, R. C., [1962], Conformations of service and retail activities: an example in lower orders of an urban hierarchy in a lesser developed area; *Lund Studies in Geography, Series B, Human Geography*, 24, 77–90.
MORRILL, R. L., [1965], The negro ghetto: problems and alternatives; *Geographical Review*, 55, 339–361.
MORRISSETT, I., [1958], The economic structure of American cities; *Regional Science Association, Papers and Proceedings*, 4, 239–256.
MOSER, C. A. and SCOTT, W., [1961], *British Towns*, (London).

MURPHY, R. E. and VANCE, J. E. JR., and EPSTEIN, B. J., [1955], Internal structure of the central business district; *Economic Geography*, 31, 21–46.

MUTH, R. F., [1962], The spatial structure of the housing market. *Regional Science Association, Papers and Proceedings*, 7, 207–20.

NELSON, H. J., [1955], A service classification of American cities; *Economic Geography*, 31, 189–210.

OLSSON, G. and PERSSON, A., [1964], The spacing of central places in Sweden; *Regional Science Association, Papers and Proceedings*, 12, 87–93.

PHILBRICK, A. K., [1957], Principles of areal functional organisation in regional human geography; *Economic Geography*, 33, 299–336.

POWNALL, L. L., [1953], The functions of New Zealand towns; *Annals of the Association of American Geographers*, 43, 332–350.

PRED, A., [1963], Business thoroughfares as expressions of urban negro culture; *Economic Geography*, 39, 217–233.

RATCLIFF, R. U., [1949], *Urban land economics*, (New York).

REILLY, W. J., [1931], *The law of retail gravitation*, (New York).

ROTERUS, V. and CALEF, W., [1955], Notes on the basic- nonbasic employment ratio; *Economic Geography*, 31, 17–20.

SEYFRIED, W. R., [1963], The centrality of urban land values; *Land Economics*, 39, 275–285.

SIMMONS, J., [1964], The changing pattern of retail location; *University of Chicago, Department of Geography, Research Paper*, 92.

SIMON, H. A., [1955], On a class of skew distribution functions; *Biometrica*, 42, 425–440.

SKINNER, G. W., [1964], Marketing and social structure in rural China, Part I; *Journal of Asian Studies*, 24, 3–43.

SMAILES, A. E., [1947], The analysis and delimitation of urban fields; *Geography*, 32, 151–161.

SMAILES, A. E. and HARTLEY, G., [1961], Shopping centres in the Greater London area; *Institute of British Geographers, Transactions and Papers*, 29, 201–213.

SMITH, P. J., [1962], Calgary: a study in urban pattern; *Economic Geography*, 38, 315–329.

SMITH, R. H., [1964], *The geographical relevance of functional town classification*, (Mimeographed).

SMITH, R. H., [1965], Method and Purpose in Functional Town Classification; *Annals of the American Association of Geographers*, 55, 539–548.

STAFFORD, H. A. JR., [1963], The functional bases of small towns; *Economic Geography*, 39, 165–175.

STEIGENGA, W., [1955], A comparative analysis and a classification of Netherlands towns; *Tijdschrift voor Economische en Sociale Geografie*, 46, 106–112.

STEWART, C. T., [1958], The size and spacing of cities; *Geographical Review*, 48, 222–245.

STINE, J. H., [1962], Temporal Aspects of Tertiary production elements in Korea; *Urban Systems and Economic Development*, 68–88, (Eugene, Oregon).

THOMAS, E. N., [1960], Areal associations between population growth and selected factors in the Chicago urbanised area; *Economic Geography*, 36, 158–170.

THOMAS, E. N., [1961], Toward an expanded central place model; *Geographical Review*, 51, 400–411.
THOMAS, E. N., [1962], The stability of distance-population-size relationships for Iowa towns from 1900–1950; *Lund Studies in Geography, Series B, Human Geography*, 24, 13–29.
THORPE, D. and RHODES, T. C., [1966], The shopping centres of the Tyneside urban region and large scale grocery retailing; *Economic Geography*, 42, 52–73.
THÜNEN, J. H. VON, [1826], *Der Isolierte Staat in Beziehung auf Landwirtschaft und Nationalökonomie*, (Hamburg).
TOBLER, W. R., [1963], Geographic area and map projections; *Geographical Review*, 53, 59–78.
TUOMINEN, O., [1949], Das Einflussgebiet der Stadt Turku im System der Einflussgebiëte S.W. Finnlands; *Fennia*, 71, 114–121.
ULLMAN, E. L., [1941], A theory of location of cities; *American Journal of Sociology*, 46, 853–864.
ULLMAN, E. L. and DACEY, M. F., [1960], The minimum requirements approach to the urban economic base; *Lund Studies in Geography, Series B, Human Geography*, 24, 121–143.
VINING, R., [1955], A description of certain spatial aspects of an economic system; *Economic Development and Cultural Change*, 3, 147–195.
VOORHEES, A., [1955], A general theory of traffic movement; *Proceedings of the Institute of Traffic Engineers*, 46–56.
WATSON, J. W., [1955], Geography: a discipline in distance; *Scottish Geographical Magazine*, 71, 1–13.
WINGO, L. JR., [1961], *Transportation and urban land*, (Washington D.C.).
WINSBOROUGH, H. H., [1962], City growth and city structure; *Journal of Regional Science*, 4, 35–49.
YEATES, M., [1965], Some factors affecting the spatial distribution of Chicago land values, 1910–60; *Economic Geography*, 41, 57–70.
ZIPF, G. K., [1949], *Human behaviour and the principle of least effort*, (Cambridge, Massachusetts).

CHAPTER TEN

Models of Industrial Location

F. E. IAN HAMILTON

The major contributions to location theory have been made by economists, mostly with a view to integrating location studies with the main body of economic theory. The very word 'location', however, implies the existence of spatial relations, inter-relationships, and patterns, so that models of industrial location are, by definition, part of geography. Of course, industrial production *is* an economic activity. Yet unless it is so integrated that the procurement of materials, the manufacturing process, and the marketing of the product all take place at the same point (as in an intermediate stage of a factory process), the spatial separation of materials and markets involves transport to overcome distance and so effect the geographical linkage of supply with demand. Geographers have published hundreds of works giving descriptions and analyses of empirical evidence concerning the reasons for, the effects of, and changes in, the location of industries. In general this evidence is unbalanced in favour of a few branches of industry, a few world regions or nations, and an idiographic approach which tends to dissipate, rather than to integrate, the body of location theory and practice. It is not difficult to see why this should be so.

The term 'industry' describes a wide range of activities. These are as diverse as the quarrying of chalk, the smelting and refining of metals, the assembly of electronic equipment, and the ubiquitous supermarket. Industries can be classified into four groups, therefore, according to the operational process that they use: respectively, extraction, processing, assembly, and service. Each group requires specific inputs (materials, labour, capital) from specific sources for its operation and provides specific outputs for purchase in specific markets and market areas. The locational requirements of industries in each of the four groups differ as a result; and within each group these requirements are as a 'theme with variations'. Beyond this level, the number of location variables increases rather than decreases. Industries are not only heterogeneous activities; each branch is susceptible to diverse means of organization. The relative importance of the location factors involved, and hence the final location chosen, is likely to vary, even for the same branch of industry, as between different scales of production and between the one-plant firm, the intra-industry multi-plant corporation, the multi-industry firm with one or several plants and the state-managed enterprise. Further interpretations of the reality of location arise from analyses that account changes in time and in technique, and that compare industrial development in the light of variations in the size of regions and nations and in their resource endowment, population and markets.

This essay does not purport to fill in gaps in empirical research. Rather it presents some general and particular conceptual models of industrial location phenomena, using and criticizing both the familiar and the unfamiliar, and suggests models of location in the context of capitalist and socialist economic systems, settlement hierarchies, history, regions, nations and continents.

THE CHANGING CHARACTER OF MODEL TECHNIQUES

We begin, however, by considering certain progressive changes which have occurred during the last century in the ways in which models have attempted to reach a solution to the location problem.

The models that pre-date the First World War were generally simple constructions in which few variables were assumed and rigid mathematical proof (usually involving geometry) was applied to determine some optimum location. Launhardt (1882), for example, developed the location triangle for a given factory in which two points represented the sources of the materials it used and the third point was the market where it sold its products. Launhardt determined the optimum location as the point within the triangle at which the shortest lines (representing distance) met from the three points. Later (1885) he presented a more complex solution, the 'pole principle' (see Palander, 1935, p. 143). In the same work Launhardt demonstrated that ideal market areas could be hexagonal (Lösch, 1954, p. 114), but he proved that in practice they tended to be irregular polygons with boundaries as curves of the fourth degree. In the twentieth century, Weber (who was deeply impressed by Launhardt's work) and Palander have made extensive use of geometry.

Alfred Weber (1909) made the first important break with geometry, however, when he introduced the 'isodapane', or a line joining points of equal cost. This technique permitted greater diagrammatic flexibility and greater approximation to reality. The retreat from geometric determinism had begun. Palander (1935) and Hoover (1937) used isodapanes (and other isolines) extensively to show the locational irregularities in procurement and distribution costs that resulted from the varied pattern of different transport media with different cost/distance ratios. These and other examples were often translated into graphs which were popular among economists.

Like geometry, however, the iso-line and graph techniques suffered severe limitations. They remained satisfactory only so long as the few variables that *were* considered were reduced to fixed points, and distance was expressed by the length of a line, by the radius of a circle, or by some iso-interval. Hardware models never seemed to be popular. Only those of Launhardt (a mechanical weight model) and Weber (similar, but developed on a Varignon frame) had any significance and they were primitive, though no doubt adaptable to more variables with effort.

The need for a new method to cope with several location variables became apparent when it was realized that neither were supply or demand point-formed nor were industries organized only in one-plant firms. Hitchcock responded to this need by devising (1941) an algebraic formula for minimizing transport in the distribution of a product from several manufacturing plants (located at different points) to a large number of market areas. This was the origin of the formula:

$$y = \Sigma a_{ij} x_{ij} \text{ summed for } i = 1, 2 \ldots m,$$
$$\text{and } j = 1, 2 \ldots n,$$

where y is the total cost, a_{ij} is the cost of transporting one ton product from the ith factory to the jth market, and x_{ij} is the number of tons to be moved. This linear programme technique is eminently applicable to industrial location problems if the optimum (maximum or minimum) solution can be expressed algebraically as a linear equation. Kuhn and Kuenne (1962, p. 21) assert that 'it is not fortuitous that the origins and development of programming techniques are so closely tied to the problem of spatial economics, and that these techniques have resulted in the advance of breakthrough proportions in obtaining solutions to formerly insoluble spatial problems'. However the application of the technique is not without its problems. Koopmans and Beckman (1957) developed a model to assign n factories to n locations in such a way as to maximize the combined profits of the plants. They found that as the number of variables considered increases, the programming method becomes too cumbersome for satisfactory solution of the 'assignment' or 'allocation' problem. Garrison (1959, p. 479), for instance, shows that 120 possible location patterns must be compared in assigning only five plants to five locations! Similarly, Bos (1965, pp. 76-84) found that some 10^9 linear programmes were required to achieve the optimum spatial dispersion of three vertically-integrated industries over ten possible locations. He concludes (Bos, 1965, p. 84) that further progress in this field requires the development of 'an operationally efficient (algorithmic) method of solution for mixed integer programmes.' An algorithm has been applied to a Weberian location problem involving $n=4$ by Kuhn and Kuenne (1962); they assert first, that it is superior to all geometrical, iso-line, and physical analogue methods and second, that even with more than four variables the calculations involved 'are well within the capability of the desk calculator for large numbers of elements' (1962, p. 21).

For reasons that are set out in the next section, however, models are becoming more elastic in their approach to location problems. Greater realism can be achieved if these methods are complemented by stochastic models which in part are based on game theory, 'random walks', Monte Carlo simulation, and Markov-chain analysis.

IN SEARCH OF REALITY

An attempt is made in Figure 10.1 to construct a model which forms the cornerstone of industrial location model-building. It is logical to start with the three kinds of entrepreneur that decide location in the real world: the private capitalist, the corporate capitalist, and the State administration (local, republic, national, or federal). The model thus clearly distinguishes three different approaches to the location problem. These vary in importance both in time and place. Private capitalists determined the location pattern during the industrial revolution or during the early stages of industrialization while capital requirements were small. Joint stock companies and corporations have become more important entrepreneurs in the capitalist sector in the twentieth century. And now the State not only determines location choice in the communist world, but it plays an important part also in location decisions in the mixed economies of the non-communist world. These entrepreneurial differences indicate that the location models of Launhardt, Weber, and Hoover need not contradict those of Lösch or the socialist school, and vice versa; rather they may be complementary.

Profit motivates decisions in the capitalist sector, while such less easily quantified motives as 'the national interest', 'social cost' or 'social benefit' colour State location choices. Our model specifically omits the word 'maximum' from either criteria since this expresses an ideal not a reality. Greenhut and Colberg (1962, pp. 43–44) argue convincingly that, given exactly the same business or plant to locate, no two private entrepreneurs would judge alternative locations by the same quantitative 'maximum' profit. One will seek a large profit; another will be content with some profit. Indeed, no two states judge social cost-benefit in exactly the same way. Thus divergence of entrepreneurial opinion allows greater flexibility of location choice by increasing the range of 'optima'. Indeed, 'most human decision-making, whether individual or organizational, is concerned with the discovery and selection of *satisfactory* alternatives; only in exceptional cases is it concerned with . . . optimal alternatives' (March and Simon, 1958, pp. 140–141). The one, however, is not necessarily inferior to the other (Odhnoff, 1965). Circumstances that are peculiar to each kind of entrepreneur also influence their

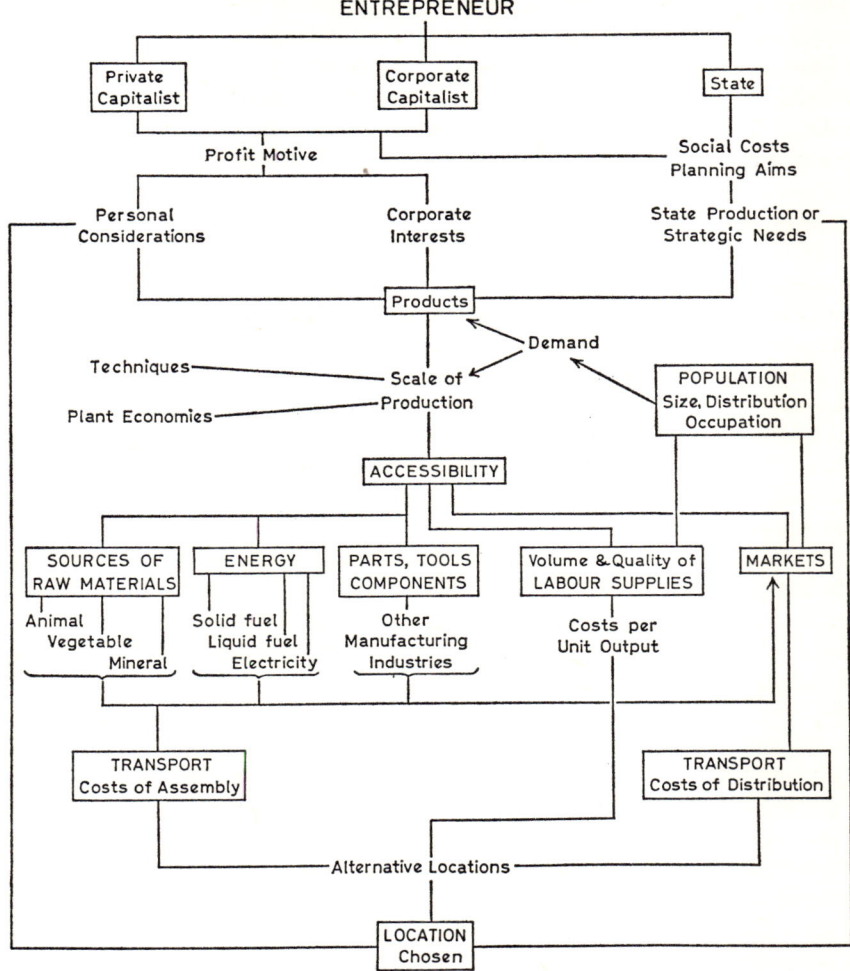

10.1 A basic model of the factors influencing industrial location decisions.

interpretations of the motive and hence their final choice of location. Personal considerations of either a partial economic nature (good financial contacts) or a purely social character (climate, hobby possibilities) are more likely to influence the private entrepreneur in choosing location than the corporation or State. More important may be his development of some small or growth industry in chance or random locations in response to a combination of favourable local conditions (inventions, laws, capital), a willingness to operate in the known environment of the 'home town', and an unwillingness to try an unknown locality elsewhere (Katona and Morgan, 1952). Evidence suggests that few private capitalists undertake comparative cost analyses to justify or refute their 'hunches' (McLaughlin and Robock, 1949, Chaps. 2–3; Alkjaer, 1953, p. 90; Klemme, 1959, pp. 71–77; Estall and Buchanan, 1961, p. 18; Goss, 1962, p. 16; Luttrell, 1962). When a firm grows into a corporation, personal factors are subordinated to the interests of shareholders and to the efficiency of plant in a highly competitive market: location decisions are influenced by corporate interests which include the efficient spatial, technical and economic relationships of new plant to existing plants, and the division of the market to ensure adequate supplies of products to existing or potential market areas. Large firms endeavour to compare location costs because of the greater risks they run on account of their scale.

As our model shows, the State can bring its social and planning policies to bear upon decisions in the capitalist sector of a mixed economy; these may take the form of negative and positive 'push-pull' forms of control (Manners, 1965, p. 153). Theoretically, the State ought to be absolved of personal influence in location policy; in practice powerful individuals or groups may steer important projects to particular areas even where planning is strictly centralized. Generally, however, the considerations that influence State interpretation of 'social benefit' or 'national interest' may include: the development of material-oriented industries to reduce the excessive export of mineral resources in raw or semi-processed form, or of manufacturing to reduce the dependence upon imports of capital-goods (Hamilton, 1962); the strategic allocation of projects both to many and to isolated areas (Hamilton, 1964A, pp. 59–61); and the location of industries to develop backward, under-privileged, or problem regions (Hamilton, 1964B, pp. 78–80).

Extraneous elements are introduced into the next stage of the model: demand (influenced by population size and occupation structure), techniques, and plant economies. These factors do not cause 'noise' because together they condition the line or range of production (and indirectly, therefore, the 'input mix'), they establish how many plants of optimum scale are required to serve the market, and they indicate how far the production process is technically divisible or indivisible and thus amenable to spatial dispersal or concentration. The scale of operation chosen is an important consideration in examining the accessibility that locations offer to supplies of materials and labour, and to markets. The importance of the range of products is now revealed: in part conditioning scale, in part conditioning cost structure where complex industries, e.g. chemicals, are concerned (Isard *et al.*, 1959). Entrepreneurs should then compare the summed totals of production and transport (procurement) costs of raw materials, fuel, power and components at, the costs of labour inputs in, and the costs of distributing the products to the market from alternative locations (Fig. 10.1). The model is comprehensive, therefore; yet *per se* it cannot indicate the relative importance of each location factor either for industries in agregate or for an industry in particular. Nevertheless, it does hint at a variety of input and output characteristics, and hence at various location relationships. The segregation of animal, vegetable and mineral raw materials suggests differing methods and sources of procurement and different degrees of 'perishability'; and the distinction between 'raw materials' and 'parts, tools and components' points up fundamental dissimilarities between primary (more nature-tied) and secondary (man-made and more 'footloose') sources of materials. The variable transportability of energy, and different labour characteristics, are also indicated. Finally, the model links the sources of materials, energy and components with 'markets' to emphasize the diversity of markets and market areas for capital goods, and it links population with 'markets' to underline the existence of distinct consumer markets. These details are essential in view of the frequently imprecise usage of the terms 'materials' and 'markets' in location literature. Their distinction, moreover, adds effectively to the beauty of the model.

As Figure 10.1 indicates, the final selection of location is made according to a comparison of economic costs, but the decision may be swayed by the individual's personal desires or by some particular State planning requirement. National or regional economic development and planning policies affect the location decision-making model in one further respect. The selection of a certain regional allocation of plants to fulfil planning needs may well have a feedback effect upon the scale and range of outputs of the plants concerned and hence their precise location. In other words, the increasingly important macro-economic approach to location choice demands more frequent reappraisal of the micro-economic approach. Once a locality is selected, a site must be found for the plant. The site that is chosen must usually ensure the availability of: adequate land for plant, stores, access facilities, and future expansion; adequate water supplies for processes or products; the right kind and quality of transport service; sufficient labour within an easily accessible area; and the preservation of amenity as stipulated by government regulations regarding site relationships with other land uses (agricultural, forestry, extractive, residential, recreational), atmospheric pollution, noise, danger or strategy.

It cannot be denied that the economics of procurement, and production distribution combined is *the* important – even deciding – location consideration. To infer that it is the only one, however, and accordingly to explain location mechanically, as many location models do, is to deny that man is human. The location patterns that emerged in the period between the industrial revolution and the application of comprehensive planning (whenever that may be) reveal that 'accident, habit, . . . and man's impulse to conquer his environment and to canalize his random impulses into orderly activities produced . . . nothing less than the empire of muddle' (Mumford, 1934, pp. 194–195). When knowledge was restricted and individualism was strong, man's gregariousness led to specialization which, under the conditions of an expanding economy, produced specialized agglomerations so that 'a variety of regional opportunities were neglected, and the amount of wasteful cross-haulage in commodities that could be produced with equal efficiency in any locality was increased . . .' (Mumford, 1934, p. 171). Moreover, man's thinking on the location of industries is formed under pressure from a vast array of human practices, prejudices, habits, laws and systems. Nationalism, imperialism and other political realities provide the true framework within which the location, distribution and spread of industries are encouraged, facilitated, restricted or prohibited (Odell, 1963; Hamilton, 1964A, pp. 46–64). Similarly, the attitude of communities, as Wallace and Ruttan (1960, p. 140), Paterson (1963, p. 130) and Hiner (1965, pp. 23–24) point out, can attract or repel or select industries on both the regional and local levels. 'Anomaly' industries or locations may result from specific ethnic distributions (Alexander, 1963, pp. 308–309). The early start of an industry in either a rational or a random location with some initial advantage (Pred, 1965), given dynamic entrepreneurship, may set in train the development of auxiliary industries, services and urban infrastructure that, at a later stage, ensures the inertia of the industry. If entry into the market is difficult for competitors, more advantageous locations may remain unexploited. Perhaps, however, the possibilities of 'sub-optimal' locations are nowhere as great as for industries manufacturing for the ultimate consumer. Here, product differentiation, as Greenhut (1964, Chap. 11) shows, distorts the shape and the extent of market areas and creates 'product islands'. Equally, advertising (Alexander, 1963, p. 310) may give extra marketing advantages (and profits) to a firm or a locality which, in cost terms, is less favourably located than rivals are. Even if the economic geographer is convinced that transport costs explain location patterns, there is no direct relationship between transport rates and geographic space (Alexander, 1963, p. 473). Special rate-fixing and pricing policies may often distort location patterns in practice.

While, therefore, the beauty of simple economic interpretation is the attraction of a model's vital statistics, the reality of her heart reminds one of the dangers of ignoring aspects which cannot be expressed in figures!

EARLY MODELS OF INDUSTRIAL LOCATION

It is against this background that one must examine early attempts at model-building. Many of the ideas at the core of location theory have their roots in the nineteenth century. To the early British economists – Adam Smith, J. S. Mill and David Ricardo – the location of industry was partly a function of the spatial distribution of agricultural surpluses for feeding industrial labourers and for processing. Writing between 1775 and 1810, Smith develops the threads of a model for industry at a time when technical innovations and canal transport were changing the face of Britain. Around mid-century, Karl Marx (1850) revealed what might be termed a 'capital localization-location' model which was framed in a world of international political inequalities. The last quarter of the nineteenth century saw a growing interest, especially among theorists in Germany, in location patterns within regions. Schäffle (1878) suggested the outline of a gravity model. According to this, industries develop chiefly in or near larger towns which, as markets, attract industry in direct proportion to the square of the distances between them:

$$M_{ij} = P_i P_j (d_{ij})^{-2}$$

for two centres i and j where M_{ij} is the market attraction of the two towns, P_i and P_j represent their populations, and d_{ij} is the distance separating them. Any significant deviation from this pattern Schäffle attributes to the location of industries near raw material or fuel resources. The model seems to explain broadly the patterns of 'ubiquitous' and of market-oriented sporadic industries; it would not necessarily function well, however, in explaining much sporadic manufacturing. In a wider context, the model implies that there is a greater localization of industry the larger is the urban population, the shorter is the distance to the market, and the closer are large cities to one another. It provides theoretical support for the more recent empirical findings of Bogue (1949) and Duncan (1959) in the United States.

Perhaps the most influential models of the nineteenth century were those developed by Launhardt.[1] He argued (Launhardt, 1882) that differences in the costs of production and prices between production centres and in the rates of transporting products modified the size and shape of the sales areas that could be supplied from those centres. Later Launhardt (1885) extended this idea to show the crucial importance of ton-mileage (weight × distance) × the transport rates at alternative locations in determining the costs of production. He concluded that entrepreneurs would develop industries in the locations with the *least costs*. It is significant, however, that Launhardt, like Schäffle before and Loria after him, never lost sight of the importance of the

[1] The work of Launhardt had a profound influence upon Weber's location theory; many of his ideas have been taken up by Palander, Lösch, and Isard more recently.

market. Thus Achille Loria (1888 and 1898) asserts that industries are located with reference to the market areas that they serve, with the exception of those industries which use bulky materials losing most of their weight in processing. The less these raw material-oriented industries needed to transport bulky materials, he argued, the larger the market area their products could command at a given price. Although Loria makes little reference to pricing policies, his model works on the premiss that entrepreneurs seek the locations that will give them *maximum profits*. This gifted Italian economist was also among the first to recognize that labour was an important location factor. He assigned industries with high labour inputs in the production process to predominantly agricultural areas. Röscher (1899) made a related observation, explaining that labour not only attracts industries to areas of high population density and localized skills, but also encourages industrial development in areas with 'workers ... (in need of) supplementary income' (Krzyżanowski, 1927, p. 279).

THE WEBER MODEL: MERITS AND DEMERITS

Although the idea of the least-cost location model originated in the work of Launhardt in the nineteenth century, the energy of Alfred Weber – in extending, modifying, and propagating it – made it all-persuasive in the twentieth century. Weber (1909) frames his model in an isolated state where natural resources for processing conform to a Thünen-ring system around a number of given market centres. These assumptions imply the existence of spatial variations in economic advantage with respect to both supply and demand. Natural resources comprise 'sporadic' materials (e.g. mineral fuels and metallic ores) and 'ubiquitous' materials (e.g. water, sands, clays, stone) which have either 'gross' weight if they lose weight in processing, or 'pure' weight if they lose none of their weight. In this way he distinguishes between more localized and more dispersed materials and between 'less mobile' and 'more mobile' materials.

Weber stresses that, within this heterogeneous environment, entrepreneurs will locate industries at the points of least cost in response to three general location factors – transport and labour (as inter-regional factors), and agglomeration or deglomeration (as intra-regional factors). He assumes that transport costs are a function of weight and distance (ton-miles). The point of least transport cost, therefore, is that at which the combined weight movements involved in assembling materials from their sources and in distributing products to their markets is at a minimum. Weber (1929, pp. 48–75) devised a 'material index', given as the weight of localized material inputs divided by the weight of the product, to indicate whether the point of 'movement minimization' (Haggett, 1965, pp. 142–148) would be near the source of materials (an index >1) or near the market (an index <1). If an industry has a high labour coefficient (according to Weber, the ratio of labour cost to the combined weights of inputs and outputs), Weber argues that labour will attract the industry to a location other than the movement minimization one if the savings in labour cost per unit output exceed the extra transport costs involved. Substantial agglomeration or deglomeration economies could encourage entrepreneurs to 'deviate' from the minimum transport and labour cost locations and to develop their industries at a third 'minimum' point. To Weber, agglomeration economies arose from local internal or external (linkage) economies of scale, labour skills, bulk buying and selling to minimize stocks, and infrastructural benefits. He stressed that, by contrast, rising land prices in growing urban areas encourage deglomeration.

Weber's model distinguishes at least fourteen hypothetical types of industry (1929, pp. 61–66) regarding weight movement, labour coefficients, and agglomeration economies; these and the 'orientations' Weber suggests are given in Table 10.1.

TABLE 10.1

Inputs, Location factors, and Least-Cost Locations for Weber's Hypothetical Industries

Industry	Ubiquitous Materials	Inputs Sporadic 'Pure' Materials	Inputs Sporadic 'Gross' Materials	Labour	Agglomeration	Deglomeration	Materials	Intermediate	Market	Labour	Agglomeration	Deglomeration
A	1								*			
B	2+								*			
C		1					?	?	?			
D	1+	1							*			
E			1				z					
F		2							*			
G			2=					*				
H			2≠ +				*					
I	1+		1				?	?	?			
J	2+											
K		unspecified		*						*		
L	,,	,,	,,	*							*	
M	,,	,,	,,	*							*	
N	,,	,,	,,		*							*

The symbols used are:

+ for 'or more' materials
= for equal weight losses for gross materials
≠ for unequal weight losses for gross materials
* for definite locations
? for alternative (equally viable) locations

This clearly reveals the main features of Weber's model. Industries A, B, and D, that process ubiquitous materials, are oriented to the market since any other location would involve the unnecessary costs of transporting ubiquitous materials in the form of a bulkier product. Weber argues that industry F, using two localized 'pure' materials, will locate at the market for the same reason. Where differentially weight-losing materials form the only inputs then entrepreneurs will locate their industries near the sources of materials, ideally, that lose most weight (industries E and H). Where two or several gross materials lose equal weight (industry G) entrepreneurs will locate at some intermediate assembly point. Weber solved this location problem by using Launhardt's location triangle, while Pick (Weber, 1929, p. 229) and Cotterill (1950, p. 67) demonstrated it respectively by the Varignon frame and the force table analogue. The location optima, however, are not so obvious either for industry C which manufactures a 'pure' sporadic material incurring (à la Weber) equal transport costs wherever the industry is located, or for industries I and J which process both localized weight-losing and ubiquitous materials. In this last case the 'minimum movement' location will depend upon the input ratio of gross materials (and their weight-losses) to ubiquitous materials. Industry K, with a high labour coefficient, is located in the cheapest labour location, industries L and M are linked and are attracted to a common centre on account of agglomeration economies, and industry N seeks out an 'out-of-town' location because of large land requirements. Clearly, the dominance of market-oriented locations in Weber's scheme refutes those who criticize him for a preoccupation with material-orientation.

By disproving the reality of Weber's basic *assumptions*, however, critics have undermined the very essence of his model. His disregard for the imponderables of human beings set out earlier is an example; this is supported in a devastating critique by Hamill based on empirical evidence from the Oregon timber industry (in Hamilton, 1964C, p. 234). To assume, for instance, that raw material sources and markets are fixed points is to ignore the geographical conditions of supply to agricultural and forest industries, and of demand generally. It also takes no account of possible spatial changes in supply or demand. Weber presupposes an unrealistic framework of perfect competition and given market conditions. This led him to analyse only cost factors and to maintain that, by definition, the least-cost location was also the maximum profit location. Production is also assumed to be given so that 'production cost has no importance and transport cost is the only factor for determining the optimum flow' (Ghosh, 1965, p. 44). Moses (1958) proves conclusively that optimum locations depend upon optimum production levels. Weber erred also in trying to reduce real differences in raw material production costs at source to a transport problem by inventing fictitious distances; this only confuses the problem. Subsequent location theorists (Hoover, 1937,

p. 52; Isard, 1956, p. 108; Fulton and Hoch, 1959, p. 52) have stressed the fallacy of the assumption that transport costs are proportional to distance. Thus Isard (1956, pp. 105–108) and Alexander (1963, pp. 473–475) show that freight rates tend to be convex and stepped with distance partly because they include fixed terminal and loading charges (Hoover, 1937) which are unaccounted for by distance. These authors emphasize, moreover, that, weight for weight, raw material freight charges are usually lower (sometimes much lower) than those for products. Yet Weber states (Friedrich, 1929, pp. 42–48) that transport costs – in addition to weight and distance – are determined by the type of carrier, the size of shipment, the nature of the region, the density of the transport network and by the nature of the goods to be carried (their quality, value, and perishability).

Dean (1938, p. 19) asserts that Weber '. . . seriously overestimated the determinate influence upon location of weight-losing materials when they were not dominant, and underestimated the attractiveness of pure materials which are never dominant.' This throws into sharp focus the lack of precise industrial input data in the model. In any case, industries A, C, and E are fictitious in the sense that no industry uses only one material input; even most simple processing activities require at least two. Smith (1955) has criticized Weber for including the use of coal in the weight index because it confuses raw-material with fuel orientation; when coal is excluded, however, the material index bears little relation to real location orientations, particularly when the weight lost is less than 75 per cent. These conclusions seem to support Dean's observations regarding the variety of influences of 'gross' and 'pure' materials on location choice.

Unreal assumptions and important inconsistencies weaken the force of Weber's labour and agglomeration analyses. Hoover (1937, p. 63) suggests that the 'labour coefficient' ought to express a relationship between labour cost and other costs (not weights) to gauge its importance. In his assumptions of unlimited labour supply and fixed regional labour cost differentials, Weber overlooks the facts of labour mobility and of labour shortage in growth areas. Sombart (1919) sees the recognition of wage differentials – the essence of capitalism – as a negation of Weber's claim to universality. Critics lay bare his misconceptions on agglomeration, pointing to the naïvity of geometric analysis and to the curious conclusion that agglomeration reinforces the attraction of cheap labour locations (Friedrich, 1929, pp. 161–162). Had Weber adopted more meaningful assumptions he might have concluded that agglomeration arises from the intensive exploitation of large and localized natural resources and major transport nodes, from urban economies, and from the growth of large city markets (Isard, 1956, p. 187). Yet labour *skill* may encourage the agglomeration of particular industries or groups of industry.

The Weber least-cost location model is, then, very 'noisy' in terms of its abstraction from real conditions. As it stands, its utility lies in its conceptual distinction of the relative importance of 'ubiquitous' and 'localized' materials, the inputs and outputs to be moved, transport and non-transport (labour, agglomeration) factors and the possible different 'orientations' of industries to materials, labour, and markets.

WEBER'S MODEL REFORMED: AN APPROACH TO REALITY

These criticisms lay the foundation for improving Weber's model by modification and extension. It appears that all *raw* materials are 'gross', although crude oil loses very little weight (Alexander, 1963, p. 345); 'pure' materials are those semi-finished industrial products (i.e. refined metals, plastics, metal components) that become the materials of fabricating, finishing, and assembly industries. In the reformed model Weber's ton-mileage transport criterion is replaced by one which compares value with weight, weight change in processing, bulk, perishability, fragility, and haulage rates. This still confirms Weber's thesis concerning ubiquitous materials since these 'abundant' materials command low – often very low – values in relation to their weight or bulk; they are thus expensive to transport. Nevertheless there may be contrasts in the types of raw material used and in the methods of mining them. These differences influence production costs directly through material procurement costs, and indirectly because they affect plant size and production processes; this gives a market advantage to low-cost producers as, for example, to manufacturers of Fletton bricks in England (Gleave, 1965). Localized materials and fuels are scarcer and command higher values, but these values vary widely according to the physical properties of, and man's uses for, such materials (e.g. iron compared with gold).

Industries, then, will tend to be located nearer the sources of their materials and fuel (or dominant input) the greater the proportion of unusable waste, the greater the bulk (perishability or fragility), the higher the freight charges, and the lower the value – weight for weight – of the raw materials compared with the products. The lower the weight loss, bulk, and the haulage rates, and the higher the value of the raw materials used, the more likely market-orientation will be. A number of location possibilities exist for industries that use 'pure' materials which neither lose nor gain weight in manufacture. Entrepreneurs will locate such industries nearer the source of the dominant material input if this has a low value/weight (bulk) ratio. Materials with high value/weight ratios are more transportable so that labour, market, or intermediate locations are more favoured for industries using them (Ross, 1896, p. 258). Industries tend to be market-oriented if they process divisible materials (oil), or assemble several 'pure' materials giving a substantial gain in weight or bulk (cf. value) in the product. Finally, the greater the gain in weight, bulk, perishability, or fragility, the higher the costs of transport, and the lower the value of the product, the greater is the tendency for entrepreneurs to locate an industry as near the centre of the market as possible to minimise distribution costs.

These principles may be modified or extended by factors which Weber did not take into account. For instance, entrepreneurs will locate industries near the cheapest adequate power-supply source if the ratio of power costs to other costs is very high. Those managing industries that depend upon labour skill, fashion, speedy contact or personal service (Smith, 1952; Martin, 1964) will choose locations where these give the greatest apparent advantages. Similarly, transport rate structures and facilities may distort the emerging patterns. Higher rates for products than for materials tip the balance further in favour of market-orientation; so does the availability of cheap (water) transport for carrying bulky low-value materials or of pipeline transport for other materials. Moreover, transport rates (Hoover, 1948) tend to 'polarize' industries at terminals, at the sources of materials or at the centres of markets, and to discourage intermediate locations unless favourable 'in-transit' rates are offered. The insistence upon cost that characterizes this 'reformed' model establishes a significant shift of emphasis from Weber's original concept of 'nearest material source' or 'nearest market' to the 'cheapest material source' and the 'most lucrative market' (Beckmann and Marschak, 1955, p. 333). This last idea provides a link with the final section, that concerns the market.

Weber laid the foundations for extending his own model, with regard to the market, in Chapter VII of *Über den Standort der Industrien* (1909). He suggests that an economy comprises a hierarchy of four 'strata' – agriculture and forestry, mining and primary processing, manufacturing, and tertiary services – that are interrelated as sources of supply and demand. This contributes also to an extension of the agglomeration concept to include industrial symbiosis and agglomeration at markets or sources of raw materials. The geographer can interpret each stratum as one market area or as a series of market areas either for industry as a whole or for particular industries. When combined, the 'strata' would give an irregular pattern of highly concentrated markets – where resources (e.g. coalfields) or nodes were localized – over a continuous and weaker market surface which reflected more extensive land uses and more dispersed settlement and population. The coalescence and overlapping of market areas with each other and with the centres or areas of materials-supply would result, as in reality. Industries, therefore, will procure their inputs from, and distribute their outputs to, a point, many points, or, most usually, an area or many areas which tend to adjoin or overlap.

The location problem here seems to resolve itself into a comparison of the importance of linear (point to point) and areal (point to area and vice versa) transport costs of procuring the materials and distributing the products (Lindbergh, 1953, p. 30). To obtain areal transport costs, Bos (1965, p. 34) shows that 'for practical purposes . . . a not too irregularly-shaped area can be approximated by calculations of cost for a circular market area of the same size'. Stefaniak (1963, p. 433) proves that a central location in a circular area minimizes distribution costs; these rise steeply if off-centre locations are chosen. Where transport cost differentials at alternative locations *are* decisive, therefore, an industry will be located at that centre where the combined areal transport costs of materials *and* products plus the linear haulage cost of transporting *either* materials *or* products from the supply area centre to the market area centre are at a minimum.[1] Naturally, an areal market (material-supply) strengthens the tendencies toward market-orientation (material-orientation), according to the principles established earlier, given point-formed material supply sources (markets). Yet the market area will always tend to have the greater attraction even where industries procure materials from an area; it will also 'pull' an industry somewhat away from materials if that industry is material-oriented. There are two reasons for this. First, areal as opposed to linear distribution magnifies the difference between the higher freight rates for products and the lower rates for materials of the same weight and bulk; the same effect results from the diseconomies of short-haul distribution in an area (especially where road transport is limited) compared with the economies of long-haul linear transport. Second, the *size* of demand in the market area is of crucial importance in determining distribution costs (Lindbergh, 1953, p. 30); but since the 'maximum profit' or 'maximum benefit' motive is more realistic than the 'least-cost' motive, industries will locate so as to tap the largest sales potential within the market area (Greenhut, 1960; Greenhut and Colberg, 1962, pp. 29–42), or to maximize some social benefit. This location may give them a large profit, or yield larger benefits, even though the costs of assembly and distribution are higher than at the least cost location. It should be possible to graft the 'market-potential' concept (Harris, 1954; Dunn, 1956; Isard *et al.*, 1960) on to this location model. In the last analysis, Weber's model can be improved if his assumption of given markets and population distribution is replaced by a dynamic approach which recognizes that relatively large-scale industrial development itself changes the geography of market potential. The location of a large, or several smaller, new plants in or near existing industrial areas speeds and strengthens the agglomerating forces and weakens the ability of unindustrialized areas to attract industry. If planning or some less direct device encourages deglomeration through the location of new plants in areas which lack industry, the process may be reversed for in time the growth of employment and industry creates a more attractive local market for consumer and capital goods in such areas, providing a far more satisfactory basis for counteracting agglomeration. Such a consideration provides a link with models of location under capitalism.

[1] Or at the point where costs *are thought to be* at a minimum.

INDUSTRIAL LOCATION UNDER CAPITALISM

Locational interdependence

Neglect of the market influence by the Weber school stimulated economists, especially in America, to consider location under the real conditions of oligopoly, duopoly, and monopoly. The entrepreneur's motive for location under these conditions is to control the largest segment (or the whole) of the market area at prices that yield him the greatest profit. The duopolist or oligopolist, therefore, must pay special attention to the location and markets of his rivals. Each firm's location is 'interdependent' (Greenhut, 1952, pp. 526–538) with those of other firms in the industry. Hotelling (1929) concludes that two firms locate their plants near the centre of the market if demand is inelastic; they will locate further apart if it is elastic. Lösch (1940), Greenhut (1957), and Devletoglou (1965) disagree; they contend that duopolists locate plants to divide, yet to gain more than half of, the market area (in sales volume); to do so they must be located at the quartiles along a line

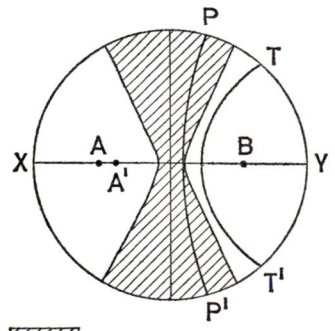

10.2 The locations chosen by duopolists in a circular market area are at A and B. Fetter's theory is introduced. If all costs are equal to both manufacturers, they will divide the market equally. Lower *production* costs at A, however, would enable the producer located there to extend his market to the line P – P'. Lower *transport* costs from A would permit expansion to the line T –T'. Theoretically, a location at A' has the same result.

through the market (locations A and B in Fig. 10.2). Yet wherever monopolistic competition occurs there are always areas (shaded in Fig. 10.2) where the market is doubtful (Greenhut, 1957, p. 86; Devletoglou, 1965, pp. 144–149), i.e. where competitors vie for the market. Even so, each manufacturer may retain a monopoly control of the areas that lie farthest away from rivals i.e. area A–X for A and B–Y for B in Figure 10.2. An entrepreneur may become a spatial monopolist if his plant (and the market it serves) is separated geographically (and economically by high transport costs) from the market areas of other firms in the industry e.g. the Pacific coast of North America (Harris, 1954). Smithies (1941) shows that the monopolist will either centralize production in relation to the market (in industries where savings in material movement or agglomeration economies far exceed the costs of distributing products); or he will divide it into equal market areas served by separate plants (where 'ubiquitous' or 'pure' materials are used or where the product bears high haulage charges on account of low value, weight, bulk, perishability, or fragility). If production is centralized, the plant may not be located at the mathematical centre (X in Fig. 10.3); it will be in a median location (A in Fig. 10.3) with access to the greatest number (or purchasing power) or buyers (Quinn, 1943). This model can be grafted on to the reformed Weber model, especially with regard to market-oriented industries.

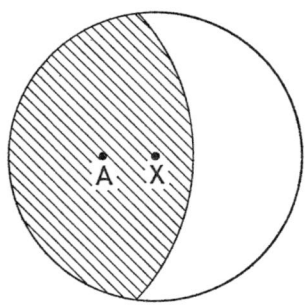

10.3 The median location. All buyers are located in the shaded area of the region concerned.

The maximum profit model

The Weber school eclipsed Loria's maximum profit concepts until Lösch published *Die räumliche Ordnung der Wirtschaft* (1940). He asserts (1954, pp. 28–29) that 'seeking the place of least cost ... is as absurd as to consider the point of largest sales as the proper location'. Lösch attempts to find the maximum profit location by comparing the costs of production at, and the market area which can be controlled from, alternative locations. Given the framework of oligopolistic competition, the location chosen may well not be the least-cost one since profits depend more upon sales revenue within an area than upon production and distribution costs (Greenhut, 1952). Unfortunately, Lösch assumes a homogeneous surface, implying ubiquitous resources with negligible variations in production and procurement costs at alternative locations. Under these conditions, he divides the market so that optimum-sized plants in each industry serve hexagonal sub-areas of equal size and that the plants of different industries serve sub-areas of different size (Lösch, 1954, pp. 124–132). He rotated the system around a common centre, transforming the homogeneous plain into a series of six alternate 'rich' and 'poor' zones. With population and settlement localized in the 'rich' zones, industries became agglomerated in the same zones to enjoy the economies of inter-industry linkage in localized market belts. Lösch shows that the location of plants at the centres of their hexagonal market zones minimizes freight costs on the product and maximizes profit. Isard (1956, p. 272) has redrawn Lösch's original concept to depict more realistic conditions; this modified system comprises increasingly large hexagons (or geometric areas) as distance from an urban agglomeration increases or population density decreases. This emphasizes agglomeration around, rather than in belts leading away from, the node.

The Löschian model is also modified by Greenhut (1952 and 1957) – to embrace cost considerations more thoroughly within an oligopolistic system. Greenhut argues that when firms seek the 'minimax' location (combining

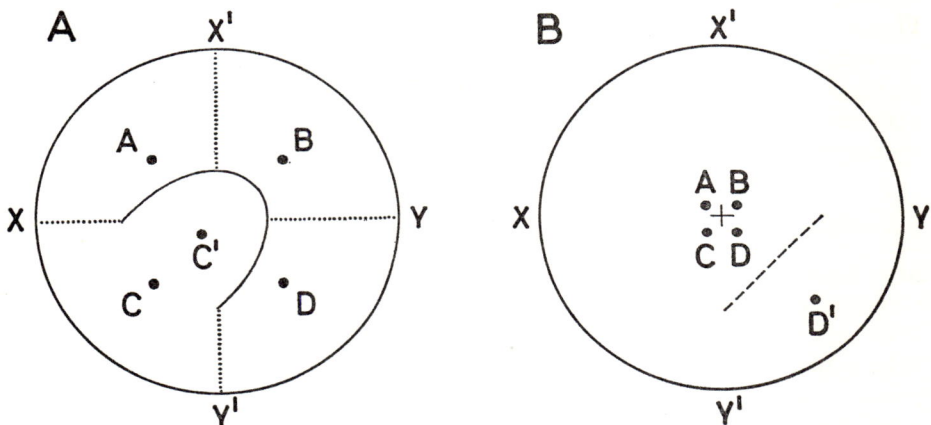

10.4 Diagrammatic representation of the problem of interdependence and maximum profit in an oligopolistic market. Four manufacturers divide a circular market area (A) by locating near the centres of their market segment at A, B, C and D, respectively. Each seeks to control a larger market by making a move to a new location as C does to C¹. Yet because each entrepreneur fears that his market may be reduced in this way, all producers locate near the centre of the total market area (B). This may encourage a new competitor to enter the market and to locate at D¹ to monopolize a sufficiently large market at the expense of D. Manufacturers must weigh up, therefore, the tactics and ability of rivals, as in a game.

minimum cost with maximum profit), the number of variables increases and the doubtful market area expands because 'uncertainty is an inherent part of the capitalist system' (Greenhut, 1957, p. 88). Such a situation arises from the existence of different costs in alternative locations, giving uncertainty about the competitiveness of rivals, and from the profit motive, giving uncertainty about the policies of rivals. Location decisions under these conditions are amenable to the use of game theory. Each competitor adopts one of several possible strategies in production and location choice; chance looms large, therefore, both in the location of individual plants and of industries as aggregates of plants. Further research is required to establish the vagaries of personal and board room decisions in more definite outline and to remove limitations to game theory application (Stevens, 1961) before 'game models' can become valuable geographically. Nevertheless, Greenhut's minimax model demonstrates that uncertainty, rivalry, and oligopolistic competition in undifferentiated products leads to industrial localization where firms think that rivals can achieve optimum sales at lowest cost. Hence man's 'follow-the-crowd' attitude leads to localized specialization. This is especially so in industries that either use fairly 'mobile' materials and manufacture valuable transportable products or serve large markets e.g. motor vehicles, textiles. If the products are differentiated, fragile or bulky, and bear high freight rates the tendency will be towards dispersal to divide the market. The problem of interdependence and maximum profit is shown for an oligopolistic market in Figure 10.4. The difficulty of establishing the minimax location scientifically has been demonstrated by Dunn (1956) who attempted to apply the 'index of location' to find the point in Florida that combines greatest market potential with minimum transport costs. This suggests that most locations are suboptimal or satisficer locations only.

Location and pricing systems

The adoption of different price systems from country to country or from industry to industry can modify resulting location patterns significantly. This section offers, therefore, further embellishment of the basic model. Where delivery prices are fixed equal for a whole market area the entrepreneur will obtain the maximum profit where his costs of selling to the market area are the lowest. This system encourages more firms to agglomerate in the largest market area as compared with other price systems (Greenhut, 1964, p. 387). The single basing-point system tends to encourage and perpetuate industrial concentration at a base point which, as a result, is an artificial least-cost point. The system effectively stifles or reduces the possibility of industrial growth in locations which may have more advantages than the base point (Stocking, 1954). Multiple basing-point systems permit spatial competition from several points so that market areas assume different shapes and sizes; market potentials vary from area to area, therefore, and may induce the selective dispersal of plants. The geographical impact of distance is only allowed full play – within the limits of transport rate policies – with the f.o.b. pricing system. It induces entrepreneurs to divide the market area by dispersing in an attempt to monopolize some segment of it (where demand is elastic); distance is the defence against competition. Again this tendency will be greater the greater the transport costs on the products. The locational relationships are possibly more complex, however, for, as Greenhut (1964, pp. 393–403) demonstrates for the U.S. paint industry, entrepreneurs may obtain their raw materials under one delivery price system but sell their products under another. This seems to be a field in which empirical research is otherwise lacking.

Too little knowledge exists about the geographical effects of business organization in a capitalist society. For instance, the location of wholesale facilities may be under-rated as a factor which 'deviates' the actual market 'centre' from the 'gravitational market centre' and induces the excessive agglomeration of market-oriented industries. This possibility is clearly underlined by Martin (1964, p. 114): 'From the manufacturer's point of view, London's effective demand for a product is disproportionate to its ultimate consumption. The market he supplies directly may consist of wholesalers and intermediaries acting as channels of demand.' Similarly, where demand for a product is fairly scattered, an entrepreneur may map out his potential market area by 'random walk' methods and choose his plant location accordingly.

LOCATION POLICY IN THE SOCIALIST WORLD

One-third of the world's population is living in countries that are managed, as planned economies, by Communist governments. Whether or not these governments apply Marxist-Leninist ideology *per se*, it is well to reflect that they do plan patterns of industrial location which differ significantly from those in 'free enterprise' economies; this is mainly because location is strictly controlled and directed according to specific guiding principles. This does not mean, however, that what is relevant to capitalism is irrelevant to socialism, or vice versa. For although 'western' theorists are described as 'bourgeois apologetic theorists for capitalism' (Balzak *et al.*, 1949, pp. 106–110; Secomski, 1956, pp. 7–20), location planners in the socialist countries do take inspiration from the work of Weber and Lösch (Saushkin, 1959, p. 47; Dziewoński, 1961, p. 7; Mrzygłód, 1962, pp. 42–45). The very facts that Weber tries to prove the raw material-orientation of industry and that Lösch centres his study on market relationships make nonsense of statements that they are trying to construe apologies for capitalist location patterns, especially when Marxist economists and geographers themselves stress (following Marx, 1850) that the distribution of capitalist industry bears little relation either to material sources or to markets. Greenhut (1957, p. 84) even asserts that 'Lösch's rational system could only be brought about by full direction from the state'.

Location under capitalism is criticized by the socialists on the grounds that it leads to disproportions in economic opportunities between regions. The increasing and abnormal agglomeration of industry in 'growth' regions is not counteracted by the disinterest of capitalists and the inadequacy of government planning measures in developing industries in backward, declining, 'stranded' or 'ruined' areas (Hoover, 1947, p. 203) where depleted resources, unemployment, and overspecialization are major problems. Socialist location policy seeks to avoid these 'contradictions' by 'paying most attention, from the very beginning, to the correct location of industry as the decisive element in the development of the whole national economy' (Secomski, 1956, p. 43).

The socialist approach to location differs from the capitalist approach in several important ways. It replaces the individual's profit motive by the State aim at raising production and income to achieve higher, more equal living standards for all everywhere. New industrial development plays the key structural and spatial role in this. The maximum social and economic benefit is required of every project. This involves a 'complex' planning approach to dovetail plant location into the whole system of spatial linkages between a given plant and other industries, settlements, transport facilities, and land uses, with an eye also on local, regional, and national repercussions. Whereas few capitalists compare alternative locations because uncertainties make it difficult for them to do so (Hague and Dunning, 1954, pp. 203–204), long-term plans and projections of future economic and social trends create a higher degree of certainty in the socialist state, so permitting comparative analyses of national, regional, and local needs and potentialities. Such analyses – when subject to defined location criteria – 'create opportunities for a wider choice of location alternatives that permits more flexible distribution (of industry)' (Secomski, 1956, p. 51). Evidence from Yugoslavia suggests that relatively exhaustive comparisons of locations are made (Hamilton, 1962, pp. 146–161; Hamilton, in preparation); an added stimulus in that state is the activation of local interest through polycentric planning (Hamilton, 1964C, p. 181). Decisions made hastily (Hamilton, 1962, pp. 191–201) or made from the viewpoint of individual plants only, as in Poland's Six-Year Plan (Mrzygłód, 1962, p. 25), can cause costly mistakes. In fact, the final location chosen often reflects the outcome of arguments between representatives of the industrial sectors, who stress the most economic location for individual plants in their particular sectors only, and the regional planning representatives whose socio-economic 'complex' approach indicates a different pattern of location as optimal (Lissowski, 1965).

The socialist countries are very varied. The vast space of the USSR is organized in huge economic and administrative regions that are larger in area than the east-central European states. East Germany, Czechoslovakia, and Poland inherited well-developed capitalist industrial regions so that postwar location patterns show considerable *unconformities*. In contrast are the Balkan states where a largely socialist industrial pattern has emerged from the beginning. And China and Yugoslavia are applying diverse planning methods. These differences, however, form the basis for testing, rather than formulating, a model which embodies principles common to all the socialist states.

Four principles of location may be noted. Industries should be located, first, near the sources of the raw materials and fuel they use, and second, near the markets for their products (Livsic, 1947). This is to minimize the transport involved 'from processing the raw materials through all subsequent semi-finishing stages, even to the marketing of the finished product' (Lenin, 1954, p. 33). These principles hint at the clear division between material- and market-oriented industries; yet spatially, the two may be linked, for as Saushkin (1959, p. 49) stresses, 'consumption increases (in mining, fuel, and power areas) as a result of industrialization and the growth of population' while 'the high level of production (in market areas) makes it possible to discover and exploit more efficiently even the second-rate raw materials and fuels'. The bases of industrial 'complexes' or agglomerations emerge, but the social frictions of overcrowding, commuting, and housing needs either encourage the dispersal of interrelated industries within a growing industrial area (Hamilton, 1964D), or prohibit new location in such inherited capitalist agglomerations as Upper Silesia (Pounds, 1959). Moreover, socialist planning removes two powerful forces which encourage agglomeration under capitalism: uncertainty and the need for close contact regarding quick changes in fashion, tastes, and consumer preference.

These considerations operate within the constraints of the third principle: to locate plants to achieve an 'even' distribution of industry (Feigin, 1954, pp. 182–183). An 'even' distribution is not meant to imply some standard ratio of industrial employment to area or to population, although it *could* be understood as that level of industrial development which, combined with other productive activities, ensures equal incomes per head of population over the whole country in the long run. Since incomes are usually unequal, this principle involes the policy 'to develop backward areas' by allocating to them proportionally more new industrial capacity than the national average (Hamilton, 1964B, p. 79). Priority is given to locating those industries in backward areas that can process materials and energy resources locally and employ much labour (Hamilton, 1963, pp. 104–105). Governments undertake geological research in such areas to establish the existence and nature of resources that would strengthen the local bases for processing industry; an example is sulphur in south-eastern Poland (Hamilton, 1964D). The importance of developing backward areas is often overemphasized; in reality the principle is to ensure that 'the development of production in one area would not hinder . . . (that) in other areas' (Saushkin, 1959, p. 49). This rather vague notion is modified and conceptualized by Dziewoński (1958) who writes: 'In view of experience . . . the principle (requires the provision of) equal possibilities . . . and opportunities for social and economic development in all parts of the country and of giving similar living standards'. Naturally a very long time is necessary to achieve this. In fact industrial growth is planned for all regions, including the most developed, to supply capital equipment for factories being constructed in the backward areas in exchange for materials and manufactures lacking elsewhere (Kidrić, 1948, p. 34). Nevertheless, the model involves a progressive dispersal of plants to all regions (defined as 'territorial administrative areas') of the country, and within each region to both the larger and the smaller towns. It also involves balanced industrial employment for both men and women locally, and balanced regional development in stressing variety as well as some degree of specialization often around some heavy industrial core area (Caesar, 1954). The fourth principle is to disperse plants in the interests of security and national defence, as has happened in the location of Yugoslav iron and steel capacities (Hamilton, 1964A, pp. 59–61).

Recently, Mrzygłód developed an industrial location model that is at once amenable to these principles and to those outlined in the 'reformed' Weber model. Socialist planning, asserts Mrzygłód (1962, pp. 35–47), must avoid the two costly extremes of over-concentration and over-dispersion by creating sub-regional industrial complexes. These comprise various plants that use local raw materials, that are 'mobile' and employ local labour and that, as branch or auxiliary plants, are linked with a new or expanded 'parent' industry – or location leader (Estall and Buchanan, 1961, pp. 167–170) – within each area. The number of plants would vary with the economic possibilities (Myrdal, 1957), but they would be assigned to different towns to achieve an even spatial distribution and in such a way as to relate their size to the labour available in each town and town hinterland. A framework of Löschian market areas is replaced by one of labour-supply areas. This would reduce commuting and migration to a minimum. The growth of industry and its effects should ensure a balance between jobs and present and future labour supply *within the region*, except where the exploitation of large and valuable resources requires migration. This implies that priority is given generally to allocating plants to those regions with a labour surplus. Mrzygłód suggests three possible variations on this theme (Fig. 10.5) involving a region A that possesses resources for processing by an industry which is divisible into spatially separated stages and which manufactures a high value product. Transport costs are small. The first case (Fig. 10.5A) assumes that the labour surplus is spread evenly among three adjacent sub-regions. The processes can be divided between these sub-regions, therefore, so that A produces and refines the raw material while sub-regions B and C manufacture semi-finished components and the final product. The second variation (Fig. 10.5B) divides the industry between two sub-regions, A as before, and B which, because of a larger labour surplus, is allocated all the fabricating and finishing stages. The last variation (Fig. 10.5C) assigns the whole industry to sub-region A where the labour surplus is very large; this variation could apply also in the case of an industry using weight-losing materials mined in sub-region A. The character, rate, and spatial arrangement of industrial growth in each sub-region depends upon the needs of the nation as a whole, interacting with the needs and potentialities of the sub-region itself and of the

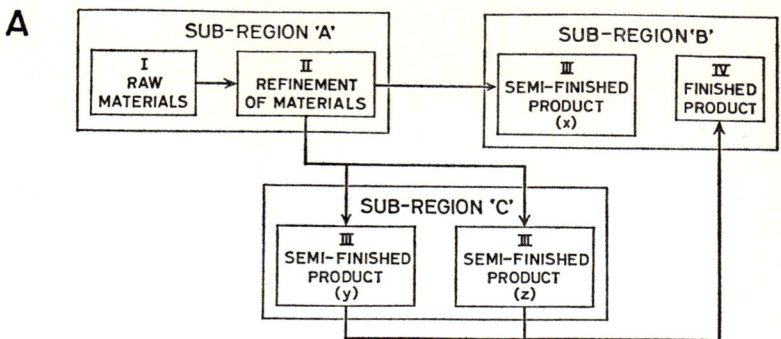

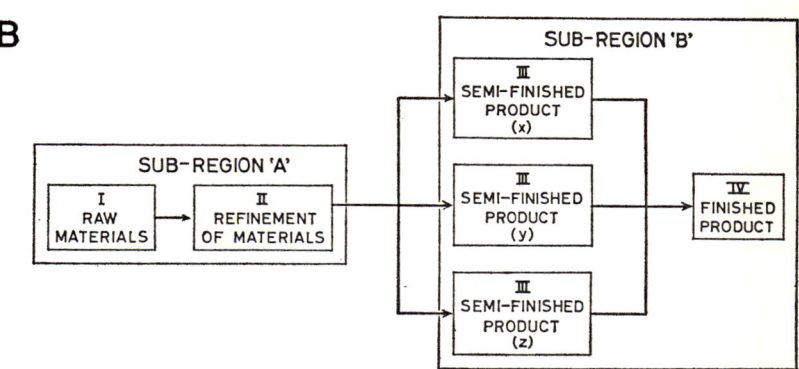

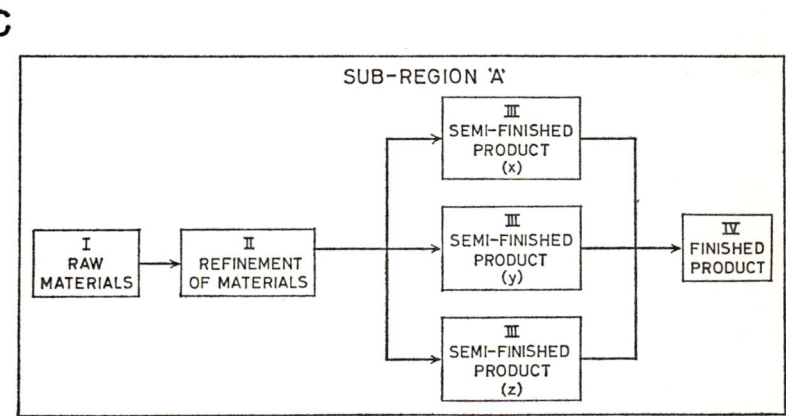

10.5 Alternative plant allocation strategies in a socialist state
(*Source: Mrzygłód, 1962, p. 40*).

adjacent sub-regions. It is obvious that neither spatially nor structurally can 'targets and capacity levels for specific commodities . . . be regarded in isolation' (Ghosh, 1965, p. 91).

The socialist models of general plant location are amenable to the theoretical and – as long as the difficulty of handling a multitude of variables is overcome – also operational, treatment embodied in the assignment or allocation-location models to be discussed in the following section (Silhána, 1964, pp. 321–328). The contributions of the Dutch economists – Tinbergen, Bos and Serck-Hansen – on spatial dispersion are also of value in this respect.

ALLOCATION-LOCATION MODELS

It is necessary to pass on from models of general plant location to consider models that assign particular plants of the same industry or of several industries to a number of locations. Significant here is the use of linear programmes and algorithms to find optimum production locations for a number of plants by allocating optimum capacities either among an equal number of locations or among a wide range of alternative locations. Both conceptual and opera-

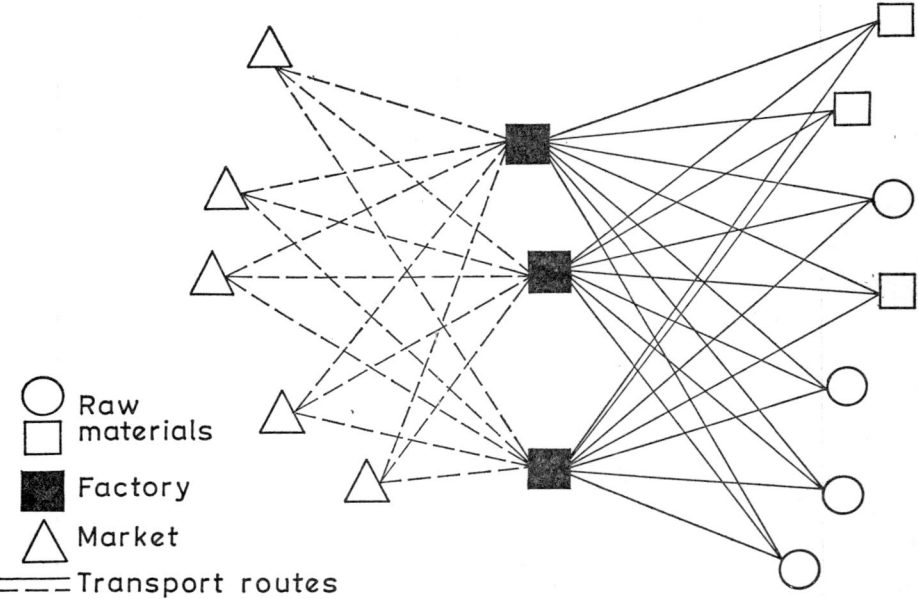

10.6 The Beckmann-Marschak assignment model. This sketches diagrammatically a problem which is to select those material and product flows which will maximize net profit to a corporation with several plants located separately, and to an industry as an aggregate of firms.

tional models are reviewed here. Beckmann and Marschak (1955) assume a diversified environment with many markets, and many potential material suppliers, each with different production costs, and with limited capacity that restricts each plant's size and saleable output. The number of locations being fixed, the task is to allocate optimum *scales* to the plants to minimize the volume and direction of flows and yet to satisfy demand. A macro-economic approach is thus required. The problem is shown diagrammatically in Figure 10.6. The model copes with situations, therefore, where the capacity of plant A exceeds the maximum output of its cheapest material's source x, so that the balance of supplies must be purchased from a dearer source y in so far as sales to the most lucrative market realize maximum profits. Goldman's conceptual model (1958) is complementary and allocates optimum iron and steel capacities among three equidistant islands which each produce one raw material and consume various quantities of steel to minimize both the flows of materials and products and also the backhauls and empty ship movements (social capital). The model also allows variation in the optimum *number* of locations (between 1 and 3), division of producers into separate blast-furnace or steel plants, and the spatial coincidence of the material sources with the markets. A Koopmans transport linear programme proves that optimum locations are a function of total demand (not cost) when transport is provided at marginal-cost rates by ships; plants may be best located, therefore, where transport costs are *above the minimum*. Indeed, the movement of empty transport capacity induces modified location patterns to provide return cargoes – witness the growth of Erie port steel industries and the Urals – Kuznetsk combine (Holzman, 1957).

More complicated are those models which assign n manufacturing plants (in one or several industries) to n locations to achieve the maximum combined profits (social or planning benefits) from the plants. Koopmans and Beckmann (1957) attack this problem in two ways. First they use the 'linear assignment' approach that treats plants individually; this method allocates plant A to the most profitable location, and then assigns plant B to the most profitable of the remaining locations, and so on. Second, they apply the 'quadratic assignment' approach that counts not only transport between plants but also the effects that the costs and profits of each plant in a given location have upon the locations, costs, and profits of all other plants. The number of variables becomes prohibitive yet the model still 'promises to be an important avenue of approach towards the precise and meaningful analysis of such interdependencies' (Cox, 1965, p. 235). The model could give greater insight into the problems of linkage, agglomeration, and socio-economic costs and benefits. The difficulties which Koopmans and Beckmann found in achieving an optimum solution may be overcome by the choice of a *satisfactory*, rather than an optimal, assignment of plants. Such a solution may be of no less 'optimal' quality (Odhnoff, 1965, p. 39).

Operational linear programme models have been applied in the United States to those industries, notably agricultural processing, in which the number of variables is small. Koch and Snodgrass (1959) construct a model of the optimum geographical distribution of the US tomato-processing industry in relation to its raw materials and markets which assumes the reality of imperfect competition. Transport costs t_{ij}[1] are analysed to include processing and advertising costs, product differentiation, consumer preference for Californian canned tomatoes, and price discrimination in 'secure' markets. This model shows that existing location and production patterns are not optimal. Stollsteimer (1963) evolves a simpler operational model for finding the optimum number, size and location of plants processing one raw material into one product. In practice, entrepreneurs extend the length of the processing season to use capacity more efficiently by processing two or more inputs, as, for example, in the frozen-foods industry (Hiner, 1965). Thus Polopolus (1965) generalizes Stollsteimer's model for a case involving three raw materials and final products (sweet potatoes, tomatoes, and okra), 25 producing origins, and 10 potential processing locations in Louisiana. Costs are adjusted for joint processing of tomatoes and okra; and costs of material supply are estimated for three truck sizes on the basis of present production patterns which are adjusted for the future geographical effects of the continued urban growth of Lafayette, the uneven incidence of pests, and the competition of vegetables with sugar-cane and cotton on farms. One multi-product plant located centrally is the optimum solution for, as plant numbers increase, far higher processing costs result from decreasing scale and offset small savings in assembly costs (Polopolus, 1965, pp. 292–295). Cox (1965), in contrast, approaches the problem chiefly from the marketing angle in a model which uses an algorithm to solve the distribution of motor-vehicle bodies from three factories to four assembly plants. More recently, Efroymson and Ray (1966) have devised a branch-bound algorithm which has been used to solve practical plant allocation problems with upwards of fifty plants. Rapid progress is, therefore, being made.

Nevertheless Cox concludes that linear programme models are defective in that they treat the places of production and consumption as points rather than as areas: whence Polopolus's use of 'origins' of supplies of agricultural produce for manufacture. The danger with linear programme models seems to lie in applying complex methods to oversimplified situations to obtain either obvious or already well-established results. An example is provided by the work of Lefeber (1958, pp. 77–87). This is inevitable, nevertheless, with embryo techniques of analysis. In part, however, difficulties arise in presenting complicated programmes with many stages in published works; much

---

[1] 'T_{ij}' is the algebraic shorthand for the costs of transporting materials or products from the ith sources to the jth destination.

computer programming is a question of laboratory method rather than of conceptualization.

INDUSTRIAL LOCATION AND SETTLEMENT HIERARCHIES

Following Christaller, geographers, until recently, have neglected the relationship of primary and secondary manufacturing to the central place hierarchy. Yet industry was, and in the developing countries still is, the chief vehicle of urbanization. The intensive nature of industrial activity explains the concentration of the labour force and auxiliary services in towns, localizing there consumer and industrial markets which attract other industries and services, and unleashing a process of 'circular and cumulative causation' (Myrdal, 1957, p. 13). It should be possible, therefore, to generalize the relationships between the size, distribution, and types of industry.

The hierarchies of settlements embodied by Christaller and Lösch in their respective models of central places and market areas form a logical framework for the distribution of market-oriented industries. According to Alexandersson's study[1] of American industry only three, out of sixteen, manufacturing industries – construction, printing and publishing, and food-processing – are ubiquitous, i.e. located in all towns with over 10,000 inhabitants and can reflect faithfully the respective regular and irregular urban size-distribution hierarchies of Christaller and Lösch. Employment in these industries is positively correlated with the size of any settlement (Alexander, 1963, pp. 296–299). Individually, the sporadic industries show no such correlation and since collectively they account for most manufacturing, they apparently weaken the hypothesis of complete industrial conformity with Christaller-Lösch systems. Evidence shows, however, that ubiquitous industries stimulate their own growth by generating more income and development locally (Pfouts, 1957; Hirsch, 1959). This can attract sporadic industries to a town when that town attains the threshold (Pred, 1965, p. 168) – the minimum population or the minimum volume of sales required to support those industries. Moreover, different industries require different optimum numbers of plants of different optimum scales to serve the same national or regional market (Bain, 1954); different industries, therefore, supply variously-sized market areas. Ideally, the model proposes that sporadic industrial plants will be located in those settlements that have thresholds that are larger than, or equal to, the plants' desired market areas. This implies that the city rank-size concept is compatible with the idea of an urban-centred hierarchy of market areas (Beckmann,

[1] Alexandersson (1956) found that of 36 broad groups of industry twenty were service activities that were located in every town with over 10,000 inhabitants; only three manufacturing industries were so dispersed, the other 13 being sporadically located.

1957; Pred, 1965). Sporadic activities that are rare in small towns are more frequent in larger towns which tend to support not only larger-scale industries but also a greater range of industries of all sizes as a result of larger population, larger thresholds, and the 'nesting' of a greater number of market areas. A synthesis of the works of Bain (1954), Alexandersson (1956), and Philbrick (1957) – regarding industrial scale, cities, and areal functional organization in the United States – suggests itself here. Accordingly, only the three ubiquitous industries would be located in the smallest towns, while medium-sized centres would contain in addition, food-manufacturing, textiles, leather and shoe, and cement industries. Larger cities would be locations for steel, oil-refining, chemicals, metal containers, and distilling, along with all industries of lower-rank towns. Such products as tyres, rayon, soap, cigarettes and farm machinery appear from factories at the metropolitan level together with all lower-rank industries. The largest-scale activities are localized in the major metropolis – non-ferrous metals, motor vehicles, tractors, and pens – or, with typewriters[1] in the primate city which contains all manufacturing industries.

In practice, some branches of industry are developed in particular locations, not for their accessibility to markets, but for their proximity to bulky or perishable raw materials. Agglomeration at localized sources of raw materials or nodes, especially in a capitalist economy, can provide exceptions to the urban-industrial hierarchy (Haggett, 1965, pp. 135–152). Technical and economic changes have been at work, however, to reduce their importance in the industrial spectrum. Even so, they point the way to two modifications of the model.

The first concerns the general distribution of industry. It involves the principle that manufacturing employment as a proportion of total employment varies directly with the population potential and with local urbanization, but inversely with distance from a metropolitan or major market centre (Duncan, 1959, p. 95). Generally, as urban population densities decline with increasing distance from a central city, there is a more rapid decrease in the intensity of manufacturing and in average plant size (Bogue, 1949). Naturally, the regularity and rates of decline differ with varying regional conditions. Nevertheless, this concept seems to fit reasonably well into Isard's modification of Lösch's market-area system (Isard, 1956, p. 275). However, there tends to be a concentration of large hinterland towns with specialist industries at distances of 30–65 miles from the metropolis. This allows for the 'off-centre' location of large enterprises (e.g. steel, oil-refining, vehicles, typewriters) away from, yet near and obtaining the benefits of the urban centre with a threshold equivalent to their market areas. This tendency is reinforced in a socialist economy by the planned spread of industry to

---

[1] Bain shows that typewriter manufacturing requires the least number of plants to serve the national market (Bain, 1954, p. 36).

achieve more equal regional opportunities. Closely-spaced settlements of relatively similar size and often with related specializations may form a 'dispersed city' (Burton, 1963) which, as a unit, contains industrial plants with market areas which equal the threshold of a single and much larger city. At greater distances from the metropolis (over 250 miles in the U.S.A.), much localized manufacturing can thrive in centres that enjoy monopolistic market conditions under the protection of the prohibitive freight-in costs that arise over long distances.

This modification underlines the probability of a more than proportional localization of sporadic industries in metropolitan centres and also in places where the access to such centres, the frequency and size of other urban centres, and the population potential, are the greatest. The explanation is simple. Ubiquitous industries are small-scale, simple activities that are closely tied to consumer markets and not to other industries (except cans and packaging for food-processing). The thirteen sporadic industries defined by Alexandersson (1956) comprise larger, more complex, and individually more specialized, activities that do demand spatial proximity to each other since 'the bulk of the composite market confronting industry is industry itself' (Kenyon, 1960, p. 168). The point is summarized succinctly by Chinitz and Vernon (1960, p. 130): 'The chain of processes between the raw materials and final products has been growing longer and longer. The tendency for any one plant in the chain to use materials which are already processed has continued to grow. As a result, and in increasing degree, plants hold down their freight-in costs by locating near other plants'. These tendencies in a capitalist system underline the validity of Lösch's 'city-rich' manufacturing belts. For even if his geometric arrangement of areas is unacceptable, the fact remains that sporadic industries (including much mining) grow and perpetuate their growth in the city-rich zones where 'the greatest number of locations coincide, the maximum number of purchases[1] can be made locally, the sum of the minimum distances between industrial locations is least, and in consequence... shipments... are reduced to a minimum' (Lösch, 1954, p. 24).

The second modification of the basic model concerns the relationship between the different types of industrial activity and the distribution and size of settlements. Rarely do towns of equal size possess equal advantages for the same industries, so that no rules can be laid down for other than ubiquitous activities. Each town specializes according to the natural resources that are available locally, and to the produce that its traders handle. Its economy, however, does not function in isolation; the growth of some kind of manufacturing in one town influences economic trends in neighbouring towns, and vice versa. Curry (1964) shows that the diffusion of industry results in specialization by towns and regions according to the arc-sine law. Industries in

[1] (and sales) – author.

town A determine about one half of the industrial structure of adjacent town B (the remainder being autonomous development), while town C, nearest neighbour of B, has one-third of its structure determined by the industries of A, one-third by those of B, while one-third is autonomous, and so on. The type of industries in towns $B, C, \ldots n$ are affected by those in the town of origin A according to the diffusion coefficient $d\text{-}k/t$, in which d and t are units of distance and time respectively. Whence the association and linkage of industries in neighbouring settlements and the evolution of areal specialization. Where the urban network is dense the precise directions of influence of the industry of a given town may reflect the operation of random human processes. In addition, Alexandersson's work shows that where towns are widely spaced there is a tendency throughout the hierarchy towards a wider range of industries; in contrast, where the network of towns is more intense there is a tendency towards a smaller range of industries and greater specialization in each town. This seems to reflect the greater spatial possibilities for inter-urban division of labour, the close spacing of towns offering inter-urban external economies of scale. It may also be explained by the tendency of industries with medium-sized plants (e.g. metal-working, textiles) to 'swarm' (Florence, 1963) in separate, though neighbouring, settlements to gain economies of close linkage. Further support is given, therefore, to the first modification of the model.

Individual branches of industry show varying relationships to settlement size. Mining, quarrying, and forest industries are more typical of smaller, more rural settlements; as the size of settlements increases so the proportion of extractive industries decreases (Winsborough, 1959). Manufacturing has a high index of urbanization, yet medium-sized settlements tend to have a higher proportion of manufacturing employment than larger centres (where services are proportionally more important). Both Winsborough and Duncan distinguish two groups of manufacturing which have divergent locational associations. Processing industries[1] have an urbanization index near to zero, indicating a closer correlation between industrial location and the distribution of the total urban and rural population; their importance shows a negative relationship with increasing settlement size. Fabricating industries[2] show a high index of urbanization and a clearly positive correlation with settlement size. The two extreme distribution patterns in the USA are provided by the scattered textile and furniture industries with low indices, and the highly localized, highly urbanized clothing and motor vehicle industries. Broad support for this model is given by Estall (1966, appendix II) through his index of relative dispersal for industries in New England. While some

---

[1] e.g. Forest products, metallurgy, food-processing, textiles, and other non-durable goods industries.

[2] Especially transport equipment (excluding motor vehicles), printing and publishing, electrical machinery, and metal products.

industries may diverge from this general pattern in other countries, there is no reason to expect any marked deviation.

MODELS OF AN IDEAL SPATIAL DISPERSION OF INDUSTRY

Akin to models of industries framed within the settlement hierarchy are those of spatial dispersion developed by the Dutch economists, Tinbergen, Bos, and Serck-Hansen. These models are treated separately since they are not based upon empiricism but suggest an ideal economic and social cost-benefit distribution of industries, accounting the different optimum sizes of plants and market areas in different industries. Tinbergen (1961) assumes a closed economy in which agricultural production and population are spread evenly. Industries are ranked into a hierarchy according to the number of optimum plants n in each so that $n_1, n_2, n_3 \ldots n_H = 1$. The model does not assign these to locations. By specifying the number of centres and their possible industrial compositions it calculates that combination of plants of various industries in various centres which would minimize production and transport costs (Tinbergen, 1961 and 1964). The hypotheses of the model are that each urban centre with an industry of a given rank h contains also all industries of lower rank, and that only one plant of the highest-ranking

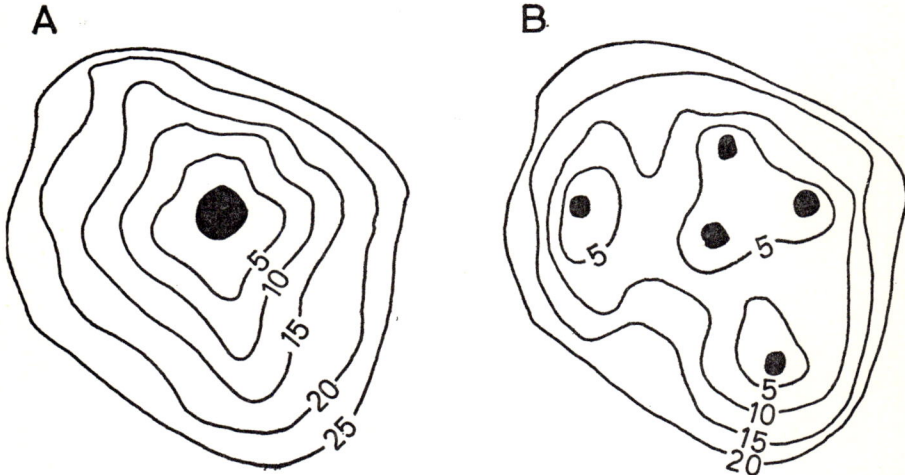

10.7 Dispersion and socio-economic costs. Isodapanes show that the costs of supplying food to one centre where industry is agglomereated may be higher (A) than to several dispersed centres with equivalent demand (B). The concept can be extended to include the social costs, for example, of commuting or of providing extra housing and social facilities in centres of large-scale immigration.

industry is located in any one settlement and that only this plant exports products to other centres and to the agricultural areas.

Bos (1965) modifies this model to include vertically-integrated industries, unevenly distributed population and demand (which are localized in either a large or a small number of settlements), and different settlement types. The aim is to minimize total location costs by dispersing industry to agricultural areas to reduce the costs of supplying agricultural produce to the industrial labour force; its conceptual implications are shown in Figure 10.7. This assumes decentralized marketing of food supplies. In part it suggests Tinbergen's idea (1962) that the full social costs of location (public services, commuting, etc.) should be paid by entrepreneurs to restrict agglomeration to that level at which social benefits equal social costs and to disperse industry so long as the social and economic benefits offset the social and economic costs of doing so. With a large number of centres, industries supply the needs of agricultural areas and they exchange products among themselves. The number of calculations becomes astronomical because plant numbers vary from industry to industry, a very large number of industrial settlement types are possible, and for each type continuous variations in their relative positions are also possible. Bos simplifies by assuming that transport costs to a large number of centres equal those to a circular market area and that plants in each industry have a certain *minimum*, rather than optimum, size. Industries are ranked as in Tinbergen's model but as their number increases the range of possible types of industrial centre increases rapidly. Further simplifications are made, therefore, by omitting the export industry and crosshauls of the same product. Bos concludes (1964, p. 69) that relatively high transport costs per ton-mile for agricultural products encourages decentralization of production over a large number of small centres as the optimum, relatively high transport costs for the products of high-rank industries (those with few plants) concentrate production in a small number of large centres as optimal, and relatively high transport costs for low-rank industry products indicates Tinbergen's system as optimal.

Further refinements are made by developing a simple programming model for several vertically-integrated industries and a small number of settlements. A regularly-shaped area is divided into equal hexagonal market areas (Fig. 10.8). The model indicates that the highest-ranking industries with one plant each will be optimally located in the most central sub-area (1). The regularity of the figure allows lower-rank industries to be located with more freedom since sub-areas 2–7 have equal accessibility to the whole market. The model is adapted here also to show the tendencies of agglomeration at central places (sub-area 1). The unequal number of industries in centres 2–7 represents the impact of unevenly distributed resources and of different optimum locations for different rank industries. It also indicates some social cost disadvantage in sub-areas 2, 4 and 6, and the disadvantage of transport costs in the outer areas (8, 9, 10). Bos stresses that the optimum national location of an industry is *not* only dependent upon transport costs for its own product and inputs, but also upon the costs of transporting the consumer goods required by the working population of the industry. 'For this reason it may *not* be optimal to locate industry 1, delivering its product only to industry 2 ... in the same centre as industry 2' (Bos, 1964, p. 85). Decentralization is thus preferable to agglomeration.

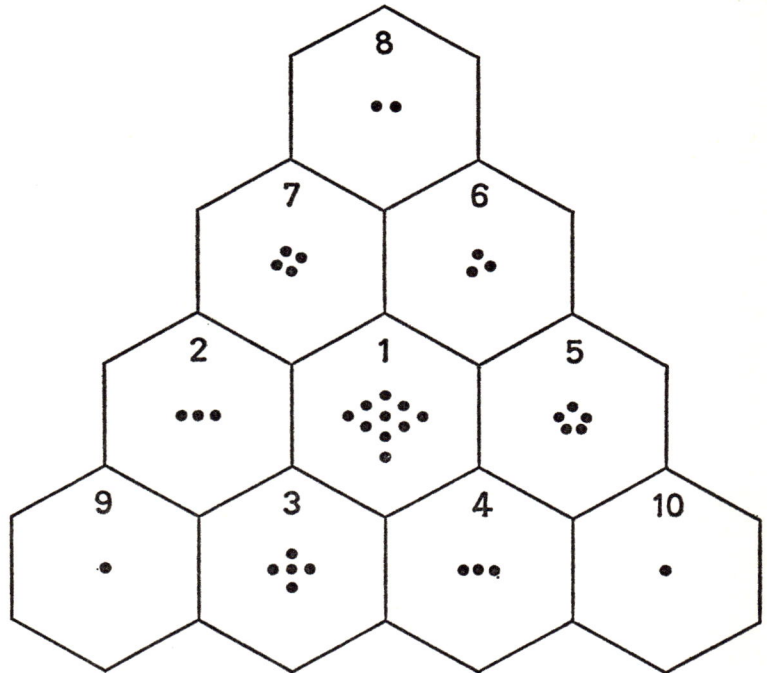

10.8 An idealized model of the spatial dispersion of industry.

Mathematical models of spatial dispersion on similar lines have been developed by Serck-Hansen (1961). This interest in decentralization mirrors the concern of the Dutch for the continuing agglomeration of industry and population in the western Netherlands between Rotterdam and Amsterdam, Tinbergen (1962), for example, asserting that there has been excessive localization of new industry in Rotterdam.

INDUSTRY THROUGH HISTORY

'Manufacture' ante-dates the 'industrial revolution' by many centuries, although most families produced their own clothing, shelter, furniture, and tools. In ancient Rome baking bread, fulling cloth, and making bricks or pottery were large-scale activities (Heaton, 1948, p. 52). A first model might show that, from ancient times until the mid-nineteenth century, the localization of world manufacturing in any era reflected the location of contemporary imperial or dominant power. Areas achieved industrial pre-eminence when they were governed strongly to give internal stability and independence, to exploit local resources and a favourable situation for trade, to expand political power to secure foreign sources of materials, slaves, or skilled labour in order to accumulate capital (Finley, 1965), and to secure adequate food for 'industrial' workers. When these advantages shifted to other areas, the industrial importance of a region waned. The rise and fall of empires and states, explaining the rise and fall of major manufacturing areas, resembles a series of waves rolling slowly across the world through time, from Mesopotamia through Rome and the North Italian city states to England. The process changes and quickens in modern times. A free-for-all is being replaced by civilization bringing the independence and integrity of all states with the knowledge to facilitate and perpetuate the economic and the geographic proliferation of industry. The world is now like a system of differently-sized canal locks or docks, each one a country protected by tariffs and regulations (the weir gates) 'trapping' industrial growth (the water level) to serve home needs (with independent water reserves representing domestic resources), yet inter-connected by channels of trade exchange (inflow and outflow) to balance supply and demand.

A second model might present the intra-regional location pattern before 1700. Capital being limited, processing equipment was usually simple, so that 'industrial' plants were small and *could* be scattered. That they *were* scattered resulted from the interplay of several factors. Energy sources (wood, wind, water, and human or animal muscles) and raw materials in use before 1700 or in the *eotechnic* phase (Mumford, 1934) (wood, agricultural produce, clays, sands, leather, skins, and wool) were ubiquitous. Labour inputs, in volume, time, and skill, were very high, and since the population was also the main market, industries were distributed according to the population and the availability of food. Transport was slow and very costly on land and this limited the area of material or food supply (or of 'putting out') to a maximum radius of a day's journey. Regional differentiation arose chiefly through the growth or lack of growth of industries that processed valuable materials (wool, metals) and produced valuable products (cloth) that were transportable by sea or from a large land area.

The 'industrial revolution' originated in England in the eighteenth century. Adam Smith gives a kaleidoscopic model of its early selective effects. Industrial progress was greater where more advantageous and cheaper transport was available (then water and sea transport). Areas which lacked these facilities, and depended upon land carriage, could develop industry only 'in proportion

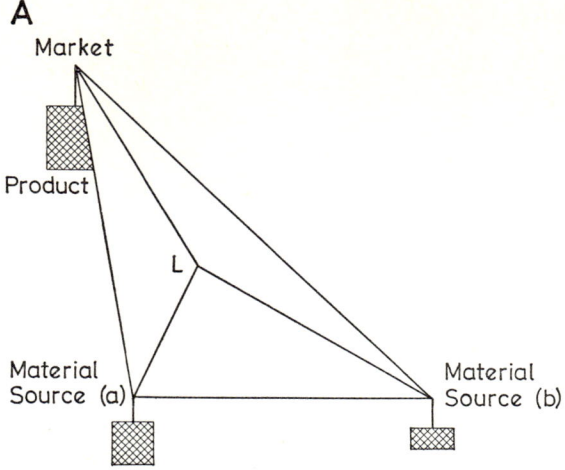
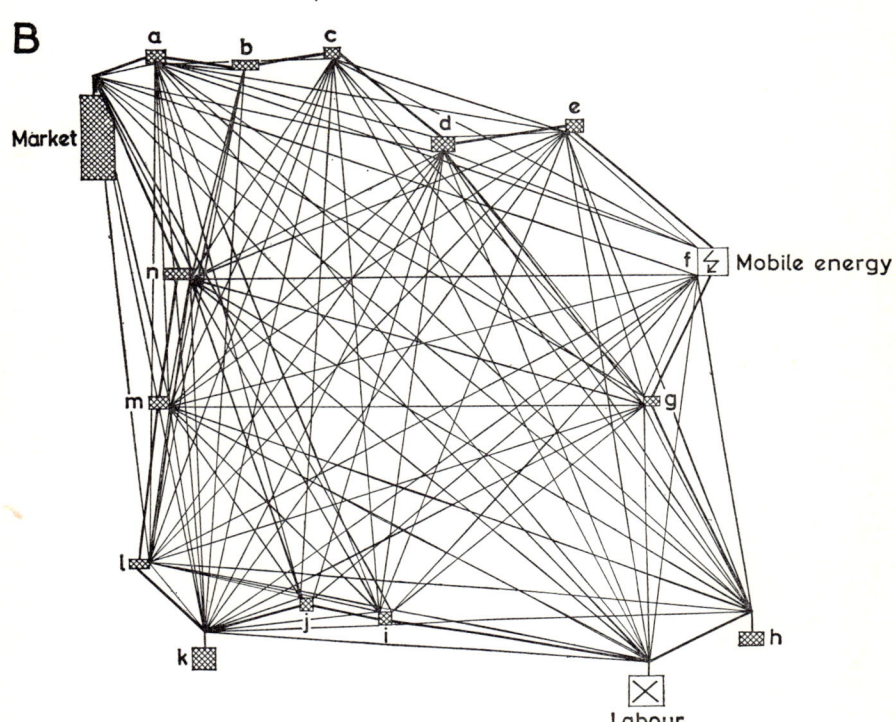

10.9 Location figures: An historical transformation. Weber's location triangle (A) is superseded by a polygon (B) which includes location factors which have become important since 1909.

to (their) riches and populousness' unless they produced goods 'whose price was very considerable in proportion to their weight' (A. Smith, 1776, p. 81). Increasing technical division of labour and improved transport were encouraging areal specialization according 'to the diverse aptitudes, capacities, and resources of different peoples and places' (1808, p. 14). And certain industries were becoming urban activities while others were tending to remain rural activities.

Radical changes occurred in the location of industry as a result of the new technology. Isard (1956) has used a modified Launhardt-Palander model to demonstrate this. The increasing use of coal as a raw material, fuel or power source made possible, and localized, the rapid expansion of existing industries,[1] on or in proximity to the coalfields. Many inventions initiated the growth of new industries[2] largely in the same areas. Weber's model describes the situation well (Fig. 10.9A) since decisions were made by small firms (Goldman, 1958, pp. 91–92) and few industries used more than three raw material inputs. As techniques were wasteful and the major input – coal – was very bulky and partly or wholly 'disappeared' in manufacture (Wrigley, 1961, pp. 3–9) the combined weights of inputs far exceeded the weight of given outputs. The power of the coalfields to localize industry was further strengthened by the occurrence of the initial sources of other raw materials in association with coal, by the expansion of local composite markets as industries and urban populations grew and by the development of infrastructure and skills which encouraged intertia. This model summarizes Mumford's *paleotechnic* phase. Unless Weber's location polygon is treated as an intraregional figure, therefore, all the weights for many industries pulled on one angle. The location figure was cast into the inter-regional context by 1909, however, as canals and railways 'extended man's mastery over space' (Launhardt, 1885, p. 206) to provide distant locations and nodes with the opportunity to use cheap coal, and as the exhaustion of local resources (other than coal) enforced a dependence upon more distant supply sources. These and other changes were beginning to favour market-orientation (Fig. 10.9A).

Continuous technological progress has shifted the balance more from the materials (especially coal as *the* source of energy) to the markets in the present century; Keir (1921) early recognized this trend in the United States. A modified location figure is presented (Fig. 10.9B) which embodies the key factors of change. These are part of the *neotechnic* phase. The larger size of the figure indicates the ease and economy of linking more distant locations as a result of cheaper and faster transport; the reduction in transport costs (especially for raw materials) has been greater than the reduction in production costs. The smaller symbols for raw materials (k and h) indicate the great reduction in the weight of raw material inputs to product outputs. The

---

[1] These included iron smelting, textiles, pottery, bricks and brewing.
[2] These included metal manufactures, machinery, and transport equipment.

differential costs of moving raw materials are now far less at alternative locations. Other symbols are introduced to denote new location factors. Symbols *a–e*, *g*, *i*, *j*, *l* and *m* represent the increasing number of intermediate 'pure' manufactured materials (metals, synthetic fibres, plastics), tools, parts, and components, which are important links in the industrial chain. Often of high value or light in weight, individually these are not important location factors, but collectively they may demand the spatial proximity of interlinked plants to each other. The corollary is the growth of diverse finishing and

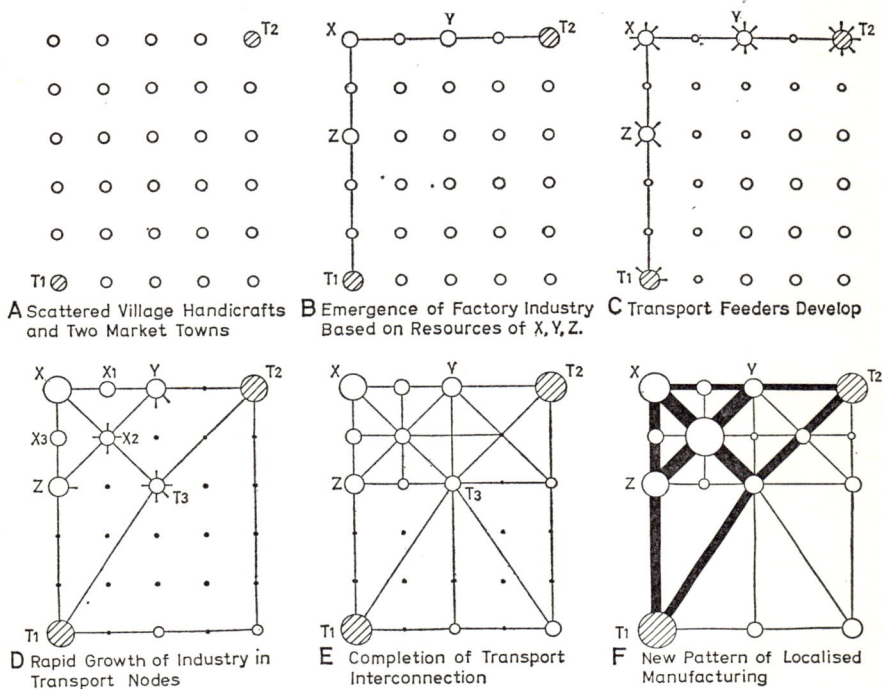

10.10 The emergence of a modern industrial location pattern.

assembly industries which use mainly power and not fuel for heat (Dales, 1953, p. 182) to produce valuable, fragile, bulky, or weight-gaining products. Generally, the freight rates charged for products are high and reflect 'what the traffic will bear'. The increasing complexity of mass-produced articles requires greater contact between producers and consumers and the provision of after-sales service both in the industrial and consumer markets (Chisholm, 1963). For these reasons, the market has become by far the most important single location factor (Fig. 10.9B). A new symbol ('mobile energy') stresses that new energy sources (oil, gas, atomic and ordinary electric power) have made coalfield locations unnecessary, and in consequence, have enhanced the locational importance of market, labour or other locations. The last new symbol (X) represents the growing importance of social factors (personal preference or planning needs), and of labour, particularly the availability of semi-skilled or high skilled personnel; in part, these factors are beginning to underline the importance of locations with pleasant amenities. This trend may be expected to continue as labour becomes more mobile. Automation is both tending to confirm these trends still more (Osborn, 1953) and to increase the *flexibility* of location choices that is indicated by the multitude and range of intersecting lines in the figure.

A model of selective locational industrialization is developed by Pred (1965) to show how progressive agglomeration of manufacturing arose in a few centres in America between 1860 and 1910. Initial advantages emerged at centres where entrepreneurs developed rational or random new industries, because these industries set in train multiplier effects which stimulated invention and attracted immigrants who brought new skills and other industries. Once these centres were thriving, three factors ensured their further growth and discouraged new development in inefficient or non-producing centres: the improvement of transport, the lowering of production costs, and the growth of large combines which could effectively prohibit new entry to an industry by would-be competitors.

The evolution of a modern spatial pattern of industry in a sequence of stages from the handicraft era is suggested in Figure 10.10. Conceptually, the model is an outgrowth of, and an analogy to, the idealized pattern of transport development in an under-developed area that was worked out by Taaffe, Morrill and Gould (1963). A series of evenly-spaced villages (Fig. 10.10A) are locations for small handicrafts in a region which supports two market towns, T_1 and T_2, where more crafts are localized. Fuel and raw material resources are discovered at X, Y. and Z (Fig. 10.10B) which, as a result of new techniques, provide the basis for growth industries. Mines are opened at X, Y and Z to supply materials and fuel to processing industries that develop at T_1 and T_2. Feeder lines develop from all five centres (Fig. 10.10C) and bring competition into the nearest villages, causing handicrafts to contract. The emerging transport network (Fig. 10.10D) intensifies the market area for fuel and minerals causing expansion at X, Y and Z, and for the products of industries located at T_1 and T_2. Manufacturing expands in all these centres as the influx of country people swells their labour forces and consumer markets, and as new industries develop to serve the growing needs of the mines, new factory processes, transport facilities, and urban population. Crafts virtually die (Fig. 10.10D), but some villages become embryo factory locations as transport connections create new nodes (e.g. T_3) and spread the effects of industrial growth around existing nuclei (X_1, X_2, X_3). Existing centres (T_1 and T_2) become more important as the transport network is completed (Fig. 10.10E), but some initial advantages combine with transport advantages to cause greater industrial growth in certain centres. A modified location pattern emerges in Figure 10.10F based on the intensive interlinkage of industries in the resource areas with fine assembly advantages leading to the rapid growth of the most modern industries at X_2, while T_1 maintains its supremacy through inertia and its adaptable labour force. Some theoretically good nodes remain unimportant because of the difficulties of entry for would-be competitors at those locations. A highly differentiated spatial pattern with localized industry therefore replaces the even distribution of activities characteristic of the pre-industrial era.

SOME OBSERVATIONS ON MONTE-CARLO AND MARKOV CHAIN MODELS

The spread of industrial inventions is a promising field of study using Monte-Carlo simulation (Hägerstrand, 1963) and Markov-Chain models (Brown, 1964; Clark, 1965). Certainly the large number of inventions that originated in Britain (and especially in the Midlands and the North of England) after 1700 gives historical confirmation of Hägerstrand's thesis that the amount of additional invention (and thus potential proliferation of industry) is proportional to the amount of existing innovation. Both models stress the prime importance of neighbourhood: an innovation tends to appear near the place where the innovation already exists while the probability of the appearance of the innovation decreases with increasing distance from the existing phenomenon. The models could be applied to trace the diffusion of the stationary steam engine, or any other invention, either within Britain or other countries, or from Britain to regions overseas. This last is the more fascinating problem since it involves certain modifications of Hägerstrand's model. The intensity and the ease of contact and trade (the information field) were far more important than distance in diffusion overseas because resistances from barriers were strong in nearby Europe whereas few barriers were encountered in America. Differences in language, culture, economy, laws, policies, and patents reduced the ability to communicate and to apply ideas and acted as multiple brakes upon invention diffusion in Europe. These barriers clearly delayed the spread of the steam engine from Britain across Europe in both time and distance. Invented by Watt in England in 1768, the efficient steam engine began operating first in Belgium mainly through English enterprise (1780's). It appeared later in France and Germany (1790's), in Austria-Hungary in 1816, and in the Balkans in 1835 where English initiative was again important (Bićanić, 1951, p. 212).

Although they are still in their infancy, Markov-chain models seem to have most potential in tracing trends in industrial location or employment (Williams, 1965) which result, for example, from progressive technological improvement changing scales (and thus distribution patterns) of industrial plants. Markov-chain analysis may also lend itself to problems of industrial migration.

STRUCTURE, PROCESS AND STAGE

Economic growth occurs as agriculture and primary industries decline in importance relative to the growth of secondary and tertiary activities. This implies not only a change in industrial structure through different stages, but also changes in the location pattern. W. M. Davis's cyclical model of landform development suggests itself here as the analogue for a controversial model of industrial structure, process and stage.

Primary activities emerge first in any area. This is the *infancy* stage when industries are pre-eminently extractive and raw material-oriented. The cycle enters *youth* with the growth of manufacturing, especially textiles, to supply the consumer market. *Adolescence* is achieved when basic industries develop to serve producer and consumer markets. Industries tend to become localized near fuel, power and raw material sources and in good nodes as bulk transport becomes important. An area reaches *maturity* when 'it has experienced large-scale development of manufacturing . . . and associated economic development over many decades' and has 'evolved a deep-rooted and highly complex system of industries and services, many of which are inter-related in a variety of important ways' (Estall, 1966, p. 3). Maturity not only suggests a broad and balanced industrial structure, it implies the existence of developed skill and infrastructure embodying the ability to adapt to changing technological and economic conditions by developing regenerative industries – engineering, chemicals, and machine tools or, as in New England, electronics and aircraft (Estall, 1963 and 1966). In Old England no better example can be found than the West Midlands, the cradle of the industrial revolution, where inventiveness has stimulated progressive adaptation to new industries. According to Kenyon (1960, p. 65) the Paterson-Passaic district (New York Metropolitan Area) 'has evolved through several complete cycles since its inception: cotton textiles, locomotive production, silk (and wool) textiles, and lately aircraft manufacture'. An area with a narrow industrial base may find spontaneous adaptation difficult; in this case the area is *immature*. The inception of marked decline indicates *old age* (McCaskill, 1962, pp. 143–169).

Some areas never pass through this cycle. The rise and fall of mining, leaving ghost settlements, is an example of *infant mortality*. Other areas have succeeded in reaching youth with basic industries – for example, the Kielce area of central Poland with metallurgy – when technological or political changes initiated their *premature death*. Nowadays, however, better knowledge and an array of economic and other stimulants (the construction of 'location leaders', communications, or redevelopment) makes industrial *rejuvenation* more likely. Moreover industrializing regions can now progress through the earlier stages of the cycle in several years rather than in several decades. A new stage is now appearing in developed countries where automation in both the secondary and tertiary sectors is initiating a shift again in favour of capital goods.

THE INTERNATIONAL DISTRIBUTION OF INDUSTRY

The world industrial location pattern is uneven and bears no close relation to the distribution of either natural resources (even coal) or population; it can be correlated only with the regional levels of economic development. This suggests a model that is an analogue of Berry's model of developed, developing and underdeveloped countries (in Ginsburg, 1962). Marx (1850) emphasized that localized capital, backed by imperial power, was the most powerful single factor in localizing industry in a small part of the world in the nineteenth century. As inventions revolutionized production and transport, capitalist economy replaced the orientation of industries in Britain to national resources and markets by increasing dependence upon material supplies (except coal) from, and export of products to, the whole world. This suggests a world interaction model, in which capital localization combined with coal in the 'mother' country to localize there industry which depended for resources or markets upon areas (colonies) which were denied manufacturing. The rise of new independent states began to duplicate the process, but industry remains confined largely to former imperial and to capital- and resource-rich countries. As a result, industry is very highly localized in the northern hemisphere, the world manufacturing belt representing a diamond that is oriented from east to west and that is split well to the west of centre by the North Atlantic Ocean.

In detail the world industrial pattern today bears the marked imprint of the nation protecting its own industries and markets by tariffs. This is promoting the international deglomeration of industry, while attempts towards greater international integration are tending to work in the opposite direction. Large states offer large markets for large-scale industries and in the absence of particular planning controls, except where huge distances make regional autonomy an economic proposition as in the U.S.S.R., industries tend to localize. In contrast, they are dispersed in a similar geographical area that is politically fragmented. Figure 10.11 demonstrates this contrast with reference to the textile industry in the U.S.A. (with bordering areas of Canada) and in a comparable area of Europe. Greater economic development (arising from enlarged markets and exchange among integrated territories) according to

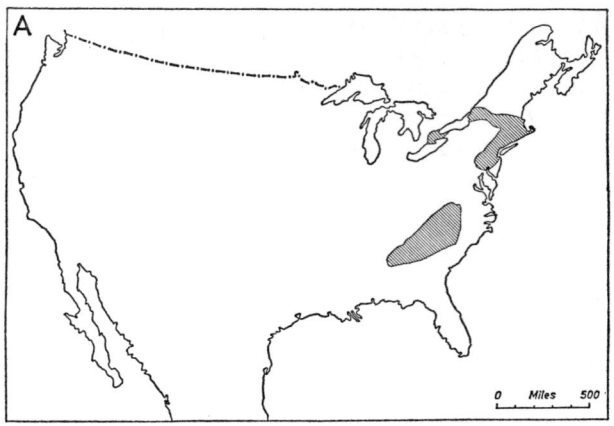

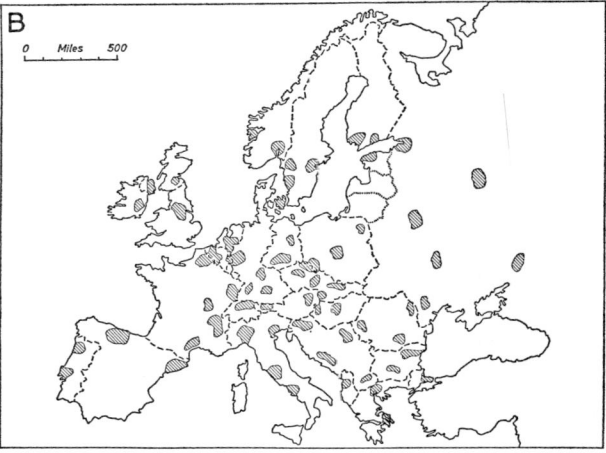

10.11 The comparative effects of political unity and of fragmentation upon industrial distributions: textile industrial areas in the United States (A) and a comparable area of Europe (B).

Myrdal's 'spread effects' means more industries and more plants that *can* be located more evenly. In practice, greater integration calls forth large-scale activities which tend, if uncontrolled, to cause greater regional differentiation than before by localizing in those places which are the most accessible to the enlarged market and which provide most linkages for modern industries. Integration in the European Economic Community, therefore, is consolidating industrial employment (Chisholm, 1962, p. 10) in a 'T-shaped complex of industrial areas that is likely to reap the greatest benefits from Common market conditions' (Wise, 1963, pp. 135–136). This is confirmed by Leviquin (1962) who shows that most of the 800 new enterprises that were developed in the European Economic Community between 1959 and 1961 were located in existing industrial agglomerations. Where economic development is planned, as in COMECON, integration can accelerate the economic 'take-off' of backward areas. The development of new industrial zones along the Vistula waterway in Poland provides an example.

INDUSTRY IN THE NATIONAL SETTING

Theorists from von Thünen to Tinbergen have framed location models in the abstract context of the isolated state. The assumption here, however, is an independent nation which has limited resources and which engages in trade to secure imports of deficient foods, materials, and manufactures. Energy and raw material resources are unevenly distributed within the country, being more contiguous than coincident with the better agricultural areas. In response, the distributions of population, settlement and transport facilities are uneven, tending to be concentrated where mineralized and good agricultural zones overlap or adjoin. Land use becomes less intensive with increasing distance from such concentrations. Given these assumptions, a number of general industrial patterns emerge in all nations.

High transport costs make perishable, bulky, or low value materials and products 'immobile' (Tinbergen, 1962), so that food processing, timber, building materials, paper, and packaging industries are located in most nations and in some regions within each nation. Sporadic industries develop to process localized domestic resources provided that suitable types and quantities of energy (which also condition industrial structure) are available (Dales, 1953). Otherwise, distributions accord with the principles established earlier and with the regional 'climaxes' to be discussed. Every country contains areas with little industry which contrast with those where sporadic industries localize to share common links or facilities, to use associated resources, and to serve associated composite markets. The nation's level of economic development will condition the size, extent, and intensity of such localizations through the degree of inter-industry integration of vertical, diagonal, convergent and indirect kinds (Florence, 1948); the transport pattern conditions their shape and alignment. Models of the shapes and sizes of sample 'manufacturing belts' are given in Figure 10.12. Two special types of industrial centre – the metropolis and the ports – may lie within, on the periphery of, or near this belt and extend or intensify its economic attraction, or they may be located well away from it and act as 'counter-magnetic poles of growth'. Whereas eccentrically-situated import/export-oriented industries

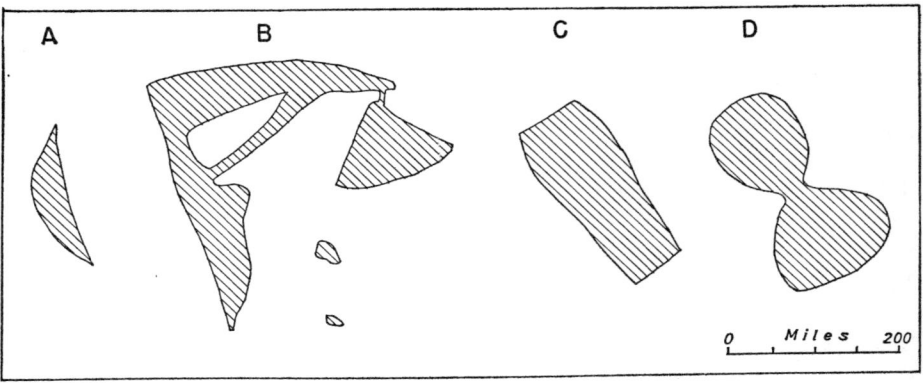

10.12 Sample manufacturing belts.
A: The 'crescent moon' in the western Netherlands, comprising an area between Eindhoven, Rotterdam and Amsterdam.
B: The German 'gallows', including the north–south Rhine axis, the west–east Ruhr–Magdeburg axis and the triangular area of industrial concentration in Saxony.
C and D: The manufacturing belt of England represented as a 'coffin' and an 'hour-glass'.

do develop at the ports where there is a large break of bulk, they do not develop at the land frontiers (of either seaboard or land-locked states) where the same transport facilities continue with no break of bulk. Strategic considerations may also discourage such frontier development. Yet border locations will attract industry if resources are localized there (e.g. coal in Western Europe) or if differential tariffs operating along the frontier offer opportunities for serving two or more national markets instead of one (e.g. the Maggi and Knorr food-processing combines along the German-Swiss frontier).

REGIONAL INDUSTRIAL 'CLIMAXES'

The reality of intricate spatial interdependencies lies at the root of an increasing emphasis in geography upon 'complex' economic regions, central place studies, and city regions. Presented here are some complementary stochastic models of regions with different spatial structures of industry. Hitherto this problem has received most attention in studies of the 'territorial-production complex' in the Soviet Union, chiefly by Kolossovsky, Saushkin, Alampiev, and Pokshishevsky. The original idea occurs in Weber's work (Friedrich, 1929, p. 196) and is developed broadly by Chardonnet (1953). Recently Isard has applied the notion to an oil-refinery, petrochemicals, and synthetic fibres complex in Puerto Rico (Isard et al., 1959). The models that follow are constructed on the premiss that, given the capital and the market demand, differences in regional industrial patterns are a function of the character, size, and variety of available natural resources and of human resourcefulness. They resemble Kolossovsky's eight 'energy-material-manufacturing' cycles (Kolossovsky, 1958). These are modified, however, to encompass variations in population distribution, labour supply, and composite market demand. None should be interpreted rigidly, either in structure or in spatial distinctiveness (Lonsdale, 1965). Rather they suggest regional 'climaxes' of sporadic industries that can be modified by external factors in the same way that vegetational 'climaxes' can be.

The simplest model depicts the scattered or localized extraction of valuable minerals in sparsely-populated regions which suffer from isolation and harsh physical environment. Only localized concentrating, refining, and power plants are important. The second model comprises temperate forest areas where dispersed saw mills and more localized pulp-paper, chemical distillation, and veneer factories are typical; or tropical areas with pharmaceutical, chemical, and rubber industries. In either variant, plants tend to localize in forest margins between the material sources and the markets. These regional models may overlap with each other and be coincident also with sources of cheap hydro-electric power and an electro-metallurgical/electro-chemical complex. If regional demand is large enough, specialist machinery industries may develop within any of these 'climaxes' to meet their demands.

The next two models relate to predominantly farming areas, although 'islands', 'inliers', or 'outliers' of other mining and manufacturing complexes may occur within them. The fourth model, for a commercial livestock region, comprises: widely scattered dairies, more localized leather tanning, meat-packing, and woollen textiles industries; and labour-intensive textile, engineering, or specialist manufactures where labour surpluses exist. The spatial structure of industry in the fifth model of a commercial mixed-farming area is more intricate, as an 'organism' of inter-related plants. Scattered factories process food crops, cotton, tobacco, and oil-seeds. Other industries develop alongside processing plants to use by-products (e.g. for livestock-feed), to supply the processing industries with cans, boxes, machinery from more localized plants, to serve farming needs in fertilizers, wire, tools, machines, and tractors (from plants that are centrally located to one or more farm regions), to serve the farm community in household goods, and to employ female labour in textile, footwear, printing, and other light industries.

The most dominantly industrial landscapes are associated with the bituminous coalfields and their related metal and non-metallic mineral resources. The sixth model, then, comprises a complex of coal-mining, coke-chemicals, gas-chemicals, electric power, heavy metallurgical, glass, and cement industries; the bulkiness of materials and of the products also ties heavy engineering industries closely to the coalfields. Less localized are industries using lighter by-products, or power from the coalfield, and employing female labour – plastics, artificial fibres, and textiles. Large fields give rise to major urban concentrations and transport networks, so attracting consumer-oriented and transport equipment (e.g. locomotives and ships) to nodes around the coalfields. Somewhat similar complexes can develop in regions with large ore-fields. Small or brown coalfields attract few industries other than electric power, gas and chemicals. The seventh model is typified by two variable structures: first, iron-ore mining, iron and steel production and manufacture, with coke-chemicals and second, non-ferrous metal-mining and processing with sulphuric acid and fertilizer chemicals integrated with smelters to use by-product gases. Oil-drilling and refining is the chief component of the eighth regional model which comprises highly localized petro-chemicals and synthetic fibres plants, associated with decentralized oil and chemical processing equipment, textile-processing, and power industries. Gas may be an important energy source for other manufacturing industries (Kortus, 1963).

The last two models concern areas where human resourcefulness is most important: the ports and the metropolis. Large ports have a triple spatial structure. Industries that process bulky, divisible, imported materials or weight-gaining products[1] are located along the waterfront. Industries that either serve, or use the products of, the first group are located near the port.[2] Light industries that are labour-oriented are located in the suburbs. The metropolis has an industrial structure which responds to its administrative, educational, and commercial functions, large labour supply and large and varied composite market: clothing, luxury goods, office and business machinery, scientific instruments, vehicles, electrical equipment, light chemicals, furniture, pharmaceuticals, and foods. If the metropolis is large, then substantial cost differentials exist between alternative locations within the city, especially with regard to land, labour, and transport. The model of the metropolitan spatial structure of industry (Fig. 10.13), therefore, comprises differing localizations of associated industries in different optimum locations (Haig, 1927). Central locations are occupied by industries (Group A

---

[1] For example, grain-milling, sugar-refining, public utilities, oil-refining, metallurgy, cables, vehicles, shipbuilding, and heavy engineering and machinery.

[2] These include food manufacture, light chemicals, furniture, newspapers, engineering, cans, Packaging, and machinery.

in Fig. 10.13) in which the need for the best access to skilled labour from the whole area (e.g. instruments, tools, printing), to the central business district (e.g. clothing, office machinery), and to the whole urban market for distribution (e.g. services, newspapers), offsets high land costs (Stefaniak, 1963; Lowenstein, 1963). The 'swarming' of closely associated activities in small

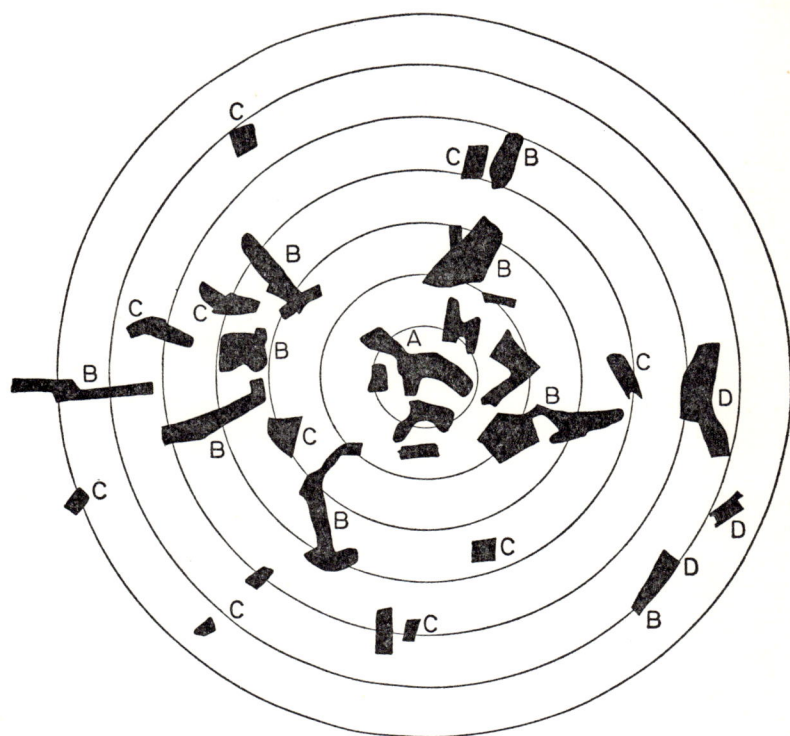

10.13 A model of the spatial industrial structure of a metropolis. This model is based on a map of the industrial areas of London in Martin (*1964, p. 122*). The radii are at intervals of 2½ miles. This should be compared with Lowenstein's (*1963*) synthesis of the intra-urban industrial patterns of sample American cities. Note the 'wedging' effect of transport lines on the distribution pattern, offering greater possibilities in the inner or middle ring for external economies and linkage.

enterprises gaining external scale economies in the central district may give rise to sharply-defined industrial quarters, e.g. the clustering of metal plating, printing machinery and parts, printing works, typesetting, lithographic plate-making and ancillary trades in the Clerkenwell–Fleet Street area of London (Martin and Hall, in Clayton, 1964, p. 34). Larger enterprises seek cheaper land, a good location for material assembly and product distribution, and access to unskilled or semi-skilled male and female suburban labour. They locate along radial or 'ring' transport arteries and include port industries (Group B in Fig. 10.13), food manufactures, electrical, engineering and light industries (Group C). On the outskirts are industries which either require large land areas for assembly-line production, for stores and for waste, or which are dangerous and obnoxious; this group (D) comprises vehicles, heavy engineering, oil-refining and heavy chemicals, metallurgical and paper industries.

Chardonnet suggests the existence of three other regional complexes: autarkic, colonial and strategic (Chardonnet, 1953, pp. 168–182).

Model urban centres possess all, or some combination of, industries appropriate to their hierarchical importance in the region in which they are situated. 'Exceptional' centres occur where random industries develop, or where two or more regional types coincide or adjoin. Set theory and Venn diagrams may be used to clarify the typological 'position' of any city.

INDUSTRIAL INERTIA AND MIGRATION

Industrial patterns are never static, but most areas provide examples of both industrial *inertia* and *migration* (Costa Santos, 1961, pp. 7–9). A model of *inertia* demonstrates that the advantages of present location far outweigh the advantages of relocation through high fixed costs of plant, the existence of a large skilled labour nucleus which facilitates high quality output and the application of new techniques *in situ*, the local development of good infrastructural facilities, and the existence of complex interlinkages with other local industries that makes the movement even of the more 'footloose' industries a major operational cost risk. 'The power of a locality to hold an industry ... greatly exceeds its original power to attract. The new locality must not only excel the old, but it must excel it by margin enough to more than offset the resisting power of the matrix' (Ross, 1896, p. 265).

While several models of population migration have been developed (Haggett, 1965, pp. 35–40), industrial counterparts are virtually lacking. The main reason lies in the different 'migration' processes involved. Rarely does industrial migration, unlike population migration, involve a physical movement from one area to another. Shifts in plant location were partly of this character in the U.S.S.R. during the Second World War, and in Yugoslavia after the war (Hamilton, 1963, p. 103). Usually 'migration' results from differential rates of industrial growth which is accentuated if stagnant or declining and expanding industries are localized in separate areas. The exhaustion of some resource may initiate the contraction of an industry in one area, while the existence of that resource elsewhere may stimulate the growth of the industry in another area. Some traditional coalfield industries, particularly iron and steel, migrated to other areas as long-distance import of low-grade materials (which were formerly available in the coalfield area) became necessary. Technological change is a major cause of migration. Once barren of industry, many coalfield areas became major industrial regions, while manufactures in forest and upland areas declined. General technological progress – at once encouraging concentration and dispersion (Gravier, 1954) – has progressively freed industry from the tight locational bonds of the water-power site or coal-mine. Canals and railways permitted industrial growth at greater distances from mines, yet they still tied factories to the canal bank, waterfront, or railway siding. In the twentieth century motor transport and electricity transmission have further unfrozen industrial concentrations. A 'centrifugal-dispersion' model is implicated here; it works against the 'centripetal-concentration' model that is induced by the economies of scale and integration.

The development of branch plants plays an important role in migration and reflects the economies that accrue in marketing expanded production from, applying new techniques in, or obtaining a stable labour force at, a

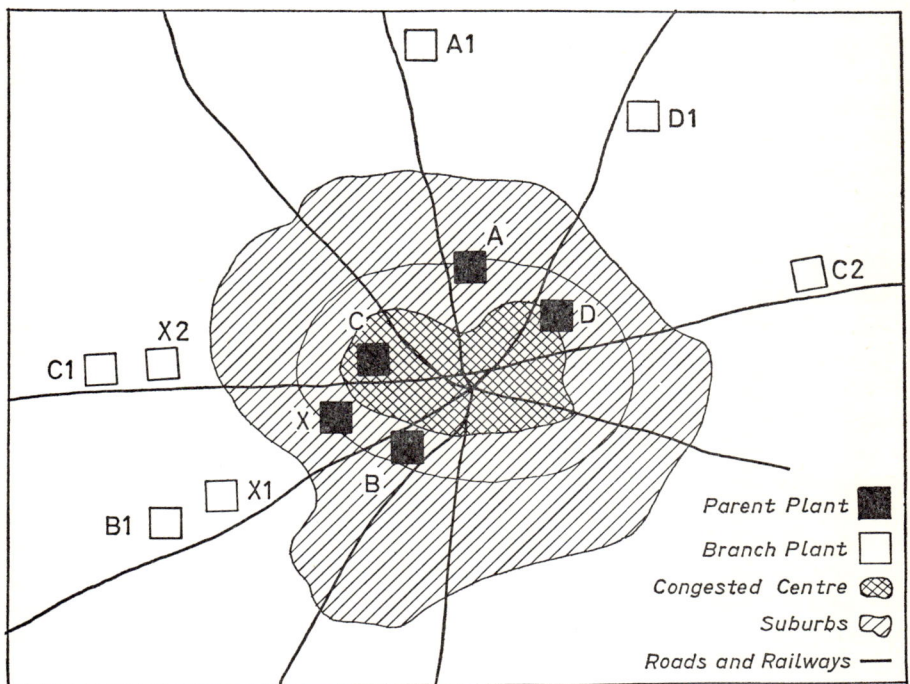

10.14 Radial migration from a large city. Branch plants are established in clearly radial fashion outside the city by firms with parent plants within the city. The foundation of a branch C2 by C is given as an example of what is *unlikely* to happen as a result of inconvenience in maintaining contact.

separate location (McLaughlin and Robock, 1949; Greenhut and Colberg, 1962). Analyses of industrial change in and around Greater London (Martin, 1964 and 1966; Keeble, 1965) show that the establishment of branch plants outside, by firms with parent plants inside the conurbation contributes significantly to centrifugal movement. This suggests a 'radial migration' model (Fig. 10.14) in which inner or outer suburban firms construct branches outside the conurbation which are accessible by the line of 'least transport effort'. This involves the shortest and most direct exit through the suburbs; branches are rarely located where links with parent plant require transport across the congested central areas. Such 'migration' may also be part of the 'suburbanization' of industry (Scott, 1963). In detail the process is more complex than these models imply. Luttrell (1962) shows how variations in transport, administrative and overhead costs between parent and branch plants condition the type of branch plant that is developed at a given distance from the main factory. A model can be constructed to distinguish three types of 'industrial distance-migration'. Branches that manufacture components for parent plants in 'footloose' industries (which serve a national market) are located, for easy control, within a radius (in Britain) of 60 miles of the main factory usually to tap available labour supplies. Intermediate plants, with some separate output, are less dependent upon the main plant and are located within a radius of 150 miles of it. Completely self-contained branch plants are established at greater distances, especially to supply growing regional markets more efficiently.

Much more research is needed before satisfactory models of inter-regional industrial migration can be presented. It seems that such models must embrace: the spatial effects of technological change; the 'regional elasticity' of industrial growth, which depends upon the importance of first, the expansion of existing plants as compared with the construction of new factories and second, the large-scale integrated plants and 'immobile' industries as compared with smaller scale or divisible plants and 'footloose' industries; comparative regional trends in population, labour and market conditions and in resource availability; and the inter-regional effects of these changes. Fuchs (1959) considers that the chief component should be the differences between regional industrial structures. Vanhove (1961), however, attaches importance to regional labour reserves, wage levels and industrial diversification, while Garrison (1960) suggests a three-region industrial migration model using Lövgren's input-output technique (1957). Markov-chain models, possibly when combined with input-output analysis, may offer a solution of the problem.

INTRA-REGIONAL CHANGE AND THE MULTIPLIER MODEL

Every factory movement and expansion sets in motion a chain reaction of corresponding scale that can be embodied in a 'multiplier' model. A useful conceptual and operational tool, the multiplier model can demonstrate the direct and indirect regional economic effects of the construction of a new, or closing of an old and important industry and the overall and selective effects on industry of general economic and social changes. The model provides an interesting example of usually positive direct and indirect relationships with positive feedback, although there may be negative side-effects where substitution is involved. Barfød (1938) showed that, by closing an oil factory employing 1,300 workers in Aarhus, some 6–10,000 workers in industries and services would become unemployed in and around the city. The total and selective regional growth effects of the location of a steel combine have been calculated by Isard and Kuenne (1953) for the New York-Philadelphia area. Pred's multiplier model (1965, p. 165) is adapted to describe graphically the long-term local impact of this project (Fig. 10.15). The same total or same

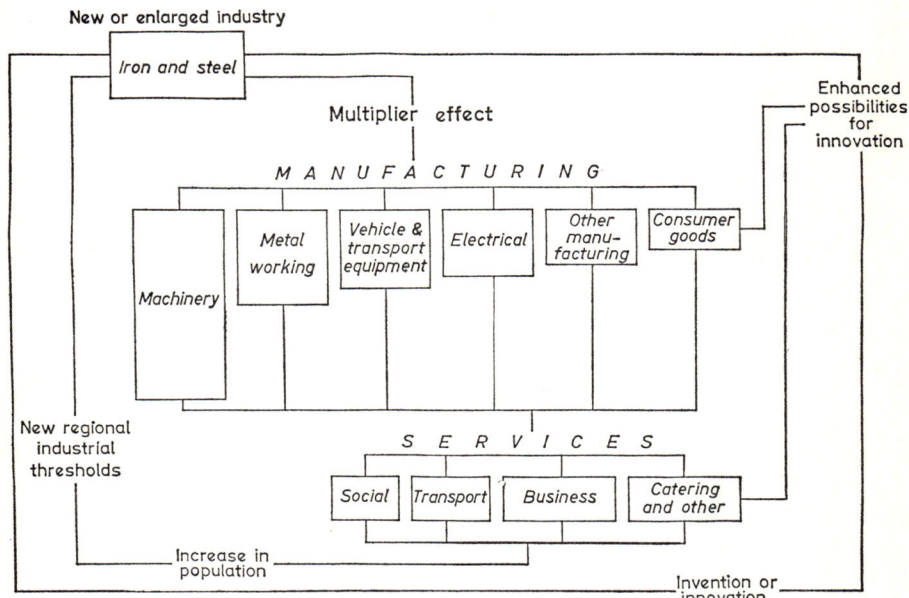

10.15 The multiplier effects of the location of an integrated steel plant in the New York-Philadelphia area. No time period is laid down for the achievement of these changes. The increases in employment (shown proportionately by the size of each box) in each activity were calculated for six rounds of chain expansions.

particular impact should not be expected in another environment, especially where capital is more scarce and less mobile and where industry and transport are less developed. The model emphasizes the importance of location leaders in shaping regional industrial structure giving greater substance, therefore, to the models of regional industrial complexes suggested earlier.

Broad socio-economic changes also induce differential changes in various industries. Hirsch (1959) asserts that a $1 million increase in final demand in the St. Louis Metropolitan area would induce four times more growth in

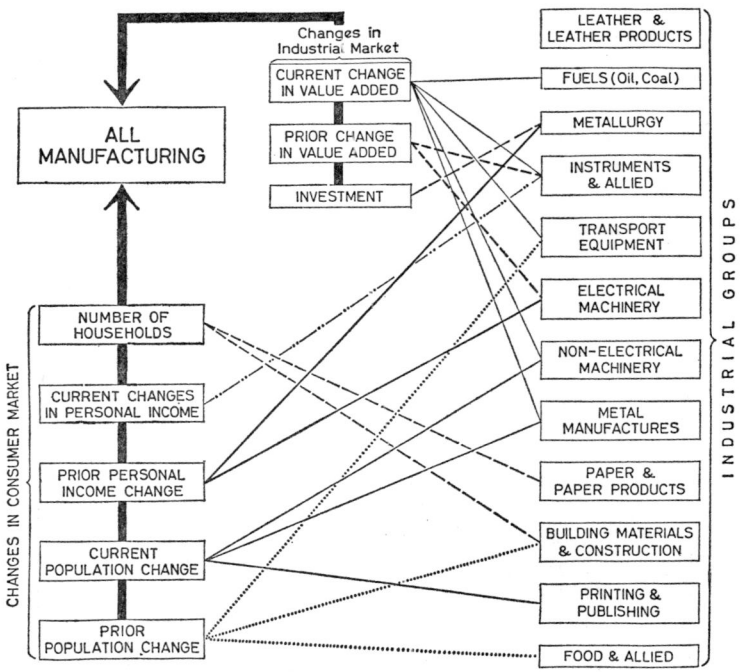

10.16 The more important relationships between eight types of change in intra-regional demand and thirteen manufacturing industries.

employment and sale turnover in printing as in petroleum industries. By using input-output techniques he demonstrates that activities with high direct employment changes per unit investment (e.g. textiles, transport equipment, printing) have also high indirect effects, while growth in timber, furniture, paper, machinery and miscellaneous manufacturing has relatively small effects on employment. Like Hirsch, Mattila and Thompson (1960) consider that intra-regional trends are more important vectors of regional change than are extra-regional factors. They analyse the effects of eight prior and present changes in State consumer and industrial markets (Fig. 10.16) upon 20 American manufacturing industries. Thirteen of these industries were strongly influenced by these changes. Figure 10.16 summarizes schematically the type of industry upon which each kind of market trend had the greatest influence. Clearly current changes in value added, for example, cause the greatest changes in the development of fuel, instruments, transport equipment, non-electrical machinery and metal manufacturing industries while changes in the number of households influence the paper and building materials industries most. Each industry is affected positively by the eight market factors with the exception of leather which seeks locations where labour supplies are readily available in existing or declining market areas. The diagram does not attempt to show more than the major effects of change upon industry, nor does it signify inter-industry relationships.

AN INDUSTRIAL EXAMPLE: IRON AND STEEL

Throughout this essay frequent references have been made to the model location affinities of various branches of processing and manufacturing industry.[1] It seems appropriate at this stage to refer to one industry in greater detail. The iron and steel industry has been chosen partly because it has become almost a model for industrial location in general, chiefly on account of the importance of changing production techniques and its apparent suitability to Weberian analysis. However, the industry is typical for other reasons. Its

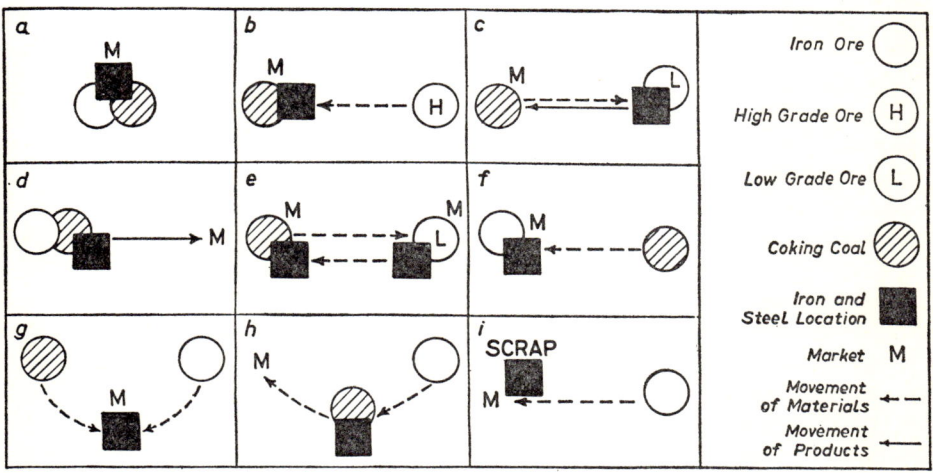

10.17 Model locations of iron and steel industries.

[1] Scale models are not in question here, for these, at best, are suitable for presenting site relationships and plant layout.

necessity to each and every national economy has involved the establishment, in a variety of locations, of various types of integrated iron and steelworks and non-integrated steel plants using a variety of processing techniques to cope with the variable inputs that are available in grade and mix. With the wide range in qualities of iron ores and coals that can be used to produce iron and, with scrap-steel and with the choice of Bessemer, Siemens-Martin, Kaldo, Tysland-Hole, electric arc and L.D. processes of steel-making, the model for the industry clearly stresses that there is no single optimum location where transport and other costs combined are at a minimum. As Moses states, 'if inputs are substitutible . . . three problems are . . . inseparable: optimum output; the optimum combination of inputs; and optimum location' (Moses, 1958, p. 272). The range of 'model' locations that can be optima is summarized diagrammatically in Figure 10.17. To these may be added the less frequent but no less viable location of non-integrated steel plants where abundant hydro-electric power or labour supplies exist, or, as location leaders, where regions are in need of development; some plants, too, may be strategically located in isolated areas (Hamilton, 1964A, pp. 62–63). These 'social' decisions may sometimes involve integrated plants, their development having been encouraged by national pride or by the activity of strongly organized regional interests. As in reality, then, the model must recognize the operation and viability of industry in sub-optimal or in satisficer locations. In this respect the iron and steel industry is more typical of the bulk of manufacturing industry than it is of industries which must be located in close proximity to their raw material supply sources.

CONCLUSION

Certain threads in model-building emerge from the complex weave of the industrial location pattern. The variety of approaches to the subject is abundantly clear. Scale is important in conditioning the degree of factual abstraction and the selection of location factors to be incorporated in the models. The variables to be handled differ as between larger and smaller scales of generalizing the organization, development and classification of industry and between different scales of spatial analysis and activity interdependence. Through time there has been a shift of emphasis which is reflected in the contrast between the models of Launhardt and Weber and those of plant assignment and spatial industrial dispersion. Such a shift is expressed in the increasing importance of macro-economic, at the expense of micro-economic, location decisions. It underlines the fact that the location decision is no longer the last act in the decision-taking process. The need to develop a given industry in a particular region generates feedback which influences the choice of the scale of plant and the nature of its output. The iron and steel industry provides a good example. The change of emphasis has been facilitated by the evolution of more refined mathematical techniques. Equally significant is the increasing importance of descriptive and stochastic models at the expense of deterministic models, especially in analysing the spontaneous evolution of the modern industrial pattern. Models of random processes and diffusion are appropriate so long as industrial location is not strictly and scientifically planned. Game theory, despite its limitation (Stevens, 1961), is applicable to the realm of oligopolistic capitalism. As industrial location becomes a matter of national and social, as well as economic interest, problems of plant assignment and dispersion take on an enhanced value. To be comprehensive, such models must embrace factors which, for lack of data, cannot be reduced at present to a common money denominator. Nevertheless, the complexity of plant allocation models – even when they use only cost and profit data – is such that 'the model-builder is torn between the ends of reality and manageability' (Koch and Snodgrass, 1959, p. 155). The latter problem is being overcome by the use of high-speed computers; but the former will remain a problem of correct programming. Many of the models that have been discussed in this essay have sought optimum location solutions – for individual plants, for an industry and for plants and industries in the framework of total spatial, social and economic interdependencies. In the light of the imperfections of reality they should seek satisficer locations. Yet the quest for optimum locations at all levels is valuable as a means of discovering alternative satisfactory solutions. As far as mortals are concerned, therefore, models that satisfy may also optimize.

REFERENCES

ALEXANDER, J. W., [1963], *Economic Geography*; (Englewood Cliffs), 661 pp.
ALEXANDERSSON, G., [1956], *The Industrial Structure of American Cities*, (Lincoln, Nebraska), 134 pp.
ALKJAER, E., [1953], *Erhvervslivets Beliggenhedsproblemer*, (Copenhagen), 90 pp.
BAIN, J. S., [1954], Economies of Scale, Concentration and the Condition of Entry in Twenty Manufacturing Industries; *American Economic Review*, 44, 15–39.
BALZAK, S. S., VASYUTIN, V. F. and FEIGIN, YA. G., [1949], *Economic Geography of the U.S.S.R.*; ed. by Harris, C. D., (New York), 620 pp.
BARFØD, B., [1938], *Local Economic Effects of a Large Scale Industrial Undertaking*, (Copenhagen), 74 pp.
BECKMANN, M., [1957–58], City Hierarchies and the Distribution of City Size; *Economic Development and Cultural Change*, 6, 243–248.
BECKMANN, M. J. and MARSCHAK, T., [1955], An Activity Analysis Approach to Location Theory; *Proceedings Second Symposium on Linear Programming*, (Washington), 331–379.

BIĆANIĆ, R., [1951], *Doba Manufakture u Hrvatskoj i Slavoniji 1750–1860*, (Zagreb), 459 p.
BOGUE, D. J., [1949], *The Structure of the Metropolitan Community: A Study of Dominance and Subdominance*; (Ann Arbor), 210 p.
BOS, H. C., [1965], *Spatial Dispersion of Economic Activity*, (Rotterdam), 106 pp.
BROWN, L., [1964], *The Diffusion of Innovation: A Markov Chain-Type Approach*; Northwestern University: Department of Geography, Discussion Paper No. 3.
CAESAR, A. A. L., [1955], On the Economic Organisation of Eastern Europe; *Geographical Journal*, 121, 451–469.
CHARDONNET, J., [1953], *Les Grands Types de Complexes Industriels*, (Paris), 196 pp.
CHINITZ, B. and VERNON, R., [1960], Changing Forces in Industrial Location; *Harvard Business Review*, 38 (1), 126–136.
CHISHOLM, M., [1962], The Common Market and British Manufacturing and Transport; *Journal of Town Planning Institute*, 1–12.
CHISHOLM, M., [1963], Tendencies in Agricultural Specialisation and Regional Concentration of Industry; *Papers and Proceedings, Regional Science Association*, 10, 157–162.
CLARK, W. A. V., [1965], Markov-Chain Analysis in Geography: An Application to the Movement of Rental Housing Areas; *Annals of the Association American Geographers*, 55, 351–359.
CLAYTON, K. M., (Ed.), [1964], *Guide to the London Excursions*, (London), 162 pp.
COSTA SANTOS, M., [1961], *Descentralizaçao Industriãl no Estado de São Paulo*, (São Paulo), 74 pp.
COTTERILL, C. H., [1950], *Industrial Plant Location: Its Application to Zinc Smelting*, (St. Louis), 155 pp.
COX, K. R., [1965], The Application of Linear Programming to Geographic Problems; *Tijdschrift voor Economische en Sociale Geographie*, 56 (6), 228–236.
CURRY, L., [1964], The Random Spatial Economy: An Exploration in Settlement Theory; *Annals of the Association of American Geographers*, 54, 138–146.
DALES, J. H., [1953], Fuel, Power and Industrial Development in Central Canada; *American Economic Review*, 43, 181–198.
DEAN, W. H., [1938], *The Theory of the Geographic Location of Economic Activities*, (Ann Arbor).
DEVLETOGLOU, N. E., [1965], A Dissenting View of Duopoly and Spatial Competition; *Economica*, 32 (126), 146–160.
DUNCAN, B., [1959], Population Distribution and Manufacturing Activity; The Non-Metropolitan United States in 1950, *Papers and Proceedings, Regional Science Association*, 5, 95–103.
DUNN, E. S., [1956], The Market Potential Concept and the Analysis of Location; *Papers and Proceedings, Regional Science Association*, 2, 183–194.
DZIEWOŃSKI, K., Translator and Editor, [1961], *Gospardarka Przestrzenna*; (Economics of Location, by A. Lösch), Warsaw, 407 pp.
EFROYMSON, M. A. and RAY, T. L., [1966], A branch-bound algorithm for plant location; *Operations Research*, Baltimore, 3, 361–368.
ESTALL, R. C., [1963], The Electronic Products Industry of New England; *Economic Geography*, 39, 189–216.

ESTALL, R. C., [1966], *New England: A Study in Industrial Adjustment*, (London), 296 pp.
ESTALL, R. C. and BUCHANAN, R. O., [1961], *Industrial Activity and Economic Geography*, (London), 232 p.
FEIGIN, YA. G., [1954], Razmeshchenie Proizvoditel'stva pri Kapitalizmie i Socializmie; *Izvestia Ekonomiki Akademii Nauk SSSR*, (Moscow), 552 pp.
FINLEY, M. I., [1965], Technical Innovation and Economic Progress in the Ancient World; *Economic History Review*, 18 (1), 29–45.
FLORENCE, P. S., [1948], *Investment, Location, and Size of Plant*, (London), 211 pp.
FLORENCE, P. S., [1953], *The Logic of British and American Industry*, (London), 368 pp.
FOLEY, D. L., [1956], Factors in the Location of Administrative Offices; *Papers and Proceedings, Regional Science Association*, 2, 318–326.
FRIEDRICH, C., [1929], *Alfred Weber's Theory of the Location of Industries*, (Chicago), 256 pp.
FUCHS, V. R., [1959], Changes in the Location of U.S. Manufacturing since 1929; *Journal of Regional Science*, 1 (2), 1–18.
FULTON, M. and HOCH, L. C., [1959], Transportation Factors Affecting Location Decisions; *Economic Geography*, 35, 51–59.
GARRISON, W., [1959–60], Spatial Structure of the Economy; *Annals of the Association of American Geographers*, 49, 471–482, and 50, 357–373.
GHOSH, A., [1965], *Efficiency in Location and Interregional Flows*, (Amsterdam), 95 pp.
GINSBURG, N., [1960], *Essays on Geography and Economic Development*; University of Chicago, Department of Geography, Research Paper, No. 62
GLEAVE, M. B., [1965], Some Contrasts in the English Brick-making industry; *Tijdschrift voor Economische en Sociale Geographie*, 56 (2), 54–62.
GOLDMAN, T. A., [1958], Efficient Transportation and Industrial Location; *Papers and Proceedings, Regional Science Assocation*, 91–106.
GORUPIĆ, D., [1954], Ekonømski Obračun Investicionog Programa; *Ekonomski Pregled*, (Zagreb) 2, 421–446.
GOSS, A., [1962], *British Industry and Town Planning*, (London), 190 pp.
GRAVIER, J. F., [1954], *Décentralisation et Progrès Technique*, (Paris), 387 pp.
GREENHUT, M. L., [1952], Integrating Leading Theories of Plant Location; *Southern Economic Journal*, 18, 526–538.
GREENHUT, M. L., [1956], *Plant Location in Theory and Practice*, New (York), 338 pp.
GREENHUT, M. L., [1957], Games, Capitalism and General Economic Theory; *The Manchester School of Economic and Social Studies*, 25, 61–88.
GREENHUT, M. L., [1960], Size of Markets versus Transport Costs in Industrial Location Surveys and Theory; *Juornal of Industrial Economica*, 8, 172–184.
GREENHUT, M. L. and COLBERG, M. R., [1962], *Factors in the Location of Florida Industry*, (Tallahalasee), 108 pp.
GREENHUT, M. L. and WHITMAN, W. T., (Eds.), [1964], *Essays in Southern Economic Development*, (North Carolina), 385–403.
HÄGERSTRAND, T., [1963], *On the Monte-Carlo Simulation of Diffusion*, (Mimeographed).
HAGGETT, P., [1965], *Locational Analysis in Human Geography*, (London), 339 pp.

HAGUE, D. C. and DUNNING, J. H., [1954], Costs in Alternative Locations – The Radio Industry; *Review of Economic Studies*, 22 (3), 203–213.
HAIG, R. M., [1927], Major Economic Factors in Metropolitan Growth and Arrangement; *Regional Plan of New York and its Environs*, (New York).
HAMILTON, F. E. I., [1962], *Recent Changes in the Location of Industry in Yugoslavia*; Unpublished Ph.D. Thesis, University of London, 338 pp.
HAMILTON, F. E. I., [1963], The Changing Pattern of Yugoslavia's Manufacturing Industry; *Tijdschrift voor Economische en Sociale Geographie*, 54 (4), 96–106.
HAMILTON, F. E. I., [1964A], Location Factors in the Yugoslav Iron and Steel Industry; *Economic Geography*, 40, 46–65.
HAMILTON, F. E. I., [1964B], The Skopje Disaster; *Tijdschrift voor Economische en Sociale Geographie*, 55 (3), 78–80.
HAMILTON, F. E. I., (Ed.), [1964C], *Abstracts of Papers*; 20th International Geographical Congress (London), 361 pp.
HAMILTON, F. E. I., [1964D], Geological Research, Planning and Economic Development in Poland; *Tijdschrift voor Economische en Sociale Geograpie*, 55, (12), 251–253.
HAMILTON, F. E. I., [1968], *Yugoslavia: Patterns of Economic Activity*; (London), 383pp.
HARRIS, C. D., [1954], The Market as a Factor in the Localisation of Industry in the U.S.; *Annals of Association of American Geographers*, 44, 315–348.
HEATON, H., [1948], *Economic History of Europe*, (New York), 792 pp.
HINER, O. S., [1965], Economic Developments in the Grimsby-Immingham Area; *Tijdschrift voor Economische en Sociale Geographie*, 56 (1), 21–32.
HIRSCH, W. Z., [1959], Inter-industry Relations of a Metropolitan Area; *Review of Economics and Statistics*, 41, 360–369.
HITCHCOCK, F. L., [1941], The Distribution of a Product from Several Sources to Numerous Localities; *Journal of Mathematics and Physics*, 20, 224–230.
HOLZMAN, F. D., [1957], The Soviet Urals-Kuznetsk Combine: A Study in Investment Criteria and Industrialisation Policies; *Quarterly Journal of Economics* 71, 368–405.
HOOVER, E. M., [1937], *Location Theory and the Shoe and Leather Industries*, (Cambridge, Mass.), 323 pp.
HOOVER, E. M., [1948], *The Location of Economic Activity*, (New York), 310 pp.
HOTELLING, H., [1929], Stability in Competition; *Economic Journal*, 39, 41–57.
ISARD, W., [1956], *Location and Space Economy*, (New York), 350 pp.
ISARD, W. and KUENNE, R. E., [1953], The Impact of Steel upon the Greater New York-Philadelphia Industrial Region: A Study in Agglommeration Projection; *Review of Economics and Statistics*, 35, 289–301.
ISARD, W. and others, [1960], *Methods of Regional Analysis*, (Cambridge, Mass.), 784 pp.
ISARD, W., SCHOOLER, E. W. and VIETORISZ, T., [1959], *Industrial Complex Analysis and Regional Development*, (Cambridge, Mass.), 294 pp.
KATONA, G. and MORGAN, J. N., [1952], The Quantitative Study of the Factors Determining Business Decisions; *Quarterly Journal of Economics*, 46, 67–90.
KEEBLE, D. E., [1965], Industrial migration from north-west London, 1940–64; *Urban Studies*, 2, 15–32.

KEIR, M., [1921], Economic Factors in the Location of Manufacturing Industry; *Annals American Academy of Political and Social Science.*
KENYON, J. B., [1960], *Industrial Localization and Metropolitan Growth: The Paterson-Passaic District*; University of Chicago, Department of Geography Research Paper No. 67, 224 pp.
KIDRIĆ, B., [1948], *Odnosi Izmedju Narodne i Privredne Politike*, (Belgrade), 146 pp.
KLEMME, R. T., [1959], Regional Analysis as a Business Tool; *Papers and Proceedings, Regional Science Association*, 5, 71–77.
KOCH, A. R. and SNODGRASS, M. M., [1959], Linear Programming Applied to Location and Product-Flow Determination in the U.S. Tomato-Processing Industry; *Papers and Proceedings, Regional Science Association*, 5, 151–166.
KOLOSSOVSKY, N. N., [1958], Proizvodstvenno-Territorial'nie Sochetanie (Kompleks) v Sovetskoi Ekonomicheskoi Geografii; *Osnovi Ekonomicheskogo Raionarovania*, (Moscow), 200 p.
KOOPMANS, T. C. and BECKMANN, M., [1957], Assignment Problems and the Location of Economic Activities; *Econometrica*, 25, 53–76.
KORTUS, B., [1963], Kompleks Przemysłowy Apszeron; *Przegląd Geograficzny*, 36 (4), 569–590.
KRYŻANOWSKI, W., [1927], Review of Literature on the Location of Industries; *Journal of Political Economy*, 35, 278–291.
KUHN, H. W. and KUENNE, R. E., [1962], An Efficient Algorithm for the Numerical Solution of the Generalized Weber Problem in Spatial Economics; *Papers and Proceedings, Regional Science Association*, 8, 21–33.
LAUNHARDT, W., [1882], Die Bestimmung des zweckmässigsten Standorts einer gewerblichen Anlage; *Zeitschrift des Vereins Deutscher Ingenieure*, 106–115.
LAUNHARDT, W., [1885], *Mathematische Begründung der Volkswirtschaftslehre*, (Leipzig), 218 pp.
LEFEBER, L., [1958], *Allocation in Space*, (Amsterdam), 151 pp.
LENIN, V. I., [1954], *Dzieła*, (Warsaw), 3 vols.
LEVIQUIN, M., [1962], *Marché commum et localisations*, (Paris-Louvain).
LINDBERGH, O., [1953], Economic-Geographical Study of the Localisation of the Swedish Paper Industry; *Geografiska Annaler*, 35, 29–49.
LISSOWSKI, W., [1965], Wpływ Układ Działowo-Gałęziowego na Układ Regionalny Planu Perspektywicznego, *Biuletyn Komitetu Przestrzennego Zagospodarowania Kraju* (Warsaw), No. 36.
LIVSIC, R. S., [1947], Nekotorie Teoriticheskie Voprosy Razmeshchenia Promishl'ennosti; *Izvestia Akademii Nauk SSSR, Otdiel Ekonomiki i Prava*, 4.
LONSDALE, R. E., [1965], The Soviet Concept of the Territorial-Production Complex; *American Slavic Review*, 24 (2), 466–478.
LORIA, A., [1888], Intorno della influenza della rendita fondaria sulla distribuzione delle industrie, *Accademia dei Lincei, Rediconti*, 4, 114–126.
LORIA., [1898], Ricerche Ulteriori della influenza della rendita fondaria sulla distribuzione delle industrie; *Rediconti*, 14, 235–243.
LÖSCH, A., [1954], *The Economics of Location*, (New Haven), 520 pp.
LÖVGREN, E., [1957], Mutual Relations Between Migration Fields: A Circulation Analysis; *Proceedings, Symposium on Migration in Sweden, Lund Studies in Geography, Series B, Human Geography*, 13, 159–169.

LOWENSTEIN, L. K., [1963], The Location of Urban Land Uses; *Land Economics*, 39, 407–420.
LUTTRELL, W. F., [1962], *Factory Location and Industrial Movement*, (London), 2 vols.
MANNERS, G., [1962], Some Location Principles of Thermal Electricity Generation, *Journal of Industrial Economics*, 10 (3), 218–230.
MANNERS, G., [1965], Areas of Economic Stress – The British Case; in Wood, W. D. and Thomas, R. S., (Eds.), *Areas of Economic Stress in Canada*, (Kingston, Ontario), 221 pp.
MARCH, J. G. and SIMON, H. A., [1958], *Organisations*, (New York).
MARTIN, J. E., [1964], The Industrial Geography of Greater London; in Clayton, R., (Ed.), *The Geography of Greater London*, (London), 111–142.
MARTIN, J. E., [1966], *Greater London: An Industrial Geography*, (London), 292 pp.
MARX, K., [1850], *Manifesto of the Communist Party*, (London), 48 pp.
MARX, K., [1909], *Capital: A Critique of Political Economy*, (Chicago), 3 vols.
MATTILA, J. M., and THOMPSON, W. R., [1960], The Role of the Product Market in State Industrial Development; *Papers and Proceedings, Regional Science Association*, 6, 87–96.
MCCASKILL, M., (Ed.), [1962], *Land Livelihood: Geographical Essays in Honour of George Jobberns*, (Christchurch), 280 pp.
MCLAUGHLIN, G. E., and ROBOCK, S., [1949], *Why Industry Moves South*; (National Planning Association, Committee of the South), 148 pp.
MOSES, L. N., [1958], Location and the Theory of Production; *Quarterly Journal of Economics*, 72, 259–272.
MRZYGŁÓD, T., [1962], *Polityka Rozmieszczenia Przemysłu w Polsce, 1946–1980*, (Warsaw), 279 pp.
MUMFORD, L., [1934], *Technics and Civlization*, (London), 495 pp.
MYRDAL, G., [1957], *Economic Theory and Underdeveloped Countries*, (London), 168 pp.
ODELL, P. R., [1963], *An Economic Geography of Oil*, (London), 219 pp.
ODHNOFF, J., [1965], On the Techniques of Optimizing and Satisficing; *Swedish Journal of Economics*, 67 (1), 24–39.
OSBORN, D. G., [1953], *Geographical Features of the Automation of Industry*; University of Chicago, Department of Geography, Research Paper No. 30, 106 pp.
PALANDER, T., [1935], *Beiträge zur Standortstheorie*, (Uppsala), 419 pp.
PFOUTS, R. W., [1957], An Empirical Testing of the Economic Base Theory; *Journal American Institute Planners*, 23, 64–69.
PHILBRICK, A. K., [1957], Principles of Areal Functional Organisation in Regional Human Geography; *Economic Geography*, 33, 299–336.
POLOPOLUS, L., [1965], Optimum Plant Numbers and Locations for Multiple Product Processing; *Journal of Farm Economics*, 47, (2), 287–295.
POUNDS, N. J. G., [1959], Planning in the Upper Silesian Industrial District; *Journal of Central European Affairs*, 18, 409–422.
PRED, A., [1965], Industrialisation, Initial Advantage, and American Metropolitan Growth; *Geographical Review*, 40 (2), 158–185.
QUINN, J. A., [1943], The Hypothesis of Median Location; *American Sociological Review*, 8, 148–156.

ROSCHER, W., [1899], *Nationalökonomik des Handels und Gewerefleisses*, (Berlin), 730 pp.
ROSS, E. A., [1896], The Location of Industries, *Quarterly Journal of Economics*, 10, 247–268.
SAUSHKIN, YU. G., [1959], *Economic Geography of the U.S.S.R.*, (Oslo), 148 pp.
SCHÄFFLE, G. F., [1878], *Bau und Leben des sozialen Korpers*, 3, (Tübingen).
SCOTT, D. R., [1963], The Location of Metropolitan Industries; *Western Australia*, 6 (4), 31–41.
SECOMSKI, K., [1956], *Wstęp do Teorii Rozmieszczenia Sił Wytwórczych*, (Warsaw), 135 pp.
SERCK-HANSEN, J., [1961], *Some Mathematical Models on the Spatial Distribution of Industry*, (Rotterdam).
SILHÁNA, V., (Ed.), [1964], *Ekonomika Průmyslu Č.S.S.R.*; (Prague), 657 pp.
SLETMO, G., [1963], *Geographical Distribution of Economic Activity in Norway*, (Bergen), 15 pp.
SMITH, A., [1776], *The Wealth of Nations*, (London), 3 vols.
SMITH, A., [1808], *The Economists Refuted*, (London).
SMITH, D. M., [1966], A Theoretical Framework for Geographical Studies of Industrial Location; *Economic Geography*, 42 (2), 95–113.
SMITH, W., [1952], *Geography and the Location of Industry*; (Liverpool), 20 pp.
SMITH, W., [1955], The Location of Industry; *Transactions, Institute of British Geographers*, 21, 1–18.
SMITHIES, A., [1941], Optimum Location in Spatial Competition; *Journal of Political Economy*, 49, 423–439.
SOMBART, W., [1919], *Der Moderne Kapitalismus*, 2 vols.
STEFANIAK, N. J., [1963], A Refinement of Haig's Theory; *Land Economics*, 4, 428–433.
STEVENS, B. H., [1961], An Application of Game Theory to a Problem in Location Strategy; *Papers and Proceedings, Regional Science Association*, 7, 143–158.
STOCKING, G. W., [1954], *Basing Point Pricing and Regional Development*, (North Carolina), 274 pp.
STOLLSTEIMER, J. F., [1963], A Working Model for Plant Numbers and Locations; *Journal of Farm Economics*, 45, 631–645.
TAAFFE, E. J., MORRILL, R. L. and GOULD, P. R., [1963], Transport Expansion in Underdeveloped countries: A Comparative Analysis; *Geographical Review*, 53, 503–529.
TINBERGEN, J., [1961], The Spatial Dispersion of Production: A Hypothesis; *Schweizerische Zeitschrift*, 97 (4), 1–15.
TINBERGEN, J., [1962], Research on the Geographical Decentralisation of Industry in the Netherlands; in *Guest Lectures in Economics*, Henderson, E. and Spaventa, L., (Eds.), (Milan), 230–242.
TINBERGEN, J., [1964], Sur une modele de la dispersion géographique de l'activité économique; *Revue d'économie politique*, 74 (1), 30–44.
VANHOVE, N. D., [1961], *De Doelmatigheid van het Regionaal-Economisch Beleid Nederland*, (Eeklo, Belgium), 157 pp.

WALLACE, L. T. and RUTTAN, V. W., [1960], The Role of the Community as a Factor in Industrial Location; *Papers and Proceedings, Regional Science Association*, 6, 133–142.

WEBER, A., [1909], *Über den Standort der Industrien, I: Reine Theorie des Standorts*, (Tübingen), 246 pp.

WEBER, A., [1914], Industrielle Standortslehre; *Grundiss der Sozialökonomik*, 4.

WILLIAMS, S. W., [1965], The Changing Character of the Fluid Milk Processing Industry in Illinois; *Illinois Agricultural Economics*, 32–39.

WINSBOROUGH, H. H., [1959], Variations in Industrial Composition with City Size; *Papers and Proceedings, Regional Science Association*, 5, 121–131.

WISE, M. J., [1963], The Common Market and the Changing Geography of Europe; *Geography*, 48 (2), 129–138.

WRIGLEY, E. A., [1961], *Industrial Growth and Population Change*, (Cambridge), 193 pp.

CHAPTER ELEVEN

Models of Agricultural Activity

JANET D. HENSHALL

A GENERAL MODEL

The study of agriculture has interested some of the most able members of our profession. Pioneer works by such men as O. E. Baker, Olaf Jonasson, Clarence F. Jones, Samuel van Valkenburg and Griffith Taylor published in *Economic Geography* in the inter-war period have been seen as the greatest contribution of that journal to our subject (Buchanan, 1959, p. 6). These studies of the agricultural regions of the world established the broad pattern, and they were followed by numerous empirical papers on agricultural land use analysing the unique causes of patterns within a specific area. According to the most recent overviews of agricultural geography (McCarty, 1954; Buchanan, 1959; Reeds, 1964) we have advanced little in our methods during the forty years since Baker's paper of 1926. The urgent need for more research is recognized by our famine frightened world, yet in 1964 it could still be said, 'Agricultural geography has not yet advanced beyond a primitive stage of development simply because many studies have been superficial investigations of extensive areas' (Reeds, 1964, p. 52). If we are to make the leap forward into maturity we must take a fresh look at our data, concepts and methods.

In the past work was often limited by a shortage of published data. Today there is a vast amount of detailed information available for most parts of the world but Coppock's (1964B) *Agricultural Atlas of England and Wales* is one of the very few examples of a geographical study utilizing this data bank. Recent work in theoretical geography has influenced our conceptual approach to agricultural problems. These developments have three major aspects: firstly there is a more theoretical approach and less concern with the 'uniqueness' of geographical distributions (Bunge, 1962); secondly there is a retreat from the deterministic interpretation of phenomena to a probabilistic and behaviourist one; thirdly the micro-geographic study is assuming a greater importance (Blaut, 1959; Brookfield, 1964). Finally our methods have benefited from the 'quantitative revolution' of the last decade and in particular the availability of computers able to handle large amounts of data has opened up new research possibilities. The wind of change in this branch of our discipline has brought with it a new awareness of the ideas other disciplines have to offer and communication with the related subjects of pedology, agroclimatology, plant ecology, rural sociology and agricultural economics in particular has increased. This chapter aims to analyse the present trends in agricultural geography with particular reference to the theoretical models developed.

The study of agriculture is concerned with individual farms having certain characteristics of area, soil, crops, livestock, etc. and complicated functional relationships based on the natural environment, the agricultural economy and the rural society. In general systems terms (von Bertalanffy, 1951) we are dealing with a set of objects (farms) with attributes (characteristics) functionally related through circulating movements (of money, labour, etc.) with energy inputs in response to the social and biological needs of the system. This system contains sub-systems, for example the plantation system, which form our agricultural regions or farming-type areas.

Models represent simplified parts of our system. As our information field becomes more complicated we turn to models for aid in understanding reality. Models of agricultural activity may be divided primarily according to whether they are based on a part of the system, that is on a farm, or on the total system or sub-system. Our experimental and analogue models are usually of the first type whilst models of agricultural land use are of the second type. Some models are *normative* in their approach (describing what ought to be under certain assumptions) whilst others are *descriptive* (describing what is actually existing). We may further divide our models on the basis of their treatment of the problem, that is whether the emphasis is on the economic (way of earning a living) or behavioural (way of life) aspects of agriculture. In practice the economic models so far developed tend to be normative, whereas the behavioural models tend to be descriptive. But the framework of the model does not define its content and some models now straddle the class boundaries.

EXPERIMENTAL MODELS

Many geographers have seen the individual farm as the focal point of our studies and the basis of our understanding of an agricultural region (Platt, 1930, 1942; Birch, 1954; Blaut, 1959; Keuning, 1964). The farm is the 'black box' of agricultural activity and the 'field' of our inquiry.

Model farms

The toy model farm with its moveable elements is probably the first model of geographic activity met by a child and as such can play an important role in his education. In the Junior School the model becomes dynamic and the toy is replaced by visits to a real farm (Farm Adoption Scheme). In this way the young child learns about the basic processes of agricultural activity. In the Secondary School childish things are put away and such models disappear from the curriculum although the Ministry of Education (1960, p. 23) does suggest the study of a single farm as an exercise in field work.

There are many ways in which model farms can be used by geographers for as 'a fundamental unit of resource utilization and organization' (Blaut, 1959, p. 81) it is a valuable tool in the task of describing and analysing the patterns of agricultural activity. Heller (1964) suggests that it may be used to describe regions, to study seasonal variations in agricultural activity and to examine the relationship between the environment and human institutions. Blaut (1953) in his study of a one-acre farm in Singapore used this actual farm as a model of resource use. The agronomist has long used experimental farms as aids to research but in the 1950's when the Imperial College of Tropical Agriculture in Trinidad set up a series of experimental 'peasant' farms they were used to illustrate a wide range of economic and social problems as well as the agronomic problems.

Developed from the 'case-study' approach was the 'representative farm' used by economists in the 1920's for type-of-farming studies. As more statistical data has become available this early method has become less popular, although it can still be useful. Stamp (1948, p. 348) suggests that in order to understand the relationship between the land use pattern and the farmer's resources, decisions, abilities, and adaptation to changes in technology and price levels, the land use map should be accompanied by a parallel study of a specimen farm. In the tropics the pattern of land use is complicated by intricate mosaics of plants and multiple crop seasons. In order to understand this pattern it has been found helpful to borrow techniques from the botanist and look not only at the individual farm but at the arrangement of plants within an individual field. On small farms in the tropics the land use pattern is made up of characteristic crop combinations often with a common spatial arrangement amongst the elements of the association, as for example, the ubiquitous hedges of pigeon pea around the fields of sugar cane in Barbados. In these cases the study of a specimen farm is essential to an understanding of the basic pattern of land use (Innis, 1961; Henshall, 1964). Occasionally such micro-study may be used to illustrate changes in land use. Beeley (1965) by mapping individual fruit trees on holdings in southern Turkey is able to show the trend away from subsistence crop mixtures towards the orderly pattern of commercial citrus orchards.

Theoretical model farms

The agricultural economists in their search for the representative farm have

[428] MODELS IN GEOGRAPHY

largely abandoned the specific farm unit. The modern tendency is to define a composite or hypothetical farm which is in some sense representative of the population being studied. The type of hypothetical farm constructed depends to some extent on the statistical information available. Often the modal farm in the frequency distribution of farms from the same universe is chosen. Butler (1960) uses this method in his study of farms and smallholdings in the North Riding of Yorkshire. The area was selected because of its apparent homogeneity but in fact revealed a variety of farming systems when a modal type was defined and the deviations from the mode examined. The modal farm was based on crop/livestock combinations as shown by the Ministry of

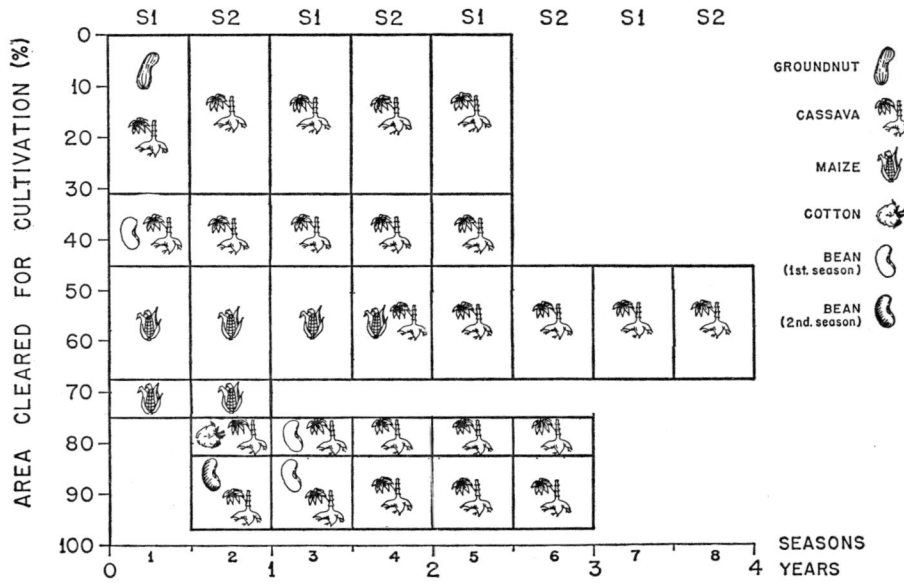

11.1 Model of the agricultural system in the *savane sableuse* of Kasai, Congo (*Source: Beguin, 1964*).

Agriculture returns and supplemented by visits to farms in different size groups. Using this approach Butler showed that a basic differentiating factor in this supposedly uniform agricultural region was the farm operator's attitude to technological innovations. In this case the use of a modal farm was very useful in illuminating the research problem.

Representative farms are a useful method of bringing micro-data to bear on macro-problems but they do involve an aggregation problem (Day, 1963). Carter (1963) criticizes modal farm studies because they are static in nature whilst the farm firm is operating in a dynamic framework, and he feels the model is most useful in studies of low income farms and highly specialized types of farming. Heller (1964) constructs model farms based on averages and includes both farm site and economic features. He suggests that these models may be used for testing the logical bases of agricultural location and for analysing shifts in agricultural patterns over longer time periods than can usually be dealt with on the basis of actual farms. In this way the overall pattern is seen in terms of its basic unit and thus some of the underlying relationships may be perceived more clearly.

Where the average farm is defined on the basis of detailed field work the background knowledge of the research worker may make the model even more useful. Johnson (1964) used his field data to define an 'average' farm and then he used this model as a basis for policy making. Beguin (1964) uses his knowledge of the Congo to develop a model illustrating tropical rotation systems (Fig. 11.1).

In an area where shifting agriculture is practised the farm is no longer the basic unit but is replaced by the area cultivated in the cycle of agriculture, as in Beguin's model (Fig. 11.1). Only rotations occupying more than 5 per cent of the area are shown. Reading horizontally the figure shows the succession of plants in the rotation whilst the abscissa shows the importance of the rotations in the total area cultivated. This type of model is an important element in the understanding of any land use pattern in the tropics.

CONCEPTUAL MODELS

Type-of-farming models

There are two main farming types for which theoretical models have been developed and used extensively: plantation and peasant. Generally these models have been derived by anthropologists or sociologists and economists and they reflect the approaches of these disciplines. The geographer has tended to accept these models as given and there has been little effort to reduce the noise level for geographical problems. But recently we have become aware that the traditional models are no longer applicable to contemporary agriculture and have begun to define new models (Gregor, 1965; Franklin, 1962 and 1965).

The characteristics considered typical of the plantation are crop and areal specialization, highly rationalized cultivation and harvesting techniques, large operating units, management centralization, labour specialization, massive production, and heavy capital investment. Perhaps the oldest explanation of the plantation sees it as a solution to the inability of the white man to do manual labour in the tropics. Research has shown since the last war that Europeans can work as well in the tropics as elsewhere but the plantation is still associated with tropical crops and monoculture. Industrialization (Waibel, 1941) and a large permanent labour force with little mechanization (Buchanan, 1958) were held to be qualities of the model limiting it to the tropics.

Gregor (1965, p. 221) in a very comprehensive study of the modern plantation has said that 'the usual classification of plantation farming as a tropical institution can no longer hold in the face of continued agricultural rationalization'. He notes that many of the characteristics formerly assumed to be the prerogative of tropical plantations are becoming associated with farms in extratropical areas. In those areas traditionally thought of as dominated by the plantation, monoculture has declined and cooler zone crops have gained in importance. The economic basis has changed and markets are no longer always in the metropolitan country or even foreign, but may be domestic as in the case of the Brazilian sugar plantations. Entrepreneurs are ethnically more diverse but may belong to the same group as the workers, and the ownership of a plantation may now be in the hands of an individual, a corporation or the state. Hutchinson (1959, p. 38) describes the modern Brazilian plantation system as 'a constellation of the corporate relations of the planter-industrialist and the private planter-supplier'. This corporate plantation he calls the 'new' plantation. Labour on the plantation is now less plentiful but has more in common with the industrial worker in its adherence to union rules than to the agricultural worker. The demographic aspects of the model have been developed by T. Lynn Smith (1959) who suggests that the population of plantation areas has certain characteristics more closely related to those of urban populations than rural populations although this has been challenged by Pico (1959) with reference to Puerto Rico and Henshall (1966) in the case of Barbados.

As the plantation has spread and changed many new types and regional variations have been recognized. New typologies have been set up such as that of Gerling (1954) based on processing complexity, or Steward (1960) based on historical stages of development and cultural variation, or Wolf (1959) identifying differential cultural adaptation of labour to the modern plantation society. The old plantation model has been replaced by a dynamic model which recognizes a continuum of change and development.

Models of peasant agriculture fall into two groups; those which use the 'way of life' approach, usually set up by anthropologists, and those which use the 'way of earning a living' approach, usually preferred by economists. As an example of the first approach Firth (1951) uses the word 'peasant' to describe any society of small producers for their own consumption, and Robert Redfield (1956, p. 18) describes peasants as people whose 'agriculture is a livelihood and a way of life, not a business for profit'. Wolf (1954) stresses that a peasant lives on land he controls and that an agriculturist who carries on agriculture for business and reinvestment, looking on the land as capital and commodity is a farmer not a peasant. Elena Padilla (1960) defines a peasant as 'an organizational type characterized by individual ownership of the land or undivided rights over the productive unit, family and kin labour, and the use of a simple technology to raise cash crops in addition to subsistence crops' (Padilla, 1960, p. 25).

The economists approach the problem in a different way. Edwards and Rees (1964, p. 73) give us a useful working description of our model listing peasant characteristics as follows: 'small scale of operation; heavy reliance on human labour provided mainly by the peasant and members of the family, and assisted in some systems by animal and mechanical power; use of traditional ("backward") techniques and a strongly conservative attitude towards innovation; individual rather than co-operative or collective cultivation of land; and a significant concentration on production for home consumption.' Anne Martin (1958, p. 88) defines the peasant as a farmer with little or no education and therefore resistant to change and technological innovation.

Our basic model seems to describe a small-scale agricultural producer dependent on family labour using simple methods but Franklin (1962) points out that this commonly accepted view of peasant life is full of paradoxes. We have for example the idea that the peasant is a good and careful farmer yet in reality his yields are often very low; peasant family life is extolled for its virtues yet many peasants will sacrifice their own lives to educate their children to leave the system; village solidarity with its appeal to the romantic is realistically compared to class or caste differences that often present strong barriers to community development. Franklin (1962, p. 3) shows that 'a fundamental cause of division within the village lies in the inequalities of land ownership . . . concentration of the ownership of draught animals, agricultural equipment and ready money intensifies the disparities. Thus we have the paradox common to many peasant societies of relatively vast inequalities in wealth amidst general poverty.' Traditionally the peasant has been described as illiterate and backward but field work in many parts of the world has shown that given the opportunity the peasant is quick to educate himself and when innovations are shown to be profitable he is swift to adopt them. The romantics maintain that peasants have a conservative and stable existence, but Franklin (1962) sees only long periods of slow evolution punctuated by sudden change. 'Archaic elements are more likely to survive within a peasant group and to remain integral parts of the culture, but it is unlikely that during the last 150 years many peasant societies have failed to experience important and perhaps significant changes, so that the study of change has become integral to the study of the modern peasantry' (Franklin, 1962, p. 4–5). Franklin calls the process responsible for much of this change *'agriculturization'*. He summarizes it thus; 'The partial incorporation of the peasantry within a market economy, the greater use of money, the appearance of usury and middlemen, the rise in rents following the increased competition for land, the weakening of communal bonds and the passing of traditional responsibilities' (1962, p. 9). In Europe this process was associated with the Industrial Revolution and is even now occurring in many underdeveloped countries. The Chinese commune and the Danish co-operative are different forms of a solution to the problem of the *paysans évolués*.

Recently Franklin (1966) has suggested that the peasant farmer should be defined on the basis of his labour commitment thus combining both social and economic aspects of previous models. The use of his own and family labour means that the peasant tends to be a satisficer giving importance to leisure as well as labour input, rather than an optimizer with regard to the output of the farm firm. Franklin (1965) uses labour as the basic differentiator in building up a model of the peasant system in relation to capitalist and socialist systems (Table 11.1). In this way regional variations and mixtures of systems can be studied and recognized on the basis of labour commitment in the various sectors of the economy (Franklin, 1965, p. 161). The concept of the peasant production system provides us with a fairly clean and noiseless model for the geographical study of peasant agriculture.

TABLE 11.1

Systems of Production

The Enterprise	*Peasant*	*Capitalist*	*Socialist*
Labour-commitment of the enterprise	Total	Non-total	Non-total
Institutional basis	Family	Family Joint stock	Combine
Control and direction	Family	Family-managerial	Managerial
Means of distribution	Barter-market	Market	Prescription-market
Media of distribution	Kind-money	Money	Money
Mechanization	Possible	Usual	Usual
Ownership (the right of . . .)			
(a) Direction	Chef d'entreprise for family	Chef d'entreprise managerial	Managerial
(b) Alienation	(1) Agnatic interdiction (2) Testamentary custom (3) Permitted	Permitted	Constitutional prohibition
Regulator	Labour supply	Market	State

Source: Franklin, 1965, p. 149.

When considering both peasant and plantation agriculture in an area, size alone may be used to discriminate between them. Edwards (1961) in Jamaica and Ooi Jin Bee (1959) in Malaya both use the figure of 25 acres as the upper limit for a peasant farm. Blaut (1961) develops a model for comparative analysis of peasant and plantation farming systems. This is based on resource materials and the amount of processing involved and takes into account the farmer's perception of his resources, his value judgements and his skill. Steward (1960) recognizes the interrelationships of peasant and plantation and states that the dispersed peasant is the counterpart of the slave plantation and the corporate peasant is the complement of the hacienda. If we consider the plantation and peasant systems as subsystems within the main system of agriculture then the concepts of general systems theory help us to understand the peasant-plantation relationship.

Environmental models

These are the basic deterministic concepts which underly much of our thinking about agricultural geography. They may be recognized at all levels. At the world scale tropical, sub-tropical and temperate areas, for example, all have been thought to have their own characteristic types of agriculture. O. E. Baker (1926), at the national level, has described American agriculture in relation to environment. In Britain we associate the wetter, western areas of Highland Britain with pastoral farming whilst the drier, eastern Lowland Britain is traditionally an area of arable farming.

These models have been widely accepted but they do tend to assume uniformity of agriculture within physically defined regions. It is often the juxtaposition of different agricultural types which is of interest to geographers. In the model of tropical agriculture (Fig. 11.2) we take as our universe that part of the world lying between the Tropics. The basic types of agriculture found within this region may be considered as sets. If we take as the basic elements of our classification firstly the types of production, arable or pastoral, and secondly the methods of production, subsistence or commercial we should then have logically four basic sets. But if we consider our model in terms of the area included in these sets then it is reasonable to combine both subsistence and commercial pastoral agriculture in one set since this type of agriculture is more common to the sub-tropics and temperate regions than to the tropics. This arrangement of sets can be shown diagrammatically by the use of a Venn diagram (Fig. 11.2).

Set A is defined as subsistence cultivation both shifting and settled. Set B is made up of commercial cultivation by both plantations and small farmers. Set C includes both nomadic and settled herding. Areas where only one of these sets is represented are becoming increasingly rare: pure subsistence cultivation is now found only in very remote places such as the upper parts of the Amazon Basin; herding with no associated cultivation is mainly found in areas of highly commercialized ranching such as northern Australia; commercial cultivation with virtually no subsistence is found in the more ad-

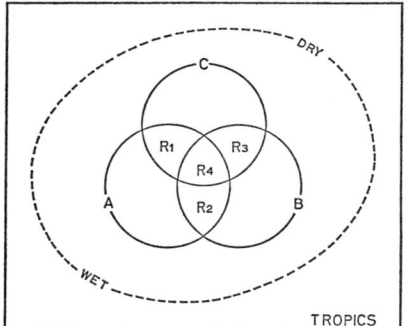

11.2 A set theory model for tropical agriculture. Set A contains subsistence cultivation; Set B commercial cultivation; and Set C herding.

vanced parts of the tropics especially where there is little room for peasants as in parts of the West Indies. By far the greater part of the tropics has a mixture of two or even all three of the basic types of agriculture. These mixtures are shown by the intersections of the sets. We can write these intersections as

$$R_1 = A \cap C$$
$$R_2 = A \cap B$$
$$R_3 = B \cap C$$
$$R_4 = A \cap B \cap C$$

R_1 is the combination of subsistence cultivation and herding, R_2 is the combination of subsistence and commercial cultivation and R_3 combines commercial cultivation and herding. R_4 covers a combination of all three basic sets and is most frequently found in conjunction with multi-racial societies. These relationships are shown in Table 11.2 (p. 435).

In most cases the intersections may be sub-divided according to the relationships between the types of agriculturist: parallel when two types of agriculture are existing side by side; dual when one group of people practise two types of agriculture; and symbiotic when two types of agriculturist have developed an interdependent system. This method of analysis does give us some idea of the complexity of agriculture in the tropics. It should be noted that as these areas become more developed the agricultural patterns become more complex.

Historical models

There are two conceptual approaches to the construction of historical models

TABLE 11.2
Tropical Agriculture

Set Intersection	Typical Environment	Sub-Division	Example
R_1	Savanna	Parallel Dual Symbiotic	Southern British Guiana Brazilian sertão Northern Nigeria – Hausa and Fulani
R_2	Humid tropics	Parallel Dual Symbiotic	Amazon – Japanese and Indian New Guinea, Yucatan West Indies
R_3	Irrigated or seasonally flooded	Parallel Dual Symbiotic	Amazon varzea Gezira Peru
R_4	Variable	—	East Africa, – European, Masai and Kikuyu North-west Argentine – European, mestizo and Amerindian

of agricultural activity. The first method looks at the problem from the point of view of the dispersal of plants and animals and culture contacts between groups of peoples. The second correlates changes in patterns of agriculture with changes in population density.

Carl O. Sauer (1952 and 1956) is an exponent of the first method. He bases his theory as to the origins of agriculture on three premises: that this new mode of life was sedentary and that it arose out of an earlier sedentary society; that planting and domestication did not start from hunger but developed in a situation in which there was both leisure and surplus food; and thirdly that primitive agriculture was located in woodlands. Sauer recognizes two ancient agricultural systems. The oldest system is known as 'hoe culture' because the main implement used is the digging stick or hoe. This method is based on mixed cropping and usually involves some bush fallowing. It is the traditional agriculture of the New World, Negro Africa and the Pacific Islands. The second basic system he identifies as a herding and sowing culture. In this system, in contrast to the hoe culture, domesticated animals play a very important role. There is little intercropping and usually only one crop season per year in the herding culture. This culture developed in the Near East and spread outwards in three directions: to the steppes of Eurasia where it became completely pastoral; to the North European plain into which the Celtic, Germanic and Slavic peoples came as cattle and horse raisers planting a few fodder crops and some rye and oats; and also along the shores of the Mediterranean. From these simple beginnings came our present complicated pattern of types of agriculture.

Malthus believed that the supply of food to the human race was inherently inelastic and this lack of elasticity was the main factor governing the rate of population growth. Thus population is seen as the dependent variable determined by preceding changes in agricultural productivity which in turn are explained by extraneous factors such as changes in technology. Ester Boserup (1965) also follows the second approach to the development of agriculture but disagrees with Malthus. She believes that population growth is the independent variable which in its turn is a major factor in determining changes in agricultural productivity. Mrs Boserup feels that her model based on changes in population density 'is conducive to a fuller understanding of the actual historical course of agriculture including the development of patterns and techniques of cultivation as well as the social structure of agrarian communities' (p. 12).

She bases her classification of systems of land use on frequency of cropping rather than on the more commonly used dichotomy of cultivated and uncultivated land, as she feels her method is more realistic when dealing with underdeveloped countries. She recognizes five types of land use as follows, in order of increasing intensity:

1 *Forest-fallow cultivation* with 20 to 25 years fallow after one or two years cultivation.
2 *Bush-fallow cultivation* with cultivation for two to as many as eight years followed by six to ten years fallow.
3 *Short-fallow cultivation* with one to two years fallow in which only wild grasses can invade the fallow land.
4 *Annual cropping*. In this system the land is left fallow for several months between the harvesting of one crop and the planting of the next. In this class may be included systems of annual rotation in which one or more of the successive crops sown is a grass or other fodder crop.
5 *Multi-cropping*. This is the most intensive system of agriculture with the same plot bearing several crops a year with little or no fallow period.

In Europe the historical development from neolithic forest-fallow to contemporary annual cropping can be recognized. In the tropics, in such countries as Nigeria where there are marked areal variations in population density, the whole range of land use systems may be seen. The most difficult step in the hierarchy is the one from long fallow to short fallow with its associated short-run decrease in output per man-hour. In the long run, however, increasing population may force changes in work habits which lead to increased productivity and increased division of labour and eventually to economic growth. Many underdeveloped countries are now at this stage of agricultural intensification and Mrs Boserup feels that her model based on past experience may help us to understand present processes.

TAXONOMIC MODELS

Classification of agricultural systems has long been a popular activity amongst geographers. Hahn (1892), Whittlesey (1936), Otremba (1950–60) and Helburn (1957) are just a few of the geographers who have worked in this field. More recently a special commission on agricultural typology has been created within the International Geographical Union. It is not intended to discuss the general problem of classification here but merely to analyse the theoretical basis of a few of the typologies set up.

Ratios and indices

In order to overcome the problem of classifying such complicated and diverse elements as farms it is necessary to develop some measure common to all. Chisholm (1964) suggests three methods of converting farm attributes to the same units based on the following indices:

1 The cash contribution of production to farm revenue
2 The cash share of inputs such as labour
3 Man/days of labour as a common index for each type of crop and livestock combination.

These indices have been much used by agricultural economists and the third index is especially useful when dealing with subsistence agriculture where few statistics are available (Clark and Haswell, 1964). There have been several other attempts to develop indices which might be used as bases for regional analysis of agriculture. One of the first of these is the distance index derived by Mather (1944). He takes farmhouses as the basic unit of his study using the formula

$$D = 1 \cdot 07 \sqrt{\frac{A}{n}}$$

where D is the average distance from one farmhouse to the nearest six farmhouses, A is the total area involved and n the total number of farmhouses. From this analysis he was able to state that farms situated to the west of the 100th meridian in the United States were more than one mile apart whilst those to the east were closer together. Livestock ratios have been used by several students to study agricultural change (Clark, 1962; Anderson, 1965). Manley and Olmstead (1965) use an index based on the monetary value of the output of the farm in a study of the geographical patterns of labour input. For studies of non-temperate agriculture Al-Maiyah (1958) uses multiple-item scaling in his study of Iraq, whilst Bhatia (1960) derives an index of crop diversification for his work in India based on the number of crops grown and the percentage of the cultivated area under different crops.

Board (1963) discusses various methods of presenting agricultural data in the form of maps of farming types and suggests the use of ratios as a basis for plotting the data. He uses Krumbein's facies maps as an analogue for his

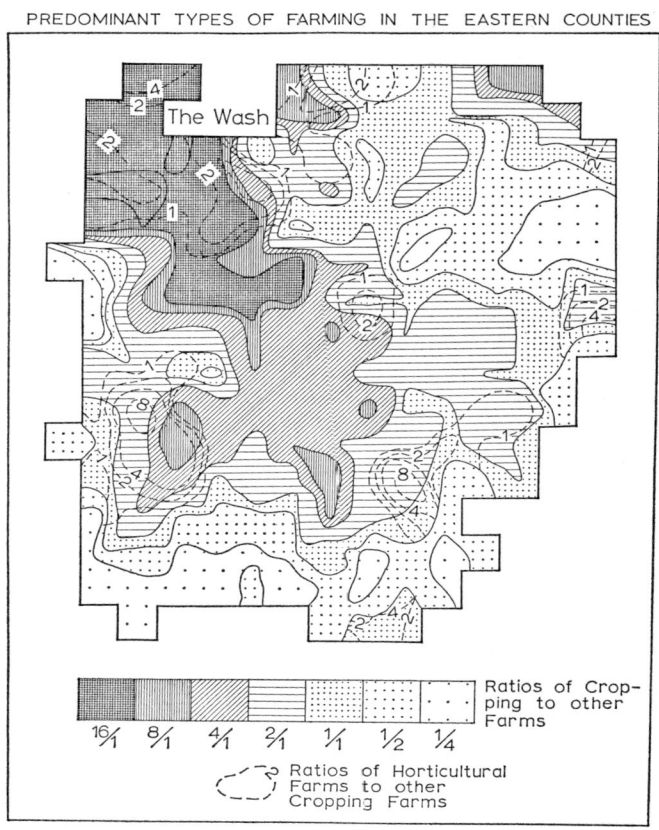

11.3 An iso-ratio map of farming types in the Eastern Counties of England (*After Board*).

cartographic presentation of farming types. Board provides us with an example based on information for Eastern England (Jackson, 1963). Using Jackson's map Board plots the ratios of cropping farms, that is farms deriving 70 per cent or more of their total gross output from crops, to all other farms (Fig. 11.3). Horticultural farms were plotted separately. The first step was to draw a square grid over the base map each square covering an area $6\frac{2}{3}$ miles by $6\frac{2}{3}$ miles. An arbitrary figure of 4 farms per grid square was chosen as the minimum number for which the ratio value for a square was calculated. The intervals chosen were as follows

$$\tfrac{1}{16} \quad \tfrac{1}{8} \quad \tfrac{1}{4} \quad \tfrac{1}{2} \quad \tfrac{1}{1} \quad \tfrac{2}{1} \quad \tfrac{4}{1} \quad \tfrac{8}{1} \quad \tfrac{16}{1}$$

which were converted for ease of plotting to decimals

·06   ·12   ·25   ·50   1·00   2·00   4·00   8·00   16·00

Board suggests that it is better to use proportions at the plotting stage and to employ ratios only at the final stage of compilation. The proportions appropriate to the above ratios are

·06   ·11   ·20   ·33   ·50   ·67   ·80   ·89   ·94

One could continue adding more iso-ratios to the basic pattern. The choice of ratios to be plotted and number of different ratios to be shown on the same map depends on the problem being analysed, the types of farming in the area being mapped and the aesthetic judgement of the cartographer. This method avoids abrupt boundaries on the map which have no basis in reality and gives a consistent cartographical method for comparison of different regions. It also provides a tool which can be used by other research workers with confidence.

Weaver's model

Underlying many classifications of agriculture is the idea of crop and livestock combinations. J. C. Weaver in the 1950's became interested in quantifying these combinations with a view to the study of change over time in the pattern of agricultural activity. He defined his objectives as follows, 'The central concern espoused is that of precise and objective measurement and pattern definition among the individual features and combinations of features of agricultural production as physical entities and economically active phenomena' (Weaver, 1954B, p. 286). In order to attain his objective Weaver set up a mathematical model for his crop-combination regions. He defined a theoretical curve based on the area of cropland being equally divided between the individual crops in the combination ranging from 100 per cent in a region of monoculture to 10 per cent in 10-crop-combination region. He then measured the actual crop percentage in the combination against his theoretical curve. Since he was interested in relative rank of deviation from the 'expected' not the actual magnitude of deviation, he used the standard deviation formula without extracting the square root. This was expressed as follows

$$\sigma = \frac{\Sigma d^2}{n}$$

where d was the difference between the actual and expected values of crop percentage and n was the number of crops in a given combination. The crop combination that showed the least deviation from the expected curve was recorded for every county. He also developed livestock-combination regions based on livestock units (Weaver, 1956). Weaver himself failed to develop his model further but it has been made use of by several other geographers notably Peter Scott (1957) in New Zealand, Thomas (1963) and Coppock (1964) in England and Wales and Singh (1965) in India.

The model Weaver set up has been criticized because it assumes that all crops are equal in any given crop-combination and fails to recognize those combinations in which one or more crops may be dominant although other crops are significant. Scott (1957) saw this problem, 'in any such application of the statistical procedure to a definition of agricultural regions the ranked percentage series in the basic crop-combinations should be retained however fragmented the resultant pattern' (p. 121). Weaver found it impossible to combine crop and livestock combinations because he was dealing with different units of measurement and thus areas where the interrelationships of crops and livestock are important, are not recognized. Not only do we have the loss of information in the Weaver model but we also have an element of subjectivity. Although the methods will give the same results with the same data if used by different people the original choice of the crops considered is dependent on subjective judgement. Finally the area in which Weaver developed his method, the Mid-West of the United States, was particularly suited to this type of classification. The areal unit on which Weaver based his calculations, the county, does not vary greatly in size as does the English parish for example, and thus certain computational problems were avoided. In the Middle West there is little areal differentiation based on physical variation in the landscape. Weaver was dealing with an area in which reality approached the ideal situation of the undifferentiated plain and thus combinations of crops were a fundamental basis for regional division. Other workers have not always recognized these underlying assumptions of Weaver's model. Despite its limitations Weaver's work is an important contribution to agricultural geography since he was one of the first people to attempt to set up a quantitative model for the classification of agricultural regions.

Factor analysis

The use of factor analysis as a means whereby the basic dimensions of a seemingly complex domain can be identified is well established in geographic research (Harman, 1962, p. 7). Kendall (1937) used it to derive a productivity index for crops and Hagood (1941) used it for defining regions based partly on agriculture, but geographers have only recently used this technique in the study of agriculture (Henshall and King, 1966; Henshall, 1966).

The rationale of modern factor analysis is to achieve parsimony of description. This is done by resolving a basic matrix of inter-relationships into a minimum, or at least a small number, of hypothetical variates or 'factors'. The geographer may consider these mathematical factors as similar to Hartshorne's element-complex (Hartshorne, 1960). It is assumed that the inter-correlations of the variables reflect certain underlying factors common to all the variables. The factors are made up broadly of two parts. One part is the general or common factor involved in all the variables; the other is the unique factor involved in each variable. The common factors often help to account for the maximum of the variance among the variables, whilst the unique factor indicates the extent to which correlations with other variables in the set do not account for the total unit variance of the variables. For a given matrix of correlation coefficients the position of the reference axes is indeterminate. In order to achieve a solution which has the greatest possible meaning and has consistency from analysis to analysis the reference axes are usually rotated to an orthogonal position where they are uncorrelated. The interpretation of the factors depends on the strength of the relationships between the variables and the factors as shown by the factor loadings. In addition the research worker must have a deep understanding of his problem and of the limitations of his input data if he is to identify the factors in a meaningful fashion.

One of our problems in dealing with agriculture is the wide number of variables we need to take account of and to analyse. Factor analysis, within the broad limits of the computer, can analyse the relationships between a large number of attributes or variables for many observations within a short space of time. This enables us to develop a classification based on a much larger number of variables than was possible before. Henshall and King (1966) tried to classify peasant agriculture in Barbados on the basis of crop-livestock combinations. Using factor analysis they were able to include 48 different types of crops and livestock for 150 observations or farms. The input data was in binary form based on presence or absence of individual attributes on each farm. Thus a matrix of phi coefficients was obtained as a measure of relationship between the items and then a conventional factor analysis was applied. Two approaches were adopted; the first, the R-mode analysis, focused attention on the observed correlations between the m variables measured over N cases, whilst the second, the Q-mode analysis, directed attention to the relationships between the farms as regards their attributes. Thus the factors recognized in the R-mode analysis were the basic crop-livestock combinations of the farms while those extracted in the Q-mode analysis were identified as farm types based on crop-livestock combinations. If the distribution of farms ranking high on each of the four factors recognized is mapped (Fig. 11.4) a regional pattern of crop-livestock combinations can be seen. The first factor which accounted for over 20 per cent of the total variation was recognized as a general factor loading highest on sugar cane and fruit trees and was strongly localized in the more isolated parts of the island. The second factor was identified as a cash-vegetable factor and is located fairly close to the main market in Bridgetown and the west coast hotels. The other two factors mainly recognized subsistence crops but they also produced distinctive patterns when mapped.

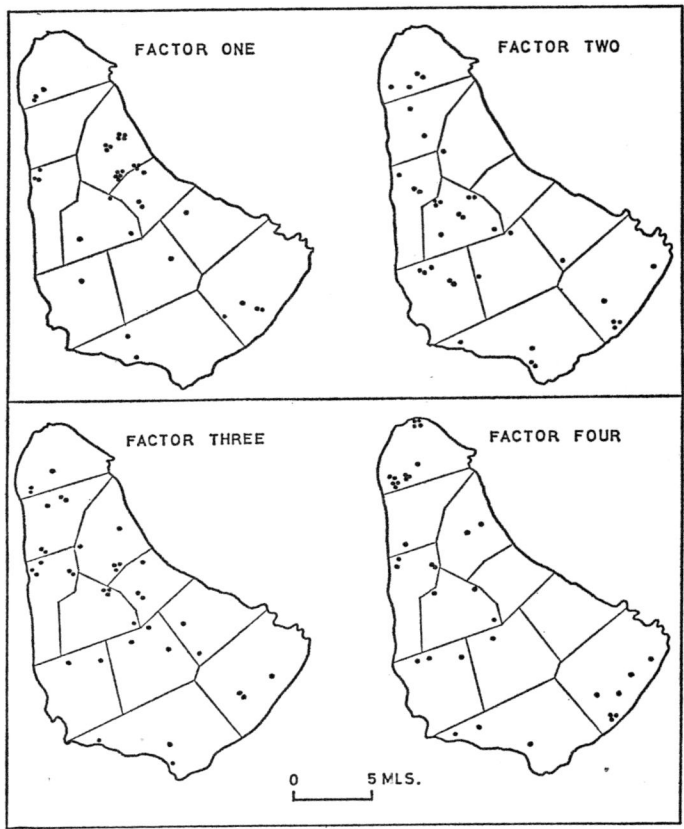

11.4 Sample farms in Barbados in the upper quartile of factor loadings for the first four factors in the Q-mode analysis.

It will be noted that the example quoted does not take into account the relative importance of individual crops or animals within the crop-livestock combination but merely their presence. However it was felt that this approach was of use in a tropical area where complicated intercropping is a feature. In addition the model set up had the built-in assumption that sugar cane was the dominant crop on virtually every farm and thus it was the secondary crops with livestock that were of interest. Of course, where the information is available, it is perfectly possible to use more detailed land use data as input for factor analysis. This method of classifying crop-livestock combinations has several advantages. A very complex combination can be handled using a computer and information is not lost in the effort to simplify the land use pattern. In addition it is possible to measure a farm's crop-livestock combination on four interval scales, these being the four factors that were extracted, and thus farms may be directly compared on the basis of their factor scores. Data collected for different areas or at different times may also be compared when analysed in a similar fashion using factor analysis.

Factor analysis has also been used to identify the relationship of a much wider range of variables to the underlying structure of agriculture and thus new agricultural regions may be recognized (Henshall, 1966). In this case the actual percentage of the holding in the various crops, livestock units and economic, social and physical characteristics of the farms were taken as attributes. Thirty-two variables were considered for 116 farms. When an R-mode analysis was carried out it was found that the 32 original variables could be collapsed into 12 new independent variates or factors. The first two factors were associated with distance variables and were identified as 'urban influence' and 'fragmentation' factors and accounted for one-third of the total variance. The next two factors accounted for one-quarter of the total variance and were identified as 'demographic' and 'motivation' factors. The situation revealed by the model showed that the basic variation in peasant agriculture is more closely related to population and settlement variables than to variations in physical environment. This is scarcely surprising in this densely populated island of gently rolling coral limestone. But if the technique is applied to other areas the order of importance of the factors may be changed or new factors revealed. Thus comparisons between areas on a common statistical basis becomes possible.

MODELS OF THE LOCATION OF AGRICULTURAL ACTIVITY

Von Thünen's model

The classical model of agricultural location outlined by von Thünen in his book *Der Isolierte Staat* (1826) is based on an econometric analysis of the estates in Mecklenburg where von Thünen farmed for forty years from 1810 until his death in 1850. Grotewold (1959) points out that only if one understands the empirical background of von Thünen's theory can one really appreciate its logic. The original model of von Thünen is inherently descriptive but later writers notably Hoover (1935), Lösch (1954), and Dunn (1954) have used the early framework as a basis for a normative model.

The underlying assumptions made by von Thünen are (1) the existence of an 'isolated state', (2) one central city as the sole market, (3) a uniform plain surrounding the city, (4) only one mode of transport, horse and cart, (5) the plain inhabited by farmers supplying the city, (6) the maximizing of profits by the farmers with automatic adjustment to the needs of the central market. However he did also consider a version of his model in which some of these assumptions of the 'isolated state' were modified by (1) introducing a navigable river on which transportation was speedier and costs only one-tenth those of land transport, (2) a minor market centre with its own trade area, (3) areal differences in the agricultural productivity of the plain around the city. Within the 'isolated state' von Thünen considered the relationship of three factors: distance of farms from the market; prices received by farmers for their goods; and finally land rent. The relationship between the first two was fairly simple. The price received by the farmer was the market price minus the cost of transportation which increased directly with distance from the market. Thus any given product was of greater value to the farmer the closer he was to the market. Land rent (Bodenrente) was defined as the return from the investment in the land. It may be derived from the expression

$$L = E(p-a) - Efk$$

where L is the locational rent per unit of land (the dependent variable), k is distance (the independent variable), E, p, f and a are constants or parameters. E is the yield per unit of land, p the market price per unit of commodity, a is the production cost per unit of commodity and f is the transport rate per unit of distance for each commodity (Dunn, 1954, p. 7). Haggett (1965, p. 161–167) discusses the substitution of a minimum-movement solution using the formula $(A/\pi)^{\frac{1}{2}}$ which simplifies the problem by omitting specific mention of the constants of market price and production cost.

The concentric rings of agricultural land use of von Thünen's hypothetical state have been well discussed by Grotewold (1959) and Chisholm (1962). The patterns seen by von Thünen 140 years ago have been radically modified by changes in transportation, new technological achievements such as refrigeration, and the replacement on the market of certain goods such as firewood by new items. In addition Lösch (1954, pp. 38–48) has pointed out that even given von Thünen's assumed conditions the formation of concentric zones was not inevitable. Yet in two spheres the nineteenth century model may still be applicable. With improvements in transportation the radius of the land use zones has become greater but the concentric zones may still be recognized on a continental scale (Backe, 1942; van Valkenburg and Held, 1952). At the other end of the scale Chisholm (1962, p. 48) considers the hamlet or farmstead in terms of von Thünen's 'isolated state' and shows that land use varies with labour input which in turn is directly related to the distance between farmhouse and field. In the less developed countries the conditions may still be similar to those of von Thünian Mecklenburg and there are several cases cited in geographical literature where land use around a settlement is directly related to distance from the settlement (for Africa by Prothero, 1957, and Steel, 1947; for India by Ahmad, 1952; for Brazil by Waibel, 1958).

The von Thünen model has probably been the most fruitful in the field of agricultural geography, but there are three respects in which the model appears in need of some theoretical revision. Firstly the model is one of partial equilibrium (although Garrison and Marble, 1957 do provide a specification of the model which can be solved by simultaneous equations). Secondly the Thunian model does not take into account the influence of non-economic factors and thus is limited in its scope. Thirdly differences in scale of the central city are not considered and Harvey (1966) suggests that the failure of many medium-sized English towns to develop distinctive land use zones in the nineteenth century may be due to economies of scale experienced by the larger towns which lead to the obliterating of the smaller market centres.

Inter-regional equilibrium models

The von Thünen model operates over space and through the use of marginal analysis shows how types of land use grade into one another over a continuum although he never discusses boundary problems. Other locational models are derived by conceptualizing areas as points. Producers, factors of production and consumers are treated as located at a series of discrete points with zero transport costs between them. An analysis of comparative advantage then 'explains' differences in production at the various points. These models are of two basic types; input-output models and spatial equilibrium models (both reviewed by Harvey, 1966, pp. 365–367).

Input-output models. These models were originally devised by Leontieff as a method of analysing national economies. They have been used for the analysis of agricultural activity by Peterson and Heady (1956), Schnittkar and Heady (1958) and Carter and Heady (1959). They have tended to concentrate on the analysis of the relationships between the various regional and commodity sectors of agriculture and the effect of economic or policy changes on production patterns. They are very difficult to operate because of the wide range of input data needed.

Spatial equilibrium models. The most 'operational' technique and probably the most popular for examining the spatial equilibrium of agricultural production patterns is linear programming. Using this method and providing sufficient data is available it is possible to determine where production should be located if certain goals are to be achieved. Early work by Fox and Tauber (1955) and Judge and Wallace (1958) has been followed by many workers especially in the United States. Outstanding in this field is the work of Heady and his colleagues at Iowa State University. It is impossible to summarize all the work done as the specifications of the various models differ according to the problem being analysed. Instead one of the more recent and important works to be produced by the Iowa team will be examined in some detail.

Egbert and Heady (1964) made a study of interregional competition and the optimal spatial allocation of crop production in the United States. They state their research problem as the need 'to bring agricultural production into greater balance with "food requirements", and to cause the interregional allocation of crops to be more consistent with differential changes in technology and factor prices by regions' (p. 374). In order to do this the United States was divided into 122 producing regions and three linear programming models were used to specify which of the regions might provide the nation's requirements for wheat, feed grains, cotton and soybeans most efficiently in 1965. The first model required that soybeans be grown in rotation, the second model allowed the least-cost mix to be used, and the third model allowed continuous cropping of soybeans in each region. The models had built into them as many as 500 restraints, an upper limit on the acreage of each crop category in each region being especially important. As usual the results are only as good as the data, a fact which the authors emphasize. Non-discrete and non-linear variables could not be handled. The use of linear programming to optimize a solution places the model as a normative one but the use of empirical data gives it a strong descriptive basis. As a problem-solving technique it has many advantages over the von Thunen model.

Decision-making models

The normative models so far discussed assume perfect knowledge and rational behaviour on the part of the farmer. Man is never motivated solely by economic considerations, and social and psychological factors play an important role in determining his decisions. Recently several models have been developed which take into account these non-economic factors and their role in determining the pattern of agricultural production. Their significance in the analysis of land-use patterns in geography has been assessed by Harvey (1966, pp. 368-373).

There are two basic approaches to this problem: one is concerned with the diffusion of information and the patterns of land use produced at different stages in the acceptance of this information; the other considers the farmer's criteria for decision making in the light of his incomplete information. The models used in the first approach are those developed by workers in the field of diffusion theory and the second approach depends mainly on game-theoretical models.

Diffusion models. There are two main constraints on the flow of information through a community. The first constraint concerns the channels of communication. If these channels are constricted in some way then the flow of information will be slower than if the channels were open in every direction. These constrictions may be of several kinds: physical such as a lake or mountain range with few passes, political such as a frontier, or socio-economic for example lack of a radio or the presence of an unpopular information officer. Where these barriers are physical it has proved possible to develop a theoretical model of the expected paths of diffusion waves (Yuill, 1965). Other barriers to communication are more difficult to quantify but there is much empirical evidence to suggest that, for example, an effective agricultural officer may generate circles of decreasing acceptance of a new technique or crop around his headquarters so that his area of influence appears as an enlightened island in a sea of inertia.

The spread of information, in an age of mass communications, may rightly be regarded as less of a constraint than in the past though it is still important in underdeveloped countries (Harvey, 1966, p. 372). But various studies have shown a difference between information availability and the acceptance of the information and this problem of acceptance forms our second constraint. In the United States where mass communications are ubiquitous it has been observed that acceptance of a new technique will vary over space. Again American studies have shown that, while mass communications provide the information, personal contact is important in the final acceptance of the idea.

Much of the early work on diffusion was done in the United States (Rogers, 1962). Agricultural economists and rural sociologists were among the first to recognize the relevance of this work and the study by Bryce and Gross (1943) on the diffusion of hybrid seed corn in Iowa is a classic in this field. Hägerstrand (1953) was the first geographer to develop a model to describe the diffusion of an innovation over space. He studied the acceptance of various new agricultural practices in an area of central Sweden and showed how the innovation spread outwards from an initial centre. He designed three models to simulate the pattern of this diffusion over space, and his third model fitted the observed data remarkably well. The model contained six working assumptions:

1 Only one person possessed the information at the start.
2 The probability of the information being accepted varied through five class of 'resistance'. These classes were established entirely arbitrarily.
3 The information is spread only by telling at pairwise meetings.
4 The telling takes place only at certain times with constant time intervals.
5 At each of these times every knower tells one other person, knower or non-knower.
6 The probability of being paired with a knower depends upon geographical distance between teller and receiver of the information.

In order to make this model operational sophisticated computational techniques are needed. These are known as Monte Carlo methods. The effectiveness of the model depends on its success in simulating patterns based on empirical data. Further work in this field has been done by Wolpert (1960) but so far Sweden is the only country which has been able to supply sufficiently detailed data for long time periods to enable such models to be run.

Game-theoretical models. The basic text on game theory by von Neumann and Morgenstern only appeared in 1944 but since then the model has been shown to have wide application. Game theory was developed to deal with the problem of optimizing decisions in the face of imperfect knowledge. Although the operational procedure is complicated the basic ideas of game theory are relatively simple. In order to present these ideas of game theory let us take an example developed by Harvey (1966, p. 369): Suppose a farmer has three possible crops he can plant on his land and that he can only use one of these (mixtures being excluded). The income from these crops varies according to weather conditions of which only four are recognized. We can then construct a matrix (known as a payoff matrix) which shows the potential return from each landuse system under each weather condition.

Crop	Weather Conditions			
	1	2	3	4
A	500	550	450	600
B	600	700	300	600
C	0	2,000	0	1,000

The entries in the cells of the matrix represent the expected income level in monetary units. Given the payoff matrix the problem then becomes one of selecting criteria for the 'best' solution. Dillon and Heady (1960) suggest seven possible criteria ranging from the maximum-minimum solution, through Simon's theory of the 'satisficer', to criteria which take into account the degree of optimism or pessimism of the farmer and his gambling instincts. The criteria thus range from purely economic to purely behavioural. If we accept the maximum-minimum solution to our problem the crop A gives the highest minimum income. If we assume that all four weather conditions occur with the same relative frequency over time then the system giving the highest income in the long run is that of crop C. Crop B gives a higher income than crop A whilst avoiding the possibility of complete crop failure and consequent absence of income. Obviously the solution will depend on a considerable background knowledge of the problem on the part of the research worker (Langham, 1963, notes the importance of political pressures on farmers) in his understanding of the farmer's decisions.

If we recognize that patterns of agricultural activity are a resultant of human decisions made by a multitude of individual farm operators then it will be seen that an understanding of the processes of decision making is basic to the improvement of our models. The 'normative' theories of the economist must be considered in the light of such concepts as learning theory and the models of the behavioural scientists.

LAND POTENTIAL MODELS

Most of these models have been developed for areas of shifting cultivation in Africa. Allan (1965) gives an exhaustive theoretical survey of these areas and defines the land use factor which forms the basis of many of these models. This factor is defined (Allan, 1965, p. 30) as 'the relationship between the duration of cultivation on each of the land or soil units used in classification and the period of subsequent rest required for the restoration of fertility'. The actual acreage of land cultivated per head of population he defines as the cultivation factor.

Gourou (1962) suggests that the average density of population in Black Africa is only 25 per square mile of total area – but 750 per square mile of cultivated area. In order to define the potential density permitted by the various types of shifting agriculture observed in Africa he suggests the following formula

$$A.C/B$$

where A equals the amount of cultivable land expressed as a proportion of the whole, B equals the total length of rotation in years (period of cultivation plus period of fallow) and C equals the number of inhabitants per acre cleared annually. If we take A at 0·8 (80 per cent), B at 8 (say one year of harvest and 7 fallow or 2 years harvest and 6 years fallow) and C at 4 inhabitants per acre cleared each year then

$A.C/B = 0·8 \times 4/8 = 0·4$ (overall potential density measured in inhabitants per acre, or 256 to the square mile).

If such a formula were applied to the whole of Black Africa it would appear that population density could increase ten times from 25 persons per square mile to 250. Obviously the application of such a formula demands precise knowledge of the existing relationships between population and cultivable land. Regional variations in soil fertility and cultivation techniques would play an important role in determining potential patterns.

Another Belgian geographer Beguin (1964) develops several more complicated formulae for assessing land carrying potential in the Kasai province of the Congo. He defines his potential production as v, the maximum agricultural production from a given area normally obtainable under the existing agricultural system. This potential varies with the agricultural system and the physical environment. If c is the years of cultivation and j is the years of fallow then $c+j$ equals the years of the agricultural cycle. Each year part of an area changes from cultivation to fallow so a given population needs $c+j$ units of area. Each year also c units of production are harvested and the annual agricultural production equals P. The production from one unit of cultivated area is

$$\frac{P}{c+j} \text{ and therefore } v=\frac{P}{c+j}.$$

Beguin states that in Kasai yields per hectare are 0·5 tons for groundnuts, 0·3 tons for beans and 3·7 tons for cassava. Therefore in a rotation of these three crops

$$P = 0 \cdot 5 + 0 \cdot 3 + 3 \cdot 7 = 4 \cdot 5 \text{ tons}$$

Land is cultivated for $2\frac{1}{2}$ years and lies fallow for $6\frac{1}{2}$ years. Thus $c+j=9$ and $v=\frac{4 \cdot 5}{9}=0 \cdot 5$ tons per hectare.

Potential for an agricultural system

In most tropical agricultural systems the number of crops is far greater than in the first model and there are usually several interlocking rotations. So for his second model Beguin includes multiple rotations. Then the model may be stated as follows

$$v = \frac{\sum_{i=1}^{l} v_i . r_i}{\sum_{i=1}^{l} r_i}$$

where l is the number of rotations, v_i is the potential of the ith rotation and r_i is the area occupied by rotation i. When this formula is applied to data from Kasai where there are six principal rotations the results are as follows

$$v_i = 0 \cdot 5, \ 0 \cdot 47, \ 0 \cdot 3, \ 0 \cdot 3, \ 0 \cdot 42, \ 0 \cdot 42 \text{ tons per hectare}$$

when weighted by the area assigned to each rotation

$$v = \frac{(0 \cdot 5)32 + (0 \cdot 47)12 + (0 \cdot 3)25 + (0 \cdot 3)6 + (0 \cdot 42)7 + (0 \cdot 42)12}{32 + 12 + 25 + 6 + 7 + 12}$$

$= 0 \cdot 41$ tons per hectare
or $= 41$ tons per square kilometre

Population density potential

This varies according to the standard of living of the population and Beguin's third model takes this into account. If u is the unit of production per capita and if the population desire one ton each per year then they would need $\frac{u}{v}$ land units. The area (s) necessary for a given population (p) so that each inhabitant could obtain a certain annual production (u) practising a certain type of agriculture on a given soil (v) would be

$$s = \frac{pu}{v}$$

Then the maximum density (d) of this population would be

$$d = \frac{p}{s} = \frac{v}{u}$$

For example if $v=60$ tons per square kilometre in an area where there are 1,000 people producing one ton each then

$$s = \frac{pu}{v} = \frac{1{,}000 \cdot 1}{60} = 16 \cdot 67 \text{ square kilometres}$$

If $s=20 Km^2$ then $p = \frac{20 \times 60}{1} = 1{,}200$ people.

If $u=1 \cdot 2$ tons then $d = \frac{60}{1 \cdot 2} = 50$ inhabitants per square kilometre.

Areal differentiation

It is more realistic to assume that the potential varies through area in response to environmental and cultural differences. If we accept that the maximum (v_M) and minimum potential (v_m) are on a continuum and vary linearly then mean potential ($\bar{v}$) equals $\frac{v_M + v_m}{2}$.

Then in a region of 100 square kilometres with a potential varying linearly from 100 tons per square kilometre to 60 tons per square kilometre, giving one ton per inhabitant per year the population maximum would be

$$p = \frac{100 \times \frac{100+60}{2}}{1} = 8{,}000 \text{ inhabitants}$$

Thus the maximum population density is 80 inhabitants per square kilometre.

Beguin develops several more refined formulae which take into account sequent occupation of land in which later occupants are forced on to land of lower potential; non-homogeneous regions; input of work per capita; and change over time as it affects population, production per capita and potential. Thus the model is both dynamic and wide ranging as to its input. Its main disadvantage is the difficulty of obtaining the empirical data on which it is based. Beguin is one of those rare people who are capable of building theoretical models on a strong basis of fieldwork. This deductive approach to model building may provide us with our most useful models for the development of agricultural geography.

FUTURE TRENDS

Perhaps the basic problem of agricultural geography is that of aggregation. This has two aspects, of scale and depth. Firstly we have a brick-laying problem, that of combining micro-studies of our basic building block, the farm, into statements about macro-areas. Secondly we have the problem of combining data concerning the physical environment with information related to the human environment.

The development of new equipment and techniques is enabling us to solve some of these long-standing problems. Modern digital computers have a vast storage capacity and can make large numbers of calculations in a very short space of time. Thus they help us to deal with the problem of combining information from many farms for an area and they make it possible to apply simulation models with their need for many iterations, to behavioural data. New statistical methods of combining quantitative and qualitative data make it possible to improve our studies in depth.

New methods are enabling us to make greater use of our information and forcing us to re-examine some of our theoretical concepts. The need for a new approach to agricultural geography has been seen (Reeds, 1964; Brookfield, 1964) and it is hoped that the models now available to us will help to bring this about.

REFERENCES

AHMAD, E., [1952], Rural Settlement Types in the Uttar Pradesh (United Provinces of Agra and Oudh); *Annals of the Association of American Geographers*, 42 (3), 223–246.

ALLAN, W., [1965], *The African Husbandman*, (London), 505 pp.

AL-MAIYAH, ALI MOHAMMED, [1958], *An Analysis of the Spatial Relationships Among Agricultural Phenomena in Iraq*, 1953; Unpublished Ph.D. dissertation, State University of Iowa, USA.

ANDERSON, J., [1965], The use of fodder and livestock units in agricultural geography: A Case Study of Soviet Land Use Policy; *Annals of the Association of American Geographers*, 55 (4), 603.

BAKER, O. E., [1926], Agricultural Regions of North America; *Economic Geography*, 2, 459–493.
BEAL, G. M. and ROGERS, E. M., [1960], The Adoption of Two Farm Practices in a Central Iowa Community; *Special Report*, No. 26, *Agricultural and Home Economics Experiment Station*, (Ames, Iowa).
BEELEY, B. W., [1965], Agricultural Change: A 'Field' Study; *Annals of the Association of American Geographers*, 55 (4), 605.
BEGUIN, H., [1964], *Modèles géographiques pour l'espace rural africain*, (Brussels).
BERTALANFFY, L. VON, [1951], An outline of general system theory; *British Journal of the Philosophy of Science*, 1, 134–165.
BHATIA, S. S., [1960], An Index of Crop Diversification; *Professional Geographer*, XII (2), 3–4.
BIRCH, J. W., [1954], Observations on the Delimitation of Farming Type Regions, with special reference to the Isle of Man; *Transactions of the Institute of British Geographers*, 20, 141–158.
BLAUT, J. M., [1953], The Economic Geography of a One-Acre Farm in Singapore: A Study in Applied Microgeography; *Malayan Journal of Tropical Geography*, 1, 37–48.
BLAUT, J. M., [1959], Microgeographic Sampling: A Quantitative Approach to Regional Agricultural Geography; *Economic Geography*, 35 (1), 79–88.
BLAUT, J. M., [1961], The Ecology of Tropical Farming Systems; *Revista Geografica*, 28 (1), 47–67.
BOARD, C., [1963], Some Methods of Mapping Farm Type Areas; Unpublished Paper.
BOSERUP, E., [1965], *The Conditions of Agricultural Growth*, (London), 124 pp.
BOWDEN, L. W., [1965], The Diffusion of the Decision to Irrigate; *University of Chicago, Department of Geography, Research Paper* No. 97.
BRANDNER, L. and KEARL, B., [1964], Evaluation for Congruence as a Factor in the Adoption Rate of Innovations; *Rural Sociology*, 29, 288–303.
BROOKFIELD, H. C., [1962], Local Study and Comparative Method: An Example from Central New Guinea; *Annals of the Association of American Geographers*, 52 (3), 242–253.
BROOKFIELD, H. C., [1964], Questions on the Human Frontiers of Geography; *Economic Geography*, 40 (4), 283–303.
BUCHANAN, R. O., [1938], A Note on Labour Requirements in Plantation Agriculture; *Geography*, 23, 156–164.
BUCHANAN, R. O., [1959], Some Reflections on Agricultural Geography; *Geography*, 44, 1–13.
BUNGE, W., [1962], Theoretical Geography; *Lund Studies in Geography, Series C, General and Mathematical Geography*, No. 1, 210 pp.
BUTLER, J. B., [1960], *Profit and Purpose in Farming*; (*A Study of Farms and Smallholdings in part of the North Riding*), Department of Economics, University of Leeds, 68 pp.
CARTER, H. C., [1963], Representative Farms – Guides for Decision Making; *Journal of Farm Economics*, 45 (5), 1449–1455.
CHISHOLM, M., [1962], *Rural Settlement and Land Use*, (London), 207 pp.

CHISHOLM, M., [1964], Problems in the Classification and use of the Farming Type region; *Transactions of the Institute of British Geographers*, 35, 91–103.

CLARK, A. H., [1962], The Sheep/Swine Ratio as a Guide to a Century's Change in the Livestock Geography of Nova Scotia; *Economic Geography*, 38 (1), 38–55.

CLARK, C. and HASWELL, M. R., [1964], *The Economics of Subsistence Agriculture*, (London), 218 pp.

COPPOCK, J. T., [1964A], Crop-livestock and enterprise combinations in England and Wales; *Economic Geography*, 40 (1), 65–81.

COPPOCK, J. T., [1964B], *Agricultural Atlas of England and Wales*, (London), 255 pp.

DAY, L. M., [1963], Use of Representative Farms in Studies of Interregional Competition and Production Response; *Journal of Farm Economics*, 45 (5), 1438–1444.

DEAN, G. W. and BENEDICTIS, DE M., [1964], A Model of Economic Development for Peasant Farms in Southern Italy; *Journal of Farm Economics*, 46 (2), 295–312.

DILLON, J. L. and HEADY, E. O., [1960], Theories of Choice in Relation to Farmer Decisions; *Agricultural and Home Economics Experiment Station, Iowa State University, Research Bulletin* 485, (Ames, Iowa).

DUNN, E. S., [1954], *The location of agricultural production*, (Gainesville), 115 pp.

EDWARDS, D., [1961], *An Economic Study of Small Farming in Jamaica*, (Mona, Jamaica), 370 pp.

EDWARDS, D. and REES, A. M. M., [1964], The Agricultural Economist and Peasant Farming in Tropical Conditions; In *International Explorations of Agricultural Economics*, (Ames, Iowa), 73–85.

EDUCATION, MINISTRY OF, [1960], Geography and Education; *Ministry of Education Pamphlet* No. 39, (London).

EGBERT, A. C., HEADY, E. O. and BROKKEN, R. F., [1964], Regional Changes in Grain Production; *Iowa Agricultural Experimental Station Research Bulletin*, No. 521, (Ames, Iowa).

ELLIOT, F. F., [1928], The 'Representative Firm' Idea Applied to Research and Extension in Agricultural Economics; *Journal of Farm Economics*, 10, 481–489.

EDMONSON, M. S., [1960], Hybrid Corn and the Economics of Innovation; *Science*, 132: 3422, 275–280.

FIRTH, R., [1951], *Elements of Social Organization*, (London).

FOX, K. and TAUBER, R., [1955], Spatial Equilibrium Models of the Livestock Feed Economy; *American Economic Review*, 45, 584–608.

FRANKLIN, S. H., [1962], Reflections on the Peasantry; *Pacific Viewpoint*, 3 (1), 1–26.

FRANKLIN, S. H., [1965], Systems of Production: Systems of Appropriation; *Pacific Viewpoint*, 6 (2), 145–166.

FRANKLIN, S. H., [1966], Personal Communication.

GARRISON, W. L. and MARBLE, D. F., [1957], The Spatial Structure of Agricultural Activities; *Annals of the Association of American Geographers*, 47, 137–144.

GERLING, W., *Die Plantage*, (Würzburg).

GOULD, P. R., [1963], Man against his environment: a game-theoretic framework; *Annals of the Association of American Geographers*, 53, 290–297.

GOUROU, P., [1962], *Agriculture in the African Tropics: the Observations of a Geographer*; Paper read at the University of Oxford.

GREGOR, H. F., [1965], The Changing Plantation; *Annals of the Association of American Geographers*, 55 (2), 221–238.
GROTEWALD, A., [1959], Von Thünen in Retrospect; *Economic Geography*, 35 (4), 346–355.
HÄGERSTRAND, T., [1952], The propogation of Innovation waves; *Lund Studies in Geography, Series B, Human Geography*, 4, 3–9.
HÄGERSTRAND, T., [1953], *Innovationsförloppet ur korologisk synpunkt*, (Lund).
HAGGETT, P., [1965], *Locational Analysis in Human Geography*, (London), 339 pp.
HAGOOD, M. J., [1943], Statistical methods for delineation of regions applied to data on agriculture and population; *Social Forces*, 21, 288–297.
HAHN, E., [1882], Die Wirtschaftsformen der Erde; *Petermann's Mitteilungen*, 38. 8–12.
HARVEY, D. W., [1966], Theoretical concepts and the analysis of agricultural land-use patterns in geography; *Annals of the Association of American Geographers*, 56, 361–374.
HARTSHORNE, R. and DICKEN, P., [1935], A classification of the Agricultural Regions of Europe and North America on a Uniform Statistical Basis; *Annals of the Association of American Geographers*, 25, 99–120.
HEADY, E. O. and EGBERT, A. C., [1964], Regional Programming of Efficient Agricultural Production Patterns; *Econometrica*, 32 (3), 374–386.
HELBURN, N., [1957], The bases for a classification of World Agriculture; *Professional Geographer*, 9, 2–7.
HELLER, C. F., [1964], The use of Model Farms in Agricultural Geography; *Professional Geographer*, 16 (4), 20–23.
HENSHALL, J. D., [1964], *The Spatial Structure of Barbadian Peasant Agriculture*; Unpublished M.Sc. Thesis, McGill University, Montreal.
HENSHALL, J. D. and KING, L. J., [1966], Some Structural Characteristics of Peasant Agriculture in Barbados; *Economic Geography*, 42 (1), 74–84.
HENSHALL, J. D., [1966], The Demographic factor in the structure of Agriculture in Barbados; *Transactions of the Institute of British Geographers*, 38, 183–195.
HOOVER, E. M., [1936], The measurement of industrial localization; *Review of Economics and Statistics*, 18, 162–171.
HUTCHINSON, H. W., [1959], Comments on E. T. Thompson's 'The Plantation as a Social System'; In *Plantation Systems of the New World*, Social Science Monography No. VII, 37–40, (Washington).
INNIS, D. Q., [1961], The Efficiency of Jamaican Peasant Land Use; *The Canadian Geographer*, V (2), 19–23.
JACKSON, B. G. et al., [1963], The Pattern of farming in the Eastern Counties; *Occasional Papers No. 8, Farm Economics Branch, School of Agriculture, Cambridge*.
JONASSON, O., [1925–26], Agricultural Regions of Europe; *Economic Geography*, 1, 277–344 and 2, 19–48.
JOHNSON, R. W. M., [1964], The Labour Economy of the Reserves; *Department of Economics, Occasional Paper No. 4, University College of Rhodesia and Nyasaland*, (Salisbury).
JONES, C. F., [1928–30], Agricultural Regions of South America; *Economic Geography*, 4, 1–30, 159–186 and 267–294; 5, 109–140, 277–307 and 390–421; 6, 1–36.

JONES, W. D., [1930], Ratio and isopleth maps in Regional Investigation of Agricultural Land Occupance; *Annals of the Association of American Geographers*, 20, 177-195.
JUDGE, G. G. and WALLACE, T. D., [1958], Estimation of spatial price equilibrium models; *Journal of Farm Economics*, 40, 801-820.
KENDALL, M. G., [1939], Geographical Distribution of Crop Productivity in England; *Journal of the Royal Statistical Society*, 102, 21-62.
KEUNING, H. J., [1964], Agrarische Geografie: Doelstelling, Ontwikkeling, Methoden; *Tijdschrift van het Koninklijk Nederlandsch Aardrijkskundig Genootschap* (Amsterdam), 81 (1), 10-19.
KRUMBEIN., W. C., [1956], Regional and local components in facies maps; *Bulletin of the American Association of Petroleum Geologists*, 40, 2163-2194.
LANHAM, W. J. and COUTU, A. J., [1964], Area Resource Adjustments for Specified Net Revenue Goals and Levels of Factor Prices on Farms in Economic Area 7, N. Carolina; *Agricultural Economics Information Series No. 109, Department of Agricultural Economics, North Carolina State of the University of North Carolina at Raleigh.*
LANGHAM, M. R., [1963], Game theory applied to a policy problem of rice farmers; *Journal of Farm Economics*, 45 (1), 151-162.
LATHAM, J. P., [1959], The Distance Relations and some other characteristics of Cropland Areas in Pennsylvania; *Technical Report No 6, Contract NONR 551 (O1) University of Pennsylvania.*
LEONTIEFF, W. W., [1953], *Studies in the structure of the American Economy*, (New York).
LÖSCH, A., [1954], *The Economics of Location*; (translated by W. W. Woglom), (New Haven).
MANLEY, V. P. and OLMSTEAD, C. W., [1965], Geographical Patterns of Labour Input as related to output indexes of scale of operation in American Agriculture; *Annals of the Association of American Geographers*, 55 (4), 629-630.
MARTIN, A., [1958], *Economics and Agriculture*, (London), 169 pp.
MATHER, E. C., [1944], A linear-distance map of farm population in the United States: *Annals of the Association of American Geographers*, 34, 173-180.
MCCARTY, H. H., [1954], Agricultural Geography; In *American Geography: Inventory and Prospect*, James, P. E. and Jones, C. F. (Eds.), (Syracuse), 258-277.
MCCLURE, J. A., [1964], The use of correlation and factor analytic techniques in comparing farmland potentials; *Annals of the Association of American Geographers*, 54 (3), 430.
MIKHEYEVA, V. S., [1963], An economic-mathematical model of the Location of Farm Production by Regions of the Soviet Union; *Soviet Geography, Review and Translation*, 4 (3), 24-29.
MILLER, B. R. and KING, R. A., [1964], Models for measuring the impact of technological change on location of marketing facilities; *Agricultural Economics Information Series No. 115. Department of Agricultural Economics, North Carolina State of the University of N. Carolina, Raleigh.*
NEUMANN, J. VON and MORGENSTERN, O., [1944], *Theory of games and economic behaviour*, (Princeton).
OOI, JIN BEE, [1959], *Land, People and Economy in Malaya*, (London).

OTREMBA, E., [1950–60], Allgemeine Agrar-und Industriegeographie; *Erde und Weltwirtschaft*, 3, 213–229 and 343–351.

PADILLA, E., [1960], Contemporary Social-Rural Types in the Caribbean Region; In *Caribbean Studies: A Symposium*, V. Rubin (Ed.), (Seattle), 22–28.

PICO, R., [1959], Comments on 'Some observations relating to population dynamics in plantation areas of the New World' by T. Lynn Smith in *Plantation Systems of the New World*. Social Science Monographs No. VII, (Washington).

PLATT, R. S., [1930], Pattern of Occupancy in the Mexican Laguna District; *Transactions of the Illinois State Academy of Science*, 22, 533–541.

PLATT, R. S., [1942], *Latin America: Countrysides and United Regions*, (New York), 564 pp.

PLAXICO, J. S. and TWEETEN, L. G., [1963], Representative farms for Policy and Projection Research; *Journal of Farm Economics*, 45 (5), 1458–1465.

PROTHERO, R. M., [1957], Land Use at Soba, Zaria Province, Northern Nigeria; *Economic Geography*, 33, 72–86.

REDFIELD, R., [1956], *Peasant Society and Culture*, (Chicago), 92 pp.

REEDS, L. G., [1964], Agricultural Geography: Progress and Prospects; *Canadian Geographer*, 8 (2), 51–63.

ROGERS, E. M., [1962], *Diffusion of Innovations*, (New York).

ROGERS, E. M., [1964], Bibliography of Research in the Diffusion of Innovations: *Research on the Diffusion of Innovations No 1, Department of Communication, Michigan State University*.

RYAN, B. and GROSS, N. C., [1943], The Diffusion of Hybrid Seed Corn in Two Iowa Communities; *Rural Sociology*, 8, 15–24.

SAUER, C. O., [1952], Agricultural Origins and Dispersals; *American Geographical Society, Bowman Memorial Lectures*, 2.

SAUER, C. O., [1956], The Agency of man on the Earth; In *Man's Role in Changing the Face of the Earth*, Thomas, W. L. Jr., (Ed.), (Chicago).

SCOTT, P., [1957], The Agricultural Regions of Tasmania; *Economic Geography*, 33, 109–121.

SINGE, H., [1965], Crop combination regions in the Malwa tract of Punjab; *Deccan Geographer*, 3 (1), 21–30.

SMITH, T. L., [1959], Some observations relating to population dynamics in plantation areas of the New World; In *Plantation Systems of the New World*, Social Science Monographs, VII, (Washington), 126–132.

STAMP, L. D., [1948], *The Land of Britain: Its Use and Misuse*, (London), 507 pp.

STEEL, R. W., FORTES, M. and ADY, P., [1947], Ashanti Survey 1945–6: An Experiment in Social Research; *Geographical Journal*, 110, 149–179.

STEWARD, J. H., [1960], Perspectives on Plantations; *Revista Geografica*, No. 52, 26 (1), 77–85.

TAKAYAMA, T. and JUDGE, G. G., [1964], An interregional activity analysis model for the agricultural sector; *Journal of Farm Economics*, 46 (2), 349–365.

TAYLOR, G., [1930], Agricultural Regions of Australia; *Economic Geography*, 6, 109–134 and 213–242.

THOMAS, D., [1963], *Agriculture in Wales during the Napoleonic Wars: A study in the geographical interpretation of historical sources*; (Cardiff).

THÜNEN, J. H. VON, [1875], *Der Isolierte Staat in Beziehung auf Landwirtschaft und*

Nationalökonomie; Third edition, (Berlin). (A first edition of Part I appeared in 1826).
VALKENBURG, S. VAN, [1931–36], Agricultural Regions of Asia; *Economic Geography*, 7, 217–237; 8, 109–133; 9, 1–18, 109–135; 10, 14–34; 11, 227–246, 325–337; 12, 27–44, 231–249.
VALKENBURG, S. VAN and HELD, C. C., [1952], *Europe*, (New York).
WAIBEL, L., [1958], *Capítulos de geografia tropical e do Brasil*, (Rio de Janeiro).
WEAVER, J. C., [1954], Changing Patterns of cropland use in the Middle West; *Economic Geography*, 30 (1), 1–47.
WEAVER, J. C., [1954], Crop-combination regions in the Middle West; *Geographical Review*, 44 (2), 175–200.
WEAVER, J. C., [1954], Crop-combinations regions for 1919 and 1929 in the Middle West; *Geographical Review*, 44 (4), 560–572.
WEAVER, J. C., [1954], Isotope and Compound: A framework for Agricultural Geography; *Annals of the Association of American Geographers*, 44 (3), 286–288.
WEAVER, J. C., HVAG, L. P. and FENTON, B. L., [1956], Livestock Units and combination regions in the Middle West; *Economic Geography*, 32, 237–259.
WEAVER, J. C., [1956], The county as a spatial average in agricultural geography; *Geographical Review*, 46 (4), 536–565.
WHITTLESEY, D., [1936], Major agricultural regions of the Earth; *Annals of the Association of American Geographers*, 26, 199–240.
WOLF, E. R., [1955], Types of Latin American Peasantry: A Preliminary Discussion; *American Anthropologist*, 57 (3), 452–471.
WOLPERT, J., [1963], *Decision making in Middle Sweden's Farming: A Spatial Behavioural Analysis*; University of Wisconsin Ph.D., University Microfilms, (Ann Arbor).
WOLPERT, J., [1964], The decision process in spatial context; *Annals of the Association of American Geographers*, 54 (4), 537–558.
YUILL, R. S., [1965], A simulation study of barrier effects in spatial diffusion problems; *Michigan Inter-University Community of Mathematical Geographers, Discussion Papers*, 5.
ZABKO-POTOPOWICZ, A., [1957], The development of the geography of agriculture since World War I; *Przeglad geograficzny*, (Warsaw), 29 (1), 21–46.

PART IV

Models of Integrated Systems

CHAPTER TWELVE

Regions, Models and Classes

DAVID GRIGG

INTRODUCTION

In the last half century there have been great changes in the aims and methods of academic geography; yet in spite of these changes geographers have remained concerned with the idea of the region and have used regional methods of investigation not only in regional geography but also in the systematic branches of the subject. Not surprisingly there have been a number of controversies over the nature of the region and some disagreement about the methods which should be used in their delimitation. Some geographers have made searching criticisms of the validity of the regional concept (Crowe, 1938, pp. 7–11; Kimble, 1951, pp. 151–174); others, whilst accepting the usefulness of regionalization as a method of inquiry, have regretted the prominence given within the field of geography to regional geography (Leighly, 1937, pp. 125–141; Ackerman, 1945, pp. 121–143 and 1953, pp. 189–197; Thornthwaite, 1961, pp. 345–356). The problems have perhaps been further confused by the fact that 'regions' are an issue outside the world of geography in the form of 'regionalism' and 'regional planning'. There are obvious interrelationships between the regional concept, methods of regional delimitation, regional geography, regionalism and regional planning (Morgan, 1939, pp. 68–88; Gilbert, 1953, pp. 345–371); but our concern here is primarily with regionalization as a *method* of investigation, which it is assumed, can be used in any branch of geography. We are also necessarily concerned with the regional concept.

There have been a number of admirable reviews of the regional concept and regional methods (Hartshorne, 1939, pp. 250–284 and 1959, pp. 129–145; Robinson, 1953, pp. 49–58; Gilbert, 1960, pp. 157–175; Whittlesey, 1954, pp. 21–68). But for the most part such reviews have been mainly concerned with ideas and methods of geographers alone; and they have also usually assumed that regionalization is a method of investigation peculiar to geography and thus having no parallel with methods in other sciences. There are at least two reasons why a broader view of the region should be attempted.

In the first place whilst geographers have done much to develop regional ideas and methods, the concept has never been confined to geography; and lately has been adopted or independently developed in a number of other sciences. We may refer, for example, to the work of some plant ecologists and phytosociologists. Both are concerned with the nature of plant communities; the latter with describing the floristic composition, structure and physiogonomy of plant communities, the former with the inter-relations between plant communities and their environment. Whilst the spatial distribution of plant communities has not been a prime concern in these sciences it is an issue which inevitably arises. Similarly whilst soil science is perhaps mainly concerned with the definition and genesis of soil types, the distribution of these types and their possible arrangement in soil regions has received some consideration. Climatology is less peripheral to the work of geographers than either pedology or plant ecology and there is little need here to draw attention to the problems of classifying climatic types and the establishment of climatic regions (Hare, 1951, pp. 111–134). Nor is the use of regional methods confined to physical geography and allied disciplines for in fields related to human geography there has been much consideration of regional problems. In cultural anthropology there has been work on the *culture areas* of primitive peoples, particularly in North America (Wissler, 1917; 1927, pp. 881–891 and 1928, pp. 894–900; Kroeber, 1931A, pp. 646–647; 1931B, pp. 248–265 and 1963). Urban sociologists have used what is essentially the regional concept in studying the *natural* and *social areas* of cities (Park, Burgess and McKenzie, 1925; Shevky and Bell, 1955). A number of rural sociologists have studied the *socio-cultural areas* of parts of the United States (Hagood et al., 1941, pp. 216–233 and 1943, pp. 287–297; Lively and Gregory, 1954, pp. 21–31) and recently there has been a growing interest among economic historians in the regional approach (Broude, 1960, pp. 588–596), particularly amongst agricultural historians (Hoskins, 1954, pp. 3–11). Even amongst economists who for long have been concerned primarily with the aggregate analysis of economies there has been of late an awareness that a regional approach may be fruitful. In this connection the work of W. Isard and his school of regional scientists is of particular note (Isard, 1960; Isard and Cumberland, 1961; Garnsey, 1956, pp. 27–39; Isard, 1956, pp. 13–26).

This by no means exhausts the manifestations of the regional concept and the use of regional methods in allied fields, but it is sufficient to indicate that in many sciences where the spatial distribution of phenomenon over the earth's surface, and particularly where these phenomena can be studied from an ecological point of view, the regional approach has been adopted or evolved independently. It is of interest to note that whilst the antecedents of the regional approach can perhaps be traced far back into the history of all these subjects, there seems to have been a marked revival of interest at the end of the nineteenth century and in the early part of the twentieth century (Kroeber, 1931B; Heberle, 1943, p. 280; Whittlesey, 1954, pp. 23–24).

But there is a second reason for looking at the region and regional methods in a broader perspective than has hitherto been customary. It can be argued that whilst the methods of different sciences are often very distinctive and appear at first sight to have little in common, yet science in general is using similar methods and that these can be reduced to a number of fundamental procedures. Once such a reduction has been achieved the problems of procedure sometimes become problems of philosophy as much as of a particular science. The field of the philosophy of science has been held to be practised by physicists who know no philosophy and philosophers who know no physics. This is unkind and untrue but indicates the dangers of trespassing in such an esoteric field. None the less it does seem worthwhile to search for parallels in the methods of different sciences, particularly where the methods used in an immature science can be shown to be similar to those in an advanced and long-established subject. Such parallels often exist, although obscured by differences in terminology and perhaps also by the blinkers of academic specialization.

Some simple procedures are common to all sciences. The observation and recording of facts, the classification of those facts, the development of inductive generalizations and the formation of deductive theories is characteristic of most fields of investigation. Here it is argued that, although in geography the classifications of allied fields are often adopted, yet the procedure which is analogous to the stage of classification in other sciences is regionalization. Such a view has of course been implicit in the work of many geographers but it has been stated more explicitly in recent years (De Jong, 1962; Bunge, 1962, pp. 14-23; Grigg, 1965, pp. 465-491). If we accept such a view then the procedures and principles of classification may be with profit applied to regionalization.

Classification is fundamental to the advance of any science, but it is generally an early stage in development. Thus it is no longer an issue of significance in say, chemistry, whilst in subjects such as geography or sociology it is still immaturely developed. But in a group of sciences – zoology and botany in particular, classification – or taxonomy – is of central importance and in the last twenty-five years has been subject to a great deal of critical examination. There is consequently a considerable literature in these subjects on the fundamental problems of classification (Gilmour and Walters, 1964, pp. 1-22; Gilmour, 1961, pp. 27-45; Sokal and Sneath, 1963; Simpson, 1961).

Now if we accept the argument that classification and regionalization are analogous procedures – and there are considerable objections to this view – we may take the argument one stage further. The fundamental procedures of taxonomy are based on the procedures of classification and division of formal logic. Classification can profitably be considered in terms of formal logic to clarify some of the issues and this is very useful when applied to a particular field (Gilmour and Walters, 1964, pp. 1–22; Cline, 1949, pp. 81–91; Grigg, 1965, pp. 465–491).

It will be as well to briefly review the argument so far. It is held that it would be beneficial to consider the regional concept in a broader perspective. First, by comparing the ideas and methods of geographers with those using the methods implicitly and explicitly in a variety of fields where the spatial distribution of phenomena is studied. Second, to seek an analogy between regionalization and classification, and particularly in the parts of those subjects where classification is of major importance but where there is no concern with spatial distribution. Third, by considering the procedures of regionalization in the light of the principles of classification and division of formal logic.

We must turn now to the region as seen by geographers.

THE DEVELOPMENT OF THE REGIONAL CONCEPT IN GEOGRAPHY

There is no agreed definition of the term 'region' when it is used unqualified by an adjective: but it seems generally to be used to mean a part of the earth's surface which is distinguished in some defined way from surrounding areas. This distinctiveness may be based upon a single criterion such as the mean annual temperature or the proportion of workers engaged in manufacturing industry; or it may be based upon a number of criteria. Thus a part of the earth's surface may be said to be an agricultural region because all the farms in the region have more than a certain acreage, grow a similar proportion of wheat and have the same range of farm equipment. Following Whittlesey (Whittlesey, 1954, p. 35) we may thus distinguish between single-feature and multiple-feature regions.

There are a great many synonyms for the word region. Thus terms such as province, division, zone, belt, locality, and district have all been used by geographers in much the same sense as 'region' save that these words have often been used to imply a particular rank in a hierarchy of regions. Thus for example Whittlesey suggested that the following terms be used to suggest regions of a different order: locality, district, province and realm (Whittlesey, 1954, pp. 48–51). Russian geographers have laid great stress on the need for a hierarchy of regions and have suggested a variety of terms to denote categories of region of a different order (Solntsev, 1962, p. 10; Grigor'yev, 1962, p. 183). In this paper the term region is used without any implication of rank.

The idea of the region has a long history, but the first systematic presentations of the concept came in the eighteenth century when geographers found political units an inadequate basis for description and sought more 'natural areas. The regional idea was much discussed in the first half of the nineteenth century, attracted less attention in the second half only to flourish again in the last decade. Regional thinking has remained prominent in geography ever since. The first two decades of this century saw the development of three distinct approaches to the regional concept.

The Pays concept

The French school of geographers of the early twentieth century produced a series of monographs on parts of France which were held to have a distinctive way of life, or *genre de vie*. These pays were for the most part rural, untouched by industrialization, still little influenced by the transport revolution and essentially local. Distinctive economies existed which were characterized by vernacular building styles, particular ways of farming and patterns of settlement. De la Blache, the leading geographer of this school, did not, as earlier geographers had argued, see these distinctive economies as a simple response to a particular environment. Rather he saw that:

'... the adjustment of each society to the peculiarities of the local physical environment, taking place over many centuries, produces local characteristics in that society which are not to be found elsewhere. Man and nature become moulded to one another over the years rather like a snail and its shell. Yet the connection is more intimate even than that, so that it is not possible to disentangle influences in one direction, of man on nature, from those in another, of nature on man. The two form a complicated amalgam' (Wrigley, 1965, p. 8).

The French writers had a profound influence on concepts of the region in other countries; but, as Kimble and Wrigley have pointed out (Kimble, 1951, pp. 167–168; Wrigley, 1965, pp. 7–13) it was a method of analysis suited to localized, agrarian societies and has been markedly less successful in dealing with modern industrial societies. Amongst French geographers themselves new concepts of the region have gained favour and have been used to deal with a now industrialized France, yet the methods of de la Blache still prove successful in territories remaining primarily rural and localized (McDonald, 1964, pp. 20–23; Annette, 1965, pp. 1–5).

The natural region

Whilst the French geographers dealt with small parts of the earth's surface and were concerned with both the physical and what came to be called the cultural environment, A. J. Herbertson in his celebrated paper of 1905 (Herbertson, 1905, pp. 300–312) tried to divide the whole world into regions on the basis primarily of its physical features. He pointed out that whilst there existed several divisions of the world upon the basis of single features such as climate or vegetation, the actual environment man lived in was composed of a great many elements; of these he claimed that climate, configuration and vegetation were the most significant. Vegetation, Herbertson assumed, corresponded closely to climate, whilst configuration he regarded as secondary. Thus his actual map of natural regions appears to be little more than a climatic classification. Although Herbertson later is said to have wished to have laid more stress on vegetation (Fleure, 1952, p. 98), his final version shows no radical changes (Stamp, 1957, pp. 204–206) from the first.

Herbertson's paper was of great importance; it pointed to the close relationships which existed between climate and vegetation, and stressed that whilst the separate elements of the physical environment may be more conveniently studied individually, man, animals and plants experience the total environment, not the elements separately. His work did not deal with the processes which link the separate elements of the environment, nor was his regional delineation any more than a provisional sketch. But it surely pointed to the most interesting possibilities for further research.

Unfortunately English speaking geographers took up a relatively minor, if attractive, feature of his work. Herbertson suggested that the natural regions might form a useful background to the study of human societies:

> By comparing the histories of the same race in two different regions; or of a succession of races in the same region, it should be possible to arrive at some knowledge of the invariable effect of a type of environment on its inhabitants and permit some estimation of the non-environmental factors in human development. (Herbertson, 1905, p. 309.)

It was this aspect of Herbertson's work, particularly after a posthumously published paper which emphasized this approach (Herbertson, 1916, pp. 147–153) that English geographers seized upon. There were few attempts to follow up his attempt at physical-geographic regionalization or to investigate the links between climate and vegetation, and climate and soil. Instead there followed a series of works which Kimble later described as making 'the natural region serve as the plaster cast of a specific kind of human economy' (Kimble, 1951, p. 153). In this his successors are perhaps not to blame, for whilst most geographers interpret Herbertson's use of the word 'natural' as meaning physical, there is some evidence in his later work that he was using it instead in the sense of a natural classification; that is to say, a classification which embraces a wide range of criteria both physical and cultural, as distinct from one or two criteria (Herbertson, 1913B, p. 205).

Herbertson was not, as he himself pointed out (Herbertson, 1913A, p. 159) the first to attempt the division of the world based upon a number of criteria of the physical environment. A number of botanists had attempted to divide the world into vegetation regions and had been struck by the similarity between climatic and vegetation types. But perhaps the most productive work was done in Russia by soil scientists. V. V. Dokuchayev and his successors defined soil types as a product largely, on a broad scale, of climate, and also emphasized the intimate relations between soils and vegetation. From this work there soon stemmed an impressive amount of work upon the natural regions of Russia, of which the books by L. S. Berg and S. P. Suslov are perhaps the best known outside the USSR (Berg, 1950; Suslov, 1961). But as Grigor'yev has pointed out, this promising line of work ran into difficulties, firstly because the boundaries of soil, climate and vegetation did not always coincide, and secondly because there was no accurate knowledge of the processes which might cause such natural differentiation. However, Russian geographers have continued to work on what is now called the problem of geographic zonality and it may be that the key to physical regionalization has been discovered in recent work on the territorial differences in the productivity of living matter and studies of the heat and moisture balance (Grigor'yev, 1961, pp. 3–16; 1962, pp. 182–187).

Single feature regions

Both the pays concept and the idea of the natural region were based on a number of features of the earth's surface. Yet there had long been attempts to divide the earth into regions on the basis of what for the moment we may call single features. In particular a number of plant geographers, of whom de Candolle, Grisebach and Schimper are perhaps the most celebrated, had attempted to establish a classification of vegetation and to map these types. The early zoogeographers had tried to establish faunal regions of the world, the work of Sclater and Wallace being of particular note, whilst Köppen's great work on climatic regions was beginning to appear at the time when Herbertson wrote. The developments in these fields have been reviewed by a number of writers (Küchler, 1951, pp. 275–283 and 1954, pp. 429–440; Dansereau, 1951, pp. 172–173; Kendeigh, 1954, pp. 152–171; Raup, 1940, pp. 319–354; Thornthwaite, 1943, pp. 233–255; Hare, 1951, pp. 111–134; Davies, 1961, pp. 412–417; George, 1962, pp. 13–14).

Most of these attempts at regionalization had a number of features in common.

First, they attempted to deal with large areas – either the whole world or a continent. Consequently little attention was paid to the exact limits of the regions described.

Second, whilst these delimitations dealt with a single feature such as animals, or vegetation formations or climate, the criteria on which the regional delimitation was made were frequently a property not of the things regionalized but of some other element of the earth's surface which was presumed to be largely responsible for the distribution studied. Thus Merriam's life zones, which divided North America into regions were delimited not by the limits or range of a particular combination of species, but by critical isotherms which were assumed to cause the differentiation of animal communities (Kendeigh, 1954, p. 163). Similarly early attempts to map soil regions defined the limits of soil regions not in terms of properties of the soil but on the basis of supposedly important isotherms or isohyets (Basinskii, 1959, pp. 14–26). Conversely Köppen assumed that the major vegetation formations were an accurate reflection of differences in climate and thus sought to delimit climatic regions by seeking isotherms and isohyets which corresponded to the boundaries of vegetation types (Hare, 1951, pp. 111–121). These early attempts then, were both genetic and deductive.

Later there was a reaction from this approach. Other workers began to deal with smaller areas where the correspondence between climate and soil type was far from clear. There grew up a school of plant ecologists who dealt with much smaller plant communities than the formation, and who were soon aware of factors other than climate, and as Raup has observed, there was a move away from the simpler climatic determinism of the late nineteenth century (Raup, 1940, p. 331).

But these early classifications had other defects. It soon became clear the single feature classifications were in fact *not* single feature classifications. Thus some plant geographers attempted regional divisions which were based upon floristic composition, whilst others were more concerned with the physiognomy of vegetation. The two classifications did not necessarily coincide. Similar difficulties arose with the classification of soil types and climatic types. The early attempts at areal classification were made when it was presumed that there could be a natural system of classification of phenomena. Recently geographers have become aware that natural classifications are difficult if not impossible to achieve. Thus, suppose we divide the world into agricultural regions; a division based upon crop combination and livestock density will not necessarily coincide with one based upon farm size and land tenure. Nor is this likely unless one powerful factor influences all the distributions of the separate phenomena such that they co-vary spatially; it follows then that natural systems of classification will necessarily show vagueness and indeterminancy at the boundaries (Gilmour, 1961, pp. 33–34; Grigg, 1965, pp. 470–471). It would now seem to be generally accepted by geographers and taxonomists that the purpose of the classification is overriding in determining the construction of areal classifications and that one classification will not serve all purposes equally well (Grigg, 1965, pp. 482–483).

Between the wars

These three trends which had developed before the First World War continued afterwards. In particular there was an increase in the number of single feature divisions, not only of the world but of much smaller areas. Thornthwaite's attempts to devise an alternative classification of climate received much attention (Thornthwaite, 1933, pp. 433–440 and 1931, pp. 633–655) and there were several attempts at physiographic regionalization (Fenneman, 1928, pp. 261–353). But perhaps the most significant trend was the effort to regionalize the cultural features of the earth's surface; of these amongst the most noteworthy were the articles on the agricultural regions of the continents published in *Economic Geography* between 1925 and 1942, whilst in 1936 Whittlesey published a system of agricultural regions of the world (Whittlesey, 1936, pp. 199–240).

Natural regions received, as has been noted, less attention. But Herbertson's suggestion that the natural regions could be used as a basis for the study of human societies was taken up by a number of writers, particularly by J. F. Unstead. As early as 1916 (Unstead, 1916, p. 232) he had suggested that natural geographical regions should be based not simply on physical features but also upon cultural features as well; a series of papers reached their culmination in a system of geographical regions (Unstead, 1916, pp. 230–249; 1926, pp. 159–170; 1932, pp. 298–317 and 1933, pp. 175—187). This system, with a hierarchy of regions with different names, was constructed in order to 'study the interactions of the various components of the great macro-organism of the earth and its inhabitants' (Unstead, 1933, p. 175) and thus whilst not avowedly deterministic, certainly pointed the way to a concept of the geographical region as being an area where society had adjusted to the local environment and where there was consequently homogeneity of both physical and cultural features (Hall, 1935, p. 215).

It was against such a view of the region that the criticisms of later years were directed. But whilst much of the regional thinking of the inter-war years was concentrated on the concept of a *geographical region*, there were important other developments. In 1937 a committee of the Geographical Association published a report upon the classification of regions and made a most useful distinction between two broad categories of regions, generic and specific. *Generic* regions were held to be those which fall into types, instances of which can be found in any part of the world, the types resembling each other in certain selected ways. Many of the regions we have discussed hitherto would fall into this category. *Specific* regions on the other hand were single areas said to have a distinct geographical individuality determined not only by the intrinsic conditions of the area but by its location in relation to other areas (Unstead, 1937, pp. 254–255).

A leading characteristic of the specific region was its location. P. M. Roxby had criticized earlier systems on the grounds that they were defined largely independent of location, and that much of the character of a region lay in its relation to other areas, a view that has much validity (Roxby, 1926, p. 378).

Other writers thought that generic systems of regions hardly had the true essence of the idea of a region within them (Hartshorne, 1959, p. 132). The importance of location in regionalization and the problems it presents will appear later (see pp. 36–37). But a perhaps much more important development in regional theory appeared in the interwar period. All the regions hitherto discussed were *uniform* regions. That is, within the defined area there was an areal uniformity in some selected criteria. Thus in an agricultural region there would be a similarity between all the farms or the fields, and this similarity would be an inherent property of the elements making up the region. *Nodal* regions, as the new type of regions were at first called, were quite different for here an area was defined in terms of interconnections between things or places. The idea of the nodal region has been attributed to a number of writers, amongst them J. C. Galpin (1915), R. E. Park (1925), R. D. McKenzie (1933) and N. S. B. Gras (1922). Certainly the idea was being put forward by many writers at this time; even Vidal de la Blache, so closely associated with the pays concept, suggested in 1917 that the most useful way of studying regional geography in the future might be by considering the hinterland of a major city and its relationships with its tributary villages (Wrigley, 1965, p. 11). The early exponents of the nodal region were essentially concerned with such interconnections between a central place and the neighbouring countryside; two early advocates of such a view were Walther Christaller (1933) and R. E. Dickinson (1930, pp. 548–557 and 1934, pp. 278–291). Since these writings the idea of a nodal region has come to mean more. 'Even if the region is not an organism', wrote P. R. Crowe in 1938 (p. 11), 'is it not probable that the regional matrix may be organization?'. The idea that a nodal region delineates areas of organization and traces the functional relationships between places has been superimposed upon the original and more simple idea, so that it is now customary to describe the nodal region as a functional region (Robinson, 1953, pp. 49–50; Hartshorne, 1959, p. 135). Certainly the differences between the two reflects a basic difference between two methods of classification; objects may be placed in the same class either because the objects have similar inherent properties *or* because there is a relationship between two dissimilar but connected objects (Hempel, 1952, pp. 5–6; Simpson, 1961, pp. 3–4; Grigg, 1965). The functional region has received a great deal of attention in recent years and some have asserted that it is a superior mode of investigation to the uniform region. It would perhaps be more accurate to say it is a different method and one more suited to the investigation of modern industrial economies.

Criticisms of the regional concept and regional methods

The period between the two World Wars saw a great growth in the attention given to regional geography, some advance in ideas about the nature of the region and a proliferation of regional studies. There was also a growth in popular interest in the idea with the attention given to regional planning in the USSR and latterly in the United States. But the end of this period saw also the beginnings of criticism within geography of both regional geography as the apex or crown of the field and also of the regional concept as then interpreted by most geographers. The criticisms can be briefly listed:

1 The region is not an entity or an organism; it follows that the earth's surface is not made up of a mosaic of regions whose delimitation it is the geographer's principal task to achieve.
2 It strains credulity to believe that all the properties of the earth's surface that geographers contend make up the totality of the environment should spatially co-vary exactly. Hence whilst there may be areas which are distinctive enough to be called *geographical regions*, there must be areas which have no particular character, and can be assigned to no particular region. It follows from this that (a) there may be more intermediate areas than there are 'regions', (b) if a region cannot be delimited accurately can it really exist?
3 The ecological approach to human communities is valuable; but too many geographers had assumed that human life is a function of environment and given too little weight to other factors. In other words, geographical regions bear too plainly the stamp of geographical determinism.
4 The regional concept is a static view of human life, in two senses:
(a) first, a regional system has validity for the moment at which it is devised and for no other moment. There is little provision for the study of change through time.
(b) second, regional studies have tended to treat the defined region as a community isolated from the rest of the world yet clearly no area or region in the modern world is independent of other parts of the world. From this two corollaries follow:
   (i) more attention should be given to movement and in particular interregional movement.
   (ii) a region should never be considered in isolation but as part of a system, which is ultimately the world.

To some extent these criticisms had been answered even at the end of the 1930's, when the attack began. But they undoubtedly served their purpose in bringing to the attention of the majority of geographers what a substantial minority already realized. In the remainder of this section it is hoped to show how these problems have arisen in allied fields where the regional concept is also used, and how there are in these fields divisions of opinion very similar to those in geography.

IS THE REGION A CONCRETE OBJECT?

The belief that regions are real units, 'genuine entities' (Hartshorne, 1939, p. 251), or an 'organism' (Herbertson, 1913, p. 212) is one which has existed since the early nineteenth century and has equally vigorously been attacked since then. In the late 1930's it was a view still held by many geographers (Hartshorne, 1939, p. 250). It is a measure of the success of the criticisms of the regional concept noticed above that by 1954 most American geographers denied this and Whittlesey could state 'Acceptance of the region as objective reality . . . is flatly rejected' (Whittlesey, 1954, p. 44). Six years later Hartshorne, reviewing the progress of methodology since the publication of *The Nature of Geography* wrote that attempts to see the region as a unitary, concrete object 'have passed into history' (Hartshorne, 1959, p. 31).

No attempt here will be made to re-open this argument. None the less some brief comments may be in order. The idea of Herbertson and others that the region is a macro-organism (Herbertson, 1913, p. 212) had some influence for a while and was then abruptly dismissed. However, the idea does merit some consideration. The word 'organism' has been used by ecologists and others in two ways, biological and philosophical (Egler, 1942, p. 246). When human communities are compared to organisms in the biological sense, as was done for example by the geo-politicians of Nazi Germany, then scepticism is perhaps merited. But when used in the philosophic sense more consideration should perhaps be given. Egler observes that organism is used by philosophers to refer to '. . . an entity in nature which acts as a distinct whole and which possesses characteristics of its own, even though it is separable into various kinds of parts. The term system is also applied in this latter and broader sense' (Egler, 1942, p. 246).

He applies the concept to the study of vegetation and observes that a plant community is more than simply the sum of its parts for there are relationships between plant and plant, and between plants and their environment. It is such an approach to this which has perhaps lead Bunge recently to suggest that regions 'do seem to exist as concrete unit objects' (Bunge, 1962, p. 25). For all this the prevailing view amongst geographers would now be that the region and regionalization is 'a device for segregating areal features' (Whittlesey, 1954, p. 44).

For many geographers, then, the issue as to whether the region is a real entity is dead. But we should take note first that some geographers still regard the region as a concrete object, and secondly that in other sciences where the regional concept is used the same controversy has raged in the past or continues to be pursued, apparently quite independently of the discussions amongst geographers. Of the geographers who believe the region to be a real entity, the foremost proponents have been Russian and East European geographers. Russian geographers have repeatedly stated that economic regions exist objectively. The idea of the region being a device or method is characterized as subjective and Western (and in particular American). The implications of the objective existence of regions are realized. Thus, M. S. Rozin, in a review of L. Y. Ziman's book on the economic regions of the United States (Rozin, 1961, p. 64), notes that there are divergences between Ziman's and other Soviet writers' regional divisions of that country ... 'in spite of the fact that Soviet authors proceed from actually constituted economic regions in the countries of the capitalistic world'. P. M. Alampiyev, in a discussion of the objective nature of economic regions argues that the idealistic view of regions as taken in the United States is not compatible with science. He goes on to quote I. G. Aleksandrov, Director of GOELRO in 1924, on the same problem (1961, pp. 66–70); 'If all the facts are known and we take a fully rationalistic approach to the problem, there can be only one real solution'. Nor are such views about the objective existence of regions confined to economic geography. Soviet geographers appear to believe that there are objectively existing physical-geographic regions (Kalesnik, 1961, p. 25) and a parallel to Aleksandrov's statement comes from N. A. Solntsev (Solntsev, 1962, p. 3): 'The concept of a mosaic structure of the earth's surface has become firmly established in geography at the present time'.

He goes on to argue that the study of these natural units is the purpose of geographical study.

But however vigorously Russian geographers may decry the subjective view of regions there are occasionally signs that the implications of this view can be embarrassing, and there may well be a change in their point of view in the near future. If in fact there was, this would follow the sequence of events in Poland. There were frequent statements in the Polish geographical journals that both economic and physical-geographic regions objectively exist (Kondracki, 1961, p. 28 and 1956, p. 60; Dziewonski, 1957, p. 739 and 1961, p. 613; Dziewonski and Lesczyckni, 1961, p. 81). But on the whole more attention has been given by Polish geographers to the Western view of regions than in Russian literature, and recently Wrobel has criticized Russian views of the real existence of regions, pointing out the difficulties of reconciling uniform and nodal concepts of the region (Wrobel, 1960, p. 136).

It would be a great mistake to see the dichotomy of views over the nature of regions as simply an East-West confrontation. Assuredly there are geographers in the Socialist countries who favour the view that the region is a device rather than an entity and similarly many geographers in the West still tend to regard regions as entities in some sense rather than simply the result of a method of areal classification. If we turn to fields other than geography it will soon be clear that a very similar dichotomy of views, although phrased in a different way, exists in other disciplines.

We may begin with sociology. There have been urban and rural sociologists, perhaps collectively who could be called human ecologists, who have been interested in the regional concept. Thus amongst urban sociologists Park, McKenzie and others have been interested in the natural areas of cities. They seem to have regarded such areas as entities in much the same way as some geographers contemporary to them regarded the geographical region as an entity. Within cities distinctive communities exist, characterized by the spatial correlation of a number of properties. Such a view has been criticized by, amongst others, Hatt (1946, pp. 423-427). Hatt investigated the natural areas of a city and argued that the marked sub-areas which could be shown by one way of mapping the data were far less evident if the data was mapped by (city) blocks. He criticized the approach of Quinn, Hawley and Park and McKenzie who 'assumed the existence of a real kind of data, real series of forces which produce real areas'; Hatt argued instead that human ecology 'is a way of looking at data without assuming any inherent qualities of those data'. A similar dichotomy of views appears to exist amongst rural sociologists in the United States. Thus Lively and Gregory (1954) in a discussion of socio-cultural areas make an assumption that if such areas are delimited by multi-factor analysis 'these areas may be regarded as social entities' (1954, p. 23). Other rural sociologists on the other hand doubt that regions can represent objective reality and consider the region simply an heuristic concept (Steward, 1955, pp. 297-298).

But of the controversies amongst social scientists over the existence of regions as real entities those which most clearly parallel the arguments between geographers as to whether the region is an object or merely a device are the disputes arising from recent attempts to divide the United States into *economic areas*. In a series of papers D. J. Bogue has outlined a system of economic regions of the United States. These are established by using counties as basic units. Similar and contiguous counties are grouped into State Economic Areas. These 'regions' are then successively grouped into Economic sub-regions, Economic Regions and Economic Provinces. In the most recent publication (Bogue and Beale, 1961), the methods of delimiting these regions is explained and the economic activity of each of the State Economic Areas described in some detail. Bogue has claimed that in the United States 'the phenomenon of general socio-economic homogeneity does exist' (Bogue and Beale, 1961, pp. xi and 1149) and that the State Economic Areas are 'distinctive communities' (Bogue, 1951, p. 1). This, and some of the other assumptions made by Bogue, has been severely criticized, most noticeably by Vining who doubts if there are any *natural areas* in the human economy and has denied that there is 'any optimum or single set of regions as distinct and operating entities and component parts of a human economy' (Vining, 1953, p. 52). Vining puts forward instead a quite different point of view. Whilst accepting that any studies of economic activity having a spatial extension must be based upon 'those finite and arbitrary spatial units for which data are available', he maintains that 'the spatial structure of a human economy should be regarded conceptually as virtually a continuum' (Vining, 1953, p. 44). Vining has attempted to demonstrate this view by extending Christaller's view of spatial structure (Vining, 1954–55, pp. 147–195).

A further example of the use of the regional concept in allied fields is drawn from plant ecology, where the use of the idea is implicit and the delimitation of floristic areas perhaps only a minor consideration. Plant geographers have long divided the world into various vegetation *formations*, and some have attempted to recognize hierarchies of plant communities smaller than the formation. But the analogy we seek to draw comes from the study of plant communities on a much larger scale. The controversies which have raged over the problems of defining plant communities and the assumptions which are made by workers with different viewpoints have been reviewed by Whittaker (1962, pp. 1–160), Grieg-Smith (1965, pp. 131–209) and Kershaw (1965, pp. 130–178). Two quotations from Kershaw may give an indication of the relevance of this work. He recognizes two schools of thought, one associated with the French plant sociologist Braun-Blanquet, and those who oppose his views amongst whom the American botanist Curtis is prominent. 'The two opposing schools of thought who on the one hand consider the climax association to be a complex organism or quasi-organism and on the other hand those who support the individualistic concept, also reflect the present-day ecological approaches to the delimitation and definition of community units. Thus the concept of the complex organism implies considerable interaction between the species which jointly modify the environment and form a distinguishable vegetational group. Conversely the individualistic concept regards no two communities as being identical, but considers that they show a continuous variation and accordingly cannot be readily delimited as definable units' (Kershaw, 1965, p. 130). He goes on to consider the methods and recent changes in the study of plant communities. He later again characterizes the two schools of thought: 'Braun-Blanquet erects a series of units, or associations, which can be classified in a hierarchical system, and are distinct, recognizable and definable entities. Curtis and his associates on the other hand regard vegetation as a multi-dimensional continuum, no two stands being the same' (Kershaw, 1965, pp. 168–169).

It can be seen that over a wide range of fields the view that distinct entities exist in the spatial distribution of phenomena is still held, but has been opposed by those who argue that such quasi-organisms do not exist and that phenomena – whether they be plant communities or economic communities – change imperceptibly as in a continuum. We will go on to consider some of the other problems of the region in allied fields but it may be worthwhile here to briefly digress and consider, in view of the fact that these opposed views occur in a number of fields, and have not entirely satisfactorily been resolved, whether such a difference does not represent, not two views, one which is wrong and the other right, but two quite different views of the world. Certainly this divergence of views is reflected in a quite different field, the philosophy of the explanation of historical events. Some have argued that events may be explained in terms of groups – such as the state, the middle class and so on. Others argue that this is illogical, and that events are the result of the decisions of many individuals, not of some supra-organism such as a class or national unit. If we substitute region for group, and space for time this conflict seems to resemble the argument in the spatial sciences between those who believe in entities and those who regard space as a continuum (Brodbeck, 1954, pp. 140–156; Watkins, 1955, pp. 58–62).

ECOLOGICAL REGIONS

At the beginning of this century many geographers regarded geography as the study of the influence of environment upon human society; this was later modified to take into account mutual relations between society and environment, and thus some geographers not unnaturally saw the subject as a branch of ecology (Barrows, 1923, pp. 1–14), for ecology seeks to portray the interrelationships between biological communities and their environment. But many of the critics of the regional concept saw that such a view had influenced ideas of the region, for, they considered, the worse. Many so-called geographical regions were no more than physical regions and it was assumed, very often erroneously, that the environment moulded human activities within the region. Some held that the essence of the region is the socio-economic adjustment of society in that area to the environment (Renner, 1935, p. 137).

Such a view is by no means confined to geographers. The early work on culture areas by American anthropologists (Wissler, 1917) was criticized by other anthropologists on the grounds that it was assumed that the life of the tribes of the culture areas was a function solely of the physical environment. Whilst these critics were prepared to accept that material culture, by which they meant the means of sustenance, was profoundly influenced by the natural environment, they could see no relationship between environment and the other cultural traits which characterized culture areas (Boas, 1938, pp. 670–671). Kroeber, the most distinguished of Wissler's followers very carefully described his approach to the culture areas of North America as ecological rather than environmental (Kroeber, 1963, p. 1). Many other workers have taken an ecological view of regions. Odum, for example, defined regionalism as the study of the relationships of man to geographical areas (Odum, 1942, p. 431), whilst Bogue in the work discussed earlier claims that his State Economic Areas, in which a distinctive economy prevails, represent the 'total adjustment which the population of that area has made to a particular combination of natural resources and other environmental factors' (Bogue, 1951, p. 1). Some regional scientists consider that regional studies should consider the relationships between society and environment (Garnsey, 1956, pp. 27-39) whilst there are still many geographers even in Socialist countries, where geographical determinism is anathema, who regard regional geography as the study of the interrelationships between economic life and environment (Barbag, 1959, p. 515).

There is no doubt that the reaction against environmentalism had initially a beneficial effect on geographer's attitude to regions, for it compelled them to define regions in terms of the things being regionalized rather than in terms of the environment which was held to influence those things. Hence agricultural regions began to be based upon properties of farming systems rather than simply being assumed to be a function of climate or soil (Whittlesey, 1936, p. 200). But at the same time the fear of being branded a determinist led many geographers to ignore the undoubted man/land interrelationships and to unconsciously reject the ecological, as well as the environmental approach to society. Some recent work by Zobler has highlighted this. Zobler (1957, pp. 83-95) has made a very useful distinction between two methods of regionalization. In the first, which he calls observational-descriptive, cultural regions are formed solely on the basis of socio-economic properties of the area regionalized and no relationship between these properties and earth features is assumed. In the second type, observational-relational, regional boundaries are formed on the basis of some earth feature such as climate or soil and then other data, about the type of farming for example, are assembled on the basis of these physical regions. This method permits the statistical testing of the relationship between the economic and the physical properties of the area. The methods Zobler used to check this relationship have been criticized (Mackay, 1959, p. 164), but there is no doubt that Zobler has made an important point and revived interest in the ecological approach to regions. It might of course be further observed that a spatial correlation between two phenomena such as soil and farming type does not prove any causal relationship, but it does suggest which processes might be further investigated.

The drawback to much of the earlier 'ecological regionalization' was that the relationship between environment and human activity was assumed and not tested; we have already seen such a view prevailed in the development of animal and vegetational regions. But we should not necessarily reject the ecological approach to regionalization, for this method is clearly valuable, as long as we recall that the areal correspondence of variables merely indicates a relationship but does not explain the nature of this relationship.

CORES AND BOUNDARIES

The boundaries of regions have always presented problems to geographers. If a region is thought to be a real entity then it must be presumed to have clear and determinable limits. Even if we take the alternative view and regard the region simply as a device then the fixing of limits still presents problems. This is particularly so when the region delimited is based on more than one criteria for it is rare to find that the isarithms – the usual method of delimitation – spatially coincide on the map, and the greater the number of criteria used in delimitation the greater the discrepancies. Such a lack of areal correlation has led many to be critical of the regional concept; and the problems are even greater when a *system* of regions is attempted.

Some have made a virtue of this difficulty. It has been argued that there are necessarily *core areas* where there is a spatial correspondence of the criteria used; but towards the edges of the region the correlation breaks down. Thus there are intermediate areas between the several cores of a system which cannot easily be assigned to any region. Others have, however, commented that the intermediate areas may well be more extensive than the cores (Crowe, 1938, p. 9).

The problem of cores and boundaries is by no means confined to geography; but whereas amongst geographers the idea of the core has been an expedient to avoid some of the dilemmas of regional delimitation, other workers have regarded the existence of cores and intermediate areas as quite logical. Thus amongst anthropologists the boundaries of culture areas have not received much attention although Kroeber has admitted they are the most dubious part of the concept (Kroeber, 1931B). Wissler however emphasized the culture *centres*, where the cultural traits were most typical, most intensively developed and most highly correlated with each other. He assumed that these features would diminish away from the centre, and indeed such an assumption was the basis of the theory of age-area which was closely allied to the culture area concept (Kroeber 1931B).

Other workers have assumed that intermediate areas are inevitable. Those plant geographers who have sought to divide the world into major vegetation regions on the basis of life-form and habitat have assumed that *formations* are a function primarily of a congerie of climatic characteristics; but as these characteristics but rarely change abruptly there will necessarily be areas between two major types of vegetation which show attributes of both types. Such intermediate or mixed areas are called *ecotones* (Weaver and Clements, 1938, p. 104). But those who have dealt with the plant communities of much smaller areas have found the problem of cores and boundaries a much greater difficulty. The attempt by the Braun-Blanquet school, which has selected stands, described the composition of the plant communities and then arranged these stands into a hierarchial classificatory system has been much criticized, particularly on the grounds that the selection of sites or stands for description has been highly subjective, and that intermediate sites where there is a much more mixed composition are ignored. Poore has pointed to this neglect of intermediate stands and communities. He argues that this gives an incomplete view of the nature of plant communities and goes on to suggest the idea of the *node*, which appears to be closely analogous to the idea of the core area (Poore, 1956, pp. 28-51).

REGIONALIZATION AND CLASSIFICATION

It will be apparent from the previous discussion that workers in other fields have had to deal with much the same problems as geographers when considering regions, and that there has been as much controversy over the solutions as there has been within geography. We turn now to consider a different approach to the problem. Many geographers have regarded regionalization as a form of classification, but it is not until recently that the analogy has been formally worked out (Bunge, 1962, pp. 14-26; De Jong, 1962; Grigg, 1965, pp. 465-480). The analogy has come to the forefront with the recent use of statistical methods in delimitation, and thus Reynolds (1956, p. 129) could write: 'The delineation of regions is essentially a classification process....'

In classification similar objects are grouped together; in regionalization until recently the most common method of delimiting regions was by the use of isarithms, and thus the obvious analogy with classification was perhaps missed.

Classification is an essential part of the biological sciences and over the centuries a number of simple verbal rules have evolved which guide the taxonomist. More recently there have been numerous attempts to apply statistical methods in classification (Sokal, 1965, pp. 337-391) and of course a number of statistical methods have already been used by geographers in regional delimitation. Further, both the rules of classification and the statistical methods are ultimately derived from formal logic. An outline of the procedures of classification and division may enable us to understand the procedures of regionalization more clearly.

Classification

Classification may be defined as the grouping of objects into classes on the basis of some similarity in either properties, or in the relationships between the objects. (References for this section in Grigg, 1965, pp. 465-480). The objects classified are called the *individuals*. All individuals have a number of *properties*, and the total number of individuals classified is called the *universe*.

[480] MODELS IN GEOGRAPHY

In classifying, one property which is possessed by all the individuals is selected for the grouping process and this property is called the *differentiating characteristic*. On the basis of this characteristic, individuals can be grouped into a number of *classes*. These classes, all being on the same level, are called either a *set* of classes or a *category*. If this initial grouping is unrewarding, the classes of the first order (see Fig. 12 1A) may themselves be grouped into a second set of classes on the basis of a second differentiating characteristic; the process may again be repeated and so a *hierarchy of classes* is formed. It

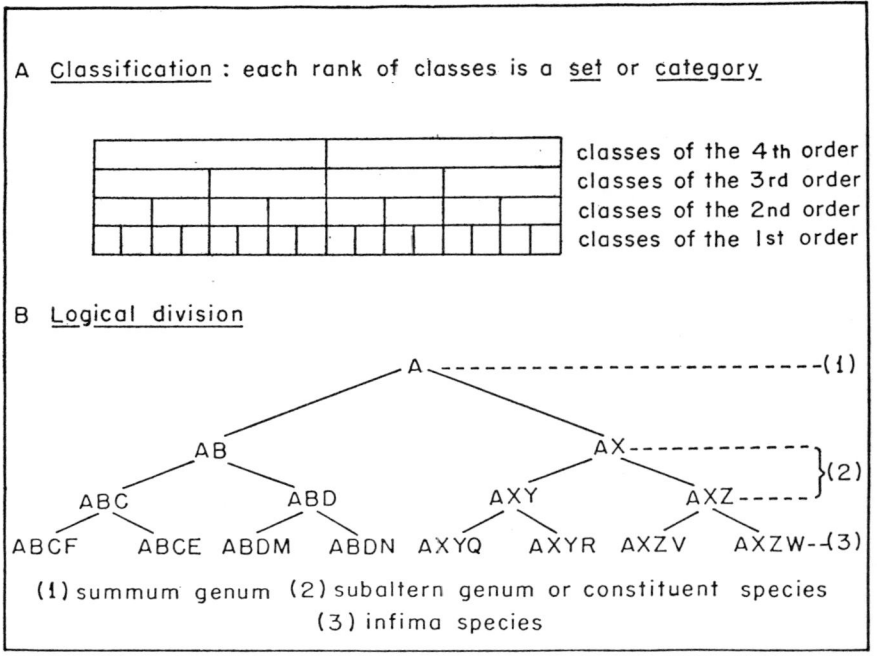

12.1 Diagrammatic representation of classification and logical division.

may be that when the initial classification is carried out to form the first set, it will be found that properties other than the differentiating characteristic vary in the same way. These are called *accessory characteristics*. A classification where the differentiating characteristic has many accessory characteristics is sometimes called a *natural classification*, one where it has few, an *artificial classification*.

Logical division

In classification we begin with a number of individuals and group them together on the basis of similarity. An allied process, logical division follows the reverse procedure. We begin with the universe, called in this case the *genus* and divide it into constituent *species* on the basis of some *principle*. The constituent species occur at the same level and are thus similar to a category of classes (Fig. 12.1B). The division can be carried further; once again a hierarchy of classes is created. But instead of seeking similarities we are looking for differences, and instead of building up we are breaking down. A special case of logical division is *dichotomous division* where the purpose is not to form a hierarchy of species but to isolate a single species. Thus, for example, land may be divided into Arable Land and not-Arable Land. Arable Land is then divided into Grain crops and not-Grain crops; a further division divides Grain crops into Wheat and not-Wheat, and we have thus isolated wheat as a species.

Are regions areal classes?

The argument then is that regions are essentially areal classes and we may then see an analogy between the basic procedures of formal logic and the methods customarily used by geographers in regionalization. The analogy perhaps needs more justification than can be given here and the reader is referred to more detailed accounts (Grigg, 1965, pp. 465–480; Bunge, 1962, pp. 14–26) elsewhere. However the argument must be briefly stated.

Classification in the broadest sense of the word may be arrived at by two processes, classification or division. These can be equated with the two basic approaches to knowledge, the inductive method and the deductive method. In the construction of many scientific classifications both methods are used to some extent and certainly the rules of division and classification apply equally well to either system. None the less many writers have recognized the two fundamental approaches. Whittlesey, for example, has stated that regional systems may be arrived at by either *aggregation* or *subdivision* (Whittlesey, 1954, pp. 38–39); Gilbert (1960, p. 60) has recognized the same distinction as has Hartshorne (1939, p. 291), whilst a number of Russian geographers have stressed the two different ways in which a hierarchy of regions may be determined (Kalesnik, 1961, p. 26). A similar divergence of method has been recognized by those dealing with plant communities (Dansereau, 1957, p. 81; Küchler, 1951, pp. 275–276; Lambert and Daly, 1964, p. 79) and soil regions (Manil, 1959, p. 8). Amongst geographers the difference has been most clearly stated by J. F. Unstead who described his own system of regions as *synthetic* and A. J. Herbertson's method, where the world is divided, as *analytical* (Unstead, 1916, pp. 236–241).

A number of important implications arise from this distinction. First, whilst either method can theoretically be used to construct areal classifications within an area of any size, classification (or synthetic regionalization) is normally restricted to relatively small areas, whilst conversely world classifications are invariably a result of division. In classification we are grouping together units or individuals on the basis of some measurable property; if the lower order categories are to have any meaning they must be based on some relatively small individual, certainly an area smaller than a State. Hence the establishment of a hierarchy of regions which began with say soil profiles or farms as the individuals yet encompassed the whole world in the highest class may be theoretically possible but would be in practice very difficult.

Thus world classifications seem invariably to be the result of division. It is important to note that division is deductive. Thus if the world is divided into vegetation types, which are then mapped, the types are assumed to exist *a priori*; this is not necessary with classification, although Küchler, in a very interesting discussion of the two methods of constructing and mapping vegetation has doubted whether *a postieri* methods are not in fact also *a priori* in practice (Küchler, 1951, pp. 275–276). Whatever the logical necessities of the two methods, it should be noted that world *divisions* are often genetic.

Our first formal analogy then is between the basic approaches to classification. It is argued that systhetic regionalization is analogous to classification, and analytical regionalization to division.

The argument may be extended further. Thus some taxonomists recognize that classification may be based upon *either* similarity between inherent properties of the objects classed *or* on the basis of a relationship between objects which may themselves be dissimilar (Simpson, 1961, p. 3). This difference seems to be analogous to that between *uniform* regions, which are based essentially on homogeneity, or similarity between parts of the region, and *nodal* regions which are defined upon the basis of relationships between parts of the region. The need for the classification of relationships is a relatively recent development in science in general. Hempel for example observed that whilst science was once concerned with the properties of objects it is now concerned more with the functions and relationships of objects (Hempel, 1952, pp. 5–6). The same development has occurred in geography, for until the 1930's regional systems invariably consisted of uniform regions. It is of particular interest to note that whereas early planning regions were essentially uniform regions, modern students have turned increasingly to nodal regions as the basis of planning in industrial economies (Friedmann, 1956) although it can be argued that uniform regions are a more suitable basis for the less industrialized areas (Fair, 1957, p. 32). A study of the remarkable changes in the policy behind Russian planning regions suggests that the Russians, whilst realizing the difference between nodal and uniform regions, have found difficulty in adjusting the two concepts to the dogmatic necessity of having one set of objectively existing regions (Wrobel, 1960, p. 136).

Generic and specific regions

If we divide or classify an area into a number of regions of the first order we create a regional system. Each of these regions will be unique in that types do not necessarily occur twice in the system, although they may do; on the other hand, the delimitation of a single region, which may be a geographical region, in order to demonstrate its individuality or in order to delimit an area for further study, does not create a regional system. What we in fact do is define the region in terms of a number of criteria; and, by definition, we assume that no other part of the world has this particular combination of criteria. Thus the universe is divided into the region and not-the-region. Thus it can be argued that the delimitation of a single region, where we are isolating a particular area from the rest of the world (universe) is analogous to *dichotomous division*. On the other hand, generic regions, where we create a system of regions (a set of co-ordinate classes), is equivalent to normal classification and division. The single region is then a special case of areal classification.

It remains then simply to compare some of the remaining terminology. If regions are areal classes, then *elements* are the equivalent of *properties*, and *place* of *individual*, *criteria* of *differentiating characteristics*, and *boundaries* of *class limits*.

The geographical individual

So far it is hoped that the analogy is convincing. But when we turn to consider the nature of the individual in regionalization, a number of problems arise. Bunge, who appears to have been the first geographer to make the analogy between classification and regionalization explicit suggests that 'place' is the equivalent of individual (Bunge, 1962, p. 16). Whilst this is true, it does not help us when we turn to actually grouping individuals into areal classes. The nature of the individual has received surprisingly little attention in geographical literature in English, although it has been much discussed by German and Russian writers. Let us begin by stating the problem in its most extreme form. Suppose we accept the idea of classification as the basis of regionalization; further let us suppose we wish to establish in an area a set of *geographical* regions. What units do we group together to form the first order regions? Let us suppose we postulate that this area can be divided into an exhaustive set of geographical regions on the basis of natural vegetation, type of farming and pattern of settlement. It is clear that the individuals we use to classify the area into a preliminary group of farms, plant communities and settlement forms will not, and in fact cannot be, the same.

Let us turn to the problem of the individual in physical geography where it has received more attention. Now when we classify plants or stones or mammals in conventional taxonomy there is no problem about distinguishing the individual. The objects classified are discrete and separate entities. But in areal classification they are not. The earth's surface, while it patently has marked areal differences, is not made up of easily identified discrete units which are irreducible; it is a continuum.

Hence the geographical individual has presented problems to all who have attempted to classify features with areal expression. In soil science the delimitation of soil regions by grouping soil profiles has been a weakness in many systems (Kellogg, 1963, p. 3) and has recently been subject to some penetrating criticisms (Jones, 1959, pp. 196–200); in plant sociology the problem has been tackled by sampling stands. But this method has not been without weaknesses, and Braun-Blanquet's system in particular has been criticized on the grounds that the 'stands' are subjectively selected. In the classification of surface morphology it has been argued that the earth's surface can be reduced to flats and slopes; and that these can then be grouped into regions on the basis of similarity. But there seems undeniably to be a subjective element in the identification of these individuals in the field (Linton, 1951, pp. 199–217). Amongst geographers who have sought to classify the totality of the physical environment even greater problems arise, and these are sometimes confused by the failure to distinguish between individuals and first order regions. This is illustrated by the controversy between some Russian physical geographers as to whether the term 'landscape' should be used to mean an individual or the equivalent of a region (Yefremov, 1961, pp. 32–43; Solntsev, 1962, pp. 3–15; Prokayev, 1962, pp. 21–29; Kalesnik, 1961, pp. 24–34).

It would seem that there is no 'natural' individual in the landscape; that is to say, there are no easily recognizable entities which form a mosaic whose interpretation and grouping is the business of regionalization. Most geographers – and others – have recognized this and instead turned to the use of what we may call 'operationally defined' individuals. Thus E. H. Hammond's classification of the landforms of North America is based on a grid system, the squares forming the individuals (Hammond, 1957 and 1958). Similarly, botanists, whilst in dispute about the nature of individuals in the study of plant communities, have reconciled themselves to the use of quadrat sampling, and pedologists although aware of the limitations of soil profiles have not overcome the difficulties. Amongst those who deal with regional differences in the cultural features of the earth's surface there has been a greater readiness to accept 'operationally defined' individuals, simply because statistics are usually only available for administrative areas.

Location and the individual

In taxonomy the position or location of the individual is of no significance; it is not a property considered in the classification system. But clearly in any geographical classification location must be considered and the means of demonstrating location is the map. Let us suppose we classify a number of farms, which are our individuals, on the basis of one differentiating characteristic. We may then group these into a number of classes on the basis of this characteristic. Now we have argued that regions are areal classes. Let us place our individual farms on a map, each individual being a dot coloured appropriate to its class. It is clear that this is a distribution map, and that we can only distinguish areal classes or regions if the individuals (farms) in the same class are contiguous. Indeed we can define regions in the terminology of classification by saying that areal classes consist of similar *and* contiguous individuals.

It is on this point that many would argue that the analogy with classification founders; even if we do not accept this it must be agreed that what has come to be called the contiguity problem presents considerable difficulties. Is there any logical necessity for similar individuals to occur together in space? It is abundantly clear that the whole concept of regionalization rests on this assumption. Further, there is an assumption that the processes which operate to cause like individuals to occur together have what we may call some spatial logic. In the case of physical geography it is possible to accept this. We may argue for example that like plants may occur in like environments (ignoring factors such as dispersal and so forth), or similarly that very unlike soil profiles will not occur adjacent to each other. It would seem then that regionalization of such phenomenon rests on an almost metaphysical assumption of what some writers have called geographic order. But when we turn to the regionalization of cultural features it is not so clear that the same type of spatial logic operates, or at least that we can so readily assume that it operates. If we return to our original example of farms, there seems no necessary reason why adjacent farms should be similar and indeed they often are not. We may of course avoid the problem by arguing that the individuals need only show an approximate likeness; and we may use ideas such as the botanists' concept of dominance to smooth over our unwanted anomalies. But the problem of spatial logic remains. To some extent the problem can be resolved by the use of various statistical methods and to these we will turn later.

SOME PRINCIPLES OF REGIONALIZATION

If it is accepted that regionalization is analogous to classification and it is remembered that the fundamentals of classification are based upon the principles of formal logic then it may be profitable to examine regional systems in the light of these principles. These are only briefly treated here; they have been dealt with in more detail elsewhere (Grigg, 1965, pp. 480–491).

1 *Classifications should be designed for a specific purpose; they rarely serve two purposes equally well.* Many geographers have emphasized the importance of *purpose* in the construction of regional systems. Purpose will determine the criteria selected and the number of regions delimited. It is equally important in the *use* of regional systems. Hence, for example, a set of soil regions based upon properties such as colour and texture will not necessarily be a reliable guide to regional variations in soil fertility. A moment's reflection shows too that this rule clarifies the difference between natural (or general-purpose classifications) and artificial (or special-purpose), classifications. Thus, a set of agricultural regions based upon a number of properties such as crop combination, land tenure, farm size and rent, would necessarily have somewhat indeterminate limits because of the implausibility of such a wide range of properties co-varying spatially exactly. On the other hand a special-purpose classification based solely on farm size could be expected to be far more accurate because only one property is considered. Although general-purpose regional systems, such as a set of geographical regions, may occasionally be very stimulating aids to interpretation, they lack precision and are perhaps, more useful in teaching than research.

2 *There exist differences in Kind between objects; objects which differ in Kind will not easily fit into the same classification.* Nineteenth-century logicians such as Mill emphasized the relevance of this rule to taxonomy (Mill, 1959, pp. 470–471) and argued that it would be illogical to include such things as different in Kind as stones and animals in the same classification system. Cline has shown that this rule may have relevance to soil classification; he suggests that organic and inorganic soils may be so different in Kind as to justify the division of the two into different systems (Cline, 1949). Bunge has briefly discussed the relevance for regional systems and points out that Land and Sea are fundamentally different in Kind and are rarely included within the same regional system (Bunge, 1962, p. 20). But the rule has a further relevance for geographers. Preston James has concluded 'that an attempt to define regions based on phenomena produced by a variety of different processes is dangerous and could lead to serious errors of interpretation. We may find ourselves trying to add things like cabbages and kings' (James, 1952, p. 204). The same author had previously observed that 'The attempt to recognize combinations of several elements often has the effect of obscuring the realities rather than throwing additional light on them' (James, 1942, p. 493). This would presumably be particularly true where the elements concerned were greatly different in Kind. Yet it is undeniable that in this century there have been an increasing number of classifications, in a number of sciences, where things greatly different in Kind are included within the same system. This is particularly true of ecological classifications, where animals or plants are classified with habitat; it is clearly true of systems of

geographical regions and equally with many of the general-purpose regional systems designed by human ecologists. Indeed the assumption made by some human ecologists is that a socio-culture area is defined by the intercorrelation of a number or variables, some of which may be very different in Kind (Lively and Gregory, 1954, pp. 22–23). Although this assumption is rarely stated so explicitly by geographers it is obviously an assumption underlying many regional systems. It is but a short step to assuming that there is a genetic relationship between the correlated properties, although this is rarely rigorously tested.

3 *Classifications are not absolute; they must be changed as more knowledge is gained about the objects under study.* An increase in understanding of the objects classified invariably leads to the revision of the classification of these objects. Jevons observed that '. . . almost every classification which is proposed in the early stages of a science will be found to break down as deeper similarities of the objects come to be observed' (Jevons, 1887, p. 691).

This is particularly true of the last thirty years as there has been a great acceleration in the understanding of the biological sciences. It has led to much criticism of existing classifications, particularly in zoology (Sokal and Sneath, 1963) and attempts to revise the methods and assumptions of taxonomy have been made. But in subjects where classifications are the basis of teaching and to a lesser extent research, this clearly presents problems, for the continued revision of existing systems may lead to confusion. Some measure of stability seems essential.

This principle has a number of implications for the interpretation of regional systems. First, it might be argued that some of the regional systems still widely used in teaching, such as Herbertson's natural regions and Whittlesey's agricultural regions, badly need revision. Second, it might be observed that the geographer has greater problems than most biologists, for not only does his increased understanding of the things classified justify revision but the things classified change much more rapidly than the aspects studied in most biological fields. Third, it might be noticed that many regional systems used by geographers are often borrowed from other sciences – this is particularly true of soil and vegetation systems – and it is not always clear that the systems used are the most recently developed or that they are necessarily suited to the geographer's purpose. Hence a map of soil regions where the classification is based upon soil genesis may not be the most suitable if the geographer is considering the relationship between soil and land use.

4 *The classification of any group of objects should be based upon properties which are properties of those objects; it follows then that differentiating characteristics should be properties of the objects classed.* No rule was more commonly

ignored at the turn of this century, not only by geographers but many others who constructed regional systems. As was noticed earlier the limits of many areal classes were determined by the use of criteria which were supposed to cause the distribution of the phenomena studied rather than criteria inherent in the phenomena. Thus agricultural regions were delimited not by properties of the farming systems studied, such as crop combinations or farm size, but by factors which were assumed to cause the observed differences in farming systems, such as soil type or climatic differences. Such a procedure is now commonly held to be invalid (Whittlesey, 1954, p. 38).

5 *When dividing, the division should be exhaustive and the classes formed should exclude each other.* If we divide the people of the world into a number of races, it is clearly not legitimate to consider only half the people; the classification is incomplete. Similarly if we construct a racial classification any one individual must fall into one class and one class alone. For example, if we make a division of the world's peoples into religious groups and arrive at four classes – Christian, Moslem, Jewish and European we have not made our classes exclude each other, for a man may be European and Muslim or European and Christian. When we apply the rule to regional systems it would appear easy to observe; thus if an area is to be divided into regions then all parts of the area must be assigned to some region and one region only. Whilst this may appear self-evident, there is no doubt that in practice the observance of this rule presents difficulties.

6 *When dividing, the division should proceed at every stage and as far as possible throughout the division upon one principle.* In any regional system the regions of the same order should be the result of a delimitation based upon the same property or some degree of expression of the same property. Thus if we find a country has been divided into four regions – the arid, the semi-arid, the humid and the super-humid we may assume that each region is based upon some variation in rainfall or evapotranspiration. But if we find that instead the same country is divided into the Dry East, the Industrial West, the Highly Populated South and the Mountainous North it is clear that the division has not proceeded at this stage upon the same principle, for one region has been delimited upon the basis of humidity, another upon relief, a third upon occupational structure and the last upon population density.

But to carry out the second part of this rule is more difficult. Some logicians argue that it is impossible to use the same principle at every stage in a division. None the less, this rule does cause us to consider the manner in which some world divisions are carried out. Thus for example; the world may be initially divided into a number of classes or regions on the basis of some property such as Relief. The classes arrived at in this manner are then subdivided upon the basis of climate and these sub-classes in turn divided upon the principle of vegetation. It may well be questioned, as Crowe (1938, pp. 7–8) has done, whether such a procedure is legitimate.

7 *The differentiating characteristic, or principle of division, must be important for the purpose of division.* In constructing a regional system some decision must be made about the criteria to be used. But it is not always clear how these criteria should be selected. Thus if we wish to divide England into type of farming regions, which of the two following lists of criteria is the more valid?

1 Proportion of land under arable, proportion of arable under wheat, density of beef cattle, proportion of farm income derived from crops.
2 Size of farm, labour per acre, rent per acre, and age of farmer.

The principles of classification give no reliable answer to this other than to suggest that the purpose of the system should determine the criteria chosen. This helps us to eliminate some of the more obvious mistakes, but upon closer examination the rule is not particularly helpful (Duncan, Cuzzort and Duncan, 1961, p. 197).

8 *Properties which are used to divide or classify in the higher categories must be more important for the purpose of the division than those used in the lower categories.* Most regional systems designed by geographers consist of regions of one order only and there are few instances of a carefully designed hierarchy of regions such as that proposed by Philbrick (Philbrick, 1957, pp. 299–336). But in world divisions where a hierarchy of regions is established, the order in which the principles are applied may well be important. The initial division should always be upon the principle which is most important to the purpose of the division. It is not unusual to find this rule neglected in elementary regional geographies. Thus we may find that western Europe is divided into a few major regions upon the basis of physiography, and then subdivided by vegetation and further subdivided on the basis of occupational structure. Such a procedure often means that the lower order regions, which are invariably the most important units of description, separate similar areas simply because they have been earlier divided upon the basis of physiography, which is largely irrelevant to the purpose of the lower order regions.

STATISTICS AND REGIONAL DELIMITATION

It is hoped that the principles of regionalization outlined in the preceding section may be of help in both the construction and the assessment of regional systems. But it must be confessed that these principles deal essentially with the strategy rather than the tactics of regionalization; nor do these principles, as they stand, help eliminate the large element of subjectivity which most methods of regionalization seem heir to. Whilst all regional systems are relative, in the sense that they depend upon the purpose of the system and the data available, and because regional differences are often differences in degree rather than kind (Gregory, 1949, pp. 59–63), none the less some attempt should be made to reduce the large number of value judgements which appear in most regionalization procedures. The first step in this direction was made by the use of descriptive statistics (Jones, 1930, pp. 177–195) in the inter-war period, and such procedures have become progressively more refined (Weaver, 1954, pp. 1–47).

Indeed by the 1930's it could be said that most geographers accepted that regional systems should be based upon some measurable property of the individuals. But the grouping of individuals was still largely a subjective procedure. In recent years many geographers have turned increasingly to statistical techniques developed in allied sciences to try and make the classification process more objective. Whilst the belief in the possibility of a really objective set of regions seems as great an illusion as the earlier belief in the possibility of a set of geographical regions, there is little doubt that the use of some statistical methods has greatly improved the validity of regional systems and hence of the generalizations that can be made from them.

The fact that the statistical methods used in regionalization have had to be adapted from other fields has had its disadvantages. Few statistical methods have been designed originally with either a spatial or ecological context in mind; whilst many methods can easily be adopted, not all the problems have been overcome. It cannot, for example, be said that the contiguity problem has been satisfactorily solved. A minor disadvantage lies in the fact that there exist few comprehensive texts on statistics for use in geography (Gregory, 1963; Duncan, Cuzzort and Duncan, 1960) and the geographer must turn for the most part to works in other fields which have somewhat incidentally a spatial and ecological interest (Grieg-Smith, 1965; Hagood and Price, 1952; Sokal, 1965, pp. 337–391; Miller and Kahn, 1965). In this section no attempt is made to give a comprehensive account of the sort of statistical methods which can be used in regionalization. Rather the aim is merely to indicate the possibilities and problems.

The individual

Most individuals used in regionalization, particularly in human geography, are administrative areas such as the parish or county, for which information is collected and published by government and other bodies. Such basic areal units may vary a great deal in size and shape, and a number of cautions should be borne in mind before they are subjected to statistical treatment.

   1 The statistics for such units are generally averages and thus may con-

ceal quite striking internal variations. For example the agricultural returns for an English parish consist of, among other things, the total acreage under each crop in the parish; this data may be converted to various ratios for easier comparison with other parishes. But such a procedure may conceal the existence of contrasting patterns of land use within each parish (Coppock, 1960, pp. 317–326). Much the same may be said of the American county (Weaver, 1956, pp. 536–565) and the problem clearly arises in case of data dealing with things other than agriculture. Hence a set of regions based upon such data may obscure some actual patterns (Hatt, 1946, pp. 423–427). Assuming that the individuals cannot be split, the problem is insoluble; and when interpreting regional systems the nature of the individual and the possible internal variations should always be borne in mind. Similar problems arise when trying to assign an individual to one of two regions. Suppose a number of individuals can be grouped to form two markedly contrasting regions; between the two regions lie a number of individuals which cannot be unambiguously assigned to either region. This may be because the marginal individuals are areal units which overlap and possess tracts of land belonging to both the contrasting regions, but the aggregate figures for the individual obscure this (Rao, 1952, p. 41).

2 The limits of farms or factories or indeed any property of the earth's surface that a geographer may wish to regionalize do not necessarily correspond with the limits of administrative units. Hence whilst, for example, a farm may lie largely within one parish, yet the crop acreages for the farm may be returned for the adjacent parish in which the farm house lies (Coppock, 1955, pp. 12–26). Similar problems arise when dealing with census data for occupations, for ambiguity may arise between place of residence and place of work.

3 Administrative units are not always of the same size and shape; this again may affect the nature of the data available for such individuals. This is best illustrated with reference to migratory data; clearly the number of emigrants from a unit area will be a function of both the size and the shape of unit. All other factors being equal, there are likely to be more commuters or migrants from a large and narrow parish or county than from a small and squarish unit (Duncan, Cuzzort and Duncan, 1960, p. 34; Chisholm, 1960, pp. 187–188).

4 The size of an individual may influence the grouping of individuals and the calculation of regional means. Robinson (1956, pp. 233–236) has shown that areal units should be weighted according to their area even when dealing with ratios; but it is not unequivocally clear how this weighting should be carried out (Duncan, Cuzzort and Duncan, 1960, p. 47; Thomas and Anderson, 1965, pp. 492–505).

Single-feature regions

Let us suppose we have a large number of individuals, for each of which there is available one measurable property and that this property is expressed in directly comparable figures. How is the area regionalized? The usual procedure is to determine the core areas where similar figures occur and to delimit these 'regions'; and then to try and assign the remaining individuals. A number of difficulties arise.

1 The similar individuals may not be contiguous; alternatively contiguous individuals may be markedly different from surrounding contiguous and similar individuals. A further possibility is that there may be single individuals which may be so different from all other individuals that they cannot be readily assigned to any region.

An early attempt to deal with the problem was made by Hagood, who used the principal component method to regionalize the southern United States. One of the factors considered was contiguity, measured by the distance in inches horizontally and vertically from an arbitrary point of origin on the base map. Thus contiguity was one of the factors considered in the factor analysis. But this was not strikingly successful and has been criticized by Gregory (1949) who has suggested an alternative method of feeding position into the factor analysis.

Some have suggested that homogeneity must necessarily be sacrificed to contiguity; that is, if an individual is adjacent to a group of similar individuals, and not similar to any other adjacent and similar group, it must be assigned to that group in spite of its lack of similarity (Gregory, 1949; Hagood, 1943). Nor can a single individual be left unassigned, for an individual by definition cannot be a class (region). Nor is it always satisfactory to create a large number of regions simply because there are a large number of small contiguous groups of similar individuals. It may be as well to decide in advance the approximate number of regions which are desired.

2 If a very large number of individuals, or basic areal units, are to be regionalized on the basis of a single property it is not uncommon to prepare a preliminary map with shadings appropriate to the magnitude of the property concerned. If this is done, care must be taken with the class limits used (Mackay, 1955, pp. 71–81).

3 A group of individuals, to form a region, must be similar and contiguous, and dissimilar from other individuals or regions. But as was observed earlier, differences may be only of degree. Thus there should be some means of deciding whether the differences between the regions are statistically significant. This issue has been discussed by Zobler (1958, pp. 140–148) and Gregory (1949, pp. 59–63). Zobler's methods have been critically reviewed by Berry (1958, pp. 301–303) and Mackay (1958, p. 164).

Multi-feature regions

Whilst single features are commonly used in geographical studies, most geographers would consider the essence of a region or regions to be found in areas where there is a similar distribution of a number of properties. This view would almost certainly be shared by workers in other fields who are concerned with the regional concept. Thus, for example, Lively and Gregory argue that one of the assumptions on which the investigation of what they call socio-cultural areas proceeds is that '. . . in terms of homogeneity the culture of the area hangs together . . . that the sum total of the intercorrelations among these traits can never be zero . . .' (Lively and Gregory, 1954). To put the matter another way, in a regional system one may expect there to be a high spatial covariance of two or more properties of the individuals under consideration. But the greater the number of properties or criteria considered the less likely, experience suggests, is it for any exactitude in spatial covariance to occur, and on a more practical level, the harder it is for a regional delimitation to be undertaken. One traditional means of determining the boundaries of regions based on a large number of properties was developed by Otto Maull (Hartshorne, 1939). Maull's girdle method is still used. For any area a number of maps of regional boundaries are obtained and then the boundaries are superimposed on one map. Where more than a given number of boundaries approximately coincide is deemed to be the boundary of a geographical region. This method can be made a little more sophisticated by measuring the degree of overlap between superimposed regions (Mowrer, 1938, pp. 86–96) but it still has deficiencies. Furthermore this is essentially an analytical method. Regions based on a large number of traits or criteria could also be constructed simply by comparing a large number of maps of single features. But it is difficult in this case to keep in mind the large number of criteria. To overcome such difficulties Hagood suggested in 1941 the use of the principal components method in regionalization, and since then an increasing number of writers have used this or related methods, such as direct factor analysis (Hagood, 1941, pp. 216–233 and 1943, pp. 287–297; Rao, 1952, pp. 33–43; Berry, 1960; Berry, 1961, pp. 263–279; Allman *et al.*, 1964, pp. 5–10; Johnson, 1965, pp. 9–12; Thompson *et al.*, 1964, pp. 1–20). In these methods the correlation between a large number of properties is measured and the significant criteria extracted from the initial properties considered. Factor analysis seems to offer at least a partial solution to the problem of multiple feature regionalization although the limitations of the method should be borne in mind (Isard, 1960, pp. 292–305).

REGIONS AS MODELS

If one looks back over the history of regionalization in the last hundred years one gains an impression of increasing precision in the methods of delimitation. There seem to be a number of significant stages in the sophistication of regionalization. First was the recognition that regions should be delimited upon the basis of properties of the individuals regionalized, and not upon the basis of some supposed 'cause' of the regions. Second was the distinction made between uniform and nodal regions, and between generic and specific regions. The third vital step in progress has been the use, both of descriptive statistics in the establishment of regional systems, and more recently of analytical statistical methods – particularly factor analysis – which have brought a greater rigour to the delimitation of regions.

This increasing precision, however, is to some extent misleading. Some would argue that the use of computers, which can handle a great number of criteria, has allowed the establishment of 'real' and 'optimum' sets of regions. Others may go further and argue that with the aid of such methods of regionalization an adequate *natural* classification of the earth's surface may be developed; that is to say, a system of regions which incorporates a very wide range of properties and also has built in an explanation of the regional differences revealed. This seems to be an over-optimistic view of the potentialities of the new methods, and it also to some extent misinterprets the purpose of regional systems. Regionalization is a means to an end, not an end in itself. A system of regions is established only as a first step in a geographical inquiry. It points the way to the study of how – and perhaps why – the regional variations described have been brought about.

With this in mind we may turn to the last section of this paper. We wish to consider in what sense regional systems may be viewed as models and to what extent regionalization is a process of model-building. It will be apparent from other chapters in this book that the term 'model' can be interpreted in a great number of ways. Nor is this variety of interpretation one confined to geographers alone. The term has been used as synonymous with 'scientific method' (Harris, 1960, p. 250), 'hypothesis', 'theory', 'law', 'explanation' (Kaiser, 1960, p. 13), and 'rule' (Skilling, 1964, p. 388). In the social sciences the term has increasingly become confused with 'hypothesis', and sometimes with 'theory', whilst many writers appear to use the word to describe any description of phenomena in mathematical terms (Black, 1962, pp. 219–243). Some philosophers have regarded this confused use of the word with asperity (Brodbeck, 1959, pp. 373–400; Black, 1962; Braithwaite, 1953, pp. 90–95; 1962, pp. 224–231; Apostel, 1960, pp. 125–161) and one writer has after a discussion of the many different uses of the word, concluded that the use of models is no more than a way of thinking. However it is hoped to show here that this may well be a profitable way of thinking.

It was stated in the opening paragraph of this section that the development of regional methods has been characterized by an increasing precision in the establishment of regional systems. But it could be argued that the measure of success of a regional system is not the exactness of the boundaries delimited but the stimulus which the system provides in explaining the regional differences revealed by the system. An example may illustrate this point. We may, by the use of computers and an elaborate factor analysis group farms of eastern England into classes and establish a system of agricultural regions, based on a great number of properties; it could be claimed that this is an optimum set of agricultural regions. Let us suppose however that we establish an alternative set of agricultural regions, in this instance we group together farms solely on the basis of two criteria, distance from major market towns and size of farm. Now the set of regions derived from this latter method offends all the rules of regionalization. Yet it might be productive of more useful generalizations about the causes of the differentiation of the agricultural landscape than the former more precise classification. It must be admitted that the example chosen is an improbable one, but there seems little doubt, for example, that modifications of Von Thünen's theory have been more productive of useful generalizations in agricultural geography than elaborate attempts to compute sets of crop combination regions (Chisholm, 1964).

The second set of agricultural regions outlined above could well be described as an example of regional model building. Let us turn to consider in what sense regional systems can be regarded as models. There appear to be four ways in which the term and the method can be applied.

Regions as abstractions of reality

At the lowest level a model is no more than an abstraction of reality. Reality is so complex that it is impossible to reproduce all its features, all its functional relationships, or the whole web of interdependences. To represent reality in words or symbols or statistics we must simplify. In the simplifying process the measure of success is the significance of the properties which we choose to represent reality. In geographical terms the map is an obvious form of this process of abstraction and simplification. A map cannot possibly show all the differences which exist over the surface of the earth or part of the earth, so that the success of the process depends upon the purpose of the map and the selection of the features to put upon it. An obvious example would be a map for the tourist. This is successful if it shows roads, towns and relief, the things most relevant to the tourists' purpose; but this is clearly not all it could show.

Reflection along these lines suggests that all regional systems are models; they must in the nature of things be based upon selection, for a regional system cannot be based upon all the possible properties of the things regionalized. Now if we consider the history of regional systems in this light it will be clear that we can regard successive attempts to classify – for example the climates of the world, or the agricultural regions of the world – as a series of approximations or models, each one which must be judged by two standards; firstly, the accuracy with which the system describes reality, and, secondly, as to how far the system is productive of new and useful generalizations. Perhaps more important is the way in which we *interpret* regional systems. If we think of regional systems as classifications of varying exactitude, then it is not difficult to come to believe that lines upon a map represent real and immovable boundaries. If on the other hand we recall that the designers of many early regional systems regarded their systems as no more than an approximation which could be later modified, then we will be far more tolerant of their limitations and perhaps more aware of the stimulus they gave to further work. The limitations of many regional systems lies not so much in the systems but in the expectations of those who interpret the system.

Regions as isomorphic systems

In the widest sense of the word model, then, we can reasonably argue that all regions are models. Let us turn to the use of the term model *in sensu stricto*. A number of philosophers have argued that the term model should be limited to its original sense which was as follows. If we are attempting to explain a system, of whatever nature, and we only partially understand it, then one way is to interpret it in terms of an isomorphic and understood system. The example often quoted is that of Faraday's comparison of the behaviour of electro-magnetism with that of fluids. The laws of a known field could then be applied to an unknown field.

There seem to be very few such regional models, and the one instance was far from productive. A number of writers in the nineteenth century considered regions to be organic entities and it was then assumed that regions could be viewed in some senses as living things. The analogy was not made very specific, but it was argued that regions grew and – most important – expanded; hence adjacent regions were in competition for space. It is not difficult to see how the idea of *lebensraum* emerged from this discussion. One of the few results of this analogy was an attempt to justify the expansion of states (considered as regional models) and their absorption of the space of other 'regions'. This model of regional systems has not received much attention, perhaps because of its association with the geopoliticians of Nazi Germany, but it does draw attention to the important and neglected topic of the nature of change in regions.

Regions as controlled models

Regional systems have two prime purposes. They are first of all descriptive such that if we wish to inquire into the nature of the agricultural geography of England, one way to undertake this task is to classify all the farms on the basis of similarity and group the farms in agricultural regions. This is a form of shorthand, for it compresses a great deal of information into a small number of categories. The classification may lead us to an inquiry into causes of the regional differentiation of agriculture. It may be possible, for example, that a map of the agricultural regions of England prompts us to investigate the importance of soil type or altitude above sea level as a factor determining agricultural differences.

A second function of regional systems is more specifically explanatory. The world of reality is extremely complex and no regional system, however rigorously designed, can incorporate all variations within itself. Nor can we always understand the operating of a system – whether it be an economic system or an organic system – by looking at it holistically. To understand the separate processes it may be necessary to look at only one, and ignore the other possible processes acting upon the system. Such a course is one frequently followed in economic theory, and it is one of great value in many geographical inquiries. Thus, for example, suppose we are trying to explain the climatic differentiation of the world. It is known that most of the major differences result from the operation of the general circulation of the atmosphere, but that the simplicity of these processes is disturbed by certain irregularly distributed features of the earth's surface such as mountains or the shape of the continents. To demonstrate how climatic regions result from the general circulation of the atmosphere it is not uncommon to construct a hypothetical continent, lacking any major mountains and standing astride the equator, and show how hypothetical regions would arise upon it. Now in explaining the actual climates of a given continent, the model regional system is taken, and the climatic regions of the continent explained in terms of the model and in the deviations from the model.

It is clear then that we can generate models of regional systems by considering only a limited number of processes and seeing how they would differentiate the earth's surface, if the other processes were ignored. Such regional models have been devised to show the impact of climate on the development of vegetation regions (Dansereau, 1954; Holdridge, 1947, pp. 367–368), whilst there seems little doubt that many maps of the soil regions of the world are arrived at in a similar manner, although this may not be specifically stated. But the construction of such regional models has been developed much more elaborately in the field of economic geography. The study of economic systems requires the consideration of a great number of factors simultaneously, and this has proved impossible to achieve. Hence many writers have turned to the consideration of the workings of a limited number of factors, other factors assumed to be constant. A model of the working of the factor in relation to a limited number of other variables is then constructed. This model must of course be internally consistent, but the final stage is to relax the initial assumptions to see if the model explains the situation in reality. Perhaps the most celebrated of these models is that constructed by Von Thünen, whose original model has been modified by a number of later workers. Briefly, von Thünen envisaged an isolated state where one city provided the only market for agricultural produce. The city lay in the centre of a level plain which possessed uniform climatic and soil characteristics. He made other assumptions about agricultural wages and means of transport. The variable analysed then was the distance of a farmer from the market. Von Thünen argued that, through the operation of agricultural rent, a concentric series of land use zones would arise around the central market. Thus a model of an agricultural regional system is generated by considering only a limited number of processes which operate to differentiate the agricultural landscape. From von Thünen's work a great number of other models of regional systems have stemmed. In passing we may note Christaller's (1933) work upon the location of cities and their spheres of influence. Lösch (1938) elaborated Christaller's ideas and tried to devise a model of economic regions, his work being as interesting for its statement of the value of this sort of approach to regions as for its substantive content. Lösch argued that neither states nor geographic nor cultural regions were an adequate means of investigating the regional structure of economic life. He wished instead to devise economic regions based upon features of economic life, not economic regions derived from geographic regions. He went on to argue:

> Even if we already knew the characteristics of economic regions – which we do not – their counterparts in the world of reality would be likely to differ more from each other than from an ideal picture. Hence studying the ideal region is both the only way to learn about the *essential*, and the first step towards investigating the *actual* structure of any real economic region. So we shall deal first with the theoretical nature of such regions, and second with their actual existence. (Lösch, 1938, p. 71.)

It is important to note that the model was only a first step in the process of investigation.

Analytical regions as models

We have considered three ways in which regions may be considered as models. We turn now to the last and perhaps the most important. Earlier we distinguished two methods of arriving at regional systems, synthetic regionalization, where similar individuals are grouped into classes, which is analogous to classification, and analytical regionalization, where an area is divided into regions by division. The latter process requires further consideration. The way in which world systems of regions are arrived at is far from clear, but there appear to be two stages. First of all a typology of the phenomena is created – soil types, or types of farming, or types of vegetation. Having created these classes, it is assumed that any part of the earth will fit into one of these classes or types. Further, the types are often both deductive and genetic, in the sense that, for example, soil types are seen primarily as a response to climate, agricultural regions as a response to climate, and natural regions to be primarily a function of climatic differences (Herbertson, 1905).

Such a process is not followed in synthetic regionalization. Admittedly the properties have to be selected before the regions are established, but there is far less *a priori* reasoning involved in the process. Hence we may be justified in seeing analytic regionalization as a form of model building, whilst synthetic regionalization is not primarily model building, except in the sense that all regional systems, being abstractions and simplifications of reality, are models.

The uses and abuses of models

There are certain advantages in regarding regional systems as models. One such is that we may then see the history of regionalization as a long process of 'eliminating noise in the model' (Chorley, 1964). If we regard, for example, Herbertson's system of natural regions as a model, two advantages result. First, Herbertson intended his system to be exploratory, to be tested by later workers, and not to be accepted as a final and complete system. The essence of models is that they are exploratory and must be tested; the sad thing about Herbertson's work is that, whilst it was stimulating and productive of many interesting generalizations, the model was not refined by later workers. We may conclude from this that many defects of model-using are not in the models themselves – provided the limitations of models are understood – but in their subsequent use by other workers. The second advantage of regarding such regional systems as models is a great gain in flexibility. If we regard regionalization as a classification process, then we immediately commit ourselves to harnessing the use of computers and the increasing precision in the establishing of similar classes, so eliminating the genetic approach, which has no role in areal classification. We end up with elegantly designed systems of regions which may have however little to offer in the way of explanations. It may well be that imperfect, deductive and genetic regional systems, however they contravene the rules of classification, may be the more stimulating vehicles of inquiry.

Having said this we must turn to inquire into the use of models, for it is often the misuse of models which may be at fault rather than the models themselves. The methodology of models has received a great deal of attention from both philosophers and scientists; they seem agreed that there are a number of pitfalls (Braithwaite, 1953, pp. 88–114) in the use of models and it is worth considering these in relation to the use of models of regional systems. Firstly, it should be remembered that a model is not a theory but a hypothesis; it is a step towards formulating a theory. A model is a means of making the unfamiliar familiar, a means of simplifying complexity; an untested model is of little value. Yet there seems little doubt that models are often confused with theories and F. Lukermann has recently exposed such misconceptions (Lukermann, 1958, 1960 and 1961). Secondly, it should be remembered that the logical necessities of a model cannot necessarily be transferred to the theory. Thus von Thünen's hypothesis may be internally consistent; but when using the hypothesis to explain land-use patterns it must be remembered that the simplifying assumptions of the model do not operate in most real situations.

The value of models

There seems little doubting the value of models of regional systems in geography. Provided the limitations are carefully borne in mind, and the dangers of using models remembered, then models can be useful for teaching and for research. We may end by noting an apparent paradox in the history of regionalization. Early regional systems such as those of Herbertson, Whittlesey and the Russian soil scientists have been shown to be deficient in many ways. Yet there is but no question that these systems were of great value at the time and opened up many avenues of inquiry. Since the early systems were developed it has been possible to harness a wide range of classificatory statistical techniques to the problems of delimiting regions. Yet, for the most part such techniques improve only the descriptive efficiency of the system. True, these systems may lead to the development of hypotheses which may explain the differences described, but they do not do so directly. On the other hand, the deductive, intuitive systems, often based on some assumed cause of regional differences – and thus liable to criticism – were at the time they were devised a great stimulus to further work. Clearly, if we regard these systems as models of regional systems – as it seems legitimate to do – then they have served their value: they have stimulated further work and produced partial explanations; they have required testing and the testing has revealed further paths to explanation. This surely justifies them. Geographical regions, natural regions and economic regions have all been stages towards a further understanding of the complexity of the differentiation of the earth's surface. As we welcome the increasing precision of modern classificatory methods of regional delimitation, we should remember that models of regional systems have an equally important role to play.

REFERENCES

ACKERMAN, E. A., [1945], Geographic training, wartime research and immediate professional objectives; *Annals of the Association of American Geographers*, 35, 121-143.
ACKERMAN, E. A., [1953], Regional research – emerging concepts and techniques in the field of geography; *Economic Geography*, 29, 189-197.
ALAMPIYEV, P. M., [1961], The objective basis of economic regionalisation and its long range prospects; *Soviet Geography*, 11, 64-74.
ALLMAN, J., COX, K. R., ERHART, R. R. and RUSSWURM, L. H., [1964], The use of standardized values in regionalisation; the example of a socio-economic spatial structure of Illinois 1960; *The Professional Geographer*, XVI, 5-10.
ANNETTE, SISTER MARY, [1965], The changing French region; *The Professional Geographer*, XVII, 1-5.
APOSTEL, L., [1960], Towards the formal study of models in the non-formal sciences; *Synthese*, XII, 125-161.
BARBAG, J., [1959], The subject and task of regional geography; *Przeglad Geograficzny*, XXXI, 495-515 (in Polish, English summary).
BARROWS, H. H., [1923], Geography as Human Ecology; *Annals of the Association of American Geographers*, 13, 1-14.
BASINSKI, J., [1959], The Russian approach to Soil Classification; *Journal of Soil Science*, 10, 14-26.
BERG, L. S., [1950], *The natural regions of the USSR*; (New York).
BERRY, B. J. L., [1958], A note concerning methods of classification; *Annals of the Association of American Geographers*, 48, 300-303.
BERRY, B. J. L., [1960], An inductive approach to the regionalisation of economic development; In Ginsberg, N. (Ed.), *Essays on Geography and Economic Development, Department of Geography Research Paper* No. 62, University of Chicago.
BERRY, B. J. L., [1961], A method for deriving multi-factor uniform regions; *Przeglad Geogriczny*, XXXIII (2), 1961, 263-279.
BOAS, F., [1938], *General Anthropology*, (Boston).
BLACK, M., [1962], *Models and Metaphors; studies in language and philosophy*, (Ithaca).
BOGUE, D. J., [1951], *State Economic Areas*, (Washington, D.C.).
BOGUE, D. J. and BEALE, C., [1961], *Economic Areas of the United States*, (Glencoe).
BRAITHWAITE, R. B., [1953], *Scientific Explanation*, (Cambridge).
BRAITHWAITE, R. B., [1962], Models in the empirical sciences; In Nagel, E., Suppes, P. and Tarski, A. (Eds.), *Logic, Methodology and Philosophy of Science*, (Stanford), 224-231.
BRODBECK, M., [1954], On the philosophy of the social sciences; *Philosophy of Science*, 21, 140-156.
BRODBECK, M., [1959], Models, meaning and theories; In Gross, L. (Ed.), *Symposium on Sociological Theory*, (New York), 373-400.
BROUDE, H. W., [1960], The significance of regional studies for the elaboration of national economic history; *Journal of Economic History*, XX, 588-596.

BUNGE, W., [1962], *Theoretical Geography*, (Lund).
CHISHOLM, M., [1960], The geography of commuting; *Annals of the Association of American Geographers*, 50, 187–188.
CHISHOLM, M., [1964], Problems in the classification and use of farming-type regions, *Transactions and Papers of the Institute of British Geographers*, 35, 91–103.
CHORLEY, R. J., [1964], Geography and analogue theory; *Annals of the Association of American Geographers*, 54, 127–137.
CHRISTALLER, W., [1933], *Die Zentralen Orte in Suddeutschland*, (Jena).
CLINE, M. G., [1949], Basic principles of soil classification; *Soil Science*, 67, 81–91.
COPPOCK, J. T., [1955], The relationship of farm and parish boundaries; *Geographical Studies*, 1, 12–26.
COPPOCK, J. T., [1960], The parish as a geographical statistical unit; *Tijdschrift voor Economische en Sociale Geografie*, 51, 317–326.
CROWE, P. R., [1938], On progress in Geography; *Scottish Geographical Magazine*, 54, 1–19.
DANSEREAU, P., [1951], Description and recording of vegetation upon a structural basis; *Ecology*, 32, 172–229.
DANSEREAU, P., [1957], *Biogeography: an ecological perspective*;
DAVIES, J. L., [1961], Aim and method in zoogeography; *Geographical Review*, 51, 412–417.
DE JONG, G., [1962], *Chorological Differentiation as the fundamental principle of geography*, (Groningen).
DICKINSON, R. E., [1930], The regional functions and zones of influence of Leeds and Bradford; *Geography*, XV, 548–557.
DICKINSON, R. E., [1934], The Metropolitan regions of the United States; *Geographical Review*, 24, 278–291.
DUNCAN, O. D., CUZZORT, R. P. and DUNCAN, B., [1961], *Statistical Geography: problems in organising areal data*, (Glencoe, Ill.).
DZIEWONSKI, K., [1957], Some problems of research for economic regionalisation of Poland; *Przeglad Geograficzny*, XXIX, 1957, 719–739, (Polish, English summary).
DZIEWONSKI, K. and LESCZYCKNI, S., [1961], Geographical studies of economic regions in Central Eastern Europe: problems and methods; In *Problems of Economic Regions*, Polish Academy of Sciences, Institute of Geography, Geographical Studies No. 27, [Warsaw].
EGLER, F. E., [1942], Vegetation as an object of study; *Philosophy of Science*, 9, 242–260.
FAIR, T. J. D., [1957], Regions for planning in South Africa; *South African Geographical Journal*, XXXIX, 26–50.
FENNEMAN, N. M., [1928], Physiographic divisions of the United States; *Annals of the Association of American Geographers*, 18, 261–353.
FLEURE, H. J., [1952], The later development of Herbertson's thought; *Geography*, 37, 97–103.
FRIEDMANN, J. R. P., [1956], The concept of a planning region; *Land Economics*, 32, 1–13.

GALPIN, J. C., [1915], The social anatomy of an agricultural community; *University of Wisconsin, Agricultural Experimental Station, Bulletin* No. 35.
GARNSEY, M., [1956], The dimensions of regional science; *Papers and Proceedings of the Regional Science Association*, 2, 27–39.
GEORGE, W., [1962], *Animal Geography*, (London).
GILBERT, E. W., [1953], Geography and regionalism; In Taylor, G. (Ed.), *Geography in the Twentieth Century*, (London), 345–371.
GILBERT, E. W., [1960], The idea of the region; *Geography*, 45, 157–175.
GILMOUR, J. S. L. and WALTERS, S. M., [1964], Philosophy and Classification; In Turril, W. B., *Vistas in Botany*, (London), IV, 1–22.
GILMOUR, J. S. L., [1961], Taxonomy; In MacLeod, A. M. and Cobley, L. S. (Eds.), *Contemporary Biological Thought*, (London), 27–45.
GRAS, N. S. B., [1922], *An introduction to economic history*, (New York).
GREGORY, C. L., [1949], Advanced techniques in the delineation of regions; *Rural Sociology*, XIV, 59–63.
GREGORY, S., [1963], *Statistical methods and the geographer*, (London).
GRIEG-SMITH, [1965], *Quantitive Plant Ecology*, (London).
GRIGG, D. B., [1965], The logic of regional systems; *Annals of the Association of American Geographers*, 55, 465–491.
GRIGOR'YEV, A. A., [1962], The present state of the theory of Geographic Zonality; In Harris, C. D., (Ed.), *Soviet Geography; Accomplishments and Tasks*; (New York), 182–187.
GRIGOR'YEV, A. A., [1961], The heat and moisture regime and geographic zonality; *Soviet Geography*, 2, 3–16.
HAGOOD, M. J., DANILEVSKY, N. D. and BEUM, C. O., [1941], An examination of the use of factor analysis in the problem of sub-regional delineation; *Rural Sociology*, 6, 216–233.
HAGOOD, M. J., [1943], Statistical methods for delineation of regions applied to data on agriculture and population; *Social Forces*, 21, 287–297.
HAGOOD, M. J. and PRICE, D., [1952], *Statistics for Sociologists*, (New York).
HALL, R. B., [1935], The geographic region; a resumé; *Annals of the Association of American Geographers*, XXV, 122–136.
HAMMOND, E. H., [1957], *Procedures in the descriptive analysis of terrain*, (Wisconsin).
HARE, F. K., [1951], Climatic classifications; In Stamp, L. D. and Wooldridge, S. W. (Eds.), *London Essays in Geography*, (London), 111–134.
HARRIS, J. E., [1960], A review of the symposium; *Symposia for the Society of Experimental Biology*, XIV, *Models and Analogues in Biology*, (Cambridge), 250–255.
HARTSHORNE, R., [1939], *The Nature of Geography*, (Lancaster, Penn.).
HARTSHORNE, R., [1959], *Perspective on the Nature of Geography*, (Chicago).
HATT, P., [1946], The concept of natural area; *American Sociological Review*, 11, 423–427.
HEBERLE, R., [1943], Regionalism; some critical observations; *Social Forces*, 21, 280–286.
HEMPEL, C. G., [1952], Fundamentals of concept formation in empirical science; *International Encyclopedia of Unified Science*, vol. 2.

HERBERTSON, A. J., [1905], The major natural regions; an essay in systematic geography; *Geographical Journal*, XXV, 300-312.
HERBERTSON, A. J., [1913A], Natural regions; *The Geographical Teacher*, VII, 158-163.
HERBERTSON, A. J., [1913B], The higher units; *Scientia*, XIV, 199-212.
HERBERTSON, A. J., [1916], Regional Environment, Heredity and Consciousness; *The Geographical Teacher*, 8, 147-153.
HOLDRIDGE, L. R., [1947], Determination of World Plant Formations from Simple Climatic Data; *Science*, 105, 367-368.
HOSKINS, W. G., [1954], Regional farming in England; *Agricultural History Review*, 11, 3-11.
ISARD, W., [1956], Regional science; the concept of the region and regional structure; *Papers and Proceedings of the Regional Science Association*, 2, 13-26.
ISARD, W., [1960], *Methods of Regional Analysis; an introduction to regional science*, (New York).
ISARD, W. and CUMBERLAND, J. C., [1961], *Regional Economic Planning: techniques of analysis*, (Paris).
JAMES, P. E., [1942], A regional division of Brazil; *Geographical Review*, XXXII, 493-495.
JAMES, P. E., [1952], Towards a further understanding of the regional concept; *Annals of the Association of American Geographers*, XLII, 195-222.
JOHNSON, R. J., [1965], Multi-variate regions: a further approach; *The Professional Geographer*, XVII, 9-12.
JONES, T. A., [1959], Soil classification – a destructive criticism; *Journal of Soil Science*, 10, 196-200.
JONES, W. D., [1930], Ratios and isopleth maps in the regional investigation of agricultural occupance; *Annals of the Association of American Geographers*, vol. 20, 177-195.
KAISER, H., [1960], Kinetic models of development and heredity; *Symposia for the society of Experimental Biology, XIV, Models and Analogue in Biology*, (Cambridge), 13-27.
KALESNIK, S. V., [1961], The present state of landscape studies; *Soviet Geography* 11, 24-34.
KENDEIGH, S. C., [1954], History and evaluation of various concepts of plant and animal communites in North America; *Ecology*, 35, 152-171.
KELLOGG, C. E., [1963], Why a new system of soil classification?; *Soil Science*, 96, 1-5.
KERSHAW, K. A., [1964], *Quantitive and Dynamic Ecology*, (London).
KIMBLE, G. H. T., [1951], The inadequacy of the regional concept; In Stamp, L. D. and Wooldridge, S. W. (Eds.), *London Essays in Geography*, (London), 151-174.
KONDRACKI, J., [1956], Natural regions of Poland; *Przeglad Geograficzny*, XXVIII, 48-60.
KONDRACKI, J., [1961], On terminology and taxonomy of regional units in Poland's physical geography; *Przeglad Geograficzny*, XXXIII, 23-28.
KROEBER, A. L., [1931A], Culture Area; *Encyclopedia of the Social Sciences*, vol. 4, 646-647.
KROEBER, A. L., [1931B], The culture area and age-area concepts of Clark Wissler; In Rice, S. A. (Ed.), *Methods in Social Science*, (Chicago).

KROEBER, A. L., [1963], *Cultural and Natural Areas of North America*, (Berkeley).
KÜCHLER, A. W., [1951], The relation between classifying and mapping vegetation; *Ecology*, 32, 275–283.
KÜCHLER, A. W., [1954], Plant geography; In James, P. E. and Jones, C. F. (Eds.), *American Geography, Inventory and Prospect*, (New York), 429–440.
LAMBERT, J. M. and DALE, M. B., [1964], The use of statistics in phytosociology; In Cragg, J. B. (Ed.), *Advances in Ecological Research*, 2, (London).
LEIGHLY, J., [1937], Some comments on contemporary geographic method; *Annals of the Association of American Geographers*, XXVII, 125–141.
LINTON, D. L., [1951], The delimitation of morphological regions; In Stamp, L. D. and Wooldridge, S. W. (Eds.), *London Essays in Geography*, (London), 199–217.
LIVELY, C. E. and GREGORY, C. C., [1954], The rural socio-cultural area as a field for research; *Rural Sociology*, 19, 21–31.
LÖSCH, A., [1938], The nature of economic regions; *The Southern Economic Journal*, 5, 71–78.
LUKERMANN, F., [1958], Towards more geographic economic geography; *The Professional Geographer*, 10, 2–13.
LUKERMANN, F., [1960], On explanation, model and description; *The Professional Geographer*, 12.
LUKERMANN, F., [1961], The role of theory in geographic enquiry; *The Professional Geographer*, 13, 1–5.
MACKAY, J. R., [1958], Chi square as a tool for regional studies; *Annals of the Association of American Geographers*, 48, 164.
MACKAY, J. R., [1955], An analysis of isopleth and chloropleth class intervals; *Economic Geography*, 31, 71–81.
MANIL, G., [1959], General considerations on the problem of soil classification; *The Journal of Soil Science*, 10, 5–13.
MCDONALD, J. R., [1964], Current controversy in French geography; *The Professional Geographer*, XVI, 1964, 20–23.
MCKENZIE, R. D., [1933], *The Metropolitan Community*, (New York).
MILL, J. S., [1959], *A System of logic*, (London).
MILLER, R. L. and KAHN, J. S., [1962], *Statistical analysis in the geological sciences*; (London).
MORGAN, F. W., [1939], Three aspects of regional consciousness; *Sociological Review*, XXVI, 68–88.
MOWRER, E. W., [1938], The isometric map as a technique of social research; *American Journal of Sociology*, 44, 86–96.
ODUM, H. W., [1942], A sociological approach to the study and practice of American regionalism; *Social Forces*, 20, 425–436.
PARK, R. E., BURGESS, E. W. and MCKENZIE, R. D., [1925], *The City*, (Chicago).
PHILBRICK, A. K., [1957], Principles of areal functional organisation in regional human geography; *Economic Geography*, 33, 299–336.
POORE, M. D., (1956), The use of phytosociological methods in ecological investigations iv. General discussion of phytosociological problems; *Journal of Ecology*, 44, 28–50.
PROKAYEV, V. I., [1962], The facies as the basic and smallest unit in landscape science; *Soviet Geography*, III, 21–29.

RAO, V. L. S. PRAKASA, [1953], Rational grouping of the districts of the Madras State; *Indian Geographical Journal*, XXVIII, 33–43.
RAUP, H. M., [1942], Trends in the development of geographic botany; *Annals of the Association of American Geographers*, XXXII, 319–354.
RENNER, G. T., (1935), The statistical approach to regions; *Annals of the Association of American Geographers*, XXV, 1935, 137–152.
REYNOLDS, R. B., [1956], Statistical methods in geographical research; *Geographical Review*, 46, 129–132.
ROBINSON, A. H., [1956], The necessity of weighting values in correlation analysis of areal data; *Annals of the Association of American Geographers*, 46, 233–236.
ROBINSON, G. W. S., [1953], The geographical region: form and function; *Scottish Geographical Magazine*, 69, 49–58.
ROXBY, P. M., [1926], The theory of natural regions; *The Geographical Teacher*, XIII, 376–382.
ROZIN, M. S., [1961], Review of L. Y. Ziman's 'Economic Regions of the United States'; *Soviet Geography*, 11, 60–65.
SHEVKY, E. and BELL, W., [1955], *Social Area Analysis*, (Stanford).
SIMPSON, G. G., [1961], *Principles of Animal Taxonomy*, (Oxford).
SKILLING, H., [1964], An operational view; *American Scientist*, 52, 388A–396A.
SOKAL, R. R., [1965], Statistical methods in Systematics; *Biological Reviews of the Cambridge Philosophical Society*, 40, 337–391.
SOKAL, R. R. and SNEATH, P. H., [1963], *Principles of Numerical Taxonomy*; (San Francisco).
SOLNTSEV, N. A., [1962], Basic problems in Soviet landscape science; *Soviet Geography*, III, 3–15.
STAMP, L. D., [1957], Major natural regions; Herbertson after Fifty Years; *Geography*, XLII, 201–216.
STEWARD, J. H., [1955], The region – a heuristic concept; *Rural Sociology*, 20, 297–298.
SUSLOV, S. P., [1961], *Physical geography of Asiatic Russia*, (San Francisco).
THOMAS, E. and ANDERSON, D. L., [1965], Additional comment on weighting values in correlation analysis of areal data; *Annals of the Association of American Geographers*, 55, 492–505.
THOMPSON, J. H., SUFRIN, S. C., GOULD, P. R. and BUCK, M., [1962], Toward a geography of economic health: the case of New York State; *Annals of the Association of American Geographers*, 52, 1–20.
THORNTHWAITE, C., [1931], The climates of North America according to a new classification; *Geographical Review*, XXI, 633–655.
THORNTHWAITE, C. W., [1933], The climates of the earth; *Geographical Review*, 23, 433–440.
THORNTHWAITE, C. W., [1943], Problems in the classification of climates; *Geographical Review*, 33, 233–255.
THORNTHWAITE, C. W., [1961], The task ahead; *Annals of the Association of American Geographers*, 51, 345–356.
TREWARTHA, G. T., [1954], *An Introduction to Climate*, (New York).
UNSTEAD, J. F., [1916], A synthetic method of determining geographical regions; *Geographical Journal*, XLVIII, 230–249.

UNSTEAD, J. F., [1926], Geographical regions illustrated by reference to the Iberian Peninsula; *Scottish Geographical Magazine*, XLII, 159-170.
UNSTEAD, J. F., [1932], The Lötschental; a regional study; *Geographical Journal*, LXXIX, 298-317.
UNSTEAD, J. F., [1933], A system of regional geography; *Geography*, XVIII, 175-187.
UNSTEAD, J. F. et al., [1937], Classification of regions of the world; *Geography* XXII, 253-282.
VINING, R., [1953], Delimitation of Economic Areas; statistical conceptions in the study of the spatial structure of an economic system; *Journal of the American Statistical Association*, 48, 44-64.
VINING, R., [1954-55], A description of certain spatial aspects of an economic system; *Economic Development and Cultural Change*, 3, 147-195.
WATKINS, J. N. L., [1955], Methodological individualism: a reply; *Philosophy of Science*, XXII, 58-62.
WEAVER, J. C., [1954], Changing patterns of cropland use in the Middle West; *Economic Geography*, 30, 1-47.
WEAVER, J. C., [1956], The county as a spatial average in agricultural geography; *Geographical Review*, 46, 536-565.
WEAVER, J. E. and CLEMENTS, F. E., [1938], *Plant Ecology*, (New York).
WHITTAKER, R. A., [1962], The classification of natural communities; *The Botanical Review*, XXVIII, 1-160.
WHITTLESEY, D., [1936], Major agricultural regions of the earth; *Annals of the Association of American Geographers*, XXVI, 199-240.
WHITTLESEY, D., [1954], The regional concept and the regional method; In James, P. and Jones, C. F. (Eds.), *American Geography: inventory and prospect*, (Syracuse), 19-68.
WISSLER, C., [1917], *The American Indian*, (New York).
WISSLER, C., [1927], The culture area concept in social anthropology; *American Journal of Sociology*, XXXII, 881-891.
WISSLER, C., [1928], The culture area as a research lead; *American Journal of Sociology*, XXXIII, 894-900.
WRIGLEY, E. A., [1965], Changes in the philosophy of geography; In Chorley, R. J. and Haggett, P., (Eds.), *Frontiers in Geographical Teaching*, (London), 3-20.
WROBEL, A., [1960], *The Warsaw Voivodship; a regional study of the Economic Regional Structure*, (Warsaw).
YEFREMEVOV, Y. K., [1961], The concept of landscape and landscapes of different orders; *Soviet Geography*, 11, 32-43.
ZOBLER, L., [1957], Statistical testing of regional boundaries; *Annals of the Association of American Geographers*, 47, 83-95.
ZOBLER, L., [1958], Decision making in regional construction; *Annals of the Association of American Geographers*, 48, 140-148.

SUPPLEMENTARY REFERENCES

The following articles and books, whilst not referred to in the text, have been found useful.

ALAMPIEV, P. M., [1962], Economic regionalisation; In Harris, C. D. (Ed.), *Soviet Geography, Accomplishments and Tasks*, (New York).

BOWEN, E. G., [1959], Les Payes de Galles; *Transactions of the Institute of British Geographers* No. 26, 1–23.

DICKINSON, R. E., [1939], Landscape and Society; *Scottish Geographical Magazine*, 55, 1–15.

DICKINSON, R. E., [1964], *City and region: a geographical interpretation*, (London).

DRYER, C. R., [1915], Natural economic regions; *Annals of the Association of American Geographers*, 5, 121–125.

DZIEWONSKI., [1961], Elements of the theory of economic regions; *Prezeglad Geograficzny*, XXXIII, 593–613.

DZIEWONSKI, K., [1961], Theoretical problems in the development of economic regions; *Papers and Proceedings of the Regional Science Association*, VIII, 43–54.

FINCH, V. C., [1939], Geographic science and social philosophy; *Annals of the Association of American Geographers*, XXIX, 1–28.

FLEURE, H. J., [1917–18], Regions in Human Geography with special reference to Europe; *The Geographical Teacher*, IX, 31–45.

JAMES, P. E., [1934], The terminology of regional description; *Annals of the Association of American Geographers*, XXIV, 78–86.

JOERG, W., [1914], The subdivision of North America into natural regions: a preliminary enquiry; IV, 55–83.

JOPE, E. M., [1963], The regional cultures of medieval Britain; In Foster, I. L. and Adcock, L., (Eds.), *Culture and Environment: Essays in Honour of Sir Cyril Fox*, (London).

KOLLMORGEN, W., [1945], Crucial deficiencies of regionalism; *Papers and Proceedings of the American Economic Review*, XXV, 377–389.

KUCHLER, A. W., [1956], Classification and purpose on vegetation maps; *Geographical Review*, XLVI, 155–167.

LEBEDEV, V. G., [1961], Principles of geomorphic regionalisation; *Soviet Geography* 2, 59–74.

LEE, SHU-TAN, [1947], Delimitation of the geographic regions of China; *Annals of the Association of American Geographers*, VII, 155–168.

LEIGHLY, J., [1938], Methodologic controversy in nineteenth-century German Geography; XXVIII, 238–258.

MACKAY, J. R., [1959], Regional geography, a quantitative approach; *Cahiers de Geographie de Quebec*, 3, 57–63.

MERRIAM, A. P., [1959], The concept of culture clusters applied in the Belgian Congo; *South Western Journal of Anthropology*, 15, 373–395.

OGBURN, W. F., [1936], Regions; *Social Forces*, 15, 6–11.

PRATT, R. S., [1957], A review of regional geography; *Annals of the Association of American Geographers*, 47, 187–190.

RIEMER, S., [1943], Theoretical aspects of regionalism; *Social Forces*, 21, 275–280.

STEVENS, A., [1939], The natural geographical region; *Scottish Geographical Magazine*, 55, 305–317.
STUART, L. C., [1954], Animal geography; In James, P. E. and Jones, C. F., (Eds.), *American Geography, inventory and prospect*, (New York), 443–451.
TEITZ, M. B., [1962], Regional theory and regional models; *Papers and Proceedings of the Regional Science Association*, 9, 35–50.
ULLMAN, E. L., [1953], Human geography and area research; *Annals of the Association of American Geographers*, XLIII, 54–66.
VANCE, R. B., [1952], The regional concept as a tool for social research; In Jensen, M., (Ed.), *Regionalism in America*, 1952, 119–140.
WHITTAKER, J. R., [1932], Regional interdependence; *Journal of Geography*, 31, 164–165.
WROBEL, A., [1962], Regional analysis and the geographic concept of region; *Papers and Proceedings of the Regional Science Association*, VIII, 37–41.

Since the above paper was written a number of articles on regionalisation have been published. They include:

BERRY, B. J. L., [1965], Identification of declining regions: an empirical study of the dimensions of rural poverty; In Wood, W. D. and Thoman, R. S., *Areas of Economic Stress in Canada*, (Kingston, Ontario).
BUNGE, W., [1966], Gerrymandering, Geography and Grouping; *Geographical Review*, 55, 256–63.
BUNGE, W., [1966], Locations are not unique; *Annals of the Association of American Geographers*, 56, 375–376.
HAGGETT, P., [1965], *Locational Analysis in Human Geography*, (London), 241–276.

CHAPTER THIRTEEN

Organism and Ecosystem as Geographical Models

D. R. STODDART

INTRODUCTION

Geography and ecology are concerned with the distribution, organization and morphology of phenomena on the surface of the earth (Slobodkin, 1962, p. 78), and both disciplines have developed similar concepts and techniques to handle similar problems (Bunge, 1964, pp. 3-4). Overt geographical interest in ecological techniques, however, has been largely confined to a small group of biogeographers, whose influence on the rest of the subject has been marginal, and to a group of American sociologists who sought for a time to restate the aims of human geography in terms of human ecology.

The influence of biological concepts in geography, however, has been both deeper and more pervasive than explicit reference might suggest. Thus, in spite of current insistence on the importance of areal differentiation as a methodological framework for geography, derived by Hartshorne from the work of von Richthofen and Hettner, much geographical work in the past hundred years has taken its inspiration directly from Darwin and the biological revolution which he began. Elsewhere I have indicated some of the main strands in the evolutionary impact on geographical thought since 1859: particularly the emphasis on changing form through time, expressed in organic analogies of ageing; the popularity of natural selection and environmental models, particularly in early American human geography; and, latterly, the application of Haeckel's concept of ecology and Tansley's of the ecosystem, which may be traced back to the third chapter of *The Origin of Species* (Stoddart, 1966).

This essay treats the biological impact on a methodological level. In Kuhn's (1962) terminology, we are concerned with the evolution of geographic *paradigms* of largely biological inspiration: paradigms, or interrelated networks of concepts on a sufficiently general level which serve to define, at least for a time, the nature of geographical goals and the conventional frameworks within which these are pursued. The paradigms examined here are those of *organism* and *ecosystem*: they are not the only ones which may have been chosen, nor necessarily the most influential, but both concern fundamental issues in the methodology of geography. The organic paradigm in particular includes several examples of smaller-scale paradigms which can be defined with greater precision: Davis's cyclic and Clements's successional concepts among them. A distinction is thus made between such generalized conceptual models, for which the term paradigm is usefully employed, and the puzzle-solving procedures (again to use Kuhn's terminology), within the paradigm framework, with which most geographic work is concerned. Ecology itself, especially in recent years, has become more concerned with empirical studies of a puzzle-solving nature, as in the problem of the definition and statistical treatment of plant and animal communities (Greig-Smith, 1964; Kershaw, 1964; Williams, 1964): and many of these techniques are of direct geographic interest at a comparable level of inquiry (Harvey, this volume). Puzzle-solving procedures of this sort, leading to the development of explanatory or predictive models for particular problems, depend on the acceptance of common standards of procedure and of common scientific aims, but they only become of methodological importance when they require the modification of the prevailing paradigm itself.

Organism and ecosystem are of interest as alternative approaches to a central theme in geographical inquiry: that of the relationship of man and environment in area. Thus Hettner states the classical position, when he writes that 'both nature and man are intrinsic to the particular character of areas, and indeed in such intimate union that they cannot be separated from each other' (Hettner, 1905, p. 554). It is also in this relationship that geography has faced two of its most difficult methodological problems, of the dualism between man and environment, and that between human and physical geography (Hartshorne, 1959, pp. 65-80). Both organic and ecosystem concepts have been used to overcome these problems, and to provide coherent frameworks for the organization of geographic data. This essay deals with the nature of these borrowings from the biological sciences, the way in which geographic interpretations have changed with developments in the biological sciences themselves, and the potential paradigmatic value of biological models in geographic methodology.

THE ORGANIC ANALOGY

Organic analogies have been used as 'explanations' of the real world since classical times. In the biological sciences in particular there has been much controversy between mechanists, vitalists and organicists over the reducibility of biological concepts and the scope and significance of organic holistic models (Loeb, 1892; Nagel, 1951; Bertalanffy, 1933, pp. 28-66). Clements's work in plant ecology may be taken as an example of the use of organic analogy in ecology itself. Clements, following Warming, developed the ideas of climax and succession in vegetation, but went beyond the empirical observation of such changes to assert that

> the developmental study of vegetation necessarily rests upon the assumption that the unit or climax formation is an organic entity. As an organism the formation arises, grows, matures, and dies. . . . Each climax formation is able to reproduce itself. (Clements, 1928, p. 3.)

The plant community is

> a complex organism, or superorganism, with characteristic development and structure. . . . It is more than the sum of its individual parts, . . . it is indeed an organism of a new order. (Clements and Shelford, 1939, p. 24.)

As with W. M. Davis's scheme of the geographical cycle (Davis, 1899), to which Clements' successional model bears close resemblance, Clements and his followers regarded the organismic idea as a revealing conceptual framework for understanding the real world: 'this concept is the "open sesame" to a whole new vista of scientific thought, a veritable *magna carta* for future progress' (Clements and Shelford, 1939, p. 24). European ecologists never wholeheartedly accepted Clements' insistence on succession or his metaphysical interpretation of community: while succession may be adequately demonstrated on a small scale (as in marshland, on sand dunes, or in pieces of dung) organismic views of community have little practical relevance at such levels; while at larger scales field workers concentrated on empirical descriptive studies in which organismic concepts had no place (Whittaker, 1953; Muller, 1958; Selleck, 1960). Similarly in pedology, the earlier organismic views of Marbut, Shaler and Whitney have been rejected by such later workers as Jenny (1941) and particularly Nikiforoff (1959).

For the early ecologists, the value of organic analogies lay in their ability to bring large quantities of discrete and often apparently unrelated data into meaningful relationship, to emphasize the organization and functional relationships existing in nature, which, once recognized, served as at least a partial explanation of their complexity. Organic analogies were particularly successful when the problems were most complex and the analytical techniques too undeveloped to produce substantive results. Hence, with the rise of descriptive ecology, with its diverse body of sophisticated descriptive and analytical techniques, vague analogy no longer served as a valid explanation of natural complexity. To take one example, while Clements emphasized the organic nature of the great vegetational units of the world, such as the selva and the coniferous forests, more detailed studies, as in East Asia, have shown that these 'do not represent the discrete units of vegetation, but are no more than the vague patterns joined together by gradual transition' (Kira and others, 1962,

p. 125). Figure 13.1 shows the overlapping distribution of 24 conifer species along a thermal gradient in Japan, and suggests that in this case the vegetation units have no discrete existence. Similar cases may be cited from animal ecology (Clark, 1946).

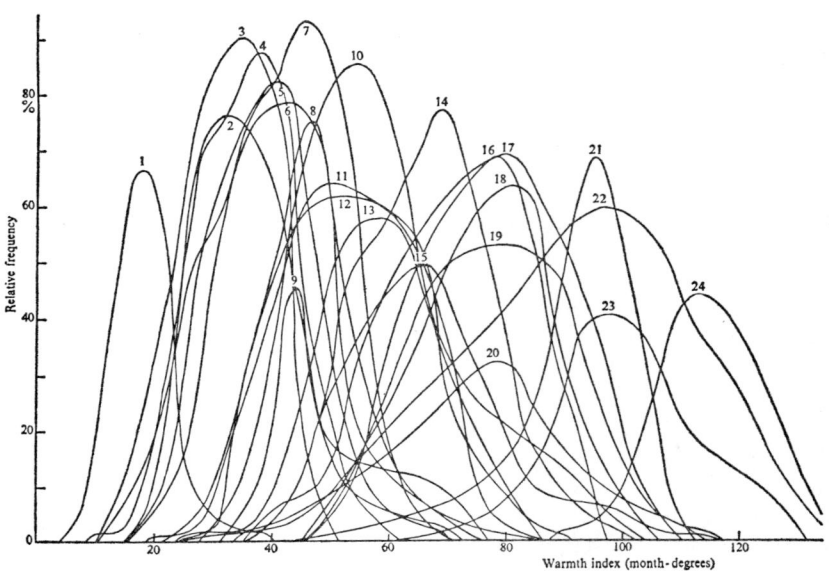

13.1 Distribution of 24 conifer species along a thermal gradient in the Central District, Japan (*After Kira, Ogawa and Yoda, 1962*).

THE ORGANIC ANALOGY IN GEOGRAPHY

This brief discussion demonstrates the appeal of organic analogy in ecological problems similar to those faced by geography. In geography itself, organic analogies are of considerable antiquity, but received fresh impetus from the Darwinian revolution, and maintained popularity until the 1930's. Carl Ritter's inspiration was in no sense biological, but for him and his followers the organic model not only provided a satisfying explanation of the relation of man and nature, but provided a religious and moral justification also. The earth for him was not 'a mere dead, inorganic planet, but an organism, a living work from the hand of a living God' (Ritter, 1866, II, p. 4), in which both animate and inanimate components form 'in a higher and comprehensive sense a cosmical life, . . . one great organism' (Ritter, 1865, p. 1). Ritter's pupil, Arnold Guyot, considered that

> Few subjects seem more worthy to occupy thoughtful minds than the contempla-

tion of the great harmonies of nature and history. The spectacle of the good and the beautiful in nature reflecting everywhere the idea of the Creator, calms and refreshes the soul. . . . Every being, every individual, necessarily forms a part of a greater organism than itself, out of which one cannot conceive its existence, and in which it has a special part to act. . . . All is order, all is harmony in the universe. (Guyot, 1850, 6, p. 85.)

Man himself, in Guyot's optimistic teleology, forms 'the bright consummate flower of this admirable organization' (Guyot, 1850, p. 297). 'The continents are made for human societies, as the body is made for the soul. The conclusion is inescapable that the entire globe is a grand organism, every feature of which is the outgrowth of a definite plan of the all-wise Creator for the education of the human family, and the manifestations of his own glory' (Guyot, in Dryer, 1924, p. 131).

More specifically, the organismic analogy has operated on three distinct levels in geographical work: those of the earth, its regions, and its states; and on each level its use long predates Darwinian evolutionary theory (Ritter and Bailey, 1928; Coker, 1910). Organic theories of the state and of the earth go back to classical and medieval times; they were revived by such philosophers as Hobbes; and were thoroughly worked out by Heinrich Ahrens in his *Organische Staatslehre* in 1850. Much of this earlier work was abstract and metaphysically teleological, as in Ritter's conception of terrestrial unity and Humboldt's cosmological philosophy, but from at least 1750 the analogy with the natural sciences of chemistry, mechanics and biology was being pursued more closely. Johann Bluntschli (1875) imbued states with the attributes of human organisms, even to details of personality and sex, while a fundamental precept of Comte's positivist philosophy was that sociology could only be understood in biological terms (Coker, 1910). Only after Darwin, however, were these somewhat metaphysical and often inchoate ideas given full expression, and social thought from about 1870 to 1900 was dominated by Darwinian thinking. In England and America Herbert Spencer and in France René Worms helped popularize organic analogies in the social sciences, and these retained vitality in geography long after they had been abandoned in other branches of human studies.

Among geographers it is to Butte (1808) and especially to Ritter that the idea of the earth as a functioning organism may best be traced. Similar views were expressed by Alexander von Humboldt and half a century later by Vidal de la Blache, who acknowledged his indebtedness to Ritter in a much quoted aphorism: 'La Terre est un tout, dont les parties sont coordonnées' (Vidal de la Blache, 1896, p. 129; also 1913, p. 289). Vidal did not of course restrict the analogy to the earth: in his own substantive work, and in that of his pupils, the theme of harmonious interrelationships reappears at the state and local-regional level (Vidal de la Blache, 1911 and 1917). Organic concepts of terrestrial unity are found in the work of Brunhes (1920, pp. 26–27 and 1952, pp. 26–28) and Lucien Febre (1925, p. 137), and for a time in the United States also: 'The earth itself is an organism,' wrote Dryer (1920, p. 13) in a more extreme statement, 'and geography is its anatomy, physiology and psychology.'

The idea of the terrestrial organism served as a unifying theme in an increasingly particularistic discipline: as such, it was a synthetic rather than an analytical concept (Batalha-Reis, 1896, p. 756). At the regional level, however, the organic analogy attracted more attention as an analytical tool, particularly in England. Herbertson raised the matter in his discussion of the 'major natural regions' in 1905, using the term 'macro-organism' for the 'complex entity' of physical and organic elements of the earth's surface: 'the soil itself the flesh, the vegetation its epidermal covering with its animal parasites, and the water the circulating life-blood automatically stirred daily and seasonally by the great solar heat. . . . If we regard the Earth as an individual, and these geographical regions, districts, localities, as representing organs, tissues and cells, we perhaps get nearest to a useful comparison' (Herbertson, 1913, p. 205). The concept of the natural region was for Herbertson a tool for assimilating multitudes of diverse facts: within this macro-organism, men are the nerve-cells, in some cases forming amorphous groups, in others organized and specialized in function. In his last paper Herbertson goes on to make the organism analogy more specific: the natural region is comparable to the botanist's plant, the zoologist's animal, or the chemist's mass of matter: natural regions are 'definite associations of inorganic and living matter with definite structures and functions, with as real a form and possessing as regular and orderly changes as those of a plant or an animal' (1913–14, pp. 158–159), and like plants and animals they can be hierarchically ranked into species, genera, orders and classes (1913–14, p. 161).

Similar thinking pervades later English work on regional methodology. Unstead, while building up a synthetic regional scheme from smaller units to larger ones, nevertheless regarded individual regions as living organisms, with distinctive morphology and physiology. He developed the analogy to include evolution, in the sense of increasing complexity, and pathology, in the sense of conditions such as floods and droughts inimical to man. While compelled to admit that regions cannot be said to die, as do biological organisms, Unstead neatly sidestepped the issue by comparing continuity of existence in the region with that of the 'germ-plasm of organisms through the successive generations' (1925, p. 168), thus endowing regions at one time with the properties of individuals, at other times with those of populations. Stevens, while criticizing the idea of a 'natural' region, allows himself to sink even deeper into the organism analogy. 'Unity within diversity, active functioning, progressive evolution are all to be associated with an organized whole, and it is this organism we have to seek. . . . The organizing agent is the human community: community and region are but aspects of the same organism . . . The unity of the natural geographical region is achieved, maintained and developed, by organization, by cohesion, and this cohesion is attained and extended by intercourse, at first within and later beyond the region, provided the organic development is an indigenous growth' (1939, pp. 308–310). The most extreme statement of the organismic concept in England came from the palaeontologist Swinnerton, who speaks of the relationship between man and land as analogous to that beyond oyster and shell. In anthropomorphic language (fighting a losing battle, adaptive response, adaptive radiation, secreting bricks and mortar), he asserts that

> those geographers who think of the world as an organism are making no fantastic comparison. Man in his relation to his environment is still a child of nature, for his reactions closely resemble those of other organisms . . . even at the limits of cultivation the cultural landscape reflects the activities of a living, throbbing organism. (1938, pp. 88–89.)

Similar, if more moderate views were expressed in America by Finch and James, and in Germany by Gradmann (1924), Hans Bluntschli (1921), and others.

Organic theories of the state go back at least to Plato, and formed the basis of Hobbes' Leviathan in pre-Darwinian times. In geographical terms, however, the concept owes its importance largely to Friedrich Ratzel, whose whole work is coloured by Darwinian and Spencerian evolutionary thinking (Steinmetzler, 1956; Wanklyn, 1961). In his *Politische Geographie*, Ratzel states that 'the earth is for us an organism, not only because it is a union of the living Volk with a rigid soil, but because the union is strengthened by the effect of one upon the other to the extent that the two can no longer be visualized as separated from each other' (1897, p. 4; quoted by Mattern, 1942, p. 96): Ratzel's first chapter is entitled 'Der Staat als bödenstandiger Organismus'. This mystical conception of the indivisibility of people and land thus goes far beyond the simple Spencerian analogy of lines of communication and arteries, seats of government and the brain, and so on, developed in *Principles of Sociology*. The state organism depends on the fundamental properties of organization and interdependence; it then assumes properties of growth and competition, and in so doing goes beyond the organismic conceptions of earth and region. In a brief but well-known paper in 1896 Ratzel developed his seven laws of the growth of states, and went on to outline the subsequently notorious idea of *Lebensraum*: 'Just as the struggle for existence in the plant and animal world always centres about a matter of space, so the conflicts of nations are in great part only struggles for territory' (Ratzel, 1897–98, p. 458). While there is undoubtedly a danger that selective quotation of this sort may do violence to Ratzel's undoubtedly scholarly position, as both Broek (1954) and Wanklyn (1961) argue, it is clear that the organic analogy for Ratzel not only provided a simple and powerful model in analytical political geography, but provided an apparently scientific justification for political behaviour. Ratzel's political geography was welcomed by others, including Vidal de la Blache (1898, pp. 108–109), but Miss Semple, in her exposition of Ratzel's work, omitted the cruder Spencerian analogies as already abandoned even in sociology (Semple, 1911, p. v). Ratzel's views served as a source for the *Geopolitik* developed in Europe between the wars in the writings of Rudolf Kjellen and Karl Haushofer. For Kjellen, the state was 'a biological manifestation or form of life', 'deeply rooted in historic and factual realities', endowed not only with morality but also with 'organic lusts' (Kjellen, 1924, p. 203; and commentary by Weigert, 1942, pp. 106–107; Fifield and Pearcy, 1944, p. 11, and Troll, 1949, pp. 128–134). Modern political geography is at pains to emphasize its empirical nature and freedom from the organismic preconceptions which stem back to Ratzel (Weigert and others, 1957).

THE ORGANIC ANALOGY: COMPONENTS AND CRITICISM

Organic analogies have thus been used by geographers at a variety of levels, and it is useful to enumerate briefly the organic properties said to be possessed by geographic areas. The fundamental criterion used by geographers is the possession of organization of constituent components into a functionally related, mutually interdependent complex, which in spite of a continuous flux of matter or energy through it, remains in an equilibrium condition. This complex in equilibrium is said to possess the properties of adaption, cohesion, reaction and recreation (Vallaux, 1929). Many writers stress, with Fleure, that 'the whole is usually more than the sum of its parts; as a whole it has functions and relations which may not be functions or relations attached to any of its parts' (Fleure, 1929, p. 13), and Anuchin (1961, p. 222) argues that

> We are fully justified in regarding as a certain unity such fundamentally different forms of matter, as dead forms, living forms, social forms, which together make up a certain whole of the material world. The whole make-up of all these forms is of course not a simple sum of these forms but more than the sum in view of the process of interaction within the whole.

One may here distinguish the influence of biological vitalism (Driesch, 1907–08), and of the holistic philosophy of Bergson (1907), Smuts (1926), and Whitehead (1952).

Secondly, and on a more empirical level, is the emphasis on morphology, stemming from Sauer's early work on landscape. With recent quantitative studies of pattern in landscape, there has been renewed interest in purely morphological studies, taking its inspiration from the work of D'Arcy

Wentworth Thompson (1917: see also Tobler, 1963). Sauer's work, however, led directly to a third organic property, that of development in the sense of change through time (Sauer, 1925, pp. 30 and 41). The work of the Berkeley school has been fundamentally of the historical type, tracing the geographic influence of culture groups on the landscape through time. Such work finds close parallel in Derwent Whittlesey's concept of sequent occupance, explicitly derived from the ecological concept of succession:

> Human occupance of area, like other biotic phenomena, carries within itself the seed of its own transformation. . . . Because of its obedience to rule analogous to that governing human organisms, the study of human occupance of area rests on secure foundations. . . . The view of geography as a succession of stages of human occupance establishes the genetics of each stage in terms of its predecessor. (Whittlesey, 1929, p. 162.)

Similar ideas are of course found in historical geography and in the frontier hypothesis of F. J. Turner (though here bound up with other Darwinian ideas), as well as in contemporary plant ecology: much of what passed for 'dynamic ecology' at this time was in fact successional or historical interpretation.

Organic analogies have thus been promoted by a wide range of geographers, to whom the subject owes its present status as an independent science. One must therefore ask why such widely held views produced so little in the way of substantive results or new lines of investigation. The debate here closely parallels that between vitalists and mechanists in biology, and similar themes were also raised in philosophy. The theme of organic unity is superficially satisfying, and has the appearance of profound insight, but as Beck states of Driesch's vitalistic views in biology, it poses no questions, and hence obtains no answers. The vitalist hypothesis cannot be tested by observation, for vitalist definitions, e.g. of entelechy, are metaphysical rather than operational in nature. The quality of 'explanation' is on a different level from that normally accepted in the physical and biological sciences. One of Vidal's own pupils, Camille Vallaux, while at first impressed by the aesthetic unity and harmony of the terrestrial whole idea, came to see that no fresh insights could be obtained from it, and that all that remained was a certain poetic attractiveness:

> Qu'on la mette, si l'on veut, au Panthéon scientifique, avec tous les honneurs funèbres; mais qu'on néglige pas de sceller sa challe; qu'elle ne soute point du tombeau. (Vallaux, 1929, p. 49.)

The vitalist position in biology was fundamentally to observe life, to fail to understand it, and to call up essentially undemonstrable, and hence unprovable, causes – the *élan vital* of Bergson and the *entelechy* of Driesch – to fill the gap. Such procedures preclude the objective formulation and testing of hypotheses, because they lie outside hypothesis (Nagel, 1951; Beck, 1961;

Caspari, 1964); and the same objection holds for organismic beliefs in geography. The use of such terms as quasi-organism or super-organism simply defers the problem of definition (Remane, 1931 and 1950).

The major objection to organic interpretations in geography, however, is that they give no assistance in actual investigation: they are synthetic, not analytical conceptions. Thus, as geographic methodology of this type became increasingly metaphysical, so it became increasingly divorced from geographical writing. With a few exceptions, such as the regional monographs of the French school, regional geography became a stereotyped catalogue of categories of regional data, while the metaphysicians talked inspiringly of an organic unity which existed only in their methodological addresses.

The vitalist position in biology on which the extreme organismic view rests has been met and resolved, largely by recent work on molecular biology, and is no longer a live issue (Beck, 1961). Even those biologists who will not admit to the more extreme mechanistic positions, such as von Bertalanffy in *Problems of Life*, take the view that an organism or whole is comprehended if not only its parts are known but also the relationships between them: whether this is a practical proposition in many branches of study at the moment does not affect the argument. Transferring this idea to geography removes the necessity for any metaphysical interpretation of organic unity. Further, as Friederichs (1958, pp. 155–156) points out, most of the qualities by which the organism is defined (such as steady-state conditions, self-regulation, and development), are not restricted to living organisms but are also possessed by non-living systems. Finally, the concept of the organism, and of organic unity, is largely, if not entirely idiographic (Siddall, 1959). As such it can make little or no contribution to an increasingly nomothetic science: though this is not to deny its importance in the development of biological thought (see especially Picken, 1960, pp. 1–22). It is of interest that it is precisely where the idiographic conception of geography has been maintained longest, as at Oxford, that the idea of organic unity of regions has had its most favourable reception. With these criticisms, the organismic idea is reduced to a metaphor of dubious value, which hinges on gross formal and functional comparisons between living matter and complexly interrelated facts in areas. Organization and interrelation cannot be denied, but there were few geographers, at least before 1939, to follow Crowe (1938, pp. 10–11) in his emphatic denial that organization implied any kind of organism. After 1939, with the eclipse of vitalism in biology and philosophy, and its disgrace in *Geopolitik*, the idea of organism has dropped out of geographical writing, except in occasional reverent mention of the work of Herbertson and Vidal de la Blache.

HUMAN ECOLOGY AND THE URBAN SOCIOLOGISTS

With the development of ecology as a formal branch study, several workers attempted to re-state the problem of the relationship between man and land less in terms of organic analogy or Davisian cause-effect determinism, than of the complexity of the interrelationships between the two and its ecological interpretation. This movement was led particularly by the Chicago group of urban sociologists, who coined the term *human ecology* (Park and Burgess, 1921) as a scientific substitute for a discredited human geography. Robert Ezra Park's programmatic statements of the scope of human ecology (Park, 1934 and 1936) deal with the web of life, the balance of nature, concepts of competition, dominance and succession, biological economics and symbiosis – all concepts derived from plant and animal ecology. For Park, human ecology investigated the processes involved in biotic balance, in which man interacts with nature through culture and technology, as demonstrated in urban life (Park, 1915; Park and others, 1925; Adajs, 1951; Evans, 1951; Hawley, 1951). McKenzie (1924; 1926 and 1934) expressed similar ideas, with a more economic bias. Human ecology developed a wide body of methodological literature (Adams, 1935; Emerson, 1943; Gerard, 1940; Hawley, 1944; Darling, 1951), some of which – dealing for example with problems of distance (McKenzie, 1929) or position (Park, 1925) – was purely geographical in nature. Cities had of course already been interpreted by geographers in organic terms (e.g. Mackinder, 1907, p.257, on London), but on a Spencerian analogical level.

The potential of human ecology as stated in purely biological terms by Park and his school, proved attractive to geographers. H. H. Barrows, in his presidential address to the Association of American Geographers in 1923, stated that:

> ... geography is the science of *human ecology*. ... Geography will aim to make clear the relationships existing between natural environments and the distribution and activities of man. Geographers will, I think, be wise to view this problem in general from the standpoint of man's adjustment to environment, rather from that of environmental influence. ... The centre of geography is the study of human ecology in specific areas. This notion holds out to regional geography a distinctive field, an organising concept throughout, and the opportunity to develop a unique group of underlying principles. (Barrows, 1923, pp. 3 and 9.)

Barrows' address aroused little support among geographers (but see Dryer, 1920, p. 16, and Carter, 1950), at least methodologically, and only the Berkeley school, following Carl Sauer, has consistently sought to interpret the relationship between man and land in ecological terms, particularly in frontier areas of the Americas. Studies from Berkeley which make explicit mention of human ecology as an aim include Aschmann's (1959) account of the history of settlement in Baja California, Gordon's (1954) on the Sinú country of Colombia, Innis (1959) on Jamaica, and Talbot (1963) on Masailand. Methodologically, however, these human ecology studies by geographers are indistinguishable from similar work by members of the Chicago school (where Sauer himself studied): compare, for example, Aschmann's work on Baja California with that on the Hawaiian Islands by Park's student Lind (1938). Both show concern with historical (successional) development within a spatial setting, weaving geography and narrative together.

Meanwhile, the sociologists themselves gradually moved away from the specifically biological position of Park. J. W. Bews, himself a botanist, followed Park closely, but beginning with Alihan's *Social Ecology* (1938), sociologists turned to community as their field of study and expressed dissatisfaction with concepts borrowed from the natural sciences (Gettys, 1940; Hawley, 1950), and with some exceptions, including a text-book by White and Renner (1936), the field delimited by Barrows and Park was abandoned by both geographers and human ecologists alike (Alihan, 1938, pp. 1–18; Schnore, 1961). For the geographer, the study of the relationship between man and environment became too specialized in the physiological sense, and though its importance was stressed by Fleure and Biasutti, there have been few substantive studies by geographers (see Monge, 1948; Lee, 1957; and compare, among others Martin, 1855; Burch and DePasquale, 1962; Newman, 1955). Geographers are generally unaware of the work which has been done on environmental influences on, for example, fertility and reproduction (Glover and Young, 1963; Grahn and Kratchman, 1963), though much relevant literature has been recently assembled by Bresler (1966). Both the geographer and the sociologist found that on any other level ecological concepts applied to human affairs only on a general and even philosophical level (Quinn, 1939; Sprout and Sprout, 1956, 1965). At the same time, the ecologists themselves were moving away from the classical statements of Clements, Wheeler and others (Wheeler, 1911 and 1928; Phillips, 1934–35), in the search for more useful analytical ideas which finally resulted in the concept of the ecosystem. Some attempts have been made in recent years to revive a 'human ecology' based on the descriptive interpretation of man-environment interrelationships in particular unique localities (Eyre and Jones, 1966), but such treatments lack methodological rigour, lean heavily on a largely discarded regional approach, appear to be mainly pedagogical in aim, and ignore developments in ecology itself over the last thirty years.

THE ECOSYSTEM AS A MODEL OF REALITY

The term *ecosystem* was formally proposed by Tansley, the plant ecologist, in 1935, as a general term for both the biome ('the whole complex of organisms –

both animals and plants – naturally living together as a sociological unit' (Tansley, 1946, p. 206)) and its habitat. 'All the parts of such an ecosystem – organic and inorganic, biome and habitat – may be regarded as interacting factors which, in a mature ecosystem, are in approximate equilibrium: it is through their interactions that the whole system is maintained' (Tansley, 1946, p. 207). Tansley's concept effectively broadens the scope of ecology itself, which is no longer purely biological in content (for a more restrictive view see Bray, 1958 and McMillan, 1956). In so doing it gives formal expression to a variety of concepts covering habitat and biome which date back at least to Forbes' use of *microcosm* in the ecosystem sense (Forbes, 1887). Friederichs (1937) termed this new ecology 'Wissenschaft von der Natur'; Thienemann (1942) distinguished it from autecology and biocenology (synecology) as *holography*; Clements termed it *bioecology* (Clements and Shelford, 1939); and Markus (1929) used the term *naturkomplex*. Friederichs himself notes that the boundary between this extended ecology and geography 'has never been sharp, for ecology deals sometimes with landscapes; . . . like geography. But the latter does not go into detail so far' (Friederichs, 1958, p. 154). In his rationale, however, Friederichs's picture of geography leans heavily on an outdated Romanticism drawn from Goethe and Alexander von Humboldt; his picture of 'the cosmos of life, . . . the great, all-life embracing holocene – the unity of nature' (1958, p. 158) accords more with Guyot and Ritter than with the subsequent development of Tansley's concept.

Fosberg (1963, p. 2) has developed Tansley's definition as follows:

> An ecosystem is a functioning interacting system composed of one or more living organisms and their effective environment, both physical and biological. . . . The description of an ecosystem may include its spatial relations; inventories of its physical features, its habitats and ecological niches, its organisms, and its basic reserves of matter and energy; the nature of its income (or input) of matter and energy; and the behaviour or trend of its entropy level.

Properties of ecosystems have been recently outlined by Sjörs (1955), Evans (1956), Whittaker (1960), Odum (1953, 1963), Duvigneaud (1962) and McIntosh (1963). Evans (1956) insists on the categorical nature of the term ecosystem, which includes a hierarchy of systems at different levels of complexity and extent. The whole terrestrial ecosystem has been termed the *ecosphere*, derived from ecosystem and biosphere, by Cole (1958).

A parallel development has been that of the Russian school of ecologists, using the terms geocenosis for the physical habitat and biocenosis for the biome, the two uniting to form the *geobiocenosis* (Sukachev, 1944, 1950 and 1958; Blydenstein, 1961). Sukachev's work has had considerable influence on Soviet 'landscape science' (Perel'man, 1961; Yefremov, 1961; Prokayev, 1962), and its use in geography has been discussed by Morgan and Moss (1965). The term geobiocenosis, however, is an unwieldy one, and fails to communicate the single most important characteristic of the ecosystem: that it is a *system* and not simply a random aggregation of discrete phenomena. Lindeman inclusively defined the ecosystem in his classic paper of 1942 as 'any system composed of physical-chemical-biological processes within a space-time unit of any magnitude' (Lindeman, 1942, p. 400), a definition which clearly includes the operational range of geography.

The ecosystem concept has four main properties which recommend it in geographical investigation. First, it is *monistic*: it brings together environment, man and the plant and animal worlds within a single framework, within which the interaction between the components can be analysed. Hettner's methodology, of course, emphasizes this ideal of unity, and some synthesis was achieved in the regional monographs of the French school, but the unity here was aesthetic rather than functional, and correspondingly difficult to define. Ecosystem analysis disposes of geographic dualism, for the emphasis is not on any particular relationship, but on the functioning and nature of the system as a whole.

Secondly, ecosystems are *structured* in a more or less orderly, rational and comprehensible way. The essential fact here, for geography, is that once structures are recognized they may be investigated and studied, in sharp contrast to the transcendental properties of the earth and its regions as

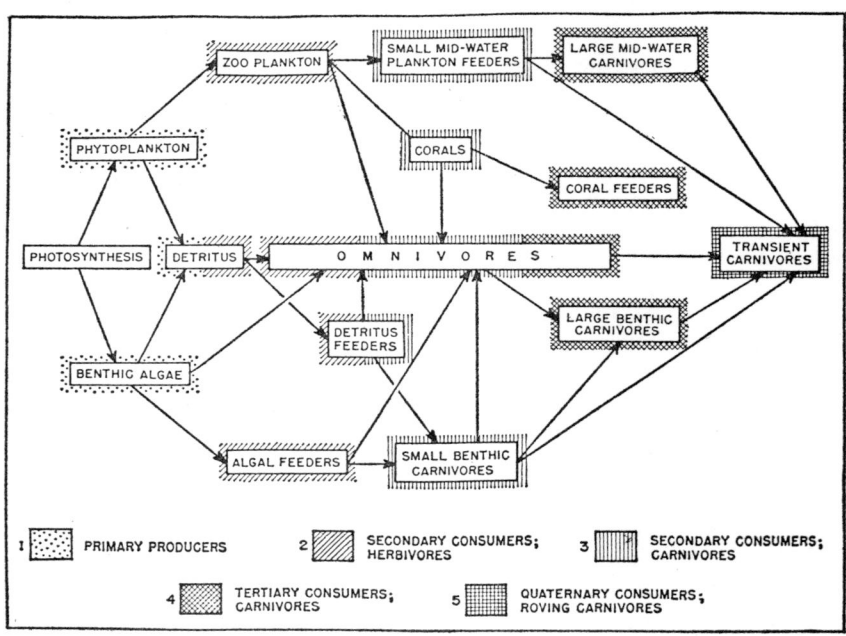

13.2 Food web of coral reefs in the Marshall Islands, showing the trophic structure in qualitative manner (*After Hiatt and Strasburg, 1960*)

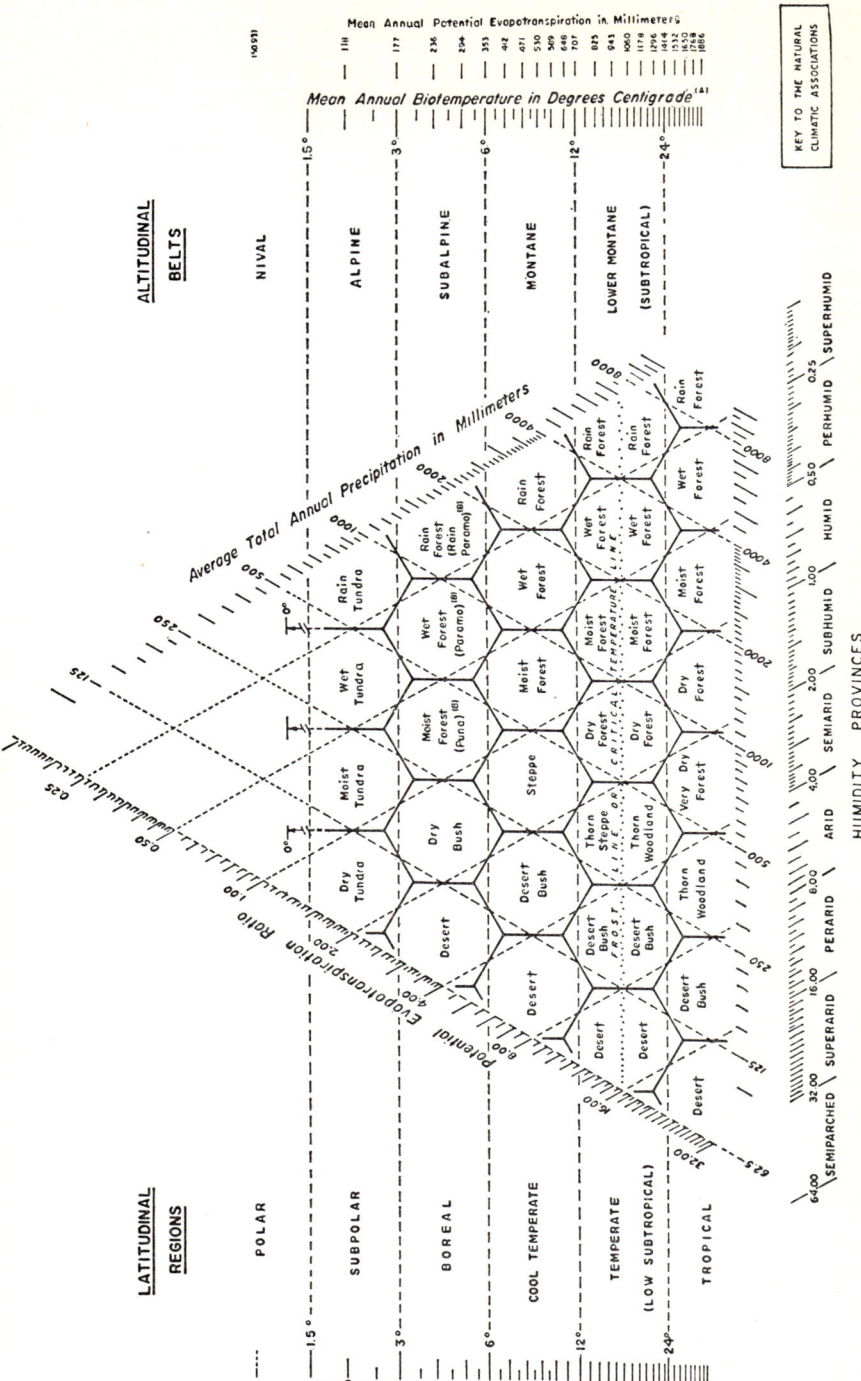

13.3 Mosaic diagram for the classification of natural life zones or ecosystems, in terms of world plant formations, based on Holdridge *After Tosi, 1964*). Mean annual biotemperature given by ($\sum$ mean monthly temperatures $>$ 0° C.)/12.

organisms or organic wholes. Much geographical work in the past has been concerned with the frameworks of systems, and the current concern with the geometry of landforms, settlement patterns and communication networks may be interpreted on this level. As an example of a structural investigation in biology, reference may be made to the work of Hiatt and Strasburg (1960) on the food web and feeding habits of over 200 species of fish in coral reefs of the Marshall Islands in the Pacific. Observation showed that the fish could be classified into five trophic groups, which were related in a rather complex manner, forming the structure shown in Figure 13.2: this includes all levels from plankton and algae to sharks and other carnivores. Among structural studies at a fairly low level of complexity made by geographers, we may cite Holdridge's work on tropical American vegetation types. Holdridge seeks to delimit ecosystem boundaries on the basis of descriptive climatic parameters with presumed physiological significance. Figure 13.3 shows the climatic limits of ecosystem units which define a mosaic of terrestrial ecosystems (Tosi, 1964), and area studies have been published on this basis for a number of Latin American states (e.g. Tosi, 1962, on Peru).

Third, ecosystems *function* (Odum, E. P., 1962 and 1963, p. 10): they involve continuous through-put of matter and energy. To take a geographical example, the system involves not only the framework of the communication net, but also the goods and people flowing through it. Once the framework has been defined, it may be possible to quantify the interactions and interchanges between component parts, and at least in simple ecosystems the whole complex may be quantitatively defined. Odum and Odum (1955) in a pioneering study, again on a Marshall Island coral reef, attempted to quantify the major trophic stages in the coral reef community – the primary producers, the herbivores, and the carnivores. Figure 13.4A shows a biomass pyramid for a measured quadrat near the seaward edge of a reef; Figure 13.4B is a mean biomass pyramid generalized from quadrats across a whole reef flat. While the details of the interpretation, particularly the trophic status of the corals, is open to question, the Odums, in this and in other studies (Odum, H. T., 1957; Odum and Pinkerton, 1955; Odum and Smalley, 1959), have certainly demonstrated the possibility of quantifying the gross structural characteristics of small ecosystems. Equally remarkable is Teal's (1962) study of a salt marsh ecosystem in Georgia. Teal constructed a food web for the salt marsh, and then measured standing crop, production and respiration for each of its components. Figure 13.5 shows in diagrammatic form the energy flow through this ecosystem, and the part played by each component of the web, with an energy input (light) of 600,000 kcal/sq.m./yr.

Fourthly, the ecosystem is a type of general system, and possesses the attributes of general systems. In general system terms, the ecosystem is an open system tending towards a steady state under the laws of open-system

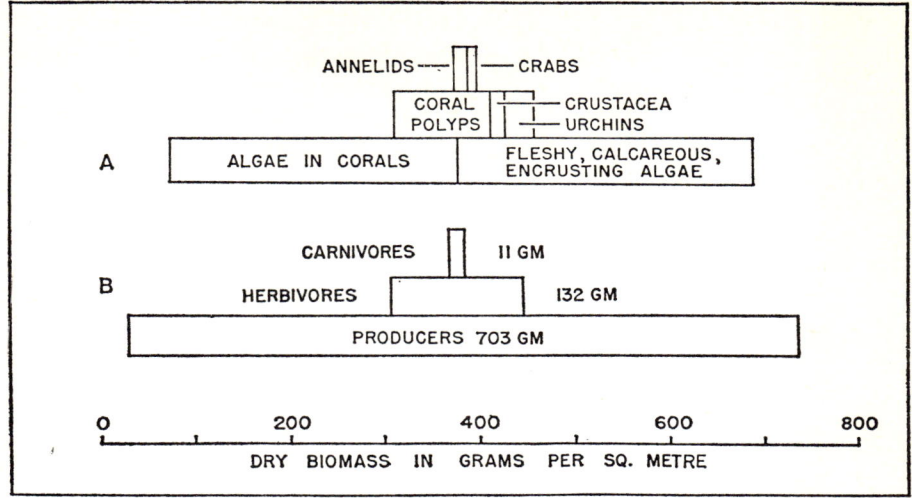

13.4 Biomass pyramids showing the dry weights of living materials in quadrats on the reef of Eniwetok Atoll in the Marshall Islands. A: a quadrat on the reef edge; B: average biomass for the reef. Gross trophic structure is here shown in a quantitative way (*After Odum and Odum, 1955*).

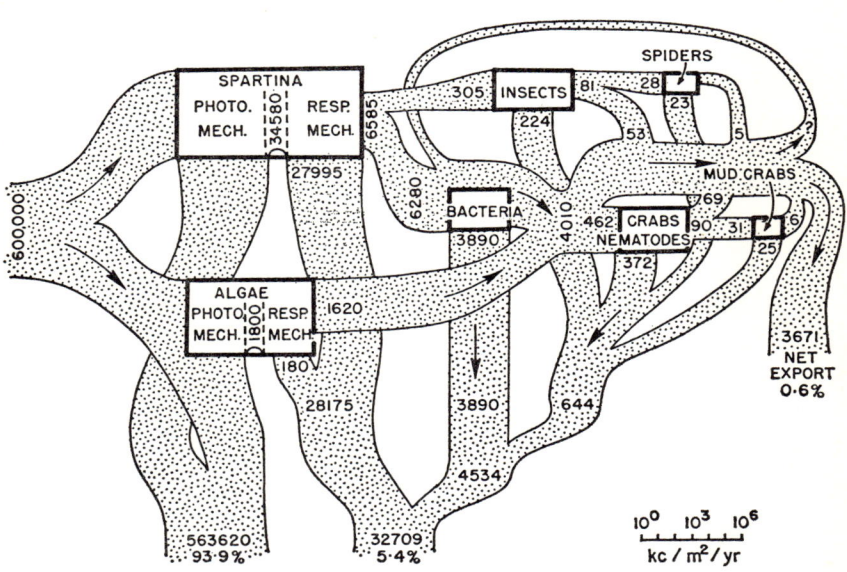

13.5 Energy-flow diagram for a salt marsh in Georgia. The numerals refer to kcal./sq.m/yr. (*After Teal, 1962*).

thermodynamics. Many of the properties of such systems have been implicitly recognized in the past – for example, the idea of climax in vegetation, of maturity in soils, and of grade in geomorphology – but most of these conceptions have been, in effect, the application of classical thermodynamic ideas to closed system situations. With the development of open system thermodynamics (Denbigh, 1951; Prigogine, 1955), many of these older ideas are being reinterpreted in a dynamic rather than a static manner. Whittaker (1953) has thus revised Clement's view on succession and climax; Jenny (1941, 1961) and more recently Nikiforoff (1959) and Auerbach (1958) have done the same for soils; and Chorley (1962) and Hack (1960) have reinterpreted landforms in open system terms.

Ecosystems in a steady state possess the property of self-regulation (action and reaction), and this is similar in principle to a wide range of mechanisms such as homeostasis in living organisms, feedback principles in cybernetics, and servomechanisms in systems engineering (Wiener, 1948; Hutchinson, 1948). Systems such as ecosystems, moreover, may be conceived on different levels of complexity, and it is the task of the geographer to search out aspects of reality which are significant at the level at which the system is conceived. Systems, in fact, possess many of the structural properties of theoretical models, and a first approximation to system structure may be reached in a model-building manner, by selection, simplification, and ordering of data at a series of levels (Chorley, 1964). Thus systems may be constructed at the framework level (e.g. settlement hierarchies or transportation nets) or as simple cybernetic systems (e.g. the mechanism of supply and demand), or at the more complex level of social systems and living organisms. Often in the case of highly complex systems, the system must be conceived at a very much lower level of complexity, in the hope of gaining insight into problems where the data are too involved or the techniques inadequate for complete understanding, or indeed where the problem itself has been insufficiently defined. In geography, for example, the study of such highly complex systems as human groups has often been carried out at the level of 'clockwork' systems, such as simple deterministic, cause-effect relationships. The potential value of a system clearly depends on the correct selection of components at the initial structuring stage, and this normally presupposes considerable experience with the problems or data involved (Boulding, 1956).

Applicability

The ecosystem concept is in origin a biological idea, and most of its applications, including those already quoted, have been from the non-human world. Some attempts have been made, however, to describe fairly complex ecosystems in which man may play some part. Fosberg (1961 and 1962), for example, after many years' work on coral atolls, attempted a general qualitative description of the coral atoll ecosystem, in terms of the media involved, the inflow of energy, primary productivity and successive elaboration, transformation and decomposition of its animal and plant community, excretion and accumulation of matter, and total turnover of matter and energy. Fosberg subsequently convened a symposium (Fosberg (ed.), 1963) to discuss the role of man in the isolated ecosystems of islands, in which the discussion ranged from man's own appraisal of his ecological status to more detailed consideration of the effects of over-population on island life. Islands in fact provide small laboratories for the testing and analysis of relatively simple and well-defined ecosystem structures. Sachet (1963) has described the effects of the introduction of pigs on the ecology of Clipperton Island: vegetation was severely checked by crabs, until men introduced pigs, which ate crabs and allowed the vegetation to grow again. The pigs have recently been killed, and the ecological readjustments are awaited. In a similar situation, Stoddart (1964) has shown how coral islands in the Caribbean, when covered with natural vegetation, are in equilibrium with major storms, and are even built up during hurricanes, but when man replaces the natural vegetation by coconut plantations the storms begin to cause catastrophic erosion. A classic study of an island ecosystem involving man is that of Thomson (1949) on the interaction of man, plants and animals in Fiji, and other works on island ecology of relevance here include those on Lind (1938) on Hawaii and Harris (1962) on the Leeward Islands.

Most ecosystems involving man are more complex than the salt marsh and coral reef systems already described (Figs. 13.3–13.5), and attempts to describe ecosystems at such complex levels are likely to be difficult until experience is gained with relatively simple or restricted systems. Fosberg's focus on islands is one way out of this problem; another, which has received considerable attention recently, is to concentrate on primitive human and sub-human groups, in the hope of obtaining insight into the structure and function of more complex organizations. Schaller's extraordinary study (1963) of the mountain gorilla, *Gorilla gorilla berengei*, its territoriality, population structure, ecology and behaviour, for example, and DeVore's of the baboon (Washburn and DeVore, 1962; DeVore and Washburn, 1964) demonstrate the intriguing possibilities of Primate geography (DeVore (ed.), 1965). Among geographers, Sauer has been pre-eminent in the study of the ecology of man in the Pleistocene, a subject currently being actively developed in relation to the South African Pleistocene (Clark, 1960; Lee, 1963) and the Australian aborigine (Birdsell, 1953); while Daryll Forde (1934), in a classic volume, studied the ecology of some two dozen modern primitive peoples. Most of these studies, however, have been conducted on traditional lines, and not within an explicit system framework: with some of the simpler groups it should be possible to delineate ecosystems with as much precision as in the non-human world.

The power of ecosystem analysis to pose new problems in geography, and hence to seek new answers, is demonstrated by Clifford Geertz's discussion of shifting cultivation and wet rice cultivation in Indonesia (Geertz, 1963). Geertz points out that most discussions of shifting cultivation emphasize its negative characteristics (Gourou, 1956), but that it is more profitably viewed in its system characteristics in relation to the tropical forest it replaces. Both are highly diverse systems, in which matter and energy cycle rapidly among the vegetation components and the topmost soil layer: the soil itself plays little part in this energy flow, and may often be impoverished. Burning is seen as a means of channelling the nutrients locked up in the vegetation into certain selected crop plants: the general ecological efficiency is lowered, but the yield to man increased. In a well-developed shifting cultivation system both structure and function are comparable to those in the tropical forest, but the equilibrium is more delicately poised. By contrast, in wet rice cultivation, the ecosystem structure is quite different, the productivity is high, and the system equilibrium more stable. The analysis is given in qualitative terms, but points the way to several lines of quantitative investigation, with clear import for land use planning and rural reform programmes.

Apart from Geertz's work, there have been few specific system-building studies in geography. The ecologist Dice, after working on natural communities, has produced a survey of ecosystem properties which may serve as a programme for human ecosystem research (Dice, 1952 and 1955): to the normal ecological feedback mechanisms of starvation, predation, disease, migration and competition, he adds (1955, pp. 82–119) public opinion, punishment and rewards, wealth, taxation, supply and demand, co-operation, and the democratic process. Dice suggests that human ecosystems may be conceived at successively larger scale intervals: tribe, homestead, village, town, city, national, and international levels (Dice, 1955, pp. 251–266). Systems theory is also being used in many branches of land use planning, for example the study of water resources (McKean, 1958); Chorley (1962) has carried systems analysis into geomorphology; Brookfield (1964) has briefly noted the potential of ecosystem studies; and Ackerman (1963), in a major paper, has pointed to system analysis as geography's great research frontier. A beginning has been made in substantive geographic work with the McGill University Savanna Research Project, explicitly organized in ecosystem terms: a first report has appeared on the Rupununi savannas of British Guiana (Eden, 1964). The ecosystem as here interpreted, however, is largely restricted to the climate-vegetation-soils complex (cf. Edwards, 1964; Eyre, 1964; and Simmonds, 1966 for similar views), and is thus narrower than the view taken in this paper.

The potential practical value of ecosystem studies in geography may be demonstrated particularly in agricultural and land use studies, and may be illustrated by a number of comprehensive recent symposia: on the biological productivity of Britain (Yapp and Watson, 1958), on the exploitation of natural animal populations (Le Cren and Holdgate, 1962), on the ecology of grazing (Crisp, 1964), and on the ecology of industrial air and water pollution, pesticides, and waste-land reclamation (Goodman, Edwards and Lambert, 1965). Ryther (1959) has discussed potential productivity in the oceans; Ovington (1962) has summarized a great deal of work on woodland productivity throughout the world; and Newbould (1963) has recently reviewed developments in production ecology. Some of the literature on production ecology is complex because of the diverse meanings attributed to terms such as productivity and efficiency (Slobodkin, 1962), but Macfadyen (1964) has proposed the use of L. Dudley Stamp's (1958) Standard Nutritional Unit (SNU), equivalent to 10^6 cal./yr. as a measure of productivity. The following table, from Macfadyen (1964, p. 10) gives sample net annual production for some terrestrial ecosystems managed by man, in SNU/hectare:

Maize, unfertilized	8·2
Maize, fertilized	24·4
British improved grazing	1·0
British farming, national average	7·0
British farming, best	13·6
Europe, wheat maximum	8·3
Sugar beet	60·5
Sugar cane, mean	65·5
Sugar cane, maximum	254·0
Cassava	280·0

Summary tables of gross and net primary productivity for a variety of natural and man-managed ecosystems are given by Odum and Odum (1959, pp. 72–73) and by Westlake (1963). It is interesting to note the extraordinarily high productivity of coral reefs by comparison with almost all other ecosystems. Measurements of gross primary productivity of Pacific reefs range from 1800–11,680 g. $C/m^2/yr.$ (cf. review in Helfrich and Townsley, 1963). At Kauai, Hawaiian Islands, where a fringing reef gave a measured productivity of 2900 gm C/m^2 yr., an adjacent sugar cane plantation, managed over the years to give maximum production, averaged only 1775 gm $C/m^2/yr.$ (Helfrich and Townsley, 1963). It is concluded that 'even under the optimum conditions and techniques employed by modern agriculturists, cultivated terrestrial areas may not be as productive as certain shallow water areas in the sea in which an abundance of benthic algae is exposed to high insolation and a continuous replenishment of nutrients' (Helfrich and Townsley, 1963, p. 50). Similarly, Ovington (1957), working in the East Anglian Breckland, found that pine plantations produced more than twice the average annual production of adjacent farmland, and in ecological terms are undoubtedly the more efficient means of fixing solar energy. Such comparative data are clearly of great importance in rational land use planning, particularly in overpopulated areas. At this empirical level geographers may make a considerable contribution to the understanding of terrestrial ecosystems (Newbould, 1964; Stamp, 1960): active exploration into the value of the ecosystem concept, especially in land use studies, is being carried out by Simmonds (1966).

Problems

It may be objected that the study of ecosystems in geography is either (a) not new, or (b) not geography. In a sense, it is true, the study of systems is implicit in most geographic work: in economics system-building goes back to Smith and Ricardo, and in human and physical geography elements of systems are even older. Mackinder himself (1930, p. 310) went so far as to state the content of geography in closed-system terms. Discussion on the geographic relevance of the ecosystem concept, however, has been tentative and vague. McMillan (1956) has argued the case for excluding geography from the ecological field, and while Rowe (1961, p. 422) believes that all ecosystems are necessarily geographical, he bases his interpretations on Hartshorne's arguments of the uniqueness of geographical phenomena, and thus, from a nomothetic point of view, vitiates his argument (Siddall, 1959).

The charge that ecosystem study is 'not geography' lies in the fact, presumably, that the ecosystem definition does not explicitly define the earth's surface as a field of operation. 'Ecology is the study of environmental relationships; geography is the study of space relationships,' states Davies (1961, p. 415), but he goes on to add that 'what is not clear is where the one stops and the other starts'. Troll (1950, pp. 170–174), among geographers, has used the ecosystem concept in his definition of *Landschaftselemente*, which he subsequently modified to *ökotop* (ecotope), an unfortunate usage since the term had already been pre-empted by Tansley. Perel'man (1961) has classified landscapes in terms of system-processes, and similar procedures are implicit in Holdridge's classifications of tropical ecosystems. The study of space relationships, however, if it is to be more than mere nominal-scale classification of areas, must involve system-building: the study of the ecosystem requires the explicit elucidation of the structure and functions of a community and its environment, with the ultimate aim of the quantification of the links between components.

Potentialities

The ecosystem is a type of general system, defined as a 'set of objects together with relationships between the objects and between their attributes' (Hall and Fagen, 1956). Partaking in general system theory, the ecosystem is potentially capable of precise mathematical structuring within a theoretical

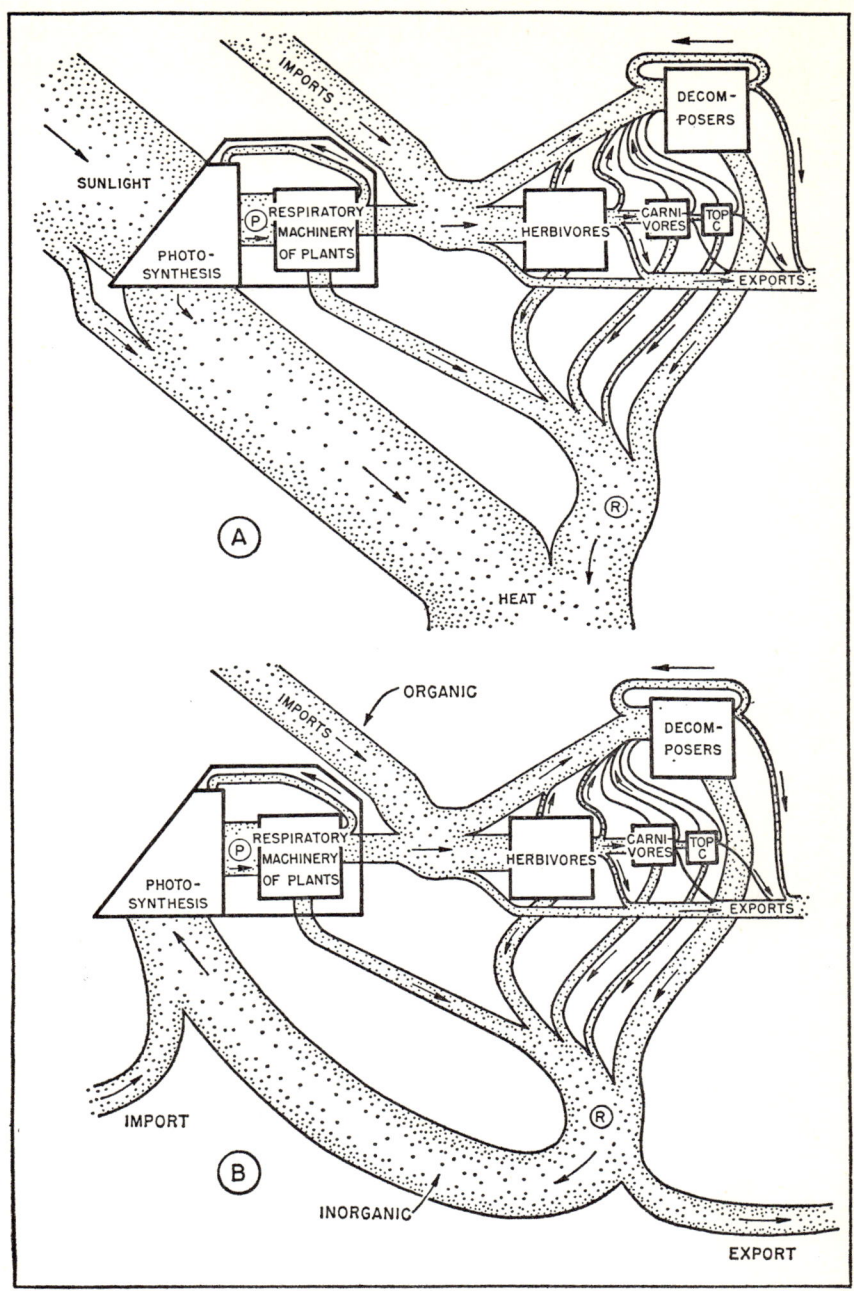

13.6 Odum's conception of (A) energy flow and (B) materials flow (the 'ecomix cycle') in a simple terrestrial ecosystem (*After Odum, 1960*).

framework, a very different matter from the tentative and incomplete descriptions of highly complex relationships which too often pass for geographical 'synthesis'. The limits of the ecosystem may be set at any desirable areal extent, and so flexible is the concept that it may be employed at any level from that of acorns (Winston, 1956) or pieces of dung (Mohr, 1943) to that of the universe itself, and is currently being used in the study of artificial ecosystems within space capsules and interplanetary rockets (Konecci, 1964). Within any areal framework the ecosystem concept will give point to inquiry, and thus highlight both form and function in a spatial setting. Simplistic ideas of causation and development, or of geographic dualism, are in this context clearly irrelevant: ecosystem analysis gives geographers a tool with which to work.

The value of systems analysis lies not only in its emphasis on organization, structure, and functional dynamics: through its general system properties, it brings geography back into the realm of the natural sciences, and allows us to participate in the scientific revolutions of this century from which the Kantian exceptionalist position excluded us (Schaefer, 1953).

A most significant implication of the ecosystem approach in geography is that systems may be interpreted in terms of cybernetics, information and communication theory, and related mathematical techniques. Quantitative interpretation of ecosystem dynamics dates back to Lindeman's paper (1942) on the energy relationships and trophic structure of the ecosystem. This approach has been developed theoretically (Odum and Odum, 1959; Odum and Pinkerton, 1955), and by the use of simulation techniques. Odum (1960) attempted to construct a simple electric analogue for the ecosystem, based on energy flow (Fig. 13.6A) and materials flow (Fig. 13.6B) in a simple ecosystem. Following Ohm's Law ($A=cv$, where A is the flow of electric current, v the voltage and c the conductivity), Odum derives an equation for force and flux in the ecosystem, of the form

$$J_e = c_e x_e$$

where J_e is the ecoflux (e.g. flow of food through a food chain circuit), c_e is the ecological conductivity of the food chain, and x_e the thermodynamic force or ecoforce. Figure 13.7 shows Odum's analogue circuit of the ecosystem, with energy input derived from batteries, food flows simulated by currents, and energy dissipation by amperage and voltage changers. Similarly, Olson has constructed more complex analogue models for the simulation of small ecosystems, incorporating both positive and negative feedback links (Neel and Olson, 1962; Olson, 1963 and 1964).

The formulation of mathematical models for simple ecosystems also suggests the possibility of digital computer simulation for the derivation of steady-state ecosystem properties (Garfinkel, 1962). Garfinkel and Sack (1964) have used an ecosystem model based on Volterra's mass-action law (that the rate at which two species interact is proportional to the product of their populations: Volterra, 1931) to derive ecosystem properties for a six species system which includes grass, bushes, trees, a small and a large herbivore, and a carnivore. The results are clearly only applicable to such simple systems, but they suggest further lines of inquiry in the simulation of more complex systems, and the techniques used may be relevant in, for example, the use of gravity models in population geography (Carrothers, 1958).

A third line of approach stems from Wiener's (1948) development of

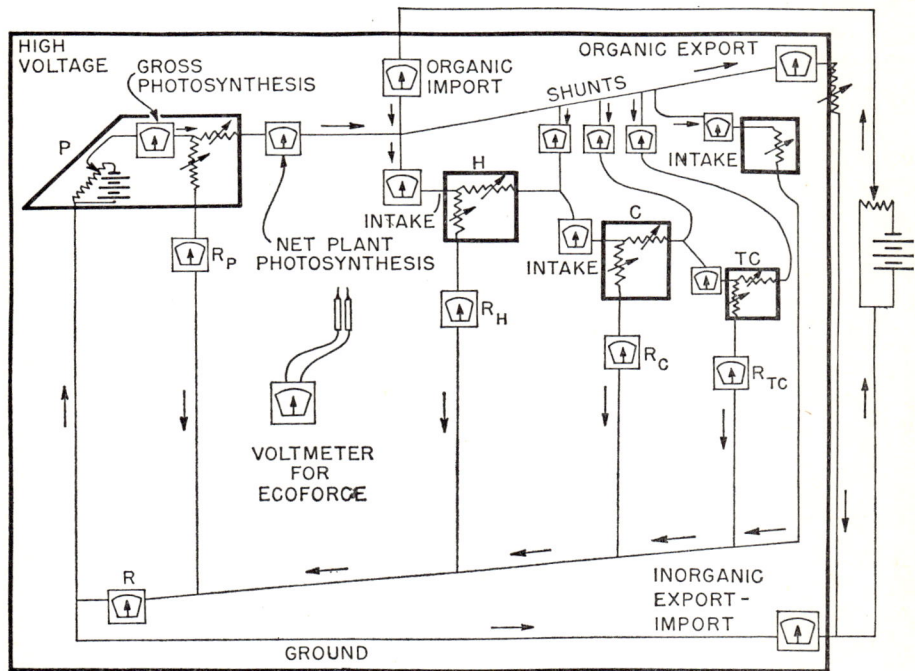

13.7 Electric analogue circuit for a steady state ecosystem modelled on the ecosystem materials flow in Figure 13.6.B (*After Odum, 1960*).

cybernetics. Ecosystems are ordered arrangements of matter, in which energy inputs carry out work. Remove the energy input and the structure will break down until the components are randomly arranged (maximum entropy), which is the most probable state. Brillouin (1962 and 1964) has shown that order, or negative entropy, in systems corresponds to information (in the information-theoretic sense: Quastler, 1958). First attempts have been made, as a result, to apply information theory to ecosystem analysis (Margalef, 1957 and 1958), and to interpret ecosystems in terms of cybernetics (Patten, 1959). Hairston (1959) and MacArthur (1960) have used these techniques in specific problems of ecological populations. Similar methods are being more extensively used in purely biological fields (Beament, 1960; Grodins, 1963; Quastler, 1953; Yockey and others, 1958; George, 1960 and 1964).

Many ecosystem models proposed are conceptually simple, and may be criticized on the grounds that they ignore the complexities of the real world. Slobodkin (1960, p. 214) points out that in the ecosystem the structure of the system itself depends on continuing energy flow, whereas in analogue systems such as Odum's electric circuit the existence of the structure is independent of energy flow. More important from a theoretical point of view, and with special geographical significance, is the fact that in complex-feedback systems, such as ecosystems (Hutchinson, 1948), Prigogine's Theorem (that steady state systems tend to a condition of minimum entropy production) is not necessarily applicable (Foster, Rappoport and Trucco, 1957). If this is indeed the case in ecosystems and geographical systems, static-structured analogue models will have little predictive power. This general argument underlies von Bertalanffy's suggested replacement of purely mechanistic models in biology with his 'organismic theory' (Bertalanffy, 1952), and Slobodkin's statement (1962, p. 82) that Odum's electric analogue model not only violates common sense, but fails to produce either substantive results or suggestions for further work (cf. Greer-Wootten, 1965).

On a general level, Maruyama's distinction between two types of mutual causal processes is especially illuminating in the geographical context. Wiener's cybernetic model is a negative feedback system, typified by such machines as the Watt's Governor and ship's automatic steering mechanisms. Negative feedback systems in geography are numerous: they include the supply and demand process (Marshall, 1890) in equilibrium economics, Malthus's model of population control (Glass, 1953), Gilbert's principle of dynamic equilibrium in river erosion (Gilbert, 1877), and the 'vicious circle' situation in tropical economic development (Nurkse, 1953, p. 4), all of which are, in effect, mechanisms for damping down fluctuations. In addition, Maruyama (1960 and 1963) distinguishes 'deviation-amplifying mutual causal processes' as the basis for a 'second cybernetics'. Deviation-amplifying or positive feedback processes are common in geography, and in ecology itself have led to the great emphasis given to problems of growth, development and succession. Modern technological development is an example of deviation-amplification on a massive scale. The recent expansion of development economics (Rostow, 1960; Meier, 1965) is based on positive feedback principles: if a factory is built or a new road constructed in a rural area, previously existing deviation-damping processes cease to operate, and the system structure is itself transformed. Problem analyses within the earlier frameworks become redundant. Maruyama's process lies at the root of geographical discontent with, for example, 'uniform plain' suppositions in location models (Nystuen, 1963); and it is clear that many ecosystems or geographical systems need to have built into them either deterministic or stochastic 'growth' processes. Geographical needs here begin to converge with those of development economics in such techniques as linear and non-linear programming, rather than with most ecosystem models currently envisaged. Positive feedback requirements thus impose severe restraints on the value of mechanical or electrical analogues (but see Olson, 1963 and 1964), and suggests that computerized mathematical model-building may be more productive (Garfinkel, 1962).

While simulation models may be constructed within given theoretical frameworks, therefore, they are largely limited by the level of sophistication of the theoretical models available. Slobodkin (1958) has suggested that the *categorization* of the theoretical models on the basis of either quantitative or formal properties will be of greater heuristic value than the construction of more and more detailed specific models: such categorization he terms a *meta-model*. The value of meta-models of this sort in geography has been discussed by Chorley (1964), and in the present state of techniques may be the most useful way in which concepts may be clarified and specific new approaches suggested.

GEOGRAPHY, THE ECOSYSTEM AND GENERAL SYSTEMS

This review of biological models in geographic methodology has demonstrated the pervasiveness of first, organic analogies, with intellectual origins far back in antiquity but deriving major impetus from Vitalist biology, and second, and more recently, of the more formal framework of ecosystem analysis. While the ecosystem concept has proved useful in several branches of geographical work, it has become apparent that its influence is seminal rather than definitive, and lies not in the ecosystem concept as such, but in its general system properties. Recent quantitative and simulation studies of ecosystems by biologists have geographical value chiefly in so far as they suggest similar lines of work: the biological emphasis on energetic and trophic structure of ecosystems, for example, is clearly of peripheral geographic significance, but the fundamental concept of system in geography is central to the development of the subject as a nomothetic science.

Paradoxically, therefore, one is led beyond the concept of the ecosystem to the recognition of the importance of system studies in geography (Ackerman, 1963; Blaut, 1962; Berry, 1964; cf. Bertalanffy, 1950; 1951 and 1956; Bertalanffy and others, 1951; Boulding, 1956). Geography is clearly concerned with systems on a multitude of levels. A preliminary attempt to develop a science of 'geocybernetics' has been made in a little-known paper by Polonskiy (1963), and several cases of system-building in geography have been discussed in this paper. The study of these *geosystems* may now replace that of ecosystems in geography, which remains concerned with precisely the same body of data, which, intuitively grasped in all its complexitiy, led to the use of simplistic organic analogies, and subsequently to Tansley's fundamental concept. Systems analysis at last provides geography with a unifying methodology, and using it geography no longer stands apart from the mainstream of scientific progress. As Berry rightly concludes, 'Geography's integrating concepts and processes concern the world-wide ecosystem of which man is the dominant part' (Berry, 1964, p. 3).

REFERENCES

ACKERMAN, E. A., [1963], Where is a research frontier?; *Annals of the Association of American Geographers*, 53, 429–440.
ADAMS, C. C., [1935], The relation of general ecology to human ecology; *Ecology*, 16 316–335.
ADAMS, C. C., [1951], The application of biological research methods to urban research problems; *Scientific Monthly*, 73, 39–40.
AHRENS, H., [1850], *Die Organische Staatslehre auf philosophisch-anthropologischer Grundlage*, (Wien).
ALIHAN, M. A., [1938], *Social ecology: a critical analysis*, (New York).
ANUCHIN, V. A., [1961], *Teoreticheskiye problemy geografii*, (Moscow).
ASCHMANN, H., [1959], The Central Desert of Baja California: demography and ecology; *Ibero-Americana*, 42, 1–282.
AUERBACH, S. I., [1958], The soil ecosystem and radioactive waste disposal to the ground; *Ecology*, 39, 522–529.
BARROWS, H. H., [1923], Geography as human ecology; *Annals of the Association of American Geographers*, 13, 1–14.
BATALHA-REIS, J., [1896], On the definition of geography as a science, and on the conception and description of the earth as an organism; *Report of the Sixth International Geographical Congress, London, 1895*, 753–766.
BEAMENT, J. W. L., [1960], Models and analogues in biology; *Symposia of the Society for Experimental Biology*, 14, 1–255.
BECK, W. S., [1961], *Modern science and the nature of life*, (London), (first pub. 1957).
BERGSON, H., [1907], *L'évolution créatrice*; (Paris), (Translation by A. Mitchell, *Creative evolution*, London, 1911).
BERRY, B. J. L., [1964], Approaches to regional analysis: a synthesis; *Annals of the Association of American Geographers*, 54, 2–11.
BERTALANFFY, L. VON, [1933], *Modern theories of development: An introduction to theoretical biology*, (London, Reissued in New York, 1962).
BERTALANFFY, L. VON, [1950], The theory of open systems in physics and biology; *Science*, 111, 23–29.
BERTALANFFY, L. VON, [1951], An outline of general system theory; *British Journal of the Philosophy of Science*, 1, 134–165.
BERTALANFFY, L. VON, [1952], *Problems of life*, (London).
BERTALANFFY, L. VON, [1956], General system theory; *General Systems*, 1, 1–10.

BERTALANFFY, L. VON, HEMPEL, C. G., BASS, R. E., and JONES, H., [1951], General system theory: a new approach to unity of science; *Human biology*, 23, 302–361.
BIRDSELL, J. B., [1953], Some environmental and cultural factors influencing the structuring of Australian aboriginal populations; *American Naturalist*, 87, 171–207.
BLAUT, J. M., [1962], Object and relationship; *Professional Geographer*, 14 (6), 1–7.
BLUNTSCHLI, H., [1921], Die Amazonasniederung als harmonischer Organismus; *Geographische Zeitschrift*, 27, 49–67.
BLUNTSCHLI, J., [1875], *Allgemeine Statslehre. Fünfte umgearbeitete Auflage der ersten Bandes der Allgemeine Statsrechts. Lehre vom Modernen Stat*; Th. 1, (Stuttgart).
BLYDENSTEIN, J., [1961], The Russian school of phytocenology; *Ecology*, 42, 575–7.
BOULDING, K., [1956], General systems theory – the skeleton of science; *General Systems*, 1, 11–17.
BRAY, J. R., [1958], Notes toward an ecologic theory; *Ecology*, 39, 770–776.
BRESLER, J. B., [1966], *Human ecology: collected readings*, (Reading, Massachusetts).
BRILLOUIN, L., [1962], *Science and information theory*, (New York), second edition.
BRILLOUIN, L., [1964], *Scientific uncertainty, and information*, (New York).
BROEK, J. O. M., [1954], *Friedrich Ratzel in retrospect*, (Mimeographed), (abstract in *Annals of the Association of American Geographers*, 44, 207).
BROOKFIELD, H. C., [1964], Questions on the human frontiers of geography; *Economic Geography*, 40, 283–303.
BRUNHES, J., [1920], *Human Geography: An attempt at a positive classification: principles and examples*, (Bowman, I. and Dodge, R. E., Eds.), (London).
BRUNHES, J., [1952], *Human geography* (abridged), (London).
BUNGE, W., [1964], Patterns of location; *Michigan Inter-University Community of Mathematical Geographers, Discussion Paper* No. 3, 1–36.
BURCH, G. E. and DEPASQUALE, N. P., [1962], *Hot climates, man and his heart*, (Springfield).
BUTTE, W., [1808], *Die Statistik als Wissenschaft*, (Landshut).
CARROTHERS, G. A. P., [1958], An historical review of the gravity and potential concepts of human interaction; *Journal of the American Institute of Planners*, 22, 94–102.
CARTER, G. F., [1950], Ecology – geography – ethnobotany; *Scientific Monthly*, 70, 73–80.
CASPARI, E., [1964], On the conceptual basis of the biological sciences; In Colodny, R. G., (Ed.), *Frontiers of science and philosophy*, (London), 131–145.
CHORLEY, R. J., [1962], Geomorphology and general systems theory; *United States Geological Survey Professional Paper* 500-B, 1–10.
CHORLEY, R. J., [1964], Geography and analogue theory; *Annals of the Association of American Geographers*, 54, 127–137.
CLARK, E. J., [1946], Studies in the ecology of British grasshoppers; *Transactions of the Royal Entomological Society of London*, 99, 173–222.
CLARK, J. D., [1960], Human ecology during Pleistocene and later times in Africa south of the Sahara; *Current Anthropology*, 1, 307–324.
CLEMENTS, F. E., [1928], *Plant succession and indicators*, (Washington).
CLEMENTS, F. E. and SHELFORD, V. E., [1939], *Bio-ecology*, (New York).

COKER, F. W., [1910], Organismic theories of the state. Nineteenth-century interpretations of the state as organism or as person; *Columbia University Studies in History, Economics, etc.*, 38 (2), 1–209.
COLE, L., [1958], The ecosphere; *Scientific American*, 198 (4), 83–92.
CRISP, D. J., (Ed.), [1964], Grazing in terrestrial and marine environments; *British Ecological Society Symposia*, 4, 1–322.
CROWE, P. R., [1938], On progress in geography; *Scottish Geographical Magazine*, 54, 1–19.
DARLING, F. F., [1951], The ecological approach to the social sciences; *American Scientist*, 39, 244–254.
DARWIN, C. R., [1859], *The origin of species by means of natural selection or the preservation of favoured races in the struggle for life*, (London), (page references to the Oxford reprint of the sixth edition, 1951).
DAVIES, J. L., [1961], Aim and method in zoogeography; *Geographical Review*, 51, 412–417.
DAVIS, W. M., [1899], The geographical cycle; *Geographical Journal*, 14, 481–504.
DENBIGH, K. G., [1951], *The thermodynamics of the steady state*, (London).
DEVORE, I., (Ed.), [1965], *Primate behavior: field studies of monkeys and apes*, (New York).
DEVORE, I. and WASHBURN, S. L., [1964], Baboon ecology and human evolution; In Howell, F. C. and Bourlière, F., (Eds.), *African ecology and human evolution* (London), 335–367.
DICE, L. R., [1952], *Natural communities*, (Ann Arbor, Mich.).
DICE, L. R., [1955], *Man's nature and nature's man: the ecology of human communities*, (Ann Arbor, Mich.).
DRIESCH, H., [1908], *The science and philosophy of the organism*; The Gifford Lectures delivered before the University of Aberdeen in 1907 and in 1908, (London), 2 vols.
DRYER, C. R., 1920, Genetic geography. The development of the geographic sense and concept; *Annals of the Association of American Geographers*, 10, 3–16.
DRYER, C. R., [1924], A century of geographic education in the United States; *Annals of the Association of American Geographers*, 14, 117–149.
DUVIGNEAUD, P., [1962], *Ecosystèmes et biosphere. L'écologie, science moderne de synthèse*, (Brussels), 2 vols.
EDEN, M. J., [1964], The savanna ecosystem, northern Rupununi, British Guiana; *McGill University, Savanna Research Series*, 1, 1–216.
EDWARDS, K. C., [1964], The importance of biogeography; *Geography*, 49, 85–97.
EMERSON, A. E., [1943], Ecology, evolution, and society; *American Naturalist*, 77, 97–118.
EVANS, F. C., [1951], Ecology and urban areal research; *Scientific Monthly*, 73, 37–38.
EVANS, F. C., [1956], Ecosystem as the basic unit in ecology; *Science*, 123, 1127–8.
EYRE, S. R., [1964], Determinism and the ecological approach in geography; *Geography*, 49, 369–376.
EYRE, S. R. and JONES, G. R. J., (Eds.), [1966], *Geography as human ecology; methodology by example*, (London).
FEBVRE, L., [1925], *A geographical introduction to history*, (London).
FIFIELD, R. H. and PEARCY, G. E., [1944], *Geopolitics in principle and practice*, (Boston).

FLEURE, H. J., [1929], *An introduction to geography*, (London).
FORBES, S. A., [1887], The lake as a microcosm; *Bulletin of the Scientific Association of Peoria*, Reprinted in *Bulletin of the Illinois Natural History Survey*, 15 (1925), 537—550.
FORDE, C. D., [1934], *Habitat, economy and society: a geographical introduction to ethnography*, (London).
FOSBERG, F. R., [1961], Qualitative description of the coral atoll ecosystem; *Atoll Research Bulletin*, 81, 1–11.
FOSBERG, F. R., [1962], Qualitative description of the coral atoll ecosystem; *Proceedings of the Ninth Pacific Science Congress*, 4, 161–167.
FOSBERG, F. R., [1963], The island ecosystem; In Fosberg, F. R., (Ed.), *Man's place in the island ecosystem*, (Honolulu), 1–6.
FOSBERG, F. R., (Ed.), [1963], *Man's place in the island ecosystem*, (Honolulu).
FOSBERG, F. R., [1965], The entropy concept in ecology; *Symposium on ecological research in humid tropics vegetation, Kuching, Sarawak*, 157–163.
FOSTER, C., RAPPOPORT, A. and TRUCCO, E., [1957], Some unsolved problems in the theory of non-isolated systems; *General systems*, 3, 9–29.
FRIEDERICHS, K., [1937], *Ökologie als Wissenschaft von der Natur oder Biologische Raumforschung*, (Leipzig).
FRIEDERICHS, K., [1958], A definition of ecology and some thoughts about basic concepts; *Ecology*, 39, 154–159.
GARFINKEL, D., [1962], Digital computer simulation of ecological systems; *Nature*, 194, 856–857.
GARFINKEL D. and SACK R. [1964], Digital computer simulation of an ecological system, based on a modified mass-action law; *Ecology*, 45, 502–507.
GEERTZ, C., [1963], *Agricultural involution: The process of agricultural change in Indonesia*, (Berkeley).
GEORGE, F. H., [1960], Models in cybernetics; *Symposia of the Society for Experimental Biology*, 14, 169–191.
GEORGE, F. H., [1964], *Cybernetics and biology*, (Edinburgh).
GERARD, R. W., [1940], Organism, society, and science; *Scientific Monthly*, 50, 340–350, 403–412 and 530–535.
GETTYS, W. E., [1940], Human ecology and social theory; *Social Forces*, 18, 469–476.
GILBERT, G. K., [1877], *Geology of the Henry Mountains*, (Washington).
GLASS, D. V., (Ed.), [1953], *Introduction to Malthus*, (London).
GLOVER, T. D. and YOUNG, D. H., [1963], Temperature and the production of spermatozoa; *Fertility and Sterility*, 14, 441–450.
GOODMAN, G. T., EDWARDS, R. W. and LAMBERT, J. M., (Eds.), [1965] Ecology and the industrial society; symposia of the *British Ecological Society*, 5.
GORDON, B. LE R., [1954], *Human geography and ecology in the Sinú country of Colombia*; Berkeley, Department of Geography, Ph.D. thesis; Ibero-Americana, (1957).
GOUROU, P., [1953], *The tropical world*, (London).
GOUROU, P., [1956], The quality of land use of tropical cultivators; In Thomas, W. L., (Ed.), *Man's role in changing the face of the earth*, (Chicago), 336–349.
GRADMANN, R., [1924], Das harmonischer Landschaftsbilde; *Zeitschrift der Gesellschaft für Erdkunde zu Berlin*, 59, 129–147.

GRAHN, D. and KRATCHMAN, J., [1963], variation in neonatal death rate and birth weight in the United States and possible relations to environmental radiation, geology and altitude; *American Journal of Human Genetics*, 15, 329–352.

GREER-WOOTTEN, B., [1965], *General systems theory – a new backbone for the 'formless' discipline of geography?*; Discussion paper, Graduate Seminar Group in Arctic Geography, McGill University, October 18, 1965, 1–18.

GREIG-SMITH, P., [1964], *Quantitative plant ecology*, (London), second edition.

GRODINS, F. S., [1963], *Control theory and biological systems*, (Columbia).

GUYOT, A., [1850], *The earth and man: Lectures on comparative physical geography*, (London).

HACK, J. T., [1960], Interpretation of erosional topography in humid temperate regions; *American Journal of Science*, 258–A, 80–97.

HAIRSTON, N. G., [1959], Species abundance and community organization; *Ecology*, 40, 404–416.

HALL, A. D. and FAGEN, R. E., [1956], Definition of system; *General Systems*, 1, 18–28.

HARRIS, D. R., [1962], The invasion of oceanic islands by alien plants: an example from the Leeward Islands, West Indies; *Transactions of the Institute of British Geographers*, 31, 67–82.

HARTSHORNE, R., [1959], *Perspective on the nature of geography*, (London).

HAWLEY, A. H., [1944], Ecology and human ecology; *Social Forces*, 22, 398–405.

HAWLEY, A. H., [1950], *Social ecology: a theory of community structure*, (New York).

HAWLEY, A. H., [1951], The approach of human ecology to urban areal research; *Scientific Monthly*, 73, 48–49.

HELFRICH, P. and TOWNSLEY, S. J., [1963], The influence of the sea; In Fosberg, F. R., (Ed.), *Man's place in the island ecosystem*, (Honolulu), 39–53.

HERBERTSON, A. J., [1905], The major natural regions: an essay in systematic geography; *Geographical Journal*, 25, 300–310.

HERBERTSON, A. J., [1913], The higher units: a geographical essay; *Scientia*, 14, 203–212, (Reprinted in *Geography*, 50 (1965), 332–342).

HERBERTSON, A. J., [1913–14], Natural regions; *Geographical Teacher*, 7, 158–163.

HETTNER, A., [1905], Das Wesen und die Methoden der Geographie; *Geographische Zeitschrift*, 11, 545–564, 615–629 and 671–686.

HIATT, R. W. and STRASBURG, D. W., [1960], Ecological relationships of the fish fauna on coral reefs of the Marshall Islands; *Ecological Monographs*, 30, 65–127.

HOLLINGSHEAD, A. B., [1940], Human ecology and human society; *Ecological Monographs*, 10, 354–363.

HUTCHINSON, G. E., [1948], Circular causal systems in ecology; *Annals of the New York Academy of Sciences*, 50, 221–246.

INNIS, D. Q., [1959], *Human ecology in Jamaica with a detailed study of peasant agriculture in the Mollison district of northern Manchester*; Berkeley, Department of Geography, Ph.D. thesis, 1–248.

JENNY, H., [1941], *Factors of soil formation: a system of quantitative pedology*, (New York).

JENNY, H., [1961], Derivation of state factor equations of soils and ecosystems; *Proceedings of the Soil Science Society of America*, 25, 385–388.

KALESNIK, S. V., [1962], About 'monism' and 'dualism' in Soviet geography; *Soviet Geography*, 3 (7), 3–16.

KERSHAW K. A., [1954], *Quantitative and dynamic ecology*, (London).
KIRA, T., OGAWA, H. and YODA, K., [1962], Some unsolved problems in tropical forest ecology; *Proceedings of the Ninth Pacific Science Congress*, 4, 124-134.
KJELLEN, R., [1924], *Der Staat als Lebensform*, (Berlin), 4th edition. (Translation by J. Sandmeier of *Staten som Lifsform*, Stockholm, 1916).
KONECCI, E. B., [1964], Space ecological systems; In Schaefer, K. E., (Eds.), *Bioastronautics*, (New York), 274-304.
KUHN, T. S., [1962], *The structure of scientific revolutions*, (London).
LE CREN, E. D. and HOLDGATE, M. W., [1962], The exploitation of natural animal populations; *Symposia of the British Ecological Society*, 2.
LEE, D. H. K., [1958], *Climate and economic development in the tropics*, (New York).
LEE, R. B., [1963], The population ecology of man in the Early Upper Pleistocene of Southern Africa; *Proceedings of the Prehistoric Society*, N.S. 29, 235-257.
LIND, A. W., [1938], *An island community: Ecological succession in Hawaii*, (Chicago).
LINDEMAN, R. L., [1942], The trophic-dynamic aspect of ecology; *Ecology*, 23, 399-418.
LOEB, J., [1892], *Organbildung und Wachsthum. Untersuchungen zur Physiologischen Morphologie der Thiere*, 2, (Wurzburg).
MACARTHUR, R., [1955], Fluctuations of animal populations, and a measure of community stability; *Ecology*, 36, 533-536.
MACARTHUR, R., [1960], On the relative abundance of species; *American Naturalist*, 94, 25-36.
MACFADYEN, A., [1964], Energy flow in ecosystems and its exploitation by grazing; In Crisp, D. J., (Ed.), Grazing in terrestrial and marine environments, *Symposia of the British Ecological Society*, 4, 3-20.
MACKINDER, H. J., [1907], *Britain and the British Seas*, (Oxford), second edition.
MACKINDER, H. J., [1930], The content of philosophical geography; *Proceedings of the International Geographical Union*, (Cambridge), (1928), 305-312.
MARGALEF, R., [1957], La teoria de la informacion en ecologia; *Memorias Real Academia, Barcelona*, 32, 373-449.
MARGALEF, R., [1958], Information theory in ecology; *General Systems*, 2, 36-71.
MARKUS, E., [1929], Naturkomplexe; *Protokoly obshchestva Estestviospytatelei pri imperatorskom Yur'evskom Universitete* (Yur'ev, Dorpat, Jurjew, or Tartu), 32, 79-94.
MARSHALL, A., [1890], *Principles of economics*, (London).
MARTIN, J. R., [1855], *Influence of tropical climates in producing the acute endemic diseases of Europeans*, (London).
MARUYAMA, M., [1960], Morphogenesis and morphostasis; *Methodos*, 12, 251-296.
MARUYAMA, M., [1963], The second cybernetics: deviation-amplifying mutual causal processes; *American Scientist*, 51, 164-179.
MATTERN, J., [1942], *Geopolitik: Doctrine of national self-sufficiency and empire*, (Baltimore).
MCINTOSH, R. P., [1963], Ecosystems, evolution and relational patterns of living organisms; *American Scientist*, 51, 246-267.
MCKEAN, R. N., [1958], *Efficiency in government through systems analysis, with emphasis on water resources development*, (New York).

MCKENZIE, R. D., [1924], The ecological approach to the study of the human community; *American Journal of Sociology*, 30, 287–301.
MCKENZIE, R. D., [1926], The scope of human ecology; *Publications of the American Sociological Society*, 20, 141–154.
MCKENZIE, R. D., [1929], Spatial distance; *Sociology and Social Research*, 13, 536–545.
MCKENZIE, R. D., [1934], Demography, human geography, and human ecology; In Bernard, L. L., (Ed.), *The fields and methods of sociology*, (New York), 52–66.
MCMILLAN, C., [1956], The status of plant ecology and plant geography; *Ecology*, 37, 600–602.
MEIER, G. M., [1965], *Leading issues in development economics*, (Oxford).
MOHR, C. O., [1943], Cattle droppings as ecological units; *Ecological Monographs*, 13 275–298.
MONGE, C., [1948], *Acclimatisation in the Andes*, (New York).
MORGAN, W. B. and MOSS, R. P., [1965], Geography and ecology: the concept of the community and its relationship to environment; *Annals of the Association of American Geographers*, 55, 339–350.
MULLER, C. H., [1958], Science and philosophy of the community concept; *American Scientist*, 46, 294–308.
NAGEL, E., [1951], Mechanistic explanation and organismic biology; *Philosophy and Phenomenological Research*, 11, 327–338.
NEEL, R. B. and OLSON, J. S., [1962], Use of analog computers for simulating the movement of isotopes in ecological systems; *Oak Ridge National Laboratory Report ORNL-*3172.
NEWBOULD, P. J., [1963], Production ecology; *Science Progress*, 51, 91–104.
NEWBOULD, P. J., [1964], Production ecology and the International Biological Programme; *Geography*, 49, 98–104.
NEWMAN, M. T., [1955], Adaptation of man to cold climates; *Evolution*, 9, 101–105.
NIKIFOROFF, C. C., [1959], Reappraisal of the soil; *Science*, 129, 186–196.
NURKSE, R., [1953], *Problems of capital formation in underdeveloped countries*, (Oxford).
NYSTUEN, J. D., [1963], Identification of some fundamental spatial concepts; *Papers of the Michigan Academy of Science, Arts and Letters*, 48, 373–384.
ODUM, E. P., [1953], *Fundamentals of ecology*, (Philadelphia).
ODUM, E. P., [1962], Relationship between structure and function in the ecosystem; *Japanese Journal of Ecology*, 12, 108–118.
ODUM, E. P., [1963], *Ecology*, (New York).
ODUM, E. P. and SMALLEY, A. E., [1959], Comparison of population energy flow of a herbivorous and a deposit-feeding invertebrate in a salt marsh ecosystem; *Proceedings of the National Academy of Science*, 45, 617–622.
ODUM, H. T., [1957], Trophic structure and productivity of Silver Springs, Florida; *Ecological Monographs*, 27, 55–112.
ODUM, H. T., [1960], Ecological potential and analogue circuits for the ecosystem; *American Scientist*, 48, 1–8.
ODUM, H. T. and ODUM, E. P., [1955], Trophic structure and productivity of a windward coral reef community on Eniwetok Atoll; *Ecological Monographs*, 25, 291–320.
ODUM, H. T. and ODUM, E. P., [1959], Principles and concepts pertaining to energy

in ecological systems; In Odum, E. P., *Fundamentals of ecology*, (Philadelphia), Chapter 3, 43–87.

ODUM, H. T. and PINKERTON, R. C., [1955], Time's speed regulator, the optimum efficiency for maximum output in physical and biological systems; *American Scientist*, 43, 331–343.

OLSON, J. S., [1963], Analog computer models for movement of nuclides through ecosystems; *Radioecology: Proceedings of the First National Symposium*.

OLSON, J. S., [1964], Gross and net production of terrestrial ecosystems; *British Ecological Society Jubilee Symposium, Supplement to Journal of Ecology*, 52, 99–118.

OVINGTON, J. D., [1957], Dry-matter production by *Pinus sylvestris* L.; *Annals of Botany*, 21, 287–314.

OVINGTON, J. D., [1962], Quantitative ecology and the woodland ecosystem concept; *Advances in Ecological Research*, 1, 103–192.

PARK, R. E., [1915], The city: suggestions for the investigation of human behaviour in the city environment; *American Journal of Sociology*, 20, 577–612.

PARK, R. E., [1925], The concept of position in sociology; *Proceedings of the American Sociological Society*.

PARK, R. E., [1934], Human ecology; *American Journal of Sociology*, 43, 1–15.

PARK, R. E., [1936], Succession as an ecological concept; *American Sociological Review*, 1, 171–179.

PARK, R. E. and BURGESS, E. W., [1921], *Introduction to the science of sociology*, (Chicago).

PARK, R. E., BURGESS, E. W. and MCKENZIE, R. D., [1925], *The city*, (Chicago).

PATTEN, B. C., [1959], An introduction to the cybernetics of the ecosystem: the trophic-dynamic aspect; *Ecology*, 40, 221–231.

PEREL'MAN, A. I., [1961], Geochemical principles of landscape classification; *Soviet Geography*, 2 (3), 63–73.

PHILLIPS, J. F. V., [1934–35], Succession, development, the climax and the complex organism: an analysis of concepts; *Journal of Ecology*, 22, 554–571; 23, 210–246 and 488–508.

PICKEN, L. E. R., [1960], *The organization of cells and other organisms*, (Oxford).

POLONSKIY, M. L., [1963], *Geokibernetkia, predmet i metod*, (Minsk), (Mimeographed).

PRIGOGINE, I., [1955], *Introduction to thermodynamics of irreversible processes*, (Springfield, Illinois).

PROKAYEV, V. I., [1962], The facies as the basic and smallest unit in landscape science; *Soviet Geography*, 3 (6), 21–29.

QUASTLER, H., (Ed.), [1953], *Information theory in biology*, (Urbana, Ill.).

QUASTLER, H., [1958], A primer on information theory; In Yockey, H. P. and others, (Ed.), *Symposium on information theory in biology*, 3–49.

QUINN, J. A., [1939], The nature of human ecology: re-examination and redefinition; *Social Forces*, 18, 161–168.

RATZEL, F., [1896], Die Gesetze des raumlichen Wachstums der Staaten. Ein Beitrag zur wissenschaftlichen politischen Geographie; *Petermanns Mitteilungen*, 42, 96–107.

RATZEL, F., [1897], *Politische Geographie*, (Munich).

RATZEL, F., [1897–98], Studies in political areas; *American Journal of Sociology*, 3, 279–313; 3, 449–463.
REMANE, A., [1939], Die Gemeinschaft als Lebensform in der Natur; *Kieler Blätter*, 2, 43–61.
REMANE, A., [1950], Ordnungsformen der Lebenden Natur; *Studium generale*, 3, 404–410.
RITTER, C., [1865], *Comparative geography*; (Translated by W. L. Gage), (Edinburgh).
RITTER, C., [1866], *The comparative geography of Palestine and the Sinaitic Peninsula*; (Translated by W. L. Gage), (Edinburgh), 4 vols.
RITTER, W. E. and BAILEY, E. W., [1928], The organismal conception: its place in science and its bearing on philosophy; *University of California Publications in Zoology*, 31, 307–358.
ROSTOW, W. W., [1960], *The stages of economic growth*, (London).
ROWE, J. S., [1961], The level-of-integration concept and ecology; *Ecology*, 42, 420–427.
RYTHER, J. H., [1959], Potential productivity of the sea; *Science*, 130, 602–608.
SACHET, M. H., [1963], History of change in the biota of Clipperton Island; In Gressitt, J. L., (Ed.), *Pacific Basin Biogeography*, (Honolulu), 525–534.
SAUER, C. O., [1925], The morphology of landscape; *University of California Publications in Geography*, 2, ii, 19–54.
SCHAEFER, F. K., [1953], Exceptionalism in geography: a methodological examination; *Annals of the Association of American Geographers*, 43, 226–249.
SCHALLER, G. B., [1963], *The mountain gorilla: ecology and behaviour*; (Chicago).
SCHNORE, L. F., [1961], Geography and human ecology; *Economic Geography*, 37, 207–217.
SELLECK, E. W., [1960], The climax concept; *Botanical Review*, 26, 534–545.
SEMPLE, E. C., [1911], *Influences of geographic environment: on the basis of Ratzel's system of anthropo-geography*, (New York).
SHALER, N. S., [1890–91], The origin and nature of soils; *U.S. Geological Survey, 12th Annual Report*, Part 1, 213–345.
SIDDALL, W. R., [1959], *Idiographic and nomothetic geography. The application of some ideas in the philosophy of history and science to geographic methodology*; Ph.D. thesis, University of Washington.
SIMMONDS I., [1966], Ecology and land use; *Transactions and Papers, Institute of British Geographers*, 38, 59–72.
SJÖRS, H., [1955], Remarks on ecosystems; *Svensk Botanisk Tidskrift*, 49, 155–169.
SLOBODKIN, L. B., [1958], Meta-models in theoretical ecology; *Ecology*, 39, 550–551.
SLOBODKIN, L. B., [1960], Ecological energy relationships at the population level; *American Naturalist*, 94, 213–236.
SLOBODKIN, L. B., [1962], Energy in animal ecology; *Advances in Ecological Research*, 1, 69–101.
SMUTS, J. C., [1926], *Holism and evolution*, (New York).
SPROUT, H. and SPROUT, M., [1956], *Man-milieu relationship hypotheses in the context of international politics*, (Princeton).
SPROUT, H. and SPROUT, M., [1965], *The ecological perspective on human affairs with special reference to international politics*, (Princeton).

STAMP, L. D., [1958], The measurement of land resources; *Geographical Review*, 48, 1–15.
STAMP, L. D., [1960], *Our developing world*, (London).
STEINMETZLER, J., [1956], Die Anthropogeographie Friedrich Ratzels und ihre ideengeschichtlichen Wurzeln; *Bonner Geographische Abhandlungen*, 19, 1–151.
STEVENS, A., [1939], The natural geographical region; *Scottish Geographical Magazine*, 55, 305–317.
STODDART, D. R., [1964], Storm conditions and vegetation in equilibrium of reef islands; *Proceedings of the Ninth Conference on Coastal Engineering*, (Lisbon, 1964), 893–906.
STODDART, D. R., [1965], Geography and the ecological approach: the ecosystem as a geographic principle and method; *Geography*, 50, 242–251.
STODDART, D. R., [1966], Darwin's impact on geography; *Annals of the Association of American Geographers*, 56, 683–698.
SUKACHEV, Y. N., [1944], On the principles of genetic classification in biocenology; *Zhür. Obshchei. Biol. (Journal of general biology)*, 5, 213–227.
SUKACHEV, V. N., [1950], Biogeozönose; *Bol'sheia Sovetskaia Entsiklopediia*, 5, 180–181.
SUKACHEV, V. N., [1958], On the principles of genetic classification in biocenology; (translated by F. Ramey and D. Daubenmire), *Ecology*, 39, 364–367.
SWINNERTON, H. H., [1938], The biological approach to the study of the cultural landscape; *Geography*, 23, 83–89.
TALBOT, L. M., [1963], *The ecology of western Masailand, East Africa*; Ph.D. thesis, Department of Geography, University of California, Berkeley.
TANSLEY, A. G., [1935], The use and abuse of vegetational concepts and terms; *Ecology*, 16, 284–307.
TANSLEY, A. G., [1946], *Introduction to plant ecology*, (London).
TEAL, J. M. JR., [1962], Energy flow in the salt marsh ecosystem of Georgia; *Ecology*, 43, 614–624.
THIENEMANN, A., [1942], Vom Wesen der Ökologie; *Biol. gen.* 15, 312–331.
THOMPSON, D'A. W., [1917], *On growth and form*, (London).
THOMPSON, L., [1949], The relations of man, animals and plants in an island community (Fiji); *American Anthropologist*, 51, 253–267.
TOBLER, W. R., [1963], D'Arcy Thompson and the analysis of growth and form; *Papers of the Michigan Academy of Science, Arts and Letters*, 48, 385–390.
TOSI, J., [1962], *Zonas de vida natural en el Peru*, (Lima).
TOSI, J., [1964], Climatic control of the terrestrial ecosystem: a report on the Holdridge model; *Economic Geography*, 40, 173–181.
TROLL, C., [1949], Geographic science in Germany during the period 1933–1945: a critique and justification; *Annals of the Association of American Geographers*, 39, 99–137.
TROLL, C., [1950], Die geographische Landschaft und ihre Erforschung; *Studium Generale*, 3.
UNSTEAD, J. F., [1926], Geographical regions illustrated by reference to the Iberian Peninsula; *Scottish Geographical Magazine*, 42, 159–170.
VALLAUX, C., [1929], La surface terrestre assimilée a un organisme; In *Les sciences géographiques*, (Paris), Chapter 2, 28–57.

VIDAL DE LA BLACHE, P., [1896], Le principe de la géographie générale; *Annales de Géographie*, 5, 129–142.
VIDAL DE LA BLACHE, P., [1898], La géographie politique à propos des écrits de M. Frédéric Ratzel; *Annales de Géographie*, 7, 97–111.
VIDAL DE LA BLACHE, P., [1911], *Tableau de la géographie de la France*; Vol. 1 of Lavisse, E., *Histoire de France illustrée depuis les origines jusqu'à la Revolution*, (Paris).
VIDAL DE LA BLACHE, P., [1913], Des caractères distinctifs de la géographie; *Annales de Géographie*, 22, 289–299.
VIDAL DE LA BLACHE, P., [1917], *La France de L'Est*, (Paris).
VOLTERRA, V., [1931], *Leçons pour la theorie mathematique de la lutte pour la vie*, (Paris).
WANKLYN, H., [1961], *Friedrich Ratzel: A biographical memoir and bibliography*, (Cambridge).
WASHBURN, S. L. and DEVORE, I., [1962], Social behavior of baboons and early man; in Washburn, S. L., (Ed.), *Social life of early man*, (London), 91–105.
WEIGERT, H. W., [1942], *Generals and geographers. The twilight of geopolitics*, (New York).
WEIGERT, H. W. et al., [1957], *Principles of political geography*, (New York).
WESTLAKE, D. F., [1963], Comparisons of plant productivity, *Biological Reviews*, 38, 385–425.
WHEELER, W. M., [1911], The ant colony as an organism; *Journal of Morphology*, 22, 307–325.
WHEELER, W. M., [1928], *Emergent evolution and the development of societies*, (New York).
WHITE, C. L. and RENNER, G. T., [1936], *Geography: an introduction to human ecology*, (New York).
WHITEHEAD, A. N., [1952], *Science and the modern world*, (London).
WHITTAKER, R. H., [1953], A consideration of climax theory: the climax as a population and pattern; *Ecological Monographs*, 23, 41–78.
WHITTAKER, R. H., [1960], Ecosystems; *McGraw-Hill Encyclopedia of Science and Technology*, 4, 404–408.
WHITTLESEY, D., [1929], Sequent occupance; *Annals of the Association of American Geographers*, 19, 162–165.
WIENER, N., [1948], *Cybernetics, or control and communication in the animal and the machine*; Actualités scientifiques et industrielles, 1053.
WILLIAMS, C. B., [1964], *Patterns in the balance of nature, and related problems in quantitative ecology*, (London).
WINSTON, P. W., [1956]. The acorn microsere with special reference to arthropods; *Ecology*, 37, 120–132.
YAPP, W. B. and WATSON, D. J., (Eds.,), [1958], The biological productivity of Britain; *Symposia of the Institute of Biology*, 7.
YEFREMOV, YU. K., [1961], The concept of landscape and landscapes of different orders; *Soviet Geography*, 2 (10), 32–43.
YOCKEY, H. P., PLATZMAN, R. L., and QUASTLER, H., (Eds.), [1958], *Symposium on information theory in biology*, (London).

CHAPTER FOURTEEN

Models of the Evolution of Spatial Patterns in Human Geography

D. HARVEY

To intercalate realities . . . is the only way to remain faithful to time; for life consists in the perpetual act of choosing and in the perpetual reservation of judgement.
LAWRENCE DURRELL

I THEORIES AND METAPHORS

History may be, in Lord Acton's words, 'as dead as the men who made it', but it is still of tremendous relevance to the living. Economic and cultural systems do not appear instantaneously at an arbitrarily determined point in time. They evolve, sometimes gradually, sometimes rapidly, sometimes scarcely at all, sometimes explosively. Nor do such systems appear arbitrarily in space. They develop at some particular place, spread, interact with other systems, become interconnected, and these connections proliferate quickly in some directions, slowly in others. Charting the changes in the spatial system over time has been the concern of many academic disciplines. Anthropology, archaeology, history, economic history, sociology, and geography, have all been deeply involved. From this vast literature two basic points emerge. The development of cultural forms over space is not a haphazard process and principles of spatial evolution *can* be developed. But no simple monolithic principle holds the key to explanation, nor, even, do few principles combined offer an adequate explanation of the tremendous complexity of change in the real world.

If we believe that the purpose of geographic research is 'to provide accurate, orderly, and rational description and interpretation of the variable character of the earth's surface' (Hartshorne, 1961, p. 21), then clearly we must be involved in the analysis of the evolution of areal patterns (Darby, 1953; Sauer, 1941; Smith, 1965). We need first of all to examine the genesis of cultural forms and this, in itself, is a matter for careful historical scholarship. But we need more than this. We need, second, to understand the *processes* of change over space and time. Scholarly studies abound in historical geography. But studies of the *processes* of change are less common. Yet these *processes* are critical to our understanding of present-day distributions; they mould and create them. A study of *process* therefore, is not the prerogative of the historical geographer alone. It concerns us all. It forms the vital link between what may seem rather obscure historical scholarship on the one hand, and penetrating studies of current spatial distributions on the other; we need above all, to dissect such all-embracing terms as 'the historical factor', 'historical momentum', 'geographical inertia', and to take a look at the multitude of *processes* subsumed under such headings.

But just as Marshall came to regard the spatial dimension as relatively unimportant to his economic system, so the 'Anglo-Saxon bias', as Isard (1956, p. 24) calls it, has led geographers to neglect the time dimension – a fault which Sauer lays firmly at Hartshorne's door (Sauer, 1963, p. 352). The importance attached to 'the historical factor' varies a great deal, but all too frequently it is regarded as a residual factor which conveniently explains away the seeming illogicalities in present-day spatial structures, which have been analysed in a static equilibrium framework. An unfortunate gap has developed, thus, between the scholarly studies of the specialist historical geographers, who frequently appear to think that the importance of their subject suddenly declines after 1800, and the analytical techniques of human geographers concerned with contemporary distributions.

Yet the importance of linking development over time and space has been keenly appreciated in the past. It was, perhaps, Ratzel's (1891; Wanklyn, 1961) greatest achievement, that he showed the close connection between the evolution of cultural forms over time and their diffusion over space. Subsequent writings in the so-called determinist school preserved this framework, even if they sometimes regard evolution, rather naïvely, as a too automatic adjustment to environment for modern tastes (Semple, 1911; Huntington, 1915 and 1945). Thus Griffith Taylor (1937) used a simple thesis of 'zones and strata' to examine the movement of races and the development of economic systems over space. The French geographers, on the other hand, preserved the time dimension by examining the symbiotic relationship between man and the land over a long period of time – an approach which yielded deep insights into the *processes* governing the mutual adjustment between man and environment (Blache, 1926; Brunhes, 1920; Sorre, 1950, 1962). German geographers took a more morphogenetic approach and tended to concentrate upon the development of landscape over time; an approach which was developed by Sauer and the 'Berkeley school' in the United States (Sauer, 1963).

Highly generalized theories of spatial evolution abound. The study of diffusion, fully developed by Ratzel, and backed up by strong influences from archaeology and anthropology, led to wide ranging studies of the movement of cultural forms over space. The movement of crops (Sauer, 1952), of town systems (Dickinson, 1951), of races (Taylor, 1937), are all examples of a general theory of diffusion looked at in a specific context. The statement of the 'frontier thesis' by Turner (1961 edition), with all of its geographic implications, led to the examination of the thesis in great detail by geographers and historians in many different environments (Wyman and Kroeber, 1957; Gulley, 1959). It led also to detailed studies of the colonization process in different environments (Bowman, 1931; Grenfell Price, 1939). Notions of distinctive cultural stages were applied to the evolution of landscape. Broek's (1932) classic study of land use change in the Santa Clara Valley, and Whittlesey's (1929) introduction of the scheme of 'sequent occupance' are cases of this. Theories concerning the importance of border regions (Cornish, 1923; Bowman, 1921) or nodal centres (Mackinder, 1917) were incorporated into a dynamic political geography, while spurious analogies between the political system and a biological organism had tragic results in the emergence of German geopolitical theory (Taylor, 1951).

Such classical writings on the evolution of spatial patterns were characterized by an informal presentation. They were, to quote Rapoport (1953), generalized theories or 'metaphors' applied to very complex situations. At best such 'metaphors' make for stimulating and exciting interpretations of whole complexes of human history. At worst they allow the subtle employment of false analogy for propaganda ends. Pregnant as such metaphors are, with concepts, ideas, and generalizations, there comes a time when the form in which such metaphors are cast seems to hinder objective judgement. The wish to be objective and scientific is somehow frustrated. And it is at this point that the student of history and geography is faced with two alternatives. He can either bury his head, ostrich-like, in the sand grains of an idiographic human history, conducted over unique geographic space, scowl upon broad generalization, and produce a masterly descriptive thesis on what happened when, where. Or he can become a scientist and attempt, by the normal procedures of scientific investigation, to verify, reject, or modify, the stimulating and exciting ideas which his predecessors presented him with. Many historians and historical geographers took the view that 'interpretation depends on scholarship', forgetting that 'without interpretation there can be no scholarship' (Barraclough, 1957). Historical scholarship cannot be conducted in an intepetative vacuum, even though many appear to attempt the feat. The first stage of scientific investigation involves the careful testing of theories and generalization, against the facts as they are determined by careful historical investigation. Modern studies draw much of their strength from precisely this process. Thus Perry's (1963) critique of the frontier hypothesis in an Australian context, Swedish studies of the expansion and retreat of rural settlement (Enequist, 1960), studies of the modern colonization process (Farmer, 1957), and studies of innovation development and regional change (Pred, 1965), examine general problems and theories of spatial development in the context of actual sets of events. Such studies have demonstrated two things. Firstly, that the simple schemes of spatial evolution put forward earlier in the century were far too crude and highly generalized to fit the real world. Reality was far more complicated than that. Second, that the actual form of presentation of the theories often lacked clarity. The informal presentation of 'metaphors' lacked the precision necessary to allow adequate testing of those theories and statements. At some stage the theory would have to be given a more formal statement, with the assumptions and hypotheses set out in a logical framework. Only if such a logical framework could be developed, would the testing, verification, and modification of hypotheses regarding the evolution of spatial systems be possible. The most suitable way to develop such a logical framework is to develop some model construct of reality which expresses the notions contained in the theory (Braithwaite, 1960; Ackoff, 1962).

II MODELS

A model may be regarded as the formal presentation of a theory using the tools of logic, set theory, and mathematics. The use of these tools allows us to identify and eliminate inconsistencies within our theory. It also allows us to use the powerful tool of algebraic analysis to make deductive statements as regards a particular system (this method is typified in classical economics), and, in some cases, to develop objective *statistical* tests of the relationship between the model we are using and the real world. To make a model *operational*, therefore, we have to develop some simple system of model building. The following is one brief schema that may prove useful (Orcutt, et al., 1961; Ackoff, 1962).

A model may specify three types of *variables* together with a set of operating characteristics (or *functions*) which link these variables. The *input* variables are independent of the model and we allow the values given to these *inputs* to vary. The *output* variables on the other hand are entirely dependent upon the model, and we attempt to show how different *outputs* result from a different pattern of *inputs*. A model may also contain *status* variables which specify certain conditions which are important, but which we keep constant throughout the operation of the model. In order to make the model operational we need to define a set of operating characteristics (*functions*) which specify the relationship between the *input*, *status*, and *output* variables. If, for example, we were developing a simple model of demand for goods and services within any community, we might regard wages as the *input* which varied according to external pressures, the demand for goods and services as the *output*, and the *status* variable as being the total population (Fig. 14.1.A). The *function* in this case would be a direct consumption function linking income with

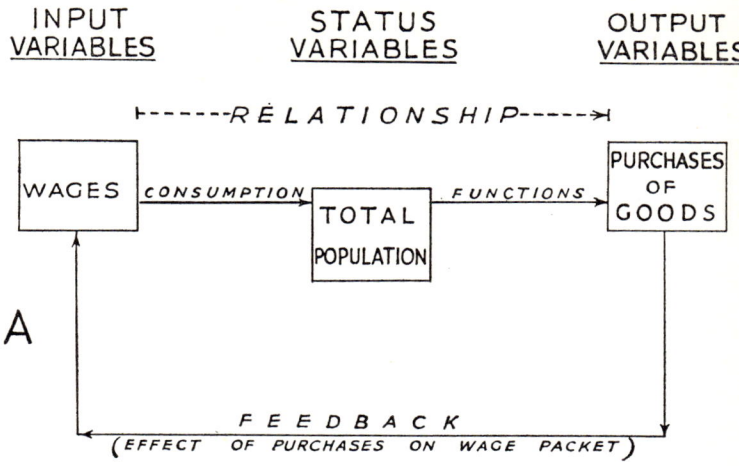

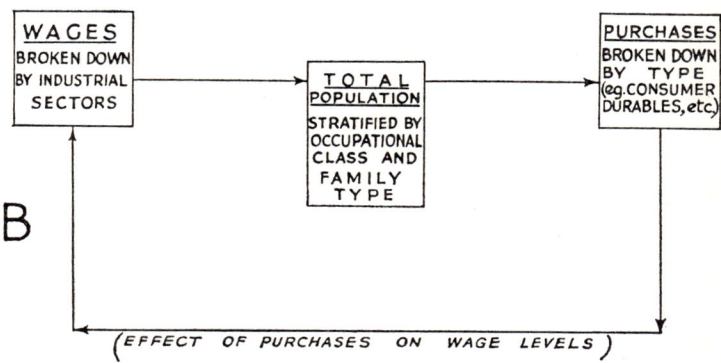

14.1 Simple model schemas for showing the relationship between wages and demand for goods and services in a community.

demand. This is a highly simplified model and we may choose to make it more realistic by breaking down the total population into different occupational classes, and further subdividing according to household characteristics which would yield a far more varied set of *status* variables (Fig. 14.1B). The *functions* in such a model need to be quantified in some way (either algebraically or statistically). The form of these *functions* may be of three distinctive types. *Deterministic* relationships specify a cause-and-effect type of relationship. *Probabilistic* relationships specify the likelihood of a particular event being followed by another event. *Functional* relationships specify how two variables co-vary without necessarily indicating any cause-and-effect relationship at all. Once we have defined all of these elements we may then 'run' our model and explore the results in relationship to the real world. In the process we may define one further element in the system. *Feedback* describes how a particular set of *outputs* may influence either the original *inputs* or, perhaps, the *status* variables. Figure 14.1.B shows how that part of the demand for goods and services which is satisfied within the community may result in a rise in the initial inputs (wages). This *feedback* effect is frequently very important.

Once we have constructed a simple model of one sector of reality, we can then link together several such individual models into a *model system*. We may, for example, collect together the outputs of several models and treat these as inputs to another model, identify feedback loops throughout the system, and so on (Steeger and Douglas, 1964). Given the complexity of most real world phenomena, it would be foolish to expect one simple model to have great explanatory power. Ultimately we must aim for a complex *model system* to represent an even more complex reality (see Section IV.ii below).

Such a model system could be developed to show how development over space and over time are related. Models of economic development, learning, population growth, and so on, express events as a function of stage and time. Clearly such models must have some spatial expression. Tentatively we might classify the kinds of spatial expression in the following four ways.

1 *Uniform*. A set of changes over time may have uniform effects over space. Climatic change may, for example, mean that the average overall temperature increases at every point on the earth's surface. If we are studying change over a relatively small area the effects of physical and human change may frequently be regarded as being uniform over the study area. A transfer in governmental power from one political party to another is a good example. Clearly, the larger the study area the less likely we will be able to identify uniform effects over space (Haggett, 1965B).

2 *Random*. A set of changes over time may be distributed randomly over space. If we are considering population growth within a country, we might find that the population growth is distributed randomly over the study area.

3 *Competitive*. An event which occurs as the result of a natural growth process (say the establishment of a town service centre with growing population and economic development), may preclude the occurrence of a similar event close by in space (the development of another service centre close by). Certain kinds of events compete for space and we may employ a relevant

spatial theory e.g. central place theory, to describe how such events are likely to be distributed over space (Morrill, 1965).

4 *Contagious*. The occurrence of a particular event as the result of a time process (say the development of a certain kind of industry in one town as a result of economic growth) may increase the probability of a similar event occurring close by in space (the development of more firms in the same industry or subsidiary firms). Again we can employ the relevant spatial theory of e.g. agglomeration, to describe the process (Isard, 1956).

This simple classification is only relative to the size of study area. Events which may be contagious on a world-wide scale may be treated as random or competitive within a very small area and vice versa. Bearing in mind this problem of scale (Haggett, 1965B) the above classification may still prove useful in formulating a model system.

The development of a model system to examine changes over both time and space also involves the whole problem of quantification. Both time and space possess metric qualities which can be misleading if not treated carefully. Any model system must, therefore, be based on a full understanding of the nature of *time* and the nature of *space*. This partly resolves itself into the technical problems of *time series analysis* and of fitting *distance functions*.

II (i) Time series Analysis

Time is a continuous variable, but most of the data available to us in human geography are discrete observations made at some point in time. Comparison between discrete sets of such observations made at different points in time then yields an estimate of how events are changing over time. Comparison of population totals between census years, annual changes in livestock numbers, are simple examples. But there are limitations to the efficiency of this procedure. Most geographic research appears curiously lax in applying an adequate methodology to the treatment of time series compared with the physical sciences and economics (Hannan, 1960; Granger, 1964).

It is convenient to distinguish between *stationary* time series and *trend* analysis. A *stationary* time series is one whose average condition is adequately described by a mean value which remains constant over time and whose fluctuating component possesses the same variance over time. We can thus expect a *stationary* time series to exhibit fluctuations, but these fluctuations will be randomly distributed around some mean value. A time *trend* on the other hand will show some change in the mean value of the series over time or some change in the variance (or both) (Granger, 1964, Chapter 2; Gregg et al., 1964).

The statistical techniques of time series analysis are concerned to isolate the different elements which make up the series of observations. Until the advent of sophisticated techniques of spectral analysis the methodology for such a separation remained intuitive rather than objective. Conventional procedure would be to decompose the series into trend component, cyclical component, seasonal component, and a randomly distributed residual (Miller and Kahn, 1962; Granger, 1964). How far these elements can be adequately distinguished depends upon the length of the time series studied. What appears as a trend over a short period may simply be part of a major oscillation when viewed over a long period. Viewing time series as a whole series of oscillations or trends superimposed upon one another is useful but poses the technical difficulty of distinguishing between the different elements involved. Initially 'it was felt that if one could determine the amplitude, period, and phase of a sine curve sufficiently accurately and subtract this from the data then the remainder ought to be an independent random series. When in fact this was done and the remainder was still found to be somewhat too smooth, it was natural to re-use the current predominant idea of the cause of the smoothness and to look for yet further sine curves to fit the data' (Granger, 1964, pp. 4–5). *Periodogram* analysis to distinguish the 'hidden periodicities' in time series suffered from the disadvantage that far too many peaks were identified and, as M. G. Kendall (1946) effectively showed, very few of these peaks could be regarded as being statistically significant. Other techniques for analysing time series were also developed. A system of *moving averages* remains a standard technique for eliminating minor oscillations, but the number of months or years used is an arbitrary choice. *Autoregressive* methods were also developed in which 'the current value x_t is assumed to have been formed from a linear sum of past values of the series together with an independent term unconnected with the past' (Granger, 1964, p. 6; Wold, 1938). *Least squares* methods have been used to describe trends, but complicated sets of changes over time have to be fitted by extremely complicated mathematical expressions, and the standard tests of goodness of fit (and statistical estimation) cannot be used because of *autocorrelation* among the residuals. The phenomenon of *autocorrelation* has proved extremely difficult to eliminate and explains why the residuals are so often distributed rather like a sine function around an already fitted curve, rather than randomly distributed (Durbin and Watson, 1950 and 1951; Wallis and Roberts, 1962; Tintner, 1952). *Spectral* techniques are far more satisfactory from a methodological point of view but are complicated to handle and compute without high-speed computers (Granger, 1964; Hannan, 1960; Rosenblatt, 1963), and in most cases the more easily computed techniques may be used, provided their dangers are fully understood.

The significance of time series analysis for geographic research does not simply lie in avoiding the dangers inherent in any oversimplified analysis of change over time – although this is very important. It has importance for geographic description and analysis.

The description of a spatial pattern or change in spatial pattern must be placed in perspective against the nature of the time series affecting the phenomena described. Consider the following hypothetical example. Suppose that the true mean density of cattle population over a long period of time is 10 per 1,000 acres and that the cattle population change may be analysed as a stationary stochastic series with a variance such that the cattle population may vary from 9·0 to 11·0 per 1,000 acres. Suppose we now examine four different counties with these characteristics and that these counties experience *stochastic* fluctuations independent of one another. We might then find a set of data for periods of time as follows:

Census Year	Cattle Density per 1,000 Acres			
	County A	County B	County C	County D
1870	10·3	11·0	9·5	9·6
1880	10·2	10·0	10·2	9·6
1890	9·7	9·3	10·6	10·7

Given the conditions we have laid down, we have no grounds for assuming that there is any *significant* difference in cattle density between counties at any one particular time, nor are there any *significant* changes over time. Yet it would be very tempting to suggest that there was decline in Counties A and B relative to Counties C and D. We might even be tempted to calculate a rate-of-rate of change! (Clarke, 1959). If, on the other hand, the variance affecting the time series was reduced so that stochastic fluctuations only ranged between 9·9 and 10·1, then the table would indicate significant differences both regionally and over time. The nature of the time processes we are examining are important for establishing the criteria of *significance* of difference in regional division (Chisholm, 1964; Blumenstock, 1953; Choynowski, 1959).

We must also consider the problem of timing of the observations on which we base spatial descriptions. Figure 14.2 shows the annual fluctuations in the hop acreage in England 1810–60. With strong cyclical fluctuation in the acreage the series is not uncharacteristic of much economic activity. Yet if we had to rely on a decennial census the kind of picture of spatial change that would emerge would be very far from the truth. In this case we fortunately possess yearly observations, and we can judge which years are relevant. Yet if we build a composite picture of land use in a set of parishes from tithe awards made between 1836 and 1845, we are forced to ignore the kind of fluctuation demonstrated here at our own peril.

The nature of time series also carries implications for geographic analysis. If we consider a time series, such as the hop acreage series, that has a long term trend and a short term set of cyclical fluctuations, then the explanation of these two different elements may be totally different. Let us suppose that the long term trend is a function of growth of population, and that the short term fluctuations are a function of changing employment conditions. If we examine change in the hop acreage between 1830 and 1840 there is an upward element in the trend, but the cyclical downswing more than compensates for this. An examination of regional change *should* separate out these two components, for the long term trend may tend to create one pattern of location while the factors governing the cyclical trend may point to a rather different

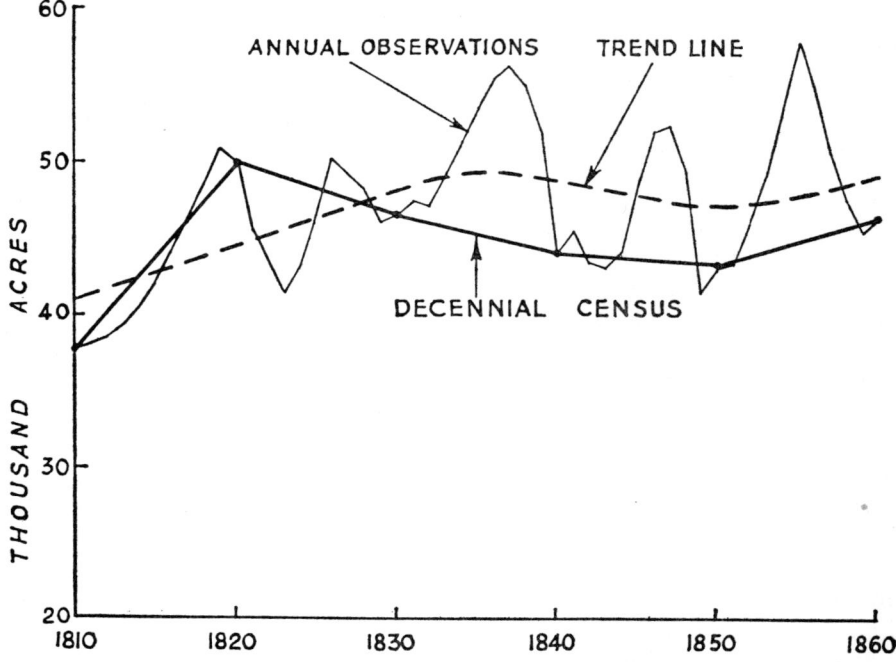

14.2 Three versions of the changing hop acreage in England and Wales, 1810–1860, showing (a) annual observations, (b) smoothed trend, and (c) change over 10 year periods.

pattern. Which is dominant may be critical for our understanding of the nature of spatial patterns. Patterns created from a whole series of short term adjustments to immediate difficulties are hardly likely to be similar to those evolved from long term adjustments.

A further implication for analysis of geographic patterns arises when we attempt to relate two variables over space which change in rather a different way over time. Curry (1962) has pointed out that the pattern of random fluctuations in climatic conditions may lead to major variations in vegetation over space. Thus chance climatic sequences favourable to a particular kind of

MODELS OF SPATIAL PATTERNS IN HUMAN GEOGRAPHY

vegetation may give rise to that vegetation which, once developed, may continue in existence for a very long period of time. In the analysis of geographic distributions, therefore, we should think in terms of the nature of the time series we are dealing with.

The problem which the treatment of time poses is not simply one of describing the pattern and form of change over time. We also need an adequate theoretical understanding of time. We tend to take for granted the metric qualities of time (Reichenbach, 1956). Yet time is only a relative and not an absolute quantity. Time only obtains its metric qualities relative to a set of events or processes, such as the motion of the earth, the length and frequency of light rays, or rates of change in matter (this is, after all, the theory behind Carbon 14 dating). In the social sciences we frequently regard astronomical time as the only convenient metric. Yet it would be perfectly feasible to invent our own time scales (civilization time, resource exhaustion time, etc.) which would free us of the predeliction to regard one period of ten years as being equivalent to another ten years. We can attempt simple 'time transformations' by using logistic curves rather than linear expressions for time periods. This problem of time as a metric is very similar to that of treating distance as a metric and a more detailed treatment of *distance functions* should indicate the kind of problem that needs to be tackled.

II (ii) Distance functions

The measurement of distance is not as simple a matter as it would seem. In economic, social, or psychological terms, 10 miles is not simply equivalent to 10 times 1 mile, and 10 miles over flat plain is rather different from 10 miles across the Alps. But in models of spatial development we frequently require relationships in which the distance variable is measured quantitatively. If we have to use a distance function in model building, then we need an adequate methodology for understanding distance.

Perhaps the most interesting argument over 'distance' has centred around the 'gravity' model. The interaction between two populations has been described as being of the form:

$$I_{ij} = \frac{P_i \cdot P_j}{d^b_{ij}}$$

where P_i and P_j are the populations at place i and j respectively, d_{ij} is the distance between the two places, and b is an exponent.

In the gravity models developed by Stewart (1947) the exponent is given the value 2, but empirical studies of behaviour of e.g. traffic flow between sectors of a city, indicate that appropriate values for the exponent may lie between 0·5 and 3·5. The higher the exponent the greater is the 'friction' of distance on interaction (Isard, 1960, Chap. 10; Olsson, 1965A). Other studies have used quadratic functions to summarize the pattern of journeys to cinema facilities in Sweden (Claesson, 1964), or to shopping facilities in a city (Lowry, 1963 and 1964).

Migration studies have also revealed a great deal about the varying impact of distance upon social activity and movement. Again, many different functions have been used to describe the observed distribution of migration distances. The pareto and exponential distributions are broadly the same as the formulation given in the gravity model, but normal, log normal, and gamma functions have all been fitted with varying degrees of success (Morrill, 1963A and 1965). Fitting such theoretical probability distributions to migration distances only has utility if the underlying theory has been explored. Thus Dacey has shown that the distribution of nearest-neighbour distances between central places arranged in a hierarchy should conform to a gamma distribution, and Olsson (1965B) has suggested that migration between places should be described by a similar function.

The difficulty of finding an appropriate function for distance in theoretical models of behaviour, has led to attempts to define distance by some social measure. Stouffer (1962) used the notion of intervening opportunity which neatly avoids the problem of physical distance by treating the number of opportunities lying between two points as a measure of the distance separating them. Journey to work data has been similarly analysed, using either the intervening opportunity model or the so-called Schneider principle (Lowry, 1964). Basically these attempts to measure distance in some other term are seeking for some adequate transformation of the distance variable. Transformations can be developed using intervening opportunity, transport cost, time of travel, direct cost, and all may be appropriate measures under certain circumstances. But models using such concepts require adequate information, and we also have to face the problem that we lose the simple Euclidean properties of space (which are easy to work with) and involve ourselves in topological problems of some complexity if we wish to map the distribution of distances between points measured in some social term (Bunge, 1962; Tobler, 1963; Getis, 1963).

The kinds of problem from which notions regarding the impact of distance have been derived are almost all 'contagious' problems in which it is assumed that a migrant or shopper attempts to minimize distance travelled while satisfying other requirements for living space, etc., at the same time. But 'competitive' activities will not yield a similar kind of function. Thus Ajo (1964A and 1964B) has fitted sine curves to describe the 'influence fields' of London and other cities.

'Distance' is, thus, a relative measure. In practice we are frequently reduced to using simple linear measures because these are easily computed and are easy to treat. But this brief review of 'distance functions' suggests that we need to examine the results of a linear treatment of distance with some caution. Not only may the nature of the distance function change over time (Hägerstrand, 1957), but it may also vary a great deal from place to place (Marble and Nystuen, 1963). Clearly, the more information we gather about this central variable in geographic analysis, distance, the quicker we shall be able to build an adequate model of spatial interaction to match models of development over time.

Distance, like time, is a continuous variable, and it is worth noting that many of the difficulties involved in analysing time series have close parallels in the analysis of data distributed over space. In the same way that much of the data available to us is in the form of a sample taken at a particular point in time, so much of the data available to us is also in the form of a spatial sample. How far that sample point is representative of the space around it, and how far there is 'autocorrelation' among points distributed over space which makes the use of normal techniques of statistical inference rather dubious, are problems of considerable importance in geographic analysis (Thomas, 1960; McCarty, 1956; Dacey, 1964D).

Observations distributed over time and over space face analogous difficulties of interpretation. Firstly these observations are but a sample (in most cases) of possible observations and the relationship between the sample and the population from which it is drawn (usually established by principles of statistical inference) cannot easily be established because of autocorrelation of observations over both time and space. Secondly, both time and space are relative metrics, and in human affairs there is no reason why we should treat the conventional metric units as being absolute values. Both time and space transformations are feasible and, in many instances, are essential to analysis of the evolution of spatial patterns over time. But the nature and form of such transformations are difficult to establish and, unfortunately, difficult to work with. Awareness of these two sets of problem seems an essential pre-requisite to the building of models of evolving spatial patterns in human geography.

III MODELS OF THE EVOLUTION OF SPATIAL PATTERNS – OPERATIONAL CONSIDERATIONS

III (i) Deterministic versus stochastic models

It is convenient to divide models according to the kind of function which is used. The basic characteristic of *deterministic* models is that the development of some system in time and space can be completely predicted provided a set of initial conditions and relationships are known. Laplace took the view, for example, that if we could successfully break down reality into its component parts then we could predict the future with certainty by solving a vast number of differential equations. *Probabilistic* or *stochastic* models, on the other hand, build random variables into their structure. Thus a *stochastic* process 'does not proceed according to some immutable law, but is at least partly dependent on random or chance factors', and *stochastic* models are also concerned with 'the corresponding wider theory of the statistics of change' (Bartlett, 1962, pp. 49–50). The random variables included in such models may be used to summarize events which are individually determined but whose aggregate effect is random, or they may be used to describe a situation where purely chance factors operate.

Stochastic models possess a greater degree of flexibility than deterministic models and in situations where there are a large number of factors to be considered or where events are highly disaggregated, a *stochastic* model often proves more realistic. Bailey (1964, p. 2) has argued, for example, that the general course of an epidemic within a population can reasonably be analysed using a *deterministic* model, but that the spread of infection within small household units can only realistically be analysed by means of a *stochastic* model. In economics a deterministic framework may be viable for examining the growth of the economy at large, but in planning output for small individual firms it is essential to build in notions of stochastic fluctuations in demand. Experiment has shown, however, that *stochastic* models of individual choice behaviour, or the spread of an epidemic among household units, yield rather different results from their deterministic counterparts when aggregated up to the same level of generality. In some cases the differences are revealing (Bailey, 1957 and 1964; Orcutt, *et al.*, 1961). One such study of marketing behaviour concluded, for example, that 'in markets where a significant amount of randomness is known to be present, the experiments with queuing formulations show that the deterministic market model yields misleading conclusions about the equilibrium values of economic variables' (Sparrow, 1965).

The choice between a *deterministic* or *stochastic* model is partly a function of ease of use. A *deterministic* model makes rigorous assumptions about behaviour but is mathematically easy to operate. On the other hand 'many of the simplest stochastic models entail considerable mathematical difficulty: even moderate degrees of reality in the model may result in highly intractable mathematics' (Bailey, 1964, p. 4). The choice may often lie, therefore, between an inflexible model of the deterministic sort in which it is possible to include several variables explicitly, and a *stochastic* model which can only treat of a few elements but treat them in a much more realistic way by summarizing deviations in terms of a random variable. The mathematical difficulty of *stochastic* models may be by-passed under certain conditions by the use of *Monte Carlo* methods (see Section III.iii.C).

The division between *deterministic* and *stochastic* models is important, but it may be possible to include both sets of relationships within a *model system*. If, for example, we develop a model to predict the distribution of population in Hertfordshire in 1985, we might use a deterministic system to predict the overall growth of population and stochastic system to distribute the population within the county. The use of one type of relationship within a model does not, therefore, exclude the use of another kind of relationship elsewhere. Bearing this point in mind, however, we will now consider the two kinds of model separately.

III (ii) Deterministic models of the evolution of spatial patterns

Deterministic models of the evolution of spatial patterns vary in the way in which they treat time and space. We can distinguish four different types of model depending upon whether the space or time dimension are included within or outside of the model itself.

A *Comparative statics.* Comparative statics can, on occasion, be a powerful tool for analysing change over space and time. The static partial spatial equilibrium models described elsewhere in this volume (Garner, Hamilton, Henshall), can be given a time dimension by altering the *inputs* as a function of time. If we possess estimates of how the *inputs* will change with time, we may then derive a succession of equilibrium solutions for different time periods. Comparison between equilibrium at different time periods may yield interesting results. Weber's model of industrial location has been usefully applied to interpret the changing location of the British iron and steel industry since 1800, and to predict the future development of the iron and steel industry in North America in a very generalized way (Isard, 1948; Isard and Capron, 1949). Spatial equilibrium models in agriculture and interregional input-output models may be similarly used with valuable results (Egbert and Heady, 1961; Chenery, 1962). The Lösch model of location of central place functions is not so easily applied to changing patterns of settlement because successive equilibrium solutions are only realistic over long time periods because settlement patterns are more fixed and rigid than most other forms of human activity. In any case the rigid and highly unrealistic assumptions of Löschian theory make it difficult to apply the model to dynamic conditions with any ease (Curry, 1962B), while Christaller's more flexible, but less rigorously formulated approach can be more easily applied to the study of change over time (Russell, 1960; Thijsse, 1963; Godlund, 1956; Stine, 1962). The changing location of retail functions might be similarly studied (Getis, 1964).

The most easily adapted of the spatial equilibrium models, however, are the von Thünen and the gravity models. Any fluctuations in the inputs to the model may be easily analysed in terms of the model. In the von Thünen case any alteration in market demand, transport cost, etc., automatically leads to an adjustment in the spatial pattern – and it is simple marginal economic analysis to show the nature of these changes. We may thus analyse the use of marginal land (Stamp, 1948; Enequist, 1960); the extension of land use systems – such as those described by Barnes (1958) and Lewthwaite (1964); the extension of industrial and economic development around nodal centres on a world-wide scale (Chisholm, 1962; Melamid, 1955), or on a more local scale (Regional Plan Association, 1964), or within a city complex (Alonso, 1964; Muth, 1961; Wingo, 1961). The whole nature of spread effects around economic growth centres as envisaged by Myrdal (1957) or indeed the advance of the frontier of settlement from a coastal centre across unsettled land, might be analysed using a dynamic von Thünen model (see Fig. 14.3). The close connection between the von Thünen model and marginal economic theory makes the model a powerful tool when applied to dynamic conditions. Although the gravity model can easily absorb changes over time in the inputs to give fairly reasonable predictions of future interaction between communities or sectors, the theoretical underpinnings of the gravity model are not entirely clear and, therefore, the application of the model to explain dynamic conditions is less powerful than the von Thünen model (Lowry, 1964; Warntz, 1959).

A major difficulty of *comparative statics* is that the time factor remains external to the model itself, so that although the model shows how a certain set of changes over time will automatically result in spatial adjustments, it does not in itself allow a truly dynamic interpretation of change over space. It assumes that there are no independent processes of change which are not somehow subsumed in the original inputs. Thus notions of growing economies of scale and specialization, agglomeration economies, or even straight conversion cost, are excluded from a von Thünen model used in a dynamic setting. Such processes may mean a mere time-lag in the achievement of a static spatial equilibrium, and therefore such processes may be unimportant in the long-run. The utility of Weber's theory for explaining long-run locational change in iron and steel production demonstrates this point. But in some cases the actual processes of change will affect the long-run spatial pattern. Processes of agglomeration and cumulative change for example, may be fundamentally important to agricultural change (Harvey, 1963) and in such cases a dynamic von Thünen model will be inappropriate. At best, therefore, *comparative statics* has only a limited application and the circumstances where such an approach may be used are relatively restricted.

B *Process models*. In general we remain rather ignorant of the nature of the processes shaping the evolution of spatial patterns in human geography and as we have already suggested above, the processes of change may, in certain cases, have a permanent effect upon the spatial pattern that emerges. Models

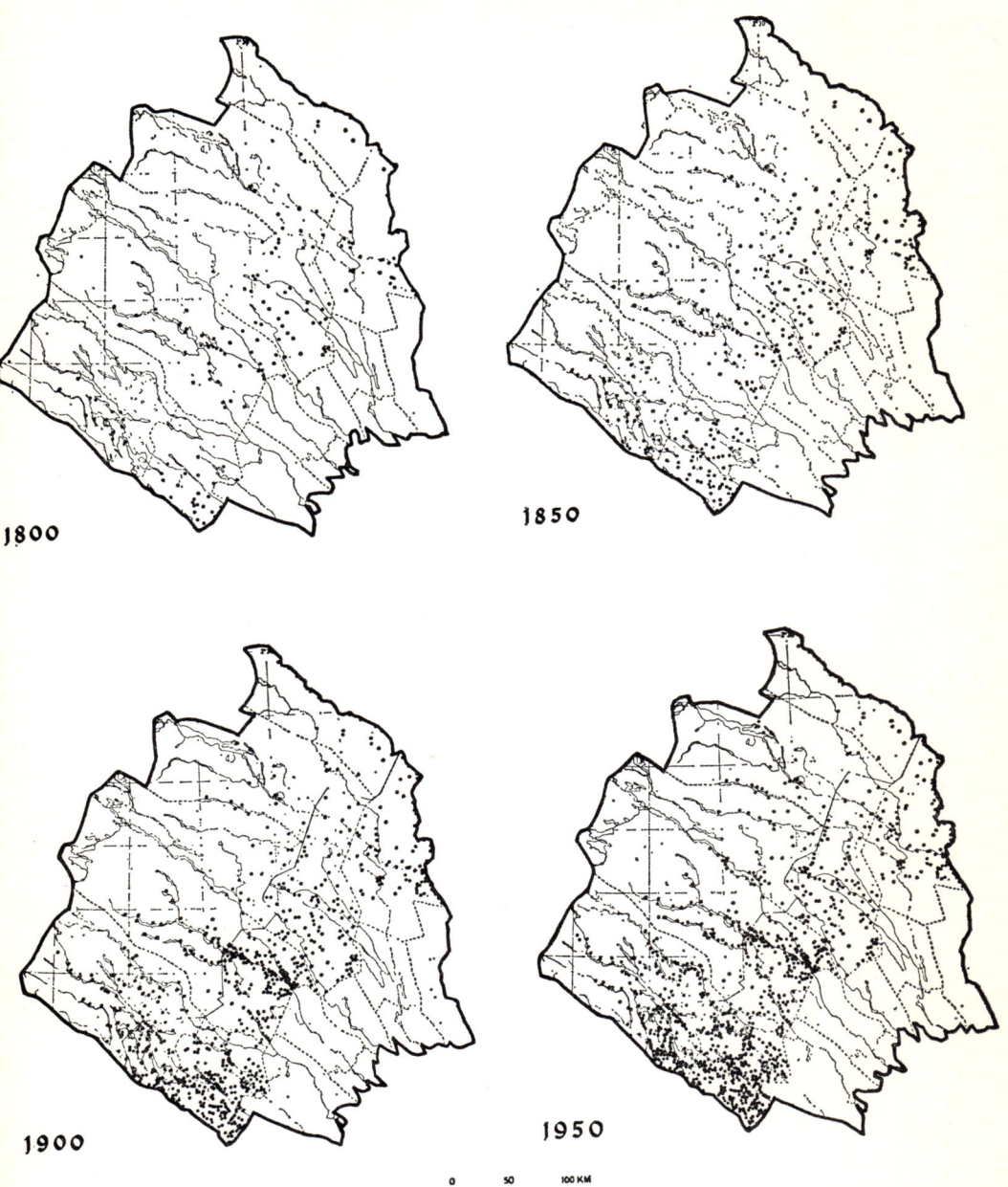

14.3 The changing pattern of inhabited settlements in inner North Sweden, 1800–1950.

exploring the implications and form of certain processes of change are thus very valuable. In most cases such models are partial analyses of a single set of processes and we are far from possessing a general model of a whole collection of processes at the same time. Some examples of process models are described below.

(i) *Migration*. The process of movement of population is a major variable affecting the evolution of spatial patterns and it is one that has been studied in some detail (Olsson, 1965A and 1965B; Morrill, 1965). The deterministic models describing the process vary in the emphasis they give to economic, behavioural and communication factors. Some models treat migration as a simple response to income differentials and therefore part of the general process of spatial interaction which leads to the elimination of inter-regional wage differentials and the regional 'equalization in factor returns' (Borts, 1960; Perloff, *et al.*, 1960). Models of this type have been developed by Nelson (1959), Sjaastad (1960) and Raimon (1962), while Porter (1956) indicates a probability model. The empirical evidence assembled to test these models suggests that the migration process cannot be explained in economic terms alone, and that it would take a very considerable period for the equalization of factor returns to come about. Nevertheless the economic variable is a powerful explanatory factor governing the migration process. Some models concentrate upon the 'information fields' which lead migrants to move, and the association between this kind of analysis and the study of communication networks and diffusion of information is strong. An intuitive understanding of the importance of this process led Ravenstein (1885, 1889) to formulate his 'laws' of migration and the study of migration from this point of view has been followed up by Swedish geographers who have employed both deterministic and probabilistic formulations for their models (Olsson, 1965A and 1965B; Lövgren, 1956; Wendel, 1953; Hägerstrand, 1957).

Studies of migration indicate the complexity of the process, and given this complexity it is hardly surprising that simple models of spatial evolution – such as the moving frontier – can only be highly aggregated accounts of what really happened. Simple and direct migration processes implied in theories of frontier movement or of achievement of economic equilibrium just do not exist.

(ii) *Colonization models*. The processes governing the development of new settlements to create a spatial pattern have been the subject of numerous historical studies. In many cases special factors, such as the land policy in the USA and the homestead acts, the political and religious pressures involved in colonization in Eastern Europe during the late Middle Ages, political decision making in Russian colonization schemes, make it seem almost impossible to set up theoretical models of settlement development. One interesting exception is the study by Bylund (1956 and 1960) of the settlement process in

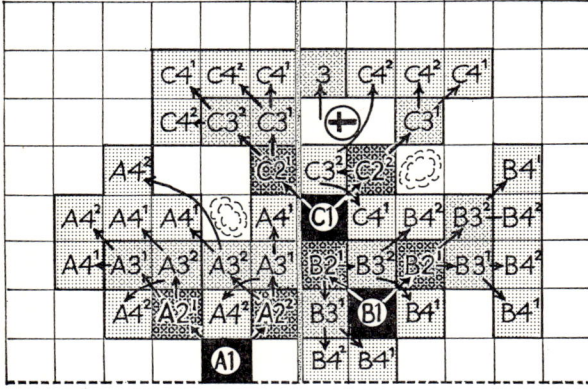

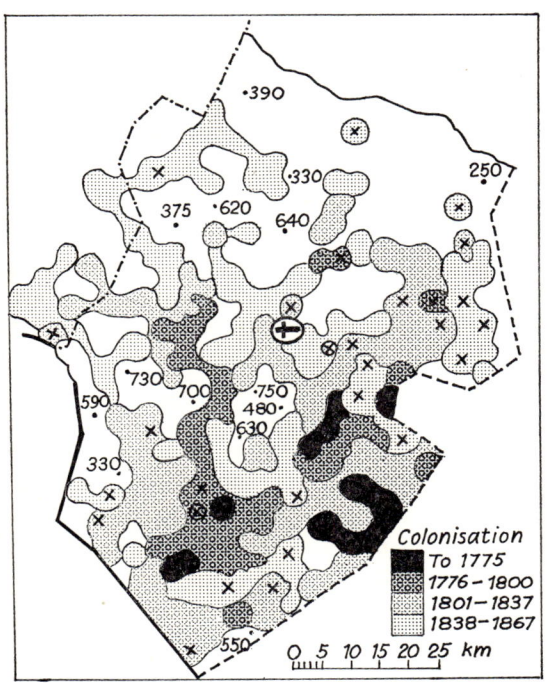

14.4 Bylund's colonization model showing (top left) values placed on proximity to mother settlement in a grid of cells, (top right) values placed on proximity to roadway (large type) and church (small type),

(centre) a simulated pattern of settlement spread around three initial centres, and

(bottom) an actual pattern of settlement colonization in north Sweden (*From Bylund, 1960*).

Lappland. The problem that Bylund set out to explain was the failure of the colonizers to locate on the best lands before bringing poorer lands into cultivation. A model was developed which contained three major features. The colonizers, it was assumed, wished to minimize distance between their new settlement and (a) their parent settlement, (b) a church, and (c) an available communication link. Using arbitrary values over a grid of equally sized cells, Bylund simulated the spread of settlement around an initial set of settlements and drew attention to the broad similarity between the simulated map and an actual map of dispersion of settlement in the area (see Fig. 14.4). This very simple deterministic model indicates how formal models of the colonization process might be developed to some advantage.

(*iii*) *Network models.* The evolution of the transport and communications network has frequently been the centre of empirical geographical studies but theoretical treatments are relatively rare. Given a certain number of points and a maximum number of links it is possible to show how transport links may be developed. If we steadily increase the number of links available over time we may show the order in which certain links should be established in order to optimize the connectivity of the network (see Haggett). The powerful tools of graph theory and network analysis indicate the potential for model building here. A simple 'stage' model of transport evolution has been usefully applied to the development of the transport network in West Africa (Taaffe *et al.*, 1963) and Kansky (1964) has used a simulation model (in fact deterministic even though a probabilistic format is used), to describe the development of the communication network in Sicily. Since analytic procedures for determining both the number of links in a network and the flows along networks are being developed in several disciplines (Berge, 1962; Ford and Fulkerson, 1962; Kleinrock, 1964), the application of these techniques to study the evolution of communications networks seems particularly promising (Garrison and Marble, 1965).

These three brief examples are not exhaustive of the kinds of *process models* which have been developed. Models which explore the processes governing town growth (Muth, 1961; Winsborough, 1962), industrial location change (Fuchs, 1962) and diffusion of information (Dodd, 1950), are further examples. The two critical processes of agglomeration and decentralization of economic activity, together with the whole problem of 'multiplier' effects, have also been given extensive theoretical treatment (Isard, 1960; Isard, Schooler and Vietorisz, 1959; Greenhut, 1963). The general aim of such models is to elucidate the processes at work. *Deterministic* models of the evolution of spatial patterns require an understanding of the processes involved, and we may therefore expect that many of these partial models of process will eventually form a set of models within a *model system* to explain spatial evolution.

C *Growth models with spatial assignment.* In the same way that *comparative statics* took the spatial equilibrium models and gave them a time dimension, so we may consider growth models and give them a spatial dimension. Various growth processes may be treated in this way, but we shall restrict consideration here to economic growth and population growth.

Clearly economic growth models can be disaggregated into regional component parts (see Keeble), either by assigning the overall growth among regions according to some specific criteria (e.g. resource availability, existing industrial structure), or by developing a set of regional growth models which can then be aggregated to give a national total. In practice the former is easier to carry out and has formed the basis for regional planning in many countries (Isard and Cumberland, 1960). The impact of national economic growth on one small area may be examined by analysing industrial structure and estimating multiplier effects (Berman, et al., 1961; Tiebout, 1962; Hirsch, 1964), or by developing a regional growth model drawing assumptions as regards inputs from a national growth model (Chenery, 1962; Perloff, 1963; Henderson and Krueger, 1965).

Models of population growth may be similarly treated. The deterministic models of growth divised by Malthus, Pearl (1939), Lotka (1945), or the more behavioural models (see Wrigley), may be given some spatial component. The assignment of population growth over space requires a close understanding of the process of migration. In general the problem has not been tackled by use of formal models, although pressing needs of planning have led to some work on the problem within metropolitan communities where the process involved in allocation is journey-to-work rather than migration (Berman, 1961). Individual community growth projections may also be totalled to give a national or regional aggregate and then compared with an overall projection. In this instance the space and time aspects are different starting points for a two-pronged study (San Francisco Bay Area, 1954).

D *Time-space models.* Ideally we might hope to include development over both time and space within the same model. If such a model is to be anything more than trivial, it would have to be exceedingly complex. It is hardly surprising, therefore, that deterministic time-space models can only be specified in a very generalized way (Berman, 1959), or for relatively small scale phenomena. Vining (1946) and Lösch (1954), for example, have examined the movement of the impact of business cycle fluctuations over space; Dodd (1950, 1953) and Rapoport (1953) have examined the spread of information over space. Another restricted example of a time space model comes from epidemiology.

Rushton and Mautner (1955) develop a simple deterministic model of the spread of an epidemic among a population. Using a simple model of growth of an epidemic that assumes that there are y *infectives* and x *susceptibles* in a homogenous population of $n-1$ individuals $(x+y=n)$, and that there is homogenous mixing in the population. The process can then be described by a simple differential equation of the form

$$\frac{dx}{dt} = -Bxy = -Bx(n-x+1)$$

This expression yields a typical 'epidemic curve' for the number of new infectives occuring at any time period (Bartlett, 1956; Bailey, 1957). If it is then assumed that there are m discrete communities each with a homogenous internal contact rate, a_i, and a homogenous mixing rate with other communities, b_{ij}, then if x_{it} denotes the number of susceptibles in the ith community at time t, and n_i is the total population in the ith community, the process may be described by the differential equation:

$$\frac{dx_i}{dt} = -x_i \left\{ a_i(n_i-x_i) - \sum_{j \neq i} b_{ij}(n_j-x_j) \right\} \quad (i = 1, 2, \ldots m)$$

Rushton and Mautner then go on to give approximate solutions. This model is difficult to solve, however, and it is a highly simplified account of a relatively uncomplicated phenomenon (uncomplicated relative to the evolution of human activity over space that is).

Deterministic study of change over space and time is thus difficult to treat within the same model and it would seem almost inevitable that we should separate out the two types of changes and treat of spatial evolution within a *model system* rather than within one single model, except in cases where very simplified treatments appear justified.

III (iii) Stochastic models of the evolution of spatial patterns

Almost all the *deterministic* models so far considered have *stochastic* counterparts. We could, therefore, examine *stochastic* models under a similar set of headings, but such a treatment would be repetitious. In any case the literature on *stochastic* models is very considerable and it would be impossible to give a comprehensive review. The works by Bartlett (1960A) and Bailey (1964) cover the ground very thoroughly as regards the role of stochastic models in the sciences. It is worthwhile, however, to attempt a brief review of the kinds of problem that have been tackled using *stochastic* models.

Population growth models have been given a *stochastic* formulation, where random variables describe birth and death probabilities and migration probabilities (Bailey, 1964, pp. 84–135; Kendall, 1948 and 1949; Bartlett, 1960B, pp. 17–42). Such models may be examined in the context of a system of regions (de Cani, 1961). Migration problems have been treated as a probability process with considerable success (Hägerstrand, 1957; Morrill, 1963B and 1965; Sforza, 1963), and in the context of small communities such a treatment seems far more realistic than the deterministic models considered in

Section III.ii.B above. Studies of epidemic processes have similarly been treated stochastically, but the formulation of a model that would extend over space and time requires considerable mathematical ingenuity. Bartlett (1956, 1957 and 1960B) has constructed models to show the spread of a measles epidemic over both time and space, but found it difficult to treat of spread among more than a six-by-six net of cells. The model is very generalized and the solution required the use of *Monte Carlo* techniques (see below Section III.iii.C). Economic growth models may similarly be given a stochastic formulation (Orcutt, *et al.*, 1961; Howard, 1960) and disaggregated to a regional growth model (Smith, 1961). Models of social interaction and diffusion of plant and animal communities, of population, of information, have similarly been treated stochastically (Rainio, 1961; Skellam, 1951; Beckmann, 1957; Landahl, 1957; Dodd, 1955; Coleman, 1964). Queuing theory (a form of stochastic model) has also been applied to the flows of messages through telephone systems over space, to traffic congestion problems, to planning water storage systems (More), and it may also have applications to understanding the location of retail activities (Curry, 1962B) and to migration processes under some constraint, such as housing shortage. But there are few examples of the use of queuing theory to analyse geographic problems (Gould, 1963), even though the technique appears most promising for understanding flows among networks, and other elements of spatial activity (Cox and Smith, 1961; Saaty, 1961; Churchman, *et al.*, 1957).

Stochastic models are extremely useful for understanding human activity because we are able to incorporate behavioural patterns and variations under an aggregate random variable. The development of *stochastic* models, however, has depended very much upon advances in the relevant branch of mathematics – i.e. the theory of stochastic processes. The techniques involved are special and the mathematical models themselves are derived from assumptions regarding mathematical processes which generate certain kinds and groups of probability distribution. These mathematical processes can, on occasion, be directly related to physical processes in the real world. For this reason it is worth taking a look at selected examples of the use of a particular kind of mathematical technique and to examine how far the processes involved can be treated as analogous to processes of development in human geography. *Quantitative ecological models* and *Markov chain models* will be considered from this point of view, and then *Monte Carlo* methods for solving complex problems will be considered as an example of how a particular technique of mathematical approximation for complex models may be applied to analyse geographic problems.

A *Quantitative ecological models*. The development of mathematical techniques for describing the distribution of points in space has been a common pursuit in several disciplines. We shall consider here only the work done by quantitative plant ecologists, because many of the models developed seem very appropriate for description in geographical research. The models are a reflection of biological theory and it would be dangerous to draw too simple an analogy between biological and geographical processes. But the techniques involved may be readily adapted to the description and analysis of spatial point patterns of any kind (Neyman and Scott, 1957).

The technique depends upon an understanding of a general class of *stochastic* processes usually termed *Poisson Processes*. If a set of events or objects are distributed randomly over space (or through time), then the probability of finding an event or object in any unit area of space (or in any interval of time) is given by the Poisson distribution. This has the general form

$$P_x = \frac{e^{-\lambda}\lambda^x}{x!}$$

where λ is the average density of points per unit area (or per time interval) and e is the base of the Napierian logarithms.

The Poisson law in itself is extremely useful, but the various 'contagious' relatives of the Poisson distribution are even more useful (Feller, 1957, pp. 146–149; Greig-Smith, 1964, Chap. 3; Coleman, 1964, Chaps. 10 and 11; Miller and Kahn, 1962, Chap. 16). The two techniques in quantitative plant ecology that utilize this property are *quadrat sampling* and *nearest neighbour analysis*.

(i) *Quadrat sampling*. A quadrat is used to denote a small areal unit for sampling within a particular study region. Having decided upon an appropriate size and shape (Greig-Smith, 1964; Morisita, 1959), the quadrats are laid down 'at random' within the study area, and the number of points which fall in any quadrat are counted and a frequency distribution of quadrats containing 0, 1, 2, ... n, points is constructed. Figure 14.5 shows a point pattern analysed by means of a systematic sample. The justification for using systematic rather than random sampling is given by Grieg-Smith (1964). The resultant observed frequency distribution may then be compared with a theoretical probability distribution generated by a particular set of processes which we suspect govern the distribution of points over space.

If we suspect that the events are randomly distributed over space either because there are many conflicting determining forces or because of genuine chance factors, then we would expect the observed distribution to be matched by a Poisson distribution. The mean number of points per quadrat in our observed distribution is put equal to λ in the Poisson distribution and theoretical probabilities of obtaining 0, 1, 2, ... n, points per quadrat are either calculated out or read off from tables (G. E. C., 1962). These probability values may then be multiplied by the total frequency of quadrats to give a theoretical frequency distribution. As Table 1 and Figure 14.5 indicate, there is a poor fit between the observed frequency distribution and the theoretical Poisson frequency distribution.

TABLE 14.1

Poisson, Negative Binomial, and Neyman Type A distributions fitted to the actual frequency distribution obtained in quadrat sampling Figure 14.5A.

Number of points in quadrat	Observed	Poisson	Negative Binomial	Neyman Type A
0	63	50·9	62·2	62·8
1	15	29·0	15·6	13·2
2	6	8·3	6·4	7·9
3	3	1·6	2·9	3·6
4	1	0·2	1·4	1·4
5	1	} 0·0	0·7	0·6
6+	1		0·8	0·5

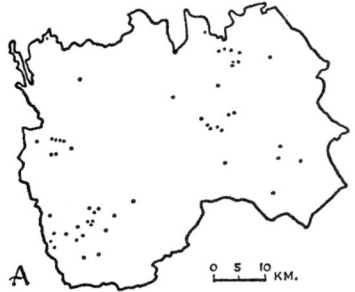
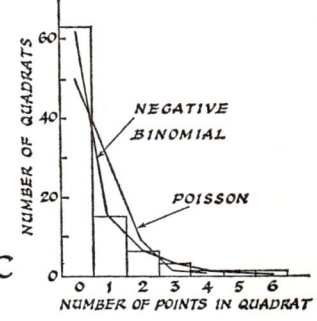

14.5 Quadrat sampling showing (A) an actual point distribution (*after Hägerstrand, 1953, Fig. 30, p. 92*), (B) a systematic sample of quadrats covering the area with points counted in each quadrat, and (C) a frequency distribution of the number of quadrats in each size class with theoretical Poisson and negative binomial distributions fitted.

We can then distinguish whether the process operating is 'contagious' or 'competitive' over space. An important property of the Poisson distribution is that its mean is equal to its variance and it has been shown that if the ratio $\frac{\text{mean}}{\text{variance}}$ is less than 1, then the pattern shows some degree of clustering and a contagious process is suggested; if the ratio is greater than 1 then the distribution is more regular than random and a competitive process is indicated. In the example of Figure 14.5 the variance is much greater than the mean and a contagious process is indicated (Greig-Smith, 1964, Chap. 3). We may then set up a model of some contagious process and test the appropriateness of this model using quadrat sampling. Skellam (1953, p. 346) illustrates the method as follows:

'If a number of points are distributed over an area in accordance with some scheme of laws (which may or may not involve notions of probability), it is possible assuming sufficient mathematical knowledge to deduce the probability distribution of the number of individuals per quadrat laid down at random.'

Conveniently, many of the ecological processes involve change over time and space at the same time, and, conveniently, the mathematical derivation of the probability distributions may also involve time-space processes. We may consider two examples.

The *negative binomial* distribution has been shown to be extremely important in quantitative plant ecology (Bliss, 1953; Anscombe, 1949, 1950; Evans, 1953), and, as we hope to show in Section IV.i, it has applications to geographic analysis. One interesting physical model which leads to the negative binomial distribution is described by Anscombe (1950, p. 360).

'If colonies are distributed randomly over an area so that the number of colonies observed in samples of fixed area has a Poisson distribution, we obtain a negative binomial distribution for the total count if the number of individuals in the colonies are distributed independently in a logarithmic distribution.'

The evaluation of the distribution is not easy, but tables are available which can easily be used to obtain theoretical probabilities using the mean and variance of the observed distribution to estimate the parameters (Williamson and Bretherton, 1964). Figure 14.5 and Table 1 show the theoretical distribution derived from tables of the negative binomial distribution and the fit is much better than the theoretical Poisson distribution.

The *Neyman Type A* distribution has a somewhat more specialized application to ecological problems. Neyman (1939, p. 36) considered the following process:

'Larvae are hatched from eggs which are being laid in so-called masses'. After being hatched they begin to travel in search of food. Their movements are slow and therefore, whenever in a given plot we find a larva, this means that the mass of eggs, from which it was hatched, must have been laid somewhere near, and this in turn means that we are likely to find in the same plot some larvae from the same litter. Of course there may be also others coming from other litters too.'

To fit this situation Neyman derived a whole class of contagious distributions which give the probability of finding a larva in any quadrat laid down at random. If we assume that the egg masses are laid randomly over space and that the number of larvae per egg mass has a Poisson distribution, then the Neyman Type A distribution may be derived (Anscombe, 1950, p. 361; Skellam, 1958). The probabilities are difficult to evaluate but Shenton (1949) and Douglas (1955) describe the method. Table 1 shows a Neyman Type A distribution fitted to the data derived from Figure 14.5.

There are various other probability distributions which have been developed to describe particular models of contagious processes (Thomas, 1949; Evans, 1953; Kendall, 1948B), while Dacey has examined a probability distribution suitable for examining processes that are competitive rather than contagious over space and which may, therefore, be applied to central place systems (Dacey, 1964A).

Quadrat sampling suffers from some defects however. Skellam (1953) points out that we cannot 'arrive at an understanding of the spatial pattern of points from a knowledge of the frequency distribution alone'. The same theoretical distribution may be generated by very different models; there are at least four appropriate models for the negative binomial distribution (Anscombe, 1950), so it only makes sense to fit a theoretical distribution if we possess other evidence as to the processes that would lead to that distribution. The problem of quadrat size is also critical because we can frequently obtain a Poisson distribution by choosing an appropriate quadrat size even though the underlying process is non-random. The kind of evidence to be derived from quadrat sampling is restricted, therefore. Failure to fit a Poisson distribution is conclusive evidence of non-randomness, and success in fitting a contagious distribution is conclusive evidence of a contagious process. But the reverse of these statements is not necessarily true. Mainly for this reason the alternative technique of *nearest neighbour* analysis has been developed.

(ii) *Nearest neighbour analysis*. Plant ecologists have paid considerable attention to the development of distance measures between points because this obviates the use of quadrats and therefore eliminates the effect of quadrat size. The literature in plant ecology is extensive and has been reviewed in Greig-Smith (1964) and Miller and Kahn (1962), while the articles by Morisita (1954, 1959), Clark and Evans (1954), Thompson (1956), Clark (1955 and 1956), Skellam (1952), and Moore (1954), warrant special mention. The application of the technique to the study of geographic problems has been explored by Dacey (1960, 1963A, 1963B, 1964B), and other examples of its uses are contained in Curry (1964) and Getis (1964).

The technique requires that we select a location within our point pattern according to some rule (Dacey, 1962; Greig-Smith, 1964) and then record measurements from that point to the 1st, 2nd, 3rd ... nth nearest neighbour. If we repeat the procedure for a number of points according to some sampling design, we may then calculate out the mean 1st order distance (and its variance), the mean 2nd order distance and so on. These mean order distances, together with the frequency distribution of each order distance, then comprise the mathematical description of pattern.

The technique has several advantages over quadrat sampling (Miller and Kahn, 1962; Clark and Evans, 1954), but is rather laborious. There are also methodological problems involved in determining the study area and overcoming 'boundary problems' (Getis, 1964, p. 391; Dacey, 1964C). The method allows us to identify 'randomness' in the population pattern and provides an accurate method of measuring the extent of the deviation from randomness. The method given by Clark and Evans (1954) is perhaps the simplest, and Dacey's classic use of the technique to test the hypotheses of spatial distribution intuitively treated by Brush demonstrates how effective the technique can be (Dacey, 1962; Brush, 1956).

As with quadrat sampling we can expect a specific model of evolution of spatial patterns to give rise to certain theoretical order distance measures. Dacey has indicated that in an ordered hierarchy of central places a gamma distribution is the appropriate theoretical distribution of order distances (Dacey, 1964C). Dacey (1964B) has also successfully tested a model of town location in Iowa which involved placing each town at the centre of a randomly selected county, and then disturbing this pattern by a random variable. As Dacey (1963B) points out, 'A central problem is: given an observed pattern estimate the probability that a specified spatial process generated that pattern'. Clearly the technique has considerable potential, but its application is limited by gaps in mathematical understanding and difficulties of defining the appropriate method.

The techniques developed in quantitative plant ecology involve an intriguing interaction between the study of processes of biological and ecological change and the derivation of theoretical probability distributions. In some cases the mathematical processes that lead to the probability distribution have a close relationship to the biological processes and because the biological processes involve development over time and space, we may expect that similar models may apply to the evolution of spatial patterns. Colonization models (such as that developed by Bylund) will clearly lead to contagious distributions of some kind, and industrial location patterns might be analysed in terms of models similar to those developed in quantitative plant ecology. If, to take a very simple example, we were interested in the location of firms involved in sub-contracting for the transport industry, we might set up a model of location which would generate a Neyman Type A distribution and then test the appropriateness of this model using quadrat sampling. Section IV.i illustrates the kind of procedure that can be relatively simply developed as an aid to our understanding of the evolution of spatial patterns.

B *Markov-chain models*. The markov process assumes dependence between an event and the immediately preceding event. In the language of probability theory we permit 'the outcome of any trial to depend upon the outcome of the directly preceding trial (and only on it)', (Feller, 1957, p. 338; Kemeny, *et al.*, 1956, p. 171). Thus specified, markov processes seem particularly useful; if we wish to examine the state of industrial location in 1966 it seems reasonable to assume that this state is a function of the pattern in 1965 plus some component of change which may be defined by a set of probabilities. The principles of the technique are relatively simple, but the construction of realistic markov chain models requires considerable mathematical ingenuity (Bharucha Reid, 1960; Feller, 1957; Doob, 1953; Bartlett, 1960A; Bailey, 1964; Täkacs, 1960). The examples that follow are very elementary, but they demonstrate the kind of problem to which markov chain techniques may be applied and the nature of the solutions that may be derived.

Let us assume a finite set of outcomes of an experiment $x_1, x_2, x_3, \ldots x_n$, and let us assume that any single outcome of the experiment, x_j, is dependent upon the preceding outcome, x_i. These outcomes of the experiment are termed *states*. Each *state* x_j is linked to a preceding *state* x_i by a probability value, p_{ij}, termed the *transition probability*. If we can define the *states* adequately, possess information regarding the *transition probabilities*, and know the initial *state*, then we can compute the behaviour of the system over time (Kemeny, *et al.*, 1956; Kemeny, *et al.*, 1960). Suppose we have two *states* x_i and x_j, then we can write in the *transition probabilities* in matrix form as follows (denoting the whole matrix by P):

$$P = \begin{matrix} \\ x_i \\ x_j \end{matrix} \begin{pmatrix} x_i & x_j \\ p_{ii} & p_{ij} \\ p_{ji} & p_{jj} \end{pmatrix}$$

In this case p_{ii} denotes the probability that *state* x_i will occur given that *state* x_i occurred on the previous trial and the values p_{ij}, p_{ji} and p_{jj}, may be similarly interpreted. If the *transition probabilities* are stable over time then further information about the development of the system over successive stages may be readily computed. We can illustrate this from a simple problem of migration (Muhsham, 1963).

Let us suppose we wish to examine the movement of household units through three different states of residence, e.g. London city, suburbs, country. We know that movement of this type is a stochastic process with household units moving in all directions at any one stage in time. Let us suppose that the probability of moving from one state to any other state over a specified period is described by the following matrix:

[578] MODELS IN GEOGRAPHY

$$\begin{pmatrix} & \text{London} & \text{Suburbs} & \text{Country} \\ & (x_1) & (x_2) & (x_3) \\ \text{London } (x_1) & 0.6 & 0.3 & 0.1 \\ \text{Suburbs } (x_2) & 0.2 & 0.5 & 0.3 \\ \text{Country } (x_3) & 0.4 & 0.1 & 0.5 \end{pmatrix}$$

Suppose an individual household starts in the country, and we wish to know the probability that that unit will be in the country after two or more specified periods. There are a number of paths which he can take and Figure 14.6

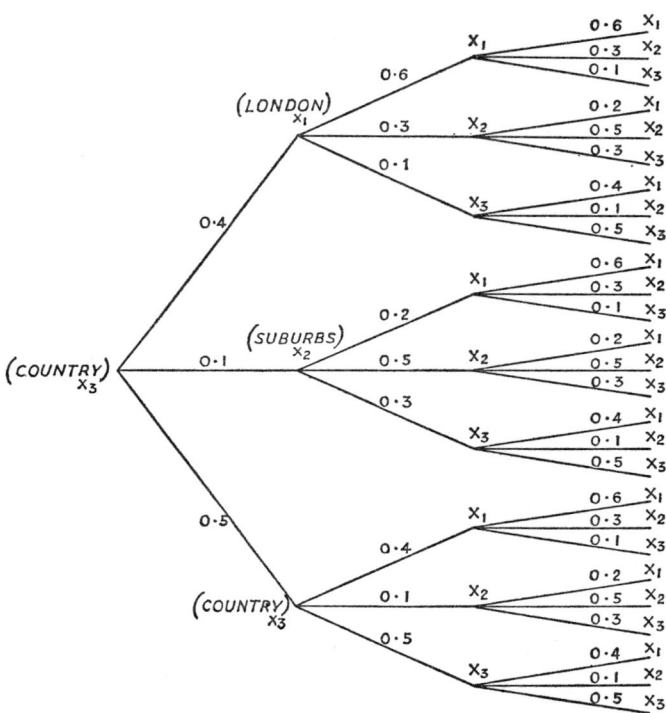

14.6 A markov chain process, showing possible migration paths for a household originating in the country, with the probability of each path assigned.

illustrates the branching process which describes these paths. The probability of being in the country after two stages is the sum of the conditional probabilities terminating in the county (x_3) after two stages which, using the multiplication rule for conditional probabilities (Hoel, 1954) yields $(0.5) \times (0.5) + (0.1) \times (0.3) + (0.4) \times (0.1) = 0.32$. It proves far quicker to analyse the problem using matrix algebra. We can calculate out a matrix, P^n, which shows the transition probabilities after n stages, by calculating out the relevant power of the matrix. Thus the square of the initial matrix shows the probability of a unit starting in any one state being in any other state after two stages. The technique of squaring the matrix is relatively simple (Kemeny, et al., 1956, Chap. 5; School Mathematics, 1960). In the example we have:

$$P^2 = \begin{matrix} & x_1 & x_2 & x_3 \\ x_1 & \\ x_2 & \\ x_3 & \end{matrix} \begin{pmatrix} 0\cdot 46 & 0\cdot 34 & 0\cdot 20 \\ 0\cdot 34 & 0\cdot 34 & 0\cdot 32 \\ 0\cdot 46 & 0\cdot 22 & 0\cdot 32 \end{pmatrix} \quad P^3 = \begin{matrix} & x_1 & x_2 & x_3 \\ x_1 & \\ x_2 & \\ x_3 & \end{matrix} \begin{pmatrix} 0\cdot 424 & 0\cdot 328 & 0\cdot 248 \\ 0\cdot 400 & 0\cdot 304 & 0\cdot 296 \\ 0\cdot 448 & 0\cdot 280 & 0\cdot 272 \end{pmatrix}$$

and from these matrices we can readily read off, for example, that a unit that starts in the country has a 0·32 probability of being in the country after 2 stages, and a 0·272 probability of being in the country after 3 stages. Using this information we can also derive information about the average state of the system after so many stages.

Let us assume that the total population in the system remains constant, and that the initial state of the system contained population distributed 50 per cent in London, 30 per cent in the suburbs, and 20 per cent in the country. This initial state of the system may be described as a probability vector of the form

$$p^{(0)} = \left(p_1^{(0)}, p_2^{(0)}, p_3^{(0)}\right) = (0\cdot 5, 0\cdot 3, 0\cdot 2).$$

It has been shown that the subsequent states of the system may be derived by multiplying the matrix of transition probabilites, P, by this vector and in general it has been shown that

$$p^{(n)} = p^{(0)}P^n = p^{(n-1)}P$$

(Kemeny, et al., 1956, p. 218). Thus the state of the system after three stages would be

$$P^{(3)} = (0\cdot 5, 0\cdot 3, 0\cdot 2) \begin{pmatrix} 0\cdot 424 & 0\cdot 328 & 0\cdot 248 \\ 0\cdot 400 & 0\cdot 304 & 0\cdot 296 \\ 0\cdot 448 & 0\cdot 280 & 0\cdot 272 \end{pmatrix} = (0\cdot 4216, 0\cdot 3113, 0\cdot 2672)$$

and the successive average states of the system calculated out in this way are

	x_1	x_2	x_3
$p^{(1)} =$	0·5	0·3	0·2
$p^{(1)} =$	0·44	0·32	0·24
$p^{(2)} =$	0·424	0·316	0·260
$p^{(3)} =$	0·4216	0·3112	0·2672

These figures show a rapid convergence towards some average state of the system and it is possible to identify that probability vector which holds the system in equilibrium. This probability vector is termed the *unique fixed point probability vector of* P and we will denote this vector by *t* (Kemeny, et al., 1956, p. 221). The values contained in this vector can be obtained by solving a set of simultaneous equations (Kemeny, et al., 1960, p. 72), and the solution in this case yields

$$t = \frac{11}{26}, \frac{4}{13}, \frac{7}{26}$$

and as a check on this we may use the relationship $tP=t$. The significant point about this equilibrium solution is that it is achieved irrespective of the original state of the system. Whatever the original distribution of population, therefore, we can say that after several stages the population distribution will be $\frac{11}{26}$ (42·3%) in London, $\frac{4}{13}$ (30·8%) in the suburbs, and $\frac{7}{26}$ (26·9%) in the country. Once this average state of the system is achieved movement between the states is compensating and a situation of statistical equilibrium is achieved.

Further information about the system can be obtained. We may wish to know, for example, the average number of stages it takes for an individual unit to move once. This value is known as the *mean first passage time* and can be readily calculated (Kemeny, et al., 1960, pp. 78–84).

This simple markov chain model is characterized by all processes being reversible. This assumption seems reasonable in a model of migration, but there are many situations where it is inappropriate. Many processes are irreversible (labour and capital turned into a road cannot be re-created by breaking up the road). A state is said to be *absorbing* if, once it is entered, it is impossible to leave it. In an absorbing markov chain 'it is certain that the process will end up in one of the absorbing states' (Kemeny, et al., 1956, p. 326). We may illustrate this process from a hypothetical example of agricultural specialization.

Let us suppose that farmers in any given region may produce one out of a choice of four products (A, B, C, D) and that states A and D require capital investment on such a large scale that they effectively represent *absorbing* states, while states B and C are *transient* states. Let us suppose that there is a trend of land conversion among the land use types but that it is easier to convert land from C to D than from B to A. We might obtain a matrix of transition probabilities among the land use types as follows:

$$P = \begin{array}{c} \\ A \\ B \\ C \\ D \end{array} \begin{array}{c} \begin{array}{cccc} A & B & C & D \end{array} \\ \left(\begin{array}{cccc} 1 & 0 & 0 & 0 \\ 0\cdot 1 & 0\cdot 6 & 0\cdot 3 & 0 \\ 0 & 0\cdot 2 & 0\cdot 4 & 0\cdot 4 \\ 0 & 0 & 0 & 1 \end{array} \right) \end{array}$$

From this matrix we can determine what proportion of the farmers in the region will end up producing crop A and crop D, and how many stages it will take, on average, before all farmers are producing either crop A or crop D; provided, that is, we know the initial state of the system. If all farmers are producing crop A to start with there will be no change in the system with time at all. But let us assume that the farmers divide their numbers among crop B and C evenly at the start. The initial state can then be described by the probability vector $P^{(0)} = (0, 0\cdot 5, 0\cdot 5, 0)$. The probability of absorption into crop A is determined in an analogous way to the determination of the *unique fixed point probability vector*. It is determined by identifying a column vector a such that $Pa = a$, putting $a_1 = 1$ and $a_4 = 0$, and the probability of absorption into D is given by identifying a column vector d such that $Pd = d$, putting $d_1 = 0$ and $d_4 = 1$, and solving a set of simultaneous equations (Kemeny, et al., 1956, pp. 326–332). The column vectors in this case are:

$$a = \begin{pmatrix} 1 \\ \frac{1}{3} \\ \frac{1}{9} \\ 0 \end{pmatrix} \quad d = \begin{pmatrix} 0 \\ \frac{2}{3} \\ \frac{8}{9} \\ 1 \end{pmatrix}$$

and combining these column vectors with the initial state of the system the number of farmers specializing in crop A is $(0)(1)+(\frac{1}{2})(\frac{1}{3})+(\frac{1}{2})(\frac{1}{9})+(0)(0)=\frac{2}{9}$ and the proportion specializing in D is $\frac{7}{9}$.

These two examples are highly simplified, but they demonstrate some of the potential uses of the markov chain technique in geographic research. In the first example we trace the movement of migrants through spaces and in this case the relevant *states* of the system are geographic locations. We could divide space into a series of cells and examine the circulation of population among such cells. We might also examine the movement of goods and services among regions and Smith (1961) has suggested how such an approach may be made to examining problems of underdeveloped regions within an economy. The movement of technical knowledge and information might similarly be studied using a markov chain approach (Brown, 1963).

In the second example we traced the movement of certain units at a fixed location among different cropping systems. In this instance we treat geographic locations as if they are consecutively sampling among certain states of economic activity. Thus Clark (1965) has examined the movement of census tracts within cities among different rental classes. It would be tempting to explore a model of regional development as if regions sample among different production systems until a satisfactory production system is found which may then be regarded as an absorbing state. A stochastic model of regional specialization becomes quite feasible using such a technique, although the model would be highly generalized. It might, therefore, be more satisfactory to examine the movement of production elements (labour, capital, etc.) among different production systems within a system of regions. A pioneer study of this type (Blumen, et al., 1955) examined the flow of workers through the industrial structure of the USA, and empirical testing of this model showed 'that the longer a person stayed in a particular job the less-likely he was to move' (Goodman, 1961, p. 841), and therefore the transition probabilities were a function of length of time in employment. The simple markov model was thus disaggregated into a 'mover-stayer' model, and Goodman (1961) has indicated that further disaggregation to take account of behavioural differences among age groups, occupational groups, race and ethnic groups, and *among regions* would improve the model markedly. Clearly, the markov chain technique has considerable potential as an aid to research into problems of evolution in human geography, since it enables us to treat of temporal dependence of events within a system of geographic locations, and to examine equilibrium as a statistical state in terms of the actual processes at work in society.

There are, of course, difficulties. We assume that there are distinctive *states* of the system in the models we have so far presented, and that change occurs in a series of discrete time intervals. But space and time are continuous variables. Similarly, the transition probabilities, which we assume to be constant with time, may vary over time. Technically it is possible to develop models which overcome all of these difficulties and diffusion models in physics (the so-called models of Brownian motion) successfully treat of movement of particles continuous in time and space. But such models are extremely complicated to operate and it will undoubtedly be some time before such models can be satisfactorily applied to economic and social development (Feller, 1957; Bharucha-Reid, 1960; Bailey, 1964). The solution of these complicated models defies conventional mathematical techniques in most cases, and solution depends upon the use of *Monte Carlo simulation* techniques, and it is to these that we shall now turn.

C *Monte-Carlo simulation.* Simulation is a term that means a variety of things to different people. Various kinds of simulation are treated elsewhere (*Morgan*), but we will be concerned only with the *Monte Carlo* method which involves 'setting up a stochastic model of a real situation and then performing sampling experiments upon it' (Harling, 1958, p. 307; Ackoff, 1962, pp. 351–363). There are three basic operations. First we identify the probability values or random variables in the stochastic model; second, we obtain a set of random numbers and use these to sample from these probability distributions according to some specified method; and third, we repeat the operation many times to get an approximate solution to the model.

We may demonstrate the technique using the simple model of migration already presented. The probability of a move from the country to London is 0·4 and we represent this quantity by the four digits 0, 1, 2, 3; we represent the probability (0·1) of a move from the country to suburb by the digit 4; and the probability of staying in the country (0·5) by the digits 5, 6, 7, 8, 9.

Movement among the other states may be similarly represented by digits. Starting, say, in the country, we can read off digits from a table of random numbers and place a unit at each stage according to the digits we have assigned to each state of the system. Table 2 shows ten households followed in this way over ten stages. We have here shown simulated migration histories of ten households and it is interesting to note the variation. One household moves eight times while another moves only twice.

TABLE 14.2

Simulated migration histories of 10 households originating in the country, over 10 stages given the migration probablities illustrated in Figure 14.6.

Stage	\multicolumn{10}{c}{Household Units}	% in country
	1	2	3	4	5	6	7	8	9	10	
0	c	c	c	c	c	c	c	c	c	c	100
1	c	s	s	c	l	l	c	l	c	c	50
2	l	s	s	l	c	l	l	l	c	c	30
3	s	s	s	l	c	l	l	c	l	s	20
4	c	s	s	s	s	l	s	c	l	s	20
5	l	l	l	s	l	l	c	l	l	s	10
6	s	l	s	l	s	l	l	l	c	10
7	s	c	s	l	c	l	s	l	l	l	20
8	s	l	c	l	l	l	s	l	s	l	10
9	s	l	c	s	l	c	s	s	s	s	20
10	l	l	c	s	s	l	c	c	s	l	30
Total No. of Moves	6	4	4	4	8	3	6	5	2	5	

c = country   s = suburbs   l = city

The property of the *Monte Carlo* solution which is most useful is the tendency to converge towards an average solution after a number of trials. Examining our first sample of ten migrants we note that 3 were located in the country after two stages. Taking more samples of ten and averaging the number found in the country after two stages we obtain a converging estimate of the theoretical value of 0·32 which we have already calculated (see Fig. 14.7). In this case we know the theoretical value, but in many cases this value is almost impossible to compute and under such circumstances we may use this *Monte-Carlo* technique to give an approximate solution. In a recent study of water resource systems (Maas, et al., 1962, p. 253) a reasonable approximation to an optimal design was found after 200 trial simulations out of a total possible number of 10^{23}. In these kinds of problem the aim is to identify or approximate to an optimal solution and special steps (such as the

Las Vagas technique) may be used to ensure that each trial is a step nearer to the solution desired (Ackoff, 1962, p. 361).

Monte Carlo techniques have been used by many disciplines. Any stochastic model which involves difficult and intractable mathematics may be approximately solved using this technique. The problems involving movement of particles and so on in atomic physics were those that first led to the full development of the technique and since this time it has found a multitude of applications both in the natural sciences (Hammersley and Handscombe, 1964) and in the behavioural sciences (Hoggatt and Balderston, 1963).

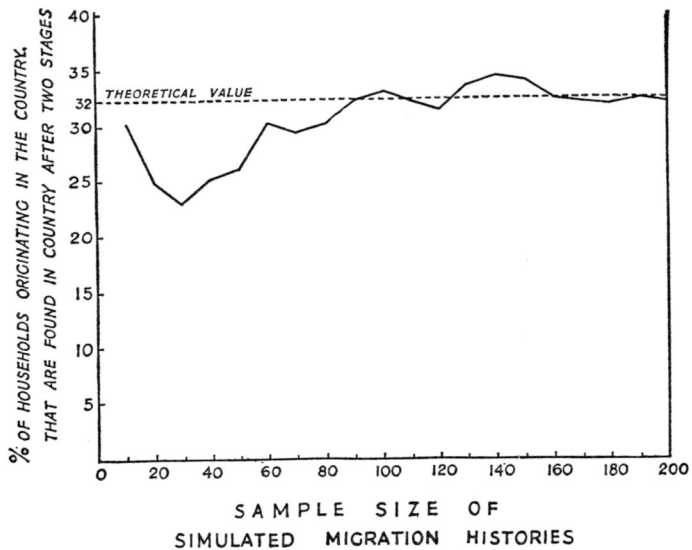

14.7 Convergence of the mean number of people found in the country after two stages, under the assumption of constant probabilities and all households originating in the country. The simulated histories show a convergence towards the theoretical value of 0.32 with increasing sample size.

Traffic control and queueing problems (Churchman *et al.*, 1957, pp. 389–475), flows in networks (Kleinrock, 1964), the progress of epidemics over time and space (Bartlett, 1960B, pp. 82–84), and water storage and allocation problems (Maas, *et al.*, 1962), have been studied using this technique. The study of the U.S. economy (Orcutt, *et al.*, 1961) uses a micro-analytical model incorporating information on individual household behaviour and then aggregates the information to obtain estimates of such features as total labour force availability for the whole U.S. economy. *Monte Carlo* techniques have also been used to simulate conditions within a static equilibrium framework (Mills, 1964). The flow of traffic in towns, journey to work, the movement of population around service facilities, may be studied as flows which have no time trend attached. The dynamic force in such models comes from an exogenous shock to the system as typified in the comparative statics models reviewed in the context of deterministic models above. Models of this type are currently being built of urban systems and, as such, they contain much that is relevant to geographic analysis (Lowry, 1964; Steeger and Douglas, 1964). In short, 'Monte Carlo methods tend to flourish on problems that involve a mass of practical complications of the sort encountered more and more frequently as applied mathematics and operational research come to grips with actualities' (Hammersley and Handscombe, 1964, p. 9). Given the complex situations with which geographic research typically deals, models of activity requiring *Monte Carlo* methods are almost bound to be of major significance.

This potential can be used in two ways. In models which have been verified by ordinary numerical methods, a *Monte Carlo* simulation may be used as a demonstration of the general form of the solution and to give some idea of the chance variations that can result from the stochastic relationships contained in the model. In highly complicated models, the technique may be basic to solution and verification.

Morrill, for example, has used the method for studying the evolution of spatial patterns over time. Assuming the spatial processes inherent in the Christaller-Lösch model of settlement location, he then shows how random disturbances cumulative over time may lead to a considerable variability of pattern even though the underlying spatial process is constant (Morrill, 1962, 1965). The model is extremely useful for pedogogic purposes but it is difficult to verify as a general predictive model (Fig. 14.8). A major difficulty here is that we have only one real world pattern to compare our model with, and the appropriateness of a more sophisticated model proposed by Morrill (1963 C and 1965) for southern Sweden is very difficult to test. This points up one of the difficulties of using *Monte Carlo* methods. Sampling on probability distributions can produce a very wide range of results, and the procedure is only meaningful if we take quite a large sample of events which are independent of one another. In our migration example we gain a reasonable estimate of the average state of the system after studying the simulated migration histories of, say, 100 household units. Problem solving, using *Monte Carlo* procedures, thus only seems reasonable where we aggregate a large number of events. One stage of Morrill's model concerns the assignment of migrants between places, and in this case the model comes close to problem solving because he is treating of the net movement of a large number of people and we may reasonably assume that a simulated model will converge towards a solution which approximates reality *if* the model is an appropriate one (Morrill, 1963B; 1963C and 1965). Thus Hägerstrand's simulated models of migration and diffusion (Hägerstrand, 1953; 1957; 1963 and forthcoming) seem to be very real attempts to solve specific problems using the *Monte Carlo* method. But even in these cases the simulated models have proved very difficult to test against reality and as Hägerstrand (forthcoming) has pointed out, all we can hope to do is to demonstrate that the general shape and form

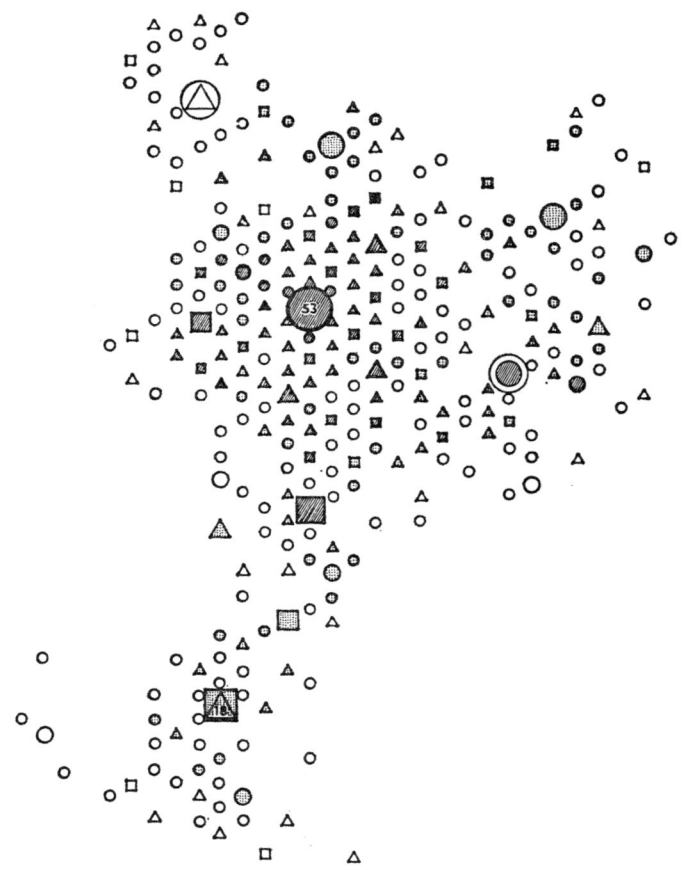

14.8 A simulated pattern of settlement spread involving a random process, but strong constraints on migration between places (*From Morrill, 1962*).

of the simulated pattern is broadly similar to patterns existing in reality for 'appropriate measurement techniques are sadly neglected.' This problem is not confined to geographic analysis, as many disciplines face the same difficulty (Bartlett, 1960B).

The *Monte Carlo* simulation technique does not seem an appropriate tool for problem solving where we are treating of a very small sample of events.

Nobody would deny that there is a chance element in the location of iron and steel works, but it would not be reasonable to set up a stochastic model of iron and steel location and then solve this model using *Monte Carlo* methods to locate just four iron and steel works. Each simulated run would give very different results, and aggregating a large number of results to give an average solution seems a very long way round to demonstrate what could easily be determined by e.g. elementary techniques of Weberian or Löschian analysis with a suitable margin allowed for imperfections in decision making and market competition (Devletoglou, 1965). Using *Monte Carlo* techniques to simulate the development of a transport network (Morrill, 1963C and 1965; Kansky, 1963) appears to pedagogic rather than problem-solving because each link is determined by a sampling procedure and once that link is chosen by a random number it is fixed. We cannot reasonably average out a large number of transport links between two points. For this reason the stage of Morrill's simulation model of settlement location which requires *Monte Carlo* simulation of the transport network seems theoretically weak (Morrill, 1963C and 1965).

Monte Carlo simulation may produce very different results on each run through of the model. It is thus very difficult to test 'whether the actual patterns could have been produced by the suggested theory' (Morrill, 1963C, p. 1). If we add to this the general point that many theoretical models can be developed, all of which could fit reality equally well (Braithwaite, 1960; Curry, 1964), then the limitations of the *Monte Carlo* technique for problem solving become all too evident.

The *Monte Carlo* method involves sampling on a stochastic model, where the random variables are already defined, in order to solve the model. If the solution shows considerable deviation from reality, we may then alter our model in some respect. In particular we may use a feedback method to identify more suitable random variables which, when sampled from, generate a solution more in accordance with reality. Such a procedure requires that we ensure that the random variables eventually defined have a valid theoretical interpretation. We may differentiate here between physical processes – which generally speaking remain constant over time – and behavioural processes – which change quite rapidly (M. G. Kendall, 1961). Given the general processes of water flow one can reasonably expect that a similar pattern of water flow in a river basin will result from a similar rainfall distribution independent of time. Defining parameters to the system and random variables by a feedback technique seems reasonable here (Maas, *et al.*, 1962). We cannot make the same assumptions as regards traffic flow where technology, cost, motivation, and behaviour, may change radically from time to time and place to place. A random variable defining propensity to migrate, for example, has been shown to vary according to size of unit and time (Hägerstrand, 1957; Olsson, 1965B). Defining parameters and random variables by feedback techniques would be dangerous here for it becomes very difficult to give these random variables any useful theoretical interpretation. Thus the probabilities which Morrill used to assign transport links (1963E and 1965) seem largely geared to fit the result. In a study of journey to work patterns in Chicago (Taaffe, *et al.*, 1963) a set of probabilities were derived from the observed data and these probabilities apply only to Chicago at that particular time and appear to have no useful theoretical interpretation. The probabilities which Hägerstrand used have a theoretical interpretation in that they relate 'information fields' derived from one set of data to 'migration fields' (Hägerstrand, 1957). It is, thus, very important to pay attention to the theoretical underpinnings of any *Monte Carlo* model. Care in this respect may make all the difference between 'a successful simulation and an incoherent attempt to programme one's own ignorance' (Harling, 1958, p. 312; Garrison, 1962, p. 100).

Monte Carlo simulation techniques must be treated with care. But if we are satisfied as to the theoretical interpretation of the random variables contained in the model, if we are treating aggregated events, and if we are certain that a simple theoretical model is inadequate to solve the problem, then the technique is an invaluable one for problem solving. As more complex stochastic models are developed by geographers the *Monte Carlo* technique will undoubtedly become a basic tool in geographic research. Its simplicity also gives it a pedagogic use, and the day may come when every classroom is equipped with a roulette wheel!

This review of stochastic models of spatial evolution has concentrated mainly on the technology involved in developing such models and has indicated that there is tremendous potential here for developing more realistic models of spatial development over time and of giving such models a quantitative mathematical expression. It remains to show in greater detail how such techniques may be applied to a particular problem.

IV THE APPLICATION OF MODELS OF SPATIAL EVOLUTION IN GEOGRAPHIC RESEARCH

Many of the general ideas indicated in Section I can be given a quantitative expression by building explicit models of development over space and time. No one single model explains reality exclusively except, of course, reality itself. We can, thus, frequently build several rather different models to explain the same reality and each might be equally good. It is this which makes model building so exciting. But the application of model building requires some demonstration.

IV (i) An empirical study – the diffiusion of an innovation

A study of diffusion would be a salutary experience for those that claim that an academic discipline should be distinguished by distinctive subject matter and distinctive problems. The problem of diffusion has been tackled in physics (the movement of particles), plant and animal ecology (the dispersal of plant and animal groups), sociology (the spread of information among social groups), archaeology and anthropology (the diffusion of physical and cultural traits), and in geography (the spread of cultural forms over space), to mention just a few. There is thus a vast literature (Rogers, 1962, cites over 500 references relevant to sociology alone while Brown, 1965A, cites 337 references in a bibliography of spatial diffusion), and many quantitative models have been developed to examine the problem. Some of these models are highly specialized and reflect the particular processes involved (this is particularly true of physics), but in other cases models developed in one discipline can be quite adequately transferred with little modification into another discipline (Brown, 1965B). We will examine two alternative models here.

A *The Hägerstrand model*. Hägerstrand (1953) examined the spread of a number of innovations among the population of a part of central Sweden. Some of these innovations concerned agricultural practices (t.b. control in cattle, pasture improvement, etc.) and some were more general (telephones, post office accounts, cars, etc.). To describe the spread of the agricultural techniques Hägerstrand built a model based on the following six assumptions (Karlsson, 1958).

1 Only one person has the information at the beginning.
2 The method is accepted at once when heard of.
3 The information is spread only by the telling at pairwise meetings.
4 The telling takes place only at certain times with constant time intervals.
5 At each of these times every knower tells one other person who may be knower or non-knower.
6 The probability of being paired with a knower depends on the geographical distance between knower and teller in a way determined by an empirical estimate derived from distance functions fitted to migration movement and telephone contacts over space.

The model is a difficult one to solve by ordinary mathematical techniques and so Hägerstrand used the *Monte Carlo* simulation technique to generate a series of 'artificial maps' of the spread of the innovation, and relied upon visual comparison between the general shape and form of the artificial series and the observed series as an indication of the appropriateness of the model

(Figs. 14.9 and 14.10). In later analysis Hägerstrand began to adjust his simple model by introducing a 'resistance' to acceptance of the information and by introducing 'barriers' to take account of lakes or other physical features which clearly disturbed the simple pattern. A computer version of this model has since been presented by Pitts (1963), and Yuile (1964) has made a special study of barrier effects. This model is a classic demonstration of how *Monte Carlo* techniques can be used to great effect to study events distributed over both space and time.

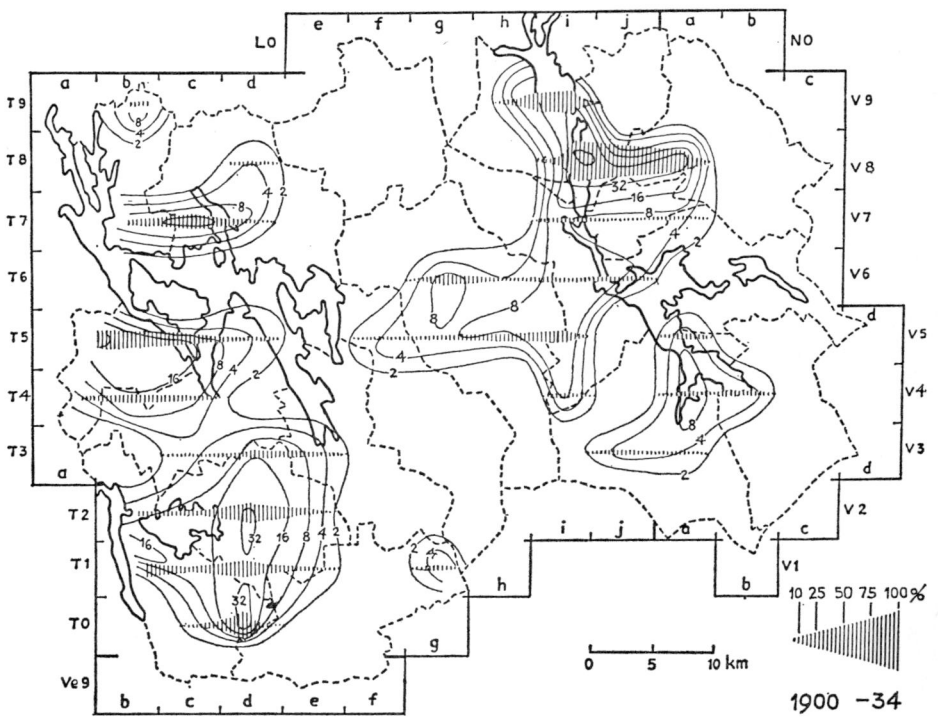

14.9 The spread of the control of t.b. in cattle in part of central Sweden, 1900–1934 (*From Hägerstrand, 1953, p. 95*).

B *An Ecological model* (Harvey, 1966). Several writers have suggested that information is spread by a 'two-step-flow' process in which the information is first accepted by a set of *leaders* (usually from mass media source of information) who then pass on the information by personal contact to other members of the community (*followers*) (Rogers, 1963; Coleman *et al.*, 1957). We might use a simple model derived from plant ecology to describe this process. Let us assume:

1 That a community contains two classes of individuals, *leaders* and *followers*.
2 That the *leaders* are distributed throughout the community randomly (i.e. according to the Poisson law) and accept the information randomly over time.
3 That the *followers* accept the information from the *leaders* or from other *followers* who know, at a constant rate, constrained over distance.
4 That the rate at which *followers* who know accumulate around a *leader* is logarithmic.

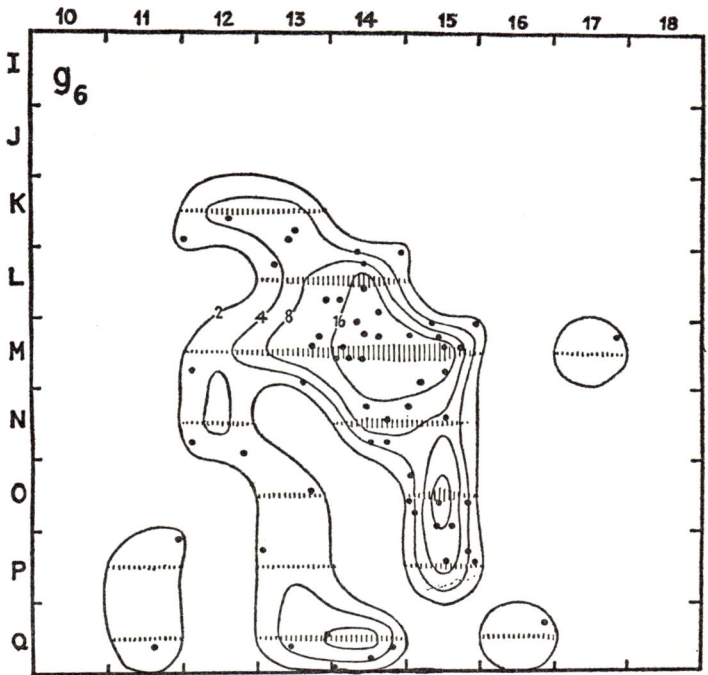

14.10 The simulated pattern of points generated from Hagerstrand's probability model of diffusion using the *Monte Carlo* technique (*From Hägerstrand, 1953, plate 3*).

Given this highly simplified model, we would expect quadrat sampling in any study area where this model operates to yield a negative binomial distribution. To test this model several of Hägerstrand's maps were quadrat sampled – Figure 14.5 is in fact one such map sampled in this way, and Table 3 shows five observed frequency distributions fitted by the negative binomial. The fairly good fits obtained suggest that the model is a reasonable one, although it is worth noting that the fits become worse as the innovation proceeds. This is understandable since the diffusion of an innovation in a

[592] MODELS IN GEOGRAPHY

finite population is best described as a logistic curve (Dodd, 1955; Mansfield, 1961).

These two examples of models fitted to examine a particular problem of diffusion demonstrate how many different models may be developed to describe reality. Each model has its defects and each its advantages. We could go further and fit a deterministic diffusion model (Rushton and Mautner, 1955), or a more theoretical model of random dispersal rather similar to the systems developed by Skellam (1951), Landahl (1953), and Beckman (1957). We might develop more complex models examining the spread of information among a network of points and bring together the powerful tools of network theory and diffusion theory (Rapoport and Rebhum, 1952; Karlsson, 1958). In fact there are a multitude of approaches to examine just this one problem and it is both interesting and instructive to examine as many of these approaches as possible. In this way we come to appreciate reality from many different viewpoints.

TABLE 14.3

Negative binomial distribution fitted to observed distributions obtained by quadrat sampling five of Hägerstrand's (1953) maps of the dispersion of innovations in central Sweden.
1 The location of acceptors of pasture improvement grants, 1928–31 (columns 1–2, Figure 15, p. 64), and 1928–33 (columns 3–4, Figure 19, p. 66).
2 The location of acceptors of t.b. control in cattle, 1899–1901 (columns 5–6, Figure 28, p. 90), and 1900–34 (columns 7–8, Figure 33, p. 94).
3 The location of acceptors of milking machines, 1944 (columns 9–10, Figure 104, p. 290).

Number of points in Quadrat	(1) Obs.	(2) Theory	(3) Obs.	(4) Theory	(5) Obs.	(6) Theory	(7) Obs.	(8) Theory	(9) Obs.	(10) Theory
0	49	47·3	27	20·0	65	51·9	44	50·9	32	29·2
1	17	15·1	15	14·1	15	14·8	13	11·2	15	16·3
2	4	8·5	9	10·8	5	6·0	11	6·3	9	11·1
3	2	5·4	4	8·5	1	2·8	6	4·3	7	8·0
4	5	3·7	3	6·8	0	1·4	6	3·1	8	5·9
5	2	2·6	4	5·5	2	0·7	1	2·4	5	4·4
6	2	1·9	4	4·4	2	0·4	1	1·9	2	3·4
7	5	1·4	2	3·6			1	1·5	3	2·6
8	1	1·0	3	2·9			1	1·2	1	2·0
9	1	0·8	3	2·4			1	1·0	1	1·5
10			3	2·0	0	0·3	0	0·9	2	1·2
11			2	1·6			0	0·7	1	0·9
12	2	2·3	4	1·3			1	0·6	1	0·7
13			1	1·1			1	0·5	3	2·8
14+			6	5·0			3	3·5		

IV (ii) A synoptic model of the evolution of spatial patterns

Consider the following statement by Sauer (1963, pp. 359-360):

'... culture is the learned and conventionalized activity of a group that occupies an area. A culture trait originates at a certain time in a particular locality. It gains acceptance – that is, it is learned by a group – and is communicated outwards, or diffuses, until it encounters sufficient resistance, as from unsuitable physical conditions, from alternative traits, or from disparity of culture level. These are processes involving time; and not simple chronologic time, but especially those moments of culture history when a group possesses the energy of invention or the receptivity to acquire new ways.'

This general account of the development and spread of culture forms contains the elements of a model of spatial evolution. It indicates that spatial patterns are the result of myriad human decisions. Yet very little attention has been paid to the decision making processes in models of spatial development. We may, thus, consider the general schema laid out by Sauer and examine it in the light of many of the advances being made in decision making and learning theory (George; Isard and Dacey, 1962).

Let us assume an individual decision maker operating in an uncertain environment under conditions of total ignorance. In order to survive he must seek out a production system which yields a living. If he aspires to some level of living above mere subsistence (this depends on 'motivation'), he must seek this out, but once he has achieved a 'satisfactory' living standard he may cease to search for a better production system (Simon, 1957; Wolpert, 1964). In any environment there are a very large number of possible production systems. Let us rank these from x_n to x_i according to how far they 'satisfy' a particular producer with x_n defined as the optimal production system. We can then group these production systems into those that do not yield even a subsistence living $(x_i - x_i)$, those that yield a subsistence living but do not 'satisfy' $(x_j - x_s)$, and those that are satisfactory $(x_t - x_n)$. An individual may search among these production systems in the hope of finding a satisfactory solution. He may do this by selecting a production system at random and continuing to select until he hits upon a satisfactory solution entirely by chance (Fig. 14.11A) (Radner, 1964; Gould, 1965). Starting at random a producer may accumulate information about the environment over time and he may use this to help choose a more favourable production system. He will do so by a process of learning, and this involves the development and statement of certain principles which may allow a producer to eliminate certain groups of poor solution (if the generalization is inaccurate it may lead to the elimination of some favourable solutions) (Fig. 14.11B). The choice which any producer makes at any one point in time is thus a function of his knowledge at a preceding point in time. It could, thus, be described as a markov process. In fact markov chain models have been widely applied by psychologists in the statistical analysis of the learning process (Bush and Mosteller, 1955; Estes, 1960 and 1963; Suppes and Atkinson, 1960).

In an uncertain environment it would be wrong to reject a solution after just one year of failure. In this case it is possible to regard *game theory* (see *Henshall*) as a descriptive theory of behaviour, but unfortunately 'it does not provide, even in schematic form, a formulation of the elementary process that

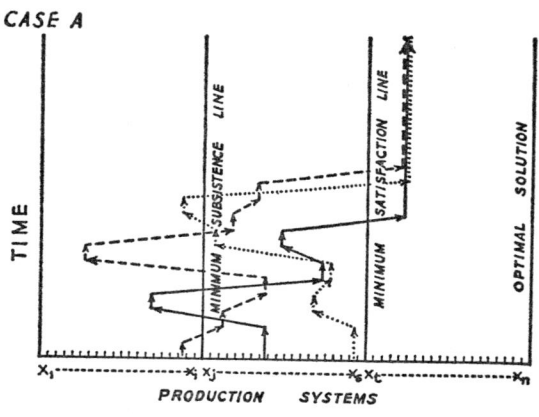

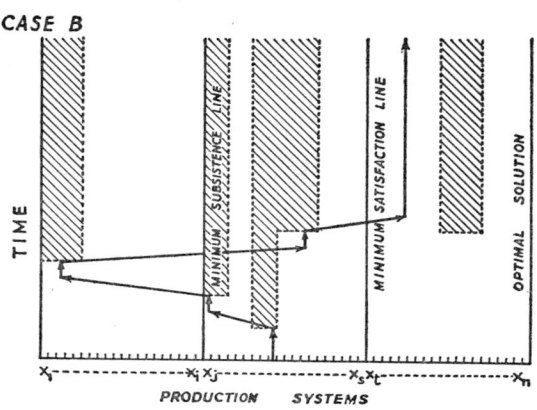

14.11 Idealized learning processes applied to the selection of a production system over time showing (A) random search with solid line representing the first successful searcher and dotted lines showing search paths of 'dependent' searchers, and (B) search process with learning an elimination of sets of production systems as generalizations about the environment are developed.

would lead an organism to select the appropriate game-theoretic choice of strategy' (Suppes and Atkinson, 1960, p. 34). It is thus possible to incorporate uncertainty into a basic learning model and use the markov chain approach to describe how individuals learn and act starting from a point in time when they are totally ignorant. We may thus view the first stage of our generalized conceptual model as involving a *stochastic* learning model which leads to the selection of a 'satisfactory' production system in the face of uncertainty.

But the process of learning and search is not simply a response to environmental stimulation (Gould, 1965). The problem of how and why technology develops is also fundamental. Typically we tend to regard technology as an exogenous variable, and yet technological development is one form of response involved in the learning process. Toynbee's notions of challenge and response are appealing in this respect, and Webb's (1931) classic study of evolving technology on the Great Plains, demonstrates the process. In advanced societies innovation tends to result from 'organizational slack' which allows the allocation of resources for experiment (Mansfield, 1961; Cyert and March, 1963; Machlup, 1962). Whether the same process operates in primitive societies is difficult to say. But we clearly need to regard technological change as one of the outputs to our learning model, and as a major factor leading to the evolution of spatial patterns (Pred, 1965).

So far we have examined learning as an individual process. We need to consider the social aspects of the process for knowledge is transferred by communication among individuals. In the process of communication 'noise' will interfere so that some information will get lost. Ignoring information loss we can begin to develop these notions of adaption and learning in terms of an elementary model of social interaction.

If we consider a group of individuals searching for an appropriate production system in exactly the same environment with perfect interchange of information and exactly similar aspiration levels, then we might expect that as soon as one producer found a satisfactory solution all other producers would resort to that solution (see Fig. 14.11A), provided there was no effect upon environmental conditions. If environmental conditions are affected (e.g. prices) then the adoption of one solution will preclude that solution for other seekers to some degree or other. We are thus back to considering a competitive or a contagious process.

But information does not move freely or easily. The speed, direction, and effectiveness of information flow depends upon the communication system available to any society (Meier, 1962; Cherry, 1957), and upon the social structure (Karlsson, 1958; Coleman, 1964). We thus need to resort to much more complex models of social interaction to understand the process of group learning and group development (Rainio, 1961). We may then add this *social interaction model* to our individual learning model as a further element in our *model system*.

Social interaction leads to the formation of groups and organizations. We cannot assume that organizations act as rather huge individuals. 'To assume that organizations go through the same process of learning as do individual human beings seems unnecessarily naïve, but organizations exhibit (as do other social institutions) adaptive behaviour over time' (Cyert and March, 1963, p. 123). 'Information is not given to the firm but must be obtained, ... and alternatives are searched for and discovered sequentially and ... the order in which the environment is searched determines to a substantial extent the decisions that will be made' (Cyert and March, 1963, p. 10). We may thus introduce a further sub-model into our *model system* which examines the behaviour of organizations over time and the learning process as it applies to organizations.

So far we have considered only learning processes and social and organizational development at a particular point in space. But individuals are spread over space and communication systems spread over space. We may, therefore, introduce a model of *communication* and *diffusion of information* over space and it is this that gives the spatial dimension to our model system; we may here introduce the kind of model considered in Section IV.1.

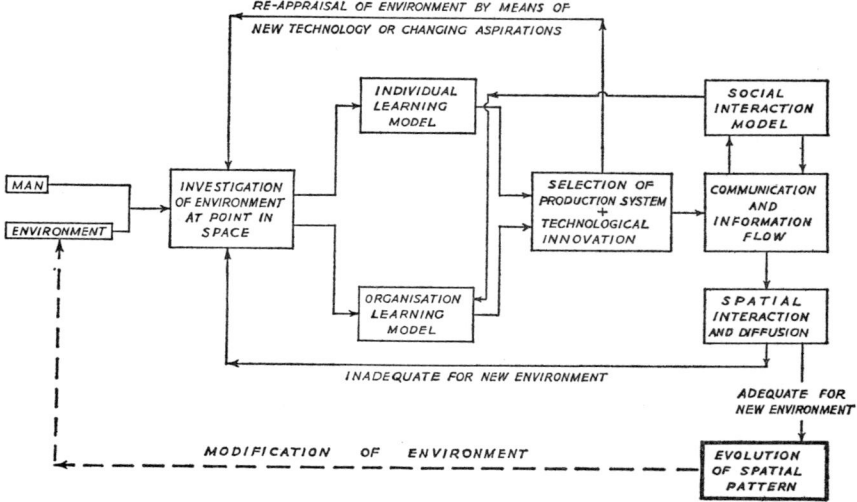

14.12 A synoptic model system to show one mode of approach to the evolution of spatial patterns in human geography.

We have thus constructed a simple model system to express the links suggested by Sauer. No one part of the model system is more important than any other part. We may consider each model as a vital link in a communications system. We may identify feedback effects. Every change in technology or change in aspiration levels will lead to a re-appraisal of the environment. Every movement of knowledge over space will need to be checked against environment and environment itself may change as the result of decisions made. Changes in the social structure may lead to changes in organizational behaviour and so on. A simple flow chart of this model system is presented in Figure 14.12.

The advantages of building a model to illustrate Sauer's statement are twofold. Firstly, we can examine the structural elements in the statement and note the interrelations and feedback effects with greater ease. Secondly, we can begin to examine individual elements in the *model system* in greater detail and begin to get at the processes involved in the evolution of a spatial pattern, and to understand the role of the psychologist, economist, sociologist, in providing us with relevant models which we can use in a more generalized system of explanation. We are not dressing up a simple elegant statement in scientific jargon, but genuinely trying to lay bare elements in reality which have for too long remained hidden from our gaze. Above all, we come to realize that we need more and yet more models to aid us in our attempt to understand the principles that govern human organization over space. A model is 'a pragmatic device, to be used freely as long as it serves its purpose, to be discarded without regrets when it fails to do so. The scientist, therefore, if he is completely a scientist, is unique among the users of metaphors in that he does not become an 'addict' of a particular way of perceiving' (Rapoport, 1953, p. 206). An understanding of the principles and potential of model building may not be a 'sufficient' condition for a *Renaissance* in geographic research; but we can be certain that without such an understanding the 'necessary' conditions for that *Renaissance* will not be fulfilled.

REFERENCES

The references are divided into sections corresponding to the sections in which they are found in the text. References that occur in more than one section are only given in full in the first section in which they occur. Thereafter only the name of the author, date, and number of the section in which the full reference may be found is given.

Section I - Theories and Metaphors

ACKOFF, R. L., GUPTA, S. K. and MINAS, J. S., [1962], *Scientific method: optimising applied research decisions*, (New York).
BARRACLOUGH, G., [1957], *History in a changing world*, (Oxford).
BLACHE, VIDAL DE LA, [1926], (English edition), *Principles of human geography*, (London).
BOWMAN, I., [1921], *The new world*, (New York).
BOWMAN, I., [1931], *The pioneer fringe*, (New York).
BRAITHWAITE, R. B., [1953], *Scientific explanation*, (London).
BROEK, J. O. M., [1932], *The Santa Clara Valley, California: a study in landscape changes*, (Utrecht).
BRUNHES, J., [1920], (English edition), *Human geography*, (London).
CORNISH, V., [1923], *The great capitals*, (London).
DARBY, H. C., [1953], On the relations of geography and history; *Transactions and Papers*, Institute of British Geographers, 19, 1–11.

DICKINSON, R. E., [1951], *The West European city*, (London).
ENEQUIST, G. and NORLING, G., (Eds.), [1960], Advance and retreat of rural settlement – papers of the Siljan Symposium; *Geografiska Annaler*, 42, 208–346.
FARMER, B. H., [1957], *Pioneer peasant colonisation in Ceylon*, (London).
GULLEY, J. L. M., [1959], The Turnerian frontier: a study in the migration of ideas; *Tijdschrift voor Economische en Sociale Geografie*, 50, 65–71 and 81–91.
HARTSHORNE, R., [1961], *Perspective on the nature of Geography*; (Association of American Geographers Monograph; Chicago).
HUNTINGTON, E., [1951], *Civilisation and climate*, (New Haven).
HUNTINGTON, E., [1945], *Mainsprings of civilisation*, (New York).
ISARD, W., [1956], *Location and the space economy*, (New York).
MACKINDER, H., [1917], *Democratic ideas and reality*, (London).
PERRY, T., [1963], *Australia's first frontier*, (Melbourne).
PRED, A., [1965], Industrialisation, initial advantage, and American metropolitan growth; *Geographical Review*, 55, 158–185.
PRICE, A. G., [1939], *White settlers in the tropics*; (American Geographical Society, New York).
RAPOPORT, A., [1953], *Operational philosophy*, (New York).
RATZEL, F., [1891], *Anthropogeographie: Der geographische Verbreitung der Menschen*, (Stuttgart).
SAUER, C. O., [1941], Foreword to historical geography; *Annals of the Association of American Geographers*, 31, 1–24.
SAUER, C. O., [1952], *Agricultural origins and dispersals*; (American Geographical Society; New York).
SAUER, C. O., [1963], *Land and life: Selected writings of Carl Sauer*, (Ed. Leighley, J.), (Berkeley and Los Angeles).
SEMPLE, E., [1911], *Influences of geographic environment*, (New York).
SMITH, C. T., [1965], The teaching of Historical Geography; In Chorley, R. J. and Haggett, P. (Eds.), *Frontiers in geographical teaching*, (London).
SORRE, M., [1950], *Les fondements de la geographie humaine, Tome 2: les fondements techniques*, (Paris).
SORRE, M., [1962], The role of historical explanation in human geography; In Wagner, P. L. and Mikesell, M. W., (Eds.), *Readings in cultural geography*, (Chicago).
TAYLOR, G., [1937], *Environment, race and nation*, (Toronto).
TAYLOR, G., (Ed.), [1951], *Geography in the 20th Century*, (London).
TURNER, F. J., [1961 edition], *Selected essays: frontier and section*, (Englewood Cliffs).
WANKLYN, H., [1961], *Freidrich Ratzel*, (Cambridge).
WHITTLESEY, D., [1929], Sequent occupance; *Annals of the Association of American Geographers*, 19, 162–165.
WYMAN, W. D. and KROEBER, C. B., [1957], *The frontier in perspective*, (Madison).

Section II - Models

ACKOFF, R. L., [1962], I.
AJO, R., [1964A], London's field response in terms of population change; *Acta Geographica*, 18, No. 2.

AJO, R., [1964B], London's field response; *Acta Geographica*, 18, No. 3.
BLUMENSTOCK, D. I., [1953], The reliability factor in the drawing of isarithms; *Annals of the Association of American Geographers*, 43, 289–304.
BUNGE, W., [1962], Theoretical geography; *Lund Studies in Geography, Series C* No. 1.
CHISHOLM, M., [1964], Problems in the classification and use of farming type regions; *Transactions and Papers, Institute of British Geographers*, 35, 91–103.
CHOYNOWSKI, M., [1959], Maps based on probabilities; *Journal of the American Statistical Association*, 54, 355–358.
CLAESSON, C. F., [1964], En korologisk publikanalys; *Geografiska Annaler*, 46, 1–130.
CLARK, A. H., [1959], *Three centuries and the Island*, (Toronto).
CURRY, L., [1962A], Climatic change as a random series; *Annals of the Association of American Geographers*, 52, 21–31.
DACEY, M. F., [1964D], Measures of contiguity for two colour maps; (mimeo.), *Department of Geography, Northwestern University*.
DURBIN, J. and WATSON, G. S., [1950–51], Testing for serial correlation in least squares regression; Part 1 and Part 2; *Biometrika*, 37, 409–428 and 38, 159–178.
GETIS, A., [1963], The determination of the location of retail activities with the use of map transformation; *Economic Geography*, 39, 14–22.
GRANGER, C. W. J., with HATANAKA, M., [1964], *Spectral analysis of economic time series*, (Princeton).
GREGG, J. V., HOSSELL, C. H. and RICHARDSON, J. T., [1964], *Mathematical trend curves: an aid to forecasting*, (Edinburgh).
HÄGERSTRAND, T., [1957], Migration and area; in Hannerberg, D., Hägerstrand, T. and Odeving, B. (Eds.), *Migration in Sweden: a Symposium*; *Lund Studies in Geography, Series B.*, No. 13.
HAGGETT, P., [1965A], Changing concepts in economic geography; in Chorley, R. J. and Haggett, P., (Eds.), *Frontiers in geographical teaching*, (London).
HAGGETT, P., [1965B], *Locational analysis in human geography*, (London).
HANNAN, E. J., [1960], *Time series analysis*, (London).
ISARD, W., [1956], I.
ISARD, W., [1960], *Methods of regional analysis*, (New York).
KENDALL, M. G., [1946], *Contributions to the study of oscillatory time series*, (Cambridge).
LOWRY, I., [1963], Location paramaters in the Pittsburgh model; *Papers of the Regional Science Association*, 11, 145–165.
LOWRY, I., [1964], A model of metropolis; *Rand Corporation*, Memorandum RM-4035-RC.
MARBLE, D. F. and NYSTUEN, J. D., [1963], Community mean information fields; *Papers of the Regional Science Association*, 11, 99–109.
MILLER, R. L. and KAHN, J. S., [1962], *Statistical analysis in the geological sciences*, (New York).
MCCARTY, H. H., HOOK, J. C. and KNOS, D. S., [1956], The measurement of association in industrial geography; *Department of Geography, State University of Iowa*.
MORRILL, R. L., [1963A], The distribution of migration distances; *Papers of the Regional Science Association*, 11, 75–84.

MORRILL, R. L., [1965], Migration and the spread and growth of urban settlement; *Lund Studies in Geography, Series B*, No. 26.
OLSSON, G., [1965A], Distance and human interaction: A bibliography and review; *Regional Science Research Institute, Bibliography Series, No. 2*.
OLSSON, G., [1965B], Distance and human interaction: A migration study; *Geografiska Annaler*, 47, Series B, No. 1, 3–43.
ORCUTT, G. H., GREENBERGER, M., KORBEL, J. and RIVLIN, A. M., [1961], *Microanalysis of socioeconomic systems – a simulation study*, (New York).
REICHENBACH, H., [1956], *The direction of time*, (California).
ROSENBLATT, M., (Ed.), [1964], *Proceedings of the symposium on time series analysis*, (New York).
STEEGER, W. H. and DOUGLAS, N. J., [1964], The simulation model; *Progress Report, No. 5*, (Community Research Program: Department of City Planning, Pittsburgh).
STEWART, J. Q., [1947], Empirical mathematical rules concerning the distribution and equilibrium of population; *Geographical Review*, 37, 461–485.
STEWART, J. Q., [1948], Demographic gravitation: evidence and applications; *Sociometry*, 11, 31–58.
STOUFFER, S. A., [1962], *Social research to test ideas*, (New York).
THOMAS, E., [1960], Maps of residuals from regression: their characteristics and uses in geographic research; *Department of Geography, State University of Iowa*.
TINTNER, G., [1952], *Econometrics*, (New York).
TOBLER, W. R., [1963], Geographic area and map projections; *Geographical Review*, 53, 59–78.
WALLIS, W. A. and ROBERTS, H. V., [1962 edition], *Statistics, a new approach*, (London).
WOLD, H., [1938], *Analysis of stationary time series*, (Stockholm).

Section III i and ii - Deterministic Models

ALONSO, W., [1964], *Location and land use*, (Cambridge, Mass.).
BAILEY, N. T. J., [1957], *The mathematical theory of epidemics*, (London).
BAILEY, N. T. J., [1964], *The elements of stochastic processes with applications to the natural sciences*, (New York).
BARNES, F. A., [1958], The evolution of the salient patterns of milk production and distribution in England and Wales; *Transactions, Institute of British Geographers*, 25, 167–195.
BARTLETT, M. S., [1962], *Essays on probability and statistics*, (London).
BERGE, C., [1962, English edition], *Theory of graphs and its application*, (London).
BERMAN, B., CHINITZ, B. and HOOVER, E. M., [1961], *Projection of a metropolis: technical supplement to the New York metropolitan region study*, (Cambridge, Mass.).
BERMAN, E. B., [1959], A spatial and dynamic growth model; *Papers of the Regional Science Association*, 5, 143–150.
BORTS, G. H., [1960], The equalisation of returns and regional economic growth; *American Economic Review*, 50, 319–347.
BYLUND, E., [1956], *Kolonisering av Pite Lappmark t. o. m. år 1867*, (Uppsala).

BYLUND, E., [1960], Theoretical considerations regarding the distribution of settlement in inner north Sweden; *Geografiska Annaler*, 42, 225–231.
CHENERY, H. B., [1962], Development policies for southern Italy; *Quarterly Journal of Economics*, 76, 515–547.
CHISHOLM, M., [1962], *Rural settlement and land use*, (London).
CURRY, L., [1962B], The geography of service centres within towns: the elements of an operational approach; In Norborg, K., (Ed.), *IGU symposium in urban geography*, (Lund).
DODD, S. C., [1950], The interactance hypothesis. A gravity model fitting physical masses and human groups; *American Sociological Review*, 15, 245–256.
DODD, S. C., [1953], Testing message diffusion in controlled experiments: charting the distance and time factors in the interactance hypothesis; *American Sociological Review*, 18, 410–416.
EGBERT, A. C. and HEADY, E. O., [1961], Regional adjustments in grain production: a linear programming analysis; *U.S. Department of Agriculture, Technical Bulletin, No. 1241*.
ENEQUIST, G., [1960], I.
FORD, L. K. and FULKERSON, D. K., [1962], *Flows in networks*, (Princeton).
FUCHS, V. R., [1962], *Changes in the location of manufacturing in the United States since 1929*, (New Haven).
GARRISON, W. L. and MARBLE, D. F., [1965], A prolegomenon to the forecasting of transportation development; *Transportation Centre Report, Northwestern University*.
GETIS, A., [1964], Temporal analysis of land use patterns with the use of nearest neighbour and quadrat methods; *Annals of the Association of American Geographers*, 54, 391–399.
GODLUND, S., [1956], The function and growth of bus traffic within the urban sphere of influence; *Lund Studies in Geography, Series B, No. 18*.
GREENHUT, M. L., [1963], *Microeconomics and the space economy*, (Chicago).
HÄGERSTRAND, T., [1957], II.
HARVEY, D. W., [1963], Locational change in the Kentish hop industry and the analysis of land use patterns; *Transactions and Papers, Institute of British Geographers*, 33, 123–144.
HENDERSON, J. M. and KRUEGER, A. O., [1965], *National growth and economic change in the Upper Midwest*, (Minneapolis).
HIRSCH, W., [1964], *Elements of regional accounts*, (Baltimore).
ISARD, W., [1948], Some locational factors in the iron and steel industry since the early nineteenth century; *Journal of Political Economy*, 56, 203–217.
ISARD, W., [1960], II.
ISARD, W. and CAPRON, W. M., [1949], The future locational pattern of iron and steel production in the United States; *Journal of Political Economy*, 57, 118–133.
ISARD, W. and CUMBERLAND, J. H., (Eds.), [1960], *Regional economic planning*, (OEEC: Paris).
ISARD, W., SCHOOLER, E. W. and VIETORISZ, T., [1959], *Industrial complex analysis and regional development*, (New York).
KANSKY, K. J., [1963], Structure of transportation networks; *University of Chicago, Department of Geography, Research Paper, No. 84*.

KLEINROCK, L., [1964], *Communication nets*, (New York).
LEWTHWAITE, G. R., [1964], Wisconsin cheese and farm type; a locational hypothesis; *Economic Geography*, 40, 95–112.
LÖSCH, A., [1954. English edition], *The economics of location*, (New Haven).
LOTKA, A., [1945], Population analysis as a chapter in the mathematical theory of evolution; In Le Gros Clark, W. E. and Medawar, P. B., *Essays on growth and form*, (Oxford).
LÖVGREN, E., [1956], The geographical mobility of labour; *Geografiska Annaler*, 38, 344–394.
LOWRY, I., [1964], II.
MELAMID, A., [1955], Some application of Thünen's model in regional analysis of economic growth; *Papers of the Regional Science Association*, 1.
MORRILL, R. L., [1965], II.
MUTH, R. F., [1961], Rural urban land conversions; *Econometrica*, 29, 1–23.
MYRDAL, G., [1957], *Economic theory and underdeveloped regions*, (London).
NELSON, P., [1959], Migration, real income and information; *Journal of Regional Science*, 1, No. 2, 43–74.
OLSSON, G., [1965A], II.
OLSSON, G., [1965B], II.
ORCUTT, G. H., et al., [1961], II.
PEARL, R., [1939], *The natural history of population*, (Oxford).
PERLOFF, H. S., DUNN, E. S., LAMPARD, E. E. and MUTH, R. F., [1960], *Regions, resources and economic growth*, (Baltimore).
PERLOFF, H. S., [1963], How a region grows; *Committee for Economic Development Supplementary Paper*, No. 17.
PORTER, R., [1956], An approach to migration through its mechanism; *Geografiska Annaler*, 38, 317–343.
RAIMON, R. L., [1962], Interstate migration and wage theory; *Review of Economics and Statistics*, 54, 428–438.
RAPOPORT, A., [1953], Spread of information through a population with a social structural bias; *Bulletin of Mathematical Biophysics*, 15, 523–546.
RAVENSTEIN, E. G., [1885 and 1889], The laws of migration; *Journal of the Royal Statistical Society*, 48 and 52, 241–305.
REGIONAL PLAN ASSOCIATION, [1962], Spread City; *Bulletin*, No. 100.
RUSHTON, S. and MAUTNER, A. J., [1955], The deterministic model of a simple epidemic for more than one community; *Biometrika*, 42, 126–132.
RUSSELL, J. C., [1960], The metropolitan city region of the middle ages; *Journal of Regional Science*, 2, 55–70.
SAN FRANCISCO BAY AREA, Office of Area Development, [1959], *Future development of the San Francisco Bay Area 1960–2020*, (Washington).
SJAASTAD, L. A., [1960], The relationship between migration and income in the United States; *Papers of the Regional Science Association*, 6.
SPARROW, F. T., Some experiments with a queueing theory model of market equilibrium; *International Economic Review*, 6, 47–64.
STAMP, L. D., [1948], *The land of Britain; its use and misuse*, (London).
STINE, J. H., [1962], Temporal aspects of tertiary production elements in Korea; In Pitts, F. R., (Ed.), *Urban systems and economic development*, (Eugene, Oregon).

TAAFFE, E., MORRILL, R. L. and GOULD, P. R., [1963], Transport development in underdeveloped countries; *Geographical Review*, 53, 503–529.

THIJSSE, J. P., [1963], A rural pattern for the future in the Netherlands; *Papers of the Regional Science Association*, 10, 133–141.

TIEBOUT, C M., [1962], The community economic base study; *Committee for Economic Development, Supplementary Paper*, No. 16.

VINING, R., [1946], The region as a concept in business cycle analysis; *Econometrica*, 14, 201–218.

WARNTZ, W., [1959], *Toward a geography of price*, (Philadelphia).

WENDEL, B., [1953], A migration schema: theories and observations; *Lund Studies in Geography, Series B.*, No. 9.

WINGO., L., [1961], *Transportation and urban land*, (Washington).

WINSBOROUGH, H. H., [1962], City growth and city structure; *Journal of Regional Science*, 4, No. 2, 35–49.

Section III iii - Stochastic Models

ACKOFF, R. L., [1962], I.

ANSCOMBE, F. J., [1949], The statistical analysis of insect counts based on the negative binomial distribution; *Biometrics*, 5, 165–173.

ANSCOMBE, F. J., [1950], Sampling theory of the negative binomial and logarithmic series distributions; *Biometrika*, 37, 358–382.

BAILEY, N. T. J., [1964], III.i and ii.

BARTLETT, M. S., [1956], Deterministic and stochastic models for recurrent epidemics; in *Proceedings of the Third Berkeley Symposium on Mathematical Statistics and Probability*, Vol. 4, (Berkeley).

BARTLETT, M. S., [1957], Measles periodicity and community size; *Journal of the Royal Statistical Society, Series A*, 120, 48–59.

BARTLETT, M. S., [1960A], *An introduction to stochastic processes*, (Cambridge).

BARTLETT, M. S., [1960B], *Stochastic population models in ecology and epidemiology*, (London).

BECKMANN, M. J., [1957], On the equilibrium distribution of population in space; *Bulletin of Mathematical Biophysics*, 19, 81–90.

BHARUCHA-REID, A. T., [1960], *Elements of the theory of markov processes and their applications*, (New York).

BLISS, C. I., [1953], Fitting the negative binomial distribution to biological data; *Biometrics*, 9, 176–200.

BLUMEN, I., KOGAN, M. and MCCARTHY, P. J., [1955], *The industrial mobility of labour as a probability process*, (Ithaca).

BRAITHWAITE, R. B., [1960], I.

BROWN, L., [1964], The diffusion of innovation: a markov chain type approach; *Department of Geography, Northwestern University, Discussion Paper*, No. 3.

BRUSH, J. E., [1953], The heirarchy of places in south western Wisconsin; *Geographical Review*, 43, 380–402.

DE CANI, J. S., [1961], On the construction of stochastic models of population growth and migration; *Journal of Regional Science*, 3, No. 2, 1–13.

CHURCHMAN, C. W., ACKOFF, R. L. and ARNOFF, E. L., [1957], *Introduction to Operations Research*, (New York).

CLARK, P. J., [1955], On some aspects of spatial patterns in biological populations; *Science*, 121, 397–398.

CLARK, P. J., [1956], Grouping in spatial distributions; *Science*, 123, 123–125.

CLARK, P. J. and EVANS, F. C., [1954], Distance to nearest neighbour as a measure of spatial relationships in populations; *Ecology*, 35, 445–453.

CLARK, W. A. V., [1965], Markov chain analysis in geography: an application to the movement of rental housing areas; *Annals of the Association of American Geographers*, 55, 351–359.

COLEMAN, J. S., [1964], *Introduction to mathematical sociology*, (New York).

COX, D. R. and SMITH, W. L., [1961], *Queues*, (London).

CURRY, L., [1962], II.

CURRY, L., [1964], The random spatial economy: an exploration in settlement theory; *Annals of the Association of American Geographers*, 54, 138–146.

DACEY, M. F., [1960], A note on the derivation of nearest neighbour distances; *Journal of Regional Science*, 2, 81–87.

DACEY, M. F., [1962], Analysis of central place and point patterns by a nearest neighbour method; In Norborg, K., (Ed.), *IGU symposium in Urban Geography*, (Lund).

DACEY, M. F., [1963A], Order neighbour statistics for a class of random patterns in multidimensional space; *Annals of the Association of American Geographers*, 53, 505–515.

DACEY, M. F., [1963B], The status of pattern analysis: identification of problems in the statistical analysis of spatial arrangement; *Regional Science Research Institute*, (mimeo.), (Philadelphia).

DACEY, M. F., [1964A], Modified Poisson probability law for point pattern more regular than random; *Annals of the Association of American Geographers*, 54, 559–565.

DACEY, M. F., [1964B], Imperfections in the uniform plane; *Michigan Inter-University Community of Mathematical Geographers, Discussion paper*, No. 4.

DACEY, M. F., [1964C], Two dimensional random point patterns: a review and interpretation; (mimeo.), *Department of Geography, Northwestern University*.

DEVLETOGLOU, N. E., [1965], A dissenting view of duopoly and spatial competition; *Economica*, 22, 140–160.

DODD, S. C., [1955], Diffusion is predictable; *American Sociological Review*, 10, 392–401.

DOOB, J. L., [1953], *Stochastic processes*, (New York).

DOUGLAS, J. B., [1955], Fitting the Neyman Type A (two parameter) distribution; *Biometrics*, 11, 149–173.

EVANS, D. A., [1953], Experimental evidence concerning contagious distributions in ecology; *Biometrika*, 40, 186–211.

FELLER, W., [1957], *An introduction to probability theory and its applications*, (2nd edition), (New York).

GARRISON, W. L., [1962], Simulation models of urban growth and development; In Norborg, K., (Ed.), *IGU symposium in urban geography*, (Lund).

GENERAL ELECTRIC COMPANY, [1962], *Tables of the individual and cumulative terms of the Poisson distribution*, (Princeton).
GETIS, A., [1964], III.i and ii.
GOODMAN, L. A., [1961], Statistical methods for the mover-stayer model; *Journal of the American Statistical Association*, 65, 841–868.
GOULD, P., [1963], Dar-es-Salaam, a geographical and queueing analysis of congestion; *East African Economics Review*, 111–120.
GREIG-SMITH, P., [1964], *Quantitative plant ecology*, (London).
HÄGERSTRAND, T., [1953], *Innovationsförloppet ur korologisk synpunkt*, (Lund).
HÄGERSTRAND, T., [1957], II.
HÄGERSTRAND, T., [1963], Geographic Measurements in Migration; in Sutter, J., (Ed.), *Human Displacements, Entretiens de Monaco*, (Monaco).
HÄGERSTRAND, T., [forthcoming], On Monte Carlo simulation of diffusion; In Garrison, W. L., (Ed.), *Quantitative Geography*, (New York).
HAMMERSLEY, J. M. and HANDSCOMBE, D. C., [1964], *Monte Carlo methods*, (London).
HARLING, J., [1958], Simulation techniques in operations research – a review; *Journal of Operations Research*, 6, 307–319.
HOEL, P., [1954 edition], *Introduction to Mathematical statistics*, (New York).
HOGGATT, A. C. and BALDERSTON, F. E., [1963], *Symposium on Simulation models*, (Cincinnati).
HOWARD, R. A., [1960], *Dynamic programming and markov processes*, (New York).
KANSKY, K. J., [1963], III.i and ii.
KENDALL, D. G., [1948A], On the generalised 'birth-and-death' process; *Annals of Mathematical Statistics*, 19, 1–15.
KENDALL, D. G., [1948B], On some modes of population growth leading to R. A. Fisher's logarithmic series distribution; *Biometrika*, 35, 6–15.
KENDALL, D. G., [1949], Stochastic processes and population growth; *Journal of the Royal Statistical Society, Series B*, 11, 230–264.
KENDALL, M. G., [1961], Natural law in the social sciences; *Journal of the Royal Statistical Society, Series A*, 124, 1–16.
KEMENY, J. G., SNELL, J. L. and THOMPSON, G. L., [1956], *Introduction to finite mathematics*, (Englewood Cliffs, New Jersey).
KEMENY, J. G. and SNELL, J. L., [1960], *Finite markov chains*, (Princeton).
KING, J. L., [1962], A quantitative expression of the pattern of urban settlements in selected areas of the United States; *Tijdschrift voor Economische en Sociale Geografie*, 53, 1–7.
KLEINROCK, L., [1964], III.i and ii.
LANDAHL, H. D., [1957], Population growth under the influence of random dispersal; *Bulletin of Mathematical Biophysics*, 19, 171–186.
LOWRY, I., [1964], II.
MAAS, A., HUFSCHMIDT, M. M., DORFMAN, R., THOMAS, H. A., MARGLIN, S. A. and FAIR, G. M., [1962], *Design of water resource systems*, (Cambridge, Mass.).
MILLER, R. L. and KAHN, J. S., [1962], II.
MILLS, G., [1964], Economics and computers; *The Technologist*, 1, No. 4, 15–22.
MOORE, P. G., [1954], Spacing in plant populations; *Ecology*, 35, 222–227.

MORISITA, M., [1954], Estimation of population density by spacing method; *Memoirs, Faculty of Science, Kyushu University*, Series E, No. 1, 187–197.
MORISITA, M., [1959], Measuring of the dispersion of individuals and analysis of the distributional patterns; *Memoirs, Faculty of Science, Kyushu University*, Series E, No. 2, 215–233.
MORRILL, R. L., [1962], Simulation of central place patterns over time; In Norberg, K., *IGU symposium in urban geography*, (Lund).
MORRILL, R. L., [1963B], The development of migration and the role of electronic processing machines; In Sutter, J., (Ed.), *Human displacements, Entretiens de Monaco*, (Monaco).
MORRILL, R. L., [1963C], The development of spatial distributions of towns in Sweden: an historical predictive approach; *Annals of the Association of American Geographers*, 53, 1–14.
MORRILL, R. L., [1965], II.
MUHSHAM, H. V., [1963], Internal migration in open populations; In Sutter, J., (Ed.), *Human displacements*; *Entretiens de Monaco*, (Monaco).
NEYMAN, J., [1939], On a new class of 'contagious' distributions, applicable in entomology and bacteriology; *Annals of Mathematical Statistics*, 10, 35–37.
NEYMAN, J. and SCOTT, E. L., [1957], On a mathematical theory of populations conceived as conglomerations of clusters; *Cold Spring Harbor Symposia on Quantitative Biology*, 22, 109–120.
ORCUTT, G. H., et al., [1961], II.
PORTER, R., [1956], III.i and ii.
RAINIO, K., [1961], A stochastic model of social interaction; *Transactions, Westermarck Society* (Helsinki), No. 7.
SAATY, T. L., [1961], *Elements of queueing theory with applications*, (New York).
SCHOOL MATHEMATICS STUDY GROUP, [1960], *Introduction to matrix algebra*, (Yale U. P. student's text, New Haven).
SFORZA, L. CAVALLI, [1963], The distribution of migration distances: models and applications to genetics; In Sutter, J., (Ed.), *Human displacements*; *Entretiens de Monaco*, (Monaco).
SHENTON, L. R., [1949], On the efficiency of the method of moments and Neyman's Type A distribution; *Biometrika*, 36, 450–454.
SKELLAM, J. G., [1951], Random dispersal in theoretical populations; *Biometrika*, 38, 196–218.
SKELLAM, J. G., [1953], Studies in statistical ecology – 1. spatial pattern; *Biometrika*, 39, 346–362.
SKELLAM, J. G., [1958], On the derivation and applicability of Neyman's Type A distribution; *Biometrika*, 45, 32–36.
SMITH, P. E., [1961], Markov chains, exchange matrices, and regional development; *Journal of Regional Science*, 3, No. 1, 27–36.
STEEGER, W. A. and DOUGLAS, N. J., [1964], II.
TAAFFE, E. J., GARNER, B. J. and YEATES, M. H., [1963], *The peripheral journey to work: a geographic consideration*, (Evanston).
TAKACS, L., [1960], *Stochastic processes*, (London).
THOMAS, M., [1949], A generalisation of Poisson's bionomial limit for use in ecology; *Biometrika*, 36, 18–25.

THOMPSON, H. R., [1956], Distribution of distances to *n*th neighbour in a population of randomly distributed individuals; *Ecology*, 37, 391–394.
WILLIAMSON, E. and BRETHERTON, M. H., [1964], *Tables of the negative binomial distribution*, (New York).

Section IV - Applications

BECKMAN, M. J., [1957], III.iii.
BROWN, L., [1965A], A bibliography on spatial diffusion; *Discussion Paper*, No. 5, Department of Geography, Northwestern University.
BROWN, L., [1965B], Models for spatial diffusion research – a review; *Technical Report* No. 3, Spatial Diffusion Study, Department of Geography, Northwestern University.
BUSH, R. R. and MOSTELLER, F., [1955], *Stochastic models for learning*, (New York).
CHERRY, C., [1957], *On human communication*, (New York).
COLEMAN, J. S., [1964], III.iii.
COLEMAN, J. S., KATZ, E. and MENZEL, H., [1957], The diffusion of an innovation among physicians; *Sociometry*, 20, 253–270.
CYERT, R. M. and MARCH, J. G., [1963], *A behavioural theory of the firm*, (Englewood Cliffs, New Jersey).
DODD, S. C., [1955], III.iii.
ESTES, W. K., [1960], A random walk model for choice behaviour; in Arrow, K. J., Karlin, S., and Suppes, P., (Eds.), *Mathematical methods in the social sciences*, (Stanford).
ESTES, W. K., [1963], Towards a statistical theory of learning; In Luce, R. D., Bush, R. R. and Galanter, E., (Eds.), *Readings in mathematical psychology, Volume 1*, (New York).
GOULD, P., [1965], A bibliography of space searching procedures for geographers; *Research Note, Department of Geography, Pennsylvania State University*.
HARVEY, D. W., [1966], Geographic processes and the analysis of point patterns: testing a diffusion model by quadrat sampling; *Transactions and Papers, Institute of British Geographers*, 40, 81–95.
HÄGERSTRAND, T., [1953], III.iii.
ISARD, W. and DACEY, M. F., [1962], On the projection of individual choice behaviour in regional analysis, Parts 1 and 2; *Journal of Regional Science*, Vol. 4, No. 1, 1–34, and Vol. 4, No. 2, 51–58.
KARLSSON, G., [1958], *Social mechanisms*, (Uppsala).
LANDAHL, H. D., [1957], III.iii.
MACHLUP, F., [1962], *The production and distribution of knowledge in the United States*, (Princeton).
MANSFIELD, E., [1961], Technical change and the rate of imitation; *Econometrica*, 29, 741–766.
MARCH, J. G. and SIMON, H. A., [1958], *Organisations*, (New York).
MEIER, R. L., [1962], *A communication theory of urban growth*, (Cambridge, Mass.).
PITTS, F. R., [1963], Problems in computer simulation of diffusion; *Papers of the Regional Science Association*, 11, 111–119.
PRED, A., [1965], I.

RADNER, R., [1964], Mathematical specification of goals for decision problems; In Shelly, M. W. and Bryan, G. L., (Eds.), *Human judgements and optimality*, (New York).
RAINIO, K., [1961], III.iii.
RAPOPORT, A. and REBHUN, L. I., [1952], On the mathematical theory of rumour spread; *Bulletin of Mathematical Biophysics*, 14, 375–383.
RAPOPORT, A., [1953], I.
ROGERS, E. M., [1962], *Diffusion of innovations*, (New York).
RUSHTON, S. and MAUTNER, A. J., [1955], III.i and ii.
SAUER, C., [1963], I.
SHELLY, M. W. and BRYAN, G. L., [1964], (Eds.), *Human judgements and optimality*, (New York).
SIMON, H. A., [1957], *Models of man*, (New York).
SKELLAM, J. G., [1951], III.iii.
SUPPES, P. and ATKINSON, R. C., [1960], *Markov learning models for multiperson interactions*, (Stanford).
WEBB, W. B., [1931], *The Great Plains*, (Boston).
WOLPERT, J., [1964], Decision making in spatial context; *Annals of the Association of American Geographers*, 54, 537–558.
YUILL, R. S., [1964], A simulation study of barrier effects in spatial diffusion problems; *Geographical Branch Office of Naval Research, Task No. 389–140, Technical Report, No. 1*.

CHAPTER FIFTEEN

Network Models in Geography
PETER HAGGETT

Handeln vom Netz, nicht von dem, was das Netz beschreibt.
L. WITTGENSTEIN, *Tractatus Logico-Philosophicus*, 1922, 6, 35.

This essay treats a wide range of 'linear' features traditionally studied within physical or human geography – rivers and railways, parishes and polygons – from the viewpoint of their basic *geometrical* characteristics. The neglected role of geometry in relation to geography has been argued in an earlier volume in the Madingley series (Haggett and Chorley: In Chorley and Haggett, 1965, Chap. 18) and this is not the place to labour this theme. Suffice to say that it follows a path blazed by Bunge (1962) amongst others who in his *Theoretical Geography* vigorously challenged the traditional dichotomy of physical and human geography. His parallels, such as those between anastomozing streams and shifting highway-courses has yet to be worked out, but mathematicians Lighthill and Whitham (1955) have already shown how continuity equations for kinematic waves may serve equally well to describe the flux and linear concentrations of motors moving along a crowded highway or clastic debris moving down a stream bed. Thus in taking Wittgenstein's advice we are looking for the basic geometrical features in a general class of 'geographical networks' that might, at first sight, appear to have rather little in common.

The organization of this chapter is simple. It is one of ascending complexity: viz., the first section treats models of the single path, the second models of the network without circuits (i.e. tree-like structures), the third models of the network *with* circuits, and the fourth models of the cellular network. There are fundamental differences in this last section and the first three in that while they are concerned with lines *along* which flows are directed, the cellular net consists of 'containing' lines or barriers, *across* which flows move. A fifth section considers the possibility of converting some naturally-occurring complex nets into simpler structures (Table 15.1).

[610] MODELS IN GEOGRAPHY

TABLE 15.1

Topological Classification of Networks

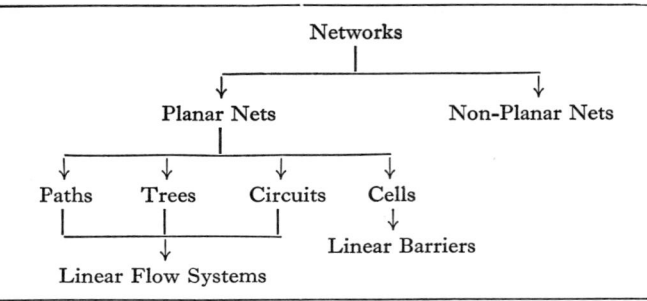

It should be emphasized at the outset that the treatment here is naïvely confined to real systems of lines on the earth's surface, i.e. geographical networks *sensu strictu*. The very wide applications of network theory to non-planar graphs is also of relevance to geographical research (e.g. in the organization of regional taxonomic units, or in critical-path analysis of field research or teaching procedures) and is treated in Haggett and Chorley (In preparation). Excellent general accounts of these aspects of network analysis are available in Avondo-Bodino (1962), Battersby (1964), Busacker and Saaty (1965), Harary, Norman and Cartwright (1965), and Cartwright (In: Haire, 1959, pp. 254–271).

I PATH GEOMETRY

The simplest component of a geographical network, the single line or path, would appear to pose few problems or provide much scope for worthwhile analysis. Yet, perhaps because of its fundamental character as the 'building block' of complex networks (Fig. 15.1), both the *location* and the *form* of the

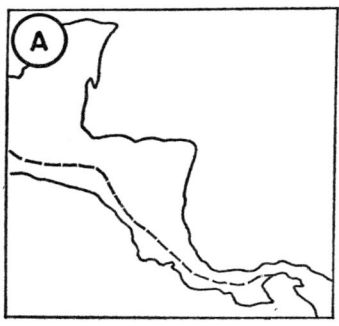
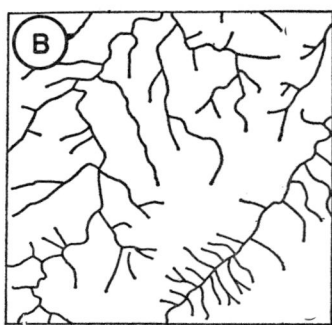

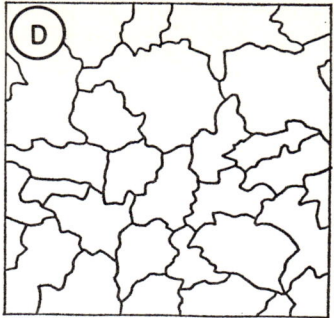

15.1 Four geographical examples of alternative topological classes of networks: (A) *paths* – the course of the Pan American Highway in Central America; (B) *trees* – stream courses, Viti Levu, Fiji; (C) *circuits* – road networks, Manawatu area, New Zealand; (D) *cells* – administrative areas, central Nigeria.

single line are surprisingly difficult to explain. Here we concern ourselves with simple examples of the locational problems posed by *direct* paths and the form problems posed by *wandering* paths. In each case our approach is expository rather than rigorously analytical; the mathematical fundamentals of much of the discussion is available in Lyusternik (1964).

1 Direct paths: locational problems

The naïve but fundamental assumption that the shortest direct distance between two points on a plane is a straight line, tends to disguise the real complexity of path-point location problems. Here we examine the extension of the problem to paths connecting three points, and go on from this to consider the more difficult N-point cases. This leads to the more complex cases posed by the location of lines running through pre-existing networks and across areas of different 'resistance'. These three different cases are geometrically similar to the problem of locating lines with respect to point-sets, line-sets and area-sets.

(a) *Point-set problems.* The simple problem of connecting three points on a plane by the shortest path has intrigued geometers since the ancient Greeks. Polya (1954) has examined the problem of the location of the shortest path between two villages which are situated near the bank of a stream which allows water to be collected *en route*. There are two solutions: (1) for a *straight* stream the solution is equivalent to that of a light-ray reflected in a mirror; (2) for an *irregular* stream the optimum water-collecting point is where the smallest ellipse that can be drawn (with the two villages as foci) touches the streambank.

Miehle (1958A, p. 232) has drawn attention to Jacob Steiner's solution of a comparable problem, that of connecting three villages by a road network of minimum total length. Steiner found that study of the angles of the triangle formed by the three villages yielded the solution. If one of the angles of the triangle is equal to or greater than 120° then the villages at the other vertices should be directly connected to this vertex. If, on the other hand, all angles in the triangle are less than 120° the solution is to connect all three villages to a junction point so located that all three roads meet at 120° angles (Fig. 15.2A).

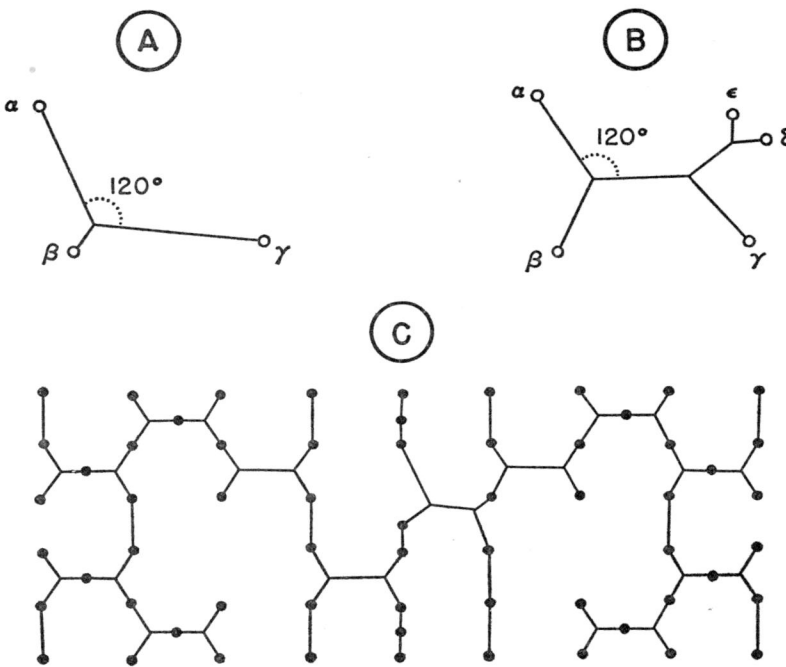

15.2 Shortest-distance paths for (A) three villages (Steiner's problem), (B) five villages, and (C) a sixty-two point weapons system (*Sources: Bunge, 1962; Miehle, 1958A*).

The general case of the shortest route between a set of N points may be studied as an extension of Steiner's problem. For N points it may be shown that the *maximum* number of junctions at which three lines meet at 120° is $N-2$. For a large set of points the mathematical solution involves the solution of sets of equations of the form

$$\frac{\partial D}{\partial x_A} = m_{A1}\frac{(x_A-x_1)}{d_{A1}} + m_{A2}\frac{(x_A-x_2)}{d_{A2}} \ldots + m_{AB}\frac{(x_A-x_B)}{d_{AB}} + \ldots = 0$$

$$\frac{\partial D}{\partial y_A} = m_{A1}\frac{(y_A-y_1)}{d_{A1}} + m_{A2}\frac{(y_A-y_2)}{d_{A2}} \ldots + m_{AB}\frac{(y_A-y_B)}{d_{AB}} + \ldots = 0$$

where x_i, y_i are the co-ordinates of point i, where m_{ij} is the weighting factor in the link connecting the two points, where D is the total distance to be minimized, and where

$$d_{ij} = \sqrt{(x_i-x_j)^2+(y_i+y_j)^2}$$

(Miehle, 1958A, p. 237). Fixed points are designated by numbers and movable junctions by letters. Since the equations are non-linear an iteration method is used where the positions of the movable junctions $(A, B, \ldots)$ converge on the minimum solution. Clearly the numerical analysis of the shortest-path problem is complex and becomes increasingly so when constraints (e.g. minimum separation distances) are placed on the junction locations: the consequent introduction of Lagrange multipliers demands advanced calculus.

As a result of the mathematical complexity a number of 'hardware' analogue models have been developed to give more direct solutions. Two of these methods the *mechanical link-length minimizer* and the *soap-film method* have been described at length by Morgan in Chapter 17 of this book (pp. 768–771; especially Plates 17.12 and 17.13). The soap-film method creates the necessary junctions and their locations by forming shapes which automatically reduce their area to a minimum. However constraints on junction spacing and weighting of links is not possible and solutions for large numbers of points are not unique. In practice therefore the mechanical method has proved more workable as, despite its bulky appearance and inconvenient operation, large systems of points can be treated with integral weights assigned to individual links and very rapid solutions of reasonable accuracy obtained. Silk (1965) made extensive use of a 33-point model in a study of optimal road designs for the county of Monmouthshire. Here fixed-pegs located on a 1/63,360 base board represented the major urban centres within the county.

All solutions, mathematical, soap-bubble, and mechanical, show the characteristic 120° junctions inherent in Steiner's three-point case. Figure 15.2 shows theoretical minimum link-length minimization for networks, including that for sixty-two weapon units in a military communications system (Miehle, 1958A, p. 236). Here an original connecting scheme had a total path length of 121·4 units. This was reduced to only 68·8 units by using the Steiner solution shown (Fig. 15.2C).

The problem of locating a route between sets of points has been studied in the practical context of railroad location by Wellington (1887). Wellington was a mining engineer who, employed on the planning of the Mexican railroad system, widened his findings to include general cases of route location. The practical problem as he saw it was to optimize the relationship between the total length of a path (i.e. the length of railroad to be constructed) and the number of points connected (i.e. the towns connected by the path). Clearly the objective in the first case was to shorten the line and in the second to connect as many places as possible (and thereby carry their traffic). From the data Wellington was able to assemble on contemporary railroad practice, he put forward three basic propositions: (1) that if all points intermediate between the two terminals of a railroad were of equal freight-generating capacity and if they were equally spaced, then the traffic varied as the square of the number of points served; (2) that if the intermediate points were 'small country towns' without competing alternative railroads, then the effect was to reduce the traffic by 10 per cent for every mile that the station was removed from the town; and (3) that if the intermediate points were 'large industrial cities' with competing railroad facilities then the loss would be still more abrupt: a reduction of 25 per cent for every mile that the station was removed from the town. Although these figures have now only antiquarian interest, and may have been quite inaccurate for contemporary conditions, they do represent an interesting early attempt to wrestle with the problem of path deviation between points. Indeed Wellington's attempt should be seen against the present state of research which still lacks a rule '... relating to the deviation of paths between major places in order to pass through intermediate places' (Garrison and Marble, 1962, p. 85). Where the ratio of fixed/variable costs is greater than one, then we might expect that the track mileage would be kept as low as possible (i.e. little deviation of the path to connect intermediate points) and vice versa. Bunge (1962, pp. 183–189) has extended this idea with reference to regional contrasts in the railroad pattern in the United States.

(b) *Line-set problems.* Perhaps the most famous puzzle in graph theory concerns a path along sets of lines. The great Swiss mathematician, Euler, began a paper on graph theory by discussing the Königsberg bridge problem in which the problem is to determine a path such that starting from any of the four parts of the city (α, β, γ and δ) one (i) crosses each river bridge only once and (ii) returns to the same point. The city of Königsberg (now Kaliningrad in East Prussia) is shown on Figure 15.3A and a schematic graph of the relationships in Figure 15.3B. Euler showed that such a path was impossible in this case and that a solution would require an even number of edges at each vertex (Ore, 1963, p. 24). Other problems followed and in 1859 an Irish mathematician, Sir William Hamilton, marketed an unusual topological puzzle. It consisted of a regular dodecahedron made of wood. Each of the twenty corners of the dodecahedron was marked with the name of an important city (Brussels, Delhi, etc.) and a large nail. The puzzle consisted in finding a route with a piece of string wound round the nails such that (i) the route began and ended at the same city and (ii) all other cities were visited once only. Although Hamilton's game must be presumed a financial failure (Ore, 1963, p. 29), his name has been preserved in the term *Hamilton line* describing a closed loop (edge train) that passes through a sequence of vertices exactly once.

Although Euler and Hamilton's puzzles may seem naïve they represent the significant forerunners to a set of highly complex problems which today occupy more computer time than any other locational analyses. For the modern analogues concern practical problems of enormous economic and military significance including the marginal costs of tanker fleets, troop convoys and a widening range of road distribution systems. The search for optimum paths is a complex and growing field of operational research: here we confine ourselves to three broad classes of locationally significant problems – shortest-route paths, maximum-flow paths, and minimum-cost paths.

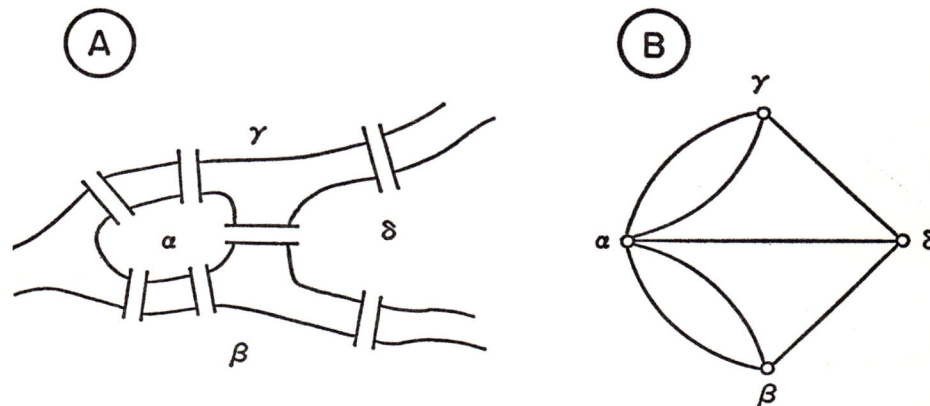

15.3 Euler's Königsberg bridge problem (*Source: Ore, 1963, pp. 23–24*).

(1) Although the *shortest-route* path through a network would seem for small graphs to be a simple matter of string and patience, it becomes surprisingly complicated and tedious where (a) very large numbers of links are involved, (b) the distance matrix is assymetric (i.e. some links are one-way only or longer in one direction than another) or (c) a number of vertices have to be joined in a single-shortest-route loop (the *travelling salesman* problem). Pollack and Wiebenson (1960) in a review of the solutions put forward to this problem draw a general distinction between computational-mathematical solutions such as the work by Moore (1959) at Harvard on the shortest paths through a maze, and analogue solutions like Rapaport and Abramson's (1959) model in which electric or electronic 'timers' are substituted for distance and the shortest route is shown by a set of illuminated links. The basic problem in both kind of studies is to reduce the long and expensive computations needed. For example, Little, Murty, Sweeney and Karel (1963) report that solutions for simple travelling salesman problems may become very costly in computer time when large numbers of cities are involved: using an IBM 7090 computer they found that while 'ten-city' paths could be derived

in from one to three seconds 'twenty-five-city' paths needed from four minutes to over one hour! Continuing research is producing still more efficient algorithms for the solution of this sort of problem (e.g. the 'Cascade' algorithm developed by Farbey, Land and Murchland (1965) at the London School of Economics and the Shen Lin algorithm developed at the Bell Telephone Laboratories). These together with the improving speed and capacity of both digital and analogue computers (i.e. improvements in both mathematical 'software' and electronic 'hardware') are converging to produce ever more rapid solutions to ever more complex shortest-route paths. Since many problems of shortest-path location remain simple in principle but tedious and time-consuming in application, this reduction in time brings some complex geographical networks into the field of possible study. Already one geographical study, that of Boye (1965) on optimum collecting routes around the Swedish town of Uddevala, has drawn on computer optimizing programmes and the potential for the future study of stubborn locational problems within geography is immense.

(2) A second group of problems is concerned with *maximum-flow paths* through networks. Here the capacities of the links in the network are constrained as in Figure 15.4A (Akers, 1960, p. 312) and the problem is to find

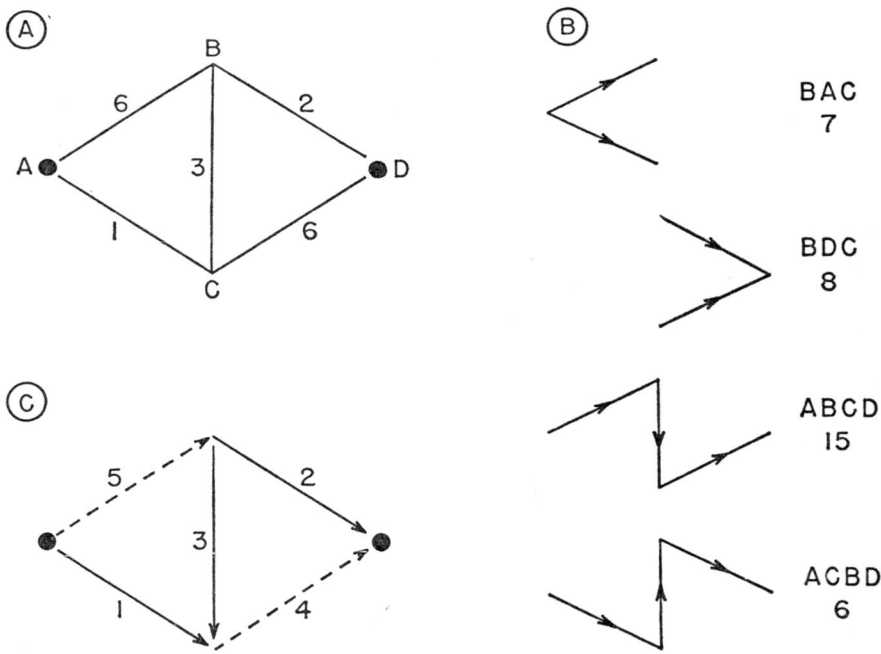

15.4 Derivation of the min-cut max-flow path for a simple network (*Source: Akers, 1960, p. 312*).

the path (or paths) that maximum flow must follow. The problem of defining maximum 'flow' was solved in 1955 by Dantzig and Fulkerson through their 'max flow – min cut' theorem which showed that the maximum flow through a network was equal to the sum of the capacity of the branches of the minimum cut. A 'cut' is any collection of branches which completely separates two terminals in the network. Thus is Figure 15.4 we can see that there are four possible cuts which separate A from D (these are plotted in Fig. 15.4B), and that the minimum cut is the line $ACBD$ with a total capacity of 6 units. This is also the maximum flow through this simple network (Fig. 15.4C).

Locational selection of the paths which the maximum flow must follow posed few problems in this simple case. All links were employed although two of them (shown by dotted lines) were carrying less than their total capacity. A number of algorithms for finding maximum-flow paths through very complex networks have been evolved (Ford and Fulkerson, 1962), but these are largely concerned with steady-state flow. Kleinrock's (1964) work has been concerned, on the other hand, with paths through networks where the flow is not steady but fluctuating and rather unpredictable (e.g. messages through telephone networks): here stochastic simulation models for alternative arrangements of the network have proved valuable. Work on paths of this kind have direct analogies with flood-routing problems on large river control networks (see Chap. 4 by More, pp. 175–179 above).

(3) A third and more complex group of line-set problems consists of *minimal-cost flow paths* through networks. Ford and Fulkerson (1962, Chap. III) in an outstanding survey, suggest that something like half the time spent on industrial and military applications of linear programming were concerned with this sub-set of problems. Because of their vast economic and military significance, a considerable body of sophisticated mathematics has been built-up to structure these studies, and practical algorithms and computer routines based on both operations research and classical electrical-network studies (e.g. the Maxwell-Kirchoff problem) have been derived. Here we do little more than point out the nature of the problem and the kind of path solutions that may be determined for them.

Figure 15.5A illustrates a simple problem in deriving minimal-cost paths through a network while obtaining maximum flow (Ford and Fulkerson, 1962, pp. 123–127). The network consists of a set of 21 links (edges) joining the two terminal points (α, β). Figure 15.5A shows the maximum flow capacity of each link; Figure 15.5B the unit shipping cost. If we assume our problem is to ship goods from α to β then it is possible to compute that a positive minimum-cost flow will get through the network to β by utilizing the paths shown by the arrows in Figure 15.5C. However for the maximum flow through the network (given the cost and capacity constraints already referred to) almost all the links are utilized (Fig. 15.5D). Two points worth noting are (i) that not all the links along the paths operate to their full capacity (links not fully 'saturated' with traffic are shown by broken lines) and (ii) that the direction of flow along some of the links may change as total flow through the network increases.

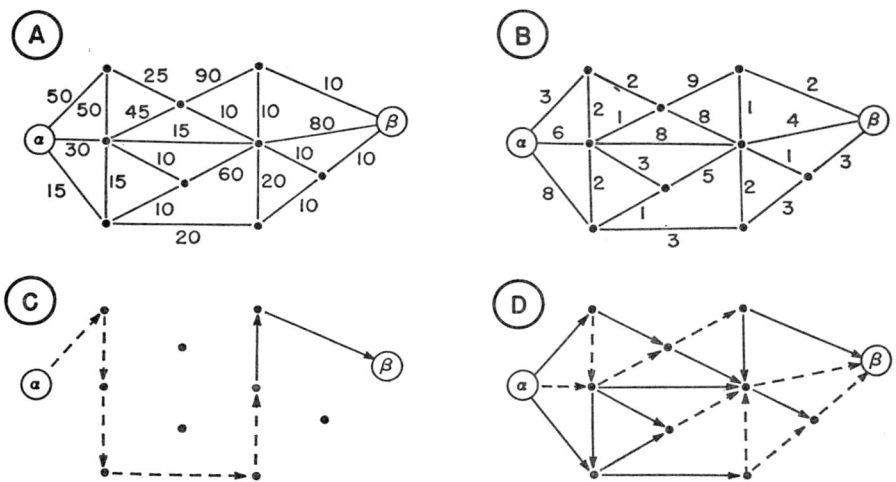

15.5 Minimal-cost flow through a complex network: (A) maximum flow capacity of each link; (B) unit shipping costs along each link; (C) initial flow pattern; (D) maximum flow pattern. Links not fully 'saturated' with flow shown by broken lines. (*Source: Ford and Fulkerson, 1962, pp. 123–127*).

(c) *Area-set problems.* The third case of path location concerns lines across areas of varying 'resistance' to movement. Wardrop (In: Herman, 1961, pp. 57–78) has examined the problem of the shortest distance between two points on a plane. If we assume that costs of movement from the first point (A) are directly proportional to distance and isotropic then the shortest distance to B is the straight line AB (Fig. 15.6). If however, a low-cost path (e.g. a motorway) is introduced (line CB), then the 'shortest' path across the plane in terms of cost will be line ACB. By making reasonable assumptions about relative costs of travel it can be shown that A lies within a region bounded by a hyperbola (Region α) where it is cheaper to use the motorway in travelling to B than using a direct overland path. A similar region (Region β), also bounded by a hyperbola, can be demarcated which indicates those points that can be most cheaply reached from A by paths using the motorway CB. Wardrop goes on to develop more complex cases on rectangular grids, and Coburn, Beesley, and Reynolds (1960) to apply more realistic analyses to the effect of the London–Birmingham motorway on optimum paths from points in the general vicinity of the new highway.

Lösch (1954, p. 184) extended the problem by considering the problem of optimum paths between two points on a plane which is divided into areas of different movement costs, e.g. land and sea, mountain and plain, etc. He noted the similarity of this problem to that of a light ray moving through media of different refractive values, and suggested that Snell's Law of Light

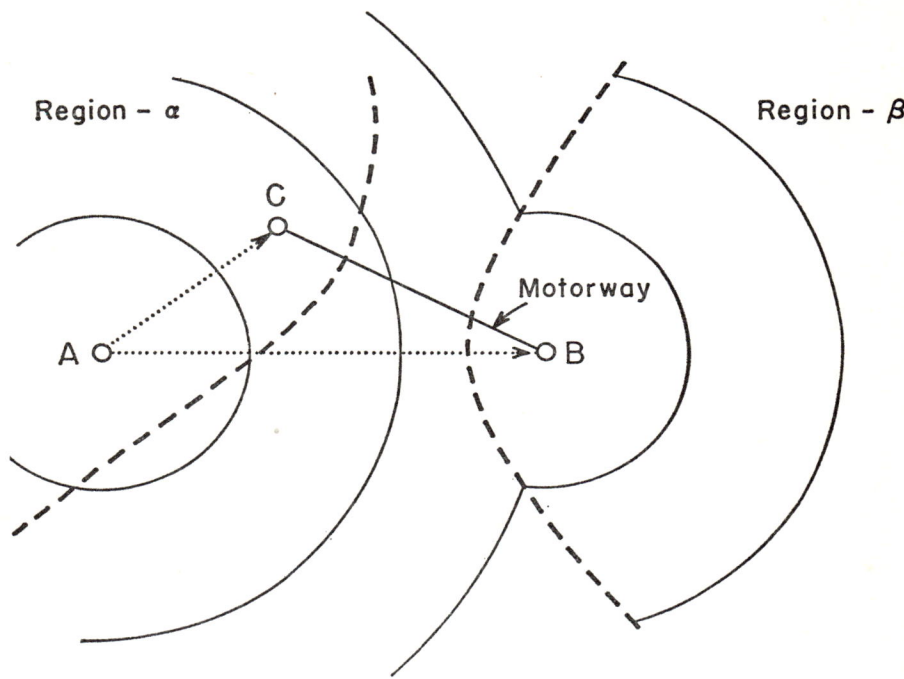

15.6 Impact of a new motorway on shortest-distance paths (*Source: Wardrop – In Herman, 1961*).

Refraction could be extended to the path location problem. For the simplest case, two adjacent areas, of different resistances, Lösch shows that the 'elbow' or inflexion point on the line joining two points will lie on the boundary between the two areas at the point where

$$f_1 \sin x - f_2 \sin y = 0$$

Here f_1 and f_2 refer to the relative resistance to movement across the two areas (i.e. freight rates) while x and y are the angles that the two sections of the path make with a line orthogonal to the boundary between the area (i.e. orthogonal to the coastline in the simple land/sea case). Haggett (1965, pp. 63–65) applies Lösch's rule to more complex cases and suggests empirical cases of routes that appear to substantiate the refractive model.

Both Wardrop's motorway problem and Lösch's refractive problem may

[620] MODELS IN GEOGRAPHY

be regarded as special cases of geodesic theory. Thus Warntz (1965) has shown the general rule for least-cost paths (geodesics) is to follow the trajectory orthogonal to the isocost lines on a surface. Figure 15.7 shows the isocost surface integrated about a Tennessee town, Murfreesboro (α), where the contours indicate the total cost of acquiring land on a route from this point to any other given point in the United States. On this basis, six least-cost routes from Murfreesboro to other parts of the country are shown. Note that the routes are orthogonal to the contours and that the paths to New York (β) and to Boston (γ) follow divergent paths. The fact that, because of variations in land acquisition costs, the least-cost paths are not straight lines

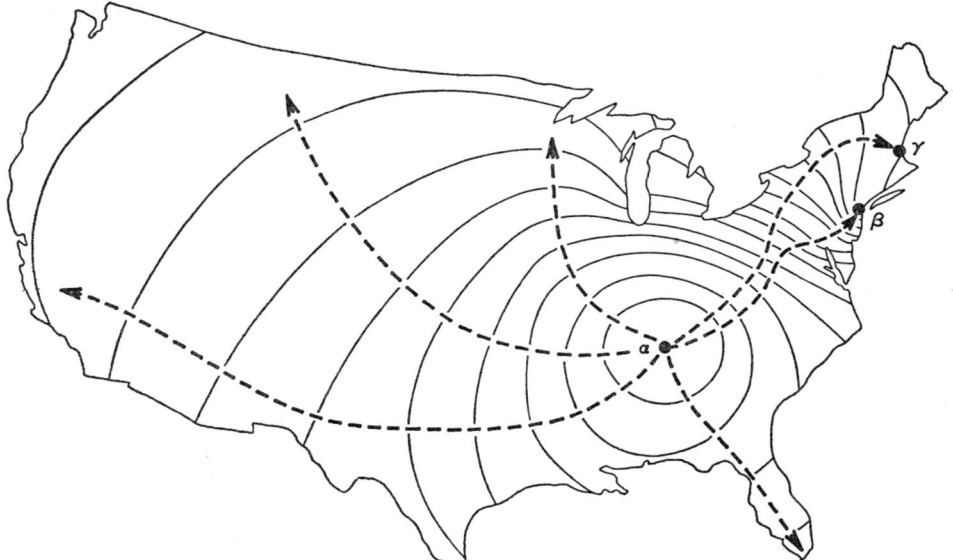

15.7 Geodesic paths on an iso-cost surface (*Source: Warntz, 1965, p. 17*).

may be demonstrated by converting the surface shown in Figure 15.7 into a three-dimensional plaster model with point α as the pit. If we place a ball on that surface it will roll 'down hill' towards Murfreesboro and (as long as the mass of the ball is sufficiently small and the gradients on the surface sufficiently gentle) its path will accurately trace the least-effort (=least cost) paths shown by the lines on the map (Warntz, 1965, pp. 15–22).

Geodesic paths have considerable practical implications. Warntz (1961) has studied the routes used by DC-8 aircraft flying across the North Atlantic from New York to cities in western Europe. Here the complex pressure systems encountered 'deflect' aircraft from the Great Circle (=least distance) paths to more complex and variable routes optimized in relation to time and fuel costs. Other notable historical studies of geodesic paths are Vance's (1961) contrast of the lines of the Oregon Trail and the Union Pacific Railroad across the Rocky Mountains, and Momsen's (1963) study of the changing role of the Serra do Mar in deflecting least-cost paths between Rio de Janeiro and São Paulo in southeast Brazil.

2 'Wandering' paths: form problems

The form of the 'rolling English road' and the meandering stream both suggest lines which vary or wander in direction rather than proceeding directly from their apparent origin to their apparent destination. For road patterns it may be possible to explain this wandering in terms of their history (e.g. the conversion of paths around the edges of fields to paved inter-village roads) and we may regard most of these as special cases of the point-set location problem. For rivers however the problem is more complex and there is growing evidence that their wandering is *periodic* in character, related perhaps to spatial 'resonance' in the flow system. This section examines some of the parameters that have been put forward to describe these wandering paths and the morphological implications of the results they show.

(a) *Descriptive parameters*. One of the simplest measures proposed to describe path wandering is to relate the length of the *observed* path (O_L) to the length of the *expected* path (E_L) where this is measured as the direct 'desire-line' distance between the two ends of the path. Thus Schumm (1963, p. 1089) has proposed descriptive categories of channel sinuosity which range from 'straight' courses ($O_L/E_L = 1\cdot 00$) through three intermediate classes – 'transitional', 'regular', and 'irregular' – to 'tortuous' courses ($O_L/E_L > 2\cdot 00$). There is, in practice, some difficulty in applying this index to river channels since it is not always clear where the terminating points for the observation should be located. For transport lines the terminal points are usually clearly defined and Kansky (1963, pp. 31–32) has proposed a general formula for a connected network of lines, viz.

$$\left\{ \sum_{i=1}^{n} \left(O_L - E_L \right)^2_i \right\} \Big/ V_n$$

where V_n is the number of vertices connected.

Studies on individual meanders and meander trains has led to investigations of the repeating elements in stream paths. Attempts have been made to define meanders in terms of their wavelength, amplitude, and radius of curvature (Leopold, Wolman and Miller, 1964, p. 295) but in practice meanders are notoriously difficult to define and slight differences in operational definition yield significant differences in the results obtained. These difficulties led Speight (1965) to investigate the application of *power-spectra* mathematics to the continuously curving form of the river. Here changes in the angle of direction of the channel measured at standard distances along the thalweg are substituted for measurements on individual 'waves'.

Spectral analysis avoids the implication of earlier meander studies that there is a single dominant wavelength of meandering discernable (even though it may be obscured by irregularities of a quasi-random nature) and substitutes the idea of a 'spectra' of oscillations measured against distance downstream. Figure 15.8A shows a characteristic thalweg trace analysed into a

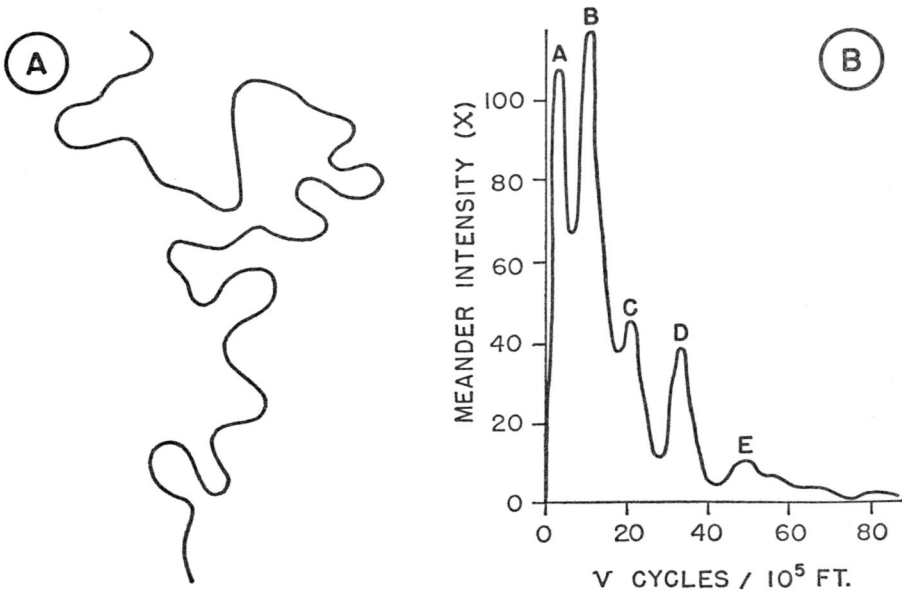

15.8 Analysis of (A) meander trains on the Angabunga river, New Guinea, in terms of (B) a power spectrum (*Source: Speight, 1965, p. 4*).

'spectral envelope' (Fig. 15.8B) containing a number of coherent oscillatory peaks. Variations in meander intensity $X(k)$ were calculated from the formula

$$X(k) = \frac{1}{n}\left[C_x(o) + \sum_{l=1}^{m-1} \left\{ C_x(l)\left(1 + \cos\frac{\pi l}{m}\right) \cos\frac{\pi k l}{m} \right\} \right]$$

where n is the number of direction angles measured (between regularly spaced points on the thalweg trace), C_x is the auto-correlation between successive angles (X), and M is the number of frequency bands, (Speight, 1965 p. 5). Values of X were plotted against the frequency in cycles per 10^5 foot (this frequency being inversely related to wavelength). We can interpret the meander spectra shown in Figure 15.8B as consisting of two intense peaks with a wavelength of about four miles (peak *A*) and two miles (peak *B*), with a series of smaller peaks at about 4/5 mile (*C*), 3/5 mile (*D*) and 2/5 mile (*E*). Although the reliability and resolving power of spectral analysis hinge on the number of direction angles measured, the spacing of these measurements, and the number of frequency bands used in producing the spectra (Blackman and Tukey, 1959) it is clear that, with the availability of digital computers, power-spectra analysis is likely to be increasingly used in the analysis of wandering paths.

(**b**) *Interpretation of morphology.* The aphorism that '. . . nature abhors a straight line' throws little interpretive light on the existence of meandering streams. The general discussion of meander theory lies outside this essay (see the summary in Leopold, Wolman and Miller, 1964, Chap. VII) but the characteristic multi-peaked spectrum of meanders suggests that ' . . . the river, like a musical instrument, may be considered as a resonant oscillatory system, the most notable difference being that the oscillations represent

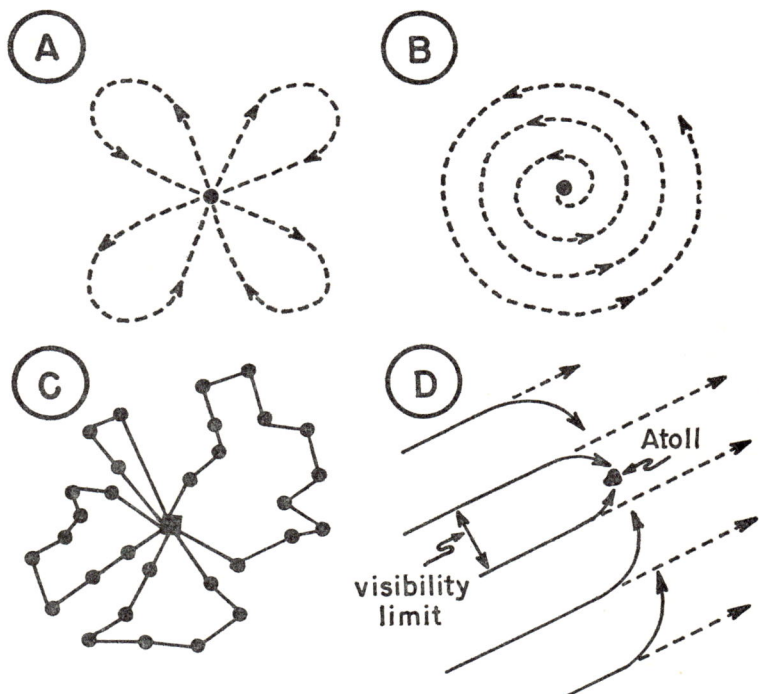

15.9 (AB) Idealized 'search' patterns. (C) Local air services around Alice Springs, Central Australia. (D) Parallel search paths of outrigger canoes moving between atolls in the Marshall Islands group, western Pacific (*Source: Bunge, 1964*).

fluctuations relative to *distance* rather than to *time*' (Speight, 1965, p. 13). Both systems require energy and it is clear that the main source of energy available to rivers is the potential energy which stems from their elevation above base level. Once a stable oscillation has been introduced into a river system it tends to be self-perpetuating, factors like bank-caving leading to the feedback of oscillatory energy into the system.

There is little evidence at the moment that the findings on meander oscillation can be transferred to other geographical networks. Certainly there are direct parallels with other *physical* phenomena (e.g. waves on polar fronts or jet-stream traces) but the oscillations in traffic flows on highways are clearly not lateral, as in meander swinging, but rather vertical, i.e. periodic changes in traffic density. Oscillations in traffic data have already been studied in terms of equations of continuity (Lighthill and Whitham, 1955; Herman, 1961) and some convergence with meander theory is possible here. Certain oscillatory paths in human geography may be aided by *search theory* developments. There is, as Bunge (1964, pp. 15–16) pointed out, some regularity discernable in the trace of hunters with the daily need to return to the central camp for food, the pattern of deliveries from a central store, and so on. Two optimal patterns are suggested in Figure 15.9 with the oscillating 'sweep' pattern about a central base (A) and the 'search' pattern from a central base with no need to return (B). Further ideas on the form of geometrical search patterns are contained in operations research journals such as the *Naval Research Logistics Quarterly* (e.g. Isbell, 1957; Gluss, 1961) and their significance for geographical networks needs vigorous exploration. A remarkably wide-ranging and useful bibliography of space-searching procedures of relevance to geographers has been compiled by Gould (1965).

II TREE GEOMETRY

The second main class of geographical networks are tree-like structures (Fig. 15.1B). These consist of sets of connected lines (rather than single lines) without any complete loops; in topological terminology they are '... connected graphs without circuits' (Ore, 1963, p. 130) and are known simply as *trees*. The most important and widely occurring tree-systems on the earth's surface are rivers. Apart from their anastomozing sections (which form closed loops) river systems have all the properties of the topologists' tree; in particular they follow the rule that trees consist simply of lines (usually known as *edges, arcs, routes* or *one-cells* in topological writing) successively added to existing lines, so that a tree with n junctions (*vertices, nodes* or *zero-cells*) has $n-1$ edges. As Figure 15.10 shows however this topological definition of a tree includes many of the forms already discussed as 'single paths' (e.g. the single line connecting a set of points is a particular kind of topological tree known as an 'open edge train') and in this section we confine our attention to those trees which, like the example in Figure 15.10D and 15.10E, have a recognizably dendritic structure.

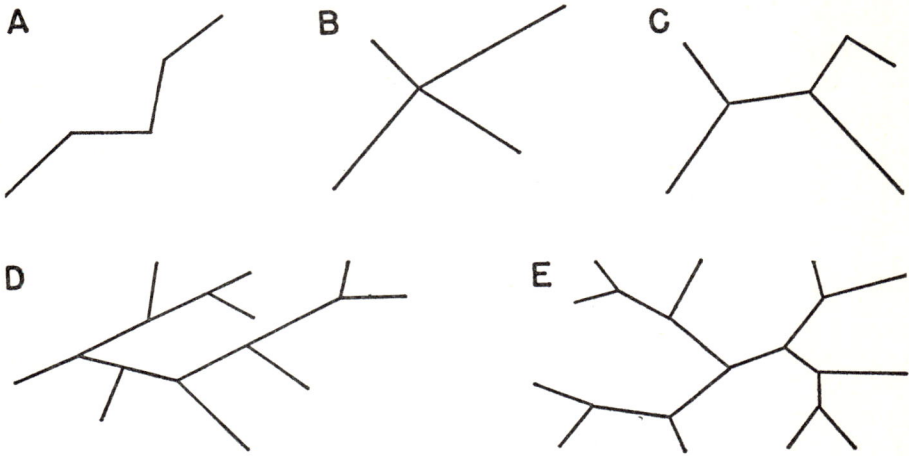

15.10 Alternative geometrical forms of 'tree' topologies (*Source: Ore, 1963*).

The problems posed by these tree-like structures are different from those posed by paths. Since there is only one path connecting any pair of vertices in a tree, 'shortest path' locational problems are here less relevant, while the form problems posed by the irregular line of stream segments (i.e. the meander problem) have already been treated. The accent shifts from location and form in this section therefore to *structure* and *growth*: here questions are directed towards the intricate concepts on network order and the interrelationship of its several parts, and towards the changes in order and balance over time.

1 Trees as ordered structures

In the recent history of geomorphology growing attention has been paid to the notion that streams are orderly and predictable phenomena. Much of this advance hinges on the ordering system put forward by Horton (1945) and from the regularities that this system showed up. Although this section is concerned with the ordering system and its implications as applied to streams, Section V of this chapter suggests Horton's work may be extended to other 'transport' systems.

(a) *Tree-ordering systems.* In the study of one major set of geographical trees, stream patterns, it was the solution of the ordering system by Horton

(1945, pp. 281–282) which provided the touchstone for a series of advances in quantitative geomorphology. Horton's method of stream ordering depended on the determination of 'parent' streams and 'tributary' streams working upstream from the mouth of the system. This decision was made on the basis of two rules: (1) that the stream joining the parent stream at the greater angle at a bifurcation point is the tributary, and (2) that if both streams at the bifurcation point join at about the same angle, the shorter is the tributary stream. Figure 15.11A shows the system applied to a hypothetical stream. Horton

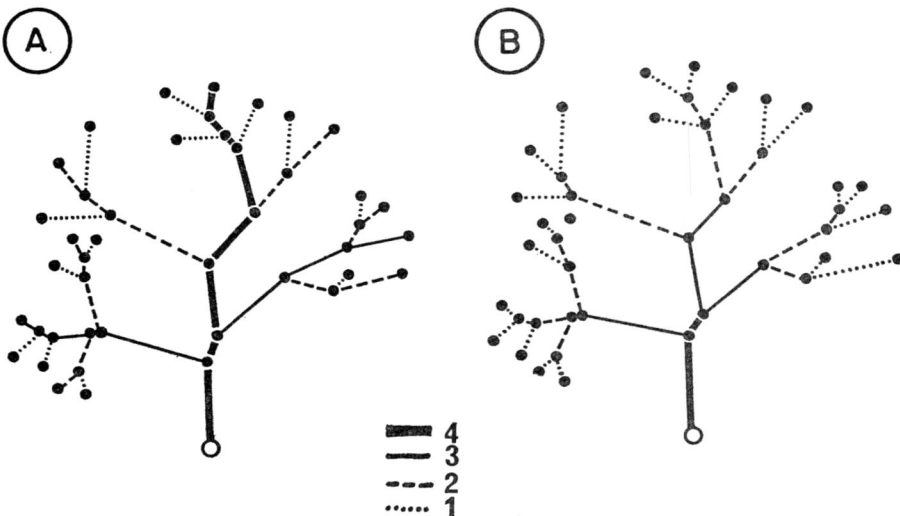

15.11 Ordering of a stream system in terms of (A) Horton's and (B) Strahler's systems (*Source: Bowden and Wallis, 1964, p. 768*).

conceded that exceptions may occur where geological controls have affected stream courses, and the ordering system was found in practice to be '... subjective and hard to replicate' (Bowden and Wallis, 1964, p. 767). To avoid such ambiguity Strahler (1952, p. 1120) redefined streams of any given order by a much simplified and easily replicable system. The application of this system is shown in Figure 15.11B. In this system unbranched fingertip tributaries are always designated as order 1. Higher order segments are created by the merger of two channels of the next lower order; they are terminated by their merger with channels of equal or higher order.

Melton (1959) has shown that channel order, according to the Strahler concept, is a simple mathematical notion that is independent of arbitrary geomorphic concepts. Order may be simply derived from the basic concepts of combinatorial analysis (Riordan, 1958, p. 110ff.) and does not depend on the existence of a single downstream direction on each channel segment. By

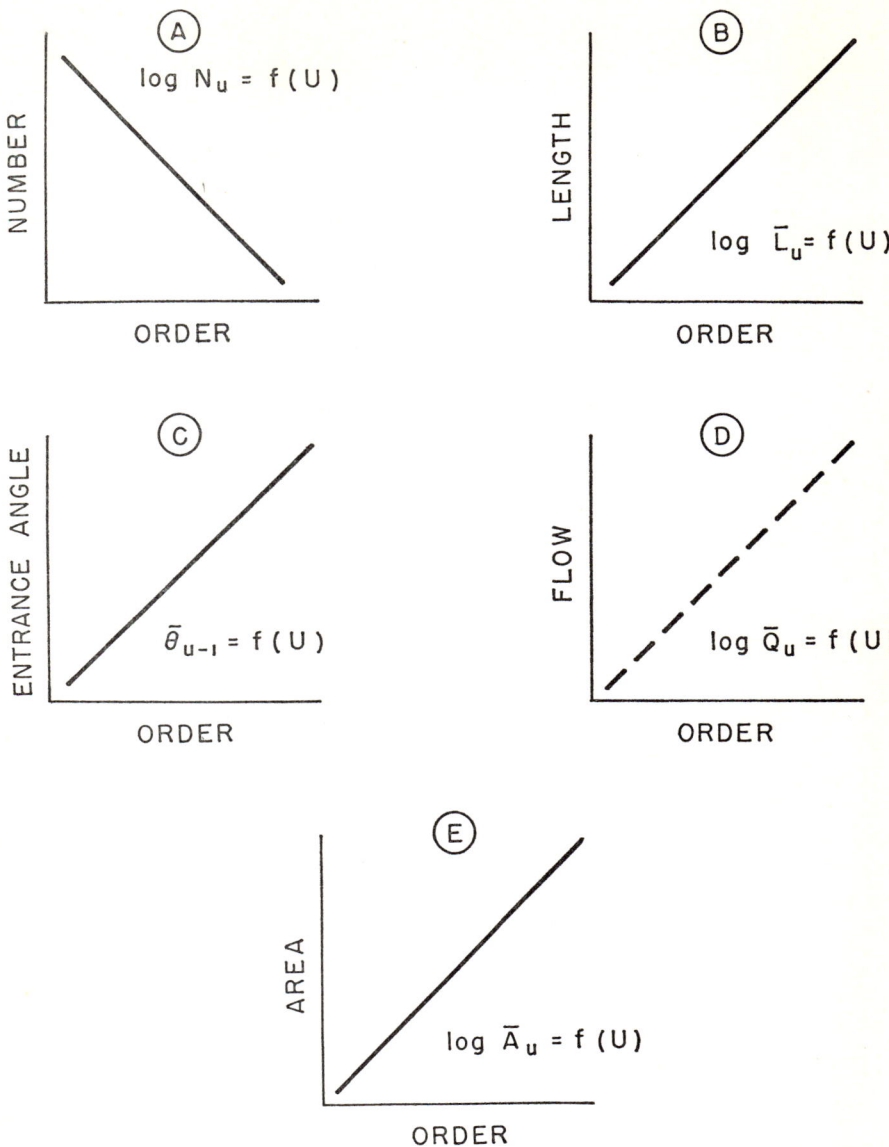

15.12 Summary of the major structural regularities within dendritic systems: (A) law of path numbers; (B) law of path lengths; (C) law of entrance angles; (D) law of path flow; (E) law of path areas.

regarding streams as a linear graph (technically a 'rooted' tree) we can extend Strahler's ordering system to other geographical phenomena (e.g. road nets) within the same topological class.

(b) *Order and structure.* The twenty years since Horton's paper have seen the detection and extended field testing of a number of structural regularities related to combinatorial order in stream systems. Strahler (In: Chow, 1965, pp. 4:39–4:76) has provided an exhaustive review with a massive bibliography of work over this period. Here we attempt to summarize five basic regularities for stream networks, confining our attention to those *planar* aspects which might be applicable to line-area relationships in other planar trees. For example the so-called 'law of stream slopes' (Horton, 1945, p. 295) is ignored since it involves a third dimension (height).

Figure 15.12 summarizes these simple relationships between network form and stream order (=*path* order) through a series of graphs. It is not suggested that all stream networks follow this pattern exactly but almost all analyses to date (conducted for a wide variety of climatic and geological conditions (Chorley, 1957)) tend to show this general pattern of linear relationships with only small deviations from the straight line. We may characterize these graphs as showing a *law of path numbers* (Fig. 15.12A) stating that the number of stream segments of each order form an inverse geometric sequence with order number (clearly a direct product of the combinatorial form of ordering); a *law of path lengths* (Fig. 15.12B) stating that the mean lengths of path segments of each successive order tend to approximate a direct geometrical sequence; a *law of entrance angles* (Fig. 15.12C) stating that the mean entrance angles of paths is directly related to the orders of the entering and receiving path; a *law of path flow* (Fig. 15.12D) stating that the mean flow along a given path is directly related to path order; and a *law of path areas* (Fig. 15.12E) stating that the mean basin areas (tributary, 'access' areas) for paths of each order form an inverse geometrical sequence with order. On all five graphs the y axis is on a logarithmic scale and the x axis (order) is on an arithmetic scale; for the direct series, the first term in the geometrical series is given by the value for first-order paths.

In discussing these findings the term 'law', originally used by Horton, must be interpreted with caution. Not all the five relationships have been equally-well validated; for stream systems, the laws of stream number, stream length, and stream area are most firmly based. The law of entrance angles hinges on work by Lubowe (1964) on dendritic networks, and that on stream flow on extrapolation by Leopold and Miller (1956). Both however accord closely with work on vascular networks (Roux, 1895) and traffic networks (Haggett, 1966) and are logical concomitants of structure within a topologically ordered tree-network.

2 Trees as optimal branching-systems

The regularities exposed in the preceding question inevitably raise questions of causation. Why should stream patterns appear to follow predictable forms? Although the answer to this question is not fully clear there are sufficient part-answers available to allow us some measure of explanation. Although discussed separately here, all involve some notion of *optimality* whether achieved (as in the static models) or in process of achievement (as in the dynamic models).

(a) *Principles of optimal systems.* Of the several attempts to consider naturally occurring phenomena as optimum mathematical designs, none have been more influential than D'Arcy Thompson's monumental treatise *On Growth and Form* (1917, 1942). Thompson was particularly intrigued by a problem that had troubled John Hunter, viz. the gradually changing angle at which successive intercostal arteries are given off from the thoracic aorta, and he summarizes Roux's rules (1895) on the branching angles of arteries. These were as follows: '... (1) If an artery bifurcates into two equal branches, these branches come off at equal angles to the main stem. (2) If one of the two branches be smaller than the other, then the main branch, or continuation of the original artery, make with the latter a smaller angle than does the smaller or 'lateral' branch. (3) All branches which are so small that they scarcely seem to weaken or diminish the main stem come off from it at a large angle, from about 70° to 90°' (Thompson, 1942, p. 129). Roux's findings have since been verified by Murray (1927) and greatly extended by Cohn (1954) as part of a general theory of vascular systems.

The point of interest here is that the vascular system may itself be viewed as a tree-like structure, and that, as Lubowe's work (1964) clearly shows, stream-systems themselves show an adjustment between flow and branching angle very comparable to those discovered by Roux. If our own regional networks belong to a general class of optimally adjusted networks then the possibilities for developing general equations for certain geographical features are greatly strengthened. Cohn's (1954) work showed that it was possible to build up, from a minimum number of parameters, a satisfactory approximation to certain real working parts of the vascular system. Starting with the problem of supplying blood to a cube divided into successively smaller parts, he was able to derive expressions for (1) the number of branches, (2) the length and resistance of each branch, (3) the total resistance of the system to flow, and to match these with clinical results. A two-dimensional analogy to Cohn's three-dimensional work is provided by the mathematician Steinhaus working on the problem of '... locating one finite area as near as possible to an infinite area'. The resulting sequence of graphs

[630] MODELS IN GEOGRAPHY

(Fig. 15.13) shows successive increases in the length of the tree-like structure as the perimeter of the finite area is increased further. Bunge (1964, p. 12) suggests therefore that the dendritic patterns represent the linear pattern which optimally satisfies the command: 'locate a set of singly connected lines of finite total length as near to an area from a given point as possible'. In his

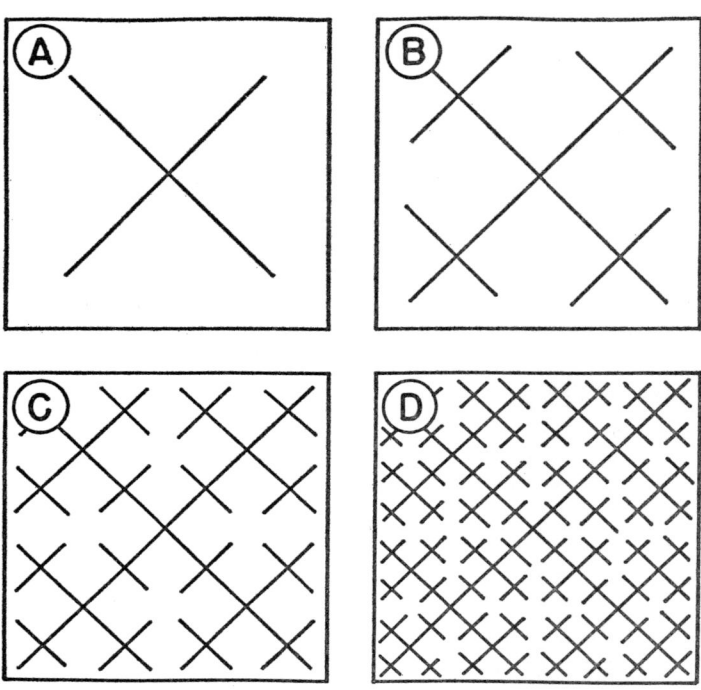

15.13 Steinhaus's solution to the optimum location of lines in respect of areas.

terms the tree is literally trying to get away from itself in the manner of self-repulsing magnets strung together, or hands stretching out to catch a baseball, or a tree stretching out its branches to get as near as possible to the available atmosphere.

(b) *Diachronic extensions: the 'state' concept.* It will be clear to the geomorphologist that the series of graphs in Steinhaus's model (Fig. 15.13) bears a very strong resemblance to Horton's (1945, p. 340) schematic diagrams for the development of a drainage net within a stream basin. Work by Ruhe (1952) on the drainage patterns developed on glacial drifts of various ages in Iowa tend to confirm Horton's schema: the older till sheets appear to be associated with patterns of greater density and integration (Fig. 15.14A).

The change over time is not however constant. Drainage density and stream frequency appear to change most rapidly in the first 20,000 years and level off in subsequent periods (Fig. 15.14B). Other observations on pattern change over time suggest that as channel length increases the number of streams also increase so that the topological order of the tree increases through time (Leopold, Wolman and Miller, 1964, p. 422).

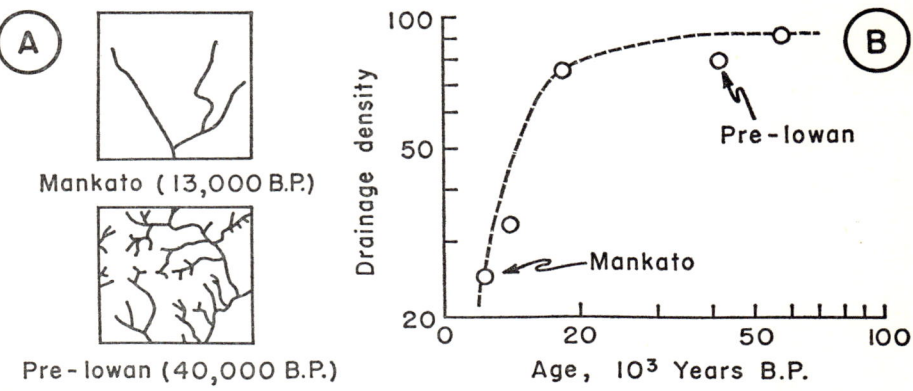

15.14 Development of dendritic systems over time: stream patterns on parts of the Des Moines lobe (*Source: Ruhe, 1952*).

There seems however no need to regard these 'diachronic' changes as special cases. Ruhe's work for example, suggests that the structural relations between the various network parameters follow similar geometrical 'laws' at each of the various time periods, i.e. the model works for different time periods in the same way as it works for different regions. We may then substitute for the idea of spatial (regional) models and temporal (periodic) models the idea of *state* models, the state being internally adjusted and theoretically independent of time and space. Similar space-time fusions have been suggested by Kansky (1963, p. 100) in his study of highway and railway networks.

(c) *Stochastic (random walk) models of branching*. Leopold and Langbein (1962) describe a theoretical approach to the characteristic dendritic forms of rivers based on concepts of statistical mechanics and entropy. They used a so-called *random-walk* simulation model to produce the most probable network forms in a structurally and lithologically homogenous region. The simulation process begins with a grid of small equal squares, each drained by a single channel which can lead in any of four directions. The direction was chosen by random number subject to the constraints that (i) reverse flows and (ii) closed loops are not permitted. (e.g. a meandering stream cannot make a complete loop in its course). When the complete graticule is filled with flow

[632] MODELS IN GEOGRAPHY

arrows a rectilinear stream pattern is generated which compares very closely with natural streams as regards both stream lengths and the number of streams of various orders.

Schenck (1963) went on to develop a programme suitable for the IBM 1620 digital computer which will generate stream patterns over an area of 600 unit cells (in fact, a rectangle of 30 by 20 units). Although the limited number of cells restricted the extent of stream development, a fair matching was achieved between the simulated patterns and sample 'natural' patterns (the United States and the Santa Fé area) (Table 15.2) and Schenck went on to show that the simulated patterns followed in broad terms Gray's (1961) law correlating the main stream length and the area drained by this stream.

TABLE 15.2

Dimensionless Numbers of Simulated and Actual Stream Numbers

Stream Order:	First	Second	Third	Fourth
Dimensionless Numbers of Streams:				
Simulation Model (Mean of 8 runs)	0·780	0·171	0·036	0·012
United States	0·777	0·174	0·040	0·009
Santa Fé	0·716	0·208	0·058	0·018

Source: Schenck, 1963, p. 5743. Dimensionless numbers of streams are defined as the number of streams of any given order divided by the *total* number in the network.

Although the computer data scattered about the regression line predicted by Gray, i.e. $L = 1\cdot4\, A^{0\cdot568}$ (where L is the main stream length and A the drained area), the results obtained do suggest that random-walk models can produce tolerably good approximations to observed regularities in dendritic structure.

One particularly interesting aspect of Schenck's model is that various boundary conditions may be substituted in the programme. Figure 15.15 shows five alternative boundary conditions with the available outlets shown as heavy lines and the main drainage lines by dashed lines. Of special significance is Figure 15.15E where the streams drain to a central 'sink' and show patterns not unlike those derived by Haggett for Portuguese road systems around small inland towns (see Fig. 15.28, below).

III CIRCUIT GEOMETRY

The third class of geographical networks are structures with closed loops or circuits (Fig. 15.1C). The most important and widely occurring circuit systems on the earth's surface are the intricate network of roads and railways which make up an important part of the total transport net of most regions. The problem of order posed by such networks are very similar to those of

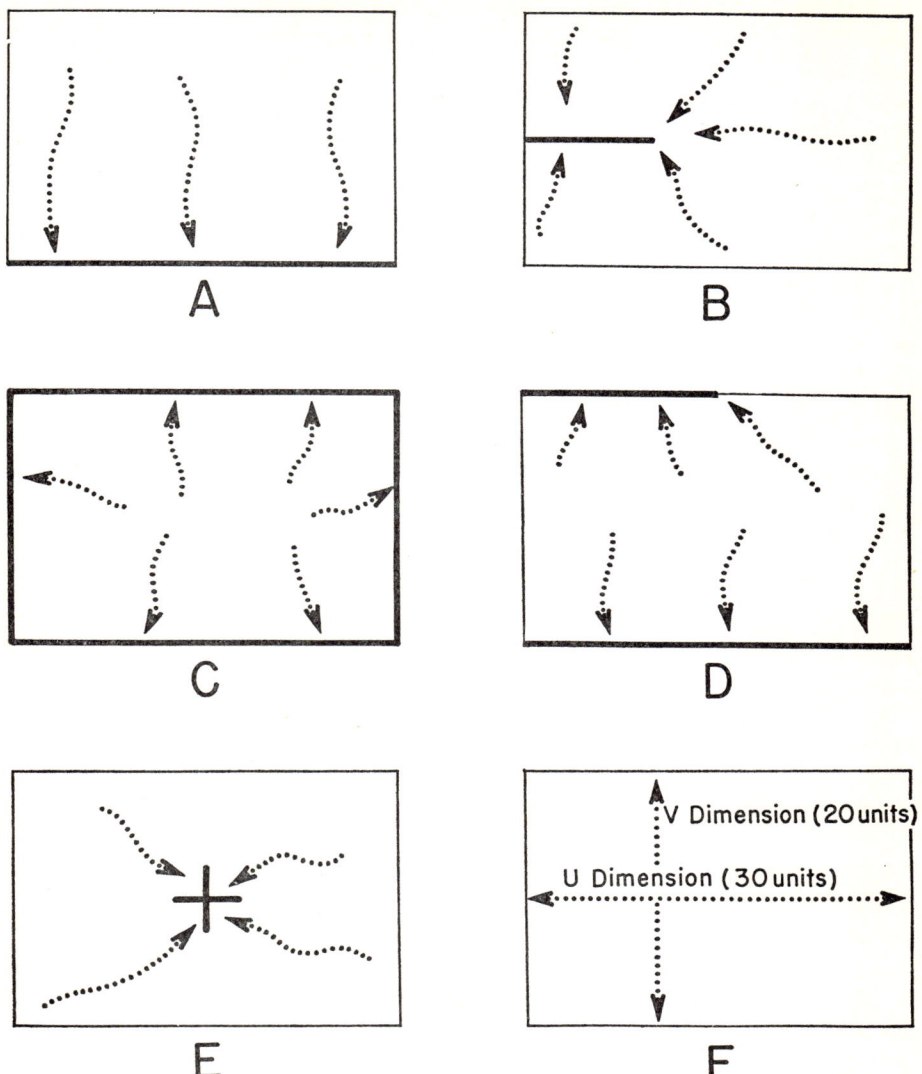

15.15 Alternative boundary conditions (heavy lines) use in the computer simulation of dendritic networks (*Source: Schenck, 1963*).

trees, but less progress has been made to date in analysing them and reducing them to predictable forms. Discussion here centres on the *structure* of such networks as revealed by the study of geometrical topological indices and the *growth models* put forward to explain their structure. The possibilities of converting circuit networks into simpler tree-like forms is considered in Section V.

1 The topological structure of circuit networks

Although the application of *graph theory* to geography has only come within the last decade it has already revolutionized our approaches to transport networks. Graph theory, a branch of topology fathered by the Swiss mathematician Euler in the eighteenth century, is closely related to algebra and matrix theory. Our discussion here will be elementary and confined to considering (i) some of the topological parameters available for describing transport networks and (ii) the results that analysis of these parameters have produced. Full discussions of graph theory with sample applications are contained in Berge (1962), and Busacker and Saaty (1965).

(a) *Topological measures of structure.* The range of topological measures developed for the description of networks is now too great to be spanned in this short chapter: Kansky (1963, pp. 5–33) and Haggett and Chorley (In preparation; Chap. 2) provide a summary of the more important ones with details of their computation. Garrison (1960) distinguishes two main groups of parameters: (i) parameters describing the whole network and (ii) parameters describing individual vertices on the network. Each group is illustrated here by sample parameters.

(1) One of the most useful of the topological measures of the *connectivity* of the whole network is the Alpha index (α). The Alpha index consists of the ratio of the observed number of fundamental circuits to the maximum number of circuits which may exist in a network (Garrison and Marble, 1962, p. 24). For planar graphs it is given by the formula:

$$\alpha = (E_n - V_n + G_n)/(2V_n - 5)$$

V_n is the number of vertices, E_n is the number of edges and G_n the number of sub-graphs in the system.

Clearly the observed number of fundamental circuits is given by the *cyclomatic number* (the dividend) while the maximum number of circuits (the divisor) is equal to the number of edges in a completely connected planar graph minus the number of edges in a complete tree. Garrison and Marble observe that multiplying the α-index by 100 gives it a range of 0 to 100 (instead of 0 to 1), allowing an interpretation of the value as 'per-cent redundancy'. A tree would clearly have zero redundancy and a completely connected planar network (i.e. a polgonal graph) 100 per cent redundancy. Figure 15.16 shows sample graphs with their appropriate redundancy values.

One further concept of importance in describing the whole graph is *diameter* (δ). The diameter is the maximum number of edges in the shortest path between the furthest pair of vertices, and is thus an index measuring the topological 'extent' of the graph (Kansky, 1963, p. 12). It is important in manipulations of matrix-algebra discussed below where determines the 'solution time' of the matrix.

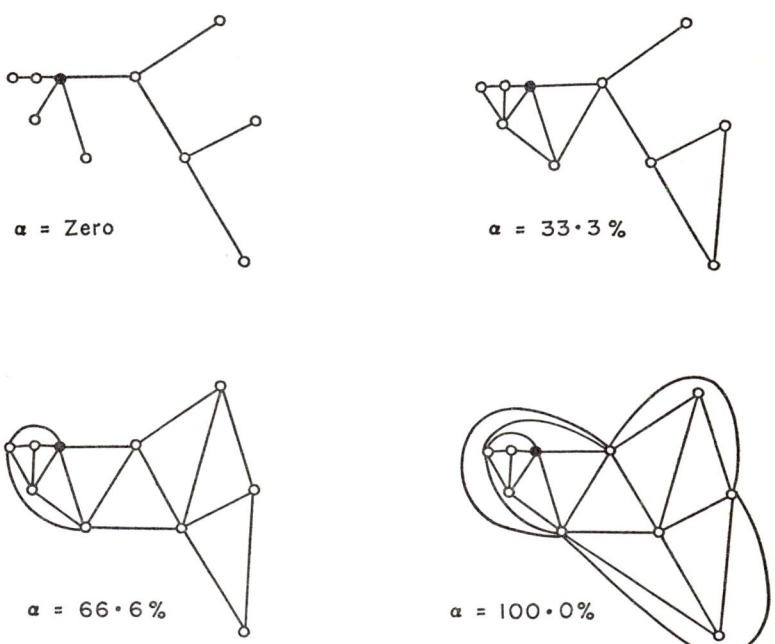

15.16 Successive changes in the redundancy index (α) with increasing connections between a set of points.

(2) Although it is possible to derive topological measures of individual vertices within a small network by direct measurement from the map, the amount of work with large networks becomes impossibly large. Solution to the problem lies in the conversion of networks to incidence or connection matrices, and their manipulation through the use of matrix algebra. Basic ideas of matrix algebra are clearly set out in Krumbein and Graybill (1965, Chap. 11).

Pitts (1965) has provided a very clear case of the use of matrix algebra in the derivation of measures describing the relative locational advantages of thirty-nine settlements in twelfth- and thirteenth-century Russia. Figure 15.17A shows the location of the settlements in relation to the rivers and trade routes and Figure 15.17B a simplified representation of this map as a planar graph. The information contained in this graph may be reduced to a

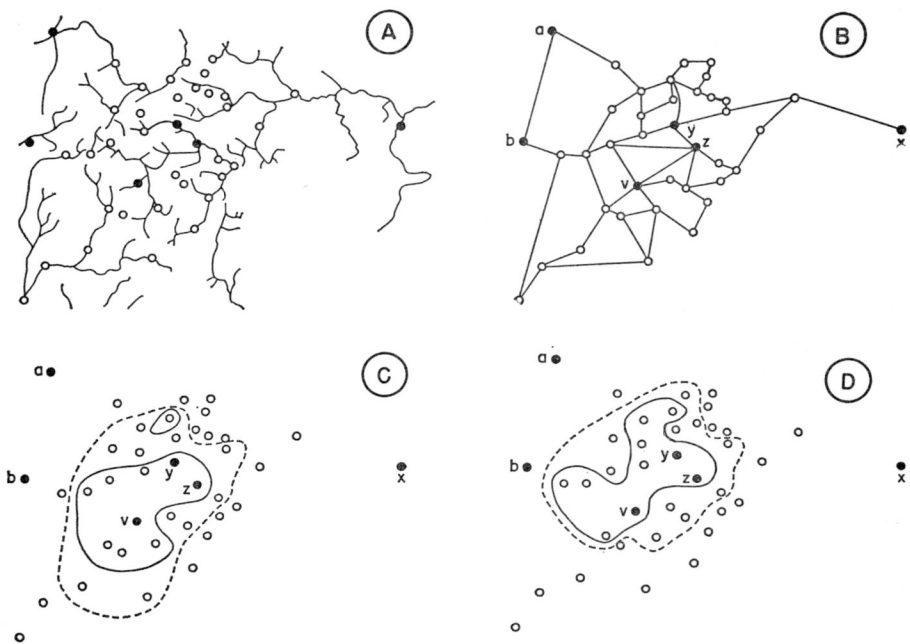

15.17 The use of graph-theoretic measures to compute the relative accessibility of Moscow (y) in twelfth- and thirteenth-century Russia. In both C and D the heavy line encloses the ten 'most connected' places and the broken line the twenty 'most connected' places (*Source: Pitts, 1965*).

one-zero array (a connection matrix) by listing all the settlements along both the rows and columns of the 39 × 39 matrix. The first four settlements in the matrix (*X*) would appear in the following form:

$$X = \begin{pmatrix} 0 & 1 & 0 & 0 & \cdot \\ 1 & 0 & 1 & 0 & \cdot \\ 0 & 1 & 0 & 1 & \cdot \\ 0 & 0 & 1 & 0 & \cdot \\ \cdot & \cdot & \cdot & \cdot & \cdot \end{pmatrix}$$

In this matrix the *ones* indicate the existence of a direct connection between pairs of places. Thus the direct link between the first settlement (*a*=Novgorod) and the second settlement (*b*=Vitebsk) is indicated by a 1 in the second cell of the top row. Absence of direct one-step connections is shown by a zero.

By raising the matrix to the power of the graph's diameter δ (eight in Pitt's example) it is possible to derive a matrix, $X(8)$; which is again shown for the first four settlements as:

$$X(8) = \begin{pmatrix} 110 & 15 & 143 & 16 & \cdot \\ 15 & 155 & 21 & 167 & \cdot \\ 143 & 21 & 580 & 32 & \cdot \\ 16 & 167 & 32 & 257 & \cdot \\ \cdot & \cdot & \cdot & \cdot & \cdot \end{pmatrix}$$

The above matrix contains the following information about the network: the diagonal entries indicate the total number of eight-step routes utilizable in going out and back from a settlement (thus Novgorod (a) has a value of 110 routes, Vitebsk (b) has 155 routes, and so on); the off-diagonal entries indicate the total number of eight-step routes going between pairs of places. To obtain the total number of alternative routes we add across the row of the $X(8)$ matrix to give a total connectivity measure for each vertex, *gross vertex connectivity*. Figure 15.17C shows the location of the 'most connected' vertex (v=Koselsk) and the 'least connected' (x=Bolgar). Moscow (y) is the fifth most-connected settlement.

Since the powered matrix ($X(8)$) contains redundant information on alternative routes, Pitts also computes a short-path matrix ($A(P)$). This has the following form for the first four settlements:

$$A(P) = \begin{pmatrix} 0 & 1 & 2 & 3 & \cdot \\ 1 & 0 & 1 & 2 & \cdot \\ 2 & 1 & 0 & 1 & \cdot \\ 3 & 2 & 1 & 0 & \cdot \\ \cdot & \cdot & \cdot & \cdot & \cdot \end{pmatrix}$$

In this matrix the diagonal entries are zeros and the off-diagonal entries indicate the number of edges traversed via the shortest path between pairs of settlements. Adding across the rows gives a total number of edges to be traversed in getting from a vertex to *all* its neighbours. Figure 15.17D shows the location of the 'most' and 'least' connected settlements respectively Kolomna (z) and Bolgar (x) on this index of *net vertex connectivity*. Moscow (y) is second on this index.

Both the indices used in Pitts' study derive from work by Shimbel (1953) on the structure of communication networks. Garrison (1960) has gone further in using a weighted measure of shortest-path accessibility, the Shimbel-Katz index (Shimbel and Katz, 1953) in a study of the highway system of the southeastern part of the United States. A scalar, s, is selected (where s is between zero and one) so that s measures the effectiveness of a one-edge connection, s^2 the effectiveness of a two-edge connection, s^3 a three-edge connection, and so on up to s^δ where δ is the graph diameter. Using Garrison's values of $s = 0\cdot 30$ and substituting them in matrix $A(P)$ would give a matrix $A(P)'$ of the form:

$$A(P)' = \begin{pmatrix} 0 & \cdot 30 & \cdot 09 & \cdot 027 & \cdot \\ \cdot 30 & 0 & \cdot 30 & \cdot 09 & \cdot \\ \cdot 09 & \cdot 30 & 0 & \cdot 30 & \cdot \\ \cdot 027 & \cdot 09 & \cdot 30 & 0 & \cdot \\ \cdot & \cdot & \cdot & \cdot & \cdot \end{pmatrix}$$

Summation across the rows gives the Shimbel-Katz index for each vertex.

Because the river routes used in Pitts' (1965) analysis are two-way connections (*symmetries*) the matrices derived from the network are symmetric about the diagonal. Prihar (1956, p. 931) has investigated the analysis of networks (particularly radio-transmitter networks) where connections may also be one way (*antimetries*) and the resulting matrices are assymetric about the diagonal. Similar matrix algebra operations may be used to derive accessibility measures for this type of circuit network and Miehle (1958B) had set these operations within a Markov framework.

One of the most intriguing approaches to network structure has been explored by Gould (1966, pp. 37-40). He used eigenvalue ratios for the orientation matrices of road links in sample quadrats in Sierra Leone and Tanzania to measure the overall orientation of the transport pattern. The less developed, peripheral areas with 'dendritic' road systems tended to have higher eigenvalue ratios; the more developed, central area lower ratios. The success of the pilot study lead Gould to speculate on the possibility of using tensor analysis to derive tensor-field maps displaying the spatial variation in the directional 'pull' of road networks. Further research on these lines is continuing at Pennsylvania State University.

(b) *Structural interrelationships.* The most extensive analysis of the structural characteristics of transport networks in topological terms was carried out at Northwestern University by Garrison and Marble (1962, pp. 27-53) and Kansky (1963, pp. 37-80). They studied a series of measures of the road and railway network of a group of fifteen countries, varying in development level from Sweden down to Angola. Figure 15.18 is an attempt to summarize their findings and relate it to other work by Ginsburg (1961). The diagram consists of two main systems, a *transport system* and an *environmental system*. These major systems are sub-divided in turn into road and rail sub-systems and socio-economic and physical sub-systems respectively. Within each sub-system 'box' are a series of parameters which describe either the form of the transport network (density, load, etc.) or the characteristics of the environ-

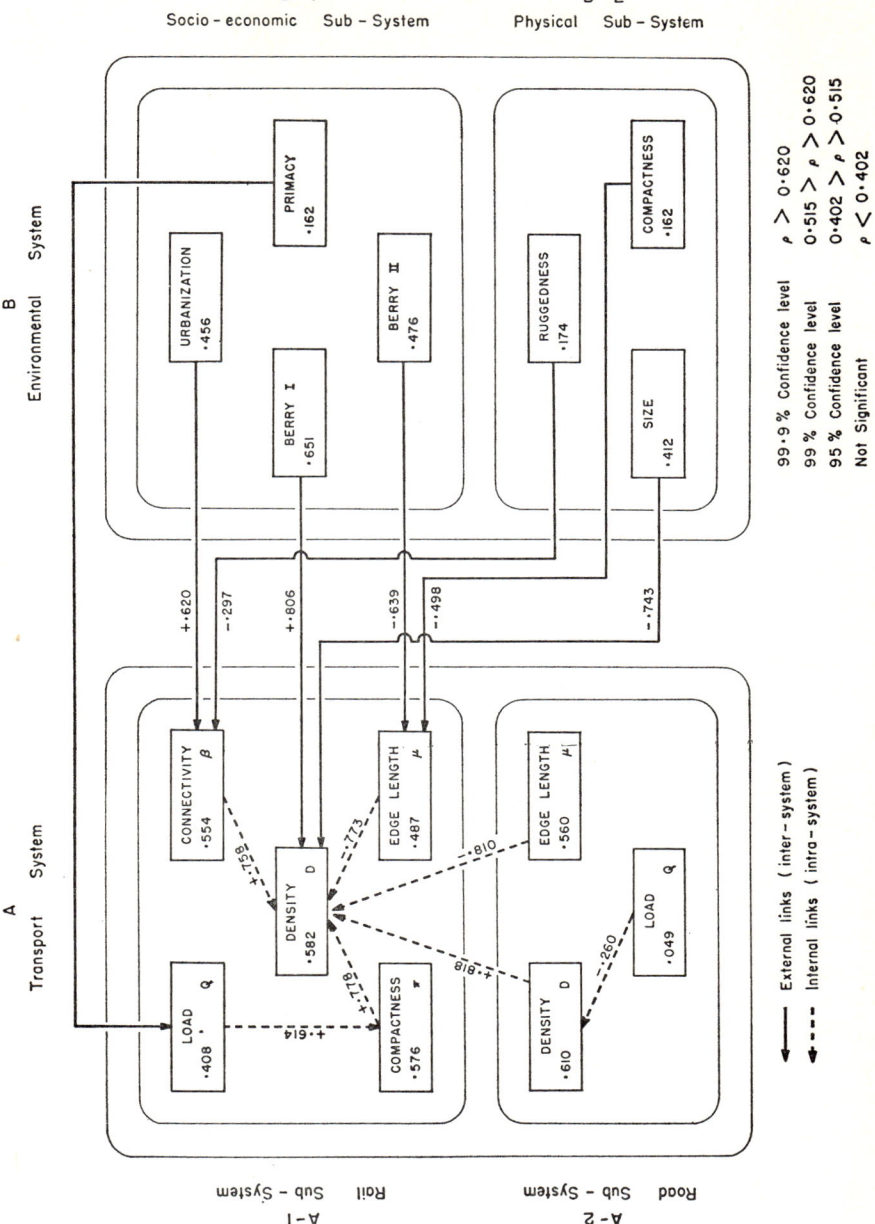

15.18 Structural interrelationships of elements of a *transport* system with its *environmental* system (*Data from Garrison and Marble, 1962, and Kansky, 1963*).

[640] MODELS IN GEOGRAPHY

ment within which that network is located (urbanization, terrain ruggedness, etc.). The parameters are largely self-explanatory: 'Berry I' refers to an index of Technical Development developed by Berry (1960) from factor analysis of 43 indices of economic development for a sample of 95 countries; 'Berry II' refers to an index of Demographic Status also determined from the same study.

Links within the system are shown by vectors. The internal links (the broken arrows) show the interlocking of the various structural parameters within the transport system; only one vector is shown for each parameter, its direction being determined by the highest correlation level (*rho* value) achieved with the other parameters within the system. The important links between the two density measures stand out with vectors clustering around the railroad-density 'box'. In general the rail sub-system was much more closely interlocked than the road system which showed more links with the rail system than with itself.

Internal links within the environmental system were not of direct interest and therefore vectors were drawn only between the environmental parameters and the transport box with which it was most closely correlated. The strength of these inter-system links are generally less than those within the transport system itself. Again the highest links are with railroad density (closely associated with the Berry I Index of general economic development). However the most striking feature of the diagram is the convergence of vectors from the environmental system on the railroad sub-system. Roads appear to be much less closely linked in with either the economy or the physical environment of the countries in which they are set.

In interpreting the diagram it should be remembered that the standard and comparability of rail data is generally better than road data, that relationships are based on a sample of only fifteen countries, and that only gross differences between countries are shown. Analysis is also hindered by the need, for the sake of clarity, to show only the dominant links in the system. However values given in the left corner of each box show the general correlation co-efficient of each parameter with all others within the transport system; these suggestions confirm the importance of the two density parameters as general guides to network structure, a point taken up in Section VI of this chapter.

An alternative approach to the latent structure in communications networks has been attempted by Garrison and Marble (1962, pp. 65–71). They begin with the airline system of Venezuela (not a planar network *sensu strictu*) which connects 59 cities (Fig. 15.19A) via a system of 104 routes (Fig. 15.19B). This is reduced topologically to a system of vertices and edges displayed in the form of a connection matrix. This matrix was analysed through component analysis to produce a set of four basic patterns of variation (Table 15.1). Factor I scaled the vertices in terms of their total number

TABLE 15.3

Factor Analysis of Latent Structure in Venezuelan Airline Routes

Components	Variation explained		Interpretation
	Per cent	Cumulative per cent	
Factor I	18·7	18·7	Size axis
Factor II	7·5	26·1	Caracas 'field' axis
Factor III	6·6	32·7	East–west 'field' axis
Factor IV	5·2	38·0	Minor regionalizing axis

Source: Garrison & Marble, 1962, p. 69.

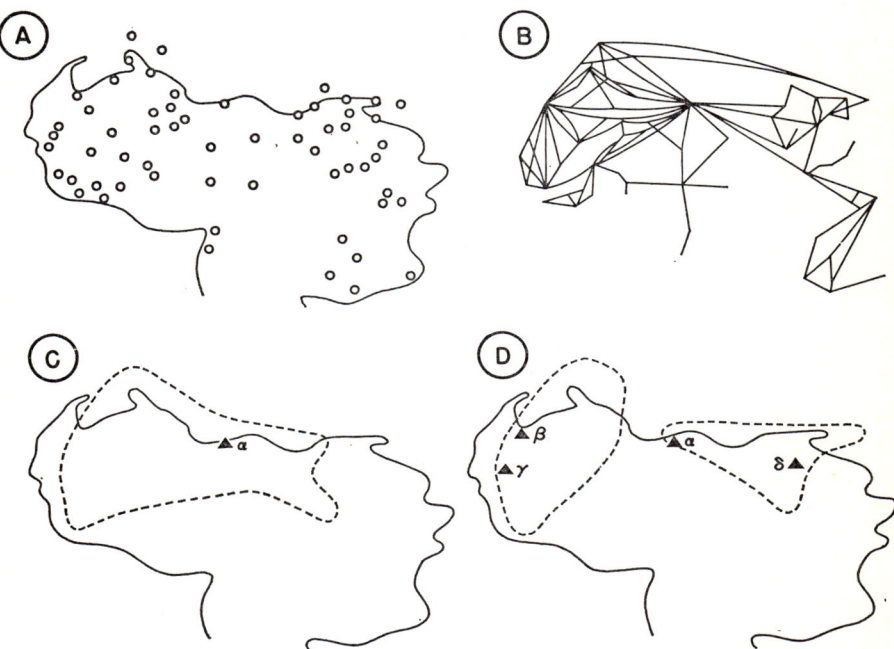

15.19 Latent structure in the pattern of internal airline links in Venezuela: (A) vertices, (B) links between vertices, (C) Factor II fields, (D) Factor III fields (*Source: Garrison and Marble, 1962*).

of direct connections with other vertices and was a nearly linear scaling of the cities by size. This 'basic' factor accounted for nineteen per cent of the total observed variation. Factor II showed up a field effect centring on the leading vertex, the city of Caracas (α) (Fig. 15.19C), and raised the level of explanation by a further seven per cent. Somewhat similar in importance was

Factor III (Fig. 15.19D) showed up a major regionalization effect – a western system centring on Maracaibo (β) and Santa Barbara (γ) stands out from the 'eastern' system centring on Caracas and Maturin (δ). The fourth factor showed a weak but detectable minor regionalization effect. Together all four factors accounted more around 38 per cent of the total variance in connectivity matrix, somewhat less than a similar component analysis applied to the structure of Argentine airline system.

2 Simulation models

There are no clear parallels in circuit networks to the models of optimal structure developed for trees in the preceding section. Certainly Lösch (1954, pp. 124–134) was concerned with the form of the road networks related to his optimal central-place pattern, and some work has been on the optimum spacing and arrangement of street plans (e.g. Prager on 'Manhattan' geometry (In: Herman, 1961, pp. 97–104)) and urban freeways (Creighton, Hoch and Schneider, 1959; Peterson, 1961). None the less, the failure to find effective parameters of internal structure seems to reflect itself in an absence of general equilibrium models of circuits. Considerably more work has been done on *change* in form and here models for the growth of networks in both developed and underdeveloped areas have been put forward (Haggett, 1965, pp. 79–86). Most, like the model of Taaffe, Morrill and Gould (1963), are based on inductive historical studies of transport growth.

In this section, attention is confined to the interesting work now being attempted on the simulation of network patterns. These models are of two kinds. The first kind, the *postdictive* model attempts to reconstruct a specific regional network at a particular stage in the past and may seek to 'fill in' subsequent stages in that evolution; the second kind, the *general* model, attempts to generate network models with less-specific historical or regional constraints.

(a) *Postdictive simulation models.* A number of limited attempts have been made to simulate the growth of complex transport networks using *Monte Carlo* methods. Garrison and Marble (1962, pp. 73–88) describe attempts to simulate the growth of the railroad system of Northern Ireland between 1830 and 1930, while Kansky (1963, pp. 132–147) and Morrill (1965, pp. 130–170) report parallel studies on the rail nets of Sicily and central Sweden. All the areas are small and their networks relatively simple in structure.

The basis of the simulation was to predict the probable *localization* of railway routes from the geographic characteristics of the area: the length, number, and rate of extension of the routes was either derived directly from the historical evidence or predicted from comparative studies of railway parameters in terms of an area's general level of economic development. Thus

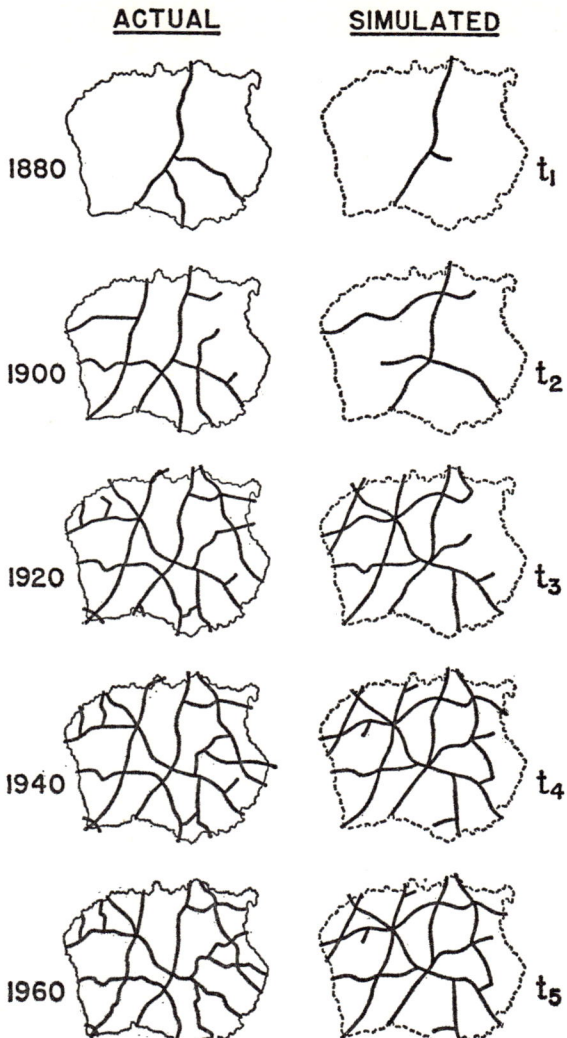

15.20 Comparison of the (A) actual growth of railway network in central Sweden with (B) the simulated growth (*Source: Morrill, 1965, pp. 130–170*).

in the case of the Sicilian railway network for 1908 the number of vertices was predicted as 16·48, the connectivity was predicted as 1·13 (β-index), and the mean edge length as 17·62 miles.

The procedure for the stochastic model may be generalized as a four-stage procedure: (1) List and weight all major settlements in the area in terms of a 'population income' score; (2) Select probable vertices from this list of settlements by a randomization process; (3) Connect the two largest vertices by a rail link; (4) Add the other edges in sequence such that '... the next largest centre joins the largest and closest centre which is already located on the network' (Kansky, 1963, p. 138). If, after all the vertices have been connected to the railroad system, un-allocated edges remain, then (5) add the edges in such a way that the circuit between the first, second, and third largest centres is completed, and continue to complete the circuits bringing in the fourth, fifth and lower-order centres in turn. The map of routes may be finally adjusted by (6) the adoption of delta-wye transformations (Akers, 1960) to simplify the links between triangles of three points, and (7) local adjustment of the routes to physical variations in the relief.

The success of the models varies considerably. The stochastic model of Northern Ireland yielded less clear results than a 'deterministic model' based on arbitrary rules of neighbourhood, regional and field effects. On the other hand, Kansky's study of Sicily provided reasonably consistent results and Morrill's work on central Sweden gives positive evidence of the possibilities of the method (Fig. 15.20).

(b) *General simulation models.* Although simulation models for the growth of complex road networks have yet to reach the sophistication of stream-pattern models (Leopold and Langbein, 1962; Schenck, 1963), roads have been included as by-products of general settlement simulation models. Thus Morrill (1965, pp. 65–82) includes a tentative road network as part of the output of his eight-generation *Monte Carlo* model of the locational evolution of a theoretical central-place hierarchy (Fig. 15.21). Although the mechanics of the migration-assignment process and the recomputing of the probability field for each time generation lie outside our discussion it is worth noting (i) that the pace of extension around the centre is uneven over both space and time and (ii) the extent of the road system is greater in the direction of early start. Spatial unevenness is related to the random element built in to the *Monte Carlo* model (Haggett, 1965, pp. 305–309) and temporal unevenness to the extending 'reach' of the migration process with improving technology: the migration probability field was successively expanded from k/D^2 in the first generation to k/D^1 in the eighth generation, where k is a constant and D the diameter of the field.

Morrill's model highlights the indirect orientation of routes with respect to the originating centre (i.e. the pattern is not rigorously stellate) and associates this with the iterative process by which roads grow by the addition of new links. When however the blotchy and uneven pattern of settlement is filled in, then new and more direct links can be formed. Comparison of the route north from the centre in the fifth and sixth generations (*E, F*) shows this realignment process in operation (Fig.15. 21).

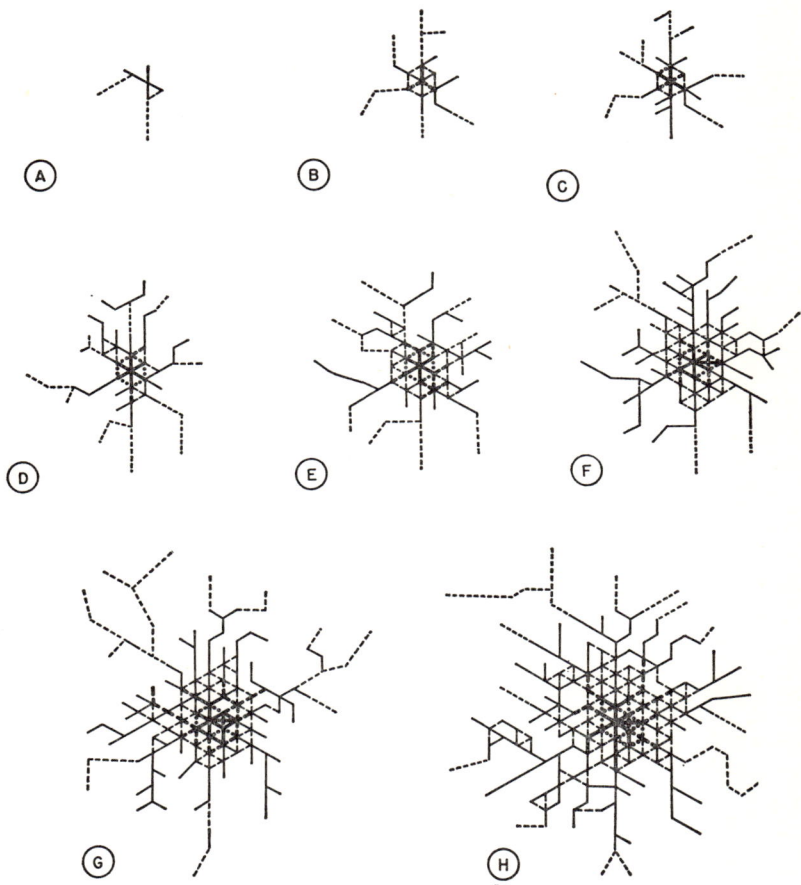

15.21 Simulation of road network around a central node by *Monte Carlo* procedures (*Source: Morrill, 1965*).

Morrill is at pains to point out (1965, p. 109) that the road net in his model is passive rather than active. That is, the road net is a product of the migration process; existing roads are *not* taken into account in the assignment of new migration routes and central places. Other artificial restrictions in the model are clear in the fixed mesh of the roads in the 'settled' area near the centre. Both the size of mesh and the triangular form are directly related to the size of the hexagonal cells in the original probability field. Where square rather than hexagonal cells are used (e.g. Morrill, 1962, p. 117), the resulting road network would also be basically rectilinear – indeed not unlike the road system over the greater part of the western United States.

IV CELL GEOMETRY

The fourth class of geographical networks, the cellular net, is different in character from the first three (Fig. 15.1D). While these consist of lines of varying complexity *along* which flows move; the cellular net consists of barrier lines *across* which flows move. Of the important and naturally-occurring cellular nets over the earth's surface, the loop of the watershed enclosing the drainage basin or the parish loop enclosing a unit of local administration typify two different but related forms. Thus cellular nets form a geometrical *dual* to the flow networks dicussed earlier in this chapter; they are areas of containment and as such have the geometrical characteristic of completely closed figures. They consist only of loops unlike the transport network which although it may consist mainly of circuit-loops also contains trees connecting to terminal vertices. We may further divide cellular nets into two geometrical classes, the first consisting of isolated loops (e.g. the closed loop of shoreline surrounding an island) and the second of contiguous loops (e.g. the loops surrounding sets of counties and forming mutual boundaries between them). This discussion centres on the second class only, treating the rather untouched topic of cell *structure*, the classical *steady-state* models of cells, and *growth* models of cells.

1 The structure of cellular nets

In contrast to stream patterns where morphometric analysis is now standard, and transport patterns where application of graph-theoretic analysis is growing apace, we have rather less comparative data on the structure of cellular networks. This lack seems due to inattention rather than measurement problems and here we briefly outline the geometrical form of cellular nets and some tentative regularities found in their structure.

(a) *Topology of cellular nets.* Cellular nets form a distinct topological class within the general group of planar graphs. They are *polygonal graphs* (Ore, 1963, p. 98) defined as a planar graph in which the edges form a set of adjoining polygons in the plane, using polygons in a general sense to indicate figures with either straight or curved boundaries. Thus a map showing the division of the USA into states would be such a polygonal graph, including as it does both straight meridional lines (the Nevada-Utah interface), curved parallel arcs (the Wyoming-Montana interface), and irregular natural lines (the Kentucky-Indiana interface).

For polygonal graphs we may apply an interesting formula first derived by Euler which states that

$$V_n - E_n + C_n = 2,$$

where V_n, E_n and C_n are the number of vertices, edges and cells respectively (Ore, 1963, p. 99). We must recall in applying this formula, known as *Eulers' polyhedral formula*, to the conterminous United States that we must include the infinite set of all points lying outside the United States (the 'outside' cell) as well as the internal cells, i.e. $C_n = 49$, not 48. Euler's formula has an interesting extension to a problem that worried Lösch (1954) in his work on the model cell for central-place hierarchies, i.e. 'what regular polygons of equal area may be used to completely fill a plane without overlapping or unused spaces?' Euler's formula may be extended to show that conditions for the establishment of these regular *tesselations* are only fulfilled when

$$1 + \frac{\rho}{\rho^*} = \frac{\rho}{2}$$

the number of edges at each vertex (ρ) are the same, and the number of edges bounding each cell (ρ^*) are the same. Only regular triangles ($\rho=6$, $\rho^*=3$), quadrangles ($\rho=\rho^*=4$) and hexagons ($\rho=3$, $\rho^*=6$) satisfy this equation (Ore, 1963, p. 108).

One classic extension of cell topology is to the 'map colouring problem'. The basic conjecture that every map can be coloured effectively (so that no two adjacent regions with a common boundary have the same colour) by the use of only *four* colours was introduced by the mathematician Cayley (1879) in the first volume of the *Proceedings of the Royal Geographical Society*. Although mathematical proofs exist for the *five-colour theorem* and partial proofs for the four-colour solution exist for limited maps (with less than 36 regions), no general *four-colour theorem* has yet been devised. Cayley's problem, so simple to demonstrate and so difficult to prove, has relevance to modern work on measures of contiguity. Dacey (1965) has discussed these measures for two and k-colour maps and has computed a table for testing hypotheses of randomness on two-colour maps.

(b) *Regularities in cellular nets.* Empirical studies of polygonal nets have concentrated on two basic regularities, *density* and *form*. Haggett (1965, pp. 53–54) examined the polygonal network of the *municipio* boundaries of Brazil, using a random sample of one hundred cells drawn from the total population of some 2,800 cells. Here the size of the cells was found to be approximately log-normal in distribution with a few very large units in the sparsely-populated areas of the upper Amazon, and many smaller cells in the more densely-populated eastern parts of Brazil. The relationship between cell size and population density was shown to be regular but inverse, a trend which is supported by examination of other systems of administrative areas in the United States and Western Europe.

Cell morphology for the Brazilian units in Haggett's (1965, pp. 50–52) study was measured in terms of (i) a shape index and (ii) the number of edges (or faces). The shape index suggested that the cells were generally elongated rather than circular with strong concentrations around three modal values characteristic of the regular triangle, square and hexagon. Although the significance of this clustering is not known it is interesting that it is just these three forms which are the regular polygons capable of completely filling a plane with a tesselation network. Study of the number of edges of the sample cells showed that, although the number of faces ranged from two to fourteen about one-third of the cells were irregular hexagons. Confirmation of the importance of the hexagonal unit was shown by the average value (5·71) for the number of faces for the whole sample. Studies of similar phenomena in other countries suggests that the hexagonal cell, albeit in an irregular form, is a rather regular feature of the structure of the sub-division systems of many countries. Similar quantitative studies on the morphology of 'natural' cell patterns are reported by Corte and Higashi (1964).

2 Steady-state models of cellular nets

Much of the literature describing cellular forms builds on simple intuitive concepts of symmetry known since the time of Greek geometers. Mathematicians like Euler, Gauss, Kepler, Plateau and Riemann were fascinated by the mathematics of cellular structures like the honeycomb, while physicists from Maxwell to Kelvin speculated on the tensional forces by which the hexagonal architecture was achieved. More recently the growing interest in equitable electoral divisions (Weaver and Hess, 1963) and optimum regional boundaries (Haggett, 1965, Chap. 9) have led to renewed interests in these regular cellular forms. Here we confine our review to the geographical implications of these steady-state models.

(a) *Unbounded nets: the 'mid-continental' problem.* Hexagons have long been recognized as the regular polygons which allow the greatest amount of packing of regular cells on a plane, allowing both the total edge length (i.e. total length of interfaces) and the accessibility (i.e. distance from the centre of the cell to all points within its boundary) to be minimized (Fig. 15.22). Both Christaller (1933) and Lösch (1954) used it as the basic modular unit in their complementary models of settlement hierarchies. Garner (Chap. 9, pp. 306–319 above) has discussed the structure of these theories and shown how modifications of the theoretical hexagonal structure may improve the 'fit' of these models with reality. Ideas of hexagonal systems have also been put forward for the structure of physical cells (e.g. ice-wedge polygons) on broadly similar grounds, and Morgan (Chap. 17, pp. 731–732) has illustrated the 'hardware' models that may be used to simulate their form.

Since these models are discussed elsewhere in this volume, the only

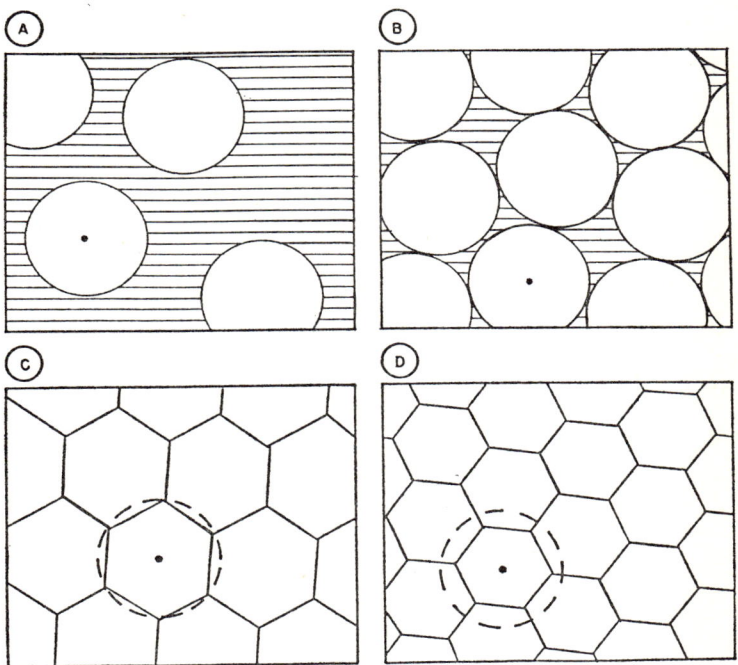

15.22 Packing of centres in the colonization of a plain to give cellular hexagonal territories (*Source: Lösch, 1954, p. 110*).

general comment we shall make on the optimum-hexagonal model is that it is essentially 'mid-continental' in application, in the sense that an infinite plane is assumed and boundary conditions are undefined. It is not without significance that in central-place studies the classic areas (e.g. Iowa or South Germany) are conveniently free from difficult boundary disturbances.

(b) *Bounded nets: the 'island' problem.* To extend the equilibrium solution from the infinite plane to finite geographical areas requires some modification. If we retain the concept of the co-equal 120° junctions of the

21*

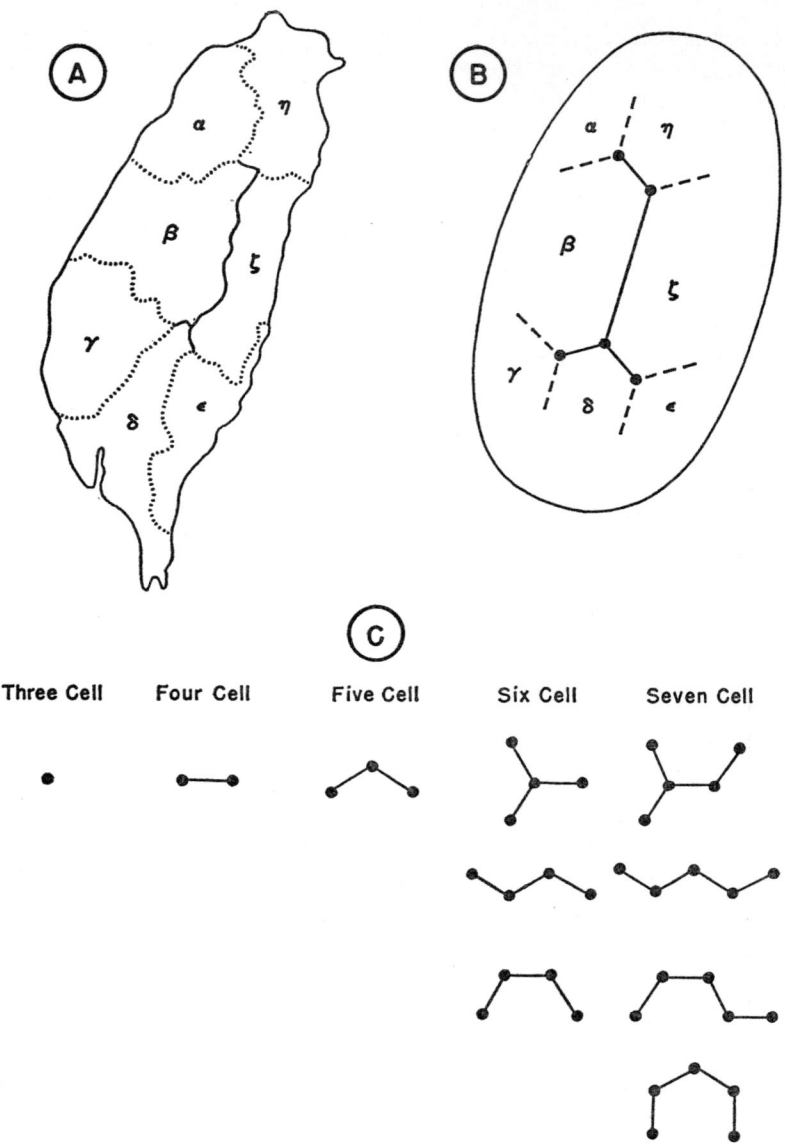

15.23 Topological character of cellular nets: (A) administrative areas of Formosa, showing 'polar furrow'; (B) generalized form of (A); (C) range of polar furrow arrangements for fixed cell numbers (*Source: Thompson, 1942, p. 516*).

hexagon but introduce the concept of a fixed boundary then we have a problem in space-partitioning not unlike that of dividing an island into N counties, all of which have a maritime boundary (i.e. are not encircled by the rest). The Island of Formosa (Fig. 15.23A) with its seven maritime provinces forms such a case. Topologically we may simplify the Formosa case by generalizing the coastline of the island as an ellipse (the boundary), regarding the seven provinces as cells and replacing the province boundaries by cell 'walls' set at co-equal 120° angles (Fig. 15.23B).

If within our island we have only three cells meeting together then, in terms of our steady state theory, we may assume they would meet at a point at equal angles. If we have four cells they need an 'internal' wall (i.e. one that does not have one end touching the coastal boundary). With five cells we need *two* such internal walls, with six cells *three*, and so on, following the rule that we require $N-3$ internal walls to separate N cells. These internal walls, designated as *polar furrows* by the cell biologist (Thompson, 1942, edn. p. 516), are of special interest to our model in that (as Fig. 15.23C) shows, they may be arranged in numbers of different ways. With the four- and five-cell island there is no ambiguity but clearly with six cells there are *three* different arrangements of the three-link polar furrow. Seven cells give four alternatives, eight cells give twelve alternatives, nine cells give twenty-seven alternatives, and beyond that the number increases rapidly with estimates by Brückner of 50,000 alternative arrangements for thirteen cells and around

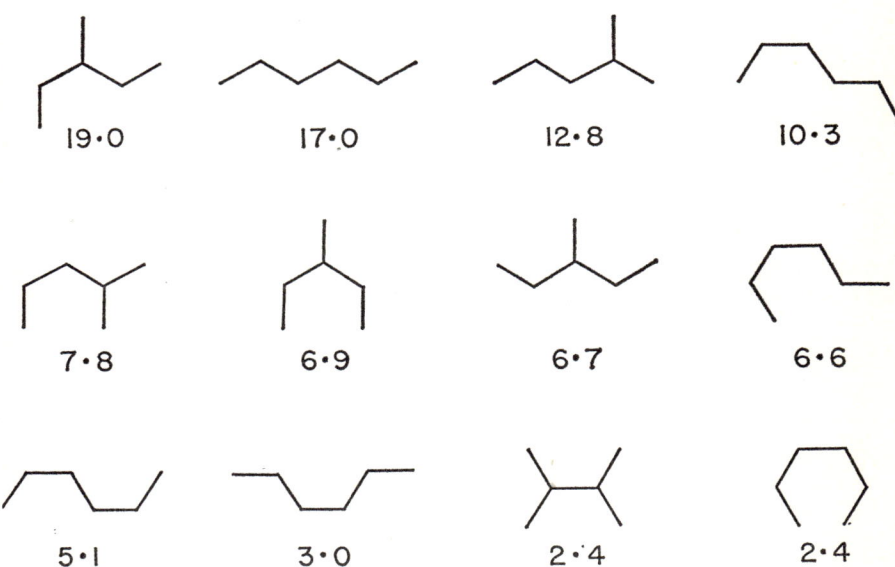

15.24 Twelve possible forms of the polar furrow for eight-cell divisions of an island, arranged in a stability sequence (*Source: Thompson, 1942, p. 607*).

30,000,000 alternatives for sixteen cells (Thompson, 1942 edn., p. 598). These higher numbers are largely of academic interest since with real geographical situations (e.g. islands divided into counties) it seems likely that internal non-maritime cells would be established in the interior; Brückner's estimates refer to cases where *all* the cells retain a maritime boundary.

The realization that the number of cell patterns is, from the topological viewpoint, strictly limited leads on to a consideration of (i) how many of the *possible* patterns actually occur and (ii) in what proportions they occur. Although the strictly geographical answers to these questions have yet to be determined, there is some evidence from the biological sciences that there may be a continuum of cell forms ranging from rather stable (and therefore common) to highly unstable (rare) arrangements. Figure 15.24 shows the twelve possible arrangements of polar furrows (each five-edges in length) for eight-cell islands arranged in a stability sequence. The frequencies are based on laboratory study of more than a thousand epiblastic cells (Thompson, 1942 edn., pp. 606–607) and show ranges from forms as common as 1:5 to as rare as 1:50. Thompson's findings suggest a rich field of research in the stability of geopolitical patterns over a wide time-space range, for areally equable divisions were not in each case a regular criterion of stability.

3 Growth models of cellular nets

Some of the characteristics of the 'steady state' models of cells, the trihedral angles and the tendency to equal-area cells, have been difficult to find in reality. Some of these difficulties may be overcome, as Haggett (1965, pp. 53–55) suggests, by suitable transformations of the base-plane (e.g. the substitution of income-space planes for distance planes in trade-cell studies) but there seem to be more fundamental problems to overcome. Steady-state models imply a state of equilibrium without stating the stages by which this equilibrium is achieved, i.e. they represent an instantaneously-achieved situation. This is often unhelpful since cellular networks are (i) usually formed by an iterative step-by-step process, and (ii) the early boundaries are not necessarily adjusted to late-stage conditions. Indeed the relaxation times of many geographical systems are so slow that early boundaries, once established, may themselves form an important reference line with respect to which later boundaries are formed.

Some of the most useful ideas on this iterative cell-division process can be generalized from detailed work on the physics of fracture (well summarized by Irwin (In: Flugge, 1958, pp. 551–590)). Here the argument rests on two points: (1) If we take a plane sheet of material then we may expect the flaws (points, lines or areas of weakness) to follow an assymetric frequency distribution, the largest flaws being least common and smaller flaws being most common. If we further assume the pattern of flaws is randomly distributed then we may expect on average, the larger flaws to be further apart and the smaller flaws closer together. (2) Each fracture of this idealized surface will create around it a buffer zone of tensional relief such that new fractures will be unlikely to form within a specific distance. This is because the fracture reduces the stress in its vicinity and increases the strength of the remaining, unfractured, surface by removing the points of greatest weakness.

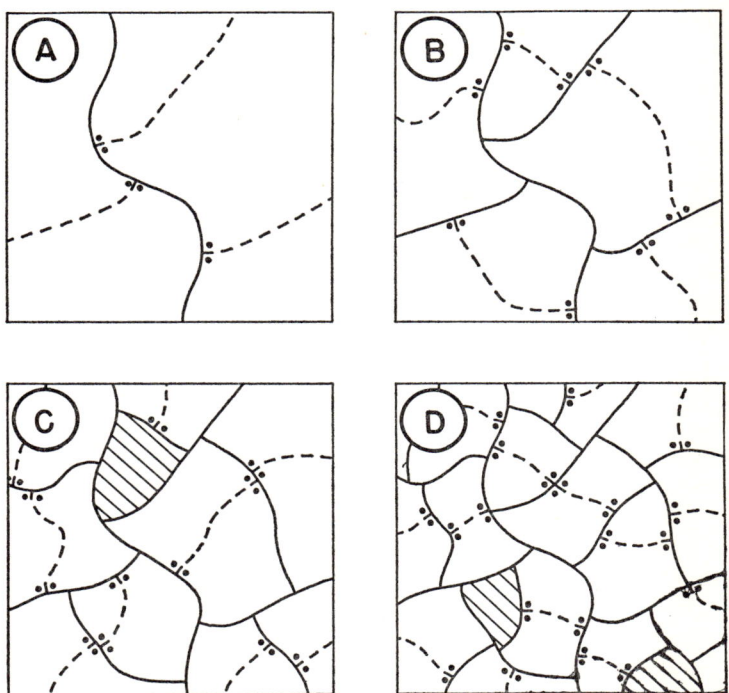

15.25 Four-stage 'fracturing' model of the sub-division of a plane with randomly distributed zones of weakness. The shaded areas indicate areas too small to be fractured under given stress, where stress is increasingly monotonically from A to D.

The general relationships between fracture, stress relief and spacing may be given by the equation

$$l = 0.7 \left(\frac{N}{\tau_o}\right)^m, \quad \tau_o < N$$

where l is the average crack spacing for a given applied tension, τ_o in terms of a flaw distribution parameter m, and N is the nominal tensile strength of a small sample (Lachenbruch, 1962, p. 43). Only where $m = \infty$ is the crack spacing determined by stress relief only, and by extension it is only in this unlikely limiting case that division of cells into a regular tesselation of hexagons is likely to occur. Smaller values of m denote more influence by flaws on crack spacing.

If we begin with a value of m greater than unity then we may assume that the earliest cracks to appear are likely to be guided by lines of weakness. The zone of stress relief caused by the early fractures is unlikely to be wide enough to achieve self-adjusting spacing and the path of such fractures may well be sinuous. However the tension relief will give greater uniformity of strength in the remaining undivided areas and raise the value of m. The fracture of these areas under increased tension is therefore likely to give more regular cell patterns with values of l more dependent on stress relief and less dependent on the location of flaws. Figure 15.25 suggests a four-stage model of cell division to illustrate the process in action.

One special feature of this diagram is that cracks tend to intersect at right angles rather than in the trihedral fashion of the steady-state model. Lachenbruch (1962, pp. 45–46) points out that the zone of stress relief around a fracture line is not a uniform band but (i) dies out asymptotically with distance normal to the fracture and (ii) is anisotropic with relief at a maximum in a direction perpendicular to the crack and at a minimum in a direction parallel with the crack. If a tensional crack extends towards an existing crack at an oblique angle it will tend to 'veer' towards the normal as it enters the stress-relief zone. Therefore we may argue that '. . . the intersection of a crack with a pre-existing one tends to be orthogonal . . . and, conversely, the orthogonal intersection suggests that one of the cracks involved predates the other' (Lachenbruch, 1962, p. 46).

In practice, fracture theory has been restricted in its geographical applications to the study of polygonal structures in frozen ground. However, the relative scarcity of trihedral junctions in other cellular networks and the high number of orthogonal intersections suggests that it might be worth exploring the extension of tension-failure models to historical sequences of cell division (Haggett, 1965, p. 52). Again the analysis of the growth of administrative area boundaries is a rich area for experimentation.

V NETWORK TRANSFORMATIONS

We may regard the foregoing four sections – on paths, trees, circuits and cells – as basically concerned with naturally-occurring topologies, i.e. the networks exist in this form and we analyse them in the *same* form in which we find them. There are however no *a priori* grounds for thinking that we must necessarily retain this convention and in this section we look at the problems and possibilities in converting one important class of geographical networks (transport circuits) into a topologically simpler form (Horton trees).

1 The problem of circuit order

Much of the significant advances in the quantitative analysis of watershed geometry over the last twenty years stem from Horton's (1945, p. 281) recognition of stream order and its modification to a simpler combinatorial system by Strahler (1952, p. 1120). Indeed, Bowden and Wallis (1964, p. 767) describe this ordering system as '. . . the touchstone by which drainage net characteristics could be related to each other and to hydrologic and erosional processes'. The success of the Horton-Strahler approach in the recognition of fundamental regularities in drainage systems suggests that the search for comparable ordering-systems for more complex networks – notably transport circuits – would be worth pursuing.

Attempts to order or categorize transport networks containing circuits have not, so far, been conspicuously successful. To be sure, useful functional classifications of railroad networks in terms of their flow characteristics have been attempted by Wallace (1958) but the recognition of the classes – 'internal', 'originating', 'terminating', 'bridge', and 'balanced traffic' lines – demands data that often are simply not available. An alternative line of approach is suggested by the recognition of circuit hierarchies (Haggett and Chorley. In preparation) but this has yet to be tested in field conditions.

Very simple transport systems may consist merely of single lines or sets of lines arranged in a tree-like fashion. Thus Rasmusson (1962, p. 80) has mapped the intricate tracery of the lanes connecting the small granite pits on the island of Malmön with the bigger roads and railways leading to the exporting ports. Other Scandinavian geographers have mapped the seasonal flow of logs down the river systems of Sweden and Finland to the Baltic ports (e.g. Hultland, 1962). In these cases the transport net is simple enough for Strahler-type ordering systems to be directly applied, and no simplification problem need be encountered.

These simple networks are however in the minority. Road systems and rail systems in almost all developed countries consist of a highly complex interlacing of circuits with trees or isolated lines as appendages. Thus Garrison and Marble (1962, p. 35) found the average cyclomatic number for a group of fifteen countries (including units as undeveloped as the Sudan and Angola) as high as 200: by comparison, the cyclomatic number for trees is *zero*. The problem to be tackled is thus a formidable one as the inspection of the road trace on any single O.S. one-inch sheet will confirm.

Direct ordering of 'natural' circuits is thus inherently more difficult than that of 'natural' trees studied by Horton, in that circuits have *alternative* paths connecting vertices on the network. It seems therefore more profitable to consider ways of simplifying circuit networks into trees, allowing thereby the direct application of the Horton-Strahler approach to the 'artificial' trees thus formed. The fact that circuits (a higher-order topologic form) are theoretically decomposable into trees (a lower-order topologic form) is clear by inspection. Interest in the number of possible trees in a fully-connected graph has been stimulated by work on maser analysis, and Ku and Bedrosian (1965) have derived general solutions for partitioning full graphs into complementary trees.

2 A circuit-order algorithm

Haggett (1966) has suggested a five-stage algorithm for the breaking down of circuit-type networks into trees. The algorithm is based on the concept of accessibility and shortest-distance paths, and is illustrated here with reference to a hypothetical transport network: an island (Fig. 15.26A) with a road network some 193 miles in length. It has 21 edges and 18 vertices of which four (I, II, III, IV) are major settlements arranged in decreasing order of size.

Stage 1: 'pole' identification. Within a drainage system, identification of an outfall point must precede the recognition of the watershed. With large watersheds the outfall point is commonly located where the river enters the sea; with smaller watersheds, where a tributary stream joins a larger one. In the same way that the choice of an outfall point may be largely arbitrary, so with the choice of a *pole* for a transport network. Any one of the eighteen vertices on the island network may be regarded as the polar centre of a small part of the network. It would be possible to rank these vertices in terms of their *vertex degree* (i.e. the number of directly connected edges (Garrison and Marble, 1962, p. 15), or in terms of other topological characteristics (such as the Shimbel-Katz index (Shimbel and Katz, 1953)). If external evidence on flows is available, vertices may be ordered in terms of the traffic moving through each vertex. Where direct flow data is not available it is possible to predict traffic generation with some accuracy from other parameters (Tanner, 1961); failing this, simple ordering of vertices in terms of population size (I, II, III, IV) will provide a very rough index of flow.

Clearly then it is possible, for any given circuit network, to (i) recognize the vertices on the network and (ii) order these vertices in terms of some size parameter from the largest pole P_1 to the smallest P_n.

Stage 2: 'polar-network' isolation. The equivalent concept to the drainage basin in circuit analysis is the *polar network*. A drainage basin may be defined as the surface area drained by the water flowing past a given outfall. Exchange of traffic flows between the various poles on a circuit network makes the direct extension of this concept impossible and we must substitute a more limiting concept, accessibility. We may define a polar network as *that part of a network nearer (in terms of distance along the shortest route on the network) to a given pole P_1 than to any other pole $P_{2, 3, \ldots n}$*. Thus in Figure 15.26C, we may split the island network into two parts, the one nearer to centre I, the other nearer to centre II. The two parts of the network are separated by a

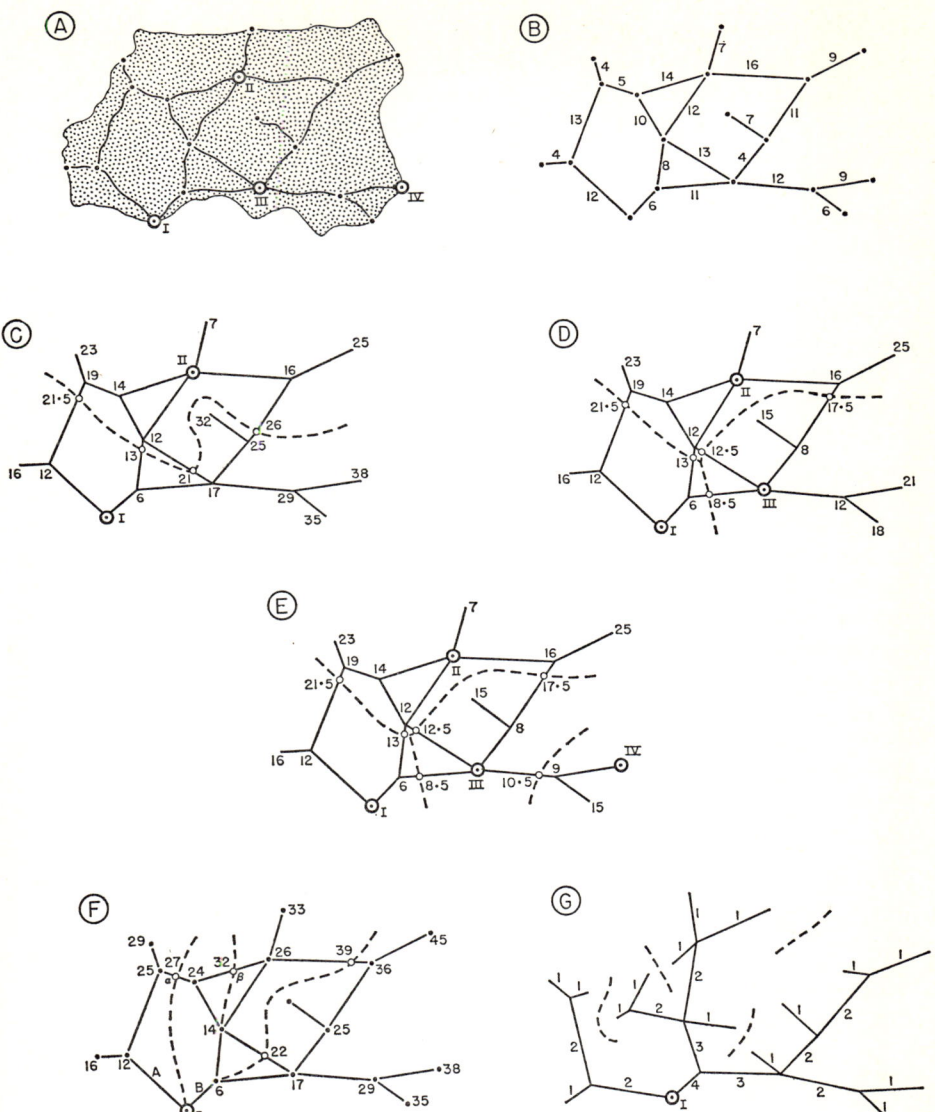

15.26 Stages in the decomposition of the road circuit of a hypothetical island into a dendritic form (*Source: Haggett, 1966*).

[658] MODELS IN GEOGRAPHY

dashed line joining the *indifference points*, i.e. points equidistant from both centres in terms of the shortest distance along the network. Exactly similar procedures may be used to split the island network into three polar networks centring on *I*, *II*, and *III* (Fig. 15.26D) or four polar networks centring on all four settlements (Fig. 15.26E).

In using the concept of nearness for our definition of polar networks we are merely extending the concept of Dirichlet regions on planes to planar graphs. *Dirichlet regions* (also called 'Voronoi polygons' or 'Thiessen polygons') consist of polygons the interior of which consists of all the points in the plane which are nearer to a particular lattice point that to any other lattice point (Dirichlet, 1850; Coxeter, 1961, pp. 53–54).

Stage 3: determination of 'network' hierarchies. We may regard small watersheds as mathematically 'nested' within larger watersheds: thus the Tennessee watershed (with its Ohio outfall) is nested within the larger Mississippi-Missouri watershed (with its outfall on the Gulf of Mexico). Nesting also occurs in the pattern of transport flows, as Philbrick (1957) has clearly shown, and some operational definition of the nesting hierarchy for polar networks is clearly needed. In the case of the island, the first centre is assumed to dominate the second, the second the third, and so on. This means that the introduction of each 'lower-order' centre adds a new tier to the number of polar networks in the manner shown in Figure 15.27.

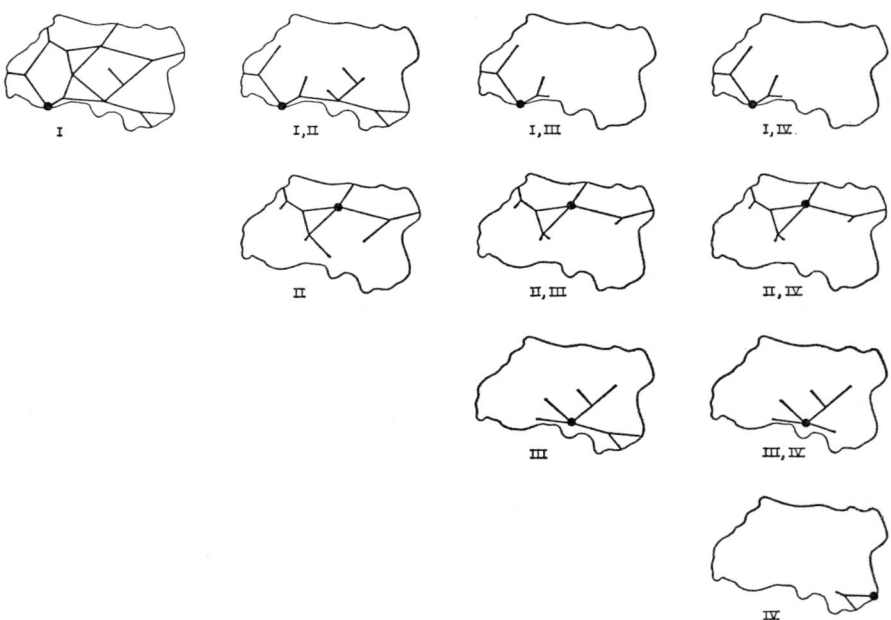

15.27 Hierarchy of road networks based on centres I, II, III and IV (see Fig. 15.26) (*Source: Haggett, 1966*).

It will be clear from this diagram that four centres create as many as *ten* polar networks. The general rule for the maximum number of polar networks is

$$\max N_n = \tfrac{1}{2}(P_n^2 + P_n)$$

where N_n are the polar networks and P_n the poles. We should note however that this is a *maximum*, and that in practice the introduction of new lower-order centres will 'disturb' only those polar networks immediately about it. Comparison of Figures 15.27D and 15.27E shows that on the island network, the introduction of centre *IV* disturbs the networks centred on *II* and *III* but *not* that centred on *I*; consequently the polar networks N (*I, III*) and N (*I, IV*) are identical. Since the smallest number of networks we can create is only twice the number of poles less one (i.e. assuming lower-order centres are introduced sequentially along a straight line at increasing distances from the highest-order centre) we know that

$$2(P_n) - 1 > N_n > \tfrac{1}{2}(P_n^2 + P_n)$$

Thus with ten centres, we know the number of polar networks will be greater than 19 but less than 55.

Stage 4: circuit decomposition. This stage of circuit simplification has no direct analogy in watershed analysis since stream patterns contain (at least in their upstream sections) no significant closed loops or circuits. Examination of the road network about centre *I* show that while there are alternative routes between this pole and other vertices on the network some routes are considerably longer than others. It is possible therefore to identify *indifference points* on the network that are equidistant from the pole by alternative routes. Thus indifference point α (on Fig. 15.26F) is 27 miles from *I* by both route *A* and route *B*; similarly indifference point β is 32 miles from *I* by alternative routes. Using these indifference points as 'breaks' on the circuit it is possible to identify a dendritic structure (Fig. 15.26G) consisting of the shortest-distance paths to a given pole.

Clearly gross distance may be in an inappropriate measure of 'nearness' and any set of numbers incorporating capacity, time, etc. could be substituted for distance without violating the indifference-point principle (Haggett, 1965, p.38).

Stage 5: Path ordering. Once the circuits have been decomposed, the ordering of the resulting paths is exactly similar to that of drainage nets. Strahler's (1952) combinatorial ordering system may be applied to the paths on the island network (Fig. 15.26G) using the procedures described in Section II (1a) above. Path ordering is thus directly due to channel ordering and we can identify second- and fourth-order paths centring on the largest settlement, *I*. The main difference between watersheds and polar networks is that whereas the former drain towards a *line* (either a larger stream or the coastline), the latter drain towards a point.

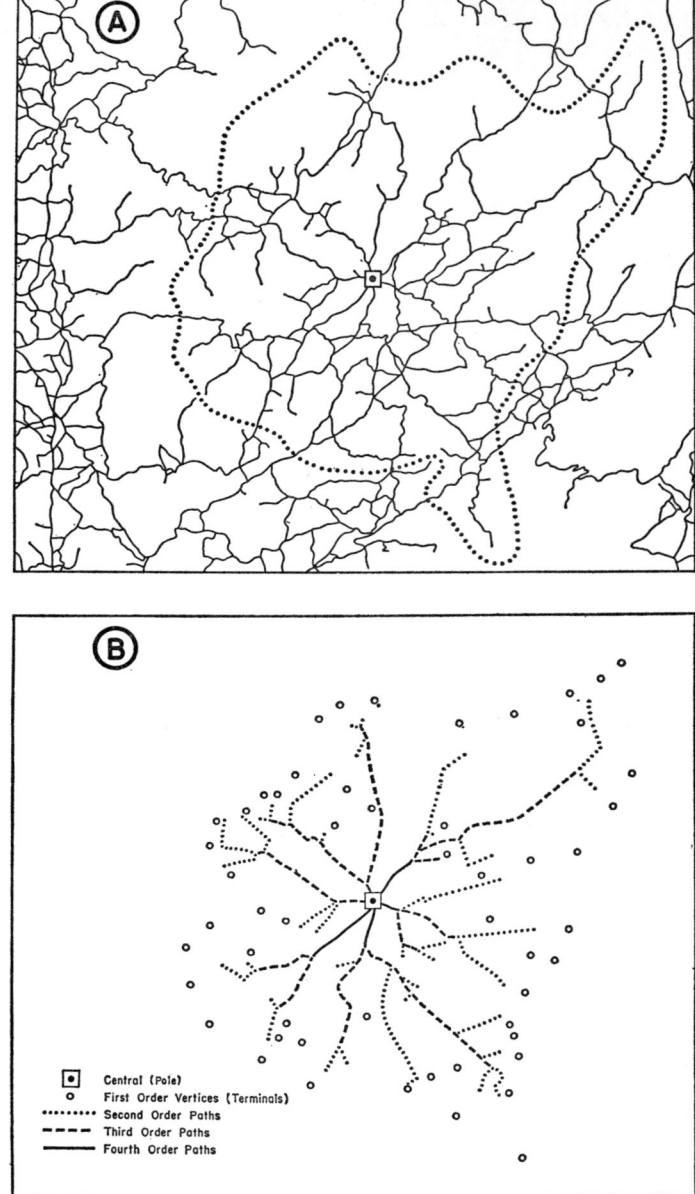

15.28 Decomposition of the road network around the town of Viseu, Portugal (A) into a dendritic form (B). First-order paths have been omitted for clarity (*Source: Haggett, 1966*).

NETWORK MODELS IN GEOGRAPHY [661]

A practical application of this conversion algorithm is shown in Figure 15.28. Here the central pole is the town of Viseu (1950, popn. 13,000) in the Beira Alta region of north-central Portugal, and the network (main and secondary roads) is shown for the region around the town in Figure 15.28A. The hierarchical position of Viseu in respect to other 'poles' is determined from population data. From this general network the polar network centred on Viseu rather than 'competing' centres is isolated (the dotted line in Fig.

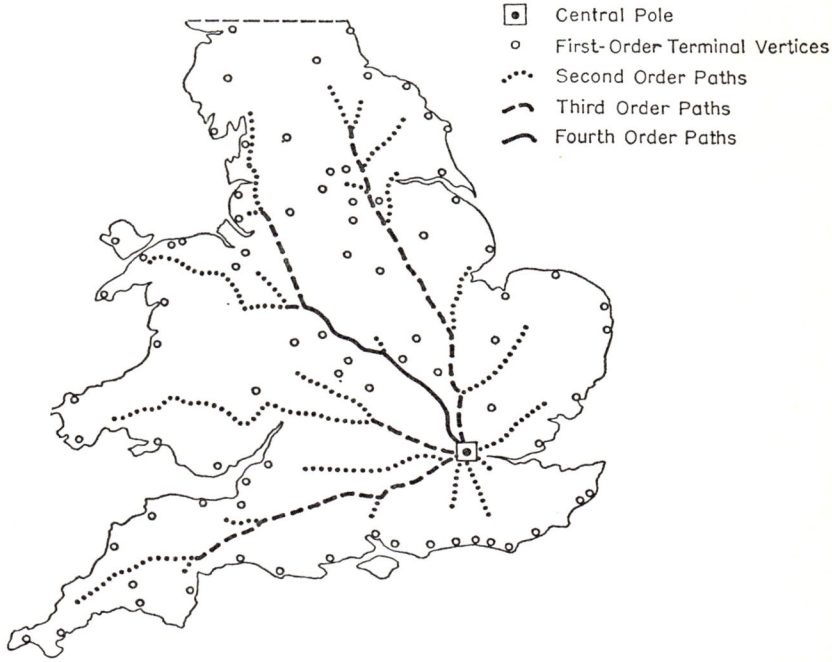

15.29 The road network of England and Wales ordered with reference to the London terminal. The map is based on the shortest route to the terminal from 210 provincial centres using Class I roads and Motorways (*Source: Haggett, 1966*).

15.28A), and dissected to give the ordered pattern of paths shown in Figure 15.28B. For simplicity the first-order paths have not been shown, but the dendritic structure of fourth-order paths centring on Viseu is striking.

A more relevant application of the ordering algorithm is shown in Figure 15.29 for leading roads (Motorways and Class A roads) in England and Wales. In this case the direct routes linking some 210 towns with the pole (i.e. London) were plotted; cross routes not on the shortest-distance path between London and these centres were not examined. The pattern shown highlights the importance of the M1 motorway from Birmingham to

[662] MODELS IN GEOGRAPHY

London; it is the only fourth-order path shown on the system. Inclusion of Scotland in the analysis would clearly increase the order of the northward path of the A1, probably making it a second fourth-order path.

3 Ordered circuits: regularities and system implications

The full testing of the ordering-system outlined above has so far been confined to one country, Portugal, where an unusually good record of traffic flows on inter-urban highways is available. Figure 15.30 summarizes the results obtained from the analysis of the northern part of the highway system centring on the city of Oporto (1950, popn. 281,000) (Fig. 15.30A). This zone, bordered on the south by the deeply-incised Douro river, on the west by the Atlantic and on much of the remaining two sides by Spain, provides an exceptionally interesting test area since the amount of traffic moving across the borders of the network (except through the pole, Oporto) is negligible; i.e. the polar network is reasonably insulated from 'overspill' flows.

What the three graphs broadly show are (i) that the number of paths is

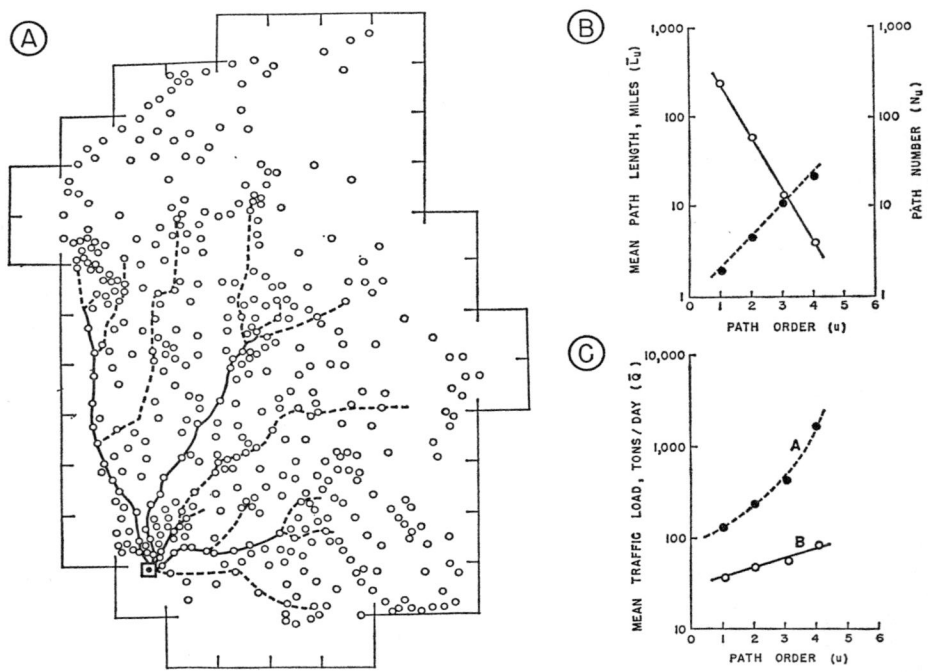

15.30 Analysis of the road network north of Oporto, Portugal. (A) Location of sampling points on the road system; (BC) Results of the analysis in terms of path number, path length, and path flow related to path order (*Source: Haggett, 1966*).

inversely related to path order (Fig. 15.30B); (ii) that the mean length of paths is directly related to order (Fig. 15.30B); and (iii) that the mean flow is directly related to order (Fig. 15.30C). Since in all graphs the y-axis is plotted on a logarithmic scale, we have a fairly exact parallel to the results shown by parallel order studies of 'natural' trees (e.g. those for stream networks shown in Fig. 15.12).

Perhaps the most interesting graph is the one which relates flow to order. Here a distinction is made between mechanical (A) and animal (B) modes of traction. The second type of traffic (largely ox-drawn wagons) show straight-line relationships directly comparable to the flow-order findings on streams; however, the relationship of motor traffic to order is concave with a proportionately higher proportion of the traffic on the high-order paths. How far these curves can be interpreted as changing equilibrium conditions within the network is under debate. One interpretation, that the network was developed in equilibrium with animal-traction flows and is now being subjected to sudden saturation of motor-traffic, is worth exploring further if only for the light it promises to throw on the form of future 'equilibrium' patterns. Certainly for drainage basins, Strahler (1958) has shown how delicate is the balance between the form of the stream network and the energy inputs and outputs; upsets in steady state (e.g. deforestation) brings rapid adjustments in drainage density of stream order.

It may then be possible to fit polar networks, with stream networks, within the general framework of general-systems analysis (Chorley, 1962). From one viewpoint we may regard the polar network simply as a series of basins draining to a common point (the central pole or 'sink') rather than a line (the coast), and as Figure 15.10 shows, programs with varying boundary conditions have already been written to include such a case (Schenck, 1963). Work at present in hand (Haggett and Chorley: In preparation) suggests that the branching angles of traffic networks are also strongly related to order, and follow in broad terms the rules of arterial branching recognized by Roux (1895). There is strong convergence in work on various types of flow networks which suggests, not that road patterns are exactly like stream patterns, still less that they are like vascular blood systems, but rather that all such branching networks are special cases of a general *optimal linear system*. The value then of the transformations discussed in this section is largely bringing to bear on familiar and peculiarly geographical problems, a very powerful and much more general mathematical-physical theory; it is clearly wasteful and unnecessary to develop special models of geographical phenomena when we can, by using appropriate transformations, regard them merely as spatial manifestations of existing general models.

VI CONCLUSION

This essay[1] has taken the basic proposition that a wide range of different geographical networks may be usefully analysed in terms of their common geometrical characteristics. Discussion throughout has centred on the *structure* of the networks, that is the geometrical pattern of the lines, their mutual relationships and their links to external controlling forces. One equally profitable approach would be to ignore the structure of networks and convert the lines to spatial *density* patterns. Haggett (1965, pp. 73–79) has shown for road and rail systems that spatial variations in the density of the lines is linked – at local, regional and international scales – to a recognizable set of economic development characteristics. Similarly, Strahler (In: Chow, 1964) has shown how regional variations in the density of streams may be linked to a set of controlling parameters which include bedrock, rainfall intensity and vegetation. Although such an approach loses much of the information contained in structural analyses there is some evidence that certain structural characteristics are density-dependent (see the cluster of vectors towards railway density in Fig. 15.18), and we may therefore argue that the density surface contains associated structural 'signals'.

Analysis of density surfaces of geographical phenomena is a separate field of current research where, as Chorley and Haggett (1965B) show, the application of mathematical models and digital computers is proving immensely successful. Surfaces, like point sets and line sets, form a distinct family of geometrical forms but it says much for the power of the mathematical-geometrical approach to geography that we can not only (1) convert different geographical phenomena into the same class (i.e. treat streams and highways as line sets) but also (2) transform geographical phenomena from one class to another (i.e. transform line sets to density surfaces). The gains that come from such increased flexibility are, as Wittgenstein saw, immense: their impact on traditional geography in the next decade may be decisive.

REFERENCES

AKERS, S. B. JR., [1960], The use of Wye-Delta transformations in network simplification; *Operations Research*, 8, 311–323.

AVONDO-BODINO, G., [1962], *Economic applications of the theory of graphs*, (New York).

[1] Since this essay was completed a number of important developments in the application of network models to geographic phenomena have taken place. At Berlin University, Werner (1966) has greatly extended the application of mathematical models to the location of transport routes on a plane, while in the United States work on stream systems as products of stochastic branching processes (e.g. Scheidegger, 1966) is likely to demand some re-casting of the interpretation of Horton's laws described above.

BATTERSBY, A., [1964], *Network analysis for planning and scheduling*, (London).
BERGE, C., [1962], *Theory of graphs and its application*, (London).
BERRY, B. J. L., [1960], An inductive approach to the regionalization of economic development; *University of Chicago, Department of Geography, Research Paper*, 62, 78–107.
BLACKMAN, R. B. and TUKEY, J. W., [1959], *The measurement of power spectra*, (New York).
BOWDEN, K. L. and WALLIS, J. R., [1964], Effect of stream-ordering technique on Horton's laws of drainage composition; *Bulletin of the Geological Society of America*, 75, 767–774.
BOYE, Y., [1965], Routing methods: principles for handling multiple travelling salesman problems; *Lund Studies in Geography, Series C, General and Mathematical Geography*, 5.
BUNGE, W., [1962], Theoretical geography; *Lund Studies in Geography, Series C, General and Mathematical Geography*, 1.
BUNGE, W., [1964], Patterns of location; *Michigan Inter-University Community of Mathematical Geographers, Discussion Paper*, 3.
BUSACKER, R. G. and SAATY, T. L., [1965], *Finite graphs and networks: an introduction with applications*; (New York).
CAYLEY, G., [1879], On the colouring of maps; *Royal Geographical Society, Proceedings*, 1, 259–61.
CHORLEY, R. J., [1957], Climate and morphometry; *Journal of Geology*, 65, 628–638,
CHORLEY, R. J., [1962], Geomorphology and general systems theory; *United States, Geological Survey, Professional Paper*, 500-B.
CHORLEY, R. J. and HAGGETT, P., (Eds.), [1965A], *Frontiers in geographical teaching*, (London).
CHORLEY, R. J. and HAGGETT, P., [1965B], Trend-surface mapping in geographical research; *Institute of British Geographers, Publications*, 37, 47–67.
CHOW, VEN TE, [1964], *Handbook of applied hydrology*, (New York).
CHRISTALLER, W., [1933], *Die zentralen Orte in Süddeutschland*, (Jena).
COBURN, T. M., BEESLEY, M. E. and REYNOLDS, D. J., [1960], The London-Birmingham motorway: traffic and economics; *Department of Industrial and Scientific Research, Road Research Technical Paper*, 46.
COHN, D. L., [1954], Optimal systems: I. The vascular system; *Bulletin of Mathematical Biophysics*, 16, 59–74.
CORTE, A. and HIGASHI, A., [1964], Experimental research on dessication cracks in soil; *U.S. Army Materiel Command, Cold Regions Research and Engineering Laboratory, Research Report*, 66.
COXETER, H. S. M., [1961], *Introduction to geometry*, (New York).
CREIGHTON, R. L., HOCH, I. and SCHNEIDER, M., [1959], The optimum spacing of arterials and expressways; *Traffic Quarterly*, 13, 447–494.
DIRICHLET, G. L., [1850], Über die Reduction der positiven quadratischen Formen mit drei unbestimmten ganzen Zahlen; *Journal für die reine und angewandte Mathematik*, 40, 209–227.
DACEY, M. F., [1965], A review of measures of contiguity for two and k-colour maps; *Office of Naval Research, Contract Nonr* 1228 (33), *Technical Report*, 2.
FARBEY, B. A., LAND, A. and MURCHLAND, J. D., [1965], The 'cascade' algorithm

for finding minimum distances on a graph; *London School of Economics, LSE-TNT-19*.
FLUGGE, S., [1958], *Handbuch der Physik*, (Berlin).
FORD, L. R. Jr. and FULKERSON, D. R. [1962], *Flows in networks*, (Princeton).
GARRISON, W. L., [1960], Connectivity of the interstate highway system; *Regional Science Association, Papers and Proceedings*, 6, 121–137.
GARRISON, W. L. and MARBLE, D. F., [1962], The structure of transportation networks; *U.S. Army Transportation Command, Technical Report*, 62–11.
GLUSS, B., [1961], An alternative solution to the 'lost at sea' problem; *Naval Research Logistics Quarterly*, 8, 117–121.
GOULD, P., [1965], *A bibliography of space-searching procedures for geographers*; Pennsylvania State University, Department of Geography, (Mimeographed Report).
GOULD, P., [1966], *On the geographic interpretation of eigenvalues*; Pennsylvania State University, Department of Geography, (Mimeographed Report).
GRAY, D. M., [1961], Interrelationships of watershed characteristics; *Journal of Geophysical Research*, 66, 1215–1223.
HAGGETT, P., [1965], *Locational analysis in human geography*, (London).
HAGGETT, P., [1966], On certain statistical regularities in the structure of transport networks, (Mimeo.).
HAGGETT, P. and CHORLEY, R. J., [In preparation], *Network models in geography: an integrated approach*, (London).
HAIRE, M., [1959], *Modern organization theory*, (New York).
HARARAY, F., NORMAN, R. Z. and CARTWRIGHT, D., [1965], *Structural models: an introduction to the theory of directed graphs*, (New York).
HERMAN, R., (Ed.), [1961], *Theory of traffic flow*; Proceedings of the symposium on the theory of traffic flow held at the General Motors research laboratories, Warren, Michigan (USA), (Amsterdam).
HORTON, R. E., [1945], Erosional development of streams and their drainage basins: hydrophysical approach to quantitative morphology; *Bulletin of the Geological Society of America*, 56, 275–370.
HULTLAND, G., [1962], Virkestransporterna i Kalix älvdal, 1951–60; *Geographica, Shrifter från Uppsala Universitets Geografiska Institution*, 27.
ISBELL J. R., [1957], An optimal search pattern; *Naval Research Logistics Quarterly*, 4, 357–359.
KANSKY, K. J., [1963], Structure of transport networks: relationships between network geometry and regional characteristics; *University of Chicago, Department of Geography, Research Papers*, 84.
KLEINROCK, L., [1964], *Communication nets: stochastic message flow and delay*, (New York).
KRUMBEIN, W. C. and GRAYBILL, F. A., [1965], *An introduction to statistical models in geology*, (New York).
KU, Y. U. and BEDROSIAN, S. D., [1965], On topological approaches to network theory; *Journal of the Franklin Institute*, 279, 11–21.
LACHENBRUCH, A. H., [1962], Mechanics of thermal contraction cracks and ice-wedge polygons in permafrost; *Geological Society of America, Special Paper*, 70.
LEOPOLD, L. B. and LANGBEIN, W. B., [1962], The concept of entropy in landscape evolution; *United States Geological Survey, Professional Paper*, 500-A.

LEOPOLD, L. B. and MILLER, J. P., [1956], Ephemeral streams: hydraulic factors and their relation to the drainage net; *United States, Geological Survey, Professional Paper*, 282-A.

LEOPOLD, L. B., WOLMAN, M. G., and MILLER, J. P., [1964], *Fluvial processes in geomorphology*, (San Francisco).

LIGHTHILL, M. J. and WHITHAM, G. B., [1955], On kinematic waves II. A theory of traffic flow on long crowded roads; *Royal Society of London, Proceedings, Series A*, 229, 317–345.

LITTLE, J. D. C., MURTY, K. G., SWEENEY, D. W. and KAREL, C., [1963], An algorithm for the travelling salesman problem; *Operations Research*, 11, 972–989.

LÖSCH, A., [1954], *The economics of location*, (New Haven).

LUBOWE, J. K., [1964], Stream junction angles in the dendritic drainage pattern; *American Journal of Science*, 262, 325–339.

LYUSTERNIK, L. A., [1964], *Shortest paths: variational problems*, (Oxford).

MELTON, M. A., [1959], A derivation of Strahler's channel-ordering system; *Journal of Geology*, 67, 345–346.

MIEHLE, W., [1958A], Link-length minimization in networks; *Operations Research*, 6, 232–243.

MIEHLE, W., [1958B], Calculation of higher transitions in a Markov process; *Operations Research*, 6, 693–698.

MOMSEN, R. P., [1963], Routes across the Serra do Mar: the evolution of transportation in the highlands of Rio de Janeiro and São Paulo; *Revista Geografica*, 32, 5–167.

MOORE, E. F., [1959], The shortest path through a maze; *Annals of the Computation Laboratory of Harvard University*, 30.

MORRILL, R. L., [1962], Simulation of central place patterns over time; *Lund Studies in Geography, Series B, Human Geography*, 24, 109–120.

MORRILL, R. L., [1965], Migration and the growth of urban settlement; *Lund Studies in Geography, Series B, Human Geography*, 26.

MURRAY, C. D., [1927], On the branching-angles of trees; *Journal of General Physiology*, 10, 725.

ORE, O., [1963], *Graphs and their uses*, (New York).

PETERSON, J. M., [1961], Freeway spacing in an urban freeway system; *American Society of Civil Engineers, Transactions*, 126, 383.

PHILBRICK, A. K., [1957], Principles of areal functional organization in regional human geography; *Economic Geography*, 33, 299–336.

PITTS, F. R., [1965], A graph theoretic approach to historical geography; *Professional Geographer*, 17 (5), 15–20.

POLLACK, M. and WIEBENSON, W., [1960], Solutions of the shortest-route problem: a review; *Operations Research*, 8, 224–230.

POLYA, G., [1954], *Induction and analogy in mathematics*, Vol. 1, (Princeton).

PRIHAR, Z., [1956], Topological properties of telecommunication networks; *Proceedings of the Institution of Radio Engineers*, 44, 929–933.

RAPAPORT, H. and ABRAMSON, P., [1959], An analogue computer for finding an optimum route through a communications network; *Institute of Radio Engineers, Transactions on Communications Systems*, CS-7, 37–42.

RASMUSSON, G., [1962], Granite quarrying and the landscape: a comparative photogeographical study of the Malmön island, the Swedish west coast; *Lund Studies in Geography, Series C, General and Mathematical Geography*, 4.

RIORDAN, J., [1958], *An introduction to combinatorial analysis*, (New York).

ROUX, W., [1895], *Ges. Abhandlungen über Entwicklungsmechanik der Organismen, Band I, Funktionelle Anpassung*, (Leipzig).

RUHE, R. V., [1952], Topographical discontinuities in the Des Moines lobe; *American Journal of Science*, 250, 46–56.

SCHEIDEGGER, A., [1966], Stochastic branching processes and the law of stream orders; *Water Resources Research*, 2, 199–203.

SCHENCK, H. Jr., [1963], Simulation of the evolution of drainage-basin networks with a digital computer; *Journal of Geophysical Research*, 68, 5739–5745.

SCHUMM, S. A., [1963], Sinuosity of alluvial rivers on the Great Plains; *Bulletin of the Geological Society of America*, 74, 1089–1100.

SHIMBEL, A., [1953], Structural properties of communication networks; *Bulletin of Mathematical Biophysics*, 15, 501–507.

SHIMBEL, A. and KATZ, W., [1953], A new status index derived from socio-metric analysis; *Psychometrika*, 18, 39–43.

SILK, J. A., [1965], Road network of Monmouthshire; *University of Cambridge, Department of Geography, B.A. Dissertation*.

SPEIGHT, J. G., [1965], Meander spectra of the Angabunga river; *Journal of Hydrology*, 3, 1–15.

STRAHLER, A. N., [1952], Hypsometric (area-altitude) analysis of erosional topography; *Bulletin of the Geological Society of America*, 63, 1117–1142.

STRAHLER, A. N., [1958], Dimensional analysis applied to fluvially eroded landforms; *Bulletin of the Geological Society of America*, 69, 279–300.

TAAFFE, E. J., MORRILL, R. L. and GOULD, P. R., [1963], Transport expansion in underdeveloped countries: a comparative analysis; *Geographical Review*, 53, 503–529.

TANNER, J. C., [1961], Factors affecting the amount of travel; *Department of Scientific and Industrial Research, Road Research Technical Paper*, 51.

THOMPSON, D'ARCY W., [1917 and 1942], *On growth and form*, (Cambridge).

VANCE, J. E. Jr., [1961], The Oregon Trail and the Union Pacific Railroad: a contrast in purpose; *Annals of the Association of American Geographers*, 51, 357–379.

WALLACE, W. H., [1958], Railroad traffic densities and patterns; *Annals of the Association of American Geographers*, 48, 352–374.

WARNTZ, W. [1961], Transatlantic flights and pressure patterns; *Geographical Review*, 51, 187–212.

WARNTZ, W., [1965], A note on surfaces and paths and applications to geographical problems; *Michigan University Inter-University Community of Mathematical Geographers, Discussion Papers*, 6.

WEAVER, J. B. and HESS, S. W., [1963], A procedure for non-partisan districting: development of computer techniques; *Yale Law Journal*, 73, 288–308.

WELLINGTON, A. M., [1887], *The economic theory of the location of railways*, (New York).

WERNER, C., [1966], Zur geometrie von Verkehrsnetzen; *Abhandlungen des 1 Geographischen Instituts der Freien Universität Berlin*, 10, 1–136.

PART V

Information Models

CHAPTER SIXTEEN

Maps as Models

C. BOARD

I don't believe in maps because it never looks like it says on the maps when you get there. From an advertisement issued by the BREWERS' SOCIETY
Maps, representing results of original surveys in visual form, constitute simple models of a 'real' world, ...   KANSKY, 1963, p. 7.
There is no such thing as a perfectly faithful model; only by being unfaithful in some *respect can a model represent its original.*
   BLACK, 1962, p. 220.

One suspects that the young ladies on the advertisement, who said they did not trust maps, were complaining more about the ability of their male companions to make something of the 'wiggly lines on the map'. Of course, no map can perfectly depict reality, but in *not* doing so it is all the more useful. The only perfectly faithful depiction would be an identical copy of reality itself. The reasons are not far to seek. Reduction in scale, the loss of the third dimension, human artifice in the creation of the representation, and the inability to read the depiction perfectly satisfactorily are the most important. Although some of the secrets of nature can be unravelled without maps, patterns over relatively large areas are often best detected and problems identified by careful studies of maps (Wooldridge and East, 1951, p. 65). The same authors observe, with envious sympathy, that 'a quick-witted mobile urchin may in one sense "know his geography" as he conducts us by tortuous by-ways from station to hotel, but neither he nor we have any adequate picture of the pattern of the town, without the benefit of maps' (1951, p. 65).

This chapter will regard maps as iconic or representational models and as conceptual models in a framework provided by the human being's struggle to communicate to his fellows something of the nature of the real world. There have been few previous attempts to generalize about maps in this way. Among the most notable are Schmidt-Falkenberg (1962), de Dainville's historical study (1964), Moles (1964) and Bunge's account of metacartography

(1962). Chorley (1964, p. 136) in a statement of the place of analogue models in geographical investigation, has pointed out that though none of them are completely successful, few are without some value. At the conclusion of a discussion which employed maps as conceptual models (Haggett, 1964, p. 380), Stamp expressed the hope that such models would be torn up if necessary, just as the representational maps would have been, once they had served a purpose or had been superseded. The map can so easily be the point of contact between 'the modern quantitative approach' and the traditional approach.

THE MAP-MODEL CYCLE: THE ARGUMENT

It is comparatively easy to visualize maps as representational models of the real world, but it is important to realize that they are also conceptual models containing the essence of some generalization about reality. In that role, maps are useful analytical tools which help investigators to see the real world in a new light, or even to allow them an entirely new view of reality.

There are then two major stages to the cycle of map making. First, the real world is concentrated in model form; secondly the model is tested against reality. In practice the scientist who makes such maps has a new view of the real world. It is also axiomatic that the cycle may begin again with the revised view of the real world. For example, a series of journeys through an area for which there is only minimal map cover may suggest that there are interesting variations in land use pattern. The next and obvious step is to make (by some suitable method) a map which records the significant elements of that land use pattern. Once completed, the map of land use is taken into the field, or is compared with reality in some other way. Speculations on the relationships between land use and physical, economic and cultural factors may be tested. In many cases such tests will involve the design and construction of new maps of both trends and relationships in an attempt to unravel some of the complex patterns of the real world. Sometimes the process of investigation starts with a map, whose elements somehow give rise to some speculation in relation to the origin of, for example, a drainage pattern or some peculiarity in a maze of property boundaries. In this case the map, which is already a model of what it portrays, is dissected as is the landscape or the environment of the real world (e.g. Conzen, 1960, Chaps. 1 and 2). After such an investigation some of the results may well be presented in map form, thus entering yet another phase of the cycle. In both model making and model testing the principles underlying the methods of map making are strikingly similar. Though the processes of abstraction, model construction and model testing may very well continue without any marked breaks, for the purposes of this essay it is convenient to start with the real world and identify distinct steps in the processes in order that the relationship between maps and models may be more clearly seen. Figure 16.1 summarizes these steps and provides a 'map' of the ensuing account.

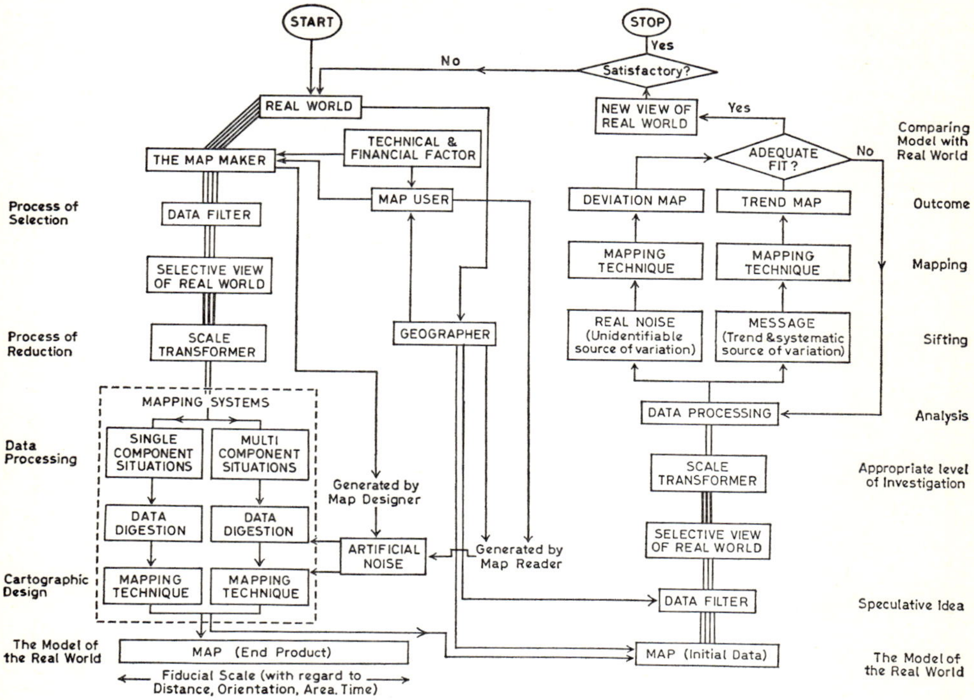

16.1 The Map-Model Cycle.

It is a truism to assert that maps are vehicles for the flow of information. Some are better vehicles than others, but the functions they perform are nevertheless similar, irrespective of their quality. It is instructive to look at the role of maps in an adaptation of a general communications system. Figure 16.2 shows such a system. The source is comparable to the real world, or the stimulus which a real world situation gives to an investigator. For example, much of the surface of the Midland counties of England is characterized by a pattern of 'ridge and furrow' (Mead, 1954; Harrison, Mead and

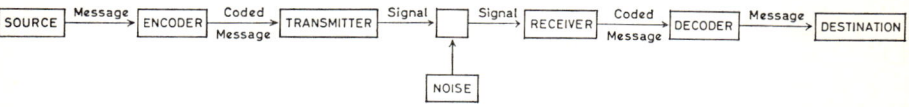

16.2 Generalized communication system (*After Johnson and Klare, 1961, p. 15*).

[674] MODELS IN GEOGRAPHY

Pannett, 1965). The curiosity of observers was excited by this phenomenon, which led to its being identified either on ground or on air photographs with a fair degree of certainty. This message generated by the cultural landscape has been simply coded (black indicates the presence of ridge and furrow) and set down in a geographical context, on a map. The stimulus of the pattern of black patches and irregular blocks provides the cartographic signal. The maps (see Fig. 16.3) are comparatively free from distracting information; they have a low noise level. Only essential names and insets are retained, and

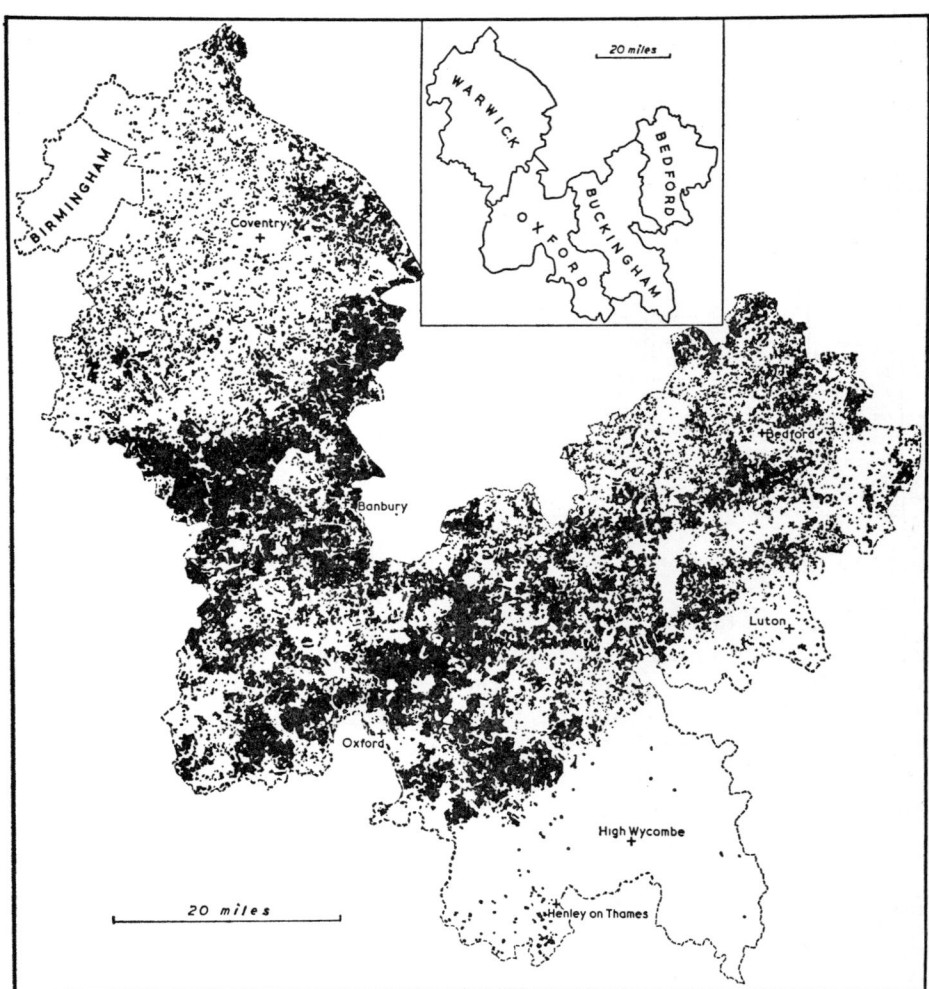

16.3 A simple cartographic signal – a ridge-and-furrow map of the Midlands of England (*Source: Harrison, Mead and Pannett, 1965, Fig. 1*).

they do not to any extent blur the signal. The stimulus, once printed and circulated in the pages of the *Geographical Journal*, is taken up through the eyes (receiver) of the readers of that periodical. The black pattern and its complementary white matrix is decoded into a pattern of distribution. The shapes are deciphered and keyed into what is already known of that fragment of the English landscape.

In other cases, for instance where it is crucial that a representation of a landscape feature such as a road junction, be firmly placed in the mind of a traveller to help him to go in the right direction, simple maps containing a maximum of information and a minimum of noise are the most effective. The provision of superfluous detail only obstructs the transmission of the message to the recipient. Advertisements such as one used by French national railways to persuade Parisians to travel at week-ends are simple messages stressing only essentials (Fig. 16.4).

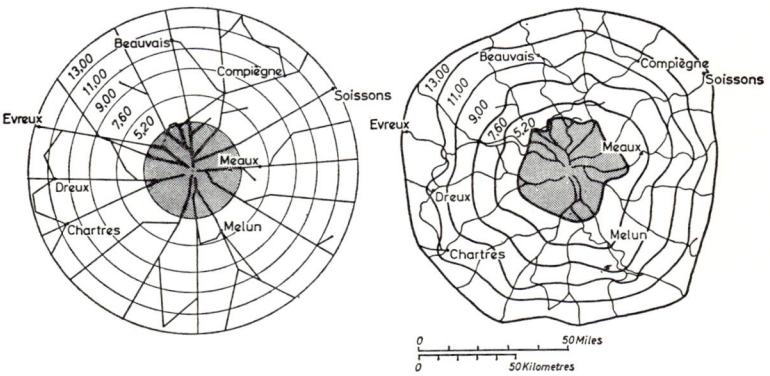

16.4 A simple cartographic message (left) indicating the price of a second class week-end ticket to different stations around Paris adapted from a leaflet advertising special fares. Contrast (right) the actual pattern of lines and places and the configuration of the price zones on the left-hand topographic map (*Source: Leaflet published by Societé Nationale des Chemins de fer francais, 1965*).

MAKING THE MODEL

The map maker

It is a commonplace that cartography, the making of maps, combines the characteristics of both a science and an art. Even the maps produced as part of the output of computers (Tobler, 1965) require the input of instructions from a designer. In fact several different sets of instructions may be desirable if a series of trial maps from a body of data is being produced as preliminary to a fair-drawn map for publication (Monmonier, 1965, p. 13). Such maps minimize the influence that human factors may have upon the final design, but they do not remove it.

Harrison (1959, pp. 29-30) has pointed out that Eckert (1908) was not wholly correct in labelling the topographic map as precise, reproducing 'facts as they exist in nature', and the thematic or geographically abstract map as artistic. However, this criticism is answered by Eckert himself:

> As long as the scale allows the objects in nature to be represented in their true proportion on the map, technical skill alone is necessary. Where this possibility ends the art of the cartographer begins. With generalization art enters into the making of maps. (Eckert, 1908, pp. 346-347.)

No map can show the objects in nature in their true proportion. The largest conceivable scale may enable the exact width of a pavement and roadway to be shown, but nothing short of a facsimile will permit the true representation of all the details of a manhole cover with the maker's name included! The very act of selection of some details rather than others for portrayal involves a decision by the cartographer and one which introduces art to the map.

In what is perhaps the most thorough account of the subjective element in maps, Wright (1942) considers scientific integrity the fundamental quality in map makers. Sometimes cartographers are tempted to depict country 'with an intricacy of detail derived largely from the imagination', especially when the quantity of known detail is scanty. Such an attitude to map making may result in a loss of information by reducing the differences between regional patterns (see Fig. 16.5). However the reverse may be true in the case of a cartographer as skilled as Robert Dawson (1776-1860), who started work as a draughtsman for the Ordnance Survey at £54 per annum in 1794. (Dictionary of National Biography, p. 678). 'Drawing in its application to maps' was not for him 'confined to delineation only, but to the whole expression of form in respect of ground . . . it is *natural history* drawing of land in full perspective perception' requiring 'the ordinary graphic qualification of the artist combined with some knowledge of some physical geography and geology' (Dawson, 1854, quoted by Harris, 1959, p. 517).

Bias in maps

The human element obtrudes further in the case of maps drawn for propaganda purposes. The aims of such persuasion may be commercial or political. Everyone is familiar with the tourist maps which are comfortably filled (or crowded) with details of the activities attractive to the potential tourist. But, it is not always so obvious that agencies, anxious to portray their territories to the best advantage, often adopt different standards for the inclusion of detail as between their territory and surrounding areas. This deliberate choice of

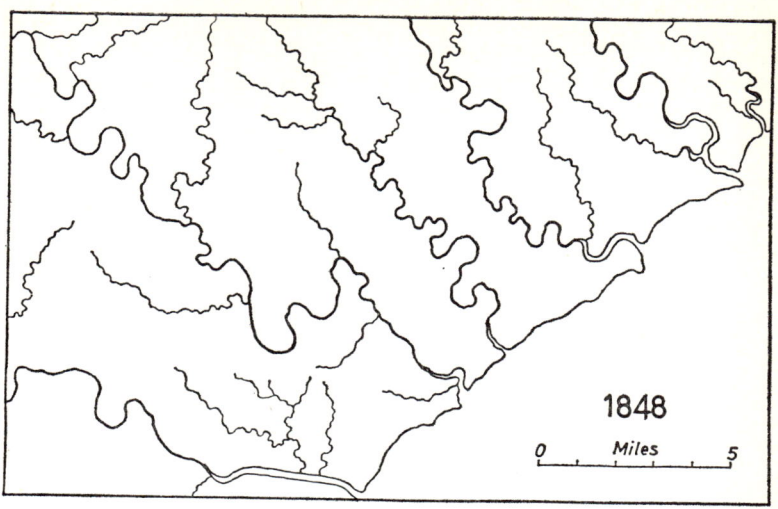

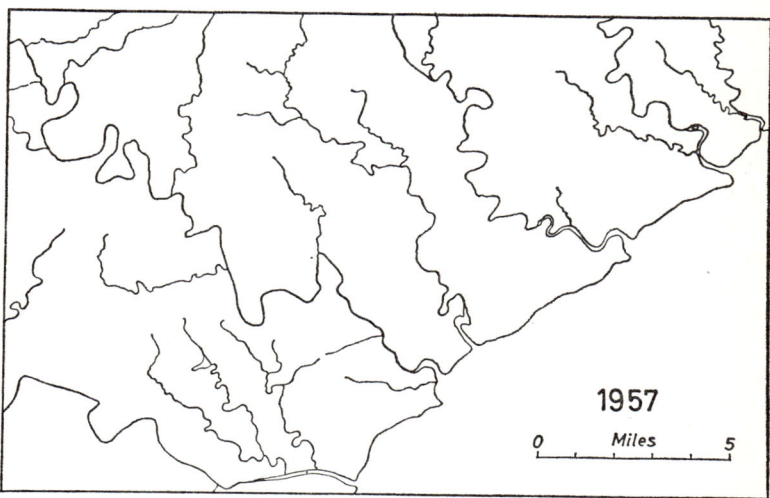

16.5 Artificial intricacy contrasted with geographical 'reality'. Rivers on Jervois' map of British Kaffraria (Eastern Cape Province) in 1848 contrasted with the river pattern on a modern map (*Trigsurvey, Pretoria, 1957, In Board, 1962, volume of maps*).

[678] MODELS IN GEOGRAPHY

detail can be even more misleading than the emphasis of detail in one area as opposed to another (Figs. 16.6 and 16.7). Occasionally, the map maker who has the task of portraying a pattern, such as a railway system with all its stations together with the connections with other systems, may have to distort

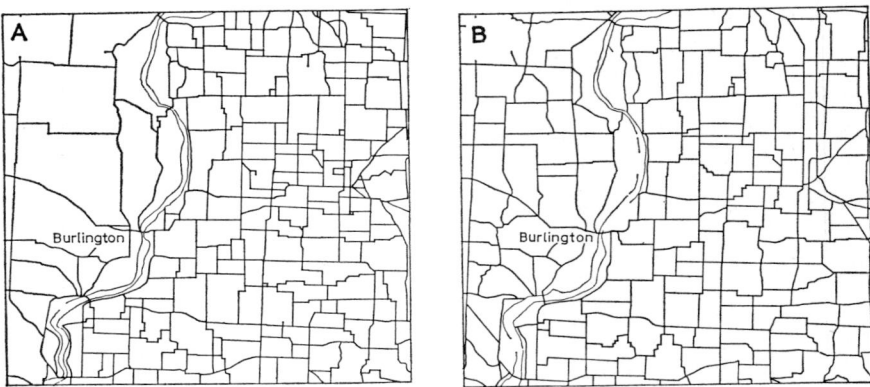

16.6 The influence of the map maker: oil company cartography contrasted with official cartography in the same areas (Illinois – Iowa).
    (A) Road pattern shown on the official highway map of Illinois.
    (B) Road pattern shown on Standard Oil Company map of Illinois.

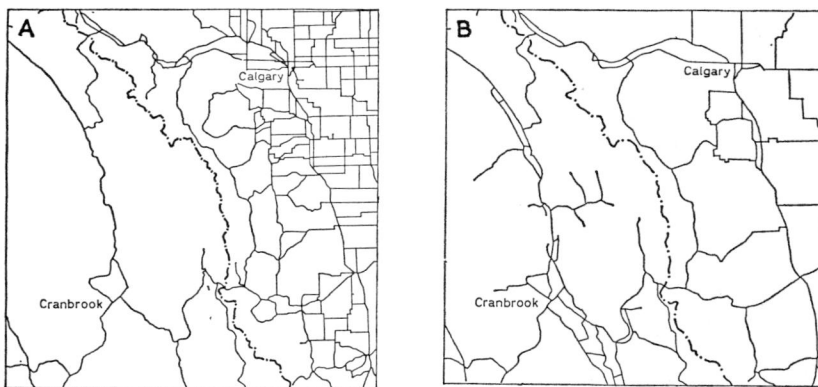

16.7 The influence of the map maker: more detail of the road pattern is given in the territory belonging to the authority producing the map.
    (A) Road pattern shown on official highway map of Alberta.
    (B) Road pattern shown on official highway map of British Columbia.

the distances and directions of lines in order to make the best use of a rectangle.

Political motives affect cartography in two main ways. In many instances, there is an official position in relation to international boundary lines and

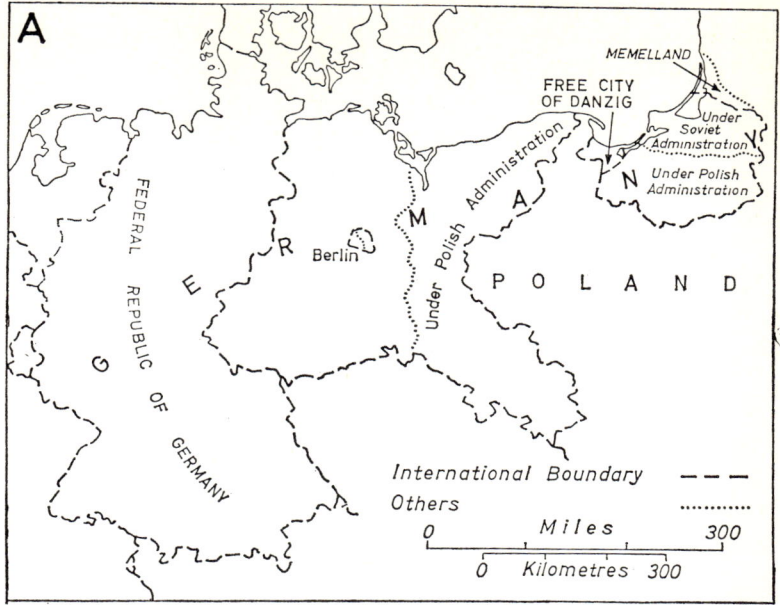

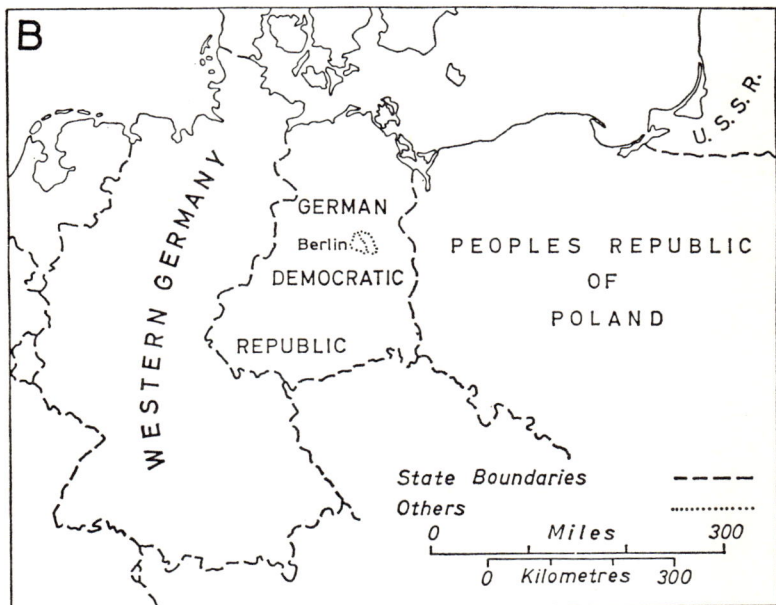

16.8 The West German view (A) contrasted with the East German view (B) of political boundaries in Central Europe (*Source: Sinnhuber, 1964, Figs. 1 and 2*).

geographical names. This is often reflected in the maps produced by a single agency, or in maps in one country which are affected by official regulations. Sometimes striking differences are created by variations in such rules in different countries. Sinnhuber (1964) has shown how the representation of the political areas and boundaries of pre-1939 Germany differs markedly in West and East German atlases, amongst others (see Fig. 16.8). In discussion Sinnhuber (1964, p. 27) also points out that place names have become so mixed up with politics that inconsistent treatment of German forms as alternatives to local forms of place names has taken place. For instance most place names in Rumania are in local forms, but names in Belgium and Italy are more frequently in German form in the *Atlas der Erdkunde* (1962).

The maker of thematic maps has a proportionately greater influence, because he is in control of the design and execution of the fair drawing as well as the processing of data to be portrayed. Typical of the bias displayed by some compilers of ethnographic maps are those by Cvijić. In particular H. R. Wilkinson has criticized his 1913 map, which indicated Macedo-Slavs in the parts of Macedonia into which Serbia hoped to expand. 'Like many other ethnographic maps of the Balkans, its ideas were dictated both by the march of events and the patriotic outlook of its author' (Wilkinson, 1951, p. 180).

Extreme cases of the flagrant use of a cartographic technique to make a particular point are fortunately relatively rare. There is an almost continuous gradation from the accidentally misleading to the deliberate attempt to distort. Maps designed for postage stamps provide interesting examples. On the one hand, the well-known 2-cent Christmas issue of the Dominion of Canada (1898) shows the British Empire in scarlet on Mercator's projection with the caption 'We hold a vaster empire than has been'. On the other hand two more recent stamps which clearly show the evidence of human artifice in mapping are those with maps of India (issued by India in 1957). These show the disputed territories of Kashmir as part of India. A contemporaneous stamp issued by Pakistan shows Kashmir as a land whose 'final status (is) not yet determined' (Kingsbury, 1964).

The map user

It would, however, be quite wrong to suggest that such aspects of map design were the product merely of the map maker's mind. Many departures from reality are perpetrated in an attempt to satisfy the requirements of map users. The most obvious illustration is the choice of map projections, especially for navigation, where Mercator or the Gnomonic projections are usual. As Robinson (1960, p. 71) emphasized, Mercator excessively exaggerates the size of the land masses in Northern latitudes. However its employment for small-scale world maps in countless atlases has been responsible for many misconceptions as to the relative sizes of different parts of the world.

Indeed Mackay (1954, p. 4) has shown that the usual aspect of Mercator does not even rate well with map users as against its oblique case when the map shape of Greenland is compared with its shape on the globe. Robinson's suggestion (1960, p. 75) that the cylindrical equal-area projection 'looks peculiar' to many people, when, provided standard parallels just below 30° are used, it has the least mean angular deformation of any equivalent world projection, demonstrates that mathematical balance may not be everything. Marschner (1943, p. 219) summarizes the position in the following manner. 'The structural property of maps on smaller scales, therefore, is a fundamental issue between the professional map users and the map makers'. But he also points out that the map users and map makers are frequently the same person. For geographical purposes Marschner (1944, p. 44) considers the property of equivalence to be most important, because so much depends on the measurement of area and the correlation of areal phenomena. Equivalence is all the more important because of the three elements (area, distance and angles) only area can be preserved as universally true on a map (Marschner, 1944, p. 45).

Map purpose

No discussion of the role of the map designer could be complete without reference to the purposes for which maps are made. Some of the more tendentious maps are of course made with the deliberate intention to deceive. They may omit details which could prove useful to some enemy power, or insert details of towns and plots of land in places devoid of habitation in order to attract settlement and land purchasers. There is abundant evidence of the latter, for example, in tract maps filed in Los Angeles and adjacent counties for towns such as Sunset, Gladstone and Richland in the late 1880's (Dumke, 1963, Chap. 14). Although maps may be made for specific purposes there is naturally enough no guarantee that they will be used in the way intended. The tract maps referred to above are a valuable source of information for the historian of urban growth in southern California. Similarly topographic maps have often been made initially for military purposes, as for example the one-inch Ordnance Survey of Britain and the 1:75,000 Austrian Staff map. But just because they provide basic information about the country, they are frequently used by non-military personnel. Indeed it is customary to use such maps as base maps for overprinting specialist information such as geology, land use or population. Linton (1948) points out that the United States Geological Survey was charged with the task of both topographical and geological survey and was able to design a topographic map 'specifically as a base for geological and other overprints'. This had a very direct influence on the design of the 1:62,500 topographic base map, which was made simpler by omitting much man-made detail and finer by the use of narrower road symbols than those usually employed on maps of comparable scales. As a result USGS geological maps are very much clearer than many others.

Another large class of maps which show clear signs of the importance of purpose in design are those intended for navigation, whether by ship, aeroplane or motor car. Ideally and in fact frequently, such maps or charts carry such information that serves to speed individuals from one place to another. The sea charts of the Marshall Islanders incorporated dominant wave directions (Lyons, 1928) and modern Admiralty charts show lighthouses, wrecks and soundings. Aeronautical charts for high-speed, low-altitude flying require prominent features at roughly 70-mile intervals to serve as checkpoints for every 4 minutes of flying time at 1,000 miles per hour. Such features are set in an extremely simplified framework with shaded relief, outlines of towns and the main lines of the transport network (Davis, 1958).

Thus we can argue that if the map user is a specialist, who has appropriate technical and financial resources at his command, maps that he commissions will tend to be 'tailor-made' to his requirements. Maps made in great quantity, for large numbers of consumers, could perhaps command the same resources by virtue of large-scale sales. On the other hand, large numbers of consumers will probably have many different requirements to make of such maps and as a result the maps themselves will represent a compromise in design. Such a difference is discernible in the design and content of truly national atlases and special editions of atlases for certain parts of the world. National atlases, because of the prestige they carry, can usually command greater financial resources than would be warranted on the grounds of numbers sold and selling price. School atlases for specific countries or regions normally include in addition to the maps to be expected in a world atlas, maps depicting special aspects of the country or region concerned. Although they are drawn specifically for that atlas, they are normally derived from more detailed maps at a larger scale. National atlas maps are frequently compiled from raw data, in order to secure consistent treatment for different topics displayed.

We shall see later how the map user can, by virtue of inbuilt or acquired restrictions on his comprehension of maps, distort the flow of information directed at him. The map maker is of course also extremely liable to distort the information by means of the particular methods he chooses to use to depict segments of the real world. Such 'artificial noise' will be considered in more detail when mapping systems are being discussed. Information about the real world is taken by the map maker, prompted by the map user, and is transformed in a number of important ways before it is presented in idealized form as a model of the real world. This information is coded in symbolic form, indeed Robinson (1960, p. 136) goes so far as to say that 'the entire map is a symbol ... and it is not quite correct to designate only certain components as symbols'. These symbols, including, conventional signs are the language of map making. Our ability to express ourselves in, or to understand this language contribute to the ease with which the cartographic message is transmitted and received. Very often trouble occurs when even the transmitter is confused about the exact nature of the message he is supposed to send. All too often in geographical texts we come across illustrations by regions of proportions of some measure related to a total amount. It is rarely clear whether the authors of pie-graph maps based on such data wish us to pick out regional patterns, estimate proportions, or estimate total figures for individual regions within the larger area. The pie-graph will not effectively do all these jobs simultaneously, judging by experiments carried out by the author.

Data selection

Having accepted the necessity for making a map, let us begin to examine how the information output of the real world is further reduced by the processes of map making. We have already seen how the very intervention of the map user and the attitude of the map maker have both combined to reduce the flow of pure information. The map maker also decides how much information can be allowed to pass through to the map. Only a selection of the innumerable bits of information can be represented, unless the map were to be at the unlikely scale of 1:1. Usually the process of data selection starts by the selection of certain classes of information and the exclusion of others. Even the most detailed topographical map usually excludes information on the precipitation or occupations of the population. Some would argue that topographic might properly exclude man-made features on the earth's surface (Stamp, 1961, p. 454). But in cartographic usage most visible features together with place names included. At this stage therefore a distinction is drawn between maps which are called thematic because they are designed to emphasize particular features or concepts and topographic maps which are more catholic in scope. The distinction is one of degree rather than of kind, but it is a useful one in that it generally reflects a fundamental difference in design. Thematic maps, being designed for comparatively small numbers of users can make use of a more esoteric language of symbolization than can topographic maps, which are usually intended to be used by a multitude of users and readers of very different capabilities. However, by selecting only part of the real world for portrayal, the map designer automatically departs from a completely faithful representation of reality. What we observe in reality is not bounded by the map's neat line.

A clear parallel to this occurs in the field of art. Gombrich (1962, p. 78) in an extremely fascinating discussion of truth and the stereotype points out that drawings from nature are not correct because they contain more detail in them. Referring to the German painter Richter's experiences in copying the Tivoli in Rome, he argues

> those who understand the notation will derive *no false information* from the drawing – whether it gives the contour in a few lines or picks out 'every blade of grass' as Richter's friends wanted to do. The complete portrayal might be the one that gives as much correct information about the spot as we would obtain if we looked from the very spot where the artist stood. . . .

Gombrich concludes 'so complex is the information which reaches us from the visible world that no picture will ever embody it all'. Significantly, Maling (1963, p. 21), in a review of quantitative methods of generalization in cartography, points out that the empirical curve which in nature is the land/sea boundary is generalized even on an air photograph. Due to limitations imposed by film grain and lens resolution the boundary is about 10 microns in width. This zone of uncertainty is from 10 to 15 times wider on any map drawn at the same scale. Because of this, irregularities narrower than twice this width cannot be shown. What Lundquist (1963, p. 35) calls 'editorial generalization', selecting *which* discrete objects should appear on a map once the number of such objects has been decided according to technical generalization, connects our discussion of data selection with generalization. It is the purpose of the map which determines which objects are to be included. For instance, a railway passenger timetable map on a relatively small scale would not normally include lines used solely for freight, however important they were. To take an extreme view, any class of information can be depicted on a map at any scale, provided that the pattern of its distribution is appropriately generalized. The Atlas of the British Flora (1962) contains maps of minute plants, whose presence is indicated within 10 kilometre squares by a black symbol in that square. The general pattern of black squares represents at that level of generalization, the distribution of a particular species.

The decision to map a particular class of object or relationship is an editorial one taken early in the process of making a map. Once the classes of features have been determined, with the exception of the minimal topographic base, the map maker will concentrate on these to the exclusion of others. The map of restaurants in France at which a good meal may have been had for 10 New Francs in 1963 (Michelin Guide, 1963, pp. 30–31) is an interesting case. Paris appears only in red as the centre of a road network serving an area devoid of such restaurants. As would be expected, many important places such as Marseille do not appear. Such maps as these present very selective views of reality. Once the selection has been made to the satisfaction of the map maker, acting sometimes in concert with map users, the next decisions revolve around the question of an appropriate scale.

Scale transformation

One seemingly uncomplicated (and as a result, overlooked) aspect of scale is the obvious one of how big an area is to be included in the map. With modern methods of printing it is perfectly possible for the whole world to be shown on a postage stamp. Such a map would not serve as a wall map to be used for teaching world political geography. There is of course an appropriate scale for any particular purpose, depending very largely upon the amount of detail the map maker wishes to include, but also upon the size of paper available. More severe restrictions operate in the case of working to a common format in an atlas or textbook, or even for a series of maps. One of the disadvantages of the Ordnance Survey's 'Third Edition' one-inch maps (published between 1901 and 1913) was the rigidity of the system of sheet lines producing a uniform size of sheet with no overlaps, taken over from earlier editions (Harley, 1962). In fact the Ordnance Survey as early as 1902 had been publishing combined sheets, where adjoining sheets contained very little land (Johnston, 1902, p. 5). Later sheet sizes were irregular, but the advantages of uniformity of format triumphed again with the Seventh Series (published from 1952 onwards). A system of overlaps between sheets ensures that map purchasers will usually get value for money. Figure 16.9 illustrates these differences for part of North Wales.

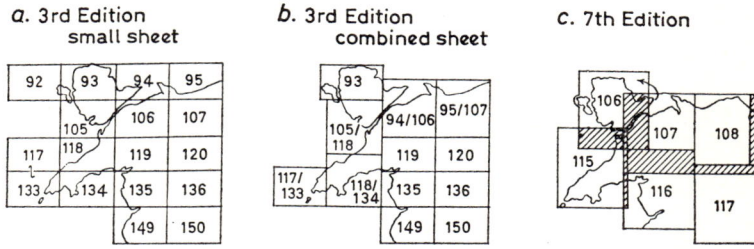

16.9 Three different arrangements of land areas within a system of sheet lines, for the same area of North Wales.
(A) Sheets same size, no overlaps.
(B) Sheets different size, no overlaps
(C) Sheets same size with overlaps.
(*Source: Ordnance Survey indexes to One Inch to the Mile sheets*) (*Crown Copyright Reserved*).

At a rather different scale, there has been much experiment to overcome some of the cartographer's basic problem, the representation of the three-dimensional globe in two dimensions. Early map projections were mainly symmetrical. Even when projections began to be interrupted to emphasize the unity of the oceans or continents respectively, they were usually symmetrical (Dahlberg, 1962). More recently further variants of interrupted projections have been developed, including Goode's arrangement of the sinusoidal in seven staggered segments. A family of azimuthal projections centred on different places has helped our understanding of the spatial relationships with respect to those places. One intractable problem remaining is that of showing all places on the globe in their true spatial relationship on a flat surface. In some cases portions of the earth's surface have been depicted twice on one map, as in the map of ocean currents in Bartholomew's Atlas of the World's Commerce (1907). In order to show the Pacific system unbroken as the relationship between the southern Indian Ocean and the Australian Bight, the longitudinal extent of nearly the whole of Australia is repeated. Repetition, again on Mercator, of some 40 degrees of longitude including Britain and western Europe at the ends of the map (Philip's University Atlas, 1946, pp. 16–17) serves to emphasize the position of those parts in relation to Eurasia and to the Atlantic without breaking the contact by sea through the Pacific Ocean.

Quite apart from questions of functional utility, the arrangement of mapped area within the neat line is subject to the vagaries of artistic taste. Robinson (1952, Chap. VIII) agrees that a map should have visual unity, eschewing centrifugal tendencies such as those inherent in flow-lines leaving the map area; and a map ought to have visual balance with elements such as land masses, title, and references balanced about the optical centre or deliberately chosen centre of interest. The use of stretches of sea for information which helps the reader to use the map, may destroy that balance.

The process of reduction

Passing information about the real world through the scale filter inevitably leads to a reduction in information. This loss of information is termed cartographic generalization and is an essential process leading to the construction of a model of the real world. Many cartographers adopt an empirical approach to generalization, using some rule of thumb such as tracing those elements of an outline which can be discerned after the original outline has been reduced to the drawing scale. The gauge of the line drawn itself determines the zone of uncertainty by obliterating indentations too small to figure independently. Maling (1963) has reviewed methods of quantitative generalization as applied to linear features, such as coasts and rivers. Imhof (1951, p. 99) points out that generalization has the effect of making the differences in patterns of features less noticeable. He also illustrates its operation through the suppression of sinuosity in rivers. In reality the Rhone from le Piz Badus to the sea is 1,320 kilometres long but measured on a map of scale 1:4 million it is about 1,000 kilometres. There is empirical evidence that the rate at which map length differs from real length is progressively reduced as map scale decreases. The Russian cartographer Volkov (Maling, 1963, p. 13) has fitted a parabola of form $y=a\sqrt{x}+b$ to Penck's measurements of the Adriatic coast on differnet maps. (y is the length of coast and x the denominator of the scale fraction) Sukhov's coefficient of general sinuosity measured for a stretch of Skärgärd coast (Maling, 1963, p. 12) also falls progressively slowly with decreasing map scale.

Lundquist (1959) in a preliminary survey of generalization sets forth several principles to be borne in mind by the map maker. These include being aware of the danger of excluding important features in areas where they may be scarce where one is applying a rigid scheme of quantitative reduction in numbers. The generalization of discrete features, such as towns and villages can be achieved in two ways. Their selection may be made according to importance until the map at the smaller scale is suitably filled with places. Or a set proportion of the places on the source map may be shown on the map of smaller scale, depending on the reduction factor. The latter is perhaps the more objective, except that the map maker still has to decide which of the

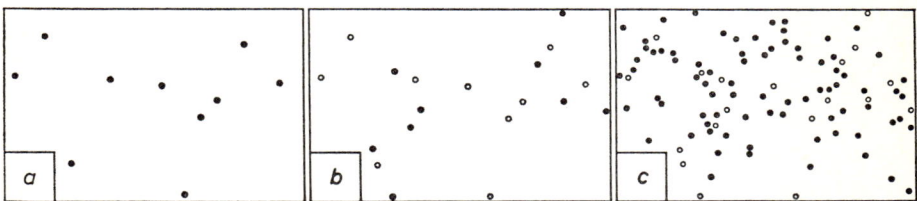

16.10 The effect of scale on the amount of detail shown. Each map shows the same part of the Witwatersrand. The places shown for the first time are black, those already on a map of a smaller scale are outlined only. (a) places shown at 1:5M, (b) places shown at 1:2.5M, (c) places shown at 1:0.5M (*Source: Times Atlas*).

smaller number of places are still to be shown. At this stage he is forced back on qualitative decision as to the relative importance of places (see Fig. 16.10) Pillewizer and Töpfer (1964) have developed a formula from a study of well-designed maps for determining the number of symbols which should be shown on maps of smaller scales generalized from those of larger scales.

$$n_F = n_A C_B C_Z \frac{\sqrt{M_A}}{M_F}$$

Where n_A is the number of symbols on the source map, of scale $1/M_A$ and n_F is the number of symbols on the derived map, of scale $1/M_F$ and C_B and C_Z are constants respectively reflecting the significance and degree of coarseness of symbol character. Maling in explanatory notes on a paper by Töpfer and Pillewizer (1966) suggested that empirical data derived from examining a number of atlas maps of Scotland broadly fitted the form of Pillewizer and Töpfer's equation.

[688] MODELS IN GEOGRAPHY

The preservation of a draft copy of the Ordnance Survey's outline map of Great Britain on the scale of 1:1,250,000 furnishes an interesting example of the process of generalizing the pattern of places on topographic maps. The earliest draft was a reduction of the Ten Mile map (1:625,000) with some features eliminated. This was redrawn to conform to the design for the smaller scale. Four sizes of town were recognized based on their population. They were distinguished by different kinds of symbols and lettering. In some parts of the map towns of 50,000 inhabitants had to be left off for fear of overcrowding it. In other areas where the density of population is low very small towns which had been left off at first were put onto the first published version (1946) on account of their local importance and others of importance were inserted. Some railway lines were also restored where they were deemed to be important through connections. A comparison of draft and published maps is given in Figure 16.11.

The generalization of symbols which cover areas, such as woodland or other land use type, involves the simplification of outline already discussed and the elimination of fragments too small to be shown at the reduced scale. Fox (1956, p. 26ff.) has suggested the minimum size of fragments which can be shown at various scales. These are based on the assumption that discrete symbols and not colours are employed. This permits a minimum map area of $\frac{1}{8}$-inch square for a separate patch of distinctive land use. Stamp (1948, p. 33) established that 6 categories could be conveniently shown at the scale of 1:633,600 if colours were employed, permitting a minimum map area of $\frac{1}{20}$-inch square depicting 155 acres. Fox therefore argues that 6 categories ought to be suitable for a map on the scale of 1:253,440 where an $\frac{1}{8}$-inch square would represent 160 acres, if symbols rather than colours were used. If the scale is doubled, twice the number of land use categories could be shown. So, not only the texture of the pattern, but the detail of classification can be related directly to changes in scale. In reducing the six basic categories of land use shown at 1:63,360 to the Ten Miles to an inch scale, Stamp (1948, p. 33) points out that although the shapes of fragments could not be preserved, the porportions of land in the different categories were maintained.

Lundquist (1963) discussing the generalization of networks of roads or railways emphasizes the importance of editorial generalization suggesting that there are relatively few problems in technical generalization. Such maps as these are usually tailored very much to the needs of particular groups of users so that the decision to retain or remove particular items of information is more critical. If a place is not sufficiently important to be shown, there is little point in inserting a road which leads to it alone.

Not only does the reduction in scale itself result in a loss of faithfulness with which reality is shown on the map, but the choice of the degree of generalization given what is possible with different techniques of printing is an important factor.

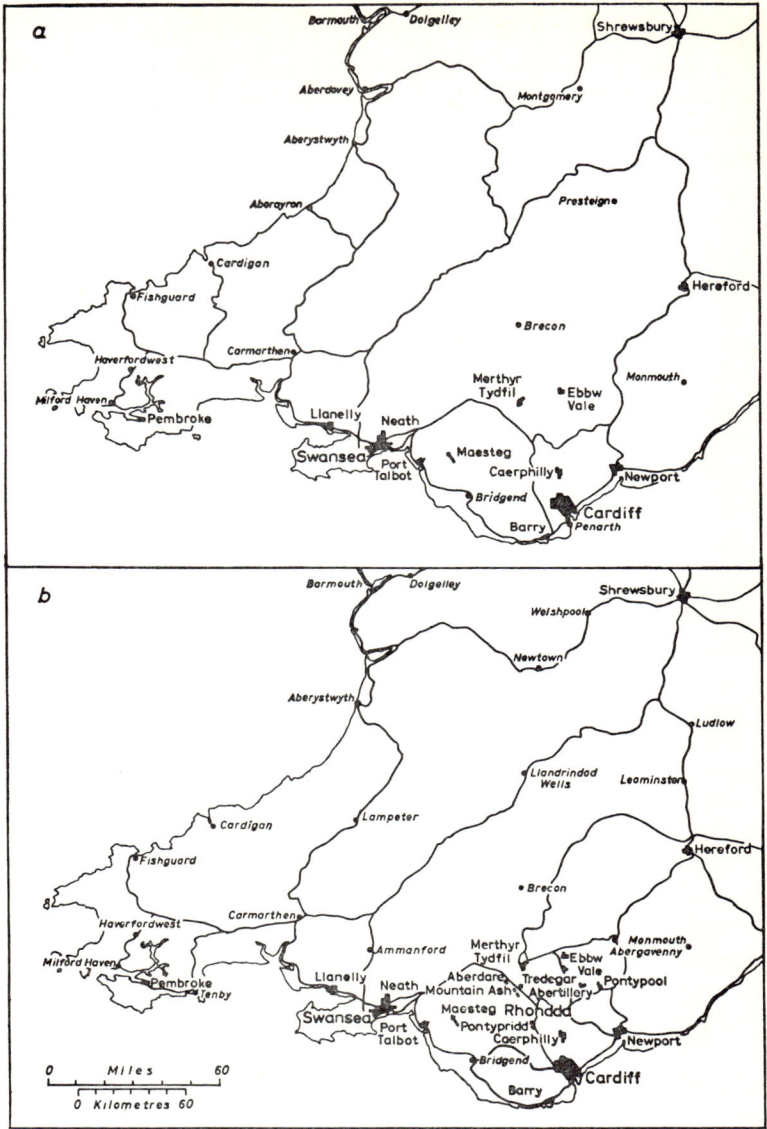

16.11 The differences between the pattern of places and railways on a draft version and the earliest published version of a small-scale map.

(*a*) draft edition, 1943: has less detail of towns and leaves out some main railways; does not sufficiently differentiate the denser pattern of features in the South Wales Coalfield.
(*b*) published edition, 1946: possesses greater verisimilitude; regional differences between the coalfield and Mid-Wales preserved (*Source: Ordnance Survey outline map on the scale of 1:1,250,000. Crown Copyright reserved*).

The mapping system

Once the basic decisions as to what to map and on which scale it should be mapped have been taken, a great range of choice of techniques is available. These are conveniently discussed under the two headings of data processing and cartographic method. Data processing, although quite vital to the end product, is not strictly a cartographic procedure. It is common to all types of description and analysis. Suffice it to point out here that no map can be better than the data from which it is compiled. The accuracy of that data may depend on observational bias, the scale of measurement and the density of measurement in relation to area being mapped.

Some maps require very little data processing before the map compilation begins. Most maps using nominal scale data and some dot maps are included in this category. Others are the end product of long and complicated calculations and are sometimes a rather insignificant part at that. Maps depicting multi-component situations clearly fall into this group. The crop-and-livestock combination maps of Weaver and others (1954 and 1956) look straightforward enough, but each enumeration district requires the computation of several sums of squares. Similarly facies maps developed by geologists (Forgotson, 1960) require considerable computation before mapping can begin. Perhaps the most complicated data processing prior to mapping is associated with the techniques of factor analysis. The effect of this form of analysis is to collapse a large number of variables upon which a series of areas is measured, for example aspects of population, income, production, mechanization, into a few basic and independent factors. Berry (1960) made choropleth maps from the loading of 95 countries on 5 components (factors). Imbrie and Purdy (1962) mapped the facies variation of carbonate rocks on the Great Bahama Bank by a similar classifying procedure. King and Henshall (1966) plotted the distribution of peasant farms in Barbados by dots according to four classes derived from a factor analysis of crops grown and kinds of livestock carried. These are essentially maps of farms of different enterprises, derived from a relatively objective procedure. A similar map, based however on subjectively defined criteria, showing enterprises of farms in Eastern England is given by Jackson, Barnard and Sturrock (1963).

Symbolization

The flow of information transmitted by the real world and filtered in the ways suggested above is now ready to be mapped. An initial distinction must be made between topographic maps and thematic maps employing the nominal and ordinal scales of measurement, and the quantitative thematic maps at higher scales of measurement. The former convey their information by means of the presence and absence of features in particular positions or areas (Fig. 16.12). Maps employing ordinal scale measurement, indicate the rela-

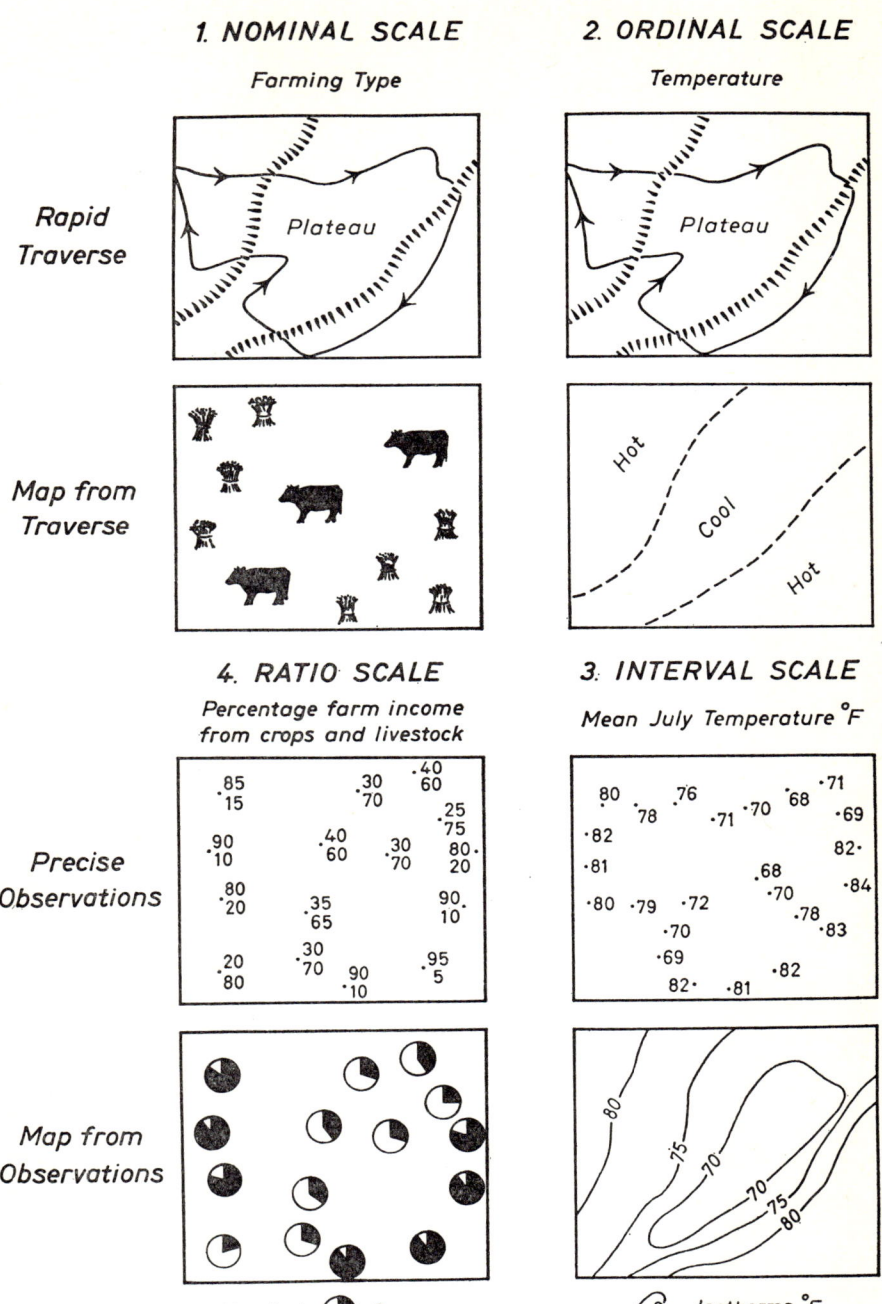

16.12 Observations and maps appropriate to the four scales of measurement.

tive size, importance or frequency of features. For such cases the techniques of generalization, particularly editorial generalization, have already determined the principal attributes of symbolization. But some choice is still left to the cartographer for varying the system of mapping. It is mainly in the realm of colour or shading, and the style of lettering where decisions still have to be taken. The exercise of choice in *this* respect is very similar to that needed for more quantitative maps, and will be mentioned when they are discussed.

For maps employing the interval and ratio scales of measurement, where statistics are being depicted, the process of handling these statistics prior to mapping furnishes much scope for further reduction in information content. Of course, this may be justified if the statistics are so unreliable that they give no more than a rough indiction of the presence of some enumerated population. Fully quantitative maps are characteristically derived from reliable statistics, available for the units of enumeration. Short of inserting a value in each of these unit-areas, the great range of values must be generalized to some extent. It is considered by Robinson (1952) that no more than 10 groups of distinctive ranges of values can be shown on a map. Jenks and Knos (1963) referring more particularly to shades of grey consider that 7 or 8 groups can be distinguished by average readers. Keates (1962) observes that from 10 to 15 separate colours may be distinguished initially, but that with training the number may rise to 50. It is usually only maps of geology, soil, vegetation and land use which incorporate *that* many colours. These are normally at the nominal or ordinal scale of measurement and have to be read carefully before they will yield information. Statistical maps on the other hand are usually intended to be read in a relatively short period of time, and so have to be correspondingly simplified. Simplification may also be achieved by combining the original units of enumeration into 'super-counties' of roughly the same size, in order to remove some of the effects of using unit-areas of widely differing size. The success of such an operation may be measured in terms of the balance between reduction in variability of size of area and the loss of detail inevitably resulting from combining areas (Haggett, 1964, p. 371). Robinson, Lindman and Brinkman (1961) have also attempted to overcome this same problem by reapportioning the values for enumeration units according to the proportion of the area of those units falling within cells of a regular hexagonal grid (see Fig. 16.13).

Class intervals

The third and most critical stage of data processing is the decision to employ a particular class interval and the base point for the gamut of intervals. Once the number of classes has been chosen, the map designer must examine the range of values (for example, densities of population) with a view to their being arranged to give adequate representation to the different parts of the range

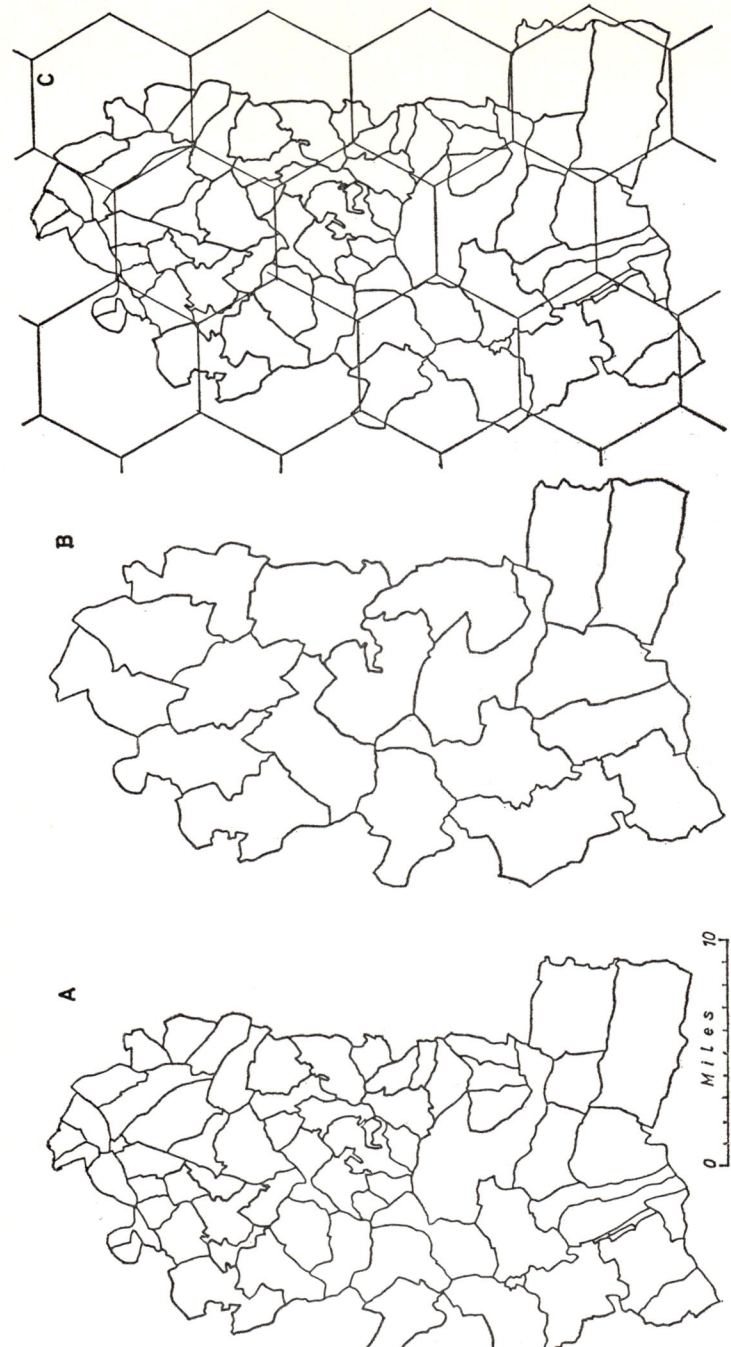

16.13 The mesh of enumeration units and their modification, an example from Hertfordshire:

(A) Pattern of parishes in the period 1896–1898 – 69 units.
(B) Combinations of parishes in (A) *(After Coppock, 1960)* – 19 units.
(C) Parishes in (A) with pattern of hexagons superimposed – 137 units.

[694] MODELS IN GEOGRAPHY

and to reproduce its characteristics. Common alternatives are steps chosen to coincide with breaks in the range of values, equal steps, steps increasing in magnitude by arithmetic increment, steps with a geometric (or logarithmic) increase in magnitude. The latter are arranged so that they normally concentrate the values in lower part of the range, and percentiles which ensure an equal number of values in each class whatever the range involved (see Fig. 16.14). Each method has its merits, but as Jenks (1963, p. 15) observes, the cartographer finds it more difficult to visualize an abstract distribution like population density and therefore does not know which method is the best. Many follow precedents set by others. One such procedure is set out by

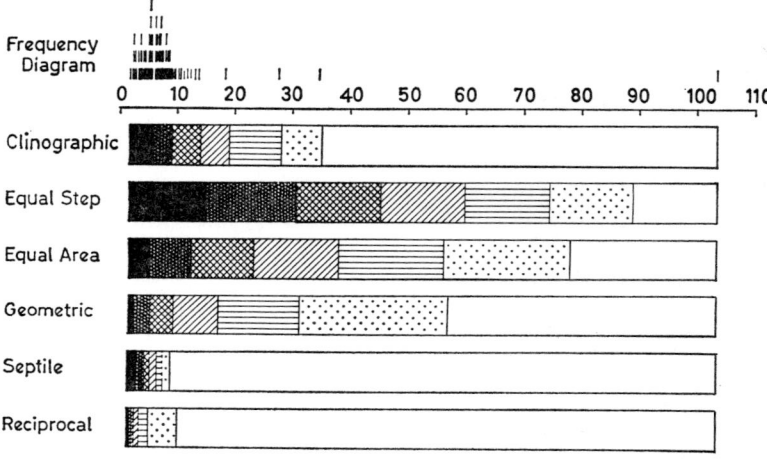

16.14 A frequency distribution (density of rural population) and six possible arrangements into seven classes showing their size and range (*After Jenks, 1963*).

Mackay (1963) and is based upon an analysis of intervals and limits selected from many geographical publications. Having permitted the cartographer to choose the number of classes and the sizes and limits of the first and last classes in the range (the latter usually being open ended), an equation determines the spacing of intervening classes. Mackay also recommends rounding off the precise values of limits obtained from the equation. Although his work, like that of others, is strictly speaking concerned with isarithms rather than statistical maps in general, the conclusions and techniques apply equally to choropleth maps.

Special problems of isarithmic maps

Further research on isarithmic maps has suggested transformation of original values in order to represent certain aspects of the distribution being mapped. For example, Krumbein (1957) discusses the advantages of transforming percentage values into angles by taking the square root of the percentage and finding the angle whose sine is equal to the square root. An arc tan. transformation can be similarly performed on ratio data giving exactly the same values (angular values) for ratios corresponding to percentages (e.g. 1:4=20%). Apart from the fact that such transformations tend to normalize the values concerned, when percentages in the middle range are bunched together on the map, those at the lower and upper ends of the scale, where percentage changes are more significant, are spread out. The opposite effect is achieved with ratio values. Such manipulations of data before mapping are applicable to multi-component situations, rather than to simple single-component cases concerning absolute values.

Blumenstock (1953) has shown that where values are derived from a sample, as for example in some agricultural and population censuses, but more commonly as meteorological observations are, observational error, sampling error and bias in some observations may all affect the reliability of isarithms. If a particular value has a relatively great chance of being incorrect by an amount which would not justify its being used to draw a detail in the isarithmic pattern, it is best ignored. Simpler, but more reliable maps result from the application of such corrections.

Another problem affecting the accuracy of isarithms based on units of enumeration is the influence of linear interpolation between the control points where the values are assumed to apply. In one case Porter (1958) showed that this procedure could result in 25 per cent of the enumeration units being misclassified, that is left on the wrong side of the isarithm. Porter was able to compare the degree of correspondence between isarithms drawn solely on a basis of statistics available for unit-areas and replotted values from supplementary information. His map shows that disagreements are related to anomalous areas of relatively small extent which the more generalized map ignored.

At the other extreme, if the units of enumeration happen to be square, so that the control points are arranged in a grid pattern, there may arise a

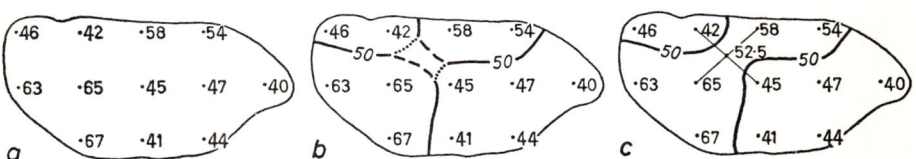

16.15 Indeterminancy in compiling isarithmic maps and its solution.
(a) the raw data arranged in grid fashion
(b) the alternative choices
(c) the solution.

situation where two alternative isarithm patterns are equally valid. Interpolating between two high values arranged diagonally can give a result differing from an interpolation between two low values arranged on the diagonal intersecting the first (Fig. 16.15). A solution is found in averaging the two pairs of values to yield a fifth control point where the diagonals cross Mackay (1953). Uhorczak (1930) saw that if the units of enumeration were arranged like bricks in a wall, the control points would be arranged in a pattern of traingles, thus overcoming the problem of indeterminacy. Czekalski (1933) followed by Mackay (1953) also recommended hexagons, which also have a triangular pattern of central control points as units for mapping. These solutions involving a reorganization of the unit-areas remain little used because of the immense labour involved in the recalculation of areas and quantities before mapping can begin (see Fig. 16.13). Robinson and others (1961) have carried out such an exercise for a portion of the Great Plains of the United States, partly with the object of avoiding cases of indeterminacy.

Point symbols

The third major class of quantitative maps are those employing a variety of point symbols. These range from the so-called dot map to the map showing quantities by proportional circles, squares, or spheres. Such maps rarely involve much preliminary data processing, and then mainly in relation to the value assigned to each symbol and the size of the symbol. Most of the calculation revolves around the values given to single dots, when it has already been decided that dots of a uniform size and value will be employed. The relative density of the pattern of dots in this case portrays the pattern of quantity over area. Robinson (1960, pp. 156–162) illustrates the effects of varying dot size and value on the appearance of the map.

With proportional symbols, such as circles, square roots are first calculated (cube roots for spheres) and a scalar applied to resulting values in order to obtain a reasonable size of symbol appropriate to the map. The representation of values by pictorial symbols (animals, men, sheaves of wheat) involves only the assignment of a range of values to a symbol of a certain size. It is unusual to find that more than four or five sizes of symbols of this kind on one map, since the map designer generally requires each size to be distinctly recognizable.

Convention and colour

In the foregoing discussion some attention has already been paid to the various kinds of symbolization available to the map designer. Data processing and symbolization are so interrelated that it would be impossible to treat the two topics separately. Nevertheless there remain some aspects of map design that can play an important part in the propagation of the message. As maps become more thematic, more specialized and more quantitative, they become more abstract. The reader will readily recall the many examples of somewhat stylized pictograms used for tourist maps. The symbols normally employed on road maps used by the motorist are only a little less pictorial. Many of the conventional signs to be found on topographic maps fall into this category too. Bagrow (1964, p I and III) shows a Mexican map that uses a line of footprints for a track and trees with a central American cast as a symbol for woodland. The Flemish and Dutch cartographers, followed by the English and German made extensive use of stylized churches and house groups on topographic maps before the end of the eighteenth century. It appears to have been the decision of the French in the years of the First Republic (1802) to alter the conventional representation of objects shown in elevation by symbols based upon the plans of objects (De Dainville, 1964). That such a change was logical is beyond question, but the mixture of symbols in both plan and elevation characterizing British and American and most European maps (with the exception of French, Swiss, Swedish and Danish) by being more conventional than logical is perhaps the more easily interpreted.

The advent of colour printing about the middle of the nineteenth century, for the first time made it possible for maps to employ extensive washes of colour as a symbol. Such colours which could be hand-painted by artists, unemployed gentlewomen or apprentices, were normally reserved for political entities in atlases. A notable exception is provided by the cadastral maps, of which single or very few copies were ever made. Indeed in England at least the convention of showing arable land in brown, woodland in green and pasture in another shade of green is of some antiquity. Some of these features appear in an estate map of Wotton Underwood at the end of the sixteenth century (Schulz, 1939 and 1954). Colour printing was used for relief maps soon after its introduction (Lyons, 1914). Layer colouring apparently originating in Germany, was to be found mainly in atlases and on wall maps. Perhaps the system which has had the longest run of popularity is von Sydow's green, white and brown in upward succession. Towards the end of the century Bartholomew's layer coloured topographic maps employed a very similar system, which has also become conventional in Britain. Both these schemes make use of the idea that the higher land ought to be darker in tone.

In other respects the conventions established in the period of hand-colouring were transferred to colour-printed maps. Blue is now almost universally used for water, both fresh and salt on topographic maps (although this has not always been so). Arab maps a millennium ago commonly showed the Red Sea in pink. Red is a colour commonly associated with towns, and this too has been employed to distinguish urban areas on German school atlases, Ordnance Survey 6 inch to the mile town maps of the 1920's, and on the maps of the Land Utilization Survey of Britain.

On thematic maps colours are less conventional, but red, by association with warmth, is frequently used on maps of population density to show the high densities. By the same token blue, a cold colour, is conventionally associated with low densities or decreases. However for maps of rainfall the blue-water association is too strong so that dark blue is conventionally reserved for very wet areas and red for the driest areas. The association of ideas is probably responsible for the popularity of shades of red for igneous rocks on geological maps (Linton, 1948, p. 143), however the imitative colouring adopted by the designers of the early maps of the British Geological Survey is also worth noting. Here the brick reds used for Old and New Sandstone systems recall the common soil colour of those outcrops, but other colour choices are less consistent with these notions. Frequently a blue to red scheme of graded colours with an intermediate band of pale tones is employed for maps intended to show a great range of values for example annual rainfall, temperature, sometimes including decreases and increases of population or to differentiate the imports and exports per caput of a commodity like timber (Bartholomew, 1907). The association of blue, blue-green or green with decrease or low values and red shades with increases or high values is now widely accepted. It is well illustrated in the *Atlas de France* (1946) and in the Ordnance Survey's 1:625,000 population maps.

Colour symbolization is also frequently found in maps which are used a lot by the general public. Transport networks, such as the London Underground or Paris Metro employ a colour code for the individual lines. Here contrasting colours aid the identification of particular routes.

By employing principles and devices such as these, most of which have some rational or at least conventional basis, the map designer helps to convey the cartographic message. By generalizing the shapes and sizes (to say nothing of the population characteristics) of towns and cities by presenting them as a series of point symbols of clearly distinguishable type, the cartographer at once diminishes the amount of information the map will convey, but hopefully he makes possible a proportional gain in legibility. The message, although it then contains less information, has a much better chance of reaching its destination. The art of cartography lies in the balance and compromise involved in the choice range between the information level and the probability of its being comprehended.

Noise

During the mapping process, both at the stage of data digestion, but rather more at the stage of mapping itself, interference of various kinds impinges upon the cartographic signal. Such unwanted disturbance is termed *noise*. Since it is generated by human agencies, it is here called *artificial noise*. Most of the *real noise* is eliminated at earlier stages in map making. It comprises information about the real world which is considered to be irrelevant to the purpose for which the map is being made. The elimination of all but the most obvious features from charts for air navigation, including the names of quite large towns, is a good illustration of the suppression of real noise. Were superfluous features to remain, the noise level of the map would be so high as to render it difficult to read in view of the high speeds of modern aircraft.

Artificial noise is of two kinds: that generated by the map designer by the methods he employs to put across the cartographic message, and that generated by the map reader who sees the elements of a map in a manner which sometimes differs from that which is intended by the designer.

Designer noise

The map designer, by making choices about class intervals, symbol values and colour or shading schemes, inserts an element of interpretation into the map. His choice is subjective, and may depend on his desire to portray a distribution in a particular way. There is no really objective arrangement of class intervals in a choropleth map. Even the employment of percentiles involves the choice of a number of classes, and such a system of class division is also directly related to the number of separate areas providing values. Wright (1942, p. 541) has observed that when the map designer relates two distributions on one map, producing 'synthetic information' from, for example, isohyets of a certain value and percentage of farmland under a particular crop, he relies more on his judgment than on unsullied information derived directly from the real world. However, as we have seen, there is a continuous range of maps from the relatively objective to the propagandist or polemic. There are many ways in which designer noise can distort the cartographic signal. Apart from the more obvious graphical possibilities already suggested, the field of lettering and titling furnishes some interesting examples. The numbering of class intervals is sometimes done in such a way that the same value appears to occur in two adjacent classes. At other times, after transformation of values to a logarithmic scale, original values are retained on the map, thus hindering the reader who may need to interpolate values. The use of awkwardly uneven or unrounded values for class limits, especially when these are based on percentiles or on equal deviations from a mean value may irritate the map reader, but it will rarely seriously inhibit comprehension.

It would clearly be impossible to convey information by maps entirely by non-verbal symbols. Maps without lettering seem unfinished and disoriented (Imhof, 1951, p. 107). Cartographic lettering often follows conventions so that the reader, once he is familiar with them, is assisted to use the map. A distinction must however be made between the style and position of lettering, and the spelling of names. Riddiford (1952) has gone so far as to say that reaction to lettering may be so violent as to prevent the user reading the map. Two extreme views are held: that lettering should be subservient to the other map detail; that lettering should be bold, or sufficiently legible to be easily read. Most discussion has centred on the issue of style versus legibility, there being a strong school of thought in favour of inconspicuous lettering. Reeves (1929, p. 437) considered that it was regrettable that any names had to appear on a map. Winterbotham (1929, p. 436) argued convincingly that:

> the art of lettering is to make it so neat and unobstrusive as not to absorb attention. Names are given to signify something in the topographical sense. I do not think one would wish to make a J or K so distinctive as to draw the eye as, for example, would a man in Bond Street in pink trousers.

Ornate and antique lettering is now usually reserved for special features, such as archaeological sites. The positioning of names on maps is frequently regarded as a craft learnt only by long apprenticeship and practice. Imhof (1962) has set down in detail, with illustrations of both good and bad work, the ground rules for positioning names. Three main principles emerge from the distillation of his great experience. The first is that names should be conveniently read in the position which the map user usually takes up. Secondly, names should be so spaced and arranged that they can be read as whole names, not appearing as two separate fragments. Thirdly, names ought clearly and unambiguously to belong to the feature they refer to. Such principles should outweigh questions of convention, such as placing river names always on the northern side of the stream. (Balchin, 1952, p. 144) In the latter case, too rigid a rule can easily introduce artificial noise to distract the reader from his task. The relationship between names and the curves of the graticule is a noisy one at best: either names, being parallel to the lines are hard to read; or if names are placed parallel to the lower neat line, their varying angle to the lines of the graticule irritates the reader.

Since the question of spelling and transliteration is not restricted to maps, it would be out of place to dwell on it at this point. Suffice it to point out that countries which use two or more languages on maps magnify the problems of cartographic lettering. Such considerations apply mainly to the reference and marginal information, which may have to be repeated in each language. The thematic maps illustrating the Tomlinson Report on the Bantu areas of South Africa (1956) provide many examples of names and descriptions in both English and Afrikaans. It is customary, at least on individual maps to reduce the possibility of confusion by placing names in one language consistently above the other. Once one is used to the idea, one learns to read alternate lines almost automatically, a habit that can sometimes be awkward. Another device is to use English and Afrikaans forms on alternate sheets of a series, as with the provisional topographical field sheets of South Africa on the scale of 1:18,000.

To flout well-established conventions, such as the depiction of the sea in blue can be a way of introducing artificial noise into the map. The Reader's Digest Atlas of the British Isles shows the sea in shades of sea-green on the grounds that the British seas do not appear blue. It takes a little time to adjust to the unfamiliar colour. More distressing perhaps is the plastic relief model of the Oxford district (1964) which shows agricultural land use in shades of light (Cambridge!) blue. It is true that the intention is to emphasize the use of land in the urban areas by employing red, orange and grey there, but the unconventional use of colour for rural land use so offends the eye at first that the message is at once distorted.

Reader noise

Cole (1964) has complained that our unwillingness to look at maps another way than with north at the top has led to 'North–South thinking', to an inability to perceive spatial relationships in other ways. Indeed one can argue that movements should be mapped so that they move away from the reader towards the top of the page. After all the coaching itineraries and motorists' routes are drawn in this fashion. Should not the expansion of settlement in the United States be plotted on a map with East at the bottom? But perhaps to follow this advice would create more noise than would be warranted in exchange for a more logical orientation, because most of us now accept that North is at the top.

The preferred orientation of maps with North at the top carries with it the corollary that they are best lit from the north-west corner. Capitalizing on this property, map designers have for some long time attempted to create an

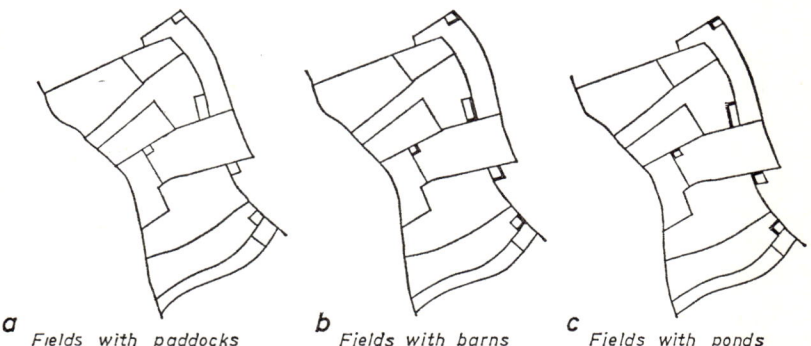

a Fields with paddocks   *b* Fields with barns   *c* Fields with ponds

16.16 An illustration of the effects of the conventional lighting of features from the North-West, or the top left-hand corner of the map.

illusion of the third dimension by thickening the boundaries of features on the shaded side, away from the north and west to indicate projections above the surface. This technique is widely employed in hill-shading. For depressions, such as ponds and pits the shaded sides lie on the north and west. In this way one can distinguish the two types of feature and identify ponds and pits with some certainty (Fig. 16.16). Reversing the map would of course destroy these illusions and would be rather like looking at air photographs the wrong way up (that is with the light not coming from the top). The discussion between Lewis (1959) and Sweeting (1958, 1959) on the interpretation of air photographs of Jamaica's Cockpit country underlines the possibility of confusion.

Another important source of noise in representational mapping is the failure of the map designer to choose shadings (or colours) to portray gradations in density or other value which look right. Map readers in some way instinctively know that one range of shadings is correct and another wrong. Skilled cartographers have for long been well aware of this phenomenon and have produced schemes of graded shades by trial and error. More recently cartographers, following the work of psychologists such as Thurstone, Ostwald and Stevens, have investigated the relationship between stimuli provided by the printed pattern and the sensation received by the map reader. Williams (1958) testing both patterns of dots and rulings found, when subjects were asked to select and place in order patterns from a wide range of choices in such a way that they gave an impression of equal gradations in density, that the percentage of white paper inked did not correspond with the Weber-Fechner Law. This states that sensation increases as the logarithm of the stimulus. This was true only of the lightly inked patterns, but for the major part of the range sensation increased arithmetically with stimulus. Similar results were obtained when using coloured patterns. Other investigators have argued that the Weber-Fechner Law implies 'that it is easier to distinguish slight differences in darker tones than in lighter tones' (Jenks and Knos, 1961, p. 323). Their empirical tests suggest that the reverse is the case. And in this they are supported by the conclusions reached by Mackay (1949), who, working with dot patterns, points out that a specific increase represents a greater proportional change when dots are fewer and farther between. It seems therefore that progressively greater increases in stimulus are required to give an impression of even increases in a sensation of density. The lack of correspondence between results of psychological tests and cartographic tests may well be put down to the very different conditions under which each type of test was carried out. Jenks and Knos (1961, p. 334) also show that the texture of a printed screen can easily disturb the map reader's sensation of a graded sequence of densities, even where the percentage area inked remains the same.

Less work has been done on the visual perception of *coloured* symbols on maps. This has not inhibited the discussion of the topic, but it is generally recognized that more research needs to be done. There are about as many views on the right system of colouring for layers on relief maps as there are colours in the spectrum. There appears to be a fundamental difference of opinion as to whether high altitudes should be lighter or darker. Imhof (1951, p. 94) suggests both are right. Starostin and Yanikov (1959) present several relief maps of the Kamchatka peninsula using different schemes of hypsographic tints. One of these is conventional in that the range passes from bright emerald green through brown to pink and white for progressively higher ground. Two others employ a green-brown range with dark brown on the mountain summits. Three make use of a range of browns, one having the dark colours for high ground and the others having light summits. The latter certainly stand out, but the lowlands, where one would expect more cultural detail, are so dark that lettering and line work there would be all but illegible. Experiments with test maps carried out by the author (1964) suggest that people who are familiar with maps tend to prefer monochrome colour schemes, with dark shades for the high points. Multi-coloured schemes are less favoured.

More research has been done on the perception of quantitative symbols of different colours. Williams (1956) found that when coloured symbols were compared with black ones of the same size and shape, only yellow ones were judged to appear more than 5 per cent larger. Considerably more is known of the way in which map readers perceive quantitative symbols printed in black. Such symbols as proportionate circles, squares, spheres and cubes are common in thematic maps used to illustrate texts on geographical topics. Williams (1956) was able to discover a general pattern from a series of experiments where students were asked to select a symbol from a large choice such that it bore some simple relationship to a standard one. For example, when asked to select a symbol which had twice the value of the standard one, the average response was to pick out one which was 1·8 times the diameter of the standard one, not one of twice the diameter, nor one which was twice the area of the standard symbol. Altogether, it seemed that for a symbol to appear x times the size of a standard one its linear dimension ought to be $x^{\cdot 8}$ larger. For the area of a circular symbol it is now conventional to make the new circle $x^{\cdot 5}$ times larger if it is intended to represent x times the value shown by the original one. Once again, it appears necessary to exaggerate the stimulus in order to convey an appropriate sensation of value. Further investigations have been carried out by Clarke (1959) and Ekman and others (1960 and 1961) to show the extent to which perception affects the interpretation of other quantitative, discrete symbols. Clarke extended work done by Croxton and Stein (1932), concluding that the 'difficulty in visual evaluation of proportional symbols increases with the number of dimensions of the symbol'. Ekman and others reworked Clarke's findings and compared them with their own experiments. They were concerned to investigate in more detail the psycho-physical 'law' that *'subjective intensity is a power function of the intensity of physical stimulation'*. Their studies confirmed that the map symbol reader operated on a scale which is a non-linear (exponential) function of the scales used in describing reality, or the variables being mapped. The values of the exponent varied from about 1 for line symbols to 0·9 for squares, 0·8 for drawn cubes, and 0·74 for drawn spheres. However the value of the exponent for Clarke's data on spheres lay between 0·5 and 0·6. Ekman and others (1961) hypothesized that if estimates of volume were essentially estimates of area for such symbols as spheres and cubes, the ratio between exponents for estimates of area and volume respectively should be 2 to 3. This was confirmed to their satisfaction because the exponent for estimates of volume were about 0·6, so that they were able to conclude that estimates of volume were almost exclusively estimates of perceived area which was in its turn not far from proportional to the geometric area covered by the symbols. Some experiments recently carried out by the author (1965) have suggested that the map reader's judgement of blacked-in segments of proportional circles, indicating the percentage of one item out of a varying total quantity, generally relate to the area blacked in rather than the angle of the segment at the centre of the circle. Because the circles varied in size, estimates of percentages were particularly affected. Since these tests were carried out in a real map situation with genuine data it would seem that the results of tests hitherto carried out only on diagrams divorced from maps *also* apply to maps. In experiments of the latter kind, von Huhn (1927, p. 34) observed that where circles of different sizes were employed, black segments were less effective for showing percentages because only angles and not arcs, chords or areas could be compared. In the author's map test general estimates of regional trend or pattern in the percentage values were difficult to make, presumably for the same reason. For relatively instantaneous impressions of spatially varying proportions the map reader needs more the angles at the centres of the circles. These difficulties are overcome only by showing proportions in circles of the same size.

The model of the real world

Thus it may be safely concluded that only a fraction of the information in the real world eventually finds its way to the map reader from the map or representational model of the real world. The ways in which maps 'perform as a device in portraying spatial properties in competition with other devices, such as photographs, pictures, graphs, language and mathematics' are encompassed in what Bunge (1962, p. 38) calls metacartography. He isolates a group of devices other than maps and mathematics as *premaps*, but comes to the conclusion that they are perhaps a subset of maps (p. 71). Bunge outlines a number of traverses to establish the boundaries between premaps and maps, successively exaggerating spatial properties of different kinds. In this way he deals with scale, shape distortion, information content versus ab-

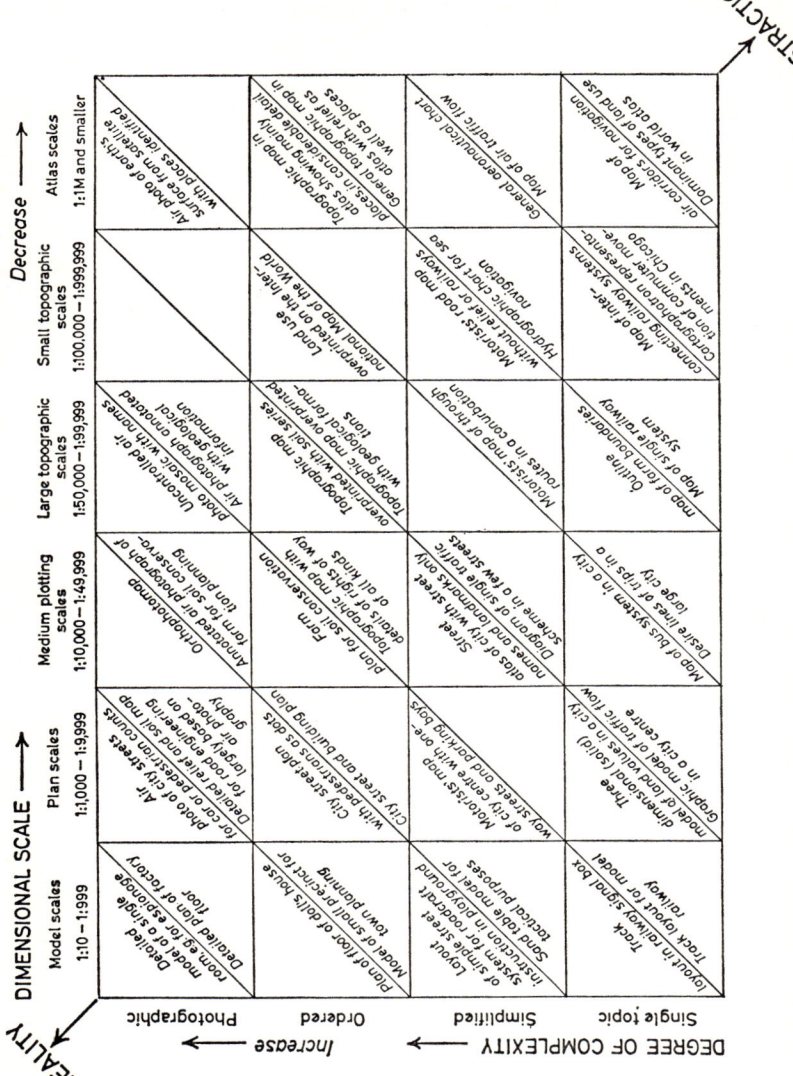

16.17 The gradient between Reality and Abstraction indicating examples of types of maps at their appropriate level of abstraction.

straction, base map data, projection angle, correspondence to the earth's surface, psychological accuracy (apparent realism), projectional conventionality and connections between places. It is not hard to perceive a number of overlaps or intersections in these traverses. Secondly Bunge outlines another set of traverses between maps and mathematics. He discusses in turn: connections between places, distance, the number of dimensions, idealization, spatial analysis, depiction of overlaps. In a final section, Bunge reviews some spatial properties for which measures have still to be adopted: homogeneity, orientation, shape, and pattern. Figure 16.17 represents another view of the relationships between maps and other devices used for portraying spatial properties. It knits together many of the topics considered separately by Bunge, bringing together maps and premaps. It regards all maps as representations of information about spatially organized features and concepts relevant to the earth's surface, in terms of a gradient between an ultimate, infinite reality and an ultimate ideal or abstraction.

Faithfulness: the gradient between reality and abstraction

In a recent paper on a general theory about the nature of knowledge, Bambrough (1964) makes extensive use of analogies with the map. The employment of maps as analogues to illustrate philosophical discussions by Bambrough, Kaplan (1964, pp. 284–285), Treisman (1966, p. 601) and Toulmin (1953, Chap. 4) suggests that they are widely and easily recognizable as models, and many of their properties furthermore are quite well known. Indeed Bambrough begins by quoting Lewis Carrol (1893) at length, reminding us of the attempt to make a really useful map which ended with one on the scale of a mile to the mile, to which the farmers objected because it would have covered the whole country and shut out the sunlight. Bambrough (1964, p. 102) commenting on methods used by metaphysicians writes:

> A direct description of the logical characteristics of our knowledge of the external world shows that each of these pictures gives undue prominence to some features of our knowledge and obscures or distorts the other features that the rival pictures emphasize. . . .
> Here again we can have either a map on the scale of one mile to the mile or we can have grasp and understanding at the cost of distortion.

It is salutary to remember that Gombrich (1962, p. 78) pointed out that the real world is so complex, so rich in detail that no picture can portray it completely and that subjectivity of the artist is not the only factor at work. He continues:

> But what matters to us is that the correct portrait, like the useful map, is an end product on a long road through schema and correction. It is not a faithful record of a visual experience but the faithful construction of a relational model.

Neither the subjectivity of vision nor the sway of conventions need lead us to deny that such a model can be constructed to any required degree of accuracy. What is decisive here is clearly the word 'required'. The form of a representation cannot be divorced from its purpose and the requirements of the society in which the given visual language gains currency.

Once we are aware that the accuracy of a map is one thing and the way it replicates reality is another, there exist possibilities for general statements about all maps in relation to the real world. The less a map is like the real world, the more abstract it is, the more it is a model of that real world. Indeed it is perfectly easy to conceive of a scale of faithfulness between the one mile to one mile map and the directional road sign, which is not only extremely abstract, but is inaccurate in terms of angles, distance and area, but (usually) up to date. To some extent the property of scale also jibes with faithfulness, the smaller the scale the less faithful a map can be to reality. However, the very obvious variation that occurs in maps of the same scale indicate that although scale may set broad limits to the faithfulness of the map, in detail other properties may be more important. Reference to Figure 16.17 makes it clear that decreasing faithfulness, or increasing abstraction has two components which are the axes of the diagram – increasing dimensional scale and decreasing complexity. Conventionally air photographs are excluded from the category *maps*. They lie some way between them and reality. They show only landscape or visible features, but depict all such features, depending on the equipment used to take them. They, like maps are also distorted in terms of area and distance but are more faithful than maps in the sense that azimuths from the centre of vertical air photographs are true. They do at least furnish a record which contains all reproducable detail at one particular point of time. In spite of their being separated from maps by convention, they are more conveniently regarded as *pseudo-maps* because of their close affinities with them. Air photographs differ from them chiefly in respect of names and invisible features. In fact, air photographs are often used as if they were maps (Wilson, 1965; Langdale-Brown and Spooner, 1963, p. 1). In the first example traffic counts were made on them; in the second seemingly uniform areas of vegetation were delimited on mosaics of photographs. A link between air photographs and maps is provided by the orthophotomap (Pumpelly, 1964), which is a photographic presentation of an area over which symbols are printed, surrounded by the marginal information usually found on a map. Horizontal displacements are eliminated and the orthographically true detail is composed of the total photographic image in a subdued colour (pale khaki) with linear features printed in dark grey. This gives it the appearance of having a third dimension. Other experimental work on the conversion of air photographs to maps is discussed by Merriam (1965), who presents the results of screening a conventional 133-line half-tone air photograph so that only the texture of the image remains. By adding tones to that image some cartographic symbolism, such as hypsometric tints, can be incorporated without destroying the detail of the original photograph. Merriam points out that the overprinting of colours on a photographic image can obscure detail. Merriam calls documents such as these *substitute maps*.

Faithfulness in terms of spatial properties

The spatial properties which underlie all others are distance, orientation and area. Any map may depart from reality in all, some or one of these characteristics. The selectivity or degree of completeness with which it represents reality which has been discussed already, is superimposed on these basic scalar qualities, but as seen in Figure 16.17, is highly dependent on them. Time furnishes a fourth factor affecting the information content of a map. Naturally the greater time lapse between survey and publication, the less

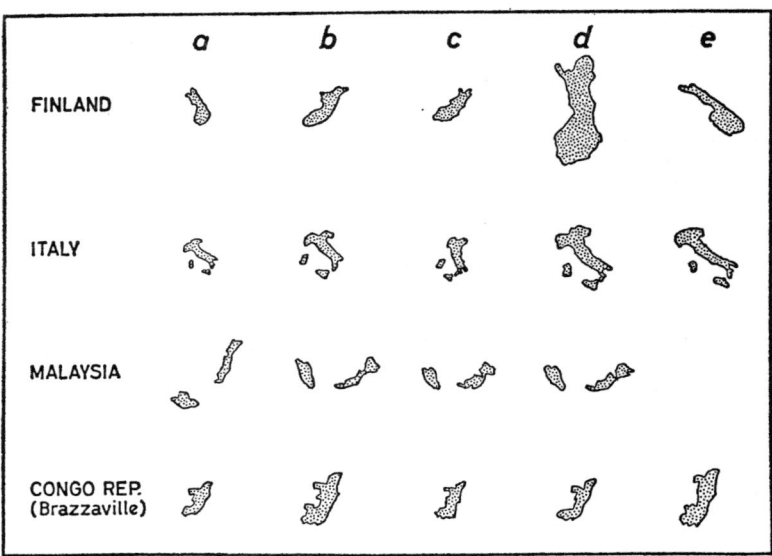

16.18 The effects on the shape and area of four countries of approximately the same size produced by the employment of different map projections.

(*a*) Oblique Azimuthal Equidistant, centred on London and its antipodal point (*Source: Cassells New Atlas, 1961, plates 4 and 5*).
(*b*) Mollweide interrupted at 20° W. and 60° E. south of the equator (*Source: Philip's University Atlas, 1960, plates 8 and 9*).
(*c*) Zenithal Equal area with hemisphere centred at 70° E. and the equator (*Source: Oxford Atlas, 1958, plate 17*).
(*d*) Mercator (*Source: The Times Atlas, 1958, volume I, plate 6*).
(*e*) Mollweide with central meridian at 40° W. in order to show the Atlantic Ocean (*Source: Faber Atlas, 1964, plate 133*).

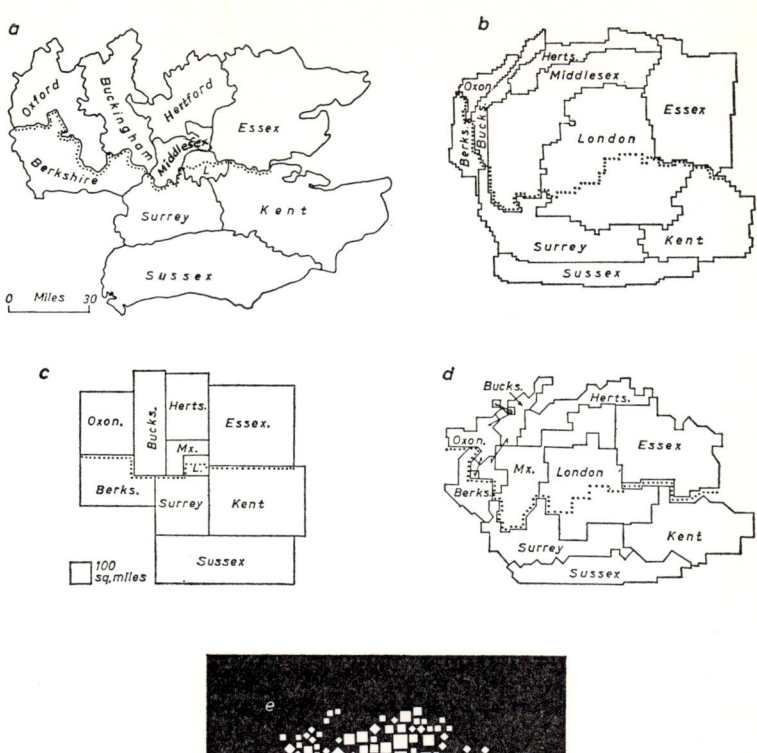

16.19 Representations of counties in south-east England.
(*a*) Conventional, from a Transverse Mercator projection.
(*b*) From Hollingsworth's map of parliamentary constituencies in *The Times*, 19 October 1964. Roughly proportional to population.
(*c*) Counties shown as rectangles whose areas are proportional to their population.
(*d*) Approximate outline of counties derived from map (*e*).
(*e*) Representation of administrative areas in south-east England, with **areas** proportional to population. Squares – urban areas; diamonds – rural districts. (*An experimental map by the Ministry of Housing and Local Government, Crown Copyright reserved.*)

likely it is that the map will be faithful to reality. Sometimes this is not so important. Consider the example afforded by one inch to one mile maps of the Geological Survey of England and Wales. Until recently these were published on a progressively ageing base map dating back to the 1890's. As long as it was possible to relate geological information to specific geographical positions the out-of-date road and settlement pattern was immaterial. However a time comes when that task becomes so difficult that a change of a base map becomes essential.

The purpose of many maps is so particular that some sacrifices of basic properties is not merely necessary but generally accepted. Topographic

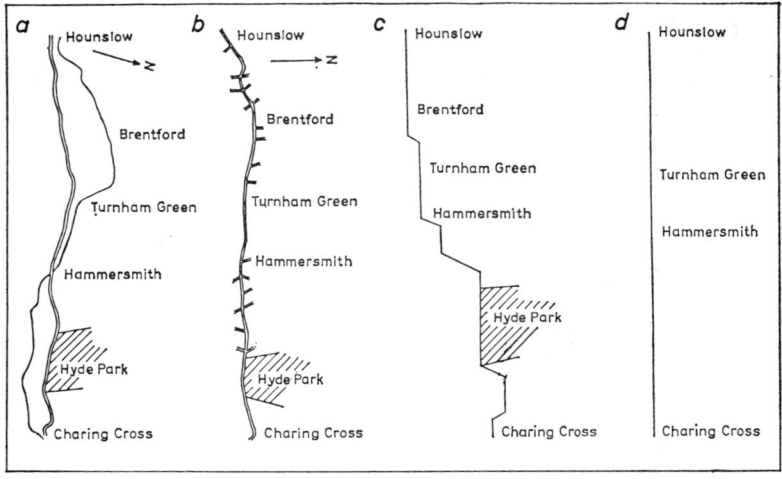

16.20 The routes between London, Charing Cross, and Hounslow, Middlesex; a distance of about 10 miles. *From:*
(a) *The Times Atlas.*
(b) Bowles' Post-chaise Companion (1778).
(c) London Transport maps of bus and trolleybus routes (1946).
(d) London Transport map of the underground railway system (1946).

maps of a fairly small area such as the British Isles, with comparatively small East–West extent and longer North–South extent sacrifice very little if drawn on a Transverse Mercator projection. Compass courses may be followed, and both area and shape are scarcely distorted near the central meridian.

By relaxing the standards of faithfulness in one property, whilst retaining the others true, one moves a little further from reality but gains from the emphasis of particular properties. Zenithal equidistant projections are often used to show distance and orientation from the centre of the map to any other point. Such maps are commonly used to show air routes from important centres and the distances to and relative positions of places from some centre, like Wellington, New Zealand, which frequently suffers from a marginal position on conventional world maps. However these maps have the major drawback of distorting both the shape and area at the margin of the map antipodal to the centre of the projection. Areal exaggeration increases away from the centre. Other kinds of map may retain areas true to scale, in order to present distributions as honestly as possible. World equal area projections are in common use for this purpose. In these, the shapes of the land masses are generally not greatly distorted (see Fig. 16.18). Another class of equal area maps shows land area true to scale, but relaxes standards of faithfulness in distance and orientation by presenting countries as rectangles in roughly their correct positions relative to each other (Kirk, 1964, p. 12) (see Fig. 16.19C). Other maps are constructed in such a way that only distance is preserved as true to scale. One of the oldest is the Peutinger table (Bagrow, 1964, p. 143) where Roman roads are shown as straight lines with places marked at the correct distances. The coaching itineraries so popular in the eighteenth and nineteenth centuries (see Fig. 16.20B) also made use of the idea. Orientation in these maps is more often than not sacrificed to distance and simplicity and area is irrelevant to their purpose.

Diagrammatic maps

A final class of map may be distinguished in which none of the three basic properties is faithful to reality. Only relative position may remain to remind us that the map is a representation of reality. These are often called cartograms but it is difficult to determine a dividing line between them and maps *sensu stricto*. Such maps usually have a very special purpose. The bus and underground railway maps of London Transport (1946, see Fig. 16.20C and D) are diagrammatic representations of the networks of services and merely show the different routes and interchange points. Distances and directions vary across these maps. Yet another group of diagrammatic maps show, not land area and topographic features, but population or volume of trade (Grotewold, 1961). The areas depicted on such maps are to scale in terms of some special property. The shapes of areas may be recognizable as in the case of Grotewold's maps or Wotyinsky's maps (1953). Recently Hollingsworth (1964) has constructed a map showing parliamentary constituencies in the United Kingdom so that their areas are roughly proportional to the electorate they contain, whilst retaining the topological properties of the real situation. Each constituency is contiguous on the map to those which border it in reality. This constraint (see Fig. 16.19B) results in a degree of distortion of shape and orientation which may well be intolerable to most map readers. An adaptation of the idea is at present being designed by the Ministry of Housing and Local Government in London (see Fig. 16.19E).

Although the areas of local authorities are proportional to the population enumerated within them, their positions relative to each other are nearer to reality because the constraint that contiguous areas should actually touch has been relaxed. Instead the areas float, as it were, in a black matrix, so that the outline of the coast is more recognizable than it is on Hollingsworth's map. The choice of two shapes of symbols – squares and diamonds – also gives the design a certain flexibility helping it to correspond more closely to reality.

The map as an end product

As an end product, a representation of reality, a map may be judged either by its artistic qualities – the fineness of line, harmony of colour and lettering and balanced layout or design, or by its usefulness. Whether a work of art or of science, such a map furnishes a new view of reality, albeit a subjective one. It also gives the map reader a view of reality not perceptible to the man on the ground (or even to one in an orbiting satellite). This it achieves by compressing and codifying reality to a comprehensible document that can be decoded and employed as a tool by which to navigate or otherwise inform the reader about the real world.

Maps such as these are representations of mental or conceptual models of spatially arranged features. Skilling (1964) reminds us that one retains mental models of places, such that, in spite of their being highly imperfect, we know that if we go in a certain direction we shall pass certain landmarks. Lynch's *The Image of the City* (1960) contains several maps of American cities drawn from such mental images of landmarks. Whether such mental models of places come from memories of travel, or from reading the map, it is important to remember that they are not infallible. In the same way the map as a model is not infallible. Nevertheless it can safely be used for prediction, for interpretation, if its properties are known and understood.

At the same time, not every feature of the model corresponds to some characteristic of its subject matter. Some features are irrelevant, such as the pink colour used for the British Empire (Kaplan, 1964, p. 285). This does not mean that the British Empire is pink because the map depicts it as such. Indeed Tom Sawyer was able to correct Huckleberry Finn when he thought their balloon ought not to be over Indiana because the ground was not pink (Twain, 1894, p. 35). Of course the states were not the same colour out of doors as they were on the map. The map did not tell lies, it used those colours for a purpose. 'It ain't to deceive you – it's to keep you from deceiving yourself' (Twain, 1894, p. 40).

Maps as representational models of the real world do need revision from time to time. The medieval T in O maps with Jerusalem at the centre of the world disc was made by craftsmen who 'made no serious attempt' to show the world as it actually was (Raisz, 1948, p. 14). As geographical knowledge grew and the cartographer became emancipated from the influence of the medieval church, this model was revised. Today changes in features on the earth's surface are brought about mainly as a result of man's activities. With the advent of rapid methods of mapping made possible by aerial photography, stereoplotters and the orthophotoscope, there is little need to accept the outdated models of the past. Revision can now proceed as fast as maps can be made and financial considerations permit. The natural landscape, within the limits set out above, can be mapped satisfactorily and needs little revision thereafter; the representation of the cultural landscape is perhaps in progressively greater need of change, as man replaces outworn structures and builds new. More frequent revision is required on the larger scale representations (Imhof, 1951, p. 131), which is partly a reflection of their complexity. It is for this reason that topographic maps, for which revision is so important, carry so little detail of an ephemeral nature. Thus, land use maps at scales larger than 1:250,000 would be likely to appear out of date far sooner than those of scales smaller than 1:1 million.

TESTING THE MODEL

Thus far we have discussed the reduction of complex reality into a codified, two dimensional form which has well-known properties. Some attention has already been paid to the purpose and utilization of the models we call maps, but this is largely incidental to the theme of representation. The distinction between thematic and general or topographic maps has already been drawn. This is paralleled by essential differences in the functions of such maps. 'Thematic maps present not only facts, but ideas, hypotheses, and the results of analysis and synthesis' (Miller and Voskuil, 1964, p. 14). Moles (1964, p. 13) clearly distinguishes between maps of two kinds. The first represent an accumulation of a reserve of information, which can be consulted rather than read. The second has the specific aim of presenting a message derived from such a reservoir of facts in as comprehensible form as possible, to the detriment if necessary, of detail, precision and accuracy. Moles observes that the average human being can effectively assimilate a message of this kind at a rate of no more than 10 bits a second. On the other hand Roberts (1962) gives an estimate that the standard, multi-coloured, large-scale topographic map of a part of the United States contains between 100 and 200 million bits of information. To assimilate all this would take well over a year of the normal map reader's time. Electronic computers, with large storage capacities may well be able to handle such stores of information by new scanning methods, but 'the map, unlike a computer memory, has the distinct advantage of permitting direct human use of its vast store of information' (Roberts, 1962, p.13). But the memory of the computer is more reliable than ours. Thus inspection of a detailed map by the human eye can yield *some immediate* return, usually in the form of a hunch or speculation about some pattern or relationship. 'La carte n'est plus un reservoir d'information à l'usage des spécialistes, elle est un outil pour faire émerger des formes et des idées dans un large public' (Moles, 1964, p. 13). This applies not only to the thematic map, but also to the less specialized, topographic map. But we must not forget that 'like carpenters' tools, maps should not be misused. More should not be expected of them than they can perform' (Wright, 1942, pp. 543–544). It is by understanding their limitations that they may be regarded as useful tools for research. By employing them in this way, the investigator is testing them against reality. In the simplest case the trained field observer goes on to the ground with a topographic or geological map. In field work the 'primary task is to make significant additions to the map. There is no better method of training in field observation: it is the sovereign cure for the attitude of the beginner to think that it is all on the map anyway; and this is to indulge in the mere incurious verification of what the map-maker has seen fit to portray' (Wooldridge and East, 1958, p. 165). But the observer does not add anything and everything, indiscriminately. To start with it may be a class of features, such as field boundary types or some aspect of current land use. The scientist, well aware of this, does not completely trust this product of human artifice to inform him of the nature of reality. We have already seen that the map maker may have seen fit to present an extremely imperfect picture of reality. The interpolation of human interpretation between reality and the scientist reminded Norbert Wiener of Einstein's dictum: 'Der Herr Gott ist raffiniert, aber boshaft Er ist nicht' (God may be artful, but He is not malicious). Nature plays fair and does not deliberately frustrate the scientist. (Wiener, 1954, p. 188).

So it may be thought that the researcher would do better to use raw data for his investigation. Having formulated a working hypothesis about the explanation of some interesting patterns, perhaps even from map evidence, the data for constructing the model can be collected directly from the real world. Indeed such a course of action avoids the interference from other sources such as maps. There are instances where maps do provide important evidence for research. In many cases long and laborious field surveys have been recorded permanently in map form. Geological, soil, vegetation and land use surveys when mapped are all documents in their own right and cannot be superseded by statistics or descriptions. It is unlikely that geological and soil surveys would be undertaken twice for the same area – at least at the same level of detail, unless for some special purpose when probably only a limited field of inquiry would be involved. Vegetation and land use surveys record relatively ephemeral features. The expense of repeating such surveys often means that they are done but once. Maps of such surveys can therefore be properly regarded as initial data with which to begin an investigation of some problem within an areal context. They are as good as raw data derived directly from observation of reality if such raw data cannot feasibly be collected.

The special role of isarithmic maps in generalizing spatial patterns

Derived maps, not based in the first place on field observations but on statistics or values calculated from sets or series treated by modern quantitative map analysis. That isarithmic maps are not often compiled and used, even as representations, let alone as an input for a geographical analysis, is certainly thanks to the tedious methods of constructing them manually. The alternative, choropleth method of depicting raw or derived values has simplicity on its side. Robinson (1961) has suggested moreover that we accept the use of isarithms for relief maps (and also for climatic distributions) by longestablished custom, but isarithmic maps for cultural distributions are avoided because they do not stress the values at particular places and they somehow transform essentially discrete data into a spatial continuum. If on the other hand it is accepted that the 'portrayal of variations from place to place is the basic aim of a geographical volume', that is some quantitative variable, the general configuration of the data is all important. Its representation by a statistical surface whose gradients may be visually assessed is most appropriate. 'When a space-series array is to be plotted on a map, only the isarithmic technique provides a graphic display of the assumption of continuous gradient, and it also smooths the data to some extent, depending upon the choice of isarithmic interval' (Robinson, 1961, p. 54). Thus the irritating and distracting vagaries of the choropleth map may be partially suppressed. The local variations that are quite likely to be emphasized in choropleth maps are probably more obvious in maps of nominally scaled data. The intricacy of land use maps even with as few categories as two – the presence of arable land and its absence for example – is well known. It is not necessary to remind generations of geographers that the task of interpreting the incredible complexity of any topographic map by eye is formidable even after years of practice. Tobler (1966) has developed techniques for use in computer mapping to generalize isarithmic patterns and reverse the process, or roughen the smoothed pattern ending with the original pattern.

No great a gap exists in geographical method than that which lies between such maps and their generalized, objective interpretation. Analytical techniques for this purpose are only now being developed, have still to be tested in a variety of conditions and are known only to a minority of investigators. The explanation of spatial patterns requires first of all the identification of pattern. The complexity of the detailed map may at first seem to reflect a random arrangement of phenomena. A search for order in this imagined chaos is therefore the first step in a scientific investigation. By some chance observation, hunch or speculation the germ of a theory may be established. Next a working model or hypothesis is developed and this tested, by experiment or comparison with the real world. Maps play a significant part in these processes. They can furnish data initially, they can suggest the hypotheses and they can be employed as tools to make and test models. In many instances they are themselves models incorporating only selected aspects of the real world reflecting some hypothetical situation.

Trend surface mapping

In a spatial context the development of methods of identifying, delineating what are called surfaces, trend-surfaces or response surfaces has greatly assisted the recognition of the role of maps in the explanation of geographical patterns. Chorley and Haggett (1965, p. 47) in a recent review of these methods state the problems facing the map interpreter in these terms: 'Because geographical problems... are characterized by areal sampling restrictions, by a multiplicity of variables, and by the interaction and simultaneous variation of most of the variables, we cannot be certain how much of the information transmitted by a map may be regarded as a "signal" and how much as random variations or "noise".' By constructing maps of the trend or message separately from maps of noise, the unexplained random variations, a more satisfactory approach to pattern interpretation may be attempted. The explanations thus given are expressed in terms of statistical relationships or as process-response models where variations in form correspond to the strength and balance of the supposed controlling factors (Chorley and Haggett, 1965, p. 48). In the same review a distinction is made between selective and objective methods of constructing trend surfaces. Since there is clearly a selective and subjective element in both methods, a distinction will here be made between mainly graphical techniques and the more rigorous mathematical techniques.

Graphical trend surface analysis

Graphical methods were for long the only ones applied to the problem of sifting the regional from the local component in spatial patterns. Graphical generalization of contours as for example used by Wooldridge (1927) to serve as a surface upon which outliers of Tertiary age could be plotted. The compilation of generalized contours suffers from variations principally due to the different ways in which individual operators interpret the ground rules. Graphical methods also include the techniques of constructing parallel and orthogonally intersecting profiles, which are themselves generalized from the available data (Krumbein, 1956). Robinson (1961) also shows that the method adopted by Tanaka Kitiro (1932), which yielded a planimetrically correct and visually pleasing impression of plastic relief, could be applied to isarithmic maps of cultural phenomena. By employing a series of uniformly spaced planes inclined at 45° the traces of which were linked to the isarithmic pattern, the general configuration of variations in population density is obtained. The advantage of this relatively objective method is that it employs an imagery familiar to those who understand the shaded relief map and it also smoothes out the local irregularities. Robinson (1961, p. 57) suggests that 'there appears to be a relatively large amount of high-frequency variation in cultural data, as compared to the purely "physical" ' possible because of the large number of factors affecting man's behaviour and a whole series of covariant factors which are less well understood in the human sciences. Furthermore isarithmic maps of cultural features do not behave in the same way as the land surface, nor are the operations of well-known causal factors readily appreciated. 'A fairly flat surface may dip gently over large areas, and, if it undulates at all, the pattern of wandering isarithms and their relation to the surrounding forms of the statistical surface may easily mask its basic flatness' (p. 58). It is therefore all the more important for geographers attempting to understand the patterns of human activity to appreciate the possibilities involved in the mapping of trends and residuals or anomalies.

More rigorous trend surface analysis

In the mathematical methods of trend surface mapping analysis before mapping is altogether more complicated. It has been tremendously aided, indeed made feasible, by the introduction of high-speed electronic computers (Krumbein and Graybill, 1966, p. 321). Methods performed objectively by the computer enable the investigator to divide each map observation into two or more parts: 'large scale' or regional trends from one edge of the map to the other and 'small scale' or local effects. By a reiterative process the original observations may be broken down into a series of trends of progressively smaller extent. The local effects, which are apparently non-systematic fluctuations, can be identified and plotted at each stage. Their pattern and nature may well suggest the interpretation of the next higher order trend surface or new variable to be built into the explanatory model being constructed. Both linear and higher order trend surfaces and residuals (the computed differences between observed values and trend values at particular points) are mapped by isarithms. It is natural at this stage to ask how well the trend fits the reality presented by observed values. Merriam and Harbaugh (1964) when fitting successively higher-order surfaces to the elevation of geological formations measured the goodness of fit in terms of the percentage reduction in the sum of squares. The surfaces are so constructed as to make the sum of the squared residual values as small as possible. In this way objectivity is achieved and a measure of correspondence with reality is obtained. Similar criteria are used in the regression models described by Chorley and Haggett (1965). In such cases the first trend is a simple linear relationship, such as between voting behaviour and proportion of the population that is rural (Thomas, 1960; see Fig. 16.21) or between cash grain farming and flat land (Hidore, 1963). The mapping of residuals from regression showed up areas where the regression under- or over-estimated the strength of voting or the amount of cash grain farming. A closer look at these areas of anomaly suggested additional variables. In the first instance, the distance from the home town of the candidate raised the level of explanation from about a third to nearly 50 per cent. In the second instance, soil quality was considered to be of importance where the regression had underestimated

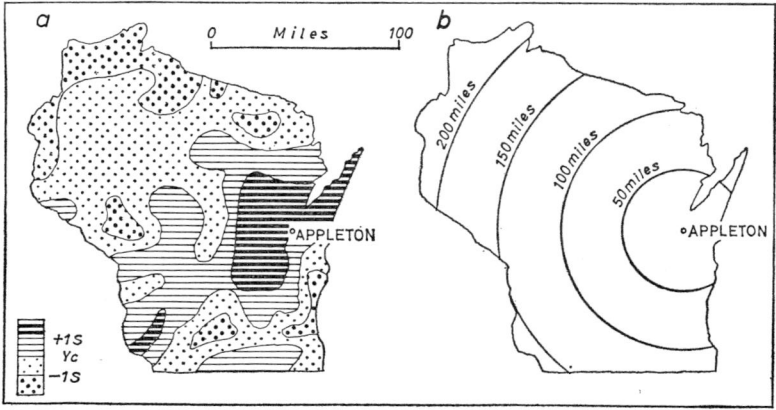

16.21 Testing the model: developing from the first hypothesis of reasons for voting behaviour, a new hypothesis from map of residuals.
(*a*) Residuals from regression $(Y_{cn} - Y_n)/S_{Y_c}$ of per cent of total vote cast for Senator McCarthy in Wisconsin on the percentage of total population that is rural. (1S 1 standard error of estimate, Y_c as predicted by the regression, $-1S$ -1 standard error of estimate). In general the regression underestimated support for the senator nearer his hometown, Appleton.
(*b*) The basis for the second hypothesis – distance from Appleton (*Source: Thomas, 1960*).

the amount of cash grain farming. In both cases, the map of residuals had real value as a means of suggesting further lines of inquiry, thus restarting the analysis. As Robinson (1961, p. 57) pointed out the different sizes of areas and hence the spacing of control points upon which isarithmic maps are based are another source of noise. Chorley and Haggett (1965, p. 61) take this argument a stage further by observing that what may be taken for noise in coarse-grained analysis, may be partly explicable in a finer grained analysis, where the mesh of control points is closer together. On the whole however it is advisable to use the finest possible mesh right from the beginning. Not to use all the available data would in fact be inefficient because trend-surface mapping itself seeks order in complexity.

CONCLUSION

Once the experimental iterations with progressively more complex models can explain no significant further variation the investigator is entitled to assume that his model, comprising a trend map with a multivariate message and a map of completely inexplicable residuals, fits the real world well enough. At this stage the cycle of research is temporarily completed because a new view of the real world is the outcome. Maps used in this way are not merely ornaments, or even portraits, they are vital tools of research. To recognize this is to return to maps their rightful place 'as pre-eminently the geographer's tool both in investigation of his problems and the presentation of his results' (Wooldridge and East, 1958, p. 64). Though placed on a level with verbal description, symbolic logic and mathematics, maps still have their peculiar disadvantages and intrinsically useful properties. 'It is too often forgotten that geographical studies are not descriptions of the real world, but rather perceptions passed through the double filter of the author's mind and his available tools of argument and representation' (Curry, 1962, p. 21). By recognizing maps as models of the real world and by employing them as conceptual models in order better to understand the real world, their central importance in geographical methodology is assured. Because the making of maps belongs properly to the profession of cartography, the geographer cannot afford to make his models without reference to cartographic practice. Many of these models will however, be made without reference to him. It is because of this that 'no-one claiming the title of geographer, however humbly, is entitled to be ignorant of how maps are made' (Wooldridge and East, 1958, p. 70).

REFERENCES

BAGROW, L., [1964], *History of cartography*, (London), 312 pp.
BALCHIN, W. G. V., [1952], Cartographic portrayal by geographers; *Indian Geographical Society (Madras)*, Silver Jubilee Volume, 141–147.
BAMBROUGH, R., [1964], Principia metaphysica; *Philosophy*, 39, 97–109.
Bartholomew's Atlas of the World's Commerce, [1907], ed. by J. G. Bartholomew, (Edinburgh), 176 plates.
BERRY, B. J. L., [1960], An inductive approach to the regionalization of economic development; In Ginsburg, N., (Ed.), *Essays on Geography and Economic Development*, University of Chicago, Department of Geography, Research Papers, No. 62.
BERRY, B. J. L., [1961], Basic patterns of economic development; In Ginsburg, N., (Ed.), *Atlas of Economic Development Chicago*, (University of Chicago Press), 110–119.

BLACK, M., [1962], *Models and Metaphors*, (Ithaca, New York), 267 pp.
BLUMENSTOCK, D. I., [1953], The reliability factor in the drawing of isarithms; *Annals of the Association of American Geographers*, 43 (4), 289–304.
BOARD, C., [1962], *The Border Region, Natural environment and land use in the Eastern Cape*; (Oxford University Press, Cape Town), 238 pp. and volume of maps.
BUNGE, W., [1962], *Theoretical Geography*; Lund Studies in Geography Series C, No. 1, 208 pp.
CARROLL, L., (pseud. Dodgson, C. L.), [1893], *Sylvie and Bruno concluded*; (Macmillan, London).
CHORLEY, R. J., [1964], Geography and analogue theory; *Annals of the Association of American Geographers*, 54 (1), 127–137.
CHORLEY, R. J. and HAGGETT, P., [1965], Trend surface mapping in geographical research; *Transactions of the Institute of British Geographers*, 37, 47–67.
CLARKE, J. I., [1959], Statistical map reading; *Geography*, 44 (2), 96–104.
COLE, J. P., [1965], *A geography of World Affairs*, (London), 348 pp.
CONZEN, M. R. G., [1960], Alnwick, a Northumberland study in town-plan analysis; *Transactions of the Institute of British Geographers*, 27, 127 pp.
COPPOCK, J. T., [1960], The parish as a geographical-statistical unit; *Tijdschrift voor sociale en economische geografie*, 51 (12), 317–326.
CROXTON, F. E. and STEIN, H., [1932], Graphic comparison by bars, squares, circles and cubes; *Journal of the American Statistical Association*, 27, 54–60.
CURRY, L., [1962], Climatic change as a random series; *Annals of the Association of American Geographers*, 52, 21–31.
CZEKALSKI, J., [1933], Mapa izarytmiczna a obraz rzeczywisty (próba analizy metody) (The isarithmic map, its method and degree of precision); *Wiadomości Służby Geograficznej*, 7 (3), 202–234, (Warsaw).
DAHLBERG, R. E., [1962], Evolution of interrupted map projections; *International Yearbook of Cartography*, 2, 36–54.
DAVIS, L. B., [1958], Design criteria for today's aeronautical charts; *Surveying and Mapping*, 18, 49–57.
DE DAINVILLE, Fr. S. J., [1964], *Le langage des géographes*, (Paris), 384 pp.
Dictionary of National Biography, [1921 edn.], volume 5, (Oxford University Press).
DUMKE, G. S., [1963], *The Boom of the Eighties in Southern California*, (San Marino, Calif.), 313 pp.
ECKERT, MAX, [1908], On the nature of maps and map logic; *Bulletin of the American Geographical Society*, 40, 344–351.
ECKMAN, G. and JUNGE, K., [1960], *Psychophysical relations in the perception of visual length, area and volume with special regard to interpretation of certain map symbols*; Paper prepared for the meeting of the International Geographical congress, Commission on a World Population Map, Stockholm, 22 pp., (Mimeographed).
EKMAN, G., LINDMAN, R. and WILLIAM-OLSSON, W., [1961], A psychophysical study of cartographic symbols; *Perceptual and Motor Skills*, 13, 355–368.
FORGOTSON, J. M. Jr., [1960], Review and classification of quantitative mapping techniques; *Bulletin of the American Association of Petroleum Geologists*, 44 (1), 83–100.

FOX, J. R., [1956], *Land-use survey general principles and a New Zealand example*, (Auckland, New Zealand), 46 pp.
GOMBRICH, E. H. J., [1962], *Art and Illusion*, (London), 388 pp.
GROTEWOLD, A., [1961], Some aspects of the geography of international trade; *Economic Geography*, 37, 309–319.
Michelin Guide, [1963], Paris, (Pneu Michelin Services de Tourisme), 997 pp.
HAGGETT, P., [1964], Regional and local components in the distribution of forested areas in South-east Brazil: a multi-variate approach; *Geographical Journal*, 130 (3), 365–380.
HARRIS, L. J., [1959], Hill shading for relief depiction in topographical maps; *Chartered Surveyor*, 91 (9), 515–520.
HARRISON, M. J., MEAD, W. R. and PANNETT, D. J., [1965], A Midland ridge-and-furrow map; *Geographical Journal*, 131 (3), 366–369.
HARRISON, R. E., [1959], Art and common sense in cartography; *Surveying and Mapping*, 19 (1), 27–38.
HARLEY, J. B., [1962], The one-inch to the mile maps of England and Wales; *Amateur Historian*, 5 (5), 130–140.
HIDORE, J. J., [1963], The relationships between cash-grain farming and land forms; *Economic Geography*, 39, 84–89.
HOLLINGSWORTH, T. H., [1964], *The Times*, 19 October 1964; (Map on p. 18).
IMBRIE, J. and PURDY, E. G., [1962], Classification of modern Bahamian carbonate sediments; In *Memoir* No. 1, *American Association of Petroleum Geologists*, Classification of carbonate rocks, a symposium, 253–272.
IMHOF, E., [1951], *Terrain et Carte*, (Zurich), 261 pp.
IMHOF, E., [1962], Die Anordnung der Namen in der Karte; *International Yearbook of Cartography*, 2, 93–129.
JACKSON, B. G., BARNARD, C. S. and STURROCK, F., [1963], *The pattern of farming in the Eastern counties*; Farm Economics Branch, School of Agriculture, Cambridge University, Occasional Papers, No. 8, 60 pp.
JENKS, G. F. and KNOS, D. S., [1961], The use of shading patterns in graded series; *Annals of the Association of American Geographers*, 51 (3), 316–334.
JENKS, G. F., [1963], Generalization in statistical mapping; *Annals of the Association of American Geographers*, 53 (1), 15–26.
JOHNSON, F. C. and KLARE, G. R., [1961], General models of communication research, a survey of the development of a decade; *Journal of Communication*, 11 (1), 13–26 and 45.
JOHNSTON, D. A., [1902], *Ordnance Survey maps of the United Kingdom, a description of their scales, characteristics, etc.*, 17 pp.
KANSKY, K. J., [1963], *Structure of transportation networks*; University of Chicago, Department of Geography Research Papers, No. 84, 155 pp.
KAPLAN, A., [1964], *The conduct of enquiry: methodology for behavioural science*, (San Francisco), 428 pp.
KEATES, J. S., [1962], The perception of colour in cartography; *Proceedings of the Cartographic Symposium*, Edinburgh, 19–28.
KING, L. J. and HENSHALL, J., [1966], Some structural characteristics of peasant agriculture in Barbados; *Economic Geography*, 42, 74–84.

KINGSBURY, R. L., [1964], The world of little maps; *Journal of Geography*, 63, 355–366.
KIRK, D., [1946], *Europe's population in the interwar years*; (Princeton, New Jersey), (for the League of Nations), 307 pp.
KITIRO, T., [1932], The orthographical relief method of representing hill features; *Geographical Journal*, 79, 213–219.
KRUMBEIN, W. C., [1956], Regional and local components in facies mapping; *Bulletin of the American Society of Petroleum Geologists*, 40, 2163–2194.
KRUMBEIN, W. C., [1957], Comparison of percentage and ratio data in facies mapping; *Journal of Sedimentary Petrology*, 27 (3), 293–297.
KRUMBEIN, W. C. and GRAYBILL, F. A., [1965], *An introduction to statistical models in geology*, (New York), 475 pp.
LANGDALE-BROWN, I. and SPOONER, R. J., [1963], *Land use prospects of Northern Bechuanaland*; Department of Technical Co-operation Directorate of Overseas Surveys, Development Study No. 1, 40 pp.
LEWIS, W. V., [1959], The karstlands of Jamaica: cockpits or rounded hills?; *Geographical Journal*, 125, 289.
LINTON, D. L., [1948], The ideal geological map; *Advancement of Science*, 5 (18), 141–149.
LUNDQUIST, G., [1959], Generalization – a preliminary survey of an important subject; *Nachrichten aus dem Karten – und Vermessungswesen*, 2 (3), 46–51.
LUNDQUIST, G., [1963], Generalization of communication net-works; Bulletin No. 4, International Cartographic Association, *Sonderdruck* from *Nachrichten aus dem Karten – und Vermessungswesen*, 5 (5), 35–42.
LYNCH, K., [1960], *The image of the city*, (Cambridge, Mass.), 194 pp.
LYONS, H. G., [1914], Relief in cartography; *Geographical Journal*, 43, 233–248 and 395–407.
LYONS, H. G., [1928], The sailing charts of the Marshall Islanders; *Geographical Journal*, 72, 325–328.
MACKAY, J. R., [1949], Dotting the dot map: an analysis of dot size, number and visual tone density; *Surveying and Mapping*, 9 (1), 3–10.
MACKAY, J. R., [1953], The alternative choice in isopleth interpolation; *Professional Geographer*, 5 (4), 2–4.
MACKAY, J. R., [1954], Geographic Cartography; *Canadian Geographer*, 4, 1–14.
MACKAY, J. R., [1963], Isopleth class intervals: a consideration in their selection; *Canadian Geographer*, 7 (1), 42–45.
MALING, D. H., [1963], Some quantitative ideas about cartographic generalization; Bulletin No. 4, International Cartographic Association, *Sonderdruck* from *Nachrichten aus dem Karten – und Vermessungswesen*, 5 (5), 6–22.
MARSCHNER, F. J., [1943], Maps and a mapping program for the United States; *Annals of the Association of American Geographers*, 33, 199–219.
MARSCHNER, F. J., [1944], Structural properties of medium- and small-scale maps; *Annals of the Association of American Geographers*, 34, 1–46.
MEAD, W. R., [1954], Ridge-and-furrow in Buckinghamshire; *Geographical Journal*, 120 (1), 34–42.
MERRIAM, D. F. and HARBAUGH, J. W., [1964], Trend-surface analysis of regional

and residual components of geologic structure in Kansas; *Kansas State Geological Survey, Special Distribution Publication* No. 11, (Lawrence, Kansas), 27 pp.

MERRIAM, M., [1965], The conversion of aerial photography to symbolised maps; *Cartographic Journal*, 2, 9–14.

MILLER, O. M. and VOSKUIL, R. J., [1964], Thematic-map generalization; *Geographical Review*, 54, 13–19.

MOLES, A. A., [1964], Théorie de l'information et message cartographique; *Sciences et Enseignement des Sciences*, (Paris), 5 (32), 11–16.

MONMONIER, M. S., [1965], The production of shaded maps on the digital computer; *Professional Geographer*, 17 (5), 13–14.

MOTT, P. G., [1964], Topographical model making using air photographs; *Cartographic Journal*, 1 (2), 29–32.

PILLEWIZER, W. and TOPFER, F., [1964], Das Auswahlgesetz: Ein Mittel zur Kartographischen Generalisierung; *Kartographische Nachrichten*, 14 (4), 117–121.

PORTER, P. W., [1958], Putting the isopleth in its place; *Proceedings of the Minnesota Academy of Science*, 25/26, 373–384.

PORTER, P. W., [1964], *A bibliography of statistical cartography*, (Minneapolis), 66 pp.

PUMPELLY, J. W., [1964], Cartographic treatments in the production of orthophotomaps; *Surveying and Mapping*, 24 (4), 567–571.

RAISZ, E., [1948], *General cartography*, (New York), 354 pp.

REEVES, E. A., [1929], in Withycombe (1929).

RIDDIFORD, C. E., [1952], On the lettering of maps; *Professional Geographer*, 4 (5), 7–10.

ROBERTS, J. A., [1962], The topographic map in a world of computers; *Professional Geographer*, 14 (6), 12–13.

ROBINSON, A. H., [1952], *The Look of maps*, (Madison, Wisconsin), 105 pp.

ROBINSON, A. H., [1960], *Elements of cartography*; 2nd edition, (New York), 343 pp.

ROBINSON, A. H., [1961], The cartographic representation of the statistical surface; *International Yearbook of Cartography*, 1, 53–63.

ROBINSON, A. H., LINDBERG, J. H. and BRINKMAN, L., [1961], A correlation and regression analysis of population densities in the Great Plains; *Annals of the Association of American Geographers*, 51, 211–222.

SCHMIDT-FALKENBURG, H., [1962], Grundlinien einer Theorie der Kartographie; *Nachrichten aus dem Karten- und Vermessungs-wesen*, 1 (22), 5–37.

SCHULZ, H. C., [1939], Elizabethan map of Wotton Underwood, Buckinghamshire; *Huntington Library Quarterly*, 3, 395–407.

SCHULZ, H. C., [1954], A Shakespeare haunt in Bucks?; *Shakespeare Quarterly*, 5 (2), 177–178.

SINNHUBER, K. A., [1964], The representation of disputed political boundaries in general atlases; *Cartographic Journal*, 1 (2), 20–28.

SKILLING, H., [1964], An operational view; *American Scientist*, 52, 388A–396A.

STAMP, L. D., [1948], *The land of Britain its use and misuse*, (London), 507 pp.

STAMP, L. D., [1961], *A glossary of geographical terms*, (London), 539 pp.

STAROSTIN, I. I. and YANIKOV, G. V., [1959], *Osnovy Topografii i Kartografii* (Principles of Topography and Cartography), (Moscow).

SWEETING, M. M., [1958], The karstlands of Jamaica; *Geographical Journal*, 124, 184–199.
SWEETING, M. M., [1959], Reply to W. V. Lewis; *Geographical Journal*, 125, 289–291.
THOMAS, E. N., [1960], *Maps of residuals from regression: their characteristics and uses in geographic research*; Department of Geography, State University of Iowa, Publication No. 2, (Iowa City), 60 pp.
TOBLER, W. R., [1965], Automation in the preparation of thematic maps; *Cartographic Journal*, 2 (1), 32–38.
TOBLER, W. R., [1966], Numerical Map Generalization; In *Discussion Paper* No. 8, Michigan Inter-University Community of Mathematical Geographers, 24 pp.
'Tomlinson Report', [1956], *Summary of the Report of the Commission for the Socio-Economic Development of the Bantu Areas within the Union of South Africa*, (Government Printer, Pretoria), U.G. 61/1955, 213 + 64 pp.
TÖPFER, F. and PILLEWIZER, W., [1966], The principles of selection, *The Cartographic Journal*, 3 (1), 10–16, translated and with explanatory notes by D. H. Maling.
TOULMIN, S. E., [1953], *The philosophy of science*, (London), 176 pp.
TRIESMAN, A., [1966], Our limited attention; *Advancement of Science*, 22 (104), 600–611.
TWAIN, M., (pseud. Clemens, S. L.), [1894], *Tom Sawyer Abroad*, (London), 208 pp.
The University Atlas, [1946], Goodall, G. and Darby, H. C., (Eds.), (George Philip, London), 96 Plates.
UHORCZAK, Fr., [1930], Metoda izarytmiczna w mapach statystycnych (The isarithmic method applied to statistical maps); *Polski Przeglad Kartograficzny*, 4, 95–129.
VON HUHN, R., [1927], Further studies in the graphic use of circles and bars: I A discussion of the Eells experiment; *Journal of the American Statistical Association*, 22, 31–39.
WEAVER, J. C., [1954], Crop-combination regions in the Middle West; *Geographical Review*, 44 (2), 175–200.
WEAVER, J. C., HOAG, L. P. and FENTON, B. L., [1956], Livestock units and combination regions in the Middle West; *Economic Geography*, 32 (3), 237–259.
WEINER, N., [1954], *Thr human use of human beings: cybernetics and society*, (Garden City, New York), 199 pp.
WILKINSON, H. R., [1951], *Maps and Politics: A Review of the ethnographic cartography of Macedonia*, (Liverpool), 366 pp.
WILLIAMS, R. L., [1956], *Statistical symbols for maps: their design and relative values*; (New Haven, Connecticut), (Yale University Map Laboratory), 115 pp.
WILSON, F. R., [1965], Collection of traffic data by aerial photographs; *Traffic Engineering and Control*, 7 (4), 258–261.
WINTERBOTHAM, H. ST. J. L., [1929], in Withycombe (1929).
WITHYCOMBE, J. G., [1929], Lettering on maps; *Geographical Journal*, 73, 429–446.
WRIGHT, J. K., [1942], Map makers are human: Comments on the subjective in maps; *Geographical Review*, 32, 527–544.

WOOLDRIDGE, S. W., [1927], The Pliocene history of the London Basin; *Proceedings of the Geologists' Association of London*, 38, 49–132.

WOOLDRIDGE, S. W. and EAST, W. G., [1958], *The spirit and purpose of Geography*; 2nd edition, (London), 186 pp.

WOYTINSKY, W. S. and WOYTINSYK, E. S., [1953], *World population and production trends and outlook*; (New York), (20th Century Fund), 1268 pp.

CHAPTER SEVENTEEN

Hardware Models in Geography

M. A. MORGAN

INTRODUCTION

It is curious that geographers who are often regarded by others if not by themselves as practical men should have paid relatively little attention to models or physical constructs either in teaching or research. Exceptions are found mainly in the field of physical geography and in particular geomorphology, but even here geologists and hydrologists have been much more productive. The comparative neglect of hardware models in geography may indicate that they have been considered inappropriate or that we have been a little tardy in realizing their value. It may be that a widespread and traditional preoccupation with the unique has diverted our attention from the problems and rewards of trying to explain the more general. The absence until recently of any rigorous mathematical or symbolic formalization in human geography and to a lesser degree in physical geography has not encouraged a search for understanding by means of analogues.

Models are of the greatest value in teaching, particularly if the students are involved in their design, construction and operation. One recalls the old saying, 'Hear and forget, see and remember, do and understand'. The educational advantages that attend students finding out for themselves are increasingly being recognized amongst science teachers. The process of making, doing, observing and measuring develops the student's intuitive capacity, and relationships and principles are often perceived long before he is able to understand a complete and rigorous explanation of what he is observing or creating.

The physical constructs however cannot and should not be isolated from conceptual models. To design a physical model we often without realizing it depend on some conceptual model; conversely observations made from a construct may lead to some mathematical or verbal formalization, in other words to a conceptual model. So the two types of models are closely related.

In the research field especially in human geography the scope for physical models seems relatively restricted. This is largely because many of the models being developed in human geography can be derived and tested with the use of a computer, which is in many respects an infinitely more subtle and flexible tool than other hardware models. In the latter too great a complexity is disastrous because the model becomes tedious to construct and is operationally difficult. The essence of a good model is simplicity and inevitably this limits the circumstances under which its use is appropriate.

This chapter has three objectives. Firstly to present a broad selection of models from a variety of disciplines, models that appear to have some bearing on the sort of topics that interest geographers; secondly to indicate in a very tentative way a few new models that could be useful and thirdly, if mainly by implication, to suggest that this is still a new field and in many respects one well worth cultivating.

STATIC MODELS

All hardware models or constructs represent some degree of abstraction of reality. The model with which most people are familiar, the three-dimensional relief model, is an abstraction. It cannot with complete fidelity represent every detail of a part of the earth's surface. Not only do we readily accept the inevitable distortions and generalizations, we frequently accept avoidable distortions probably without realizing it. Most relief models have an unnecessarily great degree of vertical exaggeration which is to be deprecated if only because it needlessly reduces the effectiveness with which surface texture can be represented. Scale factors are thoroughly examined by Imhof (1965). Techniques and materials available for relief model construction have been enhanced with the development, for example, of thermoplastics and resins (U.S. Army Map Service, 1950; Spooner, 1953). The use of computers programmed to feed taped instructions from three-dimensional co-ordinates to cutting machines which carve the final relief models is a recent and promising development (Noma and Misulia, 1959) which should do much to make a wide range of relief models more generally available.

It is not always appreciated that many other spatial variables besides relief lend themselves to modelling. The whole concept of statistical gradients and surfaces, whether physical or socio-economic in character, can very often be made more comprehensible in teaching if a model is used. Generally whenever three variables are being considered a model offers the most obvious mode of illustration. Application of models to climatological data has been discussed by Conrad and Pollak (1950) (Fig. 17.1). The difficult problem of describing the structure of a cyclone was tackled by Bjerknes (Shaw, 1934) using a wire model and Shaw (1934) made glass models of the distribution of upper air temperature. Chorley and Morgan (1962) represented two geologically similar areas by models so proportioned that each embodied to scale characteristic features associated with drainage networks and basin shapes, obtained by morphometric analysis (Fig. 17.2). Neither model represented

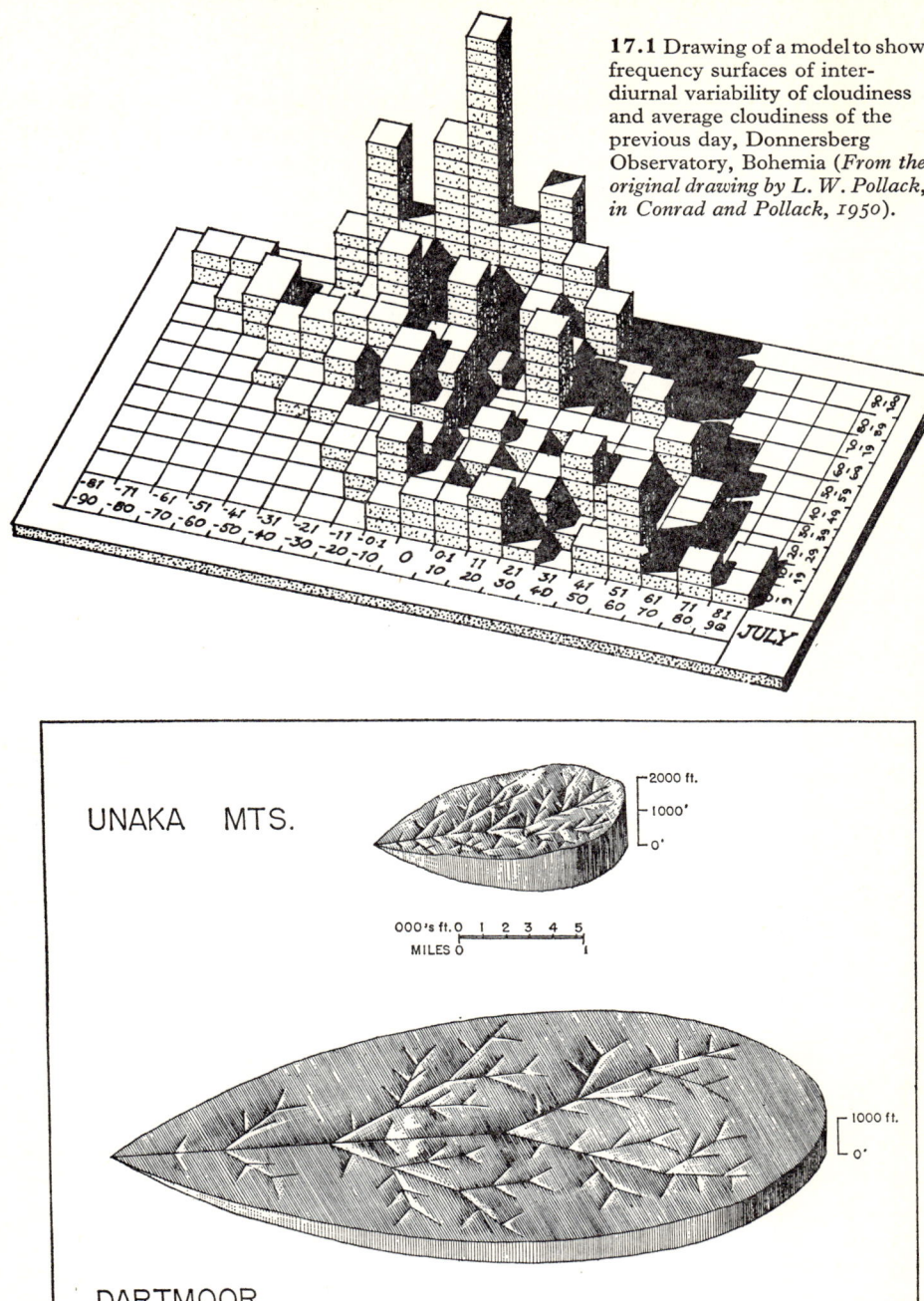

17.1 Drawing of a model to show frequency surfaces of inter-diurnal variability of cloudiness and average cloudiness of the previous day, Donnersberg Observatory, Bohemia (*From the original drawing by L. W. Pollack, in Conrad and Pollack, 1950*).

17.2 Drawing of relief models illustrating idealized drainage basins (*Source: Chorley and Morgan, 1962*).

[730] MODELS IN GEOGRAPHY

an actual part of a real landscape, but each epitomized in a quantitative manner the 'average' or 'ideal' landscape, thus permitting a relatively precise statement of regional differences. The striking contrast in drainage densities on which most of these differences hinge was explained on the basis of consistently differing rainfall intensities probably since the Miocene.

An excellent example of the use of a model to demonstrate or clarify certain statistical relationships is provided by Folk and Ward (1957) in their analysis of grain size parameters in a bar on the Brazos River in Texas. By plotting

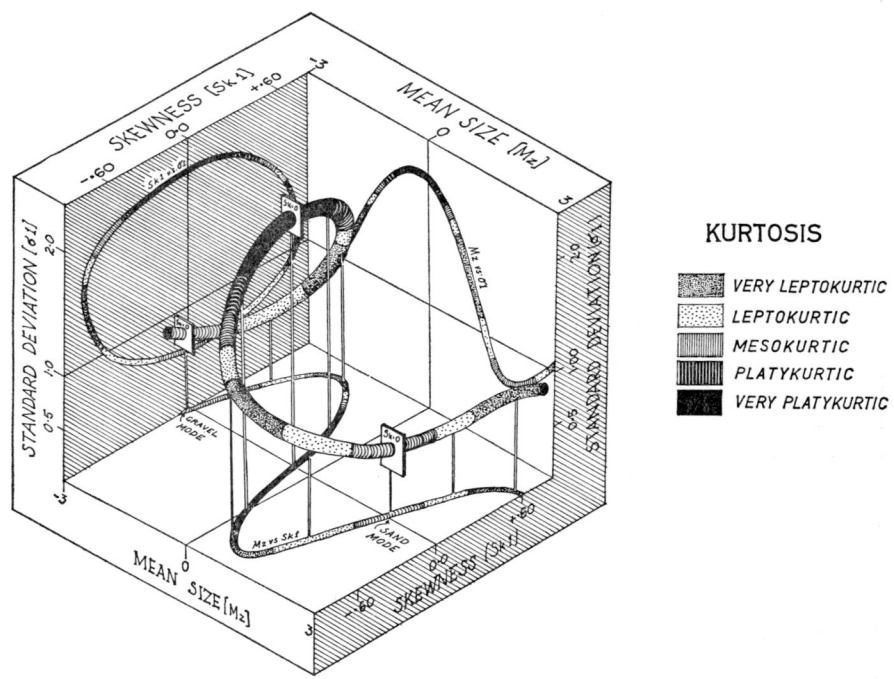

17.3 Model demonstrating a relationship between skewness, mean size, standard deviation and kurtosis (*Source: Folk and Ward, 1957*).

mean size, standard deviation and skewness against each other in turn they arrived at the three graphs shown in Figure 17.3. The degree of kurtosis associated with different parts of each graph was shown by shading. It was then noticed that the three graphs could be arranged as shown so that they were related by means of a helix, which was then constructed. Each of the three planar graphs then became a projection of the helix. The sinusoidal plots were the helix seen from above and from one side, and the circular plot was the projection of the helix seen from one end. Regular variations in kurtosis along the helix are faithfully reproduced as it is projected on each

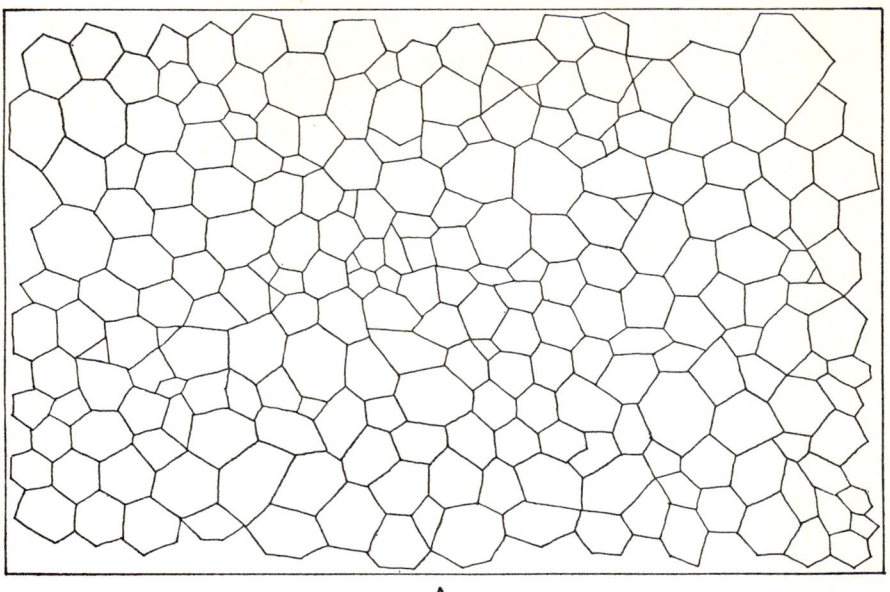

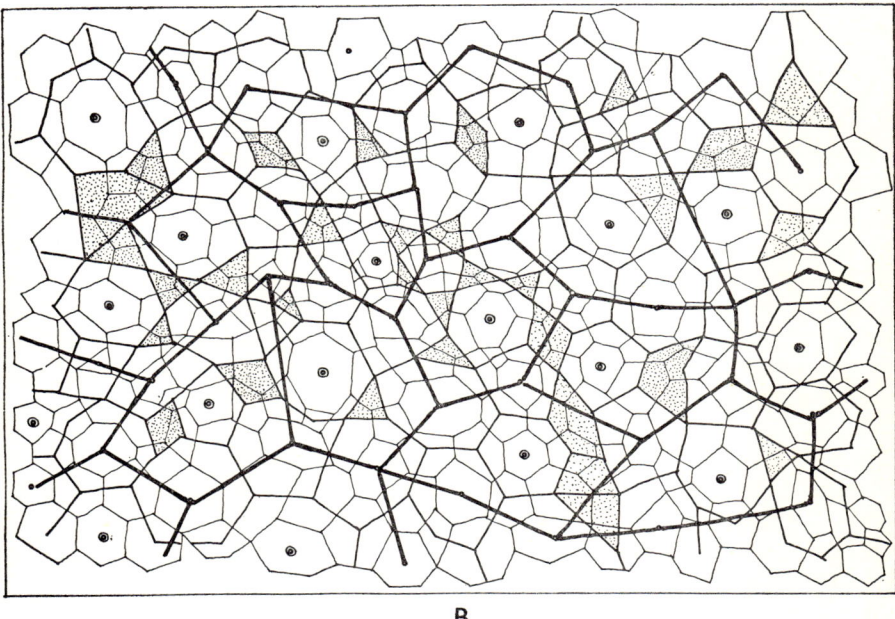

17.4 A Random network of polygons generated within expanded polystyrene.
B A sort of central place hierarchy generated from this random network. Stippled areas indicate parts of the network that cannot be incorporated within the hierarchy without considerable distortion.

surface. Between the sand and gravel modes the helix makes one complete revolution. This significant relationship is immediately evident from this drawing of the model that was made.

Balchin and Richards (1952) present a large selection of models, both static and dynamic, designed to help in teaching, and the sections on the earth's movements; the atmosphere and the oceans, and survey and map projections, should be consulted.

One should be constantly on the look-out for appropriate models for demonstration purposes and it often happens that good models are found in unexpected places. For example a piece of expanded polystyrene can be used in central place studies. This versatile material is made by forcing small roughly spherical pellets of a plastic substance into a confining mould. The pellets are packed very closely together but each retains a slightly smooth skin. If the surface of a flat piece of expanded polystyrene is carefully examined it will be seen that it consists of a network of polygons each containing a squashed and relatively soft core. When the surface is dyed the network is clearly visible. Figure 17.4A shows a characteristic pattern and an analysis of the polygons reveals that the hexagon predominates.

TABLE 17.1

Analysis of polygons in expanded polystyrene

No. of sides	3	4	5	6	7	8
No. of polygons	7	12	33	153	15	1
% total, (approx.)	3	5	15	69	7	1

Since the ultimate shape of any one polygon is a matter of pure chance the complete pattern can be regarded as a random one, and the clear predominance of the hexagon as an 'empirical regularity'. Random patterns of this sort are notoriously difficult to construct from scratch so the ready availability of such patterns in a common material is to be welcomed. It is a useful exercise to attempt to draw up an ordered hierarchy of central places on the basis of these random patterns (Fig. 17.4B) but it must be emphasized that there is no obvious unique solution. A number of attempts however would lead one fairly near to an optimum solution.

DYNAMIC MODELS IN PHYSICAL GEOGRAPHY

Perhaps not surprisingly most of the models of interest to geographers fall generally within the sphere of physical geography and specifically of geomorphology. There are good historical reasons for this emphasis. It is natural to try to create miniature replicas of parts of the earth's surface the more conveniently to study structure and process, and the growth of large scale engineering projects has stimulated work in this field since the latter half of the last century. Many models have been made by civil engineers and others to solve unique problems of great practical importance; others have been designed to explore more general or universal relationships and each to a degree has benefited from the other. Models have been used at all levels of inquiry – from the behaviour of individual grains of sand to the gross features of planetary wind circulation – to illustrate or test some hypothesis, and their success is ultimately judged by the extent to which they confirm observations about the real world.

Models of physical processes in geomorphology and structural geology

The effect of temperature changes on a homogeneous material such as clay makes a simple and instructive experiment. A shallow metal tray is filled to an even depth with a thin layer of wet clay or a thin mixture of kaolin and water. The tray is then placed in a gentle source of heat. The pattern of shrinkage cracks which develop can be photographed at various stages and the characteristics of the pattern evaluated. An interesting variation might involve freezing the material in an ice-box and then allowing it to regain room temperature. Other variants could include using non-homogeneous materials in a controlled fashion.

Van Burkalow (1945) evolved an exhaustive series of simple and repeatable experiments to evaluate the factors determining subaerial angles of repose of loose material which are of great interest to geomorphologists. With fragments less than one inch in diameter the method used to obtain a true angle of repose was to place two straight boards edge to edge and to drop a conical pile of material over the junction. When one board was moved the material left on the other slumped until the angle of repose was obtained. To produce slopes that were convex or concave in plan the boards were cut to fit together along semicircles of appropriate dimensions. For fragments between one and three inches in diameter the angle of repose was found by dropping them into a relatively tall glass sided box with a movable interior partition which allowed the width of the box to be maintained at six times the diameter of the fragments being tested. When the box was filled one side was removed and the fragments assumed a true angle of rest. In either case the angle of repose was measured with a clinometer. For uniformity artificial materials were used ranging from chilled lead shot to three inch diameter wooden blocks of various shapes. The series of experiments was designed to demonstrate general relationships rather than to solve specific geomorphological problems and in fact succeeded in formulating certain precise conclusions concerning the effect of different types of material and conditions on the angle of repose of natural slopes. The geomorphologist is frequently heavily dependent upon being able to interpret correctly the significance of transported material.

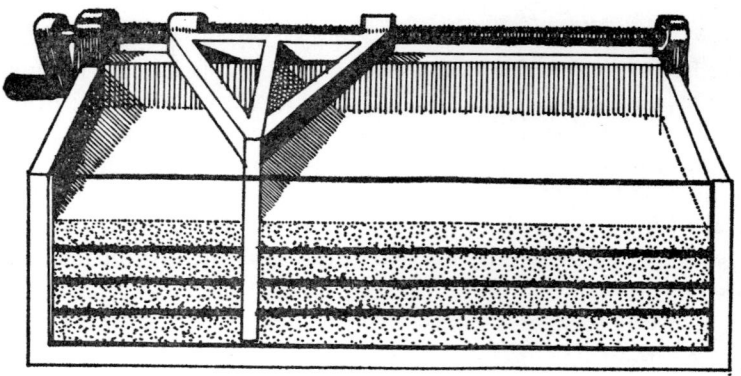
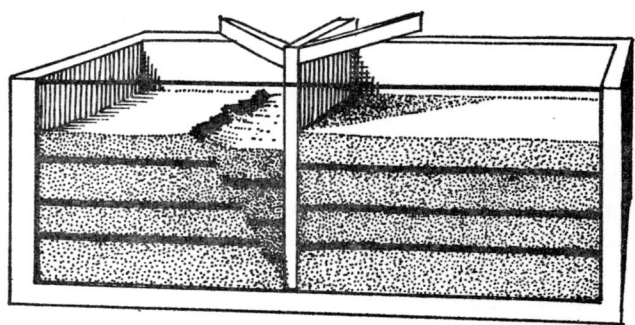
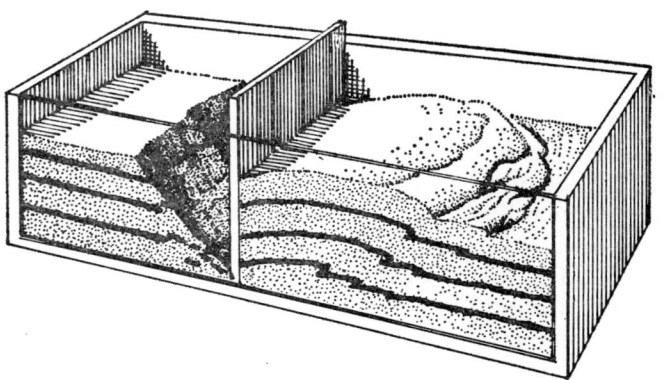

17.5 Apparatus for demonstrating normal and reverse faulting (*Source: Hubbert, 1951*).

Sands, gravels and pebbles, for example, contain in their shapes often the only clues to the history of their movement and it is consequently extremely important to be able to attribute certain shape characteristics to different modes of transport. While experimental work in the field is obviously indispensible the problem of sorting out complex variables is likely to remain a stumbling block and under these circumstances controlled laboratory experiments are suggested. Daubrée (1879) was the first of many to undertake experimental work in this field. Kuenen (1956 and 1960) has reported a series of investigations into abrasion by current and eolian action under carefully controlled conditions. The equipment used is too substantial to be easily duplicated outside a properly equipped laboratory, but his results have been valuable in clarifying a number of relationships.

While the behaviour of loose or poorly consolidated material can often be studied in the laboratory using the materials themselves it is not possible to reproduce accurately the gross behaviour of other materials such as rock or ice unless we use some other material to represent them, unless in fact we use an analogue. In terms of structural geology there are a number of physical analogues that appear satisfactory. Hubbert (1951) developed a simple model (Fig. 17.5) to demonstrate faulting. The dark layers in the model are made from any suitable contrast material with the same properties as the dry sand with which the box is filled. The middle drawing shows the development of a normal fault following the forcible shift of the dividing partition. Further movement of the divider produced reverse faulting (lower drawing) whose characteristics were found to accord well with field observation. Billings (1946) reports an experiment by Cloos in which soft mud was used successfully and it should be instructive to repeat these experiments with a variety of materials of differing competence and composition. On a much larger scale efforts have been made to simulate the effects of mountain-building, linking the observations that mountain ranges have roots penetrating into the lower part of the earth's crust with the hypothesis that sub-crustal convection currents are responsible for mountain-building. Griggs (1939) used a layer of oil loaded with sand lying upon a tank filled with water glass to simulate the surface and sub-crustal layers. Two contra-rotating drums in the water glass produced the equivalent to convection currents and deformed the oil and sand layer so that it had roots and protruberances. Kenn and Wooldridge (1965) constructed a basically similar apparatus (Fig. 17.6) in which a layer of engine lubricating oil floated upon a quantity of glycerol. Two Perspex drums in the glycerol were made to rotate in opposite directions by means of a crown-geared reversible electric motor. The lower sketch shows the patterns produced when the drums were run steadily at 81 r.p.m. As Wooldridge observed, when the drums were rotating inwards the pattern of a broad gentle uplift and extended roots recalls the classic features of epirogenetic movement. When the drums, and by analogy the sub-crustal convection

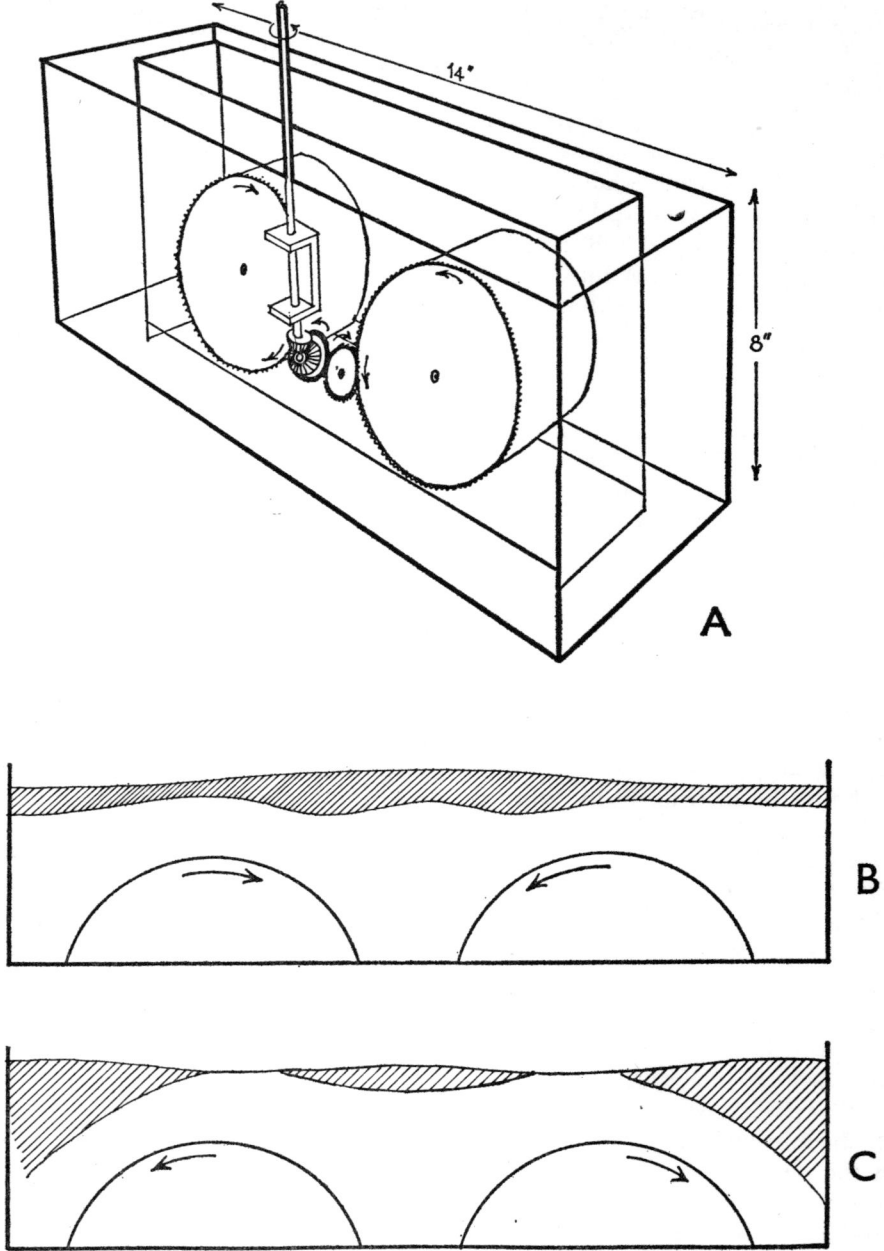

17.6 A Construction details of the mountain-building simulator.
B and C The deformation of the oil film (shaded) when the drums are rotated (*Source: Kenn, 1965*).

currents, are moving outwards the model may be thought of as demonstrating a process rather similar to continental drift.

Reiner (1959) shows how some problems of deformation and fluid motion can be represented by appropriate mechanical analogues and this has obvious bearing on the process of deformation of rocks.

The problem of simulating the movement of ice in glaciers has been dealt with by Lewis and Miller (1955). A material was needed that would exhibit the same plastic behaviour on a small scale as that of the large mass of ice in the real world that it was representing. It had necessarily to be weaker than real ice, but still capable of yielding to stress and deformation by a degree of faulting and fracturing. Eventually Lewis discovered an ideal material in the form of a mixture of two parts of kaolin (china clay) and one part of water. A relief model of a valley was made in Plaster of Paris, about 150 cms. long and 30 cms. wide, and the kaolin mixture was placed in the upper half of the valley. Plates 17.1–3 show a model roughly similar to that used by Lewis and Miller. The patterns of faults and crevasses in the model can be seen clearly. This type of model is excellent for qualitative purposes. Ideally one should arrange for a steady supply of the kaolin mixture in the cirque area, but this rather spoils the development of the bergschrund. It is usually sufficient to add quantities of kaolin as the névé becomes impoverished. If the kaolin shows a reluctance to move the valley can be greased and it may be necessary to tilt the model to as much as 20° from the horizontal. During construction a glass panel can be inserted to give a view into the cirque area. The 'glacier' can be provided with material (in the form of powdered coal, cork, etc.) to transport. Matchsticks can be placed in the surface in order to calculate speed of movement (Plate 17.2) and the general behaviour of the model will be found to accord well with field observations. Lewis and Miller deal at length with the interpretation of features developed on and in their model glacier and also discuss the possibilities of making quantitative operational models.

Chemical or electrochemical analogues have not been widely used but Muskat (1949) and Botset (1946) whose work is summarized in Karplus (1958) developed a method to assist oil engineers increase oil recovery from long-worked fields. In older fields water is sometimes injected into centrally placed wells so that the water front moves outwards and increases the oil pressure in marginal wells. It is obviously important to know the likely progress of the water front and to know the most advantageous point at which to inject water into the field. The analogue model consists of a conducting layer cut to the shape of the oil field and made of a 1 per cent agar gelatin solution with a small quantity of zinc ammonium chloride. The wells are made of plastic tubes inserted into the gelatin sheet. Injection wells are filled with a mixture of agar and a 0·1 molal deep blue copper ammonium chloride solution and the production wells contain the same material as the model field.

The injection wells are made electrically positive and supplied with potentials up to 1,000 volts and the producing wells are all made negative. When the power supply is connected a potential gradient exists between the injection and producing wells. With a flow of current blue copper ammonium ions stain the field as the 'water front' moves outwards. It is possible that the ion diffusion method may have some application in diffusion studies in human geography but the implications remain to be worked out.

Hydrological models

Models of streams, rivers and estuaries have been used for a long time, originally for demonstration purposes, but increasingly today to solve a wide range of civil engineering problems. Many of us have cause to be grateful to the late W. V. Lewis for our introduction to this peculiarly pleasurable aspect of experimental geography. Streams and rivers in miniature can be observed or created on the beach, in the gutter or garden on a rainy day. Spoil heaps of all kinds, if reasonably homogeneous can be seen to have miniature drainage patterns etched upon their surfaces. A simple river can be created upon a bed of sand in a waterproof trough and the processes of erosion and deposition can be followed. A river system can be originated on a sloping wedge of sand by gently watering the surface with an overhead sprinkler. Certain streams can be given an advantage by lowering their base level, shortening their courses or increasing their discharge. A layered basis of sand and marl can be used instead of pure sand in order to introduce an element of lithological complexity. Simple models of this sort are ideally suited to introductory courses in geomorphology in schools; they make few demands in terms of apparatus and they encourage the practice of accurate observation and recording.

When model rivers are used to solve specific problems the experimental techniques and the type of constructs have necessarily to be more sophisticated. Just how much more refined relates very much to the nature of the problem. Lewis (1944) used a stream trough four metres long and 50 cms. wide to experiment on the influence of load on gradient in the hope of throwing light on certain features associated with flood plains, in particular river terraces formed without change of sea-level. The behaviour of knickpoints in non-cohesive material has been experimentally studied by Brush and Wolman (1960) using a flume (trough) 52 feet long and 4 feet wide. Complex stratification and cross-stratification in streams and deltas (Nevin and Trainer, 1927; McKee, 1957) and meandering (Friedkin, 1945) have been studied with the use of physical models.

Civil engineering today is slowly moving away from the often inspired empiricism of its Victorian past. Vast capital outlays cannot be hazarded by inadequate information, not only concerning the structures themselves but also their effect on the site. New harbours, barrages, flood control and irrigation schemes, if only by virtue of their great size are likely to create local environmental changes. A very great deal of experimental work has been done in relation to these sorts of problems by the construction of scale models in which alternative proposals can be evaluated and their respective merits accurately assessed. One difficulty however arises from the fact that in small scale models the change of scale affects the relationships between certain properties of the model and the real world in different ways. If, for example, we make a model of the Severn Estuary at a scale of 1:10,000 the geometrical and topographical relationships can be preserved fairly easily. When we add the water though we find that an actual depth of water of say about 20 feet is represented in our model by a layer of water less than 1/40th inch thick. Not only will surface tension assume intractable proportions but it would be vir-

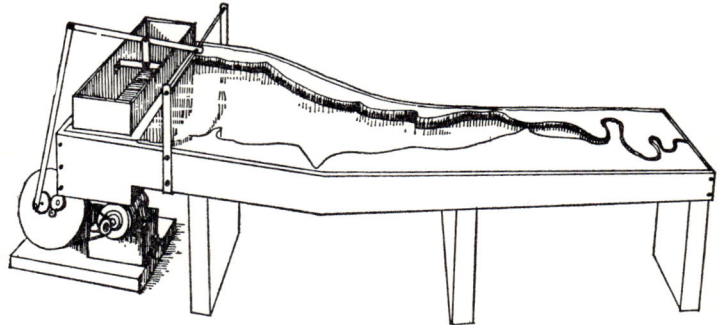

17.7 Gibson's small model of the Severn Estuary (*Source: Allen, 1947*).

tually impossible accurately to simulate tidal range and currents. Equally the sand along the bed of the river could not be reproduced to scale and even were it possible a suitable material would mostly float on the surface of the water. A coarser material in the model would represent large boulders in the real world.

In order to overcome scale difficulties it has been found possible to distort the scale of one attribute in the model in order to preserve that of another. Models of rivers and estuaries for example often have the vertical scale greatly exaggerated in order to reproduce more or less correctly the effects of turbulence in the water. Figure 17.7 shows a sketch of a model of the Severn estuary made by Professor Gibson (discussed in Allen, 1947). It has a horizontal scale of 1:40,000 and a vertical scale of 1:366. The tide in this model is produced by calculated periodic displacement of water in the reservoir by means of a float. More complicated sequences of tides are usually generated by complex linked systems of cams and plungers.

[740] MODELS IN GEOGRAPHY

For detailed experimental work empirical distortion of one attribute to preserve another is inadequate. Most models of this sort have been proportioned using the principles of dimensional analysis. Langhaar (1951) and Duncan (1953) are the principal authorities on this subject. Normally the first requirement of a model is that it shall be geometrically similar, at least in the two principal dimensions, to the prototype; the distance between any two points in the model must bear a constant ratio to the distance between the

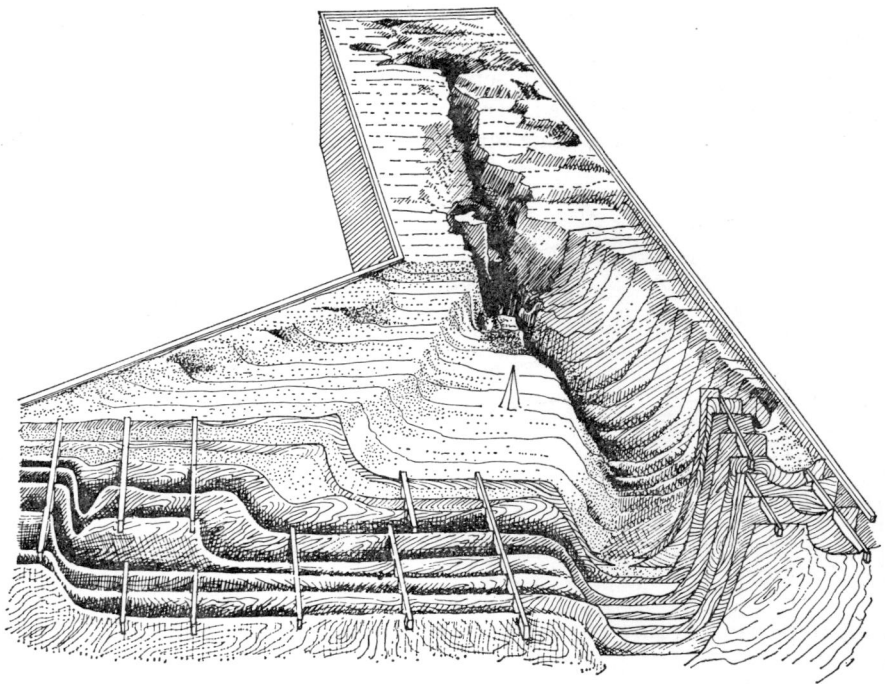

17.8 Method of constructing a scale model of an estuary. Stout serial profiles at appropriate scales are securely fixed to a rigid base. The spaces between them are filled with waterproof cement or plaster. This drawing, based on a photograph in Allen (*1947*) is of a model of the Severn Estuary constructed to examine the effects of a proposed barrage. Note the great vertical exaggeration.

corresponding two points in the real situation. In hydraulic models the similarity must usually also extend to kinematic relationships, that is the velocities and paths and patterns of movement in the water in the model should have a one-to-one correspondence with what happens in the prototype. The details of dimensional analysis and its practical implications are too complicated to be dealt with here but Pankhurst (1964) gives an excellent introduction to the subject as a whole and Allen (1947) is an outstandingly interesting and valuable source for all aspects of hydraulic models. In essence though dimensional analysis and the application of distortions calculated with the use of dimensionless numbers permits models to be made so as to preserve correctly a number of properties: to the extent that the models are made to be good analogues one can be certain that what happens in the model would happen under similar circumstances in the real world. Nevertheless it is a wise precaution constantly to test one's model against the real world because each situation contains elements of uniqueness. The most usual

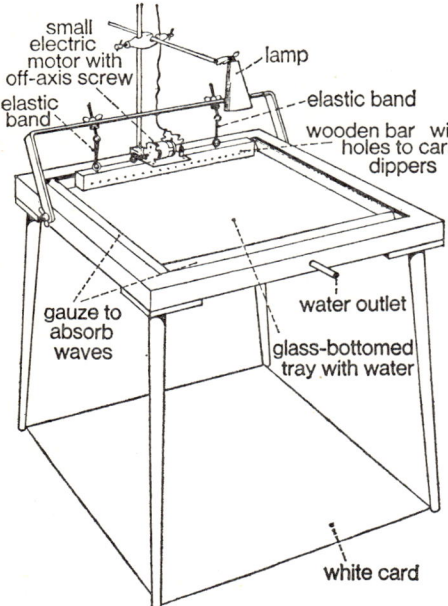

17.9 A ripple tank. The ripples are produced by vibrations of the wooden bar which in turn are induced by mounting a small off-axis screw or other weight on the shaft of the motor. The bar dipper will produce plane waves. Circular waves are made by inserting a short rod in one of the holes on the bar dipper (*Source: 'How and Why?' BBC, 1965*).

method of construction is suggested by Figure 17.8 which shows a basis of serial cross-profiles (with appropriate vertical exaggeration). The intervening areas are filled with cement or plaster of Paris (if the latter it must of course be waterproofed). The cross profiles define in this case the bedrock surface. If the model is to contain loose material over its bed than a second set of templates is made so that loose material put on to the bedrock can be accurately arranged in terms of its thickness.

Waves and coastal phenomena associated with wave action can be studied at different levels of sophistication. Simple basic experiments can be done with an easily made ripple tank, much used in physics teaching (Fig. 17.9) but in addition some sort of wave tank or flume is needed if, for example, constructive and destructive action of waves is being studied. Here, unfortunately, there is no substitute for really rugged construction and accurately designed wave generating equipment. One tends to forget the great weight of relatively small quantities of water. Stanton *et al.* (1932) and more recently Saville (1950), Johnson and Rice (1952) and McKee (1960) have accounts of work with model wave tanks which should be consulted as much from the point of view of operation and experimental design as of conclusions reached.

Until recently most hydraulic models were scale models, at least in important respects. In the past decade, however, certain electrical analogue models have been developed which have the advantage of greater speed of construction and operation and also of substantially lower cost. Glover *et al.* (1953) used such an analogue to study the effect of tidal flows superimposed upon the steady flows within a series of canals in the delta region of California. The effect of changes induced by introducing links and cut-offs between the canals was simulated electrically and the results matched well the field observations. This model technique opens up the possibility of assessing in advance of actual construction work the effect of proposed engineering works, within the system. Einstein and Harder (1961) investigated the same delta region mainly from the point of view of tidal flows and amplitudes within it. An electrical analogue of course gives virtually an instantaneous solution which can be displayed on a cathode ray tube so that parameters in the model can be adjusted rapidly and the analogue, within its limits, made to duplicate exactly the behaviour of the prototype. Harder (1963) extended the principle to simulate flood control systems with the practical objective of achieving the best phasing of the operation of reservoir sluices during a flood emergency. Harder's model related to the Kansas River basin. Flows in all tributaries and along the main channels were represented by electrical current flow and adjustable to represent a wide range of flow and rainfall conditions. The speed of response of the electrical systems to the inputs is very rapid in this case, only 0·03 seconds, so a large number of alternative plans for reservoir operation in an emergency can be evaluated in a very short period. In addition to its short term benefits the long range advantages are the comparative ease with which more permanent solutions to perennial flooding problems can be evaluated.

An unusual analogue based on heat conductance was developed by Appleby (1956). In a drainage network, whether it consists of rivers or urban sewage systems, the value of the run-off function is most difficult to calculate since it is a function both of variable input – variable in terms of time and quantity – and losses through evaporation and absorption of various kinds. Appleby's analogue is based on the fact that stream network flow and heat flow behave in the same manner and can be represented by the same differential equations. His apparatus allows a series of trial values for the run-off function to be rapidly tested against the form of an actual hydrograph. When similarity is achieved then the actual values of the run-off function can be recorded from the analogue.

Models in meteorology, climate and oceanography

Electrical analogues are beginning to be employed in meteorological research. Wallington (1961) has studied the behaviour of lee waves and Tyldesley (1965) proposed an electrical analogue for solving atmospheric diffusion equations. Non-electrical analogues have been used in evaporation and condensation studies (Turner, 1965) and tornado simulation (Turner and Lilly, 1963).

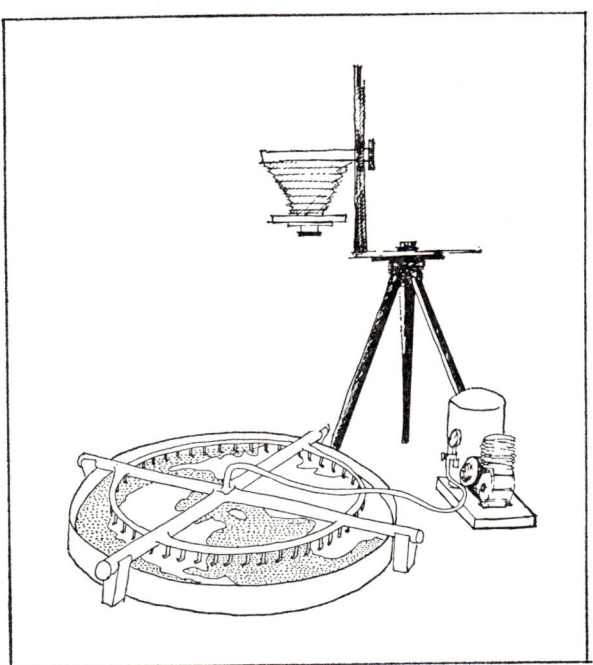

17.10 Lasareff's (1929) dishpan model for the oceanic circulation.

The problem of explaining gross planetary circulations of both wind and waters has led to an interest in simulation by model building. Lasareff (1929), then the Director of the Moscow Geophysical Institute, constructed a series of models to show the relationship between oceanic currents and trade winds which had been suggested much earlier but never experimentally verified. His apparatus (Fig. 17.10) consisted of a circular dish (later a rectangular dish was used) in which the continental masses were represented by raised plaster blocks. Wind was represented by peripheral air currents directed through adjustable nozzles over the surface of the dish which was filled with water. Powdered aluminium sprinkled on the surface revealed the pattern of water movement. The currents were recorded by time lapse methods using a vertically mounted camera. The map of the North Atlantic (Fig. 17.11A) shows not only gross features of the resulting circulation but also a number of significant secondary circulations and fits well with what is known of the real situation. It was found that small island arcs and archipelagos are of great significance in affecting circulation patterns in their immediate vicinity and occasionally over unexpectedly wide areas. Lasareff dealt in detail with the problem of choosing a suitable projection upon which to model a hemisphere. For true kinematic similarity one should use a gnomonic projection, but this is obviously impracticable since a hemisphere cannot be shown on a gnomonic projection. In practice a projection similar to the zenithal equidistant is reasonably satisfactory. (Von Arx, 1962, p. 303 gives details of spacing of parallels in the best sort of projection for this purpose.) Thermal gradients were established by fitting heating coils to the edge of the dish and with the wind generator in operation a satisfactory disposition of warm and cold currents was achieved.

Lasareff then went on to show that using models it should be possible to recreate climatic conditions of past geological periods. He assumed that the gross thermal and rotational characteristics of the earth have remained substantially unchanged and argued that consequently the main variable would have been the disposition of the continents. Given this he made models to illustrate oceanic circulation in past geological periods (Fig. 17.11B) and by extension gross palaeoclimatic conditions.

Von Arx (1957 and 1962) has developed more sophisticated models taking into account such additional factors as the rotation of the earth and certain effects caused by its curvature.

The underlying similarity between planetary circulation of water and air is emphasized by the fact that similar experimental models have been evolved to study them. Starr (1956) in a review of the work of Fultz (1956) has pointed to the defects in classical views of atmospheric circulation – specifically their failure adequately to explain the source of energy for the mid-latitude westerlies. A mathematical description of the complex dynamics of the atmosphere is extremely difficult and its interpretation would be beyond the capacity of many people interested in it. There is a strong suggestion that a combination of thermal gradients and rotational effects will explain the westerlies and Fultz has designed a model which experimentally supports this view. The apparatus consists of a flat-bottomed dishpan representing a hemisphere, filled to a depth of one inch with water. Peripheral heating and polar cooling produces a simple overturning flow pattern. When the pan is rotated from two to six revolutions a minute the simple initial circulation breaks up dramatically into eddies and vortices like the familiar sequence of cyclones and anticyclones of the mid-latitudes. From equator and pole eddies move towards the middle latitudes where westerly currents develop. The

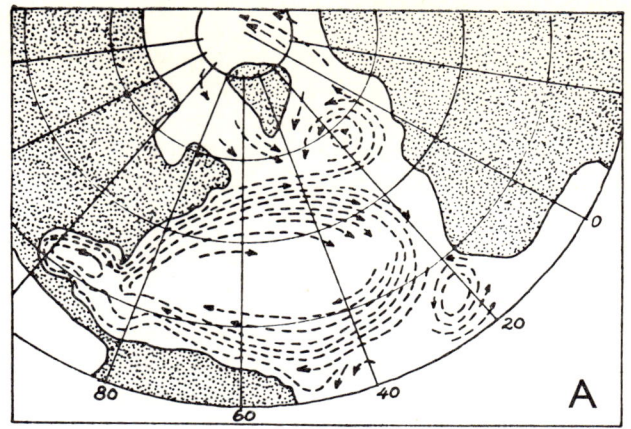

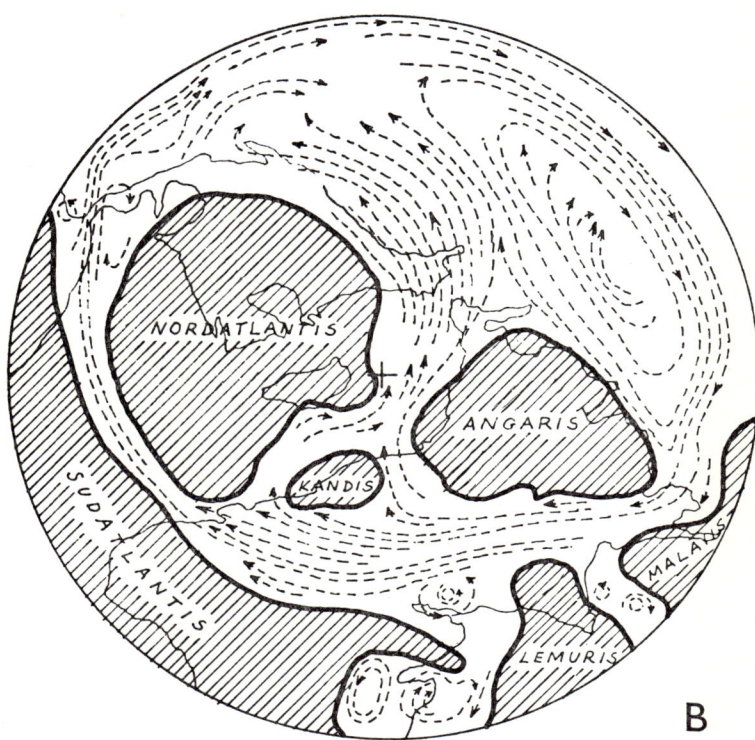

17.11 Simulated ocean currents for the North Atlantic (A) and for the northern hemisphere in the mid-Jurassic (B) (*Source: Lasareff, 1929*).

water currents in three dimensions also appear to reproduce well the vertical variations in the atmosphere. Separate eddies at the bottom of the pan are replaced nearer the surface by fast flowing undulating jet streams. A more technical account of the experimental design and results is given by Riehl and Fultz (1957) who also deal with the problem of dimensional similarity.

CONDUCTING SHEET ANALOGUES

The solution to a large variety of problems in engineering and physics depends on a knowledge of the shape of fields and the pattern of magnetic, electrical or heat flux. In very simple cases, where for example one is dealing with a single source and an infinitely large field, the solution of the resulting field pattern is comparatively simple. The problem becomes much more difficult when the boundary of the field is finite and irregular and when more than one source is considered. Under these circumstances the solution can be found by relaxation techniques which involves choosing a finite number of points or nodes in the field and estimating the potential at each node. The differential equation for the complete field is replaced in effect by a set of simultaneous equations relating the potential at each node and these are solved step by step. While the solutions are extremely accurate they are also complicated and lengthy.

An alternative is to use an analogue method and this commonly takes the form of a conducting sheet, resistance mesh or an electrolytic tank. The principle depends upon the fact that a wide variety of problems in fluid flow, heat flow, gravitation, electrical and mechanical fields are founded on the equation: Flow=driving force×acceptance. Thus a system obeying this equation can be used as a model to solve problems in any other system governed by the same equation. There is obviously an advantage in using a simple and easily constructed analogue and the conducting sheet analogue is the simplest available. Kirchoff used a thin sheet of copper for field plotting as early as 1845 (Karplus, 1958) but there are difficulties in obtaining uniform resistivity in thin sheets. The most used conducting sheet today is Teledeltos paper, originally developed for electrical recording purposes in 1948. This is made by adding carbon black (a conductor) to wood pulp. The paper has a uniform thickness and is relatively cheap; its resistivity is of the order of 2,000 ohms per square. It is somewhat anisotropic, but for most problems this can be ignored. If it is important it can be simply calculated and allowed for in experimental design.

The problem to be solved is drawn out on the sheet of Teledeltos paper. The sources and the sinks, the known equipotentials, are painted on the paper with special silver conducting paint. The plotting device consists essentially of a d.c. potential source, an accurate potential divider, a galvanometer and a probe. The output terminals of the plotter are connected, one to the source and one to the sink. When the plotter is switched on the electrical field is established over the paper. The potential divider has a scale with 100 divisions so each division is equivalent to 0·01 per cent of the potential across the terminals. The actual value in volts of this potential is unimportant for most purposes. The potential divider is set at a desired reading, say 50, and the probe moved over the surface of the paper until there is zero deflection on the galvanometer. At this probe position the potential is exactly 50 per cent of the input potential at the source. Several of these points with the same value are located and joined to form the 50 per cent equipotential line. When a sufficient number of equipotential lines have been drawn it is a relatively simple problem to sketch in the flux lines since they flow always at right angles to the equipotentials. In drawing the flux lines it is normal to space them so that they form sets of curvilinear squares each with the characteristic that mean width is equal to mean length. It is often a help to draw

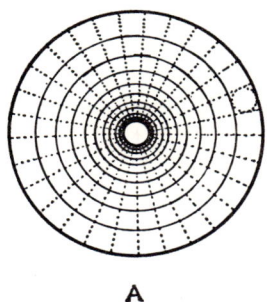

A

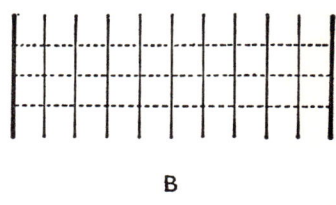

B

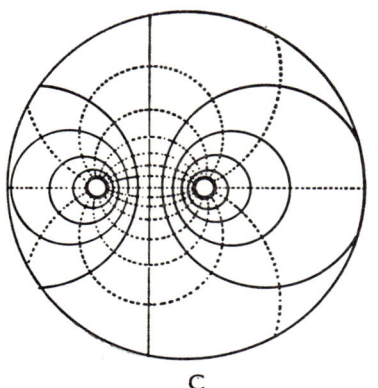

C

17.12 Characteristic field plots.

A Central source and finite circular boundary.
B Two parallel straight electrodes.
C Two sources in an infinite field.

Equipotential lines are shown by continuous lines and streamlines by dotted lines orthoganal (at right angles) to the equipotential lines. The dotted circle in A indicates one way of drawing curvilinear squares based on the plotted equipotentials.

circles whose diameter is equal to the distance between adjacent equipotential lines as a basis for inscribing the flux lines (see Fig. 17.12A). Characteristic simple plots are shown in Figures 17.12A and B. In each case boundary conditions are simple and the boundaries are finite. It should be noted that the spacing of the equipotential lines in the two examples is different. The lines about a point source are bunched closer together near the source but between two parallel plates the equipotentials are equidistant. From a point source X located in an infinitely large two-dimensional plane field the potential at a point Y at distance r from X is proportional to $\log \frac{1}{r}$ (in a three-dimensional field it is proportional to $\frac{1}{r}$). With a finite boundary, however one needs to apply Laplace's equation in two dimensions if the equipotentials are to be calculated. The field plotter, however, gives the graphic solution in a fraction of the time needed to calculate it.

It is sometimes necessary to plot fields in which the boundary or sink is at infinity. One solution is to place all the sources as near as possible to the centre of a large piece of resistance paper – around the edge of which the boundary is painted – in the hope that it is sufficiently remote not to affect the plot in the central area. For more precise work where the field is known to be symmetrical about a point the plot can be made with a finite boundary and then replotted on a transformed grid. Alternatively a method has recently been suggested by Olsen (1963) based on an adaptation of the double layer electrolytic tank described by Boothroyd, Cherry and Makar (1949). This involves cutting two circles of resistance paper about 20 inches in diameter and placing them back to back with an intervening piece of cartridge paper for insulation. The edges in contact with each other are freed of the insulated backing and finally clamped together with insulated clips. The source or sources are painted on one side and the centre of the opposite side is made a sink. The upper sheet is then able to behave as if it were part of an infinite domain. Figure 17.12C shows a characteristic plot made under these conditions.

While it is beyond question that this particular analogue is of great value in physical and mechanical sciences, its application to the sort of problems geographers concern themselves with may need some justification.

In physical terms there is an exact analogy between the pattern of an electrical field and the pattern of heat distribution and fluid flow for example, because the different physical phenomena are described by the same mathematical formalism and quantitative conclusions can be drawn about one by studying the other. In the field of physical geography true analogues are certainly feasible. But in human geography, despite the search for formal order, there is as yet no mathematical formalism, only certain 'empirical regularities' (Isard, 1956). These regularities for the time being have to serve as our models if we are to use these physical analogues, but it is as well to recognize their limitations, and to realize that precise quantitative solutions are not generally possible. On the other hand if we accept the limitations we can apply field plotting techniques in a qualitative sense to a large range of problems with considerable benefit.

We may consider possibilities first on uniform, and then on non-uniform fields. The first published account of the use of field-plotting techniques in geography is that by J. R. McKay (1965) who used an analogue field plotter to simulate the macroscopic pattern of glacier flow for large ice sheets of the Wiconsin glaciation. The experimental design is illustrated in Figure 17.13. The equipotential lines are not shown but the arrows indicate flux lines. The

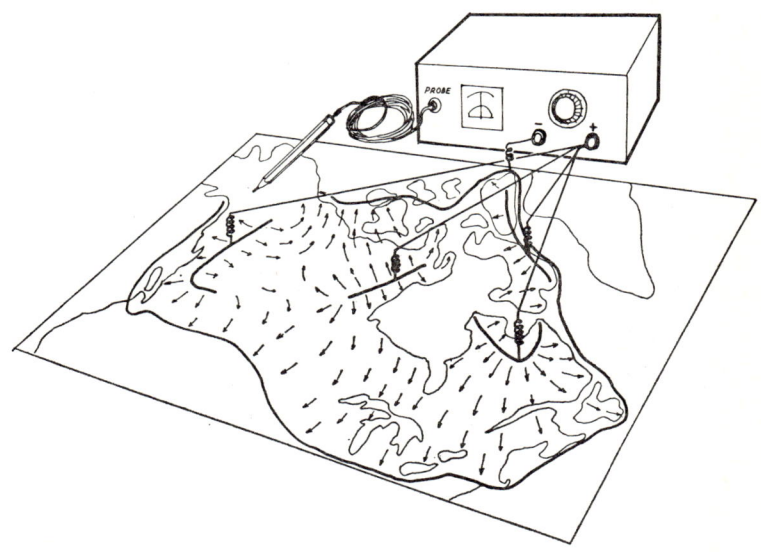

17.13 Sketch of apparatus used by MacKay (*1965*). Direction of flux lines, and (by analogy) lines of ice flow, are indicated by arrows.

sources are painted electrodes in their correct position and the boundary of the ice sheet is the painted line at zero potential. The results of the experiment were sufficiently promising (in the sense that the pattern of flux lines fitted the observed direction of ice flow under certain simplified conditions) to suggest that it would be worth developing a more refined model, which would enable different conditions to be simulated.

There are grounds for believing that the concept of 'population potential' as developed by Stewart and Warntz (1958) represents a suitable field for analogue simulation. If we equate mass with electrical potential the resulting equipotential lines from a field plot should give results comparable with those arrived at by the different method used by Stewart and Warntz. Inevitably problems of scale limit the accuracy of the final result but the method suggested should be useful both in teaching and in giving first approximations of population potentials in areas for which data is sparse or in which a crude picture is acceptable. Plate 17.4 shows the apparatus used – though the arrangement of the sources relates to another experiment. In this case a 250-volt D.C. supply was used and was fed to the resistance sheet via a number of resistors whose value was calculated to represent roughly the populations of the largest cities in Canada and USA. Very large resistances

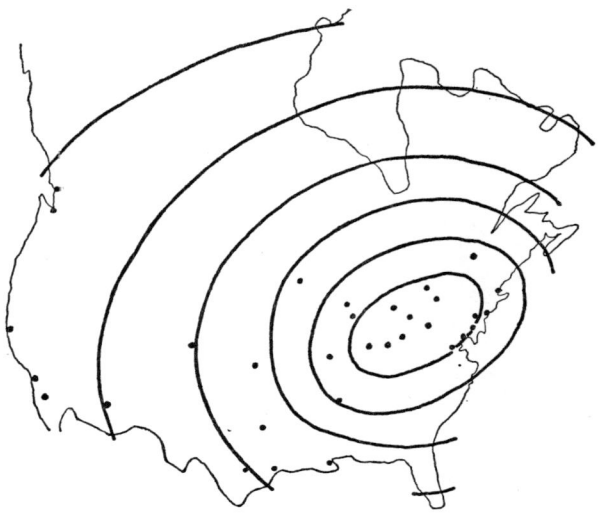

17.14 Electrical analogue to show potential population in North America. This sketch has been included merely to illustrate the order of result that may be expected using this method and is not in any sense in this form capable of yielding a precise result. The dots represent the major cities and each dot is supplied with an electrical current proportional to the population to which it refers. The resulting equipotential lines are then plotted. With careful experimental design the voltages along each equipotential line can be converted into population potential figures.

were employed to give a range of current inputs from 1 milliamp for New York to 0·06 ma. for cities with a population of from 4–500,000. The current was supplied to the resistance paper by needles correctly located and the equipotential lines were plotted (Fig. 17.14). With this apparatus actual voltages can be allocated to each equipotential line and these can be converted into population potential figures. More accurate work along these lines is currently being undertaken and it does appear to hold some promise.

There is a possibility that a similar technique could be used to simulate the territories of central places, but there are operational difficulties in applying a large number of sources with only a peripheral sink. Work is now in progress on a system with small settlements acting as sources and central places as controlled sinks. Figure 17.15 shows field plots in which the experimental design has been made progressively more complicated and could be thought of as crudely simulating the movement of a frontier of settlement from the right of each diagram to the left. In Figure 17.15B, a line of conducting paint (x) has been placed in the direction of movement of the frontier and acts as a

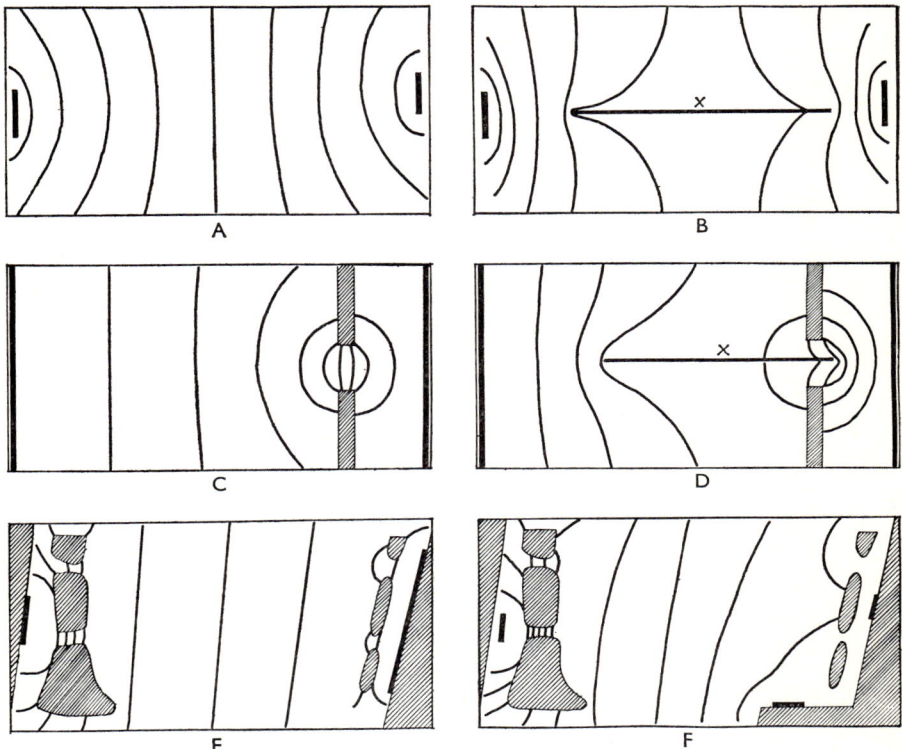

17.15 A selection of equipotential line plots. The silver paint electrodes are indicated by thick lines. 'X' represents a line of silver paint applied to the surface but not connected to the power supply. The shaded areas show where the paper has been cut away.

line of greater ease of movement. Figure 17.15C introduces a barrier with one gap (the shaded area represents the barrier which is actually cut out of the paper). Figure 17.15D places an easy route inland for some distance from the gap. The last two figures introduce two barriers with several gaps of varying width and in the final one another source of potential colonizers is added on the seaboard. The basis of the analogy is that a frontier of settlement represents a balance between pressures behind pushing it forward and attractions

[752] MODELS IN GEOGRAPHY

and difficulties in front. If these were the only factors and if they could be quantified then a simple model could be constructed. Unfortunately, if equipotential lines are equated with frontier positions the experimental conditions are such that no matter how hard or easy it is for the current to move across any part of the paper the potential difference between the source on the right and the sink on the left is always constant; so there will always be an equal number of equipotential lines between source and sink and therefore in the model the frontier reaches the final point simultaneously from all points. The difficulty can be overcome by providing a suitably distorted sink in places well beyond the shore on the left of the area. The characteristics of potential distribution within a field have some application to conformal

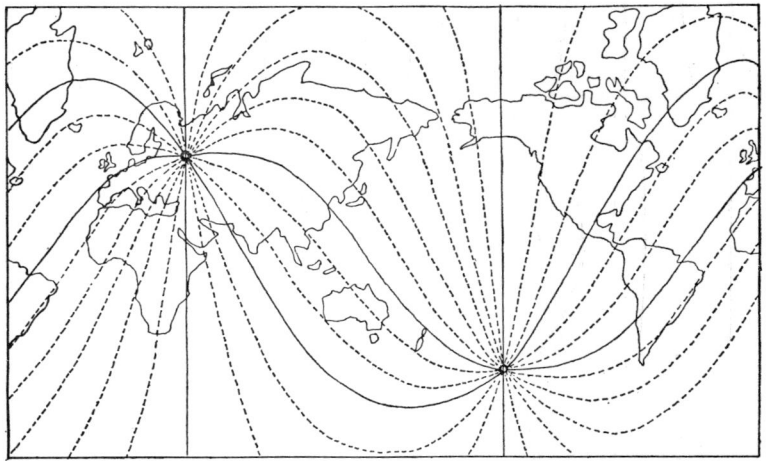

17.16 Continuous lines represent known great circle tracks on this Mercator projection. The dotted lines are equipotential lines plotted after energizing appropriate segments of the known great circle tracks.

mapping although the full implications remain to be explored. One has only to look at some field plots to become aware of the similarity between them and certain map projections, though in fact the resemblance is often only superficial. Warntz (1965), in a typically stimulating paper, has drawn attention to the geometry of surfaces and paths on conformal map projections. In one map he plots two complete great circle courses from London on a Mercator projection (equatorial case) and shows how the great circle tracks are orthogonal (i.e. at right angles) to the iso-distance lines based on London when plotted on the same graticule. Figure 17.16 develops an idea suggested by this map and uses the field plotting analogue to plot a series of great circles based on Moscow on Mercator's projection. Any great circle can be defined on a globe using a piece of taut string. Two are determined in this fashion and transferred to the Mercator graticule marked on Teledeltos paper. In fact since the meridian upon which Moscow and its antipodean point stand are also great circles only one need be plotted from the globe. Each segment of the great circles so drawn on the paper is painted with silver paint (taking care that lines at junctions are not linked up with paint) and adjacent parts of the tracks are energized in turn. The equipotential lines are drawn in and represent other great circles emanating from Moscow. If checked against a globe they will be found to be correct.

Since the Teledeltos paper can easily be cut with scissors it is amusing and instructive to play with different shapes. One can illustrate certain basic features of rivers, for example, with pieces of resistance paper. If we accept as a working hypothesis that there is at any point on a river's course, a relationship between the volume, and the width and the height above base level we can symbolize or represent a river by a triangular wedge of paper. The scale for the width is conveniently many times greater than the scale of the

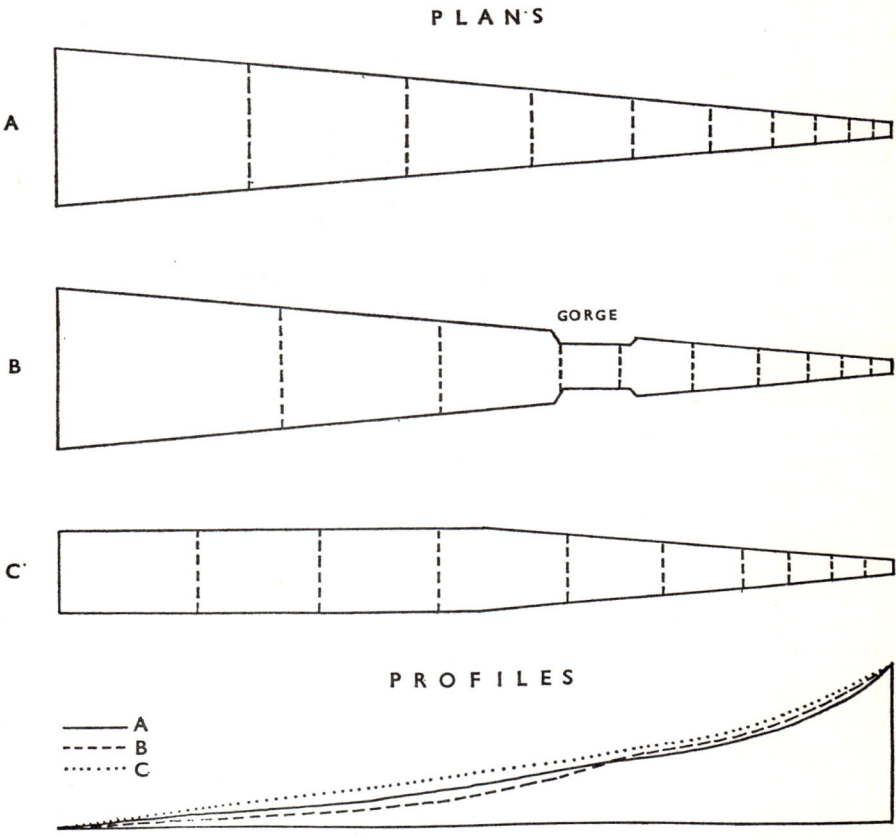

17.17 Simulation of stream characteristics by means of Teledeltos paper strips.

length. If we apply a potential at the source of the river and arrange a sink at the point where the river either enters the sea or joins another river of the same order we can plot the equipotential lines. If we equate these with height we can then derive a profile which accords fairly well with observation (Fig. 17.17). If we introduce a restriction at some point in the course by narrowing the width of our paper river we find the profile has a characteristic irregularity at that point. It is useful as an exercise not only to vary the width of the model river but to introduce tributaries, vary the flow at different points, and even perhaps attempt to simulate some features of an actual river system.

While most of the discipline in which analogue field plotting has developed are interested primarily in uniform fields, geography more often than not deals with non-uniform surfaces. The irregularity may be topographic, socio-economic or statistical for example. There are certain advantages in being able to simulate non-uniform fields but practical problems present some difficulty. On a sheet of Teledeltos paper, for example, the resistivity can be altered by cutting out a series of square holes leaving a lattice arrangement which increases the resistivity but at best this is a crude method and it is not possible to get accurate readings within the lattice because of internal fluxes. Ideally one wants a three-dimensional model made of a suitable resistance material. The resistance will vary inversely with the thickness. It has been found empirically that for crude simulations the sort of clay used by potters has a suitable resistivity if used with the standard Servomex Field Plotter F.P.144. The clay must be homogeneous and of the same consistency as when put on the potter's wheel. Plate 17.5 shows a 'relief' model to which conducting strips of copper were attached at each end. The equipotential lines are dotted in over the surface of the clay and their shape and disposition is a reflection of the topography of the surface. The conductivity of the clay is a function of the presence of interstitial water and over a period of an hour or so evaporation takes place and the resistivity changes. This is of course painfully crude and would be viewed with horror by any engineer, but it makes its point effectively and is extremely useful in the early stages of experimental design because being plastic the configurations can be altered very easily and quickly.

A more stable medium consists of a mixture of 2 parts of synthetic graphite (the conductor) to 1 part of Polyfilla (a proprietary brand of cellulose putty) which is non-conducting. These are mixed exhaustively with distilled water into a stiff paste and then modelled on a surface of clean plate glass. The terminals can be cut from copper strips and inserted into the ends of the model while it is wet. Final modelling of the surface can be accomplished by chisel and file to give the required shapes. Figure 17.18 shows the equipotential lines on a wedge of this material and the relationship between thickness and resistance can easily be seen. In a colonization or settlement model it is possible to shape the relief so that it expresses the ease or difficulty with which areas can be traversed or settled. Where the surface is thick the going is good, the relief is low, the soil is fertile, there is an absence of hostile natives, etc., etc. Merely assessing terrain potential is from a teaching point of view a valuable part of the exercise. When the potential is applied at the point of initial settlement its value is adjusted to the capacity for expansion at that time and place and the equipotential lines can then be held to reveal the rate and direction in which the frontier might be expected to move over time. The disadvantage of a single sink can be overcome by burying different sinks at various points towards the far edge of the model.

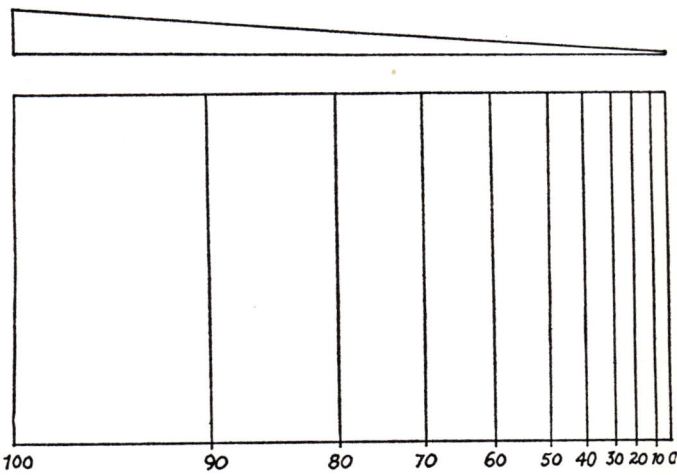

17.18 Equipotential lines within a wedge of conducting material. The cross-section is shown above and the figures indicate percentages of input potential.

It would be reasonably simple to construct a three-dimensional model of a population potential map and then to apply potentials at various points and observe the distortions that occurred because of the variable surface. A simple concentric pattern about a point would be deformed if at the point of application the surface were not uniform. Plate 17.6 shows a cardboard model of a sloping statistical surface. The rings upon it would be circular if the surface were horizontal. The distortion of the pattern is caused by the slope of the surface. There is an analogy here with Stewart and Warntz's observation that Federal Reserve Banks are displaced from the geographical centre of the area they serve towards the zone of highest population potential. This is possibly one of the most promising applications of the method and could be extended into the study of deformations over many different types of surfaces.

It is not practicable to model a continuous surface; reasonable results can be obtained by sectioning the desired surface into a number of discrete profiles. The profiles are then cut out of Teledeltos paper and the equipotential lines drawn on each in turn. They can then be made into a model and lines of equal value linked by elastic threads. Plate 17.7B shows a simple model in which its assumed movement is controlled by relief. The sections of course represent the reverse of the actual profile, i.e. areas of low resistance are high, and of high resistance are low. Plate 17.7A shows the profiles in 'topographic' terms, the mountain ranges appearing as high points, the passes and plains as low points. In each case the equipotential lines have been placed in the same relationship. Plate 17.7C shows a model in which the disposition of the equipotential lines to the far side of the model has been deliberately controlled during construction. The profiles were drawn with additional strips on the far end so that the value of each line crossing any point along the far end was predetermined. This is a more realistic situation. Naturally one can work backwards from a known situation and construct a surface upon which the equipotentials conform to a known pattern. One is then left with the problem of analysing the reasons for the shape of the resultant surface. It may well be that under such circumstances transferring the problem from one medium to another contributes nothing to its solution so one must obviously ask whether an analogue is going to be helpful before using it.

MAGNETIC AND ELECTRO-MAGNETIC ANALOGUES

These analogues are especially suitable for experiments concerned with central places. The idea that central places 'attract' custom from surrounding areas is well established and this makes the magnet analogue fairly easy to accept.

Bunge (1964) decribes a neat experiment establishing a point of some importance. An ordinary round plastic bowl about 18 inches in diameter was half filled with water. Twenty one-inch diameter corks were dumped all together and at random into the water. Each cork had a bar magnet forced through the centre with the same pole towards the top and a small lead weight at the bottom to make sure that the corks all floated the same way up. The corks represented central places and the surface of the water the uniform surface upon which the central places are disposed. The corks rapidly achieved a position of equilibrium in relation to each other based on mutual repulsion and the constraints imposed by the size of the bowl. After one hundred random tosses it was found that the corks in the middle of the bowl arranged themselves 43 times in the form of a nearly perfect hexagon with a central point, 25 times as a nearly perfect hexagon without a central point and

22 times as a nearly perfect pentagon with a central point. The remaining ten throws produced a variety of other patterns.

The simplicity of the analogue should not be allowed to blind one to its significance. While it provides some confirmation of one's intuition that there will be a unique solution (the hexagonal arrangement results in 68 per cent of the throws, cf. Table 17.1) it also demonstrates that alternative solutions to this equilibrium problem are possible. One might therefore use the experiment as an introduction to stochastic models.

Extra point is given to the use of magnetic analogues by the similarity between Reilly's Law of Retail gravitation (1931) and Coulomb's Law of Force, which is valid for practically all fields of force including magnetic and gravitational fields. Reilly's Law states that 'two cities attract retail trade from any intermediate city or town in the vicinity of the breaking point (between their two spheres of dominance) approximately in direct proportion to the populations of the two cities, and in inverse proportion to the square of the distance from these two cities to the intermediate town'. Coulomb showed in 1785 that 'the force between concentrated (point) changes varies directly with the product of the individual changes, and inversely with the square of the distance between them'. Of course Reilly's law lacks the universal validity in economic terms that Coulomb's possesses in the physical sphere, but for many purposes it can be regarded as an acceptable statement of a tendency. The inverse square law, therefore, can be expressed in magnetic terms and so a magnetic analogue is suitable for certain experiments in central place studies. In all cases where magnets are used to represent central places it is important that the polarity of the magnets should be the same at the surface over which they are disposed. Like poles repel and the flux lines between like poles form a saddle – a narrow zone in which the flux lines turn abruptly through a right angle.

Perhaps the most convenient experimental arrangement is for the magnets to be arranged with like poles upwards underneath a sheet of glass upon which a sheet of tracing paper is laid. The patterns of the field can be shown by scattering iron filings over the paper and gently tapping it (Plate 17.8). In practice friction between the iron filings and the paper limits the area over which the patterns can be made to appear, and should a collection of iron filings develop more over one magnet than the others its strength will alter in relation to the rest. A more efficient method involves use of a simple field-plotting compass. Moving from one magnet to another the needle changes direction sharply at the point of balance between them and so the individual fields can be plotted easily. The needle is more sensitive than iron filings.

Figures 17.19A–C show the results from plotting fields of equal strength permanent magnets evenly and randomly spaced. To simulate the effect of centres of different importance it is better to use electro-magnets.

In a simple electro-magnetic simulator solenoids are made with copper wire, the number of turns being made proportional to a particular property of the central places represented. This may be population, number of retail outlets, population employed in service trades or any other attribute

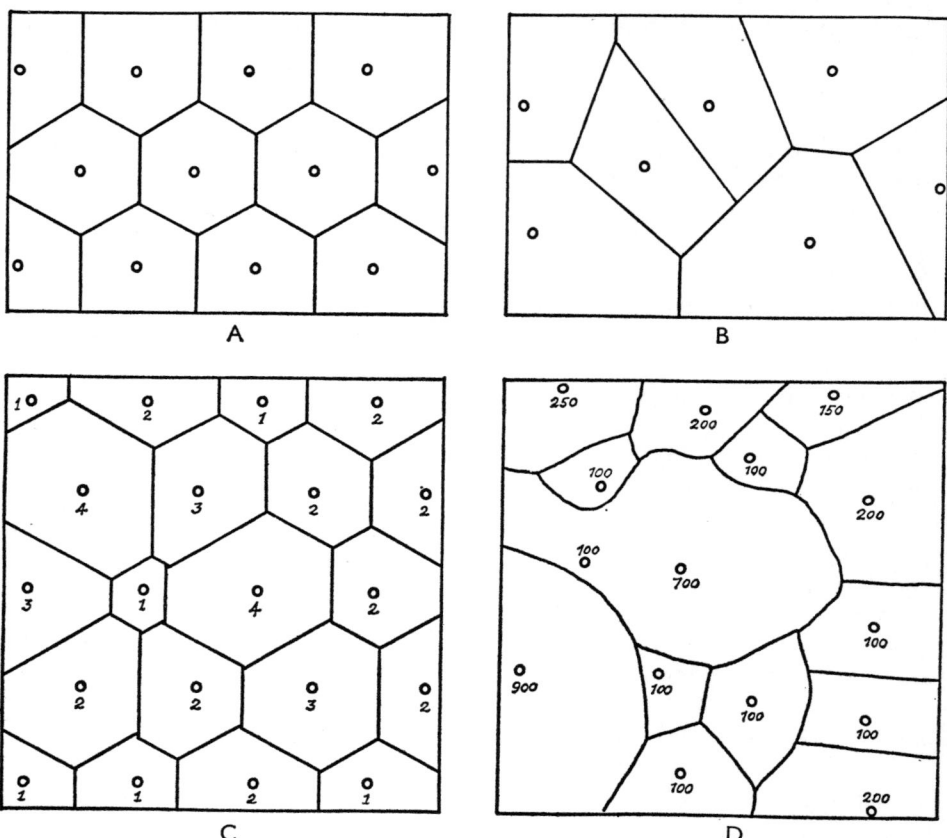

17.19 Plots of magnetic fields.

A Equal strength magnets, regular spacing.
B Equal strength magnets, irregular spacing.
C Varying strength magnets, regular spacing.
D Varying strength magnets, irregular spacing.
The numbers in C and D refer to relative strengths of the magnets used.

thought appropriate. Soft iron cores are fitted into the bobbins and the coils energized by a suitable D.C. source. The electro-magnets are placed under a sheet of glass in their proper positions and the fields outlined with a plotting compass.

Plate 17.9 shows a more sophisticated general purpose model in which the central places are represented by a series of electro-magnets to each of which is linked a rheostat by which the current entering each coil can be controlled. With coils that all have an equal number of turns variations in magnetic strength depend simply on the amount of current passing through. Rheostats can be calibrated in terms of population and the magnets whose strength they control can be disposed at will under a sheet of glass over which field plots can be made. This type of analogue from which spatial patterns can be derived has a number of applications. Many central place studies can perhaps be criticized for placing too much emphasis on the functional attributes of the centres themselves and too little on the actual areas served. Analogues of this sort can be used to simulate actual patterns of spatial organization. The inputs for each centre can be adjusted by degrees until the model situation conforms to the real situation. Then the inputs needed in the model can be related to certain attributes or combination of attributes in the real situation. The problem of creating non-uniform surfaces is always vexing in analogues of this sort and it may be that it is best to confine first experiments to real situations where a degree of uniformity is present.

BLOTTING PAPER ANALOGUES

This analogue is based on a simplified version of the technique of paper chromatography. It is a matter of common observation that a drop of water spilled on a piece of blotting paper will spread out and leave a roughly circular damp patch. If the water is supplied to the surface of a horizontal piece of blotting paper by means of a small wick the damp area will spread out until a

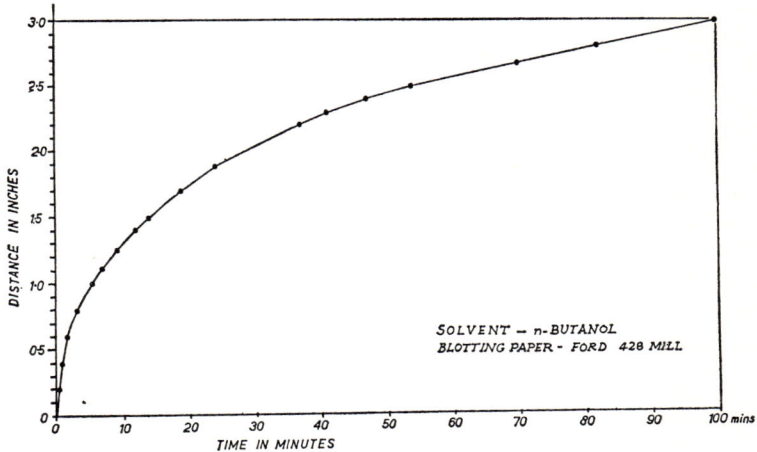

17.20 Rate of diffusion of n-butanol through blotting paper.

balance between supply and evaporation prevents further movement. In paper chromatography a compound substance is placed at the point where the wick touches the surface and as the solvent front advances outwards it carries with it at different speeds the constituents of the original compound depositing each as a ring of distinctive colour. To achieve the best separation of particles and the smoothest progression of the front requires a solvent composed of 45 parts of N-butanol, 25 parts of pyridine and 40 parts of distilled water. For many purposes, however, valuable results can be achieved by using only N-butanol as the solvent and ordinary pink blotting paper. As the solvent front advances it carries with it some of the red pigment in the blotting paper and when the process is finished and the paper dried the limits of the solvent front are marked by a thin red line. The solvent front advances very rapidly at first but soon moves more and more sluggishly. A series of experiments at normal room temperatures of about 68° F and average humidity reveal the pattern illustrated in Figure 17.20. The analogue is particularly helpful in the field of central place and diffusion studies. The circular area encompassed by the solvent front from any one point may be likened to the area of influence of a central place or market, and the analogue can easily be made to demonstrate very elegantly the reason for the hexagonal market area of a Christaller central place system (Plate 17.10). The pink blotting paper is pierced by a network of small holes, arranged at the apexes of equilateral triangles so that each hole is about two inches from its neighbour. Small wicks made of tightly rolled blotting paper are inserted tightly in each hole and rested in a dish filled with butanol. It is important that all wicks should be the same length, be made of the same amount of paper and rolled with equal tightness. It is best to fit the wicks first and then present all of them simultaneously to the butanol. Within a minute each wick will be surrounded by a fast growing circular damp patch. A little later adjacent circles will be touching and soon the whole surface will be wet as the interstitial areas fill up. As soon as the whole surface is wet the experiment should be stopped by removing the wicks. When the paper dries it will be found to be covered with a perfect network of hexagons except on the margins where the solvent has not been restrained.

The structure of market areas around central places which are not uniformly disposed over the surface can easily be resolved by using this method. If the spread of distances to nearest neighbours is very wide, however, certain difficulties can arise because the radial distance travelled by the solvent is of the order of 8–9 cms. and it cannot easily be induced to move further. On the other hand, if two wicks are placed too close together the opposing solvent fronts appear to cross each other and the pigments they carry are dispersed so that on drying out the distinctive red line is no longer present. To simulate a situation where there is great disparity in the distances between centres therefore, it is best to break the area up into smaller sections and simulate each on an appropriate scale. The result can then be compounded on one diagram. In certain cases the problem can be solved by introducing a dye into the solvent. The dye is moved more sluggishly across the paper and where the red material is absent the boundary between different centres can be determined from the disposition of the added dye.

Plate 17.11 shows a simulation of the territories of the centres of S.W. England. The model was run originally on a rectangular sheet of paper and the coastline was cut out when it had dried out. It will be apparent that since all wicks deliver the same amount of solvent the assumption is that all centres have potentially the same influence. Moreover no constraints are applied on the uniformly absorptive surface of the paper. So the illustration shows the pattern of the territories under unrealistic conditions, i.e. as it would be if all centres wielded the same influence. However, it is interesting that the irregular size and disposition of the territories arises from very simple premises. While uniform surfaces are conceptually very valuable geographers have also to be concerned with non-uniform conditions. It is not easy to introduce constraints into this particular model. However, negative areas can be created by cutting out pieces of blotting paper. A particular problem is impeding the passage of the solvent: were it not so powerful it might be possible to spray or paint on lacquers or paste on extra pieces of paper to impede the passage of the front. A promising technique of introducing variable resistance into the surface, borrowed from bio-chemistry, involves replacing blotting paper by a plastic mixture of cellulose powder and water which can be sculpted at will, and through which butanol will diffuse at different rates. Yet such merit as this analogue possesses depends on its simplicity and it is probably not worth forcing sophistication upon it.

A further series of experiments could be done to illustrate the hierarchical concepts of Christaller. By starting off with a uniform distribution of settlements as in Plate 17.10 one derives the first set of hexagons. These can then be drawn in on a fresh piece of paper at a smaller scale and wicks placed in the next lowest grade of centres (the number and arrangment depending on the factor used) from these another set of hexagons will arise defining the market areas of these more important centres. The process can be repeated several times to give the required ranks in the hierarchy, the final product being a whole landscape with all elements represented.

Another method involves using a dye composed of several different colours. Green ink, for example, will, under the action of the solvent, split into blue yellow, one colour always moving further from the wick than the other. By starting the model with centres as widely spaced as possible one will get, for example, a set of blue hexagons each containing a yellow circle inside of considerably smaller radius. One can then add new sources of supply (smaller centres) between the yellow rings eventually creating a network of yellow hexagons bearing a measurable relationship with the larger blue ones.

Experiment will produce compounds composed of a suitable proportion of mobile and reluctant colours to meet most requirements.

The analogue can also be used to represent certain diffusion features. For example Figure 17.21 represents an urban area growing near a river which can be regarded as a semi-permeable barrier. The permeability is effected by means of a blotting paper 'bridge'. If the centre of the city is represented by a wick near the bridge the bridge will permit a limited extension of the growing city on the opposite bank. There is a close parallel between the degree of movement to the paper bridge and a real bridge. The opposite bank seems to be controlled both by the width of the 'river' and the breadth of the bridge but no quantitative measurements have been done to establish relative importance. Stages through which the pattern passes in acquiring its final shape can be recorded in two ways. One can use time-lapse photography to record the various stages (bearing in mind the relation between time and speed of movement of the solvent front noted in Fig. 17.20); or one can run the model through to its final stage, dry it, replace the old wick with a new one (to provide extra dye) and run it again for a shorter time. If this process is repeated carefully a series of parallel curves is produced (Fig. 17.21F). Their distance apart is a function of relative ease or difficulty of movement. In fact, as Yuill (1964) has pointed out, it is a characteristic of river towns that they are usually better developed on one bank than another and this 'affords an excellent starting place for a solid study of city shapes and the response of the growth mechanisms to barriers'. Yuill's paper is a computer simulation study of the role of barriers in structuring human activity patterns in space, but at a very crude level the blotting paper analogue is valuable in that the method of operation is easily grasped by the mathematically innocent.

LOCATION MODELS

It must be stated at the outset that contemporary location theory has become far too complex in terms of the problems it sets itself and the techniques it uses for physical models to make any worthwhile contribution to advanced studies. But in teaching elementary location theory models have their uses. In the simplified form in which it is usually presented Weber's theory seems to offer few problems and yet the average student if asked to work out a simple example very often finds himself unable to structure the problem correctly. The two models presented here are designed to enhance understanding of the principles involved by permitting simple mechanical or graphical solutions of elementary location problems.

The simplest analogue is based on systems of pulleys and weights, an early example being quoted by Friedrich (1929). It is cheap to make and easy to operate, giving an immediate solution. Assume the problem is to find the

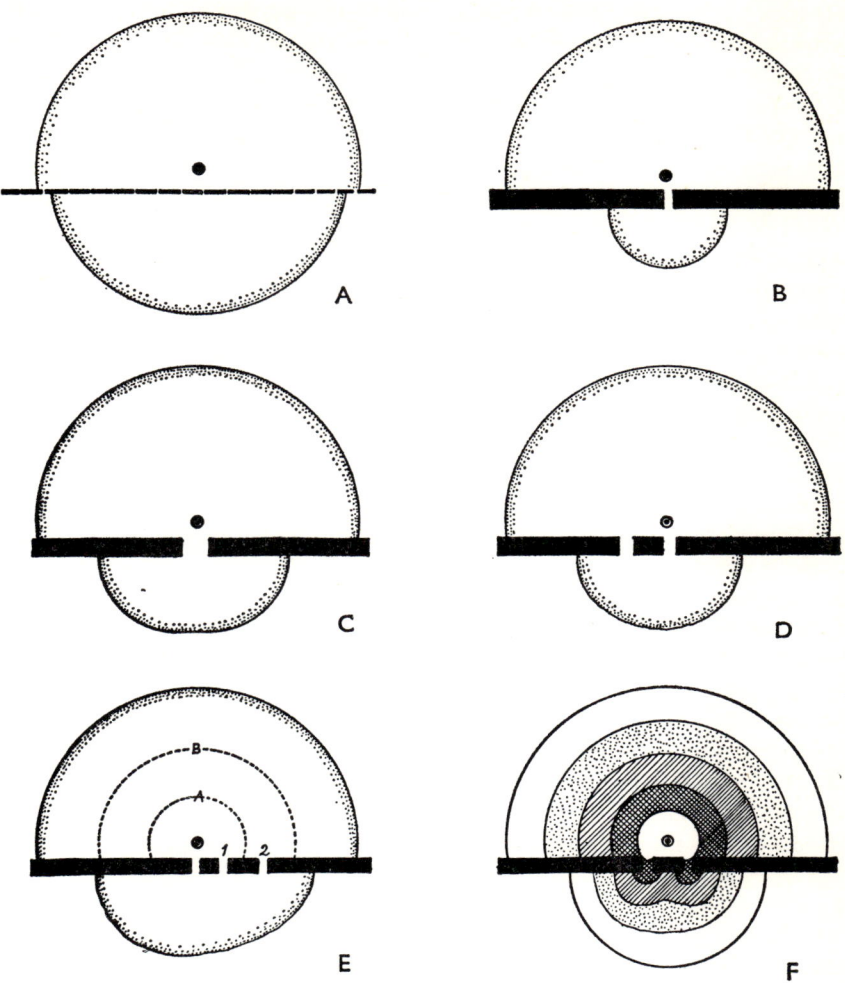

17.21 Blotting paper analogues simulating gross morphological features of river towns. A suggests that where a river can be crossed easily at many points the town will develop almost equally on both banks. B, C and D are examples of the shapes resulting from various restrictions to easy access to the opposite bank. E is a rudimentary growth model, in that when the solvent front reached the line A bridge 1 was added. Bridge 2 was added when the solvent front reached the line B. This process leads to a marked asymmetry in the shape of the part of the town on the far side of the river. F is a plot of the various stages of growth.

best location for a factory using three raw materials, A, B and C in the proportions $3:2:1$ respectively. Assume further than transport costs per unit distance are the same anywhere. We first locate on a piece of board the three raw material sources, marking their positions by free-running pulleys secured by nails. The board with pulleys attached is then placed vertically. Three pieces of thread are tied together; each end is looped over a pulley; a weight of three units is secured to the end of the thread that runs over the pulley at A, and weights of two and one unit respectively to the threads over the pulleys at B and C. When the weights are allowed to reach a point of balance the knot linking the three threads will come to rest over the best site for the factory. It is advisable to agitate the strings slightly to overcome any frictional effects and to find an average solution based on perhaps ten such agitations. A common mistake is to try to make the weights proportional to the ton-mileage component contributed by each source but these of course cannot be calculated without first knowing what the experiment is designed to prove, namely the location of the factory. If desired the market can be represented by a fourth pulley and thread to which a suitable weight is attached. The limitations of this model are firstly that one cannot allow for many sources because of friction and secondly that it is difficult to allow for handling costs. In addition the cost penalties involved in other locations cannot readily be seen.

The second model is perhaps more properly regarded as a simple aid to computation. For each raw material needed to make the final product one makes a circular disc of tracing paper or clear plastic. Around the central point are inscribed a series of concentric circles numbered sequentially from the centre. The distance between successive circles is based on how far one can transport the required weight of raw material for a given unit cost. Suppose for example that three tons of raw material A are needed to make one ton of the finished good, and assume that for one unit of expenditure these three tons can be moved one mile. Choose a suitable scale for the model, say one tenth of an inch equals one mile. Then the circles on the disc relating to A will be spaced at one tenth of an inch intervals. If to every three tons of material from A one needs only one ton from B then the circles on the disc relating to B will be three times as far apart as those around A, since one can assume in a simple case that it costs the same to send three tons one mile as to send one ton three miles. By the same argument one and a half tons of material C can be moved two tenths of an inch on the model for one unit cost.

If we now mark on a piece of paper the three raw material sources, A, B and C in their correct scale relationships we can pin to each the appropriate transparent disc. Covering them all with another piece of tracing paper we choose one of the many points where three circles of different origins intersect and add up the values ascribed to them. The total can be recorded as a 'spot height'. When enough of these points have been plotted lines of equal transport costs (isodapanes) can be drawn. The property of the isodapanes is such that the cost of assembling the raw materials in the proportions selected is the same at all points along the line of any one isodapane. The lowest value isodapane, often represented by a point, is the least cost location for the constraints imposed by the model design.

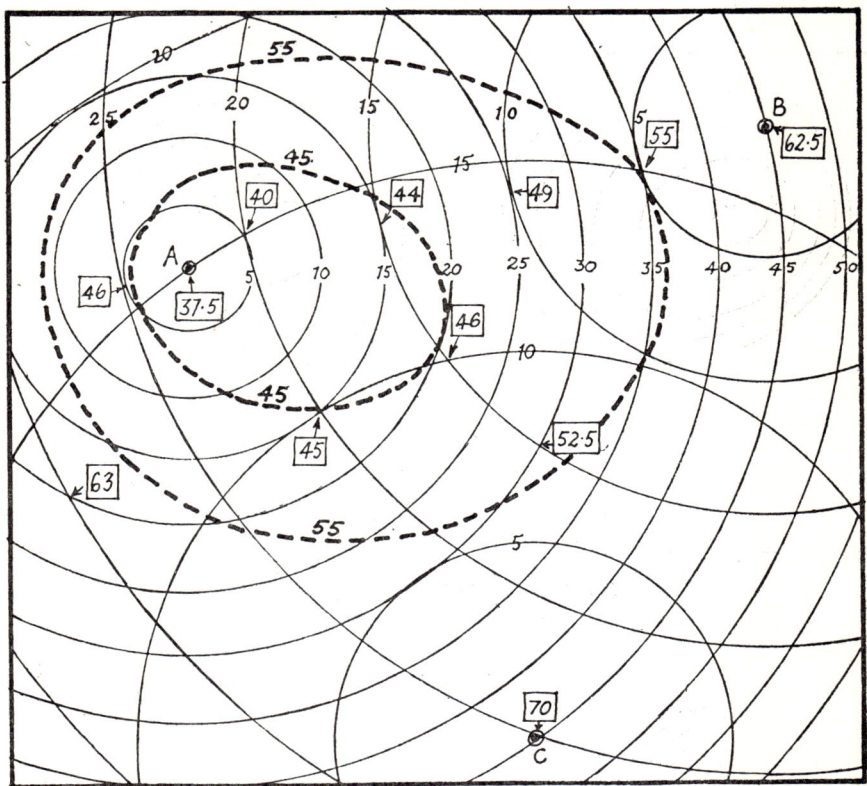

17.22 Graphical method for deriving isodapanes in simple problems. Isolines at appropriate intervals surround the three raw material sources, A, B and C. The boxed figures record the sum of the values of three sets of isolines at particular points. Given sufficient of these points the isodapanes can then be drawn. They are shown here by the thick dashed lines.

If more than three material sources are involved they need not be solved simultaneously: it is often much less confusing to deal with them in small groups and to combine the partial solutions in a final plot. Figure 17.22 should make the general procedure clear. Handling costs can be incorporated readily by converting them into distances the material could be moved and weighting the values of the relevant circles by an appropriate amount. The

[766] MODELS IN GEOGRAPHY

market can be allowed for (provided that it is regarded as existing at a single point). Figures 17.23 (II–IV) show the effect of loss and gain of weight during production both on the optimum location of the plant and on the pattern of the isodapanes. It is very easy to structure the models so that they demonstrate the basic principles of location.

Figure 17.24 shows how it is possible to build in the effect of differential freight rates. The assumption made here is that it is ten times cheaper to send

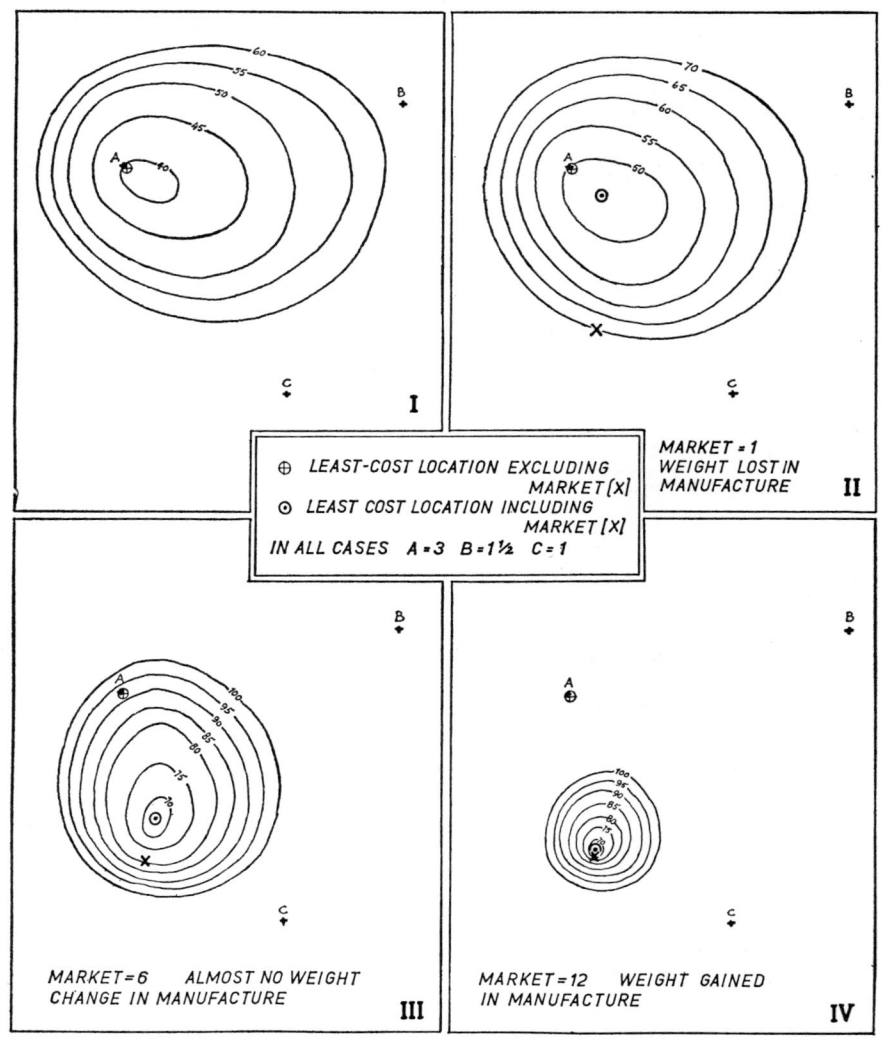

17.23 The effect of loss and gain of weight during production on the optimum location of the plant and the pattern of the isodapanes.

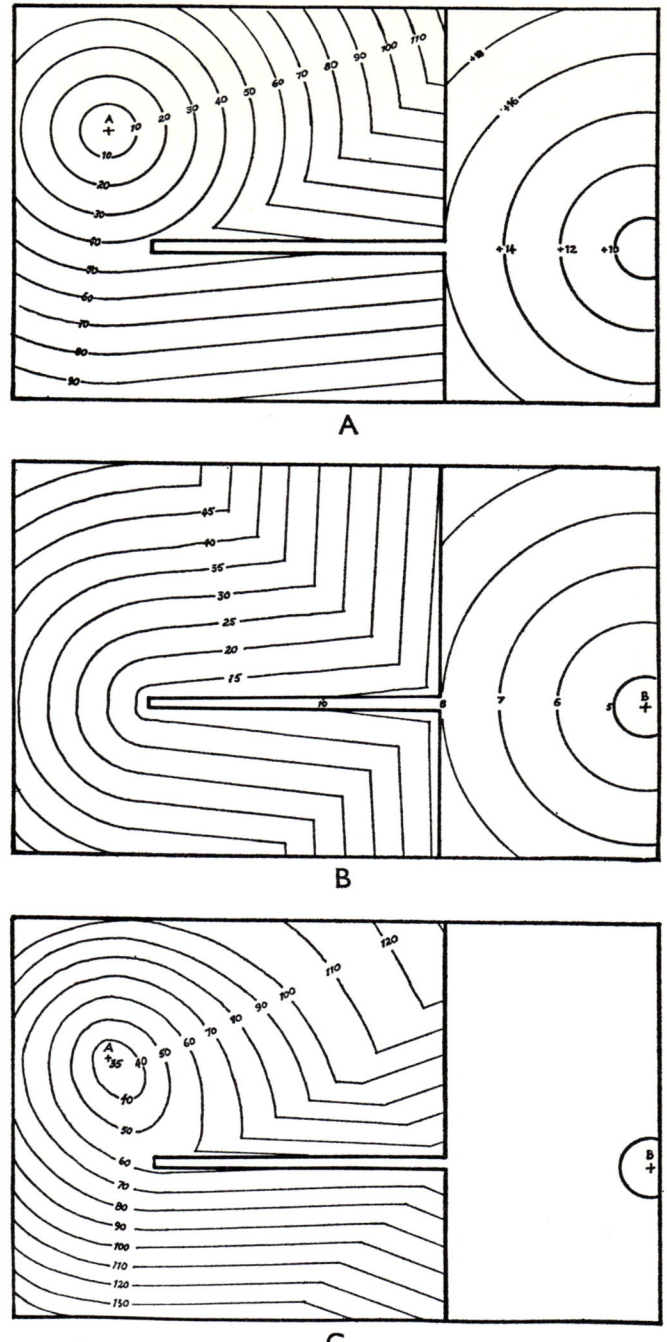

17.24 Model involving differential freight rates.

goods by sea than by land. In the model we have created an island containing raw material B, and a mainland area, penetrated by a deep estuary, and within which raw material A is found. The first two figures show the isodapanes for each material separately and the lower one the isodapanes based on the combination of the two upper patterns. While it is unlikely that this type of model could solve even a moderately sophisticated real life problem of location it should none the less prove helpful in giving the student some insights into the frictional effects of distance and in developing an intuitive feeling for space relationships. The isodapanes can be regarded as contour lines; if they are cut out of some solid material such as expanded polystyrene a three-dimensional relief model can be made to illustrate the effect of the spatial component in the cost curves. Different solutions can be compared and by cutting down through the models ready made graphs are obtainable.

NETWORKS

The study of networks, interconnections between points along more or less restricted channels, is beginning to be seen as a potentially very productive field of geographical interest (Haggett, 1965; Haggett and Chorley, In preparation), and indeed is of as much interest in physical as in human geography.

One set of problems in network analysis relates to the establishment of interconnections between points in such a way that the total length of the constituent links is minimized. If for example we wish to interconnect three villages by the shortest distance of road we find that the solution is to create a central road junction so placed that the three roads meet at an angle of 120 degrees. If any angle in the triangle formed by the villages is greater than 120 degrees the village at that vertex is connected directly to the other two. Where more than three points have to be connected in this fashion the solution can be found graphically, using movable transparent overlays on which 120 degree junctions have been drawn. Where less than about fifteen points are involved the solution is also given by a soap film. This is a particularly fascinating experiment. The points to be inter-connected are arranged in the form of pegs sandwiched between two plane surfaces. Perspex is excellent but glass or wood and glass will do as well. The parallel plates should be not more than three-quarters of an inch apart. The model is then carefully and slowly submerged in a soap solution. A 50:50 mixture of distilled water and liquid detergent behaves very well and produces a film that will with luck last up to twenty-four hours. When the model is slowly and carefully withdrawn soap films are left perpendicular to the parallel plates and they gradually contract until the minimum length links are established. All the junctions of the film

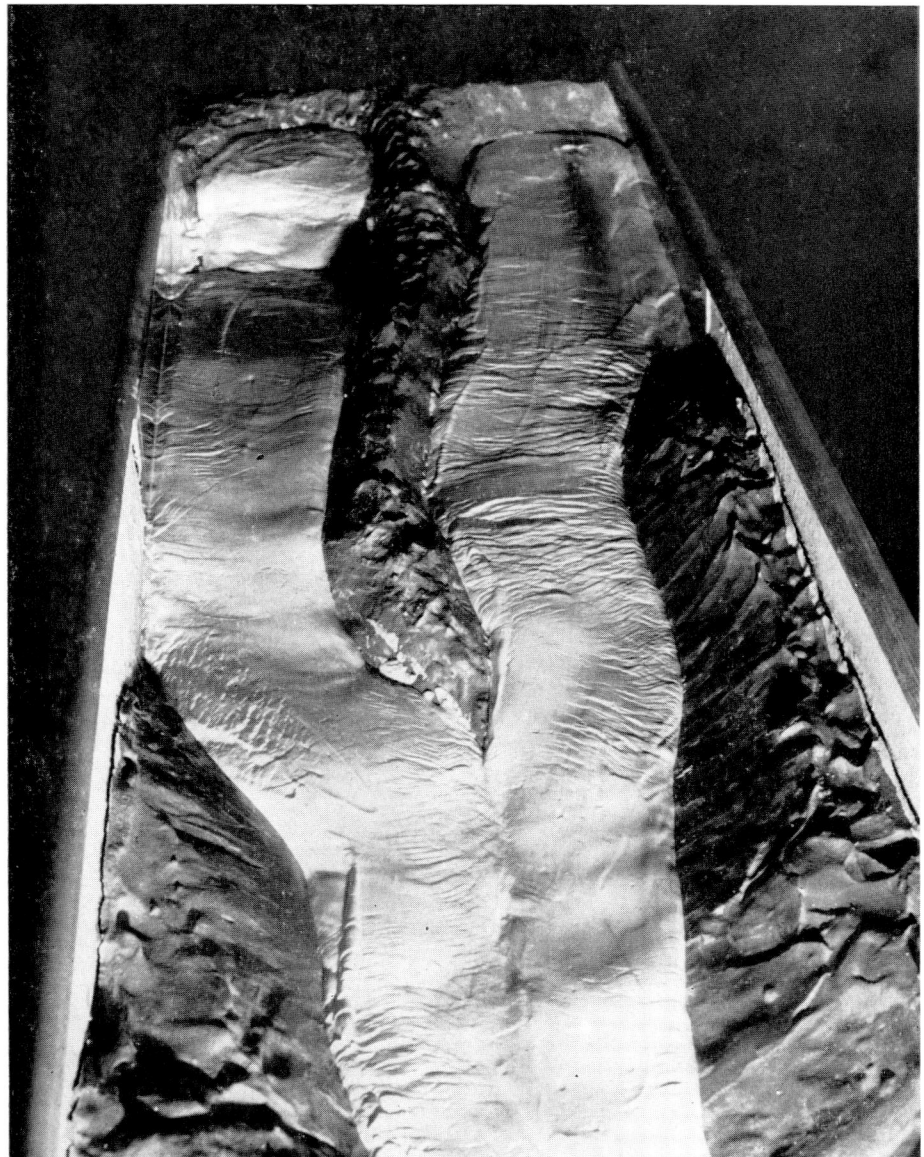

Plate 17.1 Kaolin model glacier. The left-hand valley was given a well-developed corrie that was replenished with kaolin shortly before the photograph was taken, hence the bergschrund has not yet had time to reconstitute itself. The valley on the right had no corrie and the rate of flow was significantly less.

Plate 17.2 Kaolin model glacier. Rate of flow indicated by matchsticks.

Plate 17.3 The upper reaches of the model glacier showing crevassing and step-faulting.

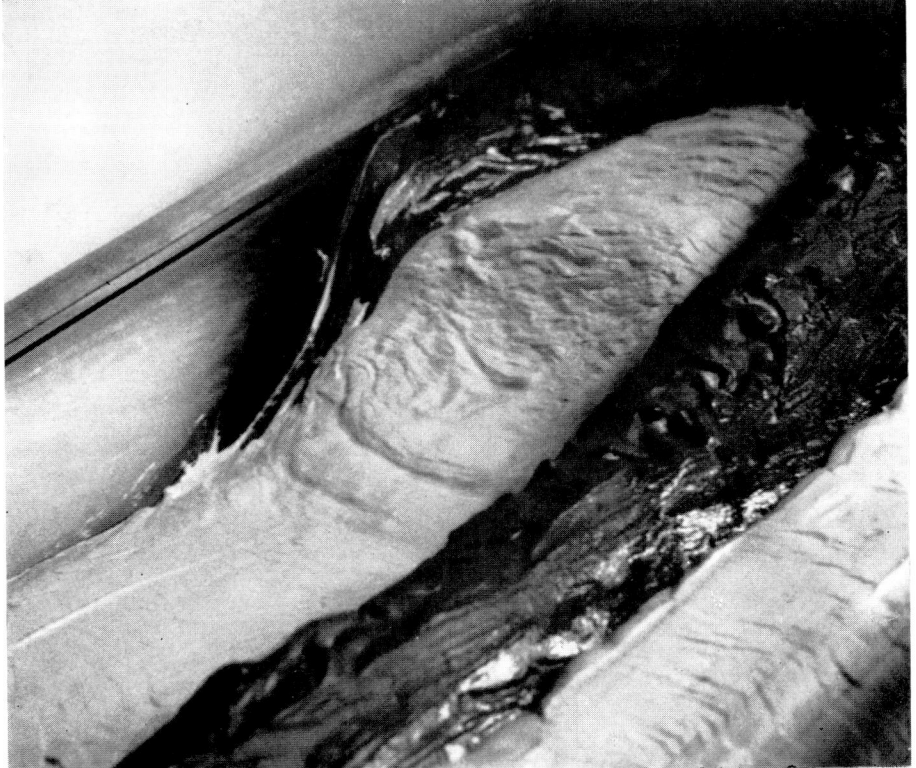

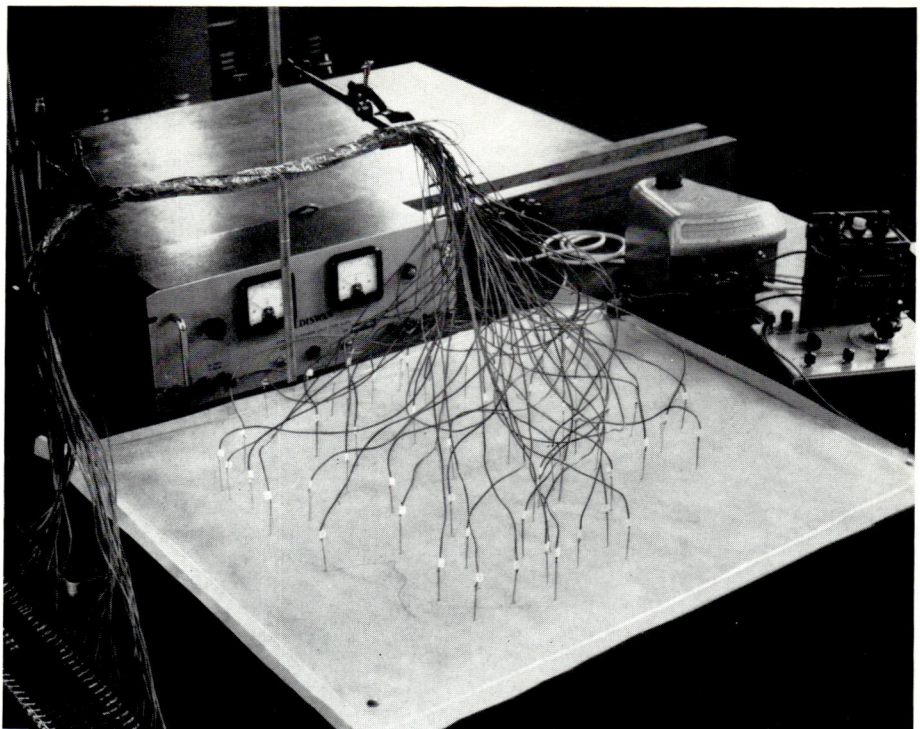

Plate 17.4 Arrangement of apparatus used in plotting population potential. The d.c. power source is in the background and feeds current to a bank of resistors (left) selected to give currents of very small magnitudes proportional to the population of the major cities. These currents are fed to the Teledeltos paper by means of needles. To the right is a potentiometer and sensitive spot galvonometer which enable the equipotential lines to be plotted. Note however that the apparatus shown here was at the time being used for another experiment. In plotting population potentials the inputs have to be placed much closer together in the centre of the paper in order to minimize boundary effects.

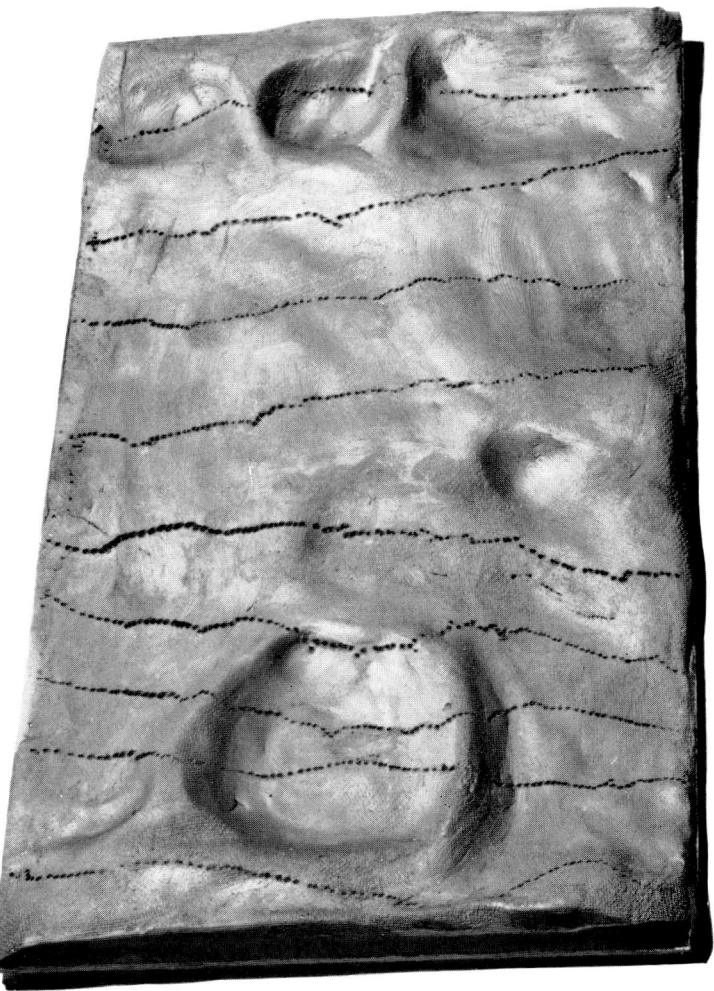

Plate 17.5 Equipotential lines directly plotted on the surface of a model moulded from pottery clay. The electrodes were fixed to the two shortest sides of the model.

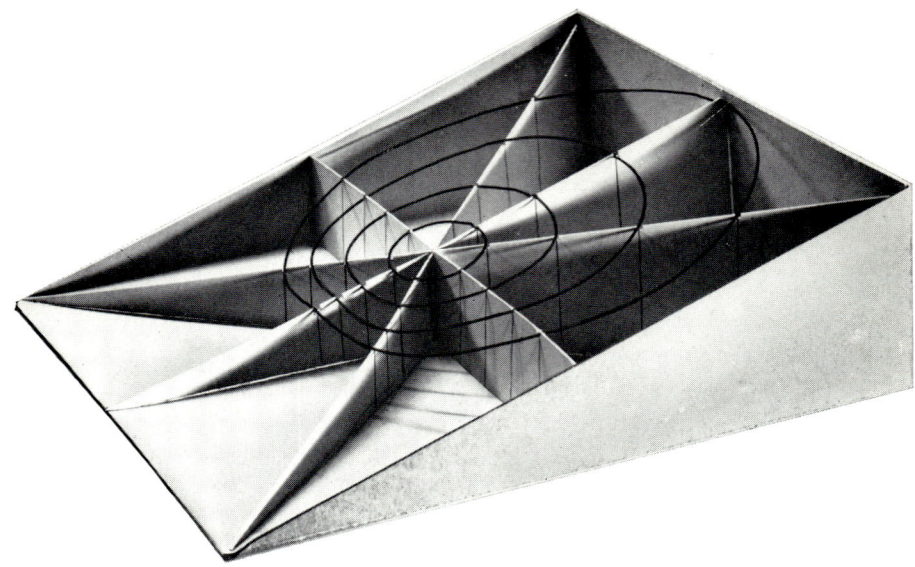

Plate 17.6 Model demonstrating the effect of a gradient on a phenomenon which varies symmetrically on a horizontal surface.

Plate 17.7 Serial profile solutions for certain three-dimensional problems. (*opposite page.*)
A – the 'topographic' profiles.
B – the resistance paper equivalents of the topographic profiles. The threads on both models represent the equipotential lines but they were of course derived only from the profiles in B.
C – shows the pattern of equipotentials resulting from varying the total resistance of each profile by the addition of extra pieces of resistance paper at the far end. These additions are not shown in the photograph.

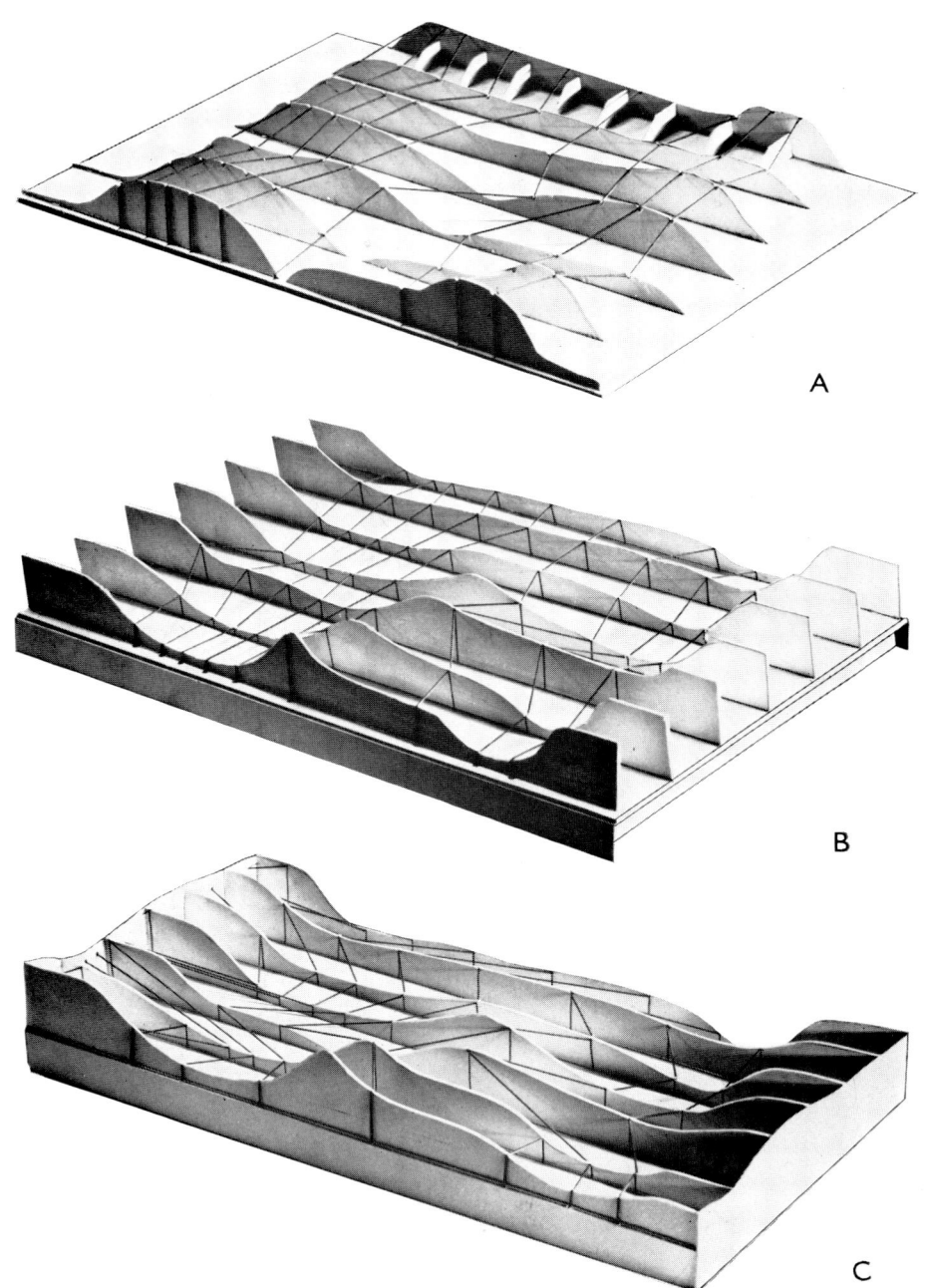

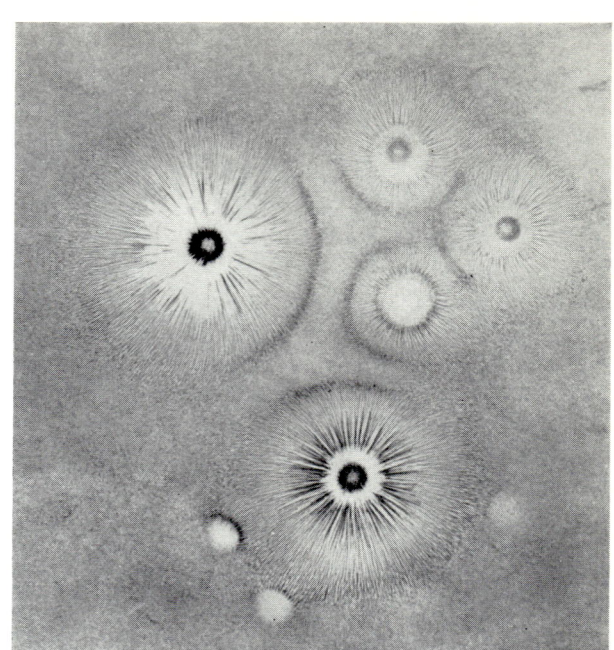

Plate 17.8 Iron filings define fields of electromagnets.

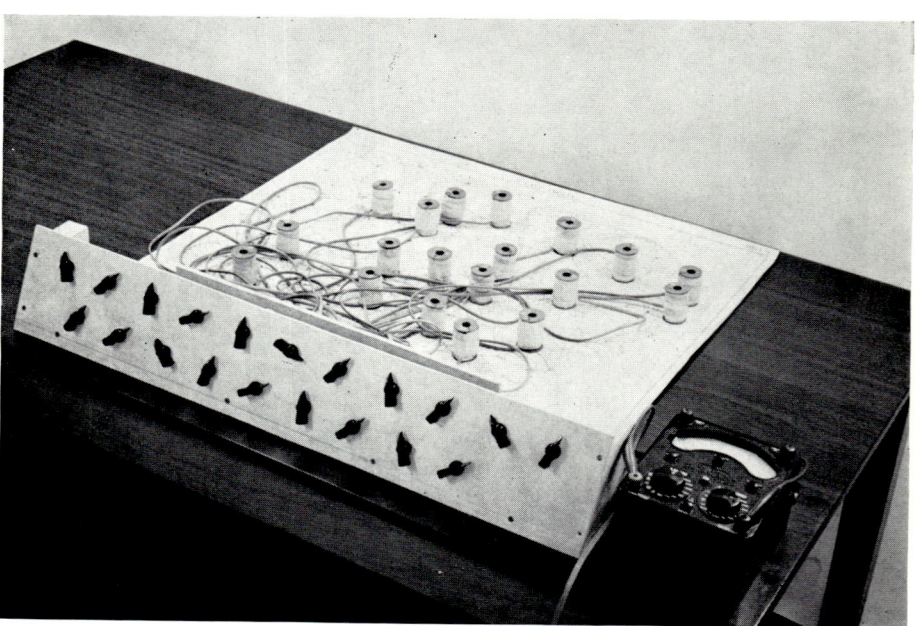

Plate 17.9 Electromagnetic analogue model for central place simulation.

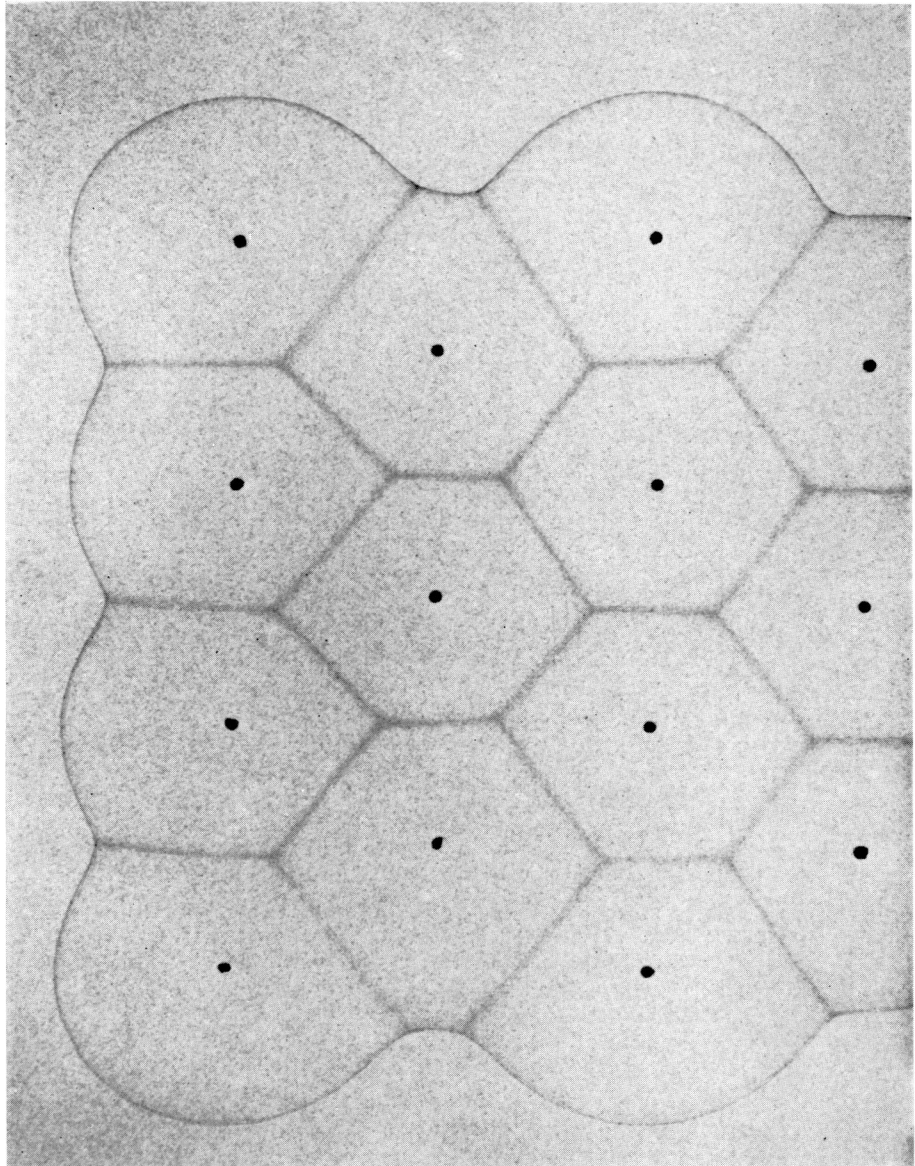

Plate 17.10 Hexagonal mesh resulting from supplying regularly-disposed wicks with equal quantities of solvent.

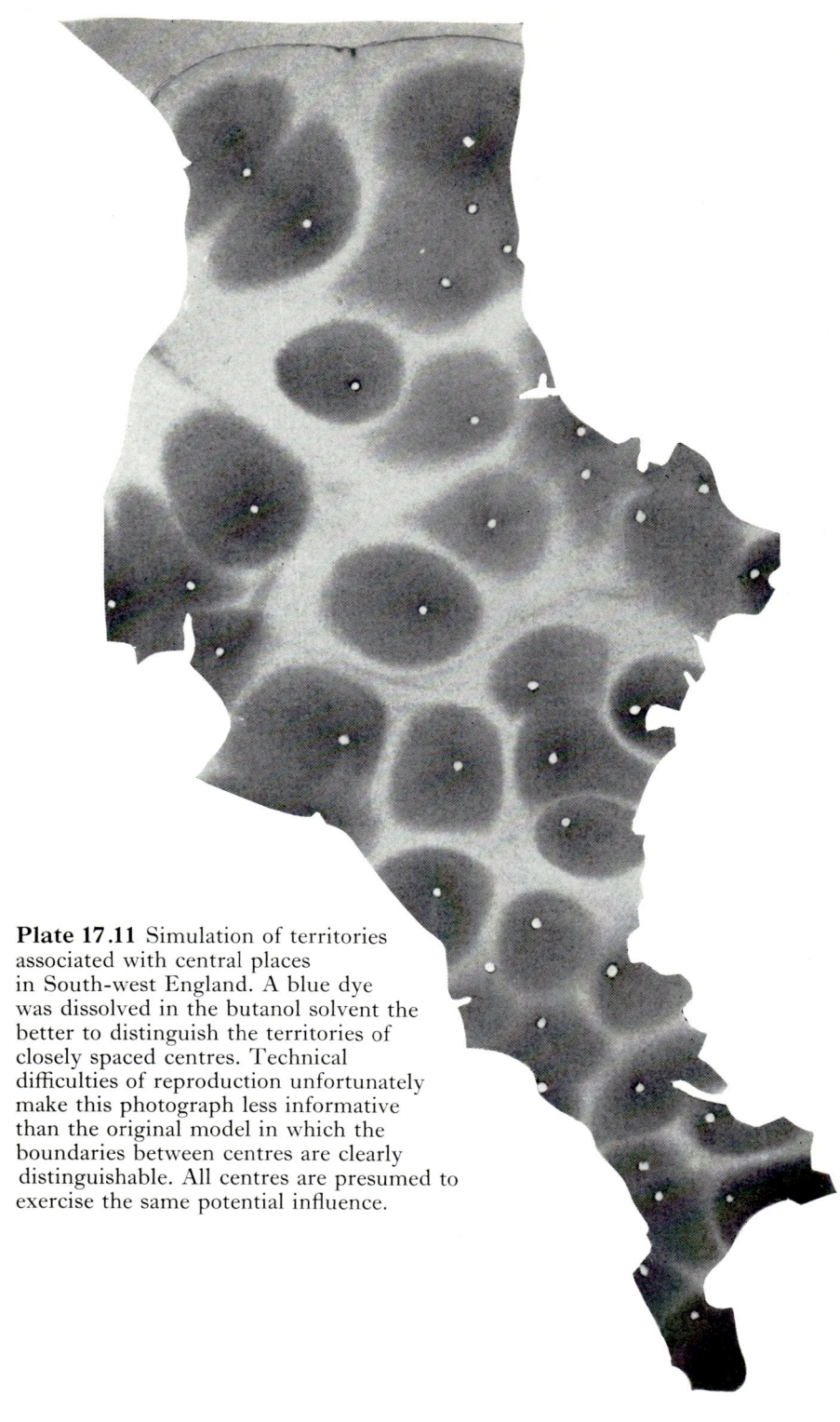

Plate 17.11 Simulation of territories associated with central places in South-west England. A blue dye was dissolved in the butanol solvent the better to distinguish the territories of closely spaced centres. Technical difficulties of reproduction unfortunately make this photograph less informative than the original model in which the boundaries between centres are clearly distinguishable. All centres are presumed to exercise the same potential influence.

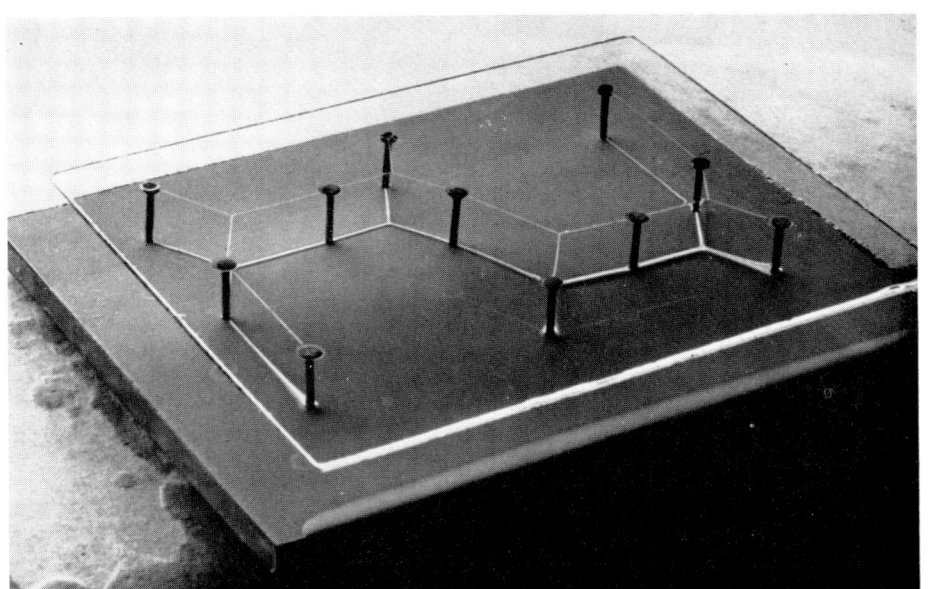
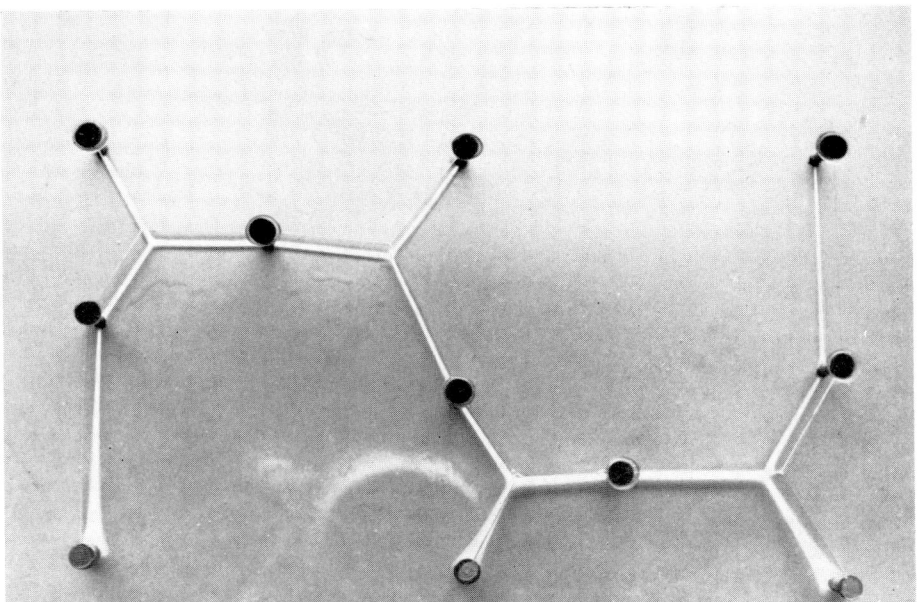

Plate 17.12 Soap film solution of link-length minimization problem involving eleven points (*Based on Miehle*, 1958).

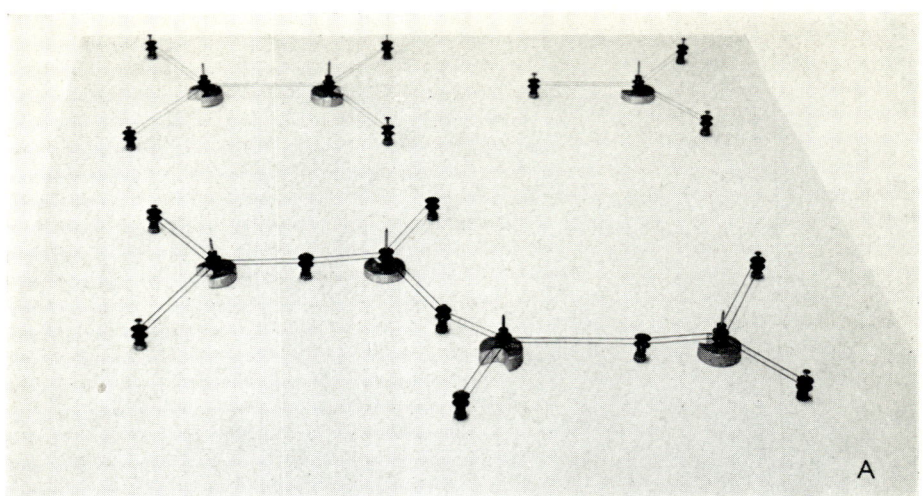

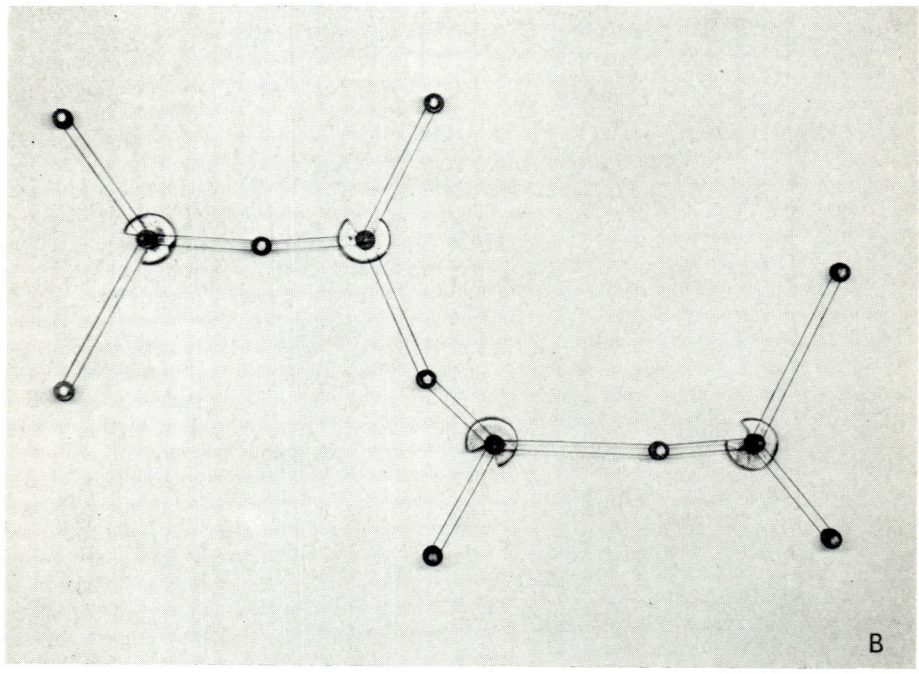

Plate 17.13 The pulley-and-thread method of link-length minimization. The pulleys on the large notched bases represent movable junctions whose positions change as the thread tightened. The notches allow junctions to be placed where necessary very close to a fixed point. The solution to the problem posed by the arrangement of points shown in the lower photograph (B) should be compared with that given by a soap film to a very similar problem (see Plate 12) (*Based on Miehle*, 1958).

are at 120 degrees. Plate 17.12 shows clearly the sort of configuration of the soap film in a typical situation. The film can be touched with a wet finger without danger of breaking. The model works because a soap film always adopts a shape which minimizes its area for the constraints imposed upon it. (Miehle, 1958; Courant and Robbins, 1941).

If too large a number of points are included a unique solution is not given by the soap film. However this disadvantage can be turned to good account. If we make a model with say about twenty-five or thirty randomly disposed pegs in it, we can dip it a number of times in the soap solution and record the pattern of interconnections established on each occasion. The minimum solution can be found by inspection, and its length compared with other solutions. If a compound drawing is made showing the frequency with which individual pairs of points are linked limited experience suggests that certain general regularities will be found, and such a procedure could form an introduction to stochastic models.

If a unique solution to a large number of points is needed it can be obtained by discretizing the network into a number of smaller parts and solving each of these in turn. The process of breaking up a large network for analysis may involve simply taking various sections in turn, but in the case of a road network for example it may be worth while first to minimize the major road net for the whole area, then the minor roads in the interstices or sectors of the major net and so on down to the local roads. These progressive changes of scale can be reconciled in a final composite map of the whole system.

Where many points are involved possibly a better solution is that proposed by Miehle (1958). Fixed points are represented to scale by pegs fixed in a board, each peg carrying a small pulley. Movable pegs, also with pulleys represent mobile junctions. A thread is looped about each peg in turn and around an appropriate number of movable pegs, the thread being finally brought back to its starting position (Plate 17.13). The loose end is then tightened. If there were no friction, and if the interconnections had been wisely chosen the first time, the system would move into equilibrium and the total length of thread used would be the least possible. In practice friction has to be overcome by easing and teasing the thread so that it is equally taut throughout the model. Faced with a large number of points and junctions one is at first bewildered by the complexity of the situation but experience helps one to decide on appropriate interconnections. Local and difficult minimizations can be solved using soap film models at first. J. A. Silk (1965) constructed a mechanical link-length minimization of the road system of Monmouthshire. The model was constructed on a one-inch O.S. map and was found to accord well with the actual road pattern.

The advantage of the mechanical minimizer is that integral weights can be assigned to individual links (by multiplying the number of linking loops of thread), constraints can be applied to the movable points and arbitrary selections of interconnections is possible, advantages not available in soap film models. Miehle in fact developed the mechanical method in order to find the answer to a problem involving the best location for a given number of control centres serving a fixed disposition of missile sites. It is not difficult to see that the method could as well be applied in central place studies, enabling one for example to judge the efficiency of the location of existing central places under certain predetermined constraining conditions. Brink and Cani (1957) in fact used an electrical analogue to determine the location of service points to serve a number of customers with a known demand pattern so as to minimize transport costs, but a link-length minimizer should be capable of giving approximately the same answer.

Networks suggest the use of electrical analogues since after all an electrical circuit is a form of a network in itself. Movement of groundwater in a three-dimensional aquifer has been simulated by means of an electrical resistance analogue (Robinove, 1962; Skibitzke, 1960), described in Chapter 5.

The application of electrical analogues to problems in spatial economics is relatively new and was elegantly illustrated by Enke (1951). He dealt with the difficult problem of determining the equilibrium prices and commodity

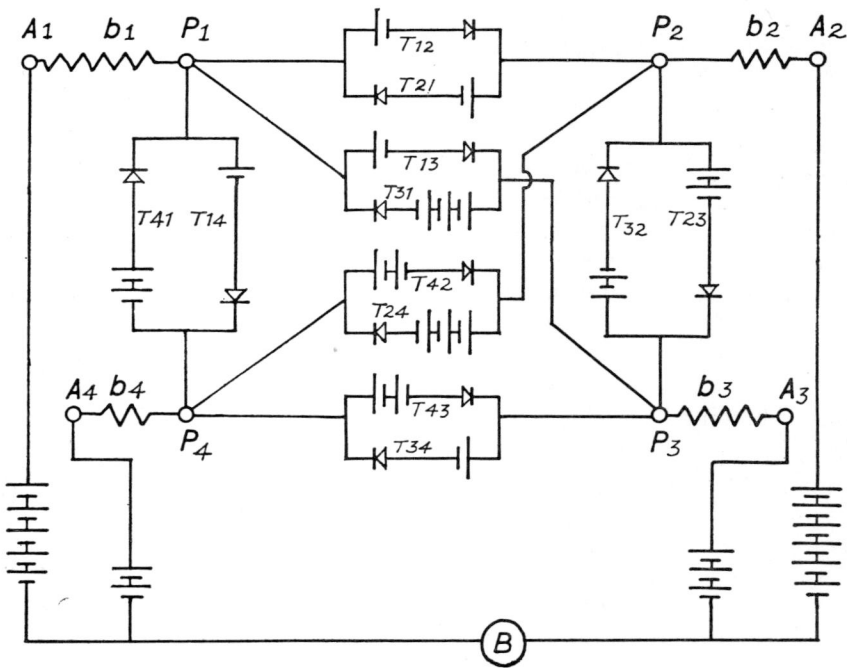

17.25 Electric circuit for determining prices and exports of a homogeneous good in spatially distinct markets (*Source: Enke, 1951*).

flows that will result when a number of interdependent trading units (four in this case) are in a position to buy or sell a particular good and where freight rates between potential trading units are significant. Each trading unit may be a seller at a relatively high price and a buyer at a low price. Transport costs are never so high as to prevent one trading unit from participating in the process of exchange. In the electrical analogue commodity flows are represented by current flow (measured in amperes) and prices by voltages. The circuit used by Enke is shown in Figure 17.25. For a detailed explanation of its design reference should be made to the original article. The analogue had particular point in 1951 because the only other way at that time of solving a problem of such complexity was through laborious iterative methods. Samuelson (1952) subsequently showed how equilibrium problems can be solved using computers, a technique made possible by the development of linear programming.

REFERENCES

ALLEN, J., [1947], *Scale models in hydraulic engineering*, (London).
APPLEBY, F. V., [1956], Run-off dynamics – a heat conduction analogue of storage flow in channel networks; *International Association of Hydrology*, Publication No. 38, (Rome, 1956), 3, 338–348.
VON ARX, W. S., [1957], *Physics and Chemistry of the Earth*; Vol. 2 (ed. by Ahrens et al.), (London), Ch. 1, 1–29.
VON, ARX, W. S., [1962], *Introduction to physical oceanography*, (Cambridge, Mass.).
BALCHIN, W. G. V. and RICHARDS, A. W., [1952], *Practical and experimental geography*, (London).
BILLINGS, M. P., [1946], *Structural geology*, (New York).
BOOTHROYD, A. R., CHERRY, E. C. and MAKAR, R., [1949], An electrolytic tank for the measurement of steady state response, transient response, and allied properties of networks; *Proceedings of the Institute of Electrical Engineers*, 96 (Pt. 1), 163–177.
BOTSET, H. G., [1946], The electrolytic model and study of recovery problems; *Transactions of the American Institute of Mechanical Engineers*, 165, 15–25.
BRINK, E. J. and DE CANI, J. S., [1957], An analogue solution of the generalised transportation problem with specific application to marketing location; *Proceedings of the First International Conference on Operational Research*, (Oxford), (English Universities Press).
BRUSH, L. M. and WOLMAN, M. G., [1960], Knickpoint behaviour in a non-cohesive material – a laboratory study; *Bulletin of the Geological Society of America*, 71, 59–74.
BUNGE, W., [1962], Theoretical geography; *Lund studies in geography, Series C, General and mathematical geography*.
BUNGE, W., [1964], Patterns of Location; *Michigan Inter-University Community of Mathematical Geographers, Discussion Paper No. 3*.

BURKALOW, A. VAN, [1945], Angle of repose and angle of sliding friction – an experimental study; *Bulletin of the Geological Society of America*, 56, 669–708. (Contains a large bibliography.)

CHORLEY, R. J. and MORGAN, M. A., [1962], A comparison of morphometric features, Unaka mountains, Tennessee and North Carolina and Dartmoor, England; *Bulletin of the Geological Society of America*, 73, 17–34.

CONRAD, V. and POLLACK, L. W., [1950], *Methods in climatology*, (Harvard), Ch. 3.

COURANT, R. and ROBBINS, H., [1941], *What is mathematics?*; (New York), Ch. 7.

DAUBREE, A., [1879], *Études synthetiques de géologie experimentale*; 2 Vols., (Paris).

DUNCAN, W. J., [1953], *Physical similarity and dimensional analysis*, (London).

EINSTEIN, H. A. and HARDER, J. A., [1961], Electric analog model of a tidal estuary; *Transactions of the American Society of Civil Engineers*, 126 (4), 855.

ENKE, S., [1951], Equilibrium among spatially separated markets – solution by electric analogue; *Econometrica*, 19, 40–47.

FOLK, R. L. and WARD, W. C., [1957], Brazos River bar; a study in the significance of grain size parameters; *Journal of Sedimentary Petrology*, 27 (1), 3–26.

FRIEDKIN, J. F., [1945], A laboratory study of the meandering of alluvial rivers; *U.S. Waterways Experimental station*, (Vicksburg, Miss.).

FRIEDRICH, C. J., [1929], *Alfred Weber's theory of the location of industries*, (Chicago).

FULTZ, D., [1956], Fluid models in geophysics; *Proceedings of the first symposium on geophysical models, Baltimore 1953*, (Washington).

GLOVER, R. E., HERBERT, D. J. and DAUM, C. R., [1953], Application to a hydraulic problem; *Transactions of the American Society of Civil Engineers*, 118, 1010.

GRIGGS, D. T., [1939], A theory of mountain building; *American Journal of Science*, 237, 611–650.

HAGGETT, P., [1965], *Locational analysis in human geography*, (London), Ch. 3.

HAGGETT, P. and CHORLEY, R. J., [in preparation], *Network models in geography, an integrated approach*, (London).

HARDER, J. A., [1963], Analog models for flood control systems; *Transactions of the American Society of Civil Engineers*, 128 (1), 993.

HOTELLING, H., [1921], *A mathematical theory of migration*; Unpublished master's thesis, University of Washington.

HUBBERT, M. K., [1951], Mechanical basis for certain geologic structures; *Bulletin of the Geological Society of America*, 62, 355–372.

IMHOF, E., [1965], *Kartographische Gelandedarstellung*, (Berlin).

ISARD, W., [1956], *Location and the space economy*, (New York).

JOHNSON, J. W. and RICE, E. K., [1952], A laboratory investigation of wind generated waves; *Transactions of the American Geophysical Union*, 33 (6), 845–854.

KARPLUS, W. J., [1958], *Analog simulation*, (New York).

KENN, M. J., [1965], Experiments simulating the effects of mountain building initiated by the late Professor S. W. Wooldridge; *Proceedings of the Geologists' Association*, 76 (1), 21–27.

KUENAN, P. H., [1956], Experimental abrasion of pebbles; 2, Rolling by current; *Journal of Geology*, 64, 336–368.

KUENAN, P. H., [1960], Experimental abrasion; 4, Eolian action; *Journal of Geology*, 68, 427–449.

LANGHAAR, H. L., [1951], *Dimensional analysis and theory of models*, (New York).

LASAREFF, P., [1929], Sur une méthode permettant de démonstrer la dépendance des courants océaniques des vents alizés et sur le rôle des courants océaniques dans le changement du climat aux époques géologiques; *Gerlands Beitrage zur geophysik*, (Leipzig), Bd. 21, 215–233.

LEWIS, W. V., [1944], Stream trough experiments and terrace formation; *Geological Magazine*, 81 (6), 240–252.

LEWIS, W. V. and MILLER, M. M., [1955], Kaolin model glaciers; *Journal of Glaciology*, 2, 533–538.

MACKAY, J. R., [1965], Glacier flow and analogue simulation; *Geographical Bulletin*, 7 (1), 1–6.

MCKEE, E. D., [1957], Experiments on the production of stratification and cross-stratification; *Journal of Sedimentary Petrology*, 27 (2), 129–134.

MCKEE, E. D., [1960], Laboratory experiments on form and structure of off-shore bars and beaches; *Bulletin of the American Association of Petroleum Geologists*, 44, 1253.

MIEHLE, W., [1958], Link-length minimisation in networks; *Operations Research*, 6, 232–243.

MUSKAT, M., [1949], *Physical principles of oil production*, (New York).

NEVIN, C. M. and TRAINER, D. W., [1927], Laboratory study in delta building; *Bulletin of the Geological Society of America*, 38, 451–458.

NOMA, A. A. and MISULIA, M. G., [1959], Programming topographic maps for automatic terrain model construction; *Surveying and Mapping*, 19, 355–366.

OLSEN, G. H., [1962], Field Plotting; *Wireless World*, 68 (2), 58. (The author has drawn attention to errors in figures 8 and 11.)

OLSEN, G. H., [1963], *Servomex field plotter manual*, (Crowborough, England).

PANKHURST, R. C., [1964], *Dimensional analysis and scale factors*, (London).

REILLY, W. J., [1931], *The law of retail gravitation*, (New York).

REINER, M., [1959], The flow of matter; *Scientific American*, 201 (6), 122–138.

RIEHL, H. and FULTZ, D., [1957], Jet streams and long waves in a steadily rotating-dishpan experiment: structure and circulation; *Quarterly Journal of the Royal Meteorological Society*, 83, 215–231.

ROBINOVE, C. J., [1962], Ground water studies and analog models; *U.S. Geological Survey, Circular* 468, (Washington), (Reprinted 1965).

SAMUELSON, P. A., [1952], Spatial price equilibrium and linear programming; *American Economic Review*, 42, 283–303.

SAVILLE, T., [1950], Model study of sand transport along an infinitely long straight beach; *Transactions of the American Geophysical Union*, 31 (4), 555–565.

SHAW, N., [1934], *Manual of Meteorology*, (Cambridge), 1, 100 and 2, 396.

SILK, J. A., [1965], *Road network of Monmouthsire*; University of Cambridge, Department of Geography, B.A. dissertation.

SKIBITZKE, H. E., [1960], International association of scientific hydrology, Publication No. 52.

SPOONER, C. S., [1953], Modernisation of terrain model production; *Geographical Review*, 43, 60–68.

STANTON, T., MARSHALL, D. and HOUGHTON, R., [1932], The growth of waves on water due to the action of the wind; *Proceedings of the Royal Society, Series A*, 137, 283–293.

STARR, V. P., [1956], The general circulation of the atmosphere; *Scientific American*, 195 (6), 40–45.
STEWART, J. Q. and WARNTZ, W., [1958], Macrogeography and social science; *Geographical Review*, 48, 167–184.
TOBLER, W. R., [1959], Automation and cartography; *Geographical Review*, 44, 536–544.
TURNER, J. S., [1963], Model experiments relating to thermals with increasing buoyancy; *Quarterly Journal of the Royal Meteorological Society*, 89, 62–74.
TURNER, J. S., [1965], Laboratory models of evaporation and condensation; *Weather*, 20, 124–128.
TURNER, J. S. and LILLY, D. K., [1963], The carbonated water tornado vortex; *Journal of Atmospheric Science*, 20, 468–471.
TYLDESLEY, B., [1965], The solution of atmospheric diffusion equations by electrical analogue methods; *Meteorological Office, Scientific Paper* No. 22, (London, HMSO).
U.S. ARMY MAP SERVICE, [1950], Refinements on production of molded relief maps and aerial photographs; *Army Map Service Bulletin* No. 29, (Washington).
WALLINGTON, C. E., [1961], An electrical analogue method of studying the behaviour of lee waves; *Weather*, 16, 222–228.
WARNTZ, W., [1959], *Towards a geography of price*, (Philadelphia).
WARNTZ, W., [1965], A note on surfaces and paths and applications to geographical problems; *Michigan Inter-University Community of Mathematical Geographers, Discussion Paper* No. 6.
WILLOUGHBY, E. O., [1946], Some applications of field plotting; *Journal of the Institute of Electrical Engineers*, 93 (3), (Communications) 275–293.
YUILL, R. S., [1965], A simulation study of barrier effects in spatial diffusion problems; *Technical Report* No. 1, *Spatial diffusion Study, Department of Geography, Northwestern University*, (Evanston, Ill.).

CHAPTER EIGHTEEN

Models of Geographical Teaching

S. G. HARRIES

INTRODUCTION

In a sense models have always been used in teaching for they are inseparable from communication. Only recently, however, has much systematic attention been given to the use and limits of the various models used in teaching geography. A consideration of this change and its implications is best viewed against the background of developments in science as a whole. Ackerman observes that the progress of science as a whole, has at any time, depended largely on the growing points in a few fertile subjects, which provided salients for the advance of other disciplines. The strategic potential of the subject-salients varied from time to time. From 1910 to the mid-1940's physics and mathematics were the most active. Chemistry, biology, geology and the social sciences were less favourably placed for advance. What of geography, which has some affinity with both the physical and social sciences? Ackerman's answer is clear. 'We have not been on the forward salients in science, nor, until recently, have we been associated closely with those who have. The reasons are not difficult to find. During the early part of this 50-year period . . . our closest associations were with history and geology. Geological study of that period, and of the thirties, was not among the inspiring growing points of science. The history and geology connections did not correct the predisposition of our scholars . . . to the deceptive simplicity of geographic determinism' (Ackerman, 1963, p. 430). As determinism became a less acceptable model of the relations of human societies to their environments, geographers turned with more interest to the social sciences and to the alternative model in possibilism. Only later, in the fifties, was this concern with the social sciences to exert its very considerable influence on geography, but not before geography had undergone an anguished search for a unique identity. Meanwhile in biology, anthropology, genetics, psychology and economics, the applications of mathematical statistics and systems analysis were actively pursued even before the advent of the electronic computer. Only in the last decade have some geographers availed themselves of this extended range of model theory, with its productive concepts and attendant techniques. It is thus not an occasion for surprise that in geography teaching old models still persist, while newer and more sophisticated models and techniques remain untried. Not that all older models are ready for the scrap heap, or that heavy teaching loads and unimaginative examinations do anything but make change difficult and slow.

In the remainder of this essay we shall attempt an appraisal of older models, before going on to examine some of the models that take their rise in modern science. Finally we shall consider some of the models of the learning process as psychologists describe them and make some tentative indications for their application.

MODELS IN GEOGRAPHY TEACHING

The use of the word 'model' in the sense that concerns us in this paper is relatively recent. Niels Bohr was among the first to use it in his investigations of the structure of the atom in 1913. Now its current use is widespread in physical, biological and social science, in history and in art. But although this sense of the word 'model' is new, its implied usage is old. Teaching is very much a conservative activity, and continues to use pre-scientific models.

An example may best serve us by way of introduction to the model most widely adopted by western medieval man. In the fourteenth century Guillaume Deguileville writes that the division between the realm of Nature and that of the Heavens was the orbit of the Moon. Something of this notion derives from Aristotle. He was interested in both biology and astronomy. He saw the characteristic of the world we inhabit was ceaseless change. Things did not happen uniformly or invariably, but 'on the whole' or 'for the most part'. The world of astronomy was different. So far as could be seen the heavenly bodies, permanently in their order, followed regular paths. The lower region of the Universe, that of change and irregularity, was Nature. The upper region of this two-part universe was the Sky. That variable element the weather was naturally included with inconstant Nature. Sky must therefore begin higher up at the orbit of the Moon. As C. S. Lewis (1964, p. 4) writes: 'It seemed reasonable to suppose that regions which differed in every observable respect were also made of different stuff'. Nature was made of the four elements, earth, water, fire, air. Air, nature and inconstancy then must end before the Sky began. In the Sky the substance was called aether. Here were the divine bodies. Below was the home of the changing and the perishable.

This example of medieval man's notion of the Universe well illustrates the bookish and scribe characteristics of the Middle Ages. When we speak of medieval man as living in an age of authority, we generally have in mind that of the Church. But in another, although partly overlapping sense, the Middle Ages was a time of many authorities whose thoughts were recorded in manuscripts. Every writer based himself on an authority – an auctour – an earlier writer, preferably in Latin. Medieval man was essentially taken with the practice of sorting out and classification of information, ideas and behaviour. It is said that of modern inventions he would have admired the card index. Supreme examples of his bringing masses of material to an ordered unity are the *Summa* of Aquinas and Dante's *Divine Comedy*. But supreme above all else in the period was the achievement of the medieval synthesis itself – the construction of a complex, harmonious mental model of the Universe from the scholarly materials of theology, science and history. This model resolved the contradictions of divergent writings and traditions, a process begun in the Graeco-Roman world and perfected in the Middle Ages. The medieval church taught on the authority of its accepted books. Although the Reformation divided Christendom, it in no way lessened the emphasis of the several churches on this model of teaching. What was taught was an authorized body of learning or knowledge. This conception of an essential body of knowledge as the stock in trade of the teacher, did not cease to be important as education became more and more a secular matter.

The notion of a body of required knowledge proved resistant to change. The scientific ideas of Bacon and Descartes on the study of nature were slow in effecting any significant change in the curriculum of schools and universities. Geography and other subjects outside the range of strictly physical science continued to be regarded as traditional bodies of knowledge, a feature well illustrated by textbooks of the eighteenth century, and re-enforced in several ways by social developments in the nineteenth century, when teachers for the growing elementary school system were recruited as pupil-teachers from the abler pupils in the elementary schools. For them geography provided convenient material to learn, teach and examine – a kind of literacy fodder. Above the elementary level geography gradually ceased to be a gloss on the scriptures and the classics. More attention was paid to the contemporary world largely as an encyclopaedic commentary on social, political and economic matters. In the new grammar schools, founded after the Education Act of 1902, geography became a more important element in the curriculum, a position it owed partly to a semi-scientific content with notions of causality on the nineteenth century determinist model, and partly to its association with history.

Only in the post-war expansion of university geography has there been evident an increasing concern for a more rigorous statistical and theoretical approach to teaching and investigation. Much of the stimulus for this has come from the social sciences. It has, however, been in some ways counterbalanced by contemporary stress on subject matter in examinations. Thus the conception of geography as a body of subject matter is still implicit in much teaching today and it is not easy to see what can effectively be done to lessen the addiction to the use of this model. Nevertheless there is available an alternative approach that may do something to reduce the over-dependence on the feeding-in-of-information model. An extract from the autobiography of R. G. Collingwood provides an introduction to its use.

Collingwood tells how his work in archaeology led him to revolt against the current logical theories of the time, rather like the revolt against scholastic logic which was effected through Bacon and Descartes, by their experience of scientific research. He writes: 'I began by observing that you cannot find out what a man means by simply studying his spoken or written statements, even though he has spoken or written with perfect command of language and perfectly truthful intention. In order to find out his meaning you must also know what the question was (a question in his own mind, and presumed by him to be in yours), to which the thing he has said or written was meant as an answer. It must be understood that question and answer, as I conceived them, were strictly correlative. A proposition was not an answer, or at any rate could not be the right answer, to any question which might have been answered otherwise. A highly detailed and particularized proposition must be the answer, not to a vague and generalized question, but to a question as detailed and particularized as itself. . . . People will speak of a savage as "confronted by the eternal problem of obtaining food". But what really confronts him is the problem, quite transitory like all things human, of spearing this fish, or digging up this root, or finding blackberries in this wood' (Collingwood, 1939, p. 31–32).

We may agree that information in books, in geography texts no less than others, is over-generalized, condensed and inert – inert in the sense of being divorced from the specific questions to which it was originally a response. Such knowledge, however, is successfully taught, learned, examined and evaluated. Some see this as a self-justifying activity; others as a mere assimilation model lacking the dimension of imagination and creation. To follow Collingwood's way is difficult. But a beginning can be made if the pupil is brought to see that the content of the average textbook, divorced from the original probing questions needed to enliven it, is, in effect, only part of the creative model, quite apart from any questions that have since been raised.

The practice of attacking problems by means of questions is very old and has been 'rediscovered' in various contexts. Inquiry by a continuous dialogue of question and response has a very long history. Aristotle tells us that it was Zeno of Elea who 'invented' the way of disputation by question and answer, or dialectic as it came to be called. Socrates used the method to expose some of the errors of contemporary belief. His pupil Plato developed it metaphysically in his theory of 'ideas' and in his search for the ultimate Idea of the Good. The christian church has long used the catechetical approach to the teaching of dogma. A similar method found favour with many eighteenth-century writers of geography textbooks.

Herbert Spencer thought that in the development of the child the same questions would emerge, as marked the progress of the race. He argued that, 'If there be an order in which the human race has mastered its various kinds of knowledge, there will arise in every child an aptitude to acquire these kinds of knowledge in the same order' (Spencer, 1861, p. 76).

Spencer, in favour of science in education, seemed to think it could be communicated in a ready-made form. He was persuaded of the need for a reformed subject-matter but not of the need of a new method. Dewey's theory of knowledge was developed in the light of recent advances in the biological sciences. Moreover, he accepted the evolutionary view of the continuity of organic life up to and including man. He realized that the child's environment could be directed and arranged so as to raise appropriate questions at the level of the child's experience. He therefore discarded the antithesis between empirical knowledge and the higher rational knowledge. For Dewey, knowledge was a mode of participation furthered and refined by the experimental method of science. He wrote of the experimental method, that, 'its significance is regarded as belonging to certain technical and merely physical matters', but he added with insight, 'it will doubtless take a long time to secure the perception that it holds equally as to the forming and testing of ideas in social and moral matters' (Dewey, 1916, p. 394).

As to geography, Dewey was alive to its possibilities and weaknesses. He saw its field as the 'connection of natural facts with social events and their consequences'. He was fully aware that to 'present specific geographical matter in its vital human bearings' required 'an informed and cultivated imagination'. Without that geography became 'a veritable rag-bag of intellectual odds and ends' (Dewey, 1916, p. 248).

A more recent attempt to use the question and answer model was the 'heuristic' method of teaching, which places the students as far as possible in the attitude of the discoverer – 'methods which involve the finding out instead of being merely told about things'. It was proposed mainly in connection with the physical sciences, but it had and still has obvious applications to geography as one of the earth sciences (Armstrong, 1903).

The question-and-answer model has an immense range of application and sophistication from the frontiers of science to detailed imprinting at the lowest levels of stimulus and response. In geography as in other fields the fruits of the method are proportional to the skill and subtlety of the user.

SCIENCE AND TEACHING MODELS

Science is a concept with much relevance to the use of models in geography.

One need not dwell for long on the history of the emergence of science as part of culture. A brief look at the achievements of science in the seventeenth century will suffice. But in passing it may be remarked that, in one sense, 'science has no beginning in human history but is as old as perception, that it begins in the evolutionary scale with the capacity to generalize in perceiving an object' (Boring, 1957, p. 5). The human organism is so constructed that its perception, like that of a number of organisms, sees the same object as the same size regardless of the distance between it and the subject. Thus, in observation both science and perception look to underlying generalities, seeing in the observed object the uniformities of nature. But whatever we may wish to make of this interesting point, modern science in the sense of a social institution did not make its appearance until the Renaissance and its accompanying revival of learning. The first century of the new scientific age was rich in achievements that are fundamental to geography as it is taught today. A brief record of items is impressive: (a) In 1593, on the threshold of the startling century, Galileo had invented the thermometer. (b) In 1600 Gilbert published his work on magnetism in Latin; the first great scientific work to be published in England. (c) Kepler laid down the three laws of planetary motion in 1609–1619. (d) Galileo used his new telescope to discover Jupiter's moons. (e) In 1628 Harvey discovered the circulation of the blood – the first great biological discovery of the new age. (f) Many of Galileo's discoveries were finally published in 1638, e.g. the law of the pendulum, the law of falling bodies, acceleration, inertia and the components of motion of a projectile etc. (g) Galileo's pupil Torricelli invented the barometer in 1643. (h) Next was found 'the spring of the air' – that gases expand to fill all free space. From this came the invention of the air pump. (i) Robert Boyle's laws of gases followed in 1660; also an important date for it marked the founding of the Royal Society for the exchange and publication of scientific communications. Newton's first communication to the Royal Society was in 1672. (j) Leeuwenhoek in 1674 used the microscope to discover micro-organisms. (k) Finally, after long delays, Newton published *Principia* in 1687. He, in his work and attitudes, reveals the essence of physical science – an insight of genius, followed by mathematical deduction and experimental verification and prediction. The model of science, at least of physical science had arrived.

The eighteenth century was less productive in scientific discovery than the seventeenth. The full implications of the revolutionary work of Kepler, Galileo and Newton needed a period of assimilation and reflection. This was as true for philosophers as for scientists; although at this time the distinction between the two was not as sharp as it is today. But it was already clear that a new method of inquiry into the processes of nature had been established. In this procedure, now known as the hypothetico-deductive method, the scientist starts with an hypothesis, he deduces therefrom a consequence that can be observed, either directly or by experiment especially arranged, and is thus enabled to verify or not the predicted observation. In this particular respect then, if he is successful, the hypothesis is confirmed. This mathematical deductive method found more acceptance in England and France; in Germany biological science made more headway. But biology at this time did not lend itself to the formulation of large generalizations like the law of gravitation, from which facts could be deducted for empirical verification. The approach of biology was phenomenological, morphological and descriptive. This form of study was equally applicable to man and his activities. Immanuel Kant was able to produce, without the experimental method, but by the power and logic of his penetrating intellect, an unrivalled account of man's cognitive and moral powers. The philosophical problems that confronted him were largely set by the advance of the physical sciences we have just been speaking about. If one may put it shortly and not too erroneously, Kant thought that the freedom of the will in man was a necessity of nature. The very notion of morality was unthinkable without free will, just as much as the conception of natural science demanded the notion of necessity as a basic postulate. There arose from this the problem of the two kinds of knowledge. It has recently appeared as the problem of the two kinds of culture. On the one side we have the realm of knowledge of nature where law and necessity prevail; on the other the knowledge of man, his activities, societies and beliefs, where we act as if man had some freedom of choice. This is the dilemma that geographers find themselves in today; they have to operate on both sides of this dichotomy. This in fact is the model of the subject over which geographers presently contend. Perhaps at this stage one should point out that for Kant himself, although he was well acquainted with this division of knowledge, the subject of geography was regarded as overwhelmingly descriptive, with the exception of a small amount of mathematical geography. Indeed, as a very cursory examination of nineteenth-century geography texts will make evident, geography remained a largely descriptive study until very recently. It is therefore unhistorical to project this dichotomy back into the nineteenth century as if it were a living issue for geographers of that time. Nevertheless, we can gain a useful insight into the nature of the subject as seen by modern geographers if we take a look at the nineteenth century attempts to bridge the gap between the study of nature and that of man and his works. This will entail a brief excursion into the background development of the social sciences in the nineteenth century.

The degree to which the methods of physical science have seemed acceptable to students of society has varied from time to time and from country to country. Hegel, when he contemplated all that lies outside the realm of nature, outside those areas where scientific law prevails, asked what logic or order of growth prevailed there. His answer was that the study of history and philosophy showed that the moving spirit was *geist* or mind. Hegel, like Kant, thought there must be a deep unity between the knower and what is known, and that knowledge is impossible without that unity. He regarded this as a unity of opposites. Hegel saw all progress, whether in the history of man or in organic life, as a series of revolutionary steps. At each such step opposites are united. This method is the dialectic, a method as old as Socrates, in which we begin with a thesis – say with man as a person, who seeks to know. To this thesis the impersonal world offers an antithesis. A conflict ensues which the synthesis resolves. This method earlier writers had applied to the work of the mind; Hegel applied it to the realities of ordinary life – to the history of institutions and to the state, and its aspirations. This is an account much simplified of what Hegel wrote but will probably serve in this context. Dilthey was against the use, in the study of the cultural sciences, of methods proper to the physical sciences. He in fact sharpened the difference between the physical sciences and the cultural sciences and history. For him each of these branches of knowledge has a distinctive method of approach, physical science aimed at explanation and cultural sciences and history at interpretation. Marx, however, saw the regulative principle of society in the material interests of classes operating through the dialectic, or conflict of interests. The Utilitarians, and later in the nineteenth century, the Positivists, held that only the natural sciences, on the model of classical mechanics could give valid empirical knowledge. The man who did more than any other to bridge the theoretical gap between the extreme positions noted above, was Max Weber, the German sociologist, much of whose thought is yet to be assimilated. He put the emphasis not on given wants, either economic or psychological, but on the total cultural pattern which made the attachment to certain classes of wants meaningful. Moreover, he assumed that the components of such a gestalt or pattern of culture should be treated as independently variable. He set out to analyse certain of these relations by the comparative method, which in terms of the logic of science is perhaps the nearest empirical equivalent to the experimental method, that is accessible to the subject matter of social science (Parsons, 1965, p. 58). An illuminating comment from an historian of science comes from a recent monograph, (Kuhn, 1962, p. x), in which he writes: 'I was struck by the number and extent of overt disagreements among social scientists about the nature of legitimate scientific problems and methods. Both history and acquaintance made me doubt that practitioners of the natural sciences possess firmer or more permanent answers to such questions than their colleagues in social sciences. Yet somehow, the practice of astronomy, physics, chemistry or biology normally fails to evoke the controversies over fundamentals that today often seem endemic among, say, psychologists or sociologists (and one may add geographers). Attempting to discover the source of that difference led me to recognize the role in scientific research of what I have since called *paradigms*. These I take to be universally recognized scientific achievements that for a time provide model problems and solutions to a community of practitioners'.

These thinkers then, reflect a growing consciousness of society throughout the nineteenth century, and a consequent development in these specialist studies of society, each seeking to establish itself on more secure theoretical bases by more precise scientific methods. It seems inevitable that those aspects of geography that are linked with the cultural sciences of economics, anthropology and sociology, should in various degrees respond to this model. Experimental as well as mathematical and conceptual models have played a part in recent advances on the physical side of the subject.

If the use of models, of whatever kind, has not effectively closed the gap between the different kinds of knowledge, it has lessened geographers' preoccupation with the unique identity of their subject, whether they base it on an integrative purpose or on a prescribed subject matter. This has come about largely by the application of research attention to concepts and techniques that transcend subject boundaries. It is significant that even in the study of history, with its regard for particularity, there is less concern with social facts as a separate type of data and more with 'social analysis as a particular set of questions which can be asked of every type of data that involves human relationships' (Hays, 1965, p. 374).

Some of these concepts have arisen in the development of modern physics and biology. Among the more important intrinsically and in their possible applications to the study of geography are: energy, feedback, information and general systems.

Energy

This is a useful concept because it has many forms that can be transformed into each other. But though it may be transformed its quantity is invariant. The concept of energy is important in biology when studying the transformations of energy in living systems. But an 'energy' concept is used in several fields of knowledge, often metaphorically rather than in its precise physical sense. 'One speaks of psychic energies, historical energies, social energies. In these senses energy is not really measurable, nor is it directly related to physical energy. Nevertheless, like physical energy, it can be released in the form of enormous physical, mental, or social activity; and, when it is, we tend to think of it as somehow "potential" in the pre-existing situation' (Brooks, 1965, p. 76). The language of energy derived from physics has proved a very useful metaphor in the study of social, political, geographical, and psychological phenomena.

Feedback

This is one of the fundamental ideas in modern engineering, particularly in automatic control and automation. 'In recent years the feedback concept has been extended still further to embrace the idea of "information feedback", which is important in biological and social phenomena as well as in the engineering of physical systems. The idea has been stated by Forrester in the following way: "An information feedback system exists whenever the environment leads to a decision that results in action which affects the environment and thereby influences future decisions". At first this may seem unrelated to amplifiers and control systems, but if we identify "environment" with "input" and "decision" with "output" we can readily see how the more general definition includes amplifiers and control systems as a special case. In the case of the amplifier the decision is completely and uniquely determined by the environment, but the concept of information feedback applies equally well when the decision is a discrete rather than a continuous function and when it is related to the environment only in a probabilistic sense' (Brooks, 1965, p. 77). 'Environment' and 'decision' used in this more general sense suggest further uses of the information feedback concept in biology, the social sciences and geography. The natural selection process in evolution and the process of learning may be regarded as information feedback systems. The teaching machine is designed to provide a feedback through the process of 'reinforcement' which helps the student to decide whether he has learned correctly. Information feedback systems and their properties have great possibilities in the analysis and interpretation of biological, cultural, economic and learning systems.

General systems theory

General systems theory is an attempt at theoretical model building somewhere between the generalizations of mathematics and the specific theories of the specialized sciences. One of its advantages is that it affords a pattern of coherence and reference to much current interdisciplinary study and research. Boulding (1965, p. 200) suggests two complementary ways in which general systems theory might be structured. First, to select certain phenomena, encountered in many disciplines, and construct general theoretical models appropriate to them. He suggests the phenomena of 'population'. In various fields 'the interaction of population can be discussed in terms of competitive, complementary, or parasitic relationships among populations of different species, whether the species consist of animals, commodities, social classes or molecules'. Other phenomena of widespread occurrence suggest themselves. A second possible approach to structure in general systems theory is a hierarchy corresponding to the complexity of the 'individuals' of the various empirical fields. This more systematic approach would lead to a system of systems. Boulding identifies a number of levels at which theoretical discourse might be conducted, extending from the level of static framework to that of the human, social and transcendental.

Only recently have social scientists and geographers taken an interest in the application of these concepts to research. The question arises whether these concepts and associated techniques are to figure in undergraduate and sixth-form teaching or are they to be regarded solely as post-graduate accomplishments? It is difficult to see how they can be so restricted. The seminars now a feature of honours geography courses already entail close reading of recent specialist papers from journals. If such papers are to be properly understood and evaluated students need some theoretical and practical acquaintance with mathematical and statistical techniques. As for the schools there is, in view of the adoption of the newer mathematics, a case for the introduction of elementary statistical skills in secondary school geography. This can only come about through a drastic re-shaping of geography teaching with appropriate changes in teacher training. Indications of a new dynamic at work in the curriculum are not lacking; despite the cramping effect of examinations, there are encouraging signs of greater theoretical rigour, higher standards of achievement, and a readier acceptance of new teaching techniques. For geography, in alliance with mathematics, the data and concepts of physical and social science afford a new and productive field of study.

LEARNING MODELS

In this section we will consider models for the learning process, a topic to which successful teaching practitioners are sometimes a little allergic. It is clear that even at the university level many teachers get by without ever raising any question about the nature of the learning process, and on what sort of models it operates. It is equally clear that they nevertheless act unconsciously as if certain models of learning were either true or at least were more acceptable than others. Perhaps the most common method by which we derive a model of learning for ourselves, and in light of which we teach others, is by trial and error. This principle is perfectly acceptable. The difficulty arises that the trial is not always a fair trial for we are seldom aware of a majority of the factors operating; equally, uncertainty prevails about errors. How can one know what is an error without a standard to judge by, and more puzzlingly, how can one get personal standards without trials and errors? This is a central and difficult problem to be worked through personally.

Before one considers models of learning, a few observations are required on two factors that even further complicate this difficult question. Learning is a process that proceeds hand in hand with *maturation*. These are the two ways in combination through which changes in students occur. They can more easily be separated conceptually than in practice. Maturation we may regard as a developmental process in which pupils manifest different characteristics, the blue-prints of which they carry in their genetic endowment. Models of learning need to be considered separately from the developmental or maturation models. Both raise in an acute form the question of the differences between the models of 'knowing that' and 'knowing how'.

The second point is that we usually consider learning models separately from cultural, and affective factors. Pupils not only learn from books and in the laboratory. They also learn about food, pop records, prejudices and other people. For this wider, and more general pattern of culture there are also a number of contending models about its nature and mode of acquirement. In the pre-scientific period these notions of culture tended to filter down as theories or models of learning. Since the seventeenth century, however, more or less systematic theories of learning have been advanced to compete with the older traditional forms. It has been estimated that it takes something between 25–75 years for a new model to be translated into classroom practice. Then it often exists alongside earlier models. It would be an interesting but professionally dangerous exercise to attempt to find out what conflicting models of learning coexist more or less peacefully within the confines of a single common room or department.

Some models of learning were current long before the twentieth century, and still have a wide vogue. I shall say something in turn about three of them; first, *Mental Discipline*; then *Natural Unfoldment* and, lastly, *Apperception*. All of these are non-experimental, were derived from a particular philosophical position, and are linked with a particular school of psychology. All were formulated by the method of introspection, much as we still construct our common-sense theories of learning. It is not then surprising that we still encounter these models in their vigorous old age. Mental discipline regards learning as the discipline of the mind, through suitable exercises in the curriculum. Behind this is a notion of the kind of knowledge best able to accomplish this end. But the idea of curriculum is less emphasized than the nature of the minds that are supposed to undergo this disciplinary process. The philosophic theory basic to it is that persons are made of mind and matter. If one holds to this mind-substance model, learning is a process of training such powers of the mind, as memory, imagination, will and thought; or, in short, a process of mental discipline. Plato thought that mental training in mathematics, particularly in geometry and philosophy, was good for those who were to conduct public affairs. The training proposed for the Guardians was of about 50 years duration. Much of this model was absorbed into Christian teaching. This interesting theme will have to be left with a bare mention.

At the Renaissance, man himself, rather than the scriptures, was taken as the measure of man's individual development. The understanding of man and his possibilities was thought to be enshrined in the classics of Greece and Rome. It was assumed that a person's direction of growth was provided and controlled from within, not by giving way to impulses, but by formulating principles which the individual set up to guide his conduct and behaviour. Learning was the harmonious development of one's inherent powers, under self-discipline, so that no faculty was under-developed. Socrates gives an authentic model in a teaching context, admittedly with smaller classes than are now customary. He sought to help students to realize what was already in their minds. In parenthesis one may remark that there is a pseudo-Socratic method to be seen today, wherein the teacher puts the idea in the student's mind with one hand and unerringly finds it with the other. Socrates thought that the environment mattered little. He did not profess to impart much information, but to draw it from the pupils by skillful questioning.

The nineteenth century was very much the century of mental discipline. Both the Arnolds, and especially Matthew, thought that Greece and Rome were the best example of the human spirit's activity. In the late nineteenth century mental discipline and the faculty theory became firmly linked and classics was regarded as the best mind-training material as well as the best repository of great truths of human experience. One of the results of this was that education was regarded solely as an art, and hence there was no point in applying the methods of science to models of learning. This attitude is readily discernible in schools and universities today.

Learning through unfoldment is usually associated with Rousseau but its influence was extended by Pestalozzi and Froebel who developed appropriate teaching methods. It stated that the natural powers of the child were good, and only needed a natural environment free from corrupting influences to allow the natural unfoldment of those powers. The duty of the teacher was to allow the child to live close to nature and to follow his natural impulses. The model of learning was one of growth or development; not the imposition of knowledge and standards. On the whole this model has been less widely accepted in England than that of mental discipline. It has perhaps more possibilities in the teaching of geography.

The last of the three older models I wish to say something about is apperception. Its origins lie in the writings of John Locke (1632–1704) and the German philosopher J. F. Herbert (1776–1841). Whereas the two earlier models assume man to be possessed of innate ideas, apperception assumes that mind is entirely a content of irreducible elements or ideas brought together by dynamic association. For Herbart the mind was an aggregate not of faculties but of ideas or mental states such as sense impressions and images. Herbart recognized in the child the stages of sense activity, memory and conceptual thinking, which he thought could best be fostered by the use in teaching of the famous *Herbartian Steps*: each lesson was to follow the sequence of Preparation, Presentation, Comparison and Abstraction, Generalization, Application. In many ways this model was a reflection of the atomistic notions in the science of that time. It has been described as 'Mental Chemistry'. It was widely used in the teaching of geography, especially in

America, where it prevailed until it was submerged by the more pragmatic methods associated with John Dewey's philosophy.

Besides these older models of learning, there exist as a result of modern experimental work newer psychologies of learning. They fall into two groups: (1) Recent developments of Associationism, known as Stimulus-Response or S-R Associationisms; one of these forms the theoretical basis of programmed learning. (2) Gestalt-field theories or models. In so far as they give theoretical support to teaching processes each of these theories has its relative strengths and weaknesses.

Stimulus-response (S-R) associationism

Early this century Herbartianism gave way gradually to a new form of associationism based not on mental but on physiological phenomena. The earlier associations were concerned with the linking or association of ideas in minds; modern physiological psychologists asserted that psychology could become a science only if it studied observable bodily processes in an experimental, verifiable manner; and ceased to base its findings on introspection. Much of their work was done on animals. Some of these 'behaviourists' studied only those aspects of animal life that were amenable to exact experimental inquiry. For them a living organism was defined as a 'self-maintaining mechanism'. S-R associationists of today may be classed as 'neo-behaviourists'. They are more interested in the observation and analysis of behaviour than in the neural mechanisms underlying it.

Gestalt-field psychology

The central idea of this school is expressed in the German word *gestalt*, which may be taken to mean an organized pattern, or an organized whole, in distinction from an aggregate of parts. Max Wertheimer, in 1912, first outlined the principles of gestalt psychology. But the notion that the universe could be explained in terms of its laws of arrangement rather than by the study of its atomistic elements, dates back to the pre-Socratic Greeks. The expansion and development of gestalt psychology was largely the work of two of Wertheimer's students, Köhler and Koffka in the United States. Both criticized the learning theory and general ideas of behaviourism. Kurt Lewin further developed the gestalt theory as a cognitive-field theory or psychology of learning, using topological and vector terms. Essentially gestalt psychology stressed the primacy of the organized whole or 'form', in distinction from the atomistic views of the associationists, who viewed the world as consisting of minute indivisible elements, endowed with their own energies. This view triumphed in the natural sciences and in physiology. The organism was conceived as a combination of small elements or cells with the reflex as the basis element of movement. This mechanical concept which was taken over by the older psychology, was contested by the gestalt psychologists.

There are variations of view within these two schools, but much more critical differences between them. Each has its own philosophical assumptions, 'laws' and technical vocabulary. A brief analysis of these will reveal points of interest and importance to geographers.

Since any system of psychology rests on some beliefs about human nature, it is difficult to separate psychology from some philosophical standpoint. Among the leading contemporary schools, the Freudian regards man as an active creature of instincts, the S-R associationists see him as essentially passive and determined by environment, while for gestalt-field psychologists man acts with purpose in his psychological environment. S-R associationism, then, has close affinities with philosophical realism or positivism. It followed on the attempts to make psychology scientific. Gestalt-field theory is more closely related to a systematic *relativism*.

Realists consider the physical world, as experienced by us, is essentially as it appears to the senses, and that its existence is independent of our knowing it. It is assumed that natural laws operate in the physical world, inevitably in terms of cause and effect. Thus the universe is seen as governed by mechanical laws. This leaves the realist with little use for other kinds of explanation. His outlook and methods are critical and empirical and grounded in verifiable facts. His approach to teaching and education is that of a determinist and environmentalist. He assumes that the environment will largely control what the students do and learn. There is a tendency for the realist to favour teaching material that has the imprimatur of authority and is likely to be of use in the contemporary world. Predictable effects are thought to reside in particular subject matter apart from student activity in the solving of problems.

Relativists neither support nor deny absolute existence. They are concerned with psychological reality or what we make or form of what comes to us. The chief notion of relativism is that a 'thing derives its qualities from its relationship to other things' (Bigge, 1964, p. 68); that knowledge is a matter of human interpretation, and not a literal description of what exists 'out there', external to man. 'The relativistic test of truth is anticipatory accuracy, not correspondence to ultimate reality'.

Both realists and relativists agree on the importance of science as a method of inquiry, but their models of science differ. The former see psychological development as a matter of learning, or a conditioning in response to external stimuli; relativists regard development as springing from a relationship between man and his culture. From his interpretations of these interactions he will form a meaningful pattern or a reality on which his thoughts and actions are based.

Since these two schools of psychology differ in these respects, we shall expect them to differ in the use of a number of terms that are of special significance to the geographer: environment (here one refers to psychological environment), perception, experience, interaction and culture. In the following outline of the differences, it must be borne in mind that the statements are necessarily general, and in any specific context would require a critical evaluation. For the S-R associationists, perception is the reading and recording of the individual's physical and social environment. It is a two-stage process in which one gives meaning to what had been previously sensed. The person's physical and psychological environments in a sense correspond. One can only sense what is there. The interaction with the environment takes place through an alternation of stimulus and response. The term experience is little used by S-R associationist psychologists. It presupposes a world of consciousness which is but a 'mentalistic copy' of the physical world. Skinner, on whose operant conditioning psychology much of the theory of teaching machines is based, writes thus: 'the private event (that is, thought or consciousness) is at best no more than a link in a causal chain, and it is usually not even that. We may think before we act in the sense that we may behave covertly before we behave overtly, but our action is not an "expression" of the covert response (that is, thought) or the consequence of it. The two are attributable to the same variables'. In S-R terms motivation is an urge to act induced by a stimulus, whether from within or without. There is no need to make the student wish to learn; it is enough to engage him in appropriate activity and use reinforcement to produce learning.

The cognitive field theory of learning which derives from the gestalt psychology, does not distinguish between the sensation of an object and its meaning. It is thought of as a highly selective, simultaneous process of sensing and making meaning. A man's environment is psychological. It is what he makes of what is around him in an interaction that is simultaneous and mutual. Motivation comes from a dynamic psychological situation characterized by a person's desire to do something about a disequilibrium within his life space. The emphasis is on the situation and not on the historical antecedents of the situation. Experience is the outcome of insightful behaviour, of acting with a purpose in the expectation of probable results. In this there are active and passive elements. This model has the advantage of allowing for the notion of 'paying attention' to the environment.

CONCLUSION

To the geography teacher, whether at school or university, the above paragraphs on psychological categories and models, may perhaps appear tedious and irrelevant to his customary activities. Such categories and models as the teacher uses in his professional activities, and these are not confined to learning models, come from varied sources; some from the conventions of his culture and upbringing, some from his specialist discipline and others from contemporary notions of sociology and psychology. Wherever derived, all have to pass the empirical test of the teaching situation. But in response to the needs of our technological society, the teaching situation, currently under strong criticism, is itself in process of change, as part of a complex of change, affecting in turn, the national structure of education, the role of the teacher and the content of the curriculum.

Quite apart from the influence of examinations and a traditional distrust of central control of the curriculum, teaching is a very conservative activity. Despite that, we have had at our command for some time research techniques as appropriate to curriculum study as to market research. But the publication of research results is only the first stage of curricular innovation. Loyalty to traditional ways has prevented the application of sufficient power and organization to the two succeeding stages in innovation; the institution of representative proving trials in schools and the diffusion of the new practices under ordinary classroom conditions. It was as recently as 1962 that a Curriculum Study Group was founded in the Department of Education and Science. Out of it grew the Schools Council, which directs, sponsors and co-ordinates educational research on curriculum and examinations from the kindergarten to university entry. The Nuffield Foundation is also engaged on curriculum research, at present mainly in mathematics and science. Their present commitments will probably prevent either body from promoting any direct research into the teaching of geography. The Foundation and the Council are, however, co-operating in a joint inquiry into courses in modern humanities or social studies, suitable for pupils of average and less than average ability, in the last two years of their schooling. In any such course on the problems of man, nature and society, relevant to the experience of the pupils, geography would have a place, but not as a body of subject matter taught in the usual academic fashion. Rather it would contribute its techniques to the co-operative treatment of selected case or sample studies. Some years ago the inclusion of geography in a comparable social studies grouping was severely criticized by geographers on the grounds that it threatened the identity of the subject, and that geography could not be effectively taught by non-geographers. Now in the light of subsequent experience of composite general courses in Sixth Forms, Further Education Colleges and Universities, interdisciplinary models of subject matter are much more acceptable. Not all, however, will go as far as Ackerman when he writes:

> We are no longer concerned about whether what we are doing is geography or not; we are concerned instead with what we contribute towards a larger goal, however infinite it may seem.

That larger goal is:

> Nothing less than an understanding of the vast, interacting system comprising all humanity and its natural environment on the surface of the earth. (Ackerman, 1963, p. 435.)

Referring more particularly to a research context, Ackerman goes on to state that systems analysis might have been ideally created as a technique for geographers. Adopted belatedly by geographers from the social and behavioural scientists, systems analysis, quite apart from its great possibilities as a technique, has the merit of re-enforcing the tendency to interdisciplinary studies.

Quantitative and analytical methods in geography are finding increasing favour at the university level. This cannot remain without influence on the teaching of geography in the grammar schools. Some of the first fruits of curriculum research in mathematics is beginning to appear in the schools. Here is a profitable field for collaboration in which geography may benefit from the application of some of these newer mathematical techniques to geographical problems.

REFERENCES

ACKERMAN, E. A., [1963], Where is a Research Frontier?; *Annals of the Association of American Geographers*, 53, 429–440.
ARMSTRONG, H. E., [1903], *The Teaching of Scientific Method*, (London), 476 pp.
BIGGE, M. L., [1964], *Learning Theories for Teachers*, (New York), 366 pp.
BORING, E. G., [1957], *A History of Experimental Psychology*, (New York), 777 pp.
BOULDING, K. E., [1956], General Systems Theory – The Skeleton of Science; *Management Science*, 2 (3), 197–208.
BROOKS, H., [1965], Scientific Concepts and Cultural Change; *Daedalus*, 94 (1), 66–83.
COLLINGWOOD, R. G., [1939], *An Autobiography*, (London), 167 pp.
DEWEY, J., [1916], *Democracy and Education*, (New York), 434 pp.
HAYS, S. B., [1965], Social Analysis of American Political History (1880–1920); *Political Science Quarterly*, 80, 373–394.
KOFFKA, K., [1935], *Principles of Gestalt Psychology*, (New York), 720 pp.
KÖHLER, W., [1947], *Gestalt Psychology*, (New York), 369 pp.
KUHN, T. S., [1962], *The Structure of Scientific Revolutions*, (Chicago).
LEWIN, K., [1936], *Principles of Topological Psychology*, (New York), 231 pp.
LEWIS, C. S., [1964], *The Discarded Image*, (Cambridge), 232 pp.
PARSONS, T., [1965], Unity and diversity in the modern intellectual disciplines: The role of the social sciences; *Daedalus*, 94 (1), 39–65.
SPENCER, H., [1861], *Education: Intellectual, Moral and Physical*, (London), 190 pp.

Index

Index

D. R. STODDART

Since this book is primarily about ideas and people rather than places, no attempt has been made to index the many incidental references to place-names in the text. The intention of this index is, first, to locate references to people and their writings, and second, to trace the main ideas common to many of the papers.
Page-references to literature citations are given in italics, thus: *236*.

Abercromby, R., 130, 133, *135*
Abortion, 200
Abrams, P., 218, *240*
Abramson, P., 615, *667*
Accessibility, 304, 337
Accuracy, in Maps, 706
Ackerman, E. A., 35, 39, *39*, 213, 243, *287*, 461, *501*, 530, 537, *538*, 775, 791, 792, *792*
Ackoff, R. L., 22, 25, *39*, 552, 582, 584, 597, 598, 603, *604*
Action, irrational, 223
Action, rational, 223
Action, Theory of, 223, 228
Action systems, 223
Acton, Lord, 549
Adams, C. C., 521, *538*
Ady, P., *457*
Age at Marriage, 199
Agglomeration, 305
Agricultural System, Classification, 437
Agriculture, 425
Agriculture, Location of, 443
Agriculture, Peasant, 430
Agriculturization, 431
Ahmad, E., 445, *452*
Ahrens, H., 515, *538*
Air Masses, 129
Air Network, 641

Ajo, R., 560, *598*, 599
Akers, S. B., Jr., 616, 644, *664*
Alampiev, P. M., *see* Alampiyev, P. M.
Alampiyev, P. M., 407, 473, *501*, *508*
Alexander, J. W., 269, *287*, 332, 333, *355*, 368, 373, 374, 389, *417*
Alexandersson, G., 332, 333, *355*, 389, 390–392, *417*
Aleksandrov, I. G., 473
Algol, 52
Algorithm, Circuit-Order, 656
Algorithm for Location Problem, 363
Alihan, M. A., 522, *538*
Alkjaer, E., 366, *417*
Allan, W., 449, *452*
Allan, W. N., 176, *183*
Allen, J., 145, *179*, 739, 740, *771*
Allen, R. A., 110, 111, *136*
Allman, J., 493, *501*
Allocation-Location Models, 386
Al-Maiyah, A. M., 437, *452*
Alonso, W., 268, 271, *291*, 336, 338, 341, 345, *355*, 564, *600*
Alpha Index, 634

Ambrose, J. W., 62, *90*
Amorocho, J., 63, 72, 78, 84, *90*, 153, 165, 167, *179*
Amorocho, J., and W. E. Hart, on Hydrological Systems, 84
Analogue, Chemical, 737
Analogue, Historical, 60
Analogue, Spatial, 62
Analogue Models, 68
Analogue System, Natural, 60
Analogy, 23
Analytical Model, in Water Resources, 176
Anderson, C. W., 267, 273, *292*
Anderson, D. L., 491, *506*
Anderson, J., 437, *452*
Anderson, R., 114, *136*
Andic, S., 270, *287*
Andrews, R. B., 278, *287*
Angle of Repose, 733
Animal Population, Behaviour of, 191
Annette, M., 465, *501*
Annulus, Hide, 106
Anscombe, F. J., 574, 575, *603*
Anthropology, Cultural, 476
Antimetry, 638
Anuchin, V. A., 518, *538*

Appleby, F. V., *179*, 742, *771*
Apostel, L., 22, 24, *39*, 494, *501*
Apperception Model, 787
Aquinas, T., 777
Areal Differentiation, 451
Area-Set Problem, 618
Aristotle, 21, 776, 777
Aristotle, on Poetry and History, 21
Armstrong, H. E., 779, *792*
Arnoff, E. L., *604*
Arnold, M., 787
Aron, R., *240*
Arx, W. S. Von., *see* Von Arx, W. S.
Aschmann, H., 521, 522, *538*
Asnani, G. C., 124, *141*
Atkinson, R. C., 594, *608*
Atmosphere, Baroclinic, 104
Atmosphere, Barotropic, 104
Atmospheric Circulation, 99
 Conceptual Models, 100
 Mathematical Models, 103
Atoll Ecosystem, 528
Atomic Structure, 776
Auerbach S. I., 528, *538*
Autocorrelation, 556
Autoregression, 556
Avondo-Bondino, G., 610, *664*

Bachmura, F. T., 257, *287*
Backwash Effects, 236
Bacon, F., 21, 777, 778
Baer, W., 261, 262, 264, *287*
Bagnold, R. A., 66, *90*
Bagrow, L., 697, 711, *719*
Bailey, E. W., 515, *546*
Bailey, N. T. J., 562, 570, 577, 582, *600*, 603
Bain, J. S., 389, 390, *417*
Baker, O. E., 425, 433, *453*
Bakker, J. P., 70
Balassa, B., 264, 282, *287*
Balchin, W. G. V., 700, *719*, 732, *771*
Balderston, F. E., 584, *605*
Baldwin, R. E., 243, 249, 266, 271, *287*, *296*
Balzak, S. S., 381, *417*
Bambrough, J. R., 21, 23, *39*, 706, *719*
Baran, P. A., 252, *287*
Barbados, 441
Barbag, J., 477, *501*

Barfod, B., 276, 277, *287*, 413, *417*
Barna, T., 255, 256, 270, *287*
Barnard, C. S., 690, *721*
Barnes, F. A., 564, *600*
Barnes, H. E., 221, *240*
Barnes, K. K., *184*
Barnum, H. G., 313, 324–326, 347, 348, 354, *355*
Barraclough, G., 551, 597
Barriers to Diffusion, 590
Barrows, H. H., 476, *501*, 521, 522, *538*
Barry, R. G., 130, *136*
Barth, J., 217
Bartholomew, J. G., 686, 698, *719*
Bartlett, M. S., 562, 570, 571, 577, 584, 586, *600*, 603
Barzanti, S., 260, *288*
Basic-Non-Basic Functions, 332
Basinski, J., 468, *501*
Baskin, C. W., *see* Christaller, W.
Bass, R. E., *539*
Batalha-Reis, J., 516, *538*
Battersby, A., 610, *665*
Baulig, H., 87, *90*
Baur, F., 134, *136*
Bayes Rule, 54
Beach Process-Response Model, 81
Beal, G. M., *453*
Beale, C., 474, *501*
Beament, J. W. L., 25, *39*, 536, *538*
Beard, L. R., 147, *180*
Beaujeu-Garnier, J., *213*
Beck, W. S., 519, 520, *538*
Becker, H., 221, *240*
Beckett, P. H. T., 63, *90*
Beckinsale, R. P., 59, 86, *91*
Beckmann, M. J., 324, *355*, 363, 375, 386, 387, 389, *417*, *421*, 571, 592, *603*, 607
Beckmann-Marschak Model, 387
Bedrosian, S. D., 656, *666*
Beeley, B. W., 427, *453*
Beers, N. R., 151
Beesley, M. E., 175, *181*, 618, *665*
Beguin, H., 281, *288*, 428, 429, 449–452, *453*
Belasco, J. E., 129, *136*
Bell, W., 462, *506*

Bénard, H., 106
Bénard Cell, 106
Bendix, R., 226, *240*
Benedictis, De M., *454*
Benson, B. T., 32, *40*, 93
Berg, L. S., 467, *501*
Berge, C., 568, *600*, 634, *665*
Bergeron, T., 97, 101, 119, 129, 133, *136*
Bergson, H., 518, 519, *538*
Berman, B., 569, *600*
Berman, E. B., 270, *288*
Bernal, J. D., 34, *39*
Berry, B. J. L., 28–30, 32, *39*, 237, 244, 247, 275, *288*, 307, 309, 313, 322, 324–326, 328, 329, 331, 337, 338, 340, 343–350, 354, *355*, *356*, 357, 403, 492, 493, *501*, 509, 537, *538*, 640, *665*, 690, *719*
Berry, B. J. L., Geographical Data Matrix, 28
Berry, F. A., Jr., 151
Beshers, J. M., *240*
Beum, C. O., *503*
Bews, J. W., 522
Bharucha-Reid, A. T., 577, 582, *603*
Bhatia, S. S., 437, *453*
Bias, in Maps, 676
Biasutti, R., 522
Bićanić, R., 401, *418*
Bifurcation Ratio, 165
Bigge, M. L., 789, *792*
Billings, M. P., 735, *771*
Binnie, G. M., 151, 152, *180*
Biology, Comparison with Geography, 34
Biome, 522, 523
Birch, A. H., 285, *293*
Birch, J. W., 426, *453*
Bird, R., 283, *288*
Birds, 192
Birdsell, J. B., 529, *539*
Birot, P., 86, *91*
Bjerknes, J., 114, 115, 119, 129, 133, *136*, *139*, 728
Black Box, 85, 426
Black Box Methods in Geomorphology, Criticisms of, 87
Black Death, 204
Black, M., 23, *39*, *240*, 494, *501*, 671, *720*
Black, R. P., 174, *182*
Blackman, R. B., 623, *665*

Blair, T. A., 123, *136*
Blalock, H. M., 81, *91*
Blaney, H. F., 158, *180*
Blaut, J. M., 425–427, 433, *453*, *537*, *539*
Bliss, C. I., 574, *603*
Blome, D. A., 312, *356*
Blotting Paper Analogue, 759
Blumen, I., 581, *603*
Blumenfeld, H., 341, 353, *356*
Blumenstock, D. I., 557, *599*, 695, *720*
Bluntschli, H., 517, *539*
Bluntschli, J., 515, *539*
Blydenstein, J., 523, *539*
Boal, F. W., 346, *356*
Board, C., 34, 438, *453*, 676, 703, 704, *720*
Boas, F., 476, *501*
Bogue, D. J., 318, 319, *356*, 369, 390, *418*, 474, 476, 477, *501*
  on Hinterland Sectors, 318
Bohr, N., 776
Bolin, B., 108, 109, 135, *136*
Bollay, E., 151
Boniface, E. S., 165, *180*
Boolean Algebra, 45
Boothroyd, A. R., 748, *771*
Boring, E. G., 780, *792*
Borts, G. H., 248, 257, 261, 267–269, *288*, 566, *600*
Bos, H. C., 254, *301*, 363, 376, 386, 393–395, *418*
Boserup, E., 190, 197, *213*, 436, *453*
Boskoff, A., *240*
Botset, H. G., 737, *771*
Bottomore, T. B., 222, *240*
Boudeville, J. R., 282, *288*
Boulding, K. E., 528, 537, *539*, 784, *792*
Boundaries, 478
Boville, B. W., 110, 114, *136*
Bowden, K. L., 626, 655, *665*
Bowden, L. W., 174, *180*, *453*
Bowen, E. G., 508
Bowman, I., 551, *597*
Boyce, R., 309, 349, 350, 354, *356*, *358*
Boyden, C. J., 118, *136*
Boye, Y., 616, *665*
Boyle, R., 780
Bracey, H. E., 313, 324, *356*
Bradbury, D. L., 112, *136*
Braham, R. T., 127

Braithwaite, R. B., 26, *39*, 43, 494, 500, *501*, 587, *597*, 603
Brandenburg, F. R., 249, 261, *288*
Brandner, L., *453*
Braun-Blanquet, J., 475, 478, 484
Braun-Blanquet School, 478
Bray, J. R., 523, *539*
Brazil, Municipios of, 309
Bresler, J. B., 522, *539*
Bretherton, M. H., 574, *607*
Bretz, J. H., 87, *91*
Bridgman, P. W., 24, *39*
Briggs, A., 270, 271, *288*
Brigham, E. F., 337, *356*
Brillouin, L., 535, *539*
Brink, E. J., 770, *771*
Brinkman, L., 692, *723*
Britain, as Imperial Power, 218
Brittan, M. R., 173, *180*
Brodbeck, M., 221, *241*, 476, 494, *501*
Broek, J. O. M., 517, *539*, 551, *597*
Brokken, R. F., 454
Brookfield, H. C., 425, 452, *453*, 530, *539*
Brooks, H., 783, 784, *792*
Broude, H. W., 462, *501*
Brown, J. A., *144*
Brown, L., *39*, 401, *418*, 581, 589, *603*, *607*
Browning, G. M., 159
Browning, K. A., 129, *136*
Brückner, M., 651, 652
Brunhes, J., 303, *356*, 515, *539*, 550, *597*
Brunner, E. De S., 312, *358*
Brush, J. E., 311–313, 324, *356*, 576, *603*
Brush, L. M., 68, *91*, 738, *771*
Bryan, L. D., *608*
Bryson, R. A., 110, *140*, *143*
Buchan, S., 165, *180*
Buchanan, R. O., 366, 384, *419*, 425, 430, *453*
Bucher, K., 249
Buck, M., 506
Büdel, J., 86, *91*
Bunge, W., 21, 33, 34, *40*, 243, *288*, 306, 307, 309, *356*, 425, *453*, 463, 472, 479, 481, 483, 486, *502*, 509, 511, *539*, 560, *599*, 609, 612, 614, 623, 624, 630, *665*, 671, 704, 706, *720*, 756, *771*
Bunge Cork Model, 756
Burch, G. E., 522, *539*
Burgess, E. W., 339, 340, *356*, 462, *505*, 521, *545*
Burley, T. M., 283, *288*
Burton, I., 175, *180*, 391
Busacker, R. G., 610, 634, *665*
Bush, R. R., 594, *607*
Butler, J. B., 428, *453*
Butte, W., 515, *539*
Butzer, K. W., 86, *91*, 131, 132, *136*, *137*
Buys Ballot, C. H. D., Wind Law, 98
Byers, H. E., 127, *137*
Bylund, E., 566–568, 576, *600*, *601*

Caesar, A. A. L., 261, *288*, 384, *418*
Cairncross, A. K., 252, 253, 261, 271, *288*
Calef, W. C., *185*, 334, *359*
Calvez, J.-Y., 260, *291*
Cameron, G. C., 263, *288*
Camp, G. D., 24, *40*
Cao-Pinna, V., 270, *289*
Capital Localization, 369
Capitalism, and Industrial Location, 377
Capron, W. M., 563, *601*
Carlston, C. A., 85, *91*
Carnap, R., 52
Carnell, F. G., 285, *293*
Carroll, L., 706, *720*
Carrothers, G. A. P., 535, *539*
Carrothers, W. J., 347, *356*
Carr-Saunders, A. M., 200, *213*
Carter, D. B., 147, *180*
Carter, G. F., 521, *539*
Carter, H. C., 428, *453*
Cartographic Signal, 674
Cartography, 32, 671
Cartography, Central Role of 34
Cartwright, D., 610, *666*
'Cascade' Algorithm, 616
Caspari, E., 520, *539*
Castle, E. N., 153, *184*
Catchment Area Units, 147
Catchment Model, 165
Causation, Circular, 389
  Cumulative, 261

Caves, R. E., 271, *288*
Caws, P., *40*
Cayley, G., 647, *665*
Cell, Bénard, 106
Cell-Division, 652
Cell Geometry, 646
Cellular Net, 646
Central Business District, 349, 350
Central Place, 563
Central Place Analogue, 760
Central Place Model, 306, 732
Central Place, Periodic, 320
Ceylon, Economic Development, 210
Chang Chai-Ch'Eng, 124, *137*
Change, 222
Change, and Functionalism, 230
Chapin, F. S., 352, *356*
Chardonnet, J., 410, *418*
Charney, J. G., 108, 109, *137*
Chatterji, M. K., 256, *289*
Chemical Analogue Model, 737
Chenery, H. B., 254–256, 263, 269, 270, 275, *289*, 563, 569, *601*
Cherry, C., 595, *607*
Cherry, E. C., 748, *771*
Chery, D. L., 154, *180*
Childs, E. C., 156, *180*
Chinitz, B., 282, *289*, 391, *418*, *600*
Chisholm, M., 243, 245, 246, 261, *289*, 399, 405, *418*, 437, 444, *453*, *454*, 495, *502*, 557, 564, *599*, *601*
Chomsky, C., 56
Chorafas, D. N., 22, 23, *40*
Chorley, R. J., 20, 24, 25, 32, 34, 35, *40*, 59, 60, 62, 73–76, 80, 85–87, 89, 90, *91*, 150, *184*, 243, 248, 253, *289*, 292, 499, *502*, 528, 530, 537, *539*, 609, 610, 628, 634, 655, 663, *665*, *666*, 672, 716–718, *720*, 728, 729, 768, *772*
Chow, Ven Te, 149, 160, 173, *180*, 628, 664, *665*
Choynowski, M., 557, *599*
Christaller, W., 37, 189, 307, 308, 312–316, 322, 324, 325, *356*, 389, 470, 475, 498, *502*, 563, 585, 648, *665*, 761

Central Place Model, 307
Churchman, C. W., 571, 584, *604*
Circuit Geometry, 632
Circulation, Atmospheric, 99
  Conceptual Models, 100
  Mathematical Models, 103
  Equatorial, 126
Ciriacy-Wantrup, S. V., *180*
City Classification, 329
City, Commercial Structure, 345
  Impact of, 237
  Internal Structure, 335
  Population Density, 343
Claesson, C. F., 560, *599*
Clapp, P. F., 109, *137*, *141*
Clark, A. H., 437, *454*, 557, *599*
Clark, B. D., 263, *288*
Clark, C., 258, *289*, 343, 345, *356*, 437, *454*
Clark, E. J., 514, *539*
Clark, J. D., 529, *539*
Clark, P. G., 255, 256, 270, *289*
Clark, P. J., 310, *356*, 575, 576, *604*
Clark, W. A. V., 309, 354, *356*, 401, *418*, 581, *604*
Clarke, J. I., 703, 704, *720*
Class Interval, 692
Classification, 463, 479, 486, 690
  in Agriculture, 437
  of Cities, 329
  Paradigms, 28
Clawson, M., 246, *289*
Clayton, K. M., 409, *418*
Clemens, S. L., *see* Twain, M.
Clements, F. E., 478, 507, 512, 513, 522, 523, 528, *539*
Clerk-Maxwell, J., *see* Maxwell, J. C.
Climate, Hardware Models, 728
Climatic Change, 131
Climatic Geomorphology, 86
Climatic Regions, 130
Climatology, Tropical, 118
Climax, Industrial, 406
Cline, M. G., 464, 486, *502*
Cloos, H., 735
Closed System, 528
Clough, S. B., 262, 264, *289*
Cluster, Regular, 312
Cluster Analysis, 81

Cluster Pattern, 306
Coale, A. J., 211, *213*
Coburn, T. M., 618, *665*
Cohen, K. J., 254, *289*
Cohn, D. L., 629, *665*
Coker, F. W., 515, *540*
Colberg, M. R., 364, 376, 412, *419*
Cole, J. P., 34, *40*, 701, *720*
Cole, L., 523, *540*
Coleman, J. S., *241*, 571, 572, 590, 595, *604*, *607*
College, Invisible, 19
Collingwood, R. G., 778, *792*
Colonization, 566
Column Vector, 28
Colyton (Devon), 199
Commandment, Second, quoted, 26
Communication, 596
Community Centre, 348
Comparative Statics, 564
Complex, Territorial-Production, 407
Computer, Digital, 167
Computer Analysis, 32
Computer Programming, 52
Computer Simulation, of Ecosystems, 534
Computers, as Dynamic Models, 49
Comte, A., 234, 515
Conducting Sheet Analogue, 746
Connectivity, of Circuits, 634
Connell, K. H., 203, *213*
Conrad, A. H., 245, 266, *289*
Conrad, V., 728, 729, *772*
Conservation of Mass, 98
Contagious Distributions, 572
Contiguity, 485
Continuity Equation, 98
Conzen, M. R. G., 672, *720*
Coombs, C. H., 52
Coppock, J. T., 425, 440, *454*, 491, *502*, 693, *720*
Coquery, M., 283, *289*
Coral Reef Productivity, 531
Core Area, 478
Coriolis Deflection, 98
Coriolis Parameter, 98
Cornish, V., 551, *597*
Corte, A., 648, *665*
Costa Santos, M., 410, *418*
Cost-Benefit Analysis, 174
Cotterill, C. H., 372, *418*
Coulomb's Law of Force, 757

Courant, R., 769, *772*
Court, A., 97, *137*
Courtheoux, J.-P., 283, *291*
Coutsoumaris, G., 263, *289*
Coutu, A. J., 456
Cox, D. R., 172, *180*, 571, *604*
Cox, K. R., 387, 388, *418*, *501*
Cox, P. R., *213*
Coxeter, H. S. M., 658, *665*
Craddock, J. M., 134, *137*
Crawford, N. H., 167, 168, *180*
Creighton, R. L., 642, *665*
Criddle, W. D., 158, *180*
Crisp, D. J., 531, *540*
Crop-Combination Region, 439
Crowe, P. R., 98, 100, 119, *137*, 461, 470, 478, 489, *502*, 520, *540*
Croxton, F. E., 703, *720*
Culling, W. E. H., 70, 72, *91*
Culture, 220
Culture Area, 462
Cumberland, J. H., 256, 270, 294, 462, *504*, 569, *601*
Curl, R. L., 72, *91*
Curry, L., 133, *137*, 348, *356*, 391, *418*, 558, 563, 571, 575, 587, *599*, *601*, *604*, 719, *720*
   on Climatic Change, 133
Curtis, J. T., 475
Cuzzort, R. P., 489, 490, 491, *502*
Cvijić, J., 680
Cybernetics, 535
Cycle, Davis, 87, 512
   of Industrial Development, 402
   of Water, 145–146
Cyclomatic Number, 634
Cyert, R. M., 254, *289*, 595, 596, *607*
Czekalski, J., 696, *720*

D^2 Analysis, 32
Dacey, M. F., 37, *40*, 310, 312, 313, 333, 334, *357*, *360*, *560*, *561*, 575, 576, *593*, *599*, *604*, *607*, 647, *665*
Dahlberg, R. E., 685, *720*
Dahrendorf, R., 227, 232, *241*
Dale, M. B., 481, *505*
Dales, J. H., 399, 405, *418*

Daly, M. C., 276, 277, *290*
Daniel, V., 68–69, *91*
Daniels, S. M., 118, *139*
Danilevsky, N. D., *503*
Dansereau, P., 467, 481, 497, *502*
Dante Alighieri, 777
Dantzig, G. B., 617
Darby, H. C., 30, *40*, 549, *597*, *724*
Darling, F. F., 521, *540*
Darwin, C. R., 90, 191, 193, 194, 196, *213*, 511, 515, 517, 519, *540*
   on Elephant, 191
   on Fertility, 193
   Atoll Theory Confirmed, 90
Darwinian Revolution, 514
Darwinism, 246
Das-Gupta, A. K., 249, 271, 273, *290*
Data, Increase in, 30
Data Bank, Size of World, 30
Data Matrix, 74
   B. J. L. Berry on, 28
   Transformation of, 28
Data Selection, in Maps, 683
Daubrée, A., 735, *772*
Daum, C. R., *772*
Davies, J. C., 234, 235, *241*
   on Revolution, 234
Davies, J. L., 467, *502*, 532, *540*
Davis, L. B., 682, *720*
Davis, S. N., 162, *181*
Davis, W. M., 86, 87, 90, *91*, 402, 512, 513, 521, *540*
   Geographical Cycle, 87, 512
   and Industrial Change, 402
Dawdy, D. R., 153, 170, 171, *181*
Dawson, R., 676
Day, L. M., 428, *454*
Dean, G. W., *454*
Dean, R. M., 279, *298*
Dean, W. H., 373, *418*
Deane, P., 254, 270, 285, *290*
De Candolle, A., 467
De Cani, J. S., 570, *603*, 770, *771*
Decision-Making, 446
Decisions, State, 366
De Dainville, S. J., 671, 697, *720*
De Geer, S., 34, *40*
Deguileville, G., 776

De Jong, G., 463, 479, *502*
Delft, University of, 145
Demographic Model, 189, 193
   Modified, 200
   Post-Industrial, 207
Denbigh, K. G., 528, *540*
Deniel, R., 199, *213*
Denudation Chronology, 62, 87
Depasquale, N. P., 522, *539*
Deprez, P., 199, 205, *214*
Descartes, R., 777, 778
Determinism, 471, 775
Deterministic Model, for Spatial Pattern, 563
   in Geomorphology, 69
Deutsch, K. W., 23, *40*
Development, Unilinear, 234
Devletoglou, N. E., 377, *418*, 587, *604*
Devore, I., 529, *540*, 548
De Vries, E., *185*
De Weist, R. J. M., 162, *181*
Dewey, J., 779, 788, *792*
Diachronic Models, 630
Diagrammatic Maps, 711
Dialectic, 778
Diameter, 635
Diamond, D. R., 283, *290*, *350*, *357*
Dice, L. R., 530, *540*
Dicken, P., 455
Dickinson, R. E., 470, *502*, *508*, 551, *598*
Diffusion, 446, 551, 585
   Barriers to, 762
   in Blotting Paper, 759
   of Innovation, 589
   Waves, 38
Dillon, J. L., 448, *454*
Dimensional Analysis, 740
Dirichlet, G. L., 658, *665*
Dirichlet Region, 658
Discard, Zone of, 354
Discriminant Analysis, 32
Dishpan Model, 106
Dispersed City, 391
Dispersion, Spatial, 393
Distance, 304
   and Population Density, 318
   Transformations of, 560
Distance Function, 559
Distance Index, Mather's, 437
Distance Measures, 304
Distribution, of Economic Development, 257

INDEX

Division of Labour in Society, 224
Dodd, S. C., 568, 569, 571, 592, *601*, *604*, 607
Dodgson, C. L., *see* Carroll, L.
Dokuchayev, V. V., 467
Donn, W. L., 131, *137*
Doob, J. L., 577, *604*
Dooge, J. C. I., 160, *181*
Döös, B. R., 109, *137*
Dorfman, R., 175, 179, *181*, *183*, *605*
Dorjahn, V. R., 199, *214*
Dorsey, H. G., Jr., 132, *138*
Douglas, J. B., 575, *604*
Douglas, N. J., 554, 585, *600*, *606*
Drainage Basin, Experimental, 149
 Model, 149
 Representative, 150
Drake, K. M., 203, 205, *214*
Drake, M., *144*
Driesch, H., 518, 519, *540*
Drummond, I., 252, *290*
Dryer, C. R., *508*, 515, 516, 521, *540*
Duckworth, W. E., 172, *181*
Duesenberry, J. S., 266, 267, *290*
Duffy, J., 219
Duke, R. L., 335, *357*
Dumke, G. S., 681, *720*
Duncan, B., 369, 390, 392, *418*, 489–491, *502*
Duncan, O. D., 238, *241*, 489–491, *502*
Duncan, W. J., 66, *92*, 740, *772*
Dunkerley, H. B., *185*
Dunn, A. J., 59, 86, *91*
Dunn, E. S., 336, *357*, 376, 380, *418*, 443, 444, *454*, *602*
Dunn, G. E., 119, *137*
Dunning, J. H., 382, *420*
Durbin, J., 556, *599*
Durkheim, E., 224, 225, 238, *241*
Durrell, L., 549
Dury, G. H., *151*
Duvigneaud, P., 523, *540*
Dwyer, D. J., 245, *290*
Dylik, J., 60, *92*
Dynamic Climatology, 97
Dynamic Equilibrium, 85, 86
Dynamic Models, Hardware, 732

Dziewoński, K., 381, 383, *418*, 473, *502*, *508*

Eady, E. T., *137*
Earth, as Organism, 515
East, W. G., 671, 714, 719, 725
Easterlin, R. A., 264, *290*
Easterly Waves, 120
Eckaus, R. S., 257, 262, *290*
Eckert, M., 676, *720*
Eckman, G., 703, 704, *720*
Eckstein, A., 254, *290*
Eckstein, O., 153, *181*, *182*
Ecological Approach, 471
Ecological Forces, 232
Ecological Model, 571, 590
Ecological Region, 476
Ecological Systems, 197
Ecology, 472, 511
 Plant, 462
Econometrics, 23
Economic Base, 332
 Analysis, 278
Economic Commission for Europe, 272
Economic Development, Literature, 244
 Non-Spatial, 248
Economic Geography, 425
Economic Growth, Typology, 247
Economic Landscape, of Lösch, 316
Economic Models, 246
Economic Region, 473
Economics, Nomothetic, 246
Ecosphere, 523
Ecosystem, 35, 511, 522, 528
Ecosystem Analogue, 534
Ecosystem, Human, 530
Ecosystem Limits, 534
Ecosystem Properties, 524
Eden, M. J., 530, *540*
Edminster, T. W., *184*
Edmonson, M. S., 454
Edwards, D., 431, 433, *454*
Edwards, K. C., 530, *540*
Edwards, R. W., 531, *541*
Effort, 304
Efroymson, M. A., 388, *418*
Egbert, A. C., 446, *454*, 455, 563, *601*
Egler, F. E., 472, *502*
Einstein, A., 714
Einstein, H. A., 742, *772*

Eisenstadt, S. N., 226, *241*
Élan Vital, 519
Eldridge, R. H., 121, *137*
Electric Analogue Model, 742
Electric Field Pattern, 748
Electrochemical Analogue Model, 737
Electrolytic Tank, 746
Element-Complex, 441
Elephant, Gestation of, 191
Eliassen, A., 104, 108, 109, 110, 134, 135, *137*
El-Kammash, M. M., 246, *290*
Elkan, W., 264, *290*
Ellermeier, R. D., 59, *92*
Elliot, F. F., 454
Emerson, A. E., 540
Emmett, W. W., 150, *181*, *182*
Enequist, G., 552, 564, *598*, *601*
Energy, 783
Enke, S., 248, 249, 252–254, *290*, 770, 771, *772*
Entelechy, 519
Entropy, 97
Environmental Model, in Agriculture, 433
Environmental Physiology, 522
Epidemic curve, 570
Epidemiology, 569, 584
Epstein, B. J., *358*
Epstein, H. A., *772*
Epstein, T. S., 236, *241*
Equatorial Circulation, 126
Equilibrium, Location, 265
Erhart, R. R., *501*
Erickson, C. O., 122, *137*
Estall, R. C., 285, *290*, 366, 384, 392, 402, *418*, *419*
Estes, W. K., 594, *607*
Estoque, M. A., 122, *137*
Estuary Model, 739
Etzioni, E., 236, *241*
Euler, L., 648
 on Königsberg Bridge Problem, 614
 Polyhedral Formula, 647
Evans, D. A., 574, 575, *604*
Evans, F. C., *356*, 521, 523, *540*, 575, 576, *604*
Evans-Pritchard, E. E., 232, 240
Evaporation Models, 156
Evaporation Pan, 156
Eversley, D. E. C., 190, *214*

Ewing, A. F., 254, *290*
Experimental Design, 73
Experimental Models, Atmospheric Circulation, 105
Export Base Model, 266
Extreme Events, E. J. Gumbel on, 152
Eyre, S. R., 522, 530, *540*

Facies Map, 438
Factor Analysis, 32, 82, 130, 440, 493, 640
Facts, Nature of, 19, 217
Fagen, R. E., 532, *542*
Fair, G. M., *183*, *605*
Fair, Periodic, 321
Fair, T. J. D., 245, *292*, 482, *502*
Fairbank, J. K., 254, *290*
Faithfull, W. G., 346, *357*
Faraday, M., 496
Farbey, B. A., 616, *665*
Farm Adoption Scheme, 427
Farm, Model, 426
Farmer, B. H., 552, *598*
Farming, Type of, Model, 429
Faulting Models, 735
Febvre, L., 516, *540*
Feedback, 78, 536, 554, 783
Fei, J. C. H., 253, *299*
Feigin, Ya G., 381, 383, *417*, *419*
Feller, W., 572, 577, 582, *604*
Fenneman, N. M., 62, *92*, 469, *502*
Fenton, B. L., *458*, *724*
Ferrel, W., 101, *137*
Fertility, 189
 Pre-Industrial Population, 198
Fiering, M. B., 173, 176, 178, 179, *181*
Fifield, R. H., 518, *540*
Finch, V. C., *508*, 517
Finley, M. I., 396, *419*
First Law of Thermodynamics, 97
Firth, R., 430, *454*
Fish, on Coral Reefs, 526
Fisher, J. L., 256, *291*
Fisher, M. R., 255, *291*
Fitzroy, R., 114, 138
Fjørtoft, R., 105
Fleagle, R. G., 100, 104, 112, *138*
Fletcher, J., *136*

Fletcher, J. O., 131, *138*
Fleure, H. J., 466, *502*, *508*, 518, 522, *541*
Flint, R. F., 132, *138*
Floating Cork Model, 756
Flöhn, H., 124-126, *138*
Flood Control Model, 742
Flood Hazard, 175
Flood Plain Management, 175
Florence, P. S., 277, *291*, 392, 406, *419*
Flugge, S., 652, *666*
Foley, D. L., *419*
Folk, R. L., 730, *772*
Follansbee, R., 166, 167, *181*
Forbes, J., 305, *357*
Forbes, S. A., 523, *541*
Ford, L. K., 568, *601*
Ford, L. R., Jr., 617, 618, *666*
Forde, C. D., 529, *541*
Fordham, P., 244, *291*
Forgotson, J. M., Jr., 690, *720*
Formalization of Theories, 43, 44
Forsdyke, A. G., 98, 112, 114, 115, 121, *138*, *144*
Fortes, M., 457
Fosberg, F. R., 523, 528, 529, *541*
Foster, C., 536, *541*
Foster, C. D., 155, 175, *181*
Foster, G. J., *357*
Fourastié, J., 283, *291*
Fourier Analysis, of Atmospheric Waves, 109
Fox, J. R., 688, *721*
Fox, K., 445, *454*
Franklin, S. H., 429, 431, 432, *454*
Free, G. R., 159
Freeman, J. C., Jr., 121, *138*
Freight rate, Differential, 767
Frenzen, P., 106, *138*
Frevert, R. K., *184*
Friction of Distance, 559
Friederichs, K., 520, 523, *541*
Friedkin, J. F., 63, 67, 68, *92*, 738, *772*
Friedmann, J. R. P., 246, 261, 262, 265, 266, 268, 271, 282, 285, *291*, 482, *502*
Friedrich, C. J., 282, *291*, 373, *419*, 762, *772*
Frisby, E. M., *138*
Froebel, J., 787
Frontier, in Geography, 19

Frontier Hypothesis, 519, 551
Frost, R., 124, *138*
Froude Number, 66
Fryer, D. W., 244, *291*
Fuchs, V. R., 412, *419*, 568, *601*
Fujita, T., 121, 127, *138*
Fulkerson, D. K., 568, *601*, 617, 618, *666*
Fulmer, J. L., 264, 268, *291*
Fulton, M., 373, *419*
Fultz, D., 106, 107, 112, 113, *138*, *139*, *142*, 744, 746, *772*, *773*
Functional Region, 305
Functional Organization, Areal, 390
Functionalism, 227
 and Change, 230
Furtado, C., 260, 262, 264, *291*
Fyot, J.-L., 260, *291*

Galileo, 780
Galloway, J. L., 114, *139*
Galpin, J. C., 470, *503*
Game Theory, 448
Gani, J., 173, *181*
Gannagé, E., 257, 281, *291*
Garbell, H. A., 119, *139*
Garfinkel, D., 534, 537, *541*
Garner, B. J., 304, 337, 338, 345-351, 353, *356*, *357*, 563, *606*, 649
Garnsey, M., 462, 477, *508*
Garrison, W., 304, 307, 322, 324-326, 326, 340, 342, 343, *355*, *357*, 363, 412, *419*, 445, *454*, 568, 588, *601*, *604*, 616, 634, 637-642, 655, 656, *666*
Gauss, C. F., 648
Geertz, C., 530, *541*
Gelernter, H., 50, 52, *56*
General Systems, 76, 326, 426, 526, 784
Generic Region, 483
Genre De Vie, 465
Geobiocenosis, 523
Geocybernetics, 537
Geodesics, 620
Geographers, Marxist, 220
Geographical Region, 469
Geography, Aim of, 549
 and Distance, 304
 Renaissance in, 597
 Values in, 219

Geometrical Properties, 34
Geometry, 609
  Circuit, 632
Geomorphic System, 76
Geomorphology, National Schools in, 59
Geopolitik, 518
George, F. H., 56, *56*, 147, *181*, 536, *541*, 593
George, W., 467, *503*
Geostrophic Wind Equation, 98
Geosystem, 35, 537
Gerard, R. W., *541*
Gerling, W., 430, *454*
Gerschenkron, A., 243, 285, 286, *291*
Gerth, H. H., 226, *241*
Gestalt Knowledge, 23
Gestalt-Field Psychology, 788
Getis, A., 304, 312, 357, 560, 563, 575, 576, 599, *601*, 605
Gettys, W. E., 522, *541*
Ghosh, A., 372, 386, *419*
Gibson, A. H., 739
Giguet, R., 153, *181*
Gilbert, E. W., 461, 481, *503*
Gilbert, G. K., 85, 86, *92*, 536, *541*
  on Dynamic Equilibrium, 85
Gilbert, W., 780
Gilchrist, B., 109, *139*
Gill, R., *56*
Gillen, F. J., 200, *215*
Gilmore, D. R., 260, *291*
Gilmour, J. S. L., 463, 464, 468, *503*
Ginsburg, L. B., 260, *291*
Ginsburg, N., 237, 243–245, *291*, *292*, 403, *419*, 638
Glacier Deformation, 737
Glade, W. P., 267, 273, *292*
Glass, D. V., 190, *214*, 536, *541*
Gleave, M. B., 374, *419*
Glover, R. E., 742, *772*
Glover, T. D., 522, *541*
Gluckman, M., 240, *241*
Gluss, B., 624, *666*
Godlund, S., 313, *357*, 563, *601*
Godson, W. L., 114, *139*
Goethe, J. W., 523
Goldman, T. A., 387, 398, *419*

Goldsmith, R. W., 243, 245, 249, 254, *292*
Gombrich, E. H. J., 683, 684, 706, 707, *721*
Goodall, G., *724*
Goode, J. P., 686
Goode, W. J., 199, *214*
Goodman, G. T., 531, *541*
Goodman, L. A., 582, *605*
Gordon, B. le R., 522, *541*
Gorilla gorilla berengei, 529
Gorupić, D., *419*
Goss, A., 366, *419*
Gottman, J., 342, *357*
Goubert, P., 190, 205, *214*
Gould, J., 226, *241*
Gould, P. R., 37, *40*, 400, *423*, 454, 506, 571, 593, 595, 603, 605, 607, 624, 638, 642, 666, 668
Gouldner, A., 232, 233, 235, *241*
Gouldner, H. P., *241*
Gourou, P., 449, 454, 530, *541*
Grade, in Rivers, 86
Graded Time, 88
Gradmann, R., 517, *541*
Grahn, D., 522, *542*
Granger, C. W. J., 555, 556, 599
Grant, E. L., 174, *181*
Graph Theory, 634
Gras, N. S. B., 249, 256, 284, *292*, 470, *503*
Gravier, J. F., *419*
Gravity Model, 369, 559
Gray, D. M., 632, *666*
Graybill, F. A., 24, 28, 32, *40*, 59, 68, 72–74, 81, 83, 93, 149, *182*, 635, *666*, 717, *722*
Great Circle, 752
Green, B. F., 50, *56*
Green, F. W. H., 129, *138*, 309, *357*
Green, L. P., 245, *292*
Greenberger, M., *600*
Greenhut, M. L., 364, 368, 376–381, 412, *419*, 568, *601*
Greer-Wootten, B., 536, *542*
Gregg, J. V., 555, 599
Gregor, H. F., 429, 430, *455*
Gregory, C. C., 462, 474, 487, 493, *505*
Gregory, C. L., 490, 492, *503*
Gregory, S., 490, *503*
Greig-Smith, P., 475, 490, 503, 512, *542*, 572, 574–576, 605
Gribaudi, F., 244, *292*
Grigg, D. B., 29, 463, 464, 468, 470, 479, 481, 485, *503*
Griggs, D. T., 735, *772*
Grigor'yev, A. A., 464, 467, *503*
Grimmer, M., 130, *139*
Grodins, F. S., 536, *542*
Gross, L., *241*
Gross Landscape, 63
Gross, N. C., 447, *457*
Grotewold, A., 443, 444, *455*, 711, *721*
Growth, Economic, 247
Growth of Information, 30
Growth Model, 569
  of Cellular Net, 652
Growth Pole, 281
Gulf Stream, Similarity to Jet Stream, 112
Gulley, J. L. M., 551, *598*
Gumbel, E. J., 152, *181*
Gunawardena, K. A., 37, *40*, 324, 325, *357*
Gupta, S. K., 22, 25, *39*, 597
Guthrie, J. A., 280, *292*
Guyot, A., 514, 515, 523, *542*

Habakkuk, J. H., 252, 253, *292*
Haberler, G., 271, *292*
Hack, J. T., 87, *92*, 165, 166, *181*, 528, *542*
Hadley, G., 100, 101, 106, 107, 113, *139*
Hadley Regime, 106, 107
Haeckel, E., *511*
Hagen, E. E., *241*, 251–253, 270, *292*
Hägerstrand, T., 38, *40*, 401, *419*, 447, 455, 561, 566, 570, 573, 585–591, 599, 601, 605, 607
Hägerstrand Model, 589
Haggett, P., 20, 32, 34–36, 38, *40*, 74, 76, 89, *91*, 237, *241*, 243, 248, 257, *289*, *292*, 304, 309, 310, 311, 315, 317, 325, 327, 328, 339, 343, 344, 357, 371, 390, 410, *419*, 444, 455, 509, 554, 555, 568, 599, 609, 610, 619, 628, 632, 634, 642, 644, 647, 648,

INDEX [803]

652, 654–664, *665*, *666*,
672, 692, 716–718, *720*,
721, 768, *772*
Hagood, M. J., 440, *455*, 462,
490, 492, 493, *503*
Hague, D. C., 382, *420*
Hahn, E., 437, *455*
Haig, R. M., 408, *420*
Haire, M., 610, *666*
Hairston, N. G., 535, *542*
Hajnal, J., 199, *214*
Hall, A. D., 532, *542*
Hall, P., 409
Hall, R. B., 469, *503*
Halley, E., 100, 101, 125
Haltiner, G. J., 134, 135, *139*
Hamill, L., 372
Hamilton, F. E. I., 366, 368,
382,–384, 410, 416, *420*,
563
Hamilton Line, 614
Hamilton's Puzzle, 614
Hammersley, J. M., 173, *181*,
584, 585, *605*
Hammond, E. H., 484, *503*
Hance, W. A., 245, *292*
Handscomb, D. C., 173, *181*,
584, 585, *605*
Hannan, E. J., 555, 556, *599*
Hanson, N. R., 20, *40*
 on Observation, 20
Haq, M. U., 249, 256, 262,
292
Hararay, F., 610, *666*
Harbaugh, J. W., 717, *722*
Harder, J. A., 742, *772*
Hardware Analogues, 613
Hardware Model, 24, 63, 727
 of Evaporation, 156
 of Rainfall Simulation, 154
Hare, F. K., 97, 110, 112, 130,
139, 462, 467, 468, *503*
Harley, J. B., 685, *721*
Harling, J., 582, 588, *605*
Harmonic Analysis, of Atmospheric Waves, 109
Harper, R. A., 285, *292*
Harris, B., 282, *292*
Harris, C. D., 304, 306, 330,
342, 357, 376, 377, *420*
Harris, D., 283, *292*
Harris, D. W., 529, *542*
Harris, J. E., 494, *503*
Harris, L. J., 676, *721*
Harris, R. A., 173, *181*
Harris, S. E., 248, 257, 267,
292

Harrison, M. J., 673, 674, *721*
Harrison, R. E., 676, *721*
Harrison, W., 82, 83, *92*
Hart, P. E., 254, 256, *293*
Hart, W. E., 63, 72, 84, *90*,
153, 165, 167, *179*
Hartley, G., 347, *359*
Hartman, G. W., 341, *357*
Hartshorne, R., 33, *40*, 243,
244, *293*, 441, *455*, 461,
470, 472, 481, 493, *503*,
511, 512, 532, *542*, 549,
550, *598*
Harvard Water Programme,
176
Harvey, D. W., 348, *357*,
445–448, *455*, 512, 564,
590, *601*, 607
Harvey, W., 780
Haswell, M. R., 437, *454*
Hatanaka, M., *599*
Hathaway, D. E., 262, *293*
Hatt, P., 474, 491, *503*
Hatt, P. K., 190, *214*
Hauser, P. M., *214*, 237, *241*
Haushofer, K., 518
Hawley, A. H., 352, *358*, 474,
521, 522, *542*
Hawthorn, G., 217
Hay, A. M., 262, *293*
Hayes, C. R., 338, *358*
Hays, S. B., 783, *796*
Hazard, of Floods, 175
Heacock, R. L., 56, *92*
Heady, E. O., 445, 446, 448,
454, 455, *601*
Heat Conductance Analogue,
742
Heaton, H., 396, *420*
Heberle, R., 462, *503*
Hegel, G., 781, 782
Helburn, N., 437, *455*
Held, C. C., 444, *458*
Hele-Shaw Model, 162
Helfrich, P., 531, *542*
Helix, 730
Heller, C. F., 427, 429, *455*
Hemming, M. F. W., 257, *293*
Hempel, C. G., 52, 227, *241*,
470, 482, *503*, 539
Hempkins, W. B., 32, *40*,
93
Henderson, J. M., 569, *601*
Henry, L., 199, *213*
Henshall, J. D., 33, 427, 436,
440, 441, 443, *455*, 563,
594, 690, *721*

Herbart, J. F., 787, 788
Herbartian Steps, 787
Herbert, D. J., *772*
Herbertson, A. J., 465–467,
469, 472, 481, 487, 499,
500, *504*, 516, 520, *542*
Herman, R., 618, 619, 624,
642, *666*
Herliny, D., 198, *214*
Hess, S. L., 98, 99, 133, *139*
Hess, S. W., 648, *668*
Hesse, M., 23, *40*
Hettner, A., 511, 512, 524,
542
Hexagon, 309, 756
Hiatt, R. W., 524, 526, *542*
Hicks, J. R., 245, 254, 260,
261, 263, 281, 285, *293*
Hicks, U. K., 285, *293*
Hide Annulus, 106
Hide, R., 106, *139*
Hidore, J. J., 718, *721*
Hierarchical Organization,
305
Hierarchies, of Settlements,
and Location, 389
Hierarchy, 489
 Blotting Paper Analogue,
761
 in Economy, 375
 of Human Ecosystems, 530
 of Natural Regions, 516
 of Settlement, 322
Higashi, A., 648, *665*
Higgins, B., 252, *293*
Highway System, 662
Hildebrand, B., 249
Hildebrand, G. H., 277, *293*
Hills, E. S., 62, *92*
Himes, N. E., *214*
Hiner, O. S., 368, 388, *420*
Hirsch, W. Z., 242, 256, 279,
293, 389, 414, *420*, 569,
601
Hirschman, A. O., 257, 260–
264, 273, 282, *293*
Historical Analogues, 61
Historical Geography, 519,
550
Historical Model, in Agriculture, 434
Historical Models, 22
History, and Industry, 395
Hitchcock, F. L., *420*
Hoag, L. P., *724*
Hobbes, T., 224, 515, 517
Hobsbawm, E. J., 252, *287*

Hoch, I., 642, *665*
Hoch, L. C., 373, *419*
Hochwald, W., 256, *293*
Hoel, P., *605*
Hoggatt, A. C., 584, *605*
Holdgate, M. W., 531, *543*
Holdridge, L. R., 497, *504*, 525, 526, 532
Holism, 518
Hollingshead, A. B., *542*
Hollingsworth, T. H., 199, *214*, 711, *721*
Holmboe, J., 114, *136, 139*
Holmes, C. D., 87, *92*
Holtan, H. N., 155
Holzman, F. D., 387, *420*
Holzner, L., 86, *92*
Hook, J. C., 578, *599*
Hoover, E. M., 211, *213*, 256, *293*, 362, 364, 372, 373, 375, 381, *420*, 443, *455*, *600*
  on Regional Growth, 256
Horton, R. E., 33, 63, 86, *92*, 156, 158–160, 165, 179, *181*, 625, 626, 628, 630, 654, 655, 664, *666*
  on Infiltration, 158
  on Morphometry, 165
Horton Ordering, 625
Horwood, E., 349, 350, *358*
Hoselitz, B. F., *240, 242*, 245, 249, 252–254, 284, 285, *293, 294*
Hoskins, W. G., 462, *504*
Hossell, C. H., 599
Hotelling, H., 377, *420, 772*
Houghton, D. H., 253, *294*
Houghton, R., *773*
Howard, A. D., 60, 77, 79, 80, 87, *92*
Howard, R. A., 571, *605*
Hoyle, B. S., 283, *294*
Hoyt, H., 334, 338, 342, 352, *358*
Hoyt, W. G., 30, *41*, 155
Hubbert, M. K., 66, *92*, 734, *772*
Hudson, J. W., *185*
Huff, D. L., 312, *358*
Hufschmidt, M. F., 175, *182, 183, 605*
Hughes, H. S., 227, *241*
Hughes, R. B., 261, 262, *294*
Hull, Theory of Learning, 43, 44
Hultland, G., 655, *666*
Human Ecology, 352, 474, 521

Humphrys, G., 260, 286, *294*
Hunter, J., 629
Huntington, E., 550, *598*
Hurd, R. M., 335, *358*
Hurricanes, 122
Hurst, H. E., 173, *182*
Huschke, R. E., 97, *139*
Hutchinson, G. E., 528, 536, *542*
Hutchinson, H. W., 430, *455*
Hutton, E. H., 24, *40*
Hutton, J., 62, *92*
Hvag, L. P., *458*
Hydraulic Geometry, 76
Hydraulic Society, 232
Hydrograph, Unit, 160
Hydrological Cycle, 84, 145–146
  Basin, 171
  Conceptual Model, 147
Hydrological Model, 738
Hydrology, Physical, 154
Hydrometeorological Analysis, 152

Ice Sheet Electric Analogue, 749
Ideal Type, of M. Weber, 225, 226
Idealization, 22
Idiographic Approach, 21, 243
Imbrie, J., 82, *92*, 690, *721*
Imhof, E., 686, 699, 700, 703, 713, *721*, 728, *772*
Individual, 490
  Geographic, 483
Inductive Generalization, 47
Industrial Inertia, 410
Industrial Location Model, 364
Industrial Migration, 410
Industrial Revolution, 31, 198, 211, 212, 396
Industry, Range of, 361
  World Distribution, 403
Inertia, Industrial, 410
Infanticide, 200
Infiltration, 158
Information, in Geography, 778
  Selection of, 23
Information field, 566
Information Flow, 595
Information Theory, 19

Inkeles, A., *241*
Innis, D. Q., 427, *455*, 522, *542*
Input, in Hydrology, 147
Input-Output Analysis, 255
Input-Output Model, 269, 445
Institut National D'Études Démographiques, 213
International Hydrological Decade, 147
Inter-Regional Analysis, 269
Inter-Tropical Convergence Zone, 119
Intervening Opportunity, 560
Invasion, 352
Invisible College, D. J. de S. Price on, 19
Ireson, W. G., 174, *181*
Iron and Steel, 415
Irwin, H., 652
Isallobaric Waves, 119
Isard, W., 246, 255, 256, 269, 270, 277, 278, 279, *294*, 306, 317–319, 327, 334, 335, 340, *358*, 367, 373, 376, 390, 398, 407, 413, *420*, 462, 493, *504*, 550, 555, 559, 563, 568, 569, 593, *598*, *599*, *601*, *607*, 748, *772*
Isarithmic Maps, 694
Isbell, J. R., 624, *666*
Iso-Cost Map, 620
Isodapane, 362, 764
Isohyets, 149

Jackson, B. G., 438, *455*, 690, *721*
Jacobs, W. C., 134, *139*
Jaffe, A. J., *214*
James, P. E., 244, 245, 283, *294*, 486, *504*, *508*, 517
Jeffreys, H., 69, *92*, 100, 102, *139*
Jenks, G. F., 692, 694, 702, *721*
Jenny, H., 513, 528, *542*
Jet Stream, and Ocean Currents, 112
  Synoptic Models, 118
Jevons, E., 240, *241*
Jevons, S., 487
Joerg, W. L. G., *508*
Johnson, A. F., 158, *183*
Johnson, A. I., 147, *182*

Johnson, D. B., 346, *356*
Johnson, D. H., 118, 126–128, *139*, *141*
Johnson, D. W., 62, 87, *93*
Johnson, F. C., 673, *721*
Johnson, H. M., 221, 222, 229, 230, *241*
Johnson, J. W., 742, *772*
Johnson, R. J., 493, *504*
Johnson, R. W. M., 429, *455*
Johnston, B. F., 252, *294*
Johnston, D. A., 685, *721*
Jonasson, O., 425, *455*
Jones, C. F., 425, *455*
Jones, E., 303, 342, *358*
Jones, G. R. J., 522, *540*
Jones, H., *539*
Jones, T. A., 484, *504*
Jones, W. D., 425, *456*, 490, *504*
Jope, E. M., *508*
Jouandet-Bernadat, R., 256, *294*
Judge, G. G., 445, *456*, 457
Judson, S., 166, *182*
Julian, P. R., 173, *182*
Junge, K., *720*

K-Value, Alexandersson's, 332, 333
  Christaller's, 308
Kahn, J. S., 73, *94*, 490, *505*, 556, 572, 575, 576, *599*, 605
Kaiser, H., 494, *504*
Kalesnik, S. V., 473, 481, 484, *504*, *542*
Kansky, K. J., 568, 587, *601*, 605, 621, 631, 634, 635, 639, 642, 644, *666*, 671, *721*
Kant, I., 781
Kaolin, 737
Kaplan, A., 19, 23, 25, 26, 38, *40*, 706, 712, *721*
Karel, C., 615, *667*
Karlsson, G., 589, 592, 595, *607*
Karplus, W. J., 738, 746, *772*
Kates, R. W., 175, *182*
Katona, G., 366, *420*
Katz, E., *607*, 637, 656, *668*
Kavesh, R., 335, *358*
Kearl, B., 453
Keates, J. S., 692, *721*
Keeble, D. E., 263, 286, *294*, 412, *420*, 569
Keir, M., 398, *420*

Keirstead, B. S., 264, *294*
Keller, F., 244, *295*
Kelley, E. J., 346, *358*
Kellogg, C. E., 484, *504*
Kelvin, Lord, see Thomson, W.
Kemeny, J. G., 577, 579–581, *605*
Kendall, D. G., 173, *182*, 605
Kendall, M. G., 440, *456*, 556, 570, 575, 587, 599, *605*
Kendeigh, S. C., 467, 468, *504*
Kendrick, M. P., 89, *95*
Kenn, M. J., 735, 736, *772*
Kennedy, B. A., 78, *93*
Kenyon, J. B., 391, 402, *421*
Kepler, J., 648, 780
Kershaw, K. A., 475, *504*, 512, *542*
Keuning, H. J., 426, *456*
Kidrić, B., 383, *421*
Kimble, G. H. T., 461, 466, *504*
King, C. A. M., 68, 82, *93*
King, J. L., *605*
King, L. J., 310, 311, 314, *358*, 440, 441, *455*, 690, *721*
King, P. F., 176, *182*
King, R. A., *456*
Kingsbury, R. L., 680, *722*
Kindleberger, C. P., 257, 270, 285, *295*
Kira, T., 513, 514, *543*
Kirchoff, G. R., 746
Kirk, D., 711, *722*
Kitiro, T., 716, *722*
Kitts, D. B., 62, *93*
Kjellen, R., 518, *543*
Klaassen, L. H., 257, 268, 269, *295*
Klare, G. R., 673, *721*
Klein, L. R., 255, *295*
Klein, W. H., 134, *140*
Kleinrock, L., 568, 584, *602*, 605, 617, *666*
Klemme, R. T., 366, *421*
Knickpoint Model, 738
Knighting, E., 135, *140*
Knos, D. S., 599, 692, 702, *721*
Knoss, D., 338, *358*
Koch, A. R., 388, 417, *421*
Koffka, K., 788, *792*
Kogan, M., *603*
Kohler, M. A., 155, 159, 169, 171, *183*
Köhler, W., 788, *792*

Kohn, A. J., 531, *542*
Kolb, C. R., 63, 65, *96*
Kolb, J. H., 312, *358*
Kolb, W. L., 226, *241*
Kollmorgen, W., *508*
Kolossovsky, N. N., 407, *421*
Kondracki, J., 473, *504*
Konecci, E. B., 534, *543*
Königsberg Bridge Problem, 614
Koopmans, T. C., 362, 387, *421*
Köppen, W., 467, 468
Korbel, J., *600*
Kortus, B., 408, *421*
Koteswaram, P., 124, *139*
Kratchman, J., 522, *542*
Krishnamurti, T. N., 107, 112, *140*
Kroeber, A. L., 462, 476, 477, *504*
Kroeber, C. B., 551, *598*
Kroft, W. C., 257, 268, *295*
Krueger, A. O., 569, *601*
Krumbein, W. C., 24, 28, 32, 40, 59, 68, 72–74, 81–83, 92, 93, 149, *182*, 438, *456*, 635, *666*, 695, 716, 717, *722*
  Experimental Design, 73
  Beach Sand Firmness, 74, 75
Krutilla, J. V., 153, *182*, *185*
Kryżanowski, W., 370, *421*
Krzywicki, L., 200, 214
Ku, Y. U., 656, *666*
Küchler, A. W., 467, 481, 482, *505*, *508*
Kuenen, P. H., 735, *772*
Kuenne, R. E., 278, 279, *294*, *295*, 363, 413, *420*, *421*
Kuhn, H. W., 363, *421*
Kuhn, T. S., 26, 27, 35, 37, 38, *41*, 511, 512, *543*, 782, *792*
  on Paradigms, 26, 27
Kuiper, E., 153, *182*
Kuiper, G. P., *92*
Kuipers, A., 22, *41*
Kuznets, S., 247, 253, 254, 264, 270, 274, 275, *295*
Kwizak, M., 110, *136*

Labour Coefficient, 371
Lachenbruch, A. H., 653, 654, *666*
Lacoste, Y., 243, *295*

[806] INDEX

Lacroix, J. L., 282, *295*
Lagrange Multipliers, 613
Lamb, H. H., 109, 130, *140*
Lambert, J. M., 481, *505*, 531, *541*
Lampard, E. E., 284, 285, *295, 602*
Land, A., 616, *665*
Land Market, Urban, 335
Land Potential, 449
Land Use, 530
   Concentric Model, 339
   Nuclei Model, 342
   Sector Model, 341
Land Value Surface, 336
Landahl, H. D., 571, 592, *605, 607*
Lane, T., 278, *295*
Lang, S. M., 147, *182*
Langbein, W. B., 30, *41*, 71–73, 79, 90, *93*, 155, 631, *644, 666*
Langdale-Brown, I., 707, *722*
Langhaar, H. L., 66, *93*, 740, *772*
Langham, M. R., 448, *456*
Language of Science, 46
Lanham, W. J., *456*
La Piere, R. T., *241*
Larimore, A., 285, *295*
Lasareff, P., 743–745, *773*
La Seur, N. E., 111, 121, *140, 142*
Lasuen, J. R., 261–263, *295*
Latham, J. P., 56, *93, 456*
Lattice, Regular, 306, 310
Laughery, K., 56
Launhardt, W., 362–364, 369, 370, 372, 398, 416, *421*
Law of Entrance Angles, 628
Law of Force, Coulomb's, 757
Law of Minimum Effort, 304
Law, of Motion, Newton's, 98
Law of Path Area, 628
Law of Path Flow, 628
Law of Path Lengths, 628
Law of Path Numbers, 628
Law of Retail Gravitation, 313, 757
Law, Wind, Buys Ballot's, 98
Laws, 24
   as Models, 24
   of Thermodynamics, 97–98
Leader-Follower Model, 596
Learning Models, 785
Learning Processes, 594
Least-Cost Location, 370, 766

Least Cost Model, 369
Least Cost Path, 620
Lebedev, V. G., *508*
Le Cren, E. D., 531, *543*
Lee, D. H. K., 522, *543*
Lee, R. B., 529, *543*
Lee, Shu-Tan, *508*
Leeds, J. R., 257, *299*
Leet, L. D., 166, *182*
Lefeber, L., 388, *421*
Le Heux, J. W. N., 70
Lehmann, O., 70
Leibenstein, H., 190, *214*
Leighly, J., 131, *140*, 461, *505, 508*
Lempfert, R. K. G., 114, *143*
Lenin, V. I., 383, *421*
Leontief, W., 255, 269, 275, *296*, 445, *456*
Leopold, L. B., 71–76, 78, 79, 90, *93*, 145, 150, 165, 166, *182*, 621, 623, 628, 631, *644, 666, 667*
Lesczycki, S., 473, *502*
Leviquin, M., 405, *421*
Levy, M. J., 190, *214*
Lewin, K., 788, *792*
Lewis, C. S., 776, *792*
Lewis, D. J., 176, *182*
Lewis, W. V., 67, 68, *93*, 702, *722*, 737, 738, *773*
Lewontin, R. C., *41*
Lewthwaite, G. R., 564, *602*
Lichty, R. W., 60, 80, 87, 88, *95*
Life Zones, Merriam's, 467
Lighthill, M. J., 609, 624, *667*
Lilly, D. K., 743, *774*
Lind, A. W., 522, 529, *543*
Lindberg, J. H., *723*
Lindbergh, O., 376, *421*
Lindeman, R. L., 524, 534, *543*
Lindman, R., 692, *720*
Linear Assignment, 387
Linear Pattern, 306
Linear Programme, Koopmans, 387
Linear Programme Models, Danger of, 388
Linear Programming, 256
Line-Set Problem, 614
Link-Length Minimization, 613, 768
Linsley, R. K., 155, 159, 167–169, 171, *180, 183*

Linton, D. L., 62, 87, *96*, 484, *505*, 681, 698, *722*
Lissowski, W., 382, *421*
List, F., 249
Little, J. D. C., 615, *667*
Lively, C. E., 462, 474, 487, *493, 505*
Livestock Ratio, 437
Livi, C., 262, 264, *289*
Livsic, R. S., 383, *421*
Location, Least Cost, 370
   Minimax, 379
   Optimum, 362
   Personal Factor, 366
   and Regions, 484
   Relativity of, 33
Location Equilibrium, 265
Location of Industry Model, 364
Location Models, Hardware, 762
Location and Settlement Hierarchy, 389
Locational Interdependence, 377
Locke, J., 787
Lockwood, D., 226, 232, 233, *241, 242*
Lockwood, J. G., 124, *140*
Locusts, Spread of, 193
Loeb, J., 512, *543*
Logical Division, 480
Lombardini, S., 285, *296*
London, 238
Long, R. R., 106, 109, 123, *140*
Long Waves, Theoretical Models, 107
Lonsdale, C., 260, *296*
Lonsdale, R. E., 407, *421*
Loria, A., 369, 370, 378, *421*
Lorimer, F., 190, 200, *214*
Lösch, A., 37, 189, 212, 282, *296*, 304, 306, 309, 313, 315–318, 322, 324, *358*, 362, 377–379, 381, 389, 391, *421*, 443, 444, *456*, 498, *505*, 563, 569, 585, 587, *602*, 618, 619, 642, 647–649, *667*
Location Model, 315
on Refraction, 618
Lotka, A. J., 190, *214*, 569, *602*
Lövgren, E., 412, *421*, 566, *602*
Lowenstein, L. K., 409, *421*

Lowry, I., 560, 564, 585, *599*, 602, 605
Lowry, R. L., 158, *183*
Lubowe, J. K., 628, 629, *667*
Ludlam, F. H., 128, 129, *136*, *140*
Lukermann, F., 500, *505*
Lund, I. A., *140*
Lundquist, G., 684, 687, 688, *722*
Luthin, J. N., 156, *183*
Luttrell, W. F., 366, 412, *422*
Lutz, V., 262, *296*
Lyle, H. H., 110, *140*
Lynch, K., 712, *722*
Lyons, H. G., 682, 697, *722*
Lyusternik, L. A., 611, *667*

Maass, A., 153, 174, *183*, 583, 584, 587, *605*
Mabogunje, A. L., 262, *296*
MacArthur, R., 535, *543*
Mace, A., 277, *293*
MacFadyen, A., 531, *543*
Machlup, F., *607*
Mackay, J. R., 477, 492, *505*, *508*, 681, 694, 696, 702, *722*, 749, *773*
Mackinder, H. J., 521, 532, *543*, 551, *598*
Maddock, T., 74–76, 78, *93*
Magee, B., 249, *296*
Magnetic Analogue, 756
Mahalanobis, P. C., 255, *296*
Makar, R., 748, *771*
Maki, W. R., 256, *296*
Maling, D. H., 684, 686, *722*
Malkus, J. S., 120–122, *140*, *142*
Malthus, T. R., 190, 191, 194, 196, 207, 212, 436, 569
Malthusian Ceiling, 198
Malthusian Precipice, 204
Malthusianism, 190
Manabe, S., 100, 105, *140*
Manil, G., 481, *505*
Manley, G., 19, *41*
Manley, V. P., 437, *456*
Manners, G., 260, 263, *296*, 366, *422*
Mansell-Moullin, M., 151, 152, *180*
Mansfield, E., 592, 595, *607*
Map Bias, 676
Map Colouring Problem, 647
Map Generalization, 687
Map of Geomorphic Activity, 60, 61
Map Making, 674
Map Projection, Choice of, 680
Map Purpose, 681
Map User, 680
Map-Model Cycle, 673
Mapping, Growth of, 30
Maps, 46, 48
  as Models, 672
  as Static Models, 48
Marble, D. F., 342, 343, *357*, 445, *454*, 561, 568, *599*, *601*, 614, 634, 638–642, 655, 656, *666*
Marbut, C. A., 513
March, J. G., 364, *422*, 595, 596, *607*
Margalef, R., 535, *543*
Marglin, S. A., *183*, 605
Market Area Analogue, 760
Market, Urban Land, 335
Markov Chain Model, 401, 571, 577
Markov Process, 72
Markus, E., 523, *543*
Maron, M. E., 51, *56*
Marriage, Age at, 199
Marriott, W., 133, *135*
Marschak, T., 375, 386, 387, *417*
Marschner, F. J., 681, *722*
Marshall, A., 224, 536, *543*, 550
Marshall, D., *773*
Martin, A., 431, *456*
Martin, F. L., 134, 135, *139*
Martin, J. E., 375, 381, 409, 412, *422*
Martin, J. R., 522, *543*
Martin, W., 286, *296*
Maruyama, M., 536, *543*
Marx, K., 234, 235, 369, 381, 403, *422*, 782
Marxism, 234
Marxist Geographers, 220
Mass, Conservation of, 98
Material Index, 371
Mathematical Models, 68
  of Development Pattern, 268
  of Atmospheric Circulation, 103
  Deterministic, 69
  of Economic Development, 254
Mather, E. C., 437, *456*
Mather, J. R., 132, *140*
Matrix Algebra, 635
Matrix, Pressure, 130
Mattern, J., 517, *543*
Mattila, J. M., 268, *301*, 414, *422*
Maturation, 785
Maull, O., 493
Maull's Girdle Method, 493
Mautner, A. J., 569, 570, 592, *602*, *608*
Max Flow-Min Cut Theorem, 619
Maximum-Flow Path, 616
Maximum Profit, 378
Maxwell, J. C., 648
Mayer, H. M., *185*, 324, 338, 346, *355*, *358*
Mayfield, R. C., 324, *358*
McCarthy, P. J., *603*
McCarty, H. H., 425, *456*, 561, *599*
McCaskill, M., 402, *422*
McCauley, J. F., 59, *94*, 95
McClellan, D. E., 114, *136*
McClure, J. A., *456*
McConnell, H., 73, *94*
McDonald, J. R., 465, *505*
McGovern, P. D., 263, *296*
McIntosh, R. P., 523, *543*
McKean, R. N., 153, 174, *183*, 530, *543*
McKee, E. D., 738, 742, *773*
McKenzie, R. D., 462, 470, 474, *505*, 521, *543*, 544, 545
McLaughlin, G. E., 263, *296*, 366, 412, *422*
McLoughlin, P. F. M., 248, *296*
McMillan, C., 523, 532, 544
McNee, R. B., 245, 246, 248, *296*
Mead, W. R., 673, 674, *721*, *722*
Meadows, P., 22, 24, *41*, *242*
Meanders, 621
Mechanical Location Model, 764
Meier, G. M., 243, 249, 251, 252, 267, 271, *296*, 536, 544
Meier, R. L., 595, *607*
Meinzer, O. E., 159
Melamid, A., 271, *296*, 564, *602*
Mellor, J. W., 252, *294*
Melton, F. A., 62, *94*

Melton, M. A., 72–74, 77, 79, 81, 82, *94*, 626, *667*
Mental Discipline, 786
Menzel, H., *607*
Merriam, A. P., 467, *508*
Merriam, D. F., 707, 717, *722*
Merriam, M., *723*
Merritt, E. S., 122, *140*
Merton, R. K., 222, 231–233, *242*
Meta-Model, 537
Metaphor, 551
Meteorological Analogue Models, 743
Metropolis, 408
Meyer, J. R., 245, 246, 256, 267, *289*, *296*
Meyers, F., 260, *297*
Meynaud, J., 243, 252, *297*
Miehle, W., 611–613, 638, *667*, 769, 770, *773*
Migration, 560
 Industrial, 410
 in Sweden, 566
Mikheyeva, V. S., *456*
Miles, M. K., 114, *140*
Milhau, T., 257, 281, *297*
Mill, J. S., 369, 486, *505*
Miller, B. R., *456*
Miller, J. P., 73, 76, *93*, 145, 165, 166, *182*, 621, 628, 631, *667*
Miller, M. M., 67, 68, *93*, 737, *773*
Miller, O. M., 713, *723*
Miller, R. L., 83, 89, *93*, *94*, 490, *505*, *556*, 572, 575, 576, *599*, *605*
 and J. M. Zeigler, Beach Process Model, 83, 89
Mills, C. W., 226, *241*, *242*
Mills, G., 254, 256, *293*, *297*, 584, *605*
Minas, J. S., 22, 25, *39*, *597*
Minimal-Cost Path, 617
Minimax Location, 379
Minimization, Link-Length, 613
Misulia, M. G., 728, *773*
Model, Abuse of, 499
 as Analogy, 23
 Defined, 22, 552
 Experimental, 426
 Internalized, 218
 Nature of, 494
 Properties of, 22
 Value of, 500

Model Drainage Basin, 149
Model for Geomorphology, 60, 61
Model of Models, R. J. Chorley on, 25
Model for Regional Systems Analysis, 36
Model Rivers, 738
Model System, 554
Model-Building, 552
 Pitfalls in, 26
Model-Testing, in Geomorphology, 89
Models, as Approximations, 23
 Dynamic, 43, 49, 52
 Experimental Circulation, 105
 Functions of, 24
 Hardware, 63, 145
 Historical, 22
 as Laws, 25
 Nature of, 21
 Scale, in Hydrology, 145
 Spatial, 22
 Static, 43, 48
 Stochastic, and Climatic Change, 133
 Structure of, 23
 Suggestive, 23
 in Teaching, 727, 775
 Types of, 25
Modernization, 236
Mohr, C. O., 534, *544*
Moles, A. A., 671, 713, 714, *723*
Momsen, R. P., 621, *667*
Monge, C., 522, *544*
Monmonier, M. S., 676, *723*
Monsoons, 123
Monte Carlo Simulation, 72, 352, 401, 571, 582, 644
Monteith, J. L., 157, *183*
Moore, E., *242*
Moore, E. F., 615, *667*
Moore, F. T., 279, *297*
Moore, P. G., 575, *605*
Moore, R. K., 56, *94*
Moore, W. E., 234–236, *242*
Moran, P. A. P., 173, *181*, *183*
More, R. J. M., *164*, 617
Morgan, F. W., 461, *505*
Morgan, J. N., 366, *420*
Morgan, M. A., 85, *91*, 582, 613, 649, 728, 729, *772*
Morgan, W. B., 523, *544*
Morgenstern, O., 448, *456*

Morisita, M., 575, *606*
Morphological Mapping, 62
Morphology, of Landscape, 518
Morphometry, 165
Morphotectonics, 62
Morrice, H. A. W., 176, *183*
Morrill, R. L., 352, 357, *358*, 400, *423*, 555, 560, 566, 570, 585–588, *599*, 600, 602, *603*, *606*, 642–646, *667*, 668
Morrissett, I., 333, *358*
Morse, C., 234, *242*
Mortality, 189
Mörth, H. T., 100, 126–128, *139*, *140*
Moser, C. A., 331, *358*
Moses, L. N., 279, *297*, 372, 416, *422*
Moss, R. P., 523, *544*
Mosteller, F., 594, *607*
Motion, Newton's Law of, 98
Mott, P. G., *723*
Mountjoy, A. B., 243, 245, *297*
Movement-Minimization, 304
Mover-Stayer Model, 582
Mowrer, E. W., 493, *505*
Mryzglód, T., 381, 382, 384, 385, *422*
Muhsham, H. V., 577, *606*
Muller, C. H., 513, *544*
Multiplier Model, 413
Multi-Purpose Scheme, 153
Mumford, L., 368, 396, *422*
Munger, E. S., 274, *297*
Murchland, J. D., 616, *665*
Murphy, F. C., *183*
Murphy, N. F., 66, *94*
Murphy, R. E., 354, *358*
Murray, C. D., 629, *667*
Murray, R., *141*
Murty, K. G., 615, *667*
Musgrave, G. W., 159
Muskat, M., 737, *773*
Muth, R. F., 344, 345, *358*, 564, 568, *602*
Myint, H., 253, 271, *297*
Myrdal, G. M., 234, 257–264, 270, 273–275, 285, *297*, 384, 389, 405, *422*, 564, *602*
Myrdal Model, 234, 258

n-Butanol, 759
Nagel, E., 21, *41*, 512, 519, *544*

INDEX [809]

Namias, J., 111, 133, 134, *136*, *141*
Nash, J. E., 167, *183*
Natural Analogue System, 60
Natural Model, in Geomorphology, 62, 63
Natural Region, 465, 469, 516
Natural Selection, 191
Natural Unfoldment, 787
Naturkomplex, 523
Navier-Stokes Equation, 162
Nearest-Neighbour Analysis, 310, 575
Needleman, L., 260, *297*
Neel, R. B., 534, *544*
Negative Binomial Distribution, 574
Negative Feedback, 78, 86, 536
Neighbourhood Centre, 348
Nelson, H. J., 346, *357*, *358*
Nelson, M. L., 176, *184*
Nelson, P., 566, *602*
Neolithic Revolution, 203
Nesting, of Birds, 192
Nets, Bounded, 649
 Unbounded, 648
Network, Cellular, 646
 Geographical, 609
Network, Linear, 34
 Types of, 609
Network Model, 568, 768
Network Transformation, 654
Neumark, D. S., 271, *297*
Nevin, C. M., 738, *773*
New Geography, G. Manley on, 19
Newbould, P. J., 531, 532, *544*
Newell, A., 50, 52, *56*
Newlyn, W. T., 285, *293*
Newman, M. T., 522, *544*
Newton, C. W., 112, *141*
Newton, I., 780
 Laws of Motion, 98
Neyman, J., 68, *94*, 572, 574, 575, *606*
Neyman Type A Distribution, 574
Nicholls, W. H., 281, 286, *297*
Nicholson, I., 261, *297*
Nikiforoff, C. C., 513, 528, *544*
Nodal Region, 305, 470
Node, 479
Noise, 20

Artificial, 698
 in Maps, 698
 Reader, 701
 in Weber Model, 373
Noma, A. A., 728, *773*
Nomothetic Approach, 21, 243
Norling, G., *598*
Norman, R. Z., 610, *666*
Normative Function of Models, 24, 25
Normative Models, in Agriculture, 426
Normative Systems, 223
North, D. C., 253, 266–268, 280, 282, *297*
Nucleii Models, of Land Use, 342
Nuptiality, 189
Nurkse, R., 274, *297*, 536, *544*
Nystuen, J. D., *357*, 536, *544*, 599

Ocean Current Model, 743
O'Connor, A. M., 258, 261, *297*
Odell, P. R., 368, *422*
Odhnoff, J., 364, 387, *422*
O'Donnell, T., 147, 153, 160, 167, 170, 171, *181*, *183*
Odum, E. P., 523, 526, 527, 531, 534, *544*
Odum, H. T., 526, 527, 531, 533–536, *544*
Odum, H. W., 476, *505*
Ogawa, H., 513, 514, *543*
Ogburn, W. F., *508*
Ohlin, G., 249, 253, *297*
O'Kelly, J. J., *183*
Okun, B., 262, 263, 274, *298*
Olivier, H., 158, *183*
Olmstead, C. W., 437, *456*
Olsen, G. H., 748, *773*
Olson, J. S., 534, 537, *544*, 545
Olsson, G., 37, *41*, 304, 312, 313, *359*, 559, 560, 566, 587, 600, 602
Ooi Jin-Bee, 245, *298*, 433, 456
Open System, 528
Optimal System, 629
Optimization, Automatic, in Hydrology, 170
Optimum Path, 615
Orchard, J. E., 245, *298*

Orcutt, G. H., 246, 254, *298*, 552, 562, 571, 584, 600, 602, 606
Order, Hobbes' Problem of, 224
 Kinds of, 20
 in Spatial Behaviour, 304
Ore, O., 614, 615, 624, 625, 646, 647, *667*
Oregon Trail, 621
Organic Analogy, in Geography, 512
Organic Relationships, 23
Organic Theory, of States, 515
Organization, 470
Organism, 471, 472, 551
 of Industrial Plants, 407
 in Plant Ecology, 513
 and Social Structure, 227
Origin of Species, 511
Osborn, D. G., 400, *422*
Ostwald, W., 702
Otremba, E., 437, *457*
Output, in Hydrology, 147
Ovington, J. D., 531, *545*
Oxford School, 520

Paauw, D. S., 246, 249, 257, 261, *298*
Padilla, E., 431, *457*
Pahl, R. E., 238–240, *242*, 286, *298*
Paish, F. W., 257, *298*
Palander, T., 362, 398, *422*
Paleoclimatic Model, 744
Paleoclimatology, 131
Palmén, E., 103, 108, 113, *141*
Palmer, C. E., 118, 119, 121, *141*
Pankhurst, R. C., 740, *773*
Pannett, D. J., 674, *721*
Paradigm, 35, 511, 782
 Model-Based, 33
 Nature of, 26
 Recognition of, 37
 Classificatory, 28
Parameter Definition, 587
Pareto, V., 224
Park, R. E., 462, 470, 474, *505*, 521, 522, *545*
Parr, J. B., 262, 283, *298*
Parsons, T., 221–229, 232–237, *242*, 782, *792*
 Theory of Action, 223, 228
Partial Systems, 84

Path, Direct, 611
  Shortest, 615
  Wandering, 621
Path Geometry, 610
Path Ordering, 659
Patten, B. C., 535, 545
Pattern, of Economic Development, 257
  Evolution of Spatial, 561
  of Population, 189
  of Rainfall, 149
  of Settlement, 306
  of World Industry, 403
Pattern Recognition, 20
Pattern Seeking, 23
Paulhus, L. H., 155, 159, 169, 171, 183
Pays, 465
Peacock, A. T., 270, 287
Pearcy, G. E., 518, 540
Pearl, R., 569, 602
Peasant Agriculture, Model, 430
Pédelaborde, P., 119, 141
Pedgley, D. E., 114, 141
Pedology, 513
Peel, D. A., 176, 182
Peixoto, J. P., 110, 143
Peltier, L. C., 86, 94
Pen, J., 243, 246, 248, 298
Penck, A., 687
Penck, W., 87, 94
Penman, H. L., 147, 148, 156–158, 183
Penner, C. M., 114, 141
Perel'man, A. I., 523, 532, 545
Periodogram Analysis, 556
Perloff, H. S., 261, 262, 266, 268, 280, 298, 566, 569, 602
Perring, F. H., 32, 41
Perroux, F., 281–283, 298
Perry, T., 551, 598
Personal Factor, in Location, 366
Persson, A., 304, 312, 313, 359
Persson, A. V., 112, 141
Perth Amboy, 66
Pestalozzi, J. H., 787
Petersen, J. W., 279, 297
Peterson, J. M., 642, 667
Pettersson, S., 98, 99, 114, 115, 141
Pfister, R. L., 279, 280, 298
Pfouts, R. W., 389, 422
Philbrick, A. K., 305, 359, 390, 422, 489, 505, 659, 667

Philip, G., 686
Phillips, J. F. V., 522, 545
Phillips, N. A., 105, 112, 141
Philosophy, Mediaeval, 776
Physical Science, Growth of, 780
Physical Systems, in Geomorphology, 63
Physiology, Environmental, 522
Picken, L. E. R., 520, 545
Pico, R., 430, 457
Piexoto, J. P., 74, 95
Pigou, A. C., 174, 183
Pillewizer, W., 687, 723, 724
Pinkerton, R. C., 526, 534, 545
Pisharoty, P. R., 124, 141
Pitts, F. R., 33, 590, 607, 635–638, 667
Planning, 382
  in Britain, 220
  Resource, 152
Plant Ecology, 513
Plant Location, 386
Plantation, 429
Plateau, F., 648
Plato, 517, 778, 786
Platt, R. S., 426, 457
Platzman, R. L., 548
Plaxico, J. S., 457
Point Symbols, 696
Point-Set Problems, 611
Point-Set Theory, J. P. Cole on, 34
Poisson Processes, 572
Polar Furrows, 651
Pole, Growth, 281
Political Divisions, as Cells, 648
Political Geography, F. Ratzel on, 517
Pollack, L. W., 728, 729, 772
Pollack, M., 615, 667
Pollock, N. C., 283, 298
Polonskiy, M. L., 537, 545
Polopolus, L., 388, 422
Polya, G., 611, 667
Polycentric Planning, 382
Polygons, Clay, 733
Polystyrene, 732
Polystyrene, 732
Ponsioen, J. A., 242
Poore, M. D., 479, 505
Popper, K., 52
Population, 189
  Animal, 191

Population Density, and Distance, 318
  Urban, 343
Population Density Potential, 450
Population Explosion, 193
Population Lapse-Rate, 318
Population Model, 576
Population Optimum, 196
Population Potential, 749
Population Pressure, 197
Porter, P. B., 20, 41
Porter, P. W., 695, 723
Porter, R., 566, 602, 606
Positive Feedback, 78, 536
Postan, M. M., 198, 204, 214
Postdictive Simulation Model, 642
Pounds, N. J. G., 383, 422
Power-Spectra Analysis, 621
Pownall, L. L., 330, 359
Prager, W., 642
Pratt, R. S., 508
Pratten, C., 279, 298
Prebisch, R., 271, 273, 298
Precipitation Simulation, 154
Pred, A., 258, 261, 262, 280, 282, 298, 307, 309, 339, 348, 355, 359, 368, 389, 390, 400, 413, 422, 552, 595, 598, 607
Pred Model, 400
Prediction, 46
  Hydrological, 150
  Weather, 133
Pressat, R., 214
Pressure, Atmospheric, Tendency, 98
Prest, A. R., 174, 183
Price, A. G., 551, 598
Price, D., 490, 503
Price, D. J. De S., 19, 30, 41
Price, W. A., 89, 95
Pricing Systems, 380
Priestley, C. H. B., 102, 142
Prigogine, I., 528, 536, 545
Prihar, Z., 638, 667
Primate City, 326
Principle of Least Effort, 304
Process of Change, 550
Process Model, in Spatial Pattern, 564
Process-Response Model, 63, 76, 83
Production Ecology, 531
Productivity, Biological, 526
Profit Maximization, 370, 378

INDEX [811]

Prokay'ev, V. I., 484, 505, 523, 545
Propositional Calculus, 44
Protestant Ethic, 225
Prothero, R. M., 445, 457
Pumpelly, J. W., 707, 723
Purdy, E. G., 690, 721

Quadrat Sampling, 572
Quadratic Assignment, 387
Quastler, H., 535, 536, 545, 548
Queney, P., 109, 142
Queue Theory, 133, 584
  in Hydrology, 153
Quinn, J. A., 378, 422, 474, 522, 545

Radcliffe-Brown, A. R., 227, 242
Radner, R., 593, 608
Railroad Location, 613
Raimon, R. L., 566, 602
Rainfall Dispersal, 155
Rainfall Simulation, 154
Rainio, K., 571, 595, 606, 608
Raisz, E., 712, 723
Ramakrishnan, K. P., 124, 125, 142
Ramanthan, K. R., 124, 142
Ramaswamy, C., 124, 142
Randall, L., 262, 298
Random Walk Model, 73, 631
Ranis, G., 253, 299
Rank-Size Regularity, 326
Rao, V. K. R. V., 270, 271, 299
Rao, V. L. S. P., 491, 493, 506
Rapaport, H., 615, 667
Rapoport, A., 551, 569, 592, 597, 598, 602, 608
Rappoport, A., 536, 541
Rasmusson, G., 655, 667
Ratcliff, R. U., 335, 359
Ratzel, F., 517, 518, 545, 546, 550, 551, 598
Raup, H. M., 467, 468, 506
Ravenstein, E. G., 566, 602
Ray, T. L., 388, 418
Real World, 704
Rebhun, L. I., 592, 608
Recurrence Interval, 152
Redfield, R., 430, 457
Reeds, L. G., 425, 452, 457
Rees, A. M. M., 431, 454

Reeves, E. A., 700, 723
Region, Crop-Combination, 439
  Generic, 483
  Industrial, 406
  as Model, 494
  Multi-Feature, 493
  Natural, 496
  Single-Feature, 467, 492
  Specific, 483
  Uniform, 470
Regional Centre, 348
Regional Concept, 461, 464
Regional Geography, 248
Regional Income Inequality, 257
Regional Models, 29
Regional Multiplier Concept, 275
Regional Planning, 220
Regional Science, 477
Regional Systems, Model for Analysis of, 36
Regionalization, Methods, 477
  Principles, 485
Regions, as Areal Classes, 481
  Physiographic, 62
Regular Cluster, 312
Regular Lattice, 306, 310
Reichenbach, H., 52, 559, 600
Reilly, W. J., 313, 359, 773
Reilly's Law of Retail Gravitation, 757
Reiner, M., 737, 773
Reiner, T., 246, 294
Reiter, E. R., 112, 142
Relaxation Time, 79
Relief Model, 728
Remane, A., 520, 546
Remote Sensing, 59, 147
Renaissance, in Geography, 597
Renner, G. T., 476, 506, 522, 548
Rent, 305
  of Land, 444
Rent Curve, 340
Research Frontier, in Geography, 19
Resistance to Diffusion, 590
Resistance Mesh, 746
Resource Planning, 153
Retail Gravitation, Law of, 313
Revelle, R., 177
Revolution, Industrial, 396
  Neolithic, 203

Theory of, 234
Rex, D. F., 132, 142
Rex, J., 217, 221, 227–229, 242
Reynolds, D. J., 618, 665
Reynolds, O., 102
Reynolds, R. B., 479, 506
Reynolds' Number, 66
Rhodes, T. C., 347, 359
Ricardo, D., 368, 532
Rice, E. K., 742, 772
Richards, A. W., 732, 771
Richardson, J. T., 599
Richardson, L. F., 134, 142
Richardson, R. W., 262, 263, 274, 298
Riddiford, C. E., 699, 723
Riehl, H., 103, 106, 111, 112, 118–123, 131, 140, 142, 746, 773
Riemann, B., 648
Riemer, S., 508
Riordan, J., 626, 668
Ripple Tank, 741
Ritter, C., 30, 31, 514, 515, 523, 546
Ritter, W. E., 515, 546
River Profile Simulation, 753
Rivlin, A. M., 600
Robbins, H., 769, 772
Roberts, H. V., 556, 600
Roberts, J. A., 713, 723
Robinove, C. J., 770, 773
Robinson, A. H., 470, 506, 680–682, 686, 692, 696, 715–718, 723
Robinson, E. A. G., 247, 254, 299
Robinson, G. W. S., 461, 506
Robock, S., 256, 262–264, 296, 299, 366, 412, 422
Rockwood, D. M., 176, 184
Rodwin, L., 283, 299
Rogers, E. M., 447, 453, 457, 589, 590, 608
Romus, P., 260, 299
Roscher, W., 370, 422
Rose, A. M., 238, 242
Rosenblatt, M., 556, 600
Rosenblueth, A., 25, 41
Ross, E. A., 374, 410, 423
Rossby, C.-G., 101, 102, 105, 108, 110–113, 119, 134, 142, 143
Rossby Regime, 106, 107
Rossby Waves, 108

Rostow, W. W., 38, 234, 245, 249–254, 256, 257, 281, 299, 536, 546
Rostow Model, 234, 249
Criticism, 252
Rotating Dishpan, 106
Roterus, V., 334, 359
Rouse, J. W., 56, 95
Rousseau, J.-J., 787
Roux, W., 628, 629, 663, 668
Row Vector, 28
Rowan, L. C., 56, 95
Rowe, J. S., 546
Roxby, P. M., 469, 506
Rozin, M. S., 473, 506
Ruhe, R. V., 630, 631, 668
Run-off Analogue Model, 742
Rural Sociology, 474
Rushton, S., 569, 570, 592, 602, 608
Russell, B., 44
Russell, J. C., 563, 602
Russia, Centrality in, 636
Russian Geography, 473
Russwurm, L. H., 501
Ruttan, V. W., 184, 253, 281, 286, 299, 368, 423
Ryan, B., 457
Ryther, J. H., 531, 546

Saaty, T. L., 571, 606, 610, 634, 665
Sabbagh, M. E., 110, 143
Sachet, M.-H., 529, 546
Sack, R., 534, 541
Salinity Control, 176
Salt Marsh, Productivity, 526
Saltzmann, B., 95, 110, 143
Samuelson, P. A., 771, 773
Sandee, J., 256, 299
Sanders, F., 97–99, 131, 144
Sansom, H. W., 114, 143
Santos, M. C., see Costa Santos, M.
Satisficer-Optimizer Model, 364
Sauer, C. O., 434, 457, 518, 519, 521, 529, 546, 549, 550, 551, 593, 596, 598, 608
Saushkin, Yv. G., 381, 382, 407, 423
Sauvy, A., 190, 196, 214
Savigear, R. A. G., 62, 64, 95
Saville, T., 68, 95, 742, 773
Sawyer, J. S., 114, 123, 131, 137, 143

Sawyer, L. R., 166, 167, 181
Scale, Problems of, 66
Time, in Climatology, 100
Scale Coverage, 246
Scale Economies, 305
Scale Index, 248
Scale of Information, 20
Scale Models, 66
in Hydrology, 145
Scale Transformation, 685
Schaefer, F. K., 534, 546
Schäffle, G. F., 369, 423
Schaller, G. B., 529, 546
Scheidegger, A. E., 69, 70, 73, 95, 664, 668
Schenck, H., Jr., 73, 95, 632, 633, 644, 663, 668
Schimper, A. F. W., 467
Schmidt-Falkenburg, H., 671, 723
Schmudde, T. H., 285, 292
Schneider, M., 642, 665
Schneider Principle, 560
Schnore, L. F., 237, 238, 241, 522, 546
Schooler, E. W., 420, 568, 601
Schuh, G. E., 257, 299
Schultz, T. W., 281, 286, 299
Schulz, H. C., 697, 723
Schumm, S. A., 60, 66, 80, 87, 88, 90, 95, 621, 668
and R. W. Lichty, on Time, 87–89
Schurr, S. H., 275, 299
Schwab, G. O., 153, 184
Schwartzberg, J. E., 211, 214
Science, Emergence of, 780
Scientific Method, 45, 463
Sclater, P. L., 467
Scorer, R. S., 99, 122, 143
Scott, B., 260, 297
Scott, D. R., 412, 423
Scott, E. L., 68, 94, 572, 606
Scott, P., 440, 457
Scott, W., 331, 358
Search Procedure, 623
Search Theory, 37, 624
Secomski, K., 381, 382, 423
Second Law of Thermodynamics, 97
Sector Model, of Land Use, 341
Sediment Abrasion, Experiments, 735
Sediment Parameters, 730
Selleck, E. W., 513, 546

Semple, E. C., 518, 546, 550, 598
Sequent Occupance, 551
Serck-Hansen, J., 386, 393, 395, 423
Servomex Field Plotter, 754
Set Characteristics, 21
Set Theory Model, 433
Settlement, 303
Settlement Hierarchy and Location, 389
Settlement Pattern, 306
Seyfried, W. R., 338, 359
Sforza, L. C., 570, 606
Shaler, N. S., 513, 546
Shape, of Industrial Areas, 406
Sharp, A. L., 155
Shaw, D. M., 137
Shaw, J. C., 50, 52, 56
Shaw, N., 728, 773
Shaw, W. N., 114, 131, 143
Sheaffer, J. R., 185
Shelford, V. E., 513, 523, 539
Shelly, M. W., 608
Shen, J., 172, 184
Shen Lin Algorithm, 616
Shenton, L. R., 575, 606
Sheppard, P. A., 109, 143
Sherman, L. K., 152, 156, 160, 161, 184
Shevky, E., 462, 506
Shifting Cultivation, 198, 530
Shils, E. A., 231–233, 242
Shimbel, A., 637, 656, 668
Shimbel-Katz Index, 637, 656
Shoemaker, E. M., 92
Shoemaker, L. A., 176, 182
Siddall, W. R., 520, 532, 546
Siegel, S., 81, 95
Significance of Difference, 557
Sigwart, C., 20, 304
Silhána, V., 386, 423
Silk, J. A., 613, 668, 769, 773
Simaika, Y. M., 174, 182
Simmonds, I., 530, 532, 546
Simmonds, R., 50, 52, 56
Simmons, J. W., 337, 338, 343–348, 354, 356, 359
Simon, H. A., 50, 56, 326, 359, 364, 422, 448, 593, 607, 608
Simonett, D. S., 56, 92
Simpson, G. C., 131, 132, 143
Simpson, G. G., 62, 95, 463, 470, 482, 506

Simpson, R. B., 56, *95*
Simulation Model, in Water Resources, 176
of Networks, 642
Singer, M., 261, *300*
Singh, H., 440, *457*
Sinnhuber, K. A., 679, 680, *723*
Sisler, D. G., 257, *300*
Sjaastad, L. A., 566, *602*
Sjörs, H., 523, *546*
Skellam, J. G., 571, 574, 575, 592, *606*, 608
Skibitzke, H. E., 162, *184*, 770, *773*
Skilling, H., 21–23, 27, *41*, 494, *506*, 712, *723*
Skinner, G. W., 321, 322, *359*
Slaymaker, H. O., 74, 89, *91*, 150, *184*
Sletmo, G., *423*
Slobodkin, L. B., 511, 531, 536, 537, *546*
Slope Development, 69, 70, 72
Smagorinsky, J., 100, 105, 109, *140*, *143*
Smailes, A. E., 309, 347, *359*
Smalley, A. E., 526, *544*
Smelser, N. J., 222, 234, 236, *242*
Smith, A., 369, 396, 398, *423*, 532
Smith, C. T., 549, *598*
Smith, D. M., *423*
Smith, P. E., 269, *300*, 571, 581, *606*
Smith, P. J., 341, 342, 352, *359*
Smith, R. H., 331, *359*
Smith, R. H. T., 262, *293*
Smith, S. C., 153, *184*
Smith, T. L., 430, *457*
Smith, W., 262, *300*, 373, 375, *423*
Smith, W. L., 172, *180*, *604*
Smithies, A., 377, *423*
Smolensky, E., 246, 256, 264, 270, *294*, *300*
Smoller, G., 249
Smuts, J. C., 518, *546*
Sneath, P. H. A., 463, 487, *506*
Snell, J. L., *605*
Snell's Law of Light Refraction, 619
Snodgrass, M. M., 388, 417, *421*
Snyder, D. E., 283, *300*
Soap-Film Analogue, 613, 768
Social Accounting Model, 255
Social Anthropology, 227
Social Contract, 224
Social Interaction Model, 595
Social Overhead Capital, 250
Social System, 230
Socialism, and Industrial Location, 382
Sociological Models, 220
Sociology, 221, 473
Socrates, 778, 787
Soil Science, 462, 467
Sokal, R. R., 62, *95*, 463, 479, 487, 490, *506*
Solberg, H., 114, 115, 133, *136*
Solntsev, N. A., 464, 473, 484, *506*
Sombart, W., 373, *423*
Sorre, M., 550, *598*
Space, as Amenity, 239
as Constraint, 239
Spacecraft, 147
Spain, Economic Development of, 219
Sparrow, F. T., 562, *602*
Spatial Analogues, 62
Spatial Description, and Time, 557
Spatial Equilibrium, 445
Spatial Models, 22
Spatial Pattern, 715
Evolution of, 561, 593
Spatial Properties, in Maps, 708
Spectral Analysis, 556, 622
Speight, J. G., 74, *95*, 621, 622, 624, *668*
Spencer, B., 200, *215*
Spencer, H., 234, 515, 517, 521, 779, *792*
Spencer, J. E., 261, *300*
Spengler, J. J., 234, 246, *300*
Spengler, O., 234
Spooner, C. S., 728, *773*
Spooner, R. J., 707, *722*
Sprout, H., 522, *546*
Sprout, M., 522, *546*
Sreenivasaiah, B. N., 125, *142*
Stafford, H. A., Jr., 324, 325, *359*
Stamp, L. D., 244, *300*, 427, 457, 466, *506*, 531, 532, 547, 564, *602*, 672, 683, 688, *723*
Stanford Watershed Model, 168, 169
Stanton, T., 742, *773*
Starostin, I. I., 703, *723*
Starr, V. P., 102, *143*, *144*, 744, *774*
State, Influence on Location, 366
of Systems, 77
State Concept, 630
State Economic Area, 474
Static Models, Hardware, 728
Statistical Models, 24
Statistics, 489
of Extreme Events, 152
Steady State, 78
Cell-Model, 648
Steady Time, 89
Steeger, W. H., 554, 585, *600*, 606
Steel, R. W., 243, *300*, 445, *457*
Stefaniak, N. J., 376, 409, *423*
Steigenga, W., 330, *359*
Stein, H., 703, *720*
Stein, J. L., 248, 257, 261, 268, 269, *288*
Stein, M., 242
Steiner, D., 32, *41*, 130, *144*
Steiner's Solution, 611
Steinhaus's Problems, 629
Steinmetzler, J., 517, *547*
Stephenson, P. H., 124, *138*
Stevens, A., 409, 516, *547*
Stevens, B., 270, *300*
Stevens, B. H., 379, 417, *423*
Stevens, S. S., 702
Steward, J. H., 430, 433, *457*, 474, *506*
Stewart, C. T., 327, 328, *359*
Stewart, J. Q., 559, *600*, 749, 755, *774*
Stimulus-Response Associationism, 788
Stine, J. H., 320, 321, *359*, 563, *602*
Stochastic Model, of Spatial Pattern, 570
Branching, 631
in Geomorphology, 72
Stocking, G. W., 380, *423*
Stoddart, D. R., 31, 35, *41*, 74, 89, 90, *91*, *95*, 248, 292, 511, 529, *547*

INDEX

Stollsteimer, J. F., 388, *423*
Stolper, W., 284, 285, *300*
Stone, R., 254, 255, *300*
Storm Model, 129
Stouffer, S. A., 560, *600*
Strahler, A. N., 60, 66, 69, 70, 73–76, 85, 86, *95*, *96*, 155, 160, 165, 166, *184*, 626, 628, 655, 663, 664, *668*
Strahler Ordering, 626
Strasburg, D. W., 524, 526, *542*
Stream Ordering, 626
Strickler, R. F., *140*
Structural Geomorphology, 62
Structure of Social Action, The, 223
Stuart, L. C., *509*
Sturmthal, A., 261, 264, *300*
Sturrock, F., 690, *721*
Subdivision of Holdings, 197
Subsystems, 78
Succession, 352
Sufrin, S. C., *506*
Sukachev, Y. N., 523, *547*
Sukhov, V. I., 687
Sukhov's Coefficient, 687
Supersystems, 78
Suppes, P., 23, 25, *41*, 594, *608*
Supple, B. E., 245, *300*
Suslov, S. P., 467, *506*
Sutcliffe, R. C., 98, 100, 107, 109, 115, *144*
Sutton, O. G., 98, 99, 134, 135, *144*
Sweeney, D. W., 615, *667*
Sweeting, M. M., 702, *724*
Swinnerton, H. H., 517, *547*
Symbolization, 690
Symbols, Colour, on Maps, 702
Synoptic Climatology, 129, 130
Synoptic Map, 130
Synthetic Systems, in Geomorphology, 80
System Properties, of Ecosystem, 524
System Synthesis Investigation, 153
Systems, Isomorphic, 496
  Optimal Branching, 629
  Partial, 84
  Physical, in Geomorphology, 63
  Recognition of, 22
Systems Analysis, 35
Systems Hydrology, 153
Systems Theory, 530

Taaffe, E. J., 400, *423*, 568, 588, *603*, *606*, 642, *668*
Takacs, L., 577, *606*
Takayama, T., 457
Take-Off, 211, 250
Talbot, L. M., 522, *547*
Tangri, S., 285, *300*
Tanner, J. C., *668*
Tansley, A. G., 511, 523, 532, 538, *547*
Tattersall, J. N., 268, 280–282, *300*
Tauber, R., 445, *454*
Taylor, R., 217
Taylor, T. G., 425, *457*, 550, 551, *598*
Teal, J. M., Jr., 526, 527, *547*
Tectonic Models, 735
Teitz, M. B., *509*
Teledeltos Paper, 746, 752, 753
Tennant, R. J., 313, 324–326, 337, 338, 343–348, 354, 355, *356*
Terrain Analogues, 63
Terrain Factors, 63
Territory, of Tits, 192
Testing, of Geomorphic Models, 89
Teweles, S., *95*
Thematic Maps, 680
Theobald, D. W., 24, *41*
Theory of Action, 228
Theory of Modernization, 236
Theory of Revolution, 234
Thermal Wind Equation, 98
Thermodynamics, Laws of, 97
Thienemann, A., 523, *547*
Thiessen, A. H., *184*
Thiessen Polygon, 149, 658
Thijsse, J. P., 563, *603*
Thirsk, J., 205, *215*
Thoman, R. S., 260, *302*
Thomas, D., 440, *457*
Thomas, E., 491, *506*, 561, *600*
Thomas, E. N., 312, 313, 324, 352, *359*, 718, *724*
Thomas, F. H., 285, *292*
Thomas, H. A., *183*, *605*

Thomas, M., 575, *606*
Thomas, M. D., 266, *301*
Thompson, D'A. W., 519, *547*, 629, 650–652, *668*
Thompson, G. L., *605*
Thompson, H. R., 575, *607*
Thompson, J. H., 493, *506*
Thompson, L., 529, *547*
Thompson, W. R., 268, *301*, 414, *422*
Thomson, W., 648
Thorn, R. B., 153, *184*
Thornthwaite, C. W., 158, *184*, 461, 467, 469, *506*
Thorpe, D., 347, *359*
Thrall, R. M., 52
Thurston, L. L., 702
Tiebout, C. M., 248, 267, *301*, 569, *603*
Time, and Industry, 395
  Kinds of, 87–89
  Nature of, 559
Time Dimension, 550
Time Series Analysis, 555
Time-Space Model, 569
Tinbergen, J., 254, 270, *301*, 386, 393–395, 405, *423*
Tintner, G., 556, *600*
Titow, J., 198, *214*
Tits, Territoriality in, 192
Tobler, W., 32, *41*, 312, *360*, 519, *547*, 560, *600*, 675, 715, *724*, 774
Todd, D. K., 162, *184*
Tomato Processing, 388
Tomlinson Report, *724*
Töpfer, F., 687, *723*, *724*
Topological Properties, 34
Topology, 610
Torricelli, E., 780
Tosí, J. A., 244, *301*, 525, 526, *547*
Toulmin, S. E., 23, *41*, 706, *724*
Town, River, Location Model, 763
Townsley, A. J., 531, *542*
Townspace, 303
Toynbee, A. J., 234, 595
Traffic Flow, 624
Trainer, D. W., 738, *773*
Transport Minimization, 363
Transport Network, 587, 638
Transition Probability, 577
Travelling Salesman Problem, 615
Tree Geometry, 624

Tree-Ordering, 625
Trend, Regional, 32
Trend Surface, 83, 130
   Graphical Analysis, 716
   on Breckland Sands, 74
   Mapping, 716
Trewartha, G. T., 118, *144*, *506*
Tricart, J., 62, 86, *96*
Troll, C., 518, 532, *547*
Tropical Climatology, Models of, 118
Tropical Cyclones, 122
Trucco, E., 536, *541*
Truth, of Models, 24
Tu, Yien-I, 256, *296*
Tucker, G. B., 103, 113, *144*
Tukey, J. W., 623, *665*
Tuominen, O., 313, *360*
Turner, F. J., 519, 551, *598*
Turner, J. S., 743, *774*
Turvey, R., 174, *183*
Twain, M., 712, *724*
Tweeten, L. G., *457*
Tyldesley, B., 743, *774*

Ubbink, J. B., 26, *41*
Uhorczak, Fr., 696, *724*
Ullman, E. L., 260, 261, *301*, 306, 307, 333, 334, 342, 357, *360, 509*
Uniform Pattern, 306
Unit Hydrograph, 160
United Nations Bureau of Social Affairs, 213
Unstead, J. F., 469, 481, *506, 507*, 516, *547*
Urban Fringe, 238
Urban Geography, 237
Urban Hinterland, 318
Urban Land Market, 335
Urban Sociology, 474, 521
Urbanization, Index of, 392
Urey, H. C., *92*
Usher, D., 270, *301*

Vajda, S., *185*
Valavanis-Vail, S., 255, *301*
Vallaux, C., 518, 519, *547*
Values, in Geography, 219
Van Burkalow, A., 733, *772*
Van Duijn, P., 38, *41*
Van Leeuwenhoek, A., 780
Van Lopik, J. R., 63, 65, *96*
Van Sickle, J. V., 248, 263, 264, *299, 300*

Van Valkenburg, S., 425, 444, *458*
Vance, J. E., Jr., 354, *358*, 621, *668*
Vance, R. B., *509*
Vanhove, N. D., 412, *423*
Variable Systems, 81
Vascular System, 629
Vasyutin, V. F., 381, *417*
Vegetation Formation, 475
Venkiteshwaran, S. P., 125, *142*
Verburg, M. C., 262, 273, *301*
Vernon, R., 391, *418*
Vidal de la Blache, P., 465, 470, 515–520, *548*, 550, 597
Vidich, A., *242*
Vietorisz, T., *420*, 568, *601*
Vigil Network, 150
Vining, R., 277, *301*, 324, 327, *360*, 474, 475, *507*, 569, 603
Vinski, I., 257, 273, *301*
Vitalism, 518
Vector Analysis, Weakness of, 31
Vectors, Row and Column, 28
   Transformation of, 29
Vegetation, 467, 513
Venezuelan Airlines, 641
Voertman, R. F., 244, *301*
Volk, D. J., *185*
Volterra, V., 534, 535, *548*
Von Arx, W. S., 744, *771*
Von Bertalanffy, L., 77, *96*, 426, *453*, 512, 520, 536, 537, 538, *539*
Von Ficker, H., 119
Von Helmholtz, H., 114, *144*
Von Huhn, R., 704, *724*
Von Humboldt, A., 30, 31, 39, 515, 523
Von Neumann, J., 448, *456*
Von Richthofen, F., 511
Von Thünen, J. H., 33, 271, 336, *360*, 405, 443–446, 457, 495, 498, 500, 563, 564
Von Thünen Model, 443, 563, 564
Voorhees, A., 304, *360*
Voronoi Polygon, 658
Vortices, 122
Vorticity Equation, 98
Voskuil, R., 257, 268, 295, 713, *723*
Vuorela, L. A., 103, *141*

Waibel, L., 430, 445, *458*
Waite, W. P., *95*
Walker, D. S., 219
Wallace, A. R., 467
Wallace, L. T., 368, *423*
Wallace, T. D., 445, *456*
Wallace, W. H., 655, *668*
Wallingford, Hydraulics Research Station, 145
Wallington, C. E., 743, *774*
Wallis, J. R., 626, 655, *665*
Wallis, W. A., 556, *600*
Walters, R. L., *95*
Walters, S. M., 32, *41*, 463, 464, *503*
Wanklyn, H., 517, *548*, 550, *598*
Ward, W. C., 730, *772*
Wardrop, J. G., 618, 619
Warming, E., 513
Warntz, W., 564, *603*, 620, *668*, 749, 752, 755, *774*
Washburn, S. L., 529, *540, 548*
Water Balance, Mathematical Model, 147
Water Resource Study, 175
Watkins, J. N. L., 476, *507*
Watson, D. J., 531, *548*
Watson, G. S., 556, *599*
Watson, J. W., 304, *360*
Watt, J., 401
Watt, K. E. F., 192, *215*
Wave Disturbances, in Trades, 119
Wave Tank, 741
Waves, in City Expansion, 353
Weather Forecasting, 133
Weaver, G. D., 86, *92*
Weaver, J. B., 648, *668*
Weaver, J. C., 439, 440, *458, 507*, 690, *724*
Weaver, J. E., 478, 490, 491, *507*
Web of Reality, 23
Webb, W. P., 595, *608*
Weber, A., 362–364, 370–378, 381, 397, 398, 415, 416, 424, 563, 564, 587, *762*
   Assumptions of, 372
Weber, M., 223–227, *242, 782*
Weber Model, Extension of, 374
Webster, R., 63, *90*
Weigert, H. W., 518, *548*
Weiss, S. F., Jr., 352, *356*

[816] INDEX

Welfare Economics, 174
Wellington, A. M., 613, 614, 668
Wendel, B., 566, *603*
Werner, C., 664, *668*
Wertheimer, M., 788
Westerlies, 107
Westermarck, E., 200, *215*
Westlake, D. F., 531, *548*
Westoff, C. F., 190, *214*
Wheeler, W. M., 522, *548*
WHIRLPOOL, 32, 82
Whitaker, E. A., *92*
Whitaker, J. K., 254, 256, *293*
White, C. L., 522, *548*
White, G. F., 153, 175, *185*
White, R. M., 102, *143, 144*
Whitehead, A. N., 44, 518, *548*
Witham, G. B., 609, 624, *667*
Whitman, W. T., *419*
Whitney, M., 513
Whittaker, J. R., *509*
Whittaker, R. H., 475, *507*, 513, 523, 528, *548*
Whitten, E. H. T., 76, 83, *96*
Whittlesey, D., 437, *458*, 461, 462, 464, 469, 472, 477, 481, 487, 488, 500, *507*, 519, *548*, 551, *598*
Wiebenson, W., 615, *667*
Wiener, D., 528, 535, *548*
Wiener, N., 25, *41*, 714, 724
Wiesner, C. J., 151, 152, *185*
Wiin-Nielsen, A., 110, *144*
Wilkinson, H. R., 680, *724*
Wilhelm, W., 267, *301*
Willett, H. C., 97–99, 111, 129, 131, 132, 134, *136*, *143, 144*
William-Olsson, W., *720*
Williams, C. B., 512, *548*
Williams, R. L., 702, 703, *724*
Williams, S. W., 401, *424*

Williamson, E., 574, *607*
Williamson, J. G., 247, 257, 258, 262, 264, *301*
Willoughby, E. O., *774*
Wills, L. J., 62, *96*
Wilson, E. B., 66, *96*
Wilson, F. R., 707, *724*
Wilson, H., 231
Wilson, T., 260, *301*
Wind, 98
Wingo, L., Jr., 261, 262, 266, 268, 280, *298*, 305, 336, 341, *360*, 564, *603*
Winsborough, H. H., 344, *360*, 392, *424*, 568, *603*
Winston, P. W., 534, *548*
Winterbotham, H. St. J. L., 700, *724*
Wise, M. J., 405, *424*
Wissler, C., 462, 476, 478, *507*
Withycombe, J. G., *724*
Wittfogel, K. A., 232, *242*
Wittgenstein, L., 609
Wold, H., 556, *600*
Wolf, A. K., 56
Wolf, E. R., 430, *458*
Wolf, P. O., 150, 153, 164, *185*
Wolfenden, H. H., *215*
Wolman, J. P., 621, 623, 631, *667*
Wolman, M. G., 68, 76, *91*, *93*, 145, *182*, 738, *771*
Wolpert, J., 448, *458*, 593, *608*
Wong, S. T., 82, *96*
Wonnacott, R. J., 261, 264, *301*
Wood, W. D., 260, *302*
Wooldridge, S. W., 62, 87, *96*, 617, 714, 716, 719, *725*, 735
Worms, R., 515
Woytinsky, E. S., 711, *725*
Wright, J. K., 676, 699, 714, *724*
Wright, M., 284–286, *302*

Wrigley, E. A., 30, 31, *41*, 189, 197, 199, 200, 204, 209, 212, *215*, 287, *302*, 398, *424*, 465, 470, *507*, 569
Wrobel, A., 473, 482, *507, 509*
Wurm, A., 66, *96*
Wyman, W. D., 551, *598*
Wynne-Edwards, V. C., 192, 193, *215*

Yanai, M., 122, 123, *144*
Yang, L. S., 254, *290*
Yanikov, G. V., 703, *723*
Yapp, W. B., 531, *548*
Yates, P. L., 260, *302*
Yeates, M. H., 338, 342, 351, *360, 606*
Yefremov, Yu. K., 484, *507*, 523, *548*
Yeh, T., *142*
Yin, M. T., 123, *144*
Yockey, H. P., 536, *548*
Yoda, K., 513, 514, *543*
Young, D. H., 522, *541*
Yuill, R. S., 447, *458*, 590, *608*, 762, 774

Zabko-Potopowicz, A., *458*
Zeigler, J. M., 83, 89, *94*
Zeno of Elea, 778
Ziman, L. Y., 473
Zipf, G. K., 304, 327, 328, *360*
Zobler, L., 477, 492, *507*
Zollschan, G. K., *242*
Zonal Indices, of Atmospheric Circulation, 110
Zonality, 467
Zone of Discard, 354
'Zones and Strata' Concept, 550
Zoo Geography, 467